Josef Ballmann
Rolf Jeltsch (Eds.)

Nonlinear Hyperbolic Equations – Theory, Computation Methods, and Applications

Josef Ballmann
Rolf Jeltsch (Eds.)

Nonlinear Hyperbolic Equations – Theory, Computation Methods, and Applications

Proceedings of the Second International Conference on Nonlinear Hyperbolic Problems, Aachen, FRG, March 14 to 18, 1988

Friedr. Vieweg & Sohn Braunschweig / Wiesbaden

CIP-Titelaufnahme der Deutschen Bibliothek

Nonlinear hyperbolic equations: theory, computation methods, and applications; proceedings of the 2nd International Conference on Nonlinear Hyperbolic Problems, Aachen, FRG, March 14 to 18, 1988 / Josef Ballmann; Rolf Jeltsch (eds.). – Braunschweig; Wiesbaden: Vieweg, 1989
(Notes on numerical fluid mechanics; Vol. 24)
ISBN 3-528-08098-1

NE: Ballmann, Josef [Hrsg.]; International Conference on Nonlinear Hyperbolic Problems ⟨02, 1988, Aachen⟩; GT

Manuscripts should have well over 100 pages. As they will be reproduced photomechanically they should be typed with utmost care on special stationary which will be supplied on request. In print, the size will be reduced linearly to approximately 75 per cent. Figures and diagramms should be lettered accordingly so as to produce letters not smaller than 2 mm in print. The same is valid for handwritten formulae. Manuscripts (in English) or proposals should be sent to the general editor Prof. Dr. E. H. Hirschel, Herzog-Heinrich-Weg 6, D-8011 Zorneding.

Vieweg is a subsidiary company of the Bertelsmann Publishing Group.

Produced by W. Langelüddecke, Braunschweig
Printed in Germany

ISSN 0179-9614

ISBN 3-528-08098-1

FOREWORD

On the occasion of the International Conference on Nonlinear Hyperbolic Problems held in St. Etienne, France, 1986 it was decided to start a two years cycle of conferences on this very rapidly expanding branch of mathematics and it's applications in Continuum Mechanics and Aerodynamics. The second conference took place in Aachen, FRG, March 14-18, 1988. The number of more than 200 participants from more than 20 countries all over the world and about 100 invited and contributed papers, well balanced between theory, numerical analysis and applications, do not leave any doubt that it was the right decision to start this cycle of conferences, of which the third will be organized in Sweden in 1990.

This volume contains sixty eight original papers presented at the conference, twenty two of them dealing with the mathematical theory, e.g. existence, uniqueness, stability, behaviour of solutions, physical modelling by evolution equations. Twenty two articles in numerical analysis are concerned with stability and convergence to the physically relevant solutions such as schemes especially deviced for treating shocks, contact discontinuities and artificial boundaries. Twenty four papers contain multidimensional computational applications to nonlinear waves in solids, flow through porous media and compressible fluid flow including shocks, real gas effects, multiphase phenomena, chemical reactions etc.

The editors and organizers of the Second International Conference on Hyperbolic Problems would like to thank the Scientific Committee for the generous support of recommending invited lectures and selecting the contributed papers of the conference.

The meeting was made possible by the efforts of many people to whom the organizers are most grateful. It is a particular pleasure to acknowledge the help of Riikka Tuominen for preparing the abstract book and Bert Pohl for his dedicated help organizing the conference. It is also a pleasure to thank Sylvie Wiertz, Angela Schneider, Gabriele Goblet and Thomas Hoerkens for preparing these proceedings. Finally the organizers are indebted to the host organizations Rheinisch Westfälische Technische Hochschule Aachen and the city of Aachen and to those organizations which provided the needed financial support for the conference: Control Data GmbH, Cray Research GmbH, Deutsche Forschungsgemeinschaft, Diehl GmbH & Co., Digital Equipment GmbH, FAHO Gesellschaft von Freunden der Aachener Hochschule, IBM Deutschland GmbH, Mathematisch-Naturwissenschaftliche Fakultät der RWTH, Ministerium für Wissenschaft und Forschung des Landes Nordrhein-Westfalen, Office of Naval Research Branch of London, Rheinmetall GmbH, US Air Force EOARD, US Army European Research Office of London, Wegmann GmbH & Co.

Aachen, September 1988
Josef Ballmann
Rolf Jeltsch

Contents

Page

NON-OSCILLATORY SCHEMES FOR MULTIDIMENSIONAL EULER CALCULATIONS WITH UNSTRUCTURED GRIDS

PAUL ARMINJON

Université de Montréal, Dépt de Mathématiques et Statistiques, C.P. 6128 Succ. A, Montréal, Québec (CANADA), H3C3J7

ALAIN DERVIEUX, LOULA FEZOUI, HERVE STEVE

INRIA, 2004 Route des Lucioles, Sophia-Antipolis 1 et 2, 06565 VALBONNE (FRANCE)

BRUNO STOUFFLET

AMD-BA, DGT-DEA B.P. 300, 78 Quai M. Dassault, 92214 SAINT-CLOUD (FRANCE)

The purpose of this paper is to present a synthesis of our recent studies related to the design of multi-dimensional non-oscillatory schemes, applying to non-structured finite-element simplicial meshes (triangles, tetrahedra). While the direct utilization of 1-D concepts may produce robust and accurate schemes when applied to non-distorted structured meshes, it cannot when non-structured triangulations are to be used. The subject of the paper is to study the adaptation of the so-called TVD methods to that context. TVD methods have been derived for the design of hybrid first-order/second-order accurate schemes which present in simplified cases monotonicity properties (see, for example, the review [2]). A various collection of first-order accurate schemes can be used, they are derived from an artificial viscosity model or from an approximate Riemann solver. However, the main feature in the design is the choice of the second-order accurate scheme ; this choice can rely either on central differencing or on upwind differencing.

1 GALERKIN AND UPWIND FINITE-ELEMENT SCHEMES FOR TRIANGLES.

Let us consider the following scalar model problem

$$\begin{cases} u_t + \vec{V} \,.\, \vec{\nabla} u &= 0 \ in \ I\!R^2 \\ u(x,0) &= u_0(x), \end{cases} \tag{1}$$

and a finite-element triangulation of $I\!R^2$; the generic element is denoted by T, and the P_1-Galerkin basis function related to a vertex i is written ϕ_i: ϕ_i is continuous, affine by element, $\phi_i = 1$ at vertex i, 0 at all other vertices. Then the P_1-Galerkin approximation scheme for (1) reads :

$$\left(\frac{\partial u}{\partial t}, \phi_i\right) - (u\vec{V}, \vec{\nabla}\phi_i) = 0 \tag{2}$$

or

$$\left(\frac{\partial u}{\partial t}, \phi_i\right) + \sum_{T \ neighbour \ of \ i} area(T)(u\vec{V})|_T(\vec{\nabla}\phi_i)|_T = 0 \tag{3}$$

with

$$\left\{ \begin{array}{lll} (u\vec{V})|_T & = & \text{some average of } u\vec{V} \text{ on } T \\ (\vec{\nabla}\phi_i)|_T & = & \left(\int_{I_1 G I_2} \vec{n}\, d\sigma\right) / \; area(T) \end{array} \right. \tag{4}$$

where I_1, I_2 are mid-sides and G the centroid of T as sketched in Fig. 1 ; the sum is taken over triangles T having i as a vertex.

The conservation properties of scheme (2) may not be clear at first glance ; it is then interesting to introduce a second scheme, that is a variant of (2) : Let $cell_i$ be the polygon around vertex i that is limited by medians as sketched in Fig 2. Then we shall call "Finite-Volume Galerkin" the following scheme:

$$area(cell_i)\frac{\partial u_i}{\partial t} + \sum_{j \text{ neighbour of } i} (u\vec{V})|_I \int_{G_1 I G_2} \vec{n}\, d\sigma = 0 \tag{5}$$

where I is the middle of the side ij and G_1, G_2 the centroids of the two triangles having ij as common side ; the sum is taken over the vertices j that are extremities of sides having i as other extremity.

Lemma 1 :*Schemes (2) and (5) are identical if (i) Mass matrix lumping by line-summing is applied to (2) and (ii) The following numerical quadratures are applied :*
- for (2) : $(u\vec{V})|_T \simeq \frac{1}{3}(u_i\vec{V}_i + u_j\vec{V}_j + u_k\vec{V}_k)$ *where* i, j, k *are the vertices of triangle* T
- for (5) : $(u\vec{V})|_I \simeq \frac{1}{2}(u_i\vec{V}_i + u_j\vec{V}_j)$.
The schemes of the above family (2), (5) do not satisfy the Maximum Principle (case V = const.) ; however, BABA and TABATA [4] proposed an upwind version of (5), that, in the case where V is constant, can be written :

$$\begin{aligned} area(cell_i)\frac{\partial u_i}{dt} &+ \sum_{j \text{ neighbour of } i} \left(\frac{u_i + u_j}{2}\right) \int_{G_1 I G_2} \vec{V} \,.\, \vec{n} d\sigma \\ &= \frac{1}{2} \sum_{j \text{ neighbour of } i} (u_j - u_i) | \int_{G_1 I G_2} \vec{V} . \vec{n} d\sigma | \end{aligned} \tag{6}$$

where the left-hand side is the Finite-Volume Galerkin term, and the right-hand side a numerical viscosity. This scheme satisfies the Maximum Principle for V = const. and preserves positiveness in the general case ; however, it is only first-order accurate and we shall discuss several ways to recover second-order accuracy locally.

2 TVD-LIKE SCHEMES FOR SYSTEMS.

The extension to second-order schemes can be performed starting from Lax-Wendroff schemes (with triangles : one-step, two-step Taylor-Galerkin schemes, [1] [8]) ; this extension can rely on an FCT approach [11] [10] [13] or a symmetric TVD one [3] [13] ; a description of all these schemes is out of the scope of this paper and we restrict ourselves to a family of MUSCL methods [15], extended to unstructured triangulations following [9].

2.1 Extension of the first-order scheme

Vijayasundaram proposed in [16] the extension of the BABA-TABATA scheme to hyperbolic systems by introducing an (approximate) Riemann solver or a flux splitting as follows : the Euler system

$$W_t + F(W)_x + G(W)_y = 0 \tag{7}$$

is discretized by

$$area(cell_i)\frac{\partial W_i}{\partial t} + \sum_{j\ neighbour\ of\ i} \Phi(W_i,\ W_j,\ \vec{\eta}_{ij}) = 0 \tag{8}$$

with

$$\left\{ \begin{array}{lll} \vec{\eta}_{ij} & = \displaystyle\int_{\partial cell_i \cap \partial cell_j} \vec{\eta}\ d\sigma \equiv (\eta^x, \eta^y) \\ \Phi & : \text{flux splitting or Riemann Solver,} \end{array} \right. \tag{9}$$

for example :

$$\Phi(W_i,\ W_j,\ \vec{\eta}) = \frac{H(W_i) + H(W_j)}{2} - \frac{1}{2}|P\left(\frac{W_i + W_j}{2}\right)|(W_j - W_i) \tag{10}$$

with

$$\begin{array}{ll} H(W) & = \eta^x F(W) + \eta^y G(W) \\ P(W) & = \frac{\partial H}{\partial W}(W) = \eta^x A(W) + \eta^y B(W) \end{array} \tag{11}$$

and

$$P = T \wedge T^{-1},\ |P| \equiv T| \wedge |T^{-1},\ \wedge\ diagonal\ . \tag{12}$$

2.2 Extension to a second-order accurate scheme.

One way to construct a MUSCL second-order extension is to introduce nodal approximate gradients [9]:

$$\vec{grad}W(i) = \left(\int\int \phi_i\ \vec{grad}W dx\right) / \int\int \phi_i dx \tag{13}$$

in order to extrapolate mid-side values :

$$\left\{ \begin{array}{ll} W_{ij} & = W_i + \frac{1}{2}\vec{grad}W(i)\ .\ \vec{ij} \\ W_{ji} & = W_j - \frac{1}{2}\vec{grad}W(j)\ .\ \vec{ij} \end{array} \right. \tag{14}$$

and then introduce them in the flux function :

$$area(cell_i)\frac{\partial W_i}{\partial t} + \sum_{j\ neighbour\ of\ i} \Phi(W_{ij},\ W_{ji},\ \vec{\eta}_{ij}) = 0. \tag{15}$$

This construction results in a scheme which is (spatially) second order accurate but may present over/undershoots in solutions.

We now study several approaches to recover (more or less surely) monotonicity.

- Limiters with nodal gradients. In order to construct a hybrid between the second-order scheme and the first-order one, the TVD approach necessitates the knowledge in the direction of the flux of four successive values of the dependent variables, let us call them

$$W_{i-1},\ W_i,\ W_j,\ W_{j+1}. \tag{16}$$

While these can be nodal values in the context of a structured grid, the values W_{i-1} and W_{j+1} have to be fictitious in the unstructured triangulation one. To derive these fictitious values, we can use the nodal gradients :

$$\left\{ \begin{array}{ll} W_{i-1} & = W_i - 2\nabla W(i)\ .\ \vec{ij} + (W_j - W_i) \\ W_{j+1} & = W_j - 2\nabla W(j)\ .\ \vec{ji} + (W_i - W_j) \end{array} \right. \tag{17}$$

We then compute, following [15], the "average" values of variations of W :

$$\begin{cases} dW_i &= ave\ (W_j - W_i,\ W_i - W_{i-1}) \\ dW_j &= ave\ (W_i - W_j,\ W_j - W_{j+1}) \end{cases} \tag{18}$$

with ($\epsilon \geq 0$) :

$$ave\ (a,b) = \begin{cases} \dfrac{a(b^2+\epsilon^2)+b(a^2+\epsilon^2)}{a^2+b^2+2\epsilon^2} & if\ a.b > 0 \\ 0 & else \end{cases} \tag{19}$$

and finally extrapolate limited values

$$\begin{cases} W_{ij}^{lim} &= W_i + \frac{1}{2}dW_i \\ W_{ji}^{lim} &= W_j + \frac{1}{2}dW_j, \end{cases} \tag{20}$$

that are introduced in the flux function Φ instead of (W_{ij}, W_{ji}). With this approach, the solutions are oscillation-free in most transonic cases, but high Mach calculations produce negatives pressures. We want then to go further in the prevention of oscillations.

- Element-by-element slope limitation. One explanation for the lack of monotonicity of the above scheme is that the nodal gradient is a mean value of element-wise gradients, that may not allow for an accurate detection of oscillations. A first way to circumvent this phenomenon is to consider, from a pessimistic point of view, each element-wise gradient for the construction of the nodal gradient ; we propose the following limited nodal gradients :

$$\left[\frac{\partial W}{\partial \xi}(i)\right]^{lim} = \min_{T\ neighbour\ of\ i} \bmod \left(\frac{\partial W}{\partial \xi}|_T\right) \tag{21}$$

and same for η, with

$$\min\ \mathrm{mod}(a_1 \ldots a_n) = \mathrm{sign}\ (a_1) \times \prod_{k=2}^{n} |\left(\frac{\mathrm{sign}\ (a_1) +\ \mathrm{sign}\ (a_k)}{2}\right)| \times \min_{k=(1,\cdots,n)} |a_k| \tag{22}$$

and with :

$$\begin{cases} \text{either} & (\xi,\eta) = (x,y) \\ \text{or} & (\xi,\eta) = (\text{direction of the local gradient, its orthogonal}). \end{cases} \tag{23}$$

This approach is very robust but rather dissipative and non-smooth.

- Upwind element formulation. We lastly propose a formulation which is inspired from the 1-D case. We return to the (16-20) context and propose a different way to define the fictitious values W_{i-1} and W_{j+1} : instead of using the nodal gradients $\nabla W(i)$, we use the usual triangle-wise Galerkin gradient $\nabla W|_T$ that we compute on the so-called upwind elements T_{ij}, T_{ji} w.r.t. the considered segment ij ; they are defined as follows : for any small enough positive number λ,

$$\begin{cases} i & +\lambda\vec{ji}\ \epsilon\ T_{ij} \\ j & +\lambda\vec{ij}\ \epsilon\ T_{ji}. \end{cases} \tag{24}$$

Then we put

$$\begin{cases} W_{i-1} &= W_i - \nabla W|_{Tij} \cdot \vec{ij} \\ W_{j+1} &= W_j - \nabla W|_{Tji} \cdot \vec{ji} \end{cases} \tag{25}$$

the rest of the calculation is as in (18)-(20). Since this construction is done after the side ij is considered, the limitation can be applied either to the primitive variables $\hat{W} = (\rho, u, v, p)$ or to characteristic variables, which are defined as follows :

$$\hat{W} = T^{-1}\left(\frac{W_i + W_j}{2}\right)(\rho, \rho u, \rho u, E) \tag{26}$$

where T is defined in (12).

- Central-difference MUSCL variant. Lastly we describe a variant of the MUSCL scheme that on second order central differencing : this scheme is obtained by replacing the usual non limited MUSCL interpolations W_{ij}, W_{ji} by the following ones :

$$\begin{aligned} \tilde{W}_{ij}^S &= \frac{\tilde{W}_i + \tilde{W}_j}{2} + k_{ij}\frac{\tilde{W}_i - \tilde{W}_j}{2} \\ \tilde{W}_{ji}^S &= \frac{\tilde{W}_i + \tilde{W}_j}{2} + k_{ij}\frac{\tilde{W}_j - \tilde{W}_i}{2} \end{aligned} \tag{27}$$

where k_{ij} is defined from four consecutive (partly fictitious) values of $\tilde{W}$:

$$k_{ij} = k_{ij}(\tilde{W}_{i-1},\ \tilde{W}_i,\ \tilde{W}_j,\ \tilde{W}_{j+1}) \tag{28}$$

following the method of symmetric TVD design [5,17] ; in the experiment presented in the sequel, the following limiter is applied :

$$\begin{aligned} r_{ij}^+ &= \frac{\tilde{W}_{i-1} - \tilde{W}_i}{\tilde{W}_i - \tilde{W}_j}, \quad r_{ij}^- = \frac{\tilde{W}_j - \tilde{W}_{j+1}}{\tilde{W}_i - \tilde{W}_j} \\ r_{ij} &= \begin{cases} \min\left(r_{ij}^+,\ r_{ji}^-,\ \frac{1}{r_{ij}^+},\ \frac{1}{r_{ji}^-}\right) & \text{if sign } (r_{ij}^+) = \text{ sign } (r_{ij}^-) \\ 0 & \text{else} \end{cases} \\ k_{ij} &= 1 - \frac{2r_{ij}}{1 + r_{ij}}\ . \end{aligned} \tag{29}$$

Furthermore, the limitation is applied separately on each primitive variable (ρ, u, v, p).

3 NUMERICAL EXPERIMENTS

3.1 Blunt body comparisons.

We present a sample of experiments performed with the simple test case of a flow past a halfly-circular blunt body with Mach at infinity equal to 8 and zero angle of attack. Although the grid used here is structured, (2000 nodes, Fig.3) both the alternation of 4-neighbours / 8-neighbours molecules ("Union Jack grid") and the bad alignment with the shock to be captured make this problem rather typical of the difficulties arising with unstructured arbitrary grids.

The upwinding of the solution is obtained by using Osher's approximate Riemann solver [12]. The solutions are computed with an implicit unfactored scheme with a first-order accurate linear operator (using the Steger-Warming splitting). For the central-difference scheme, this solver is not adequate at high CFL numbers, as shown in [7].

Application of scheme (16)-(20) : we did not get convergence of this scheme (negative pressures appear and do not disappear) for any initial state. This is explained by the non efficient detection of over-shoots/under-shoots by the averaged nodal gradients.

The second scheme, referred to as Hermitian limited or element-limited, yields the entropy contours in Fig.3a. We think that the shock is captured in a stable enough manner but much numerical viscosity is involved, in particular near the stagnation point, as one can see when considering the entropy level along the body.

The third scheme experimented is the "upwind-element" formulation, with limiters applied to the primitive variables, while in the fourth the limiters are applied to the characteristic values ; we emphasize that, due to a less viscous limitation than in the second scheme both results are more satisfactory from the standpoint of internal viscosity. Nevertheless, the characteristic limitation seems to improve both stability and accuracy. There remains a low frequency error in the entropy generation through the shock which slighly pollutes the contours.

Lastly the "central difference-MUSCL" scheme is applied ; because of the more severe limiter used here, the numerical entropy generation is larger than when the upwind schemes are applied but still rather acceptable (Fig. 3.d).

3.2 3-D applications.

For the illustration of the 3-D extensions, we refer to [6] for scheme (20), to [14] for scheme (23), and we present here a result obtained with scheme (21) : the interaction of two supersonic jets in a combustion chamber is computed with 3000 and 20,000 node-half geometries ; the maximum Mach number is close to 5 ; the second-order accurate results are in good agreement with each other(Fig. 4) while the first-order scheme applied to the 20,000 nodes grid produces a rather poor result.

4 CONCLUSION.

Several methods are described and compared for the construction of TVD-like schemes applying to arbitrary simplicial finite-element triangulations. Several schemes are robust enough for the capturing of strong shocks arising in high Mach flows ; they extend to 3-D, allowing for a large set of applications involving reentry simulations and stiff internal flows.

Acknowledgement:

We thank the SIMULOG team for yielding the grid of the combustion chamber.

5 REFERENCES

[1] ANGRAND F., DERVIEUX A., BOULARD V., PERIAUX J., VIJAYASUNDARAM G., Transonic Euler simulations by means of Finite Element explicit schemes, AIAA Paper 83-1924 (1983).

[2] ARMINJON P., Some aspects of high resolution numerical methods for hyperbolic systems of conservation laws with applications to gas dynamics, INRIA Report 520, (1986).

[3] ARMINJON P., DERVIEUX A., Schémas TVD en Eléments Finis triangulaires, in preparation.
[4] BABA K., TABATA M., On a conservative upwind finite element scheme for convective diffusion equations, RAIRO, Vol.15,1,pp 3-25(1981).
[5] DAVIS S.F., TVD finite difference schemes and artificial viscosity, ICASE Report 84-20, (1984).
[6] DERVIEUX A., FEZOUI L., STEVE H., PERIAUX J., STOUFFLET B., Low storage implicit upwind-FEM schemes for the Euler equations, IC11NMFD, Williamsburgh(USA), 1988.
[7] DESIDERI J.A., Iterative convergence of a class of implicit upwind schemes, to appear.
[8] DONEA J.,A Taylor-Galerkin method for Convective transport problems, Int. J. Numer. Meths. Eng. 20(1984) 101-120
[9] FEZOUI F., Résolution des équations d'Euler par un schéma de van Leer en éléments finis, INRIA Report 358 (1985).
[10] MORGAN K., LOEHNER R., JONES J.R., PERAIRE J., VAHDATI M., Finite element FCT for the Euler and Navier-Stokes equations, Proc. of the 6th Int. Sympos. Finite Element Methods in Flow Problems, Antibes (1986).
[11] PARROT A.K., CHRISTIE M.A., FCT applied to the 2-D Finite Element solution of tracer transport by a single phase flow in a porous medium, Proc. of the ICFD Conf. on Numerical Methods in Fluid Dynamics, Reading (1985).
[12] OSHER S., CHAKRAVARTHY S., Upwind schemes and boundary conditions with applications to Euler equations in general geometries, J. of Comp. Physics, Vol. 50, 3, 447-81 (1983).
[13] SELMIN V., Finite Element solution of hyperbolic equations, II, two-dimensional case, INRIA Report 708 (1987).
[14] STOUFFLET B., PERIAUX J., FEZOUI F., DERVIEUX A., Numerical simulation of 3-D Hypersonic Euler Flows around space Vehicles using adapted finite Elements, AIAA Paper 87-0560 (1987).
[15] VAN LEER B., Computational methods for ideal compressible flow, von Karman Institute for Fluid Dynamics, Lecture Series 1983-04.
[16] VIJAYASUNDARAM G., Transonic flow simulations using an upstream centered scheme of Godunov in Finite Element, J. Comp. Phys., vol.63, 416-433(1986).
[17] YEE H., Upwind and Symmetric Shock-capturing schemes, NASA TM 89464 (1987).

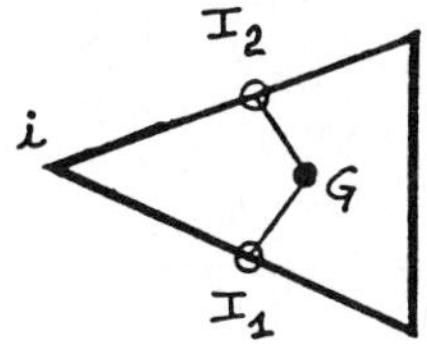

Figure 1: Calculation of $\vec{\nabla}\Phi_i|_T$

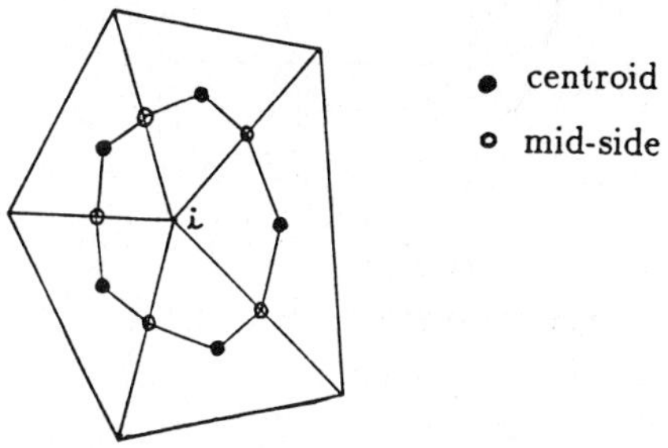

Figure 2: ***cell**$_i$*, **around** ***i***

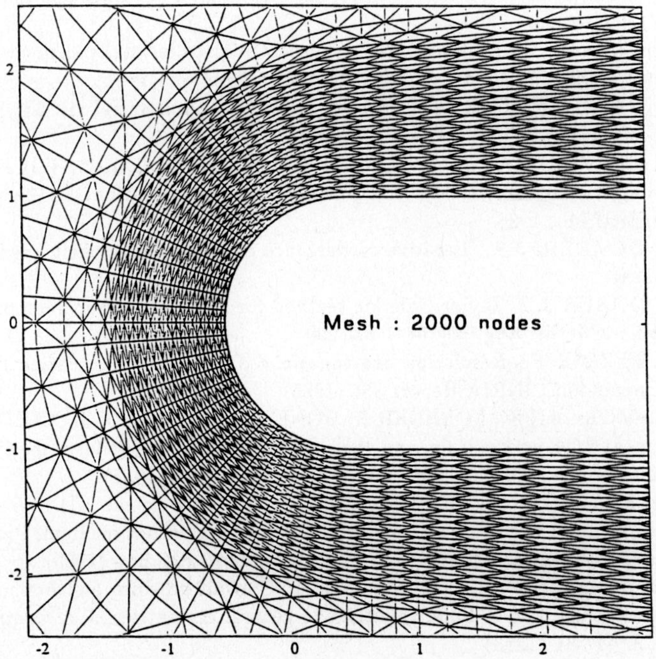

(a) Element-by-element limiters (upwind)

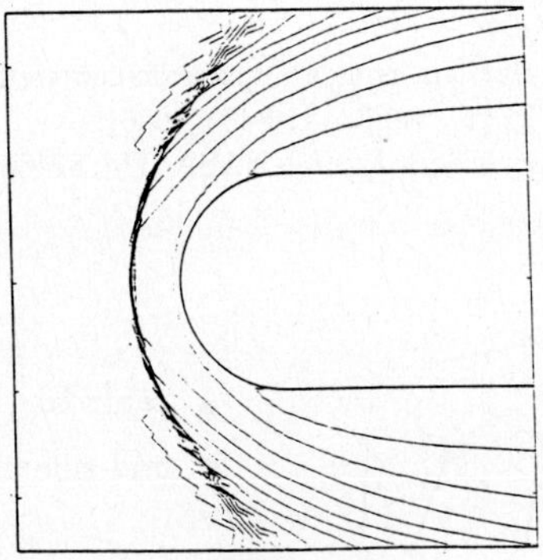

$\Delta \Sigma = .5$

Entropy contours

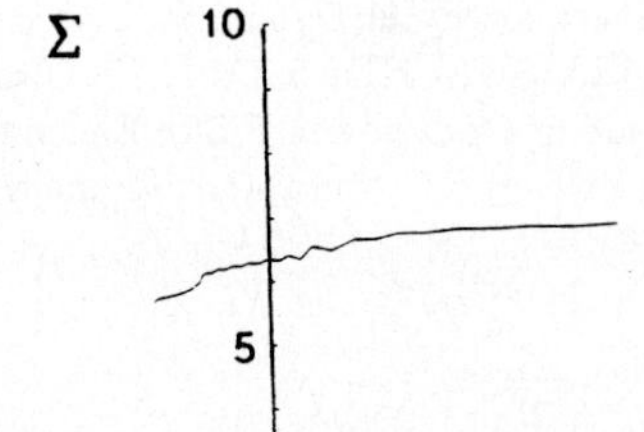

Entropy distribution along the axis and body

FIGURE 3: BLUNT BODY FLOW
(MACH AT INFINITY = 8)

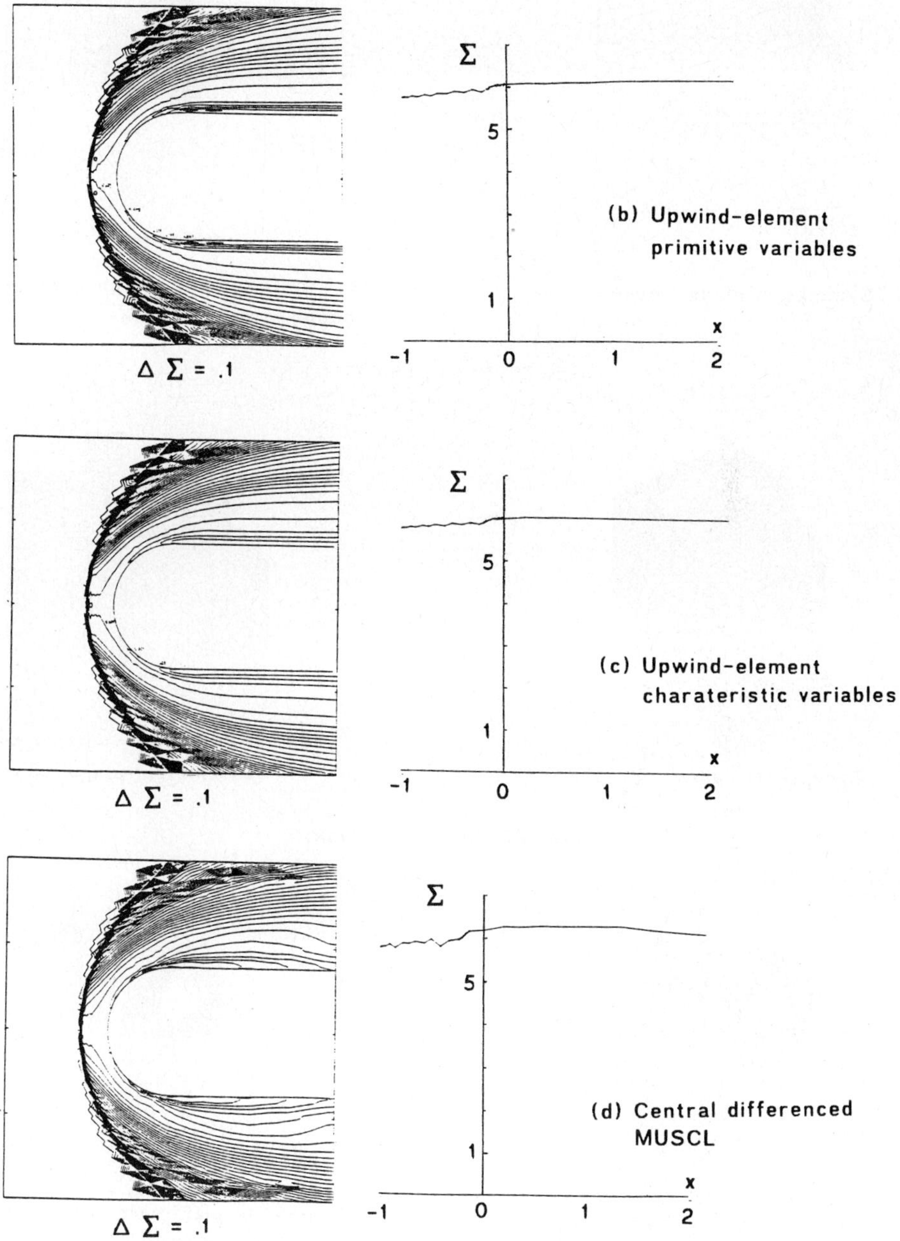

FIGURE 3, CONT'D

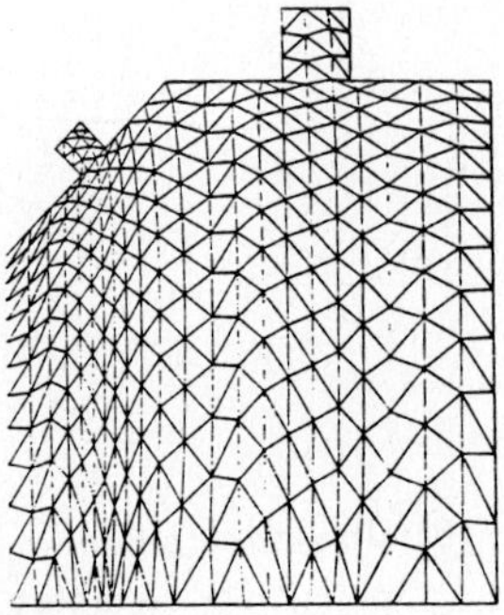

Symmetric plane: mesh

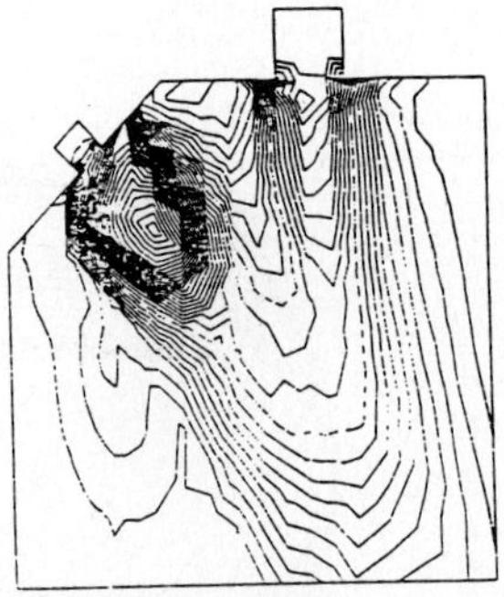

Symmetric plane: Mach contours
second-order scheme

(a) 3000-node mesh

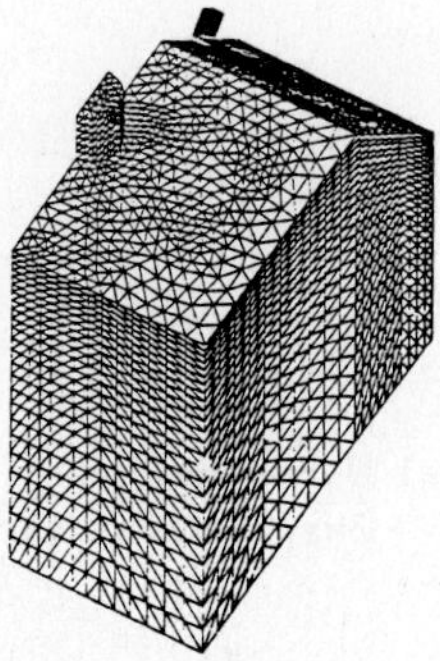

Perspective view

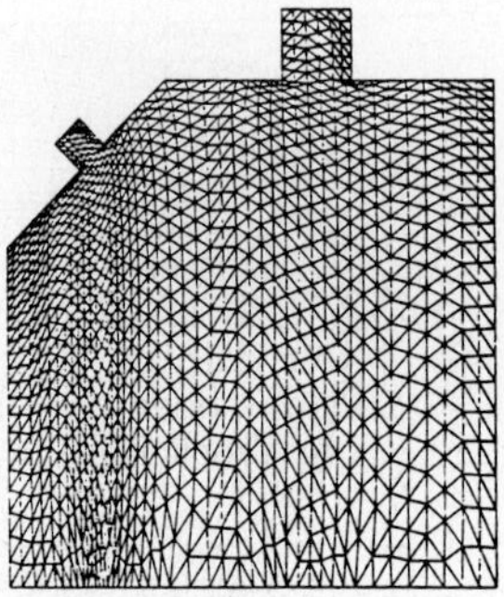

Symmetric plane

(b) 20,000-node mesh:

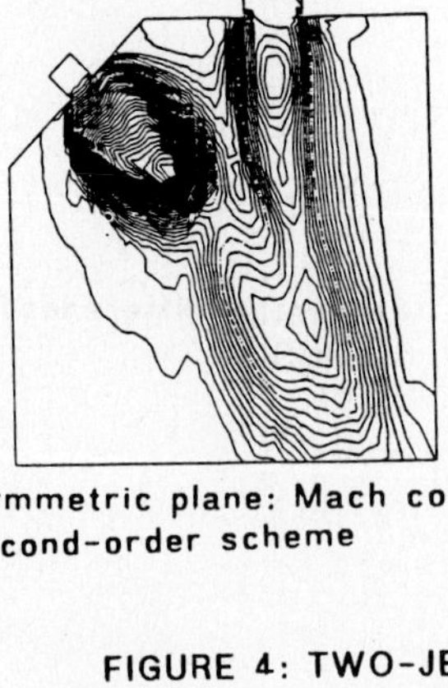

Symmetric plane: Mach contours
second-order scheme

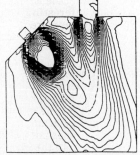

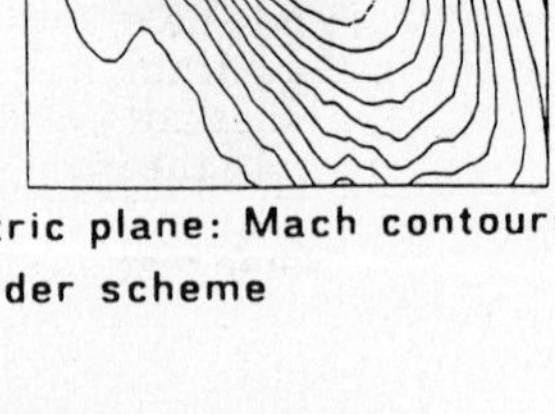

Symmetric plane: Mach contours
first-order scheme

FIGURE 4: TWO-JET 3-D FLOW IN A COMBUSTION CHAMBER

FINITE-DIFFERENCE SCHEMES WITH DISSIPATION CONTROL JOINED TO A GENERALIZATION OF VAN LEER FLUX SPLITTING

Germain BILLET

Office National d'Etudes et de Recherches Aérospatiales
29, avenue de la Division Leclerc
92320 CHATILLON, France

SUMMARY

A class of flux splitting explicit second-order finite-difference schemes S_{FS} is set up. It depends on a single parameter. The adaptation of the value of this parameter enables us to control the dissipative error included in these schemes. A generalization of Van LEER flux splitting makes possible an improvement in the numerical solution in the regions where the Mach number is relatively weak. 1D and 2D transonic flows are presented.

INTRODUCTION

The recent appearance of the flux-splitting method [1,2] used in solving the hyperbolic system of conservation laws $U_t+F(U)_x=0$ has permitted the setting up of a class of flux-splitting S_{FS} schemes that depends on the single parameter α. These schemes are stable up to CFL=2. The equivalent third-order system (ETOS) of this family has been obtained and the "optimal" value of α ($\alpha=2.5$), that minimizes the amplitude of the peaks of the numerical solution near the shock, has been determined by using the numerical tests of shock-tube flow and moving shock in 1D space [3]. The STEGER-WARMING splitting allows one to study the ETOS such as the ETOS associated with upwind scheme S_U and downwind schemes S_D defined in [4]. This study shows that the scheme S_{FS} with $\alpha=2.5$ (noted $S_{FS}{}^{OPT}$) is in fact more dissipative than with $\alpha=1$ in the case of a compression wave or a shock. The scheme S_{FS} with this last value corresponds to the STEGER-WARMING scheme used in [1]. When Van LEER flux splitting is used, the study of the ETOS appears more difficult because the jacobian matrices associated with the total flux and the partial fluxes have not the same eigenvalues and in this case, the method taken into account above cannot be applied. Nevertheless, the ETOS has been studied for the one-dimensional isothermal flow and some interesting results concerning the dispersive and dissipative properties of S_{FS} schemes were brought to light.

In this paper, the S_{FS} schemes are associated with a generalization of Van LEER splitting. This generalization, which keeps the whole properties of Van LEER splitting, has been possible because in a great number of applications, the Mach number does not reach the values -1 and 1 in a same flow. In this case, some conditions that are linked to Van LEER splitting for one of these two values can be cancelled, and the subconditioned system is correctly solved at that time by the insertion of a parameter ε in the partial fluxes. This parametric splitting (called ε-splitting) makes possible an improvement in the numerical solution, especially in the zones where the Mach number is less than 0.6.

Some results obtained for 1D shock-tube problems, a 2D steady flow in a nozzle are presented. They show that S_{FS} schemes associated with ε-splitting are well-adapted to compute such flows.

STUDY OF THE ETOS OF THE SCHEMES S_{FS}

Let the initial-value problem for hyperbolic 1D system in conservation form be:

$$U_t + F_x^+(U) + F_x^-(U) = 0, \quad U(x,0) = U^0(x), \quad -\infty < x < +\infty \, . \tag{1}$$

Here $U(x,t)$ and the fluxes F^+ and $F^- \in \mathbf{R}^m$. Let $U_i^{\,n}$ be the numerical solution of (1) at the mesh point $x_i = i\Delta x$ and at time $t = n\Delta t$. F^+ and F^- represent the partial fluxes associated with the forward and the backward moving waves respectively. Any non-linear system (1) is discretized with a second-order accuracy in time and space by the following predictor-corrector schemes S_{FS} [3] :

$$\widetilde{U}_i = U_i^{\,n} - \alpha\sigma\left(F_{i+1}^- - F_i^- + F_i^+ - F_{i-1}^+\right)^n$$

$$U_i^{\,n+1} = U_i^{\,n} + \frac{\sigma}{2\alpha}\Bigg| \alpha\left(F_{i+2}^- - F_{i-2}^+\right)^n + (1-4\alpha)\left(F_{i+1}^- - F_{i-1}^+\right)^n + (3\alpha-1)\left(F_i^- - F_i^+\right)^n \tag{2}$$

$$- \widetilde{F}_{i+1}^- + \widetilde{F}_i^- - \widetilde{F}_i^+ + \widetilde{F}_{i-1}^+ \Bigg|$$

where $\widetilde{F}_i^{\pm} = F^{\pm}(\widetilde{U}_i)$ and $\sigma = \Delta t/\Delta x$.

A way to study the properties of S_{FS} schemes is to make an analysis of the ETOS; i.e., of the system (1) approximated by the S_{FS} [5]. In this paper, we study the ETOS in the case of the 1D isothermal flow:

$$U = \begin{bmatrix} U1 \\ U2 \end{bmatrix} = \begin{bmatrix} \rho \\ \rho M \end{bmatrix} \qquad F = F^+ + F^- = \begin{bmatrix} \rho M \\ \rho(M^2+1) \end{bmatrix} \tag{3}$$

(the unities have been chosen such that $p/\rho = 1$). ρ, p *and* M represent the density, the static pressure and the Mach number respectively. We analyse the different error terms that appear in the ETOS when we choose to use either STEGER-WARMING splitting (SW) or the Van LEER splitting (VL). In both cases, we suppose $0 \leq M \leq 2$. The ETOS can be written:

$$(U_\ell)_t + (F_\ell^+)_x + (F_\ell^-)_x = \Delta x^2 \sum_{j=1}^{2} \left| \mathcal{E}_\ell^j(\sigma,M)(U_j)_{xxx} + E_\ell^j(\sigma,M,\alpha)M_x(U_j)_{xx} + e_\ell^j(\sigma,M,\alpha)(U_j)_x^3 \right| \tag{4}$$

$$+ 0(\Delta x^3); \quad \ell = 1,2 \, .$$

We can make a study of the nature of the partial differential equation (4) by considering the effect of the odd and even higher derivatives.

The terms $\mathcal{E}_\ell^{\,j}(\sigma,M)(U_j)_{xxx}$ are of a dispersive nature and lead to dispersive oscillations in the numerical solution. Fig. 1.a shows the evolution of $\mathcal{E}_1^{\,j}$ against M for two values of the Courant number. For both CFL values, the dispersive errors remain weak and the choice of the flux splitting does not have a great effect on their evolutions. The curves are continuous with VL splitting, whereas, with SW splitting, it appears to be a discontinuity of $\mathcal{E}_1^2$ due to the non-continuity of $dF^{\pm}/dU$ at M=1 with this splitting.

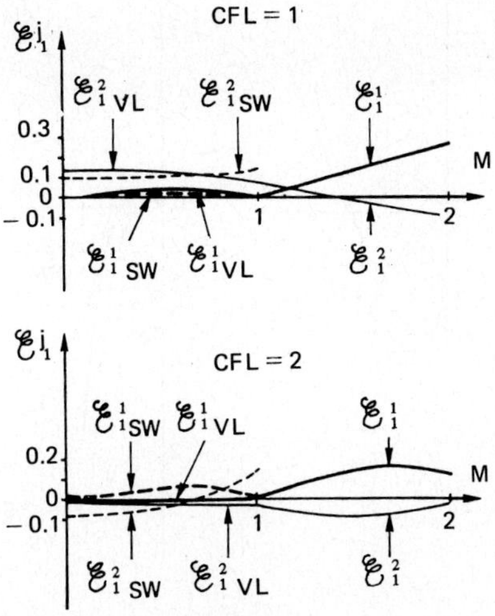

CFL = 1
M
0.3
0.1
0
– 0.1
1
2
CFL = 2
0.2
0
– 0.1
1
2

Fig. 1a

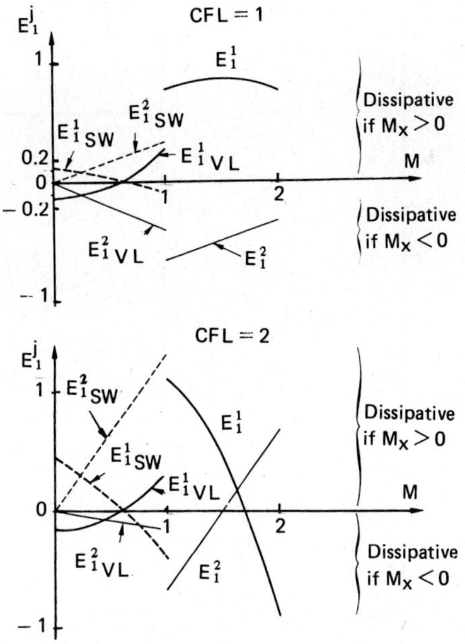

CFL = 1
1
0.2
0
– 0.2
– 1
M
1
2
Dissipative if $M_x > 0$
Dissipative if $M_x < 0$
CFL = 2
1
0
– 1
M
1
2
Dissipative if $M_x > 0$
Dissipative if $M_x < 0$

Fig. 1b

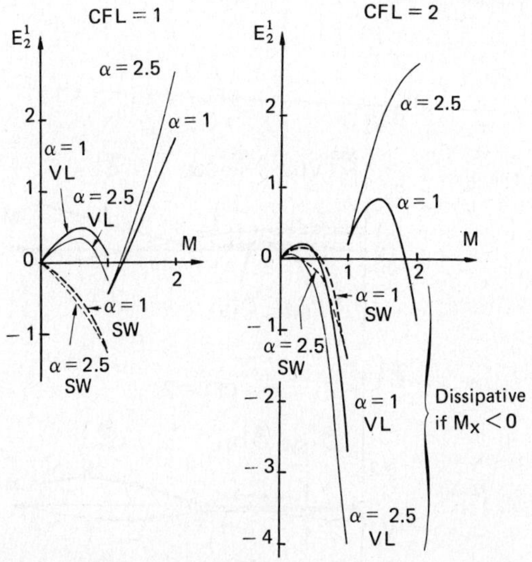

Fig. 1c

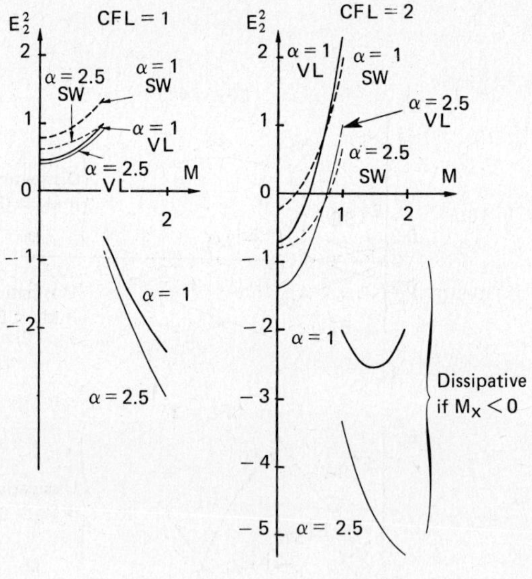

Fig. 1d

FIG. 1. Evolution of the different dispersive and dissipative error terms.
VL = Van LEER, SW = STEGER-WARMING

The terms $T_\ell = E_\ell{}^j(\sigma,M,\alpha)M_x(U_j)_{xx}$ can be defined as dissipative terms if $E_\ell{}^j(\sigma,M,\alpha)M_x>0$ and antidissipative if $E_\ell{}^j(\sigma,M,\alpha)M_x<0$. Therefore, according to the sign of $E_\ell{}^j M_x$, these terms T_ℓ can have a good behaviour by damping the contingent oscillations created by the dispersion errors or a bad effect by setting up new oscillations. $E_\ell{}^j$ are drawn on fig. 1b, 1c and 1d. Whatever the splitting (VL or SW), they are discontinuous at $M=1$, because higher order derivatives arise in these terms, and in this case even VL splitting is not capable of having the continuity of $E_\ell{}^j$ at M=1. This particularity explains the difficulties that can appear at the sonic point with these splittings. But, generally, the strength of the discontinuity is weaker with VL splitting.

$E_1{}^j$ do not depend on α. It is interesting to note that the use of VL or SW splitting for the first equation of (1) (3) gives opposite dissipative properties for a given acoustic phenomena. For example, in the case of a compression or shock wave ($M_x<0$) the scheme is, in the agregate, dissipative with VL splitting and antidissipative with SW splitting. In the case of a rarefaction ($M_x>0$), it is the contrary.

The terms $E_2{}^j$ depend on α. Whatever the splitting (VL or SW), the scheme is always more dissipative when the value of α increases (for example $\alpha=2.5$) in the case of a compression or shock wave. Contrary to this result, the scheme is more dissipative when α tends to zero for a rarefaction.

The third terms $e_\ell{}^j(\sigma,M,\alpha)(U_j)_x{}^3$ are differentials of an order lower than the previous terms and their effects are generally negligible.

To conclude this chapter, we can state that the choice of the flux splitting and the selection of the value of α have a great weight on the dissipative qualities of the schemes.

GENERALIZATION OF VAN LEER FLUX SPLITTING

This generalization has been possible thanks to the following remark. It is relatively rare to have a Mach number that reaches the values -1 and 1 in a flow. Generally, we have a main flow where $0\le M<\infty$ with or without some secondary flows where the Mach number is limited in the lower values by M_{inf} such as $-1<M_{inf}$ (in a great number of flows, $|M_{inf}|$ is small compared to 1). In these conditions, it is possible to define the following parametric ε-splitting when the one-dimensional Euler equations are considered:

$M_{inf} \le M \le 1$

$$\left\{\begin{aligned} F_1^- &= -\frac{\rho c}{4}(1+\varepsilon)(M-1)^2 \\ F_2^- &= -\frac{c}{\gamma}F_1^-[2-(\gamma-1)M] \\ F_3^- &= \frac{\gamma^2}{2(\gamma^2-1)}\frac{F_2^{-2}}{F_1^-} \end{aligned}\right. \qquad \left\{\begin{aligned} F_1^+ &= F_1 - F_1^- = \rho c M - F_1^- \\ F_2^+ &= F_2 - F_2^- = \rho c^2\left(M^2+\frac{1}{\gamma}\right) - F_2^- \\ F_3^+ &= F_3 - F_3^- = \frac{\rho c^3 M}{2(\gamma-1)}[2+(\gamma-1)M^2] - F_3^- . \end{aligned}\right. \tag{5}$$

When $M\ge 1$ $F^+=F$ and $F^-=0$.

In these expressions, c and γ represent the sound speed and the specific heat ratio. The retained decomposition verifies the following conditions:

for $M_{inf} \le M < \infty$
(a) $F = F^+ + F^-$
(b) All eigenvalues of dF^+/dU are ≥ 0
All eigenvalues of dF^-/dU are ≤ 0
(c) F^+ and F^- continuous
with $F^+ = F$ for $M \ge 1$
(d) $dF^\pm/dU$ are continuous everywhere
(e) dF^-/dU has one eigenvalue that vanishes for $M < 1$
(f) $F^\pm$ must be a polynomial in M with the lowest possible degree.

It is difficult to show that the condition (b) is respected for the Euler equations because of the complexity of the calculations. Nevertheless, in the case of the isothermal flow ($\gamma = 1$), it is possible to demonstrate that this condition is respected for $0 \le \varepsilon \le 5/3$ (when $M_{inf} = 0$). It is thought that this result can be extended to the case $\gamma = 1.4$. When $\varepsilon = 0$, the parametric flux splitting degenerates to Van LEER splitting. The evolution of the eigenvalues of SW, VL and ε-splittings is drawn in fig. 2 for $\gamma = 1$. Figure 3 shows the evolution of the components of F, F^+ and F^- with STEGER-WARMING, Van LEER and the parametric splittings ($\varepsilon = 0.2$) when $\gamma = 1.4$.

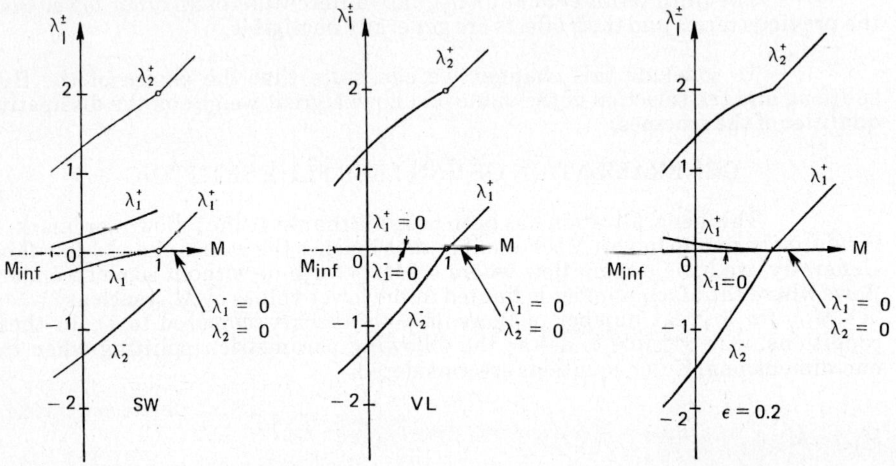

FIG. 2. Evolution of the eigenvalues of the partial fluxes against Mach number ($\gamma = 1$).

With this generalization, the merit of the continuity of $dF^\pm/dU$ for $M = 1$ is kept. But more particularly, it becomes possible to define the decomposition that gives better numerical results in the zones where the Mach number is weaker (≤ 0.6). The ε-splitting keeps the symmetry with respect to M. This property is verified easily if we take the following splitting when $-1 \le M \le -M_{inf}$:

$$\begin{cases} F_1^+ = \dfrac{\rho c}{4}(1+\varepsilon)(M+1)^2 \\ F_2^+ = \dfrac{c}{\gamma} F_1^+ [2-(\gamma-1)M] \\ F_3^+ = \dfrac{\gamma^2}{2(\gamma^2-1)} \dfrac{F_2^{+2}}{F_1^+} \end{cases} \quad and \quad F_\ell^- = F_\ell - F_\ell^+.$$

In the special flows where the Mach number can be less than -1 and greater than 1, it is possible nevertheless to define a parametric splitting that degenerates to VL splitting as well when $\varepsilon=0$ (this splitting is not presented here).

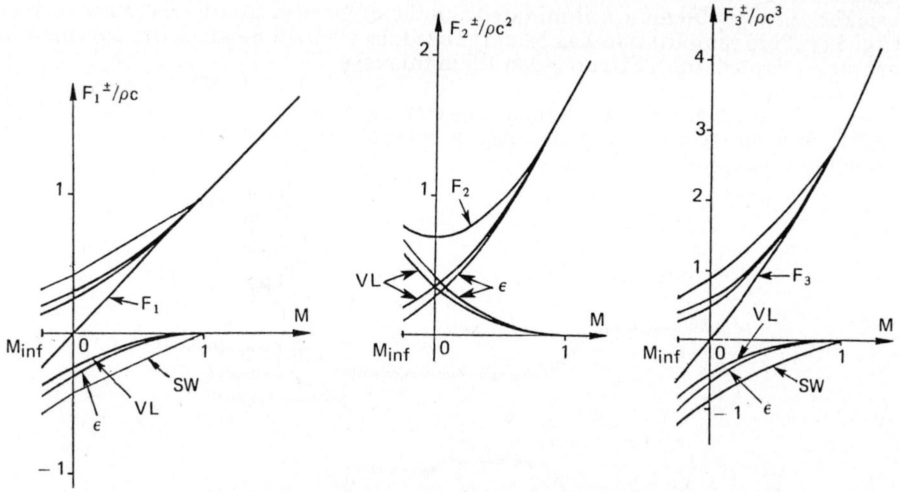

FIG. 3. Evolution of the partial fluxes. Euler equations
VL = Van LEER, SW = STEGER-WARMING, $\varepsilon=0.2$.

APPLICATIONS TO 1D AND 2D FLOWS

The new decomposition (5) associated with the $S_{FS}{}^{OPT}$ scheme ($\alpha=2.5$) has been compared with other methods:

$\alpha=1$, $\varepsilon=0$ (STEGER-WARMING scheme with Van LEER splitting)
$\alpha=2.5$, $\varepsilon=0$ ($S_{FS}{}^{OPT}$ scheme with Van LEER splitting)
$\alpha=2.5$, $\varepsilon=0.22$ ($S_{FS}{}^{OPT}$ scheme with parametric ε-splitting).

We have studied the shock-tube problem with a pressure ratio equal to 2.8 [6]. This case enables us to have a Mach number that remains relatively weak ($0 \leq M \leq 0.4$). The numerical solution is presented in fig. 4 with CFL = 1. On one hand, the solution is improved near the shock when $\alpha=2.5$ (figs. 4.a and 4.b), because the scheme S_{FS} is more dissipative with this value. But, on the other hand, the solution of the expansion wave is lightly damaged for the opposite reason: when we adapt the value of α to have a more dissipative scheme for $u_x<0$ (u represents the velocity), automatically the scheme becomes more antidissipative when $u_x>0$. In the present case, we solve this problem by using the two parameters (α,ε). The value of the first parameter α is adjusted to have a good solution when $u_x<0$ (shock or compression wave) and the second parameter ε enables us to have a correct representation of the rarefaction (Fig. 4.c).

The conjoining of $S_{FS}{}^{OPT}$ scheme with parametric flux-splitting has been applied to 2D steady flow inside a nozzle (fig. 5.a). Two parameters ε_1 and ε_2 appear. They are linked to the fluxes $\hat{F}^{\pm}$ and $\hat{G}^{\pm}$ respectively which are defined in the computational domain (see [7] for example). As in [8], some problems appear near the wall with ε_1 and $\varepsilon_2=0$, in particular strong oscillations of the numerical solution arise on the wall where the slope is strongest. The computation diverges rapidly (12th time step) (fig. 5.b). This is probably due to the strong gradient of the partial fluxes near the wall that are in this instance, sensitive to the different numerical treatments applied to the boundary mesh point and the following mesh points close to the wall. If the values of ε_1 and ε_2 are adjusted (mainly the parameter ε_2 included in the transverse fluxes), so that the gradients become weaker, these problems are eliminated and the numerical solution becomes correct (fig. 5.c). This computation has been realized for CFL = 0.8 (when the parameters are well-adapted, the CFL can reach 1.3 in this case).

To conclude this chapter, we can say that the generalization of Van LEER flux splitting makes possible a sensible improvement of 1D, 2D and probably 3D numerical solutions.

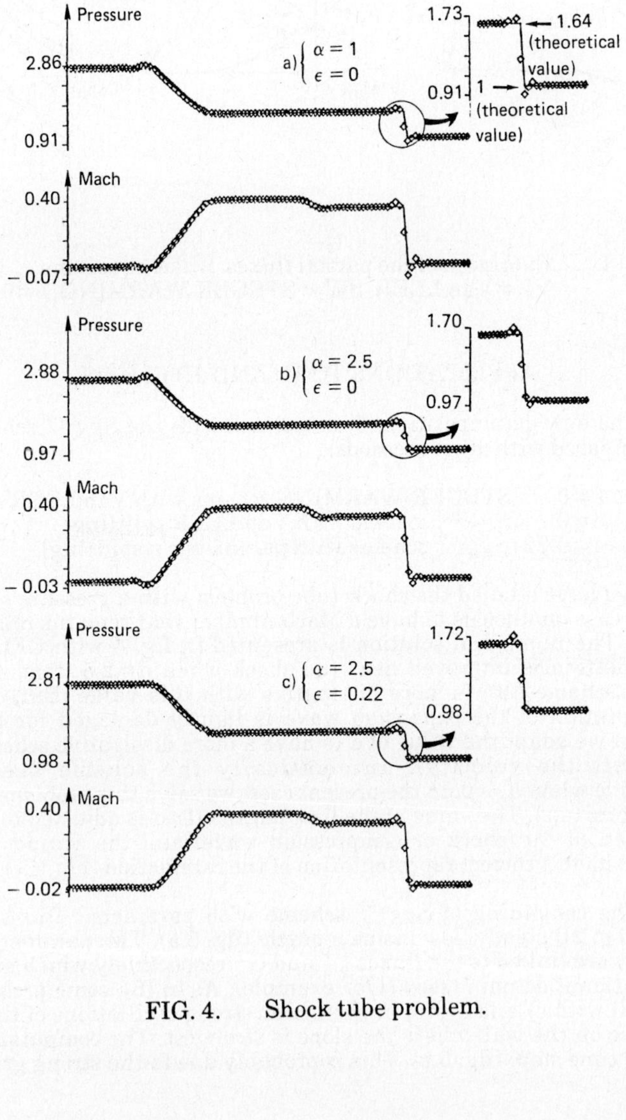

FIG. 4. Shock tube problem.

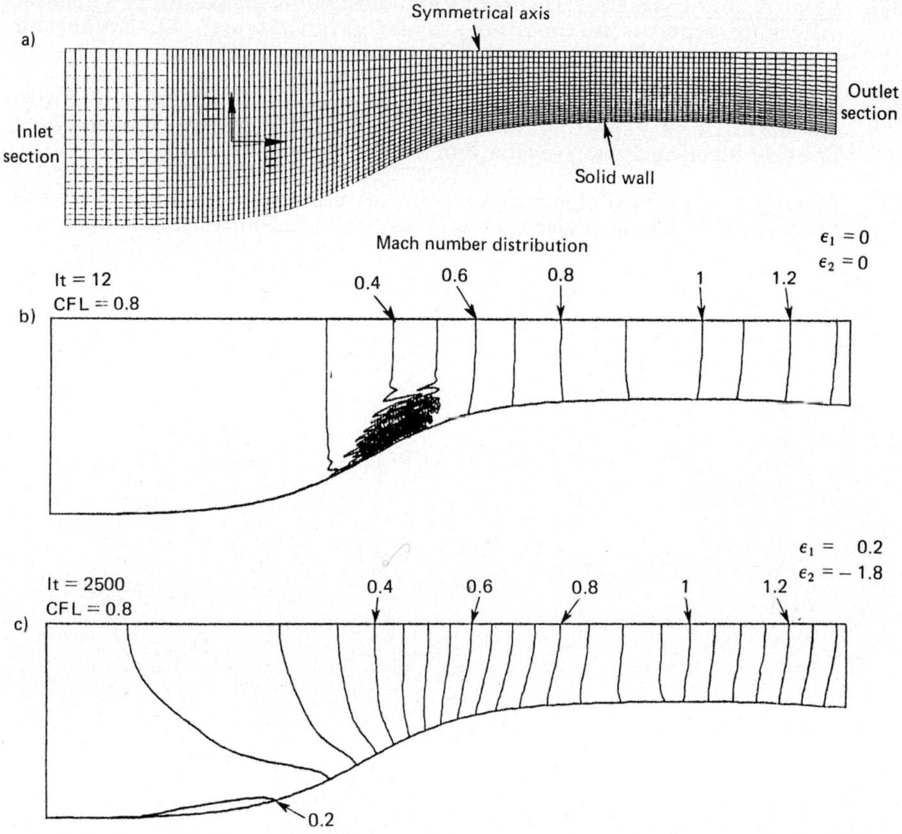

FIG. 5. Flow in a nozzle - Mach number distribution.

REFERENCES

[1] Steger J.L. and Warming R.F., "Flux vector splitting of the inviscid gasdynamic equations with application to finite-difference methods". J. Comp. Phys., Vol. 40, 1981, pp. 263-293.

[2] Van Leer B., "Flux vector splitting for the Euler equations". Von Karman Institute, Rhode St Genèse, 1983.

[3] Billet G., "A class of flux-splitting explicit second-order finite-difference schemes and application to inviscid gasdynamic equations". Eng. Comp. 1986, Vol. 3, pp. 312-316.

[4] Laval P. and Billet G., "Recent developments in finite-difference methods for the computation of transient flows". In Numerical Methods for Transient and Coupled Problems. Chapter 5, John Wiley and Sons Ltd., 1987.

[5] Peyret R. and Taylor T.D., "Computational methods for fluid flow". Springer Series in Computational Physics, 1983.

[6] Lerat A. and Peyret R., "Dispersive and dissipative properties of a class of difference schemes for nonlinear hyperbolic systems". La Recherche Aérospatiale n° 1975-2, pp. 61-79. ESA translation TT 205.

[7] Anderson W.K., Thomas J.L. and Van Leer B., "A comparison of finite volume flux vector splittings for the Euler equations". AIAA 23rd Aerospace Sciences Meeting, Reno, Nevada, Jan. 1985.

[8] Marx Y. , Etudes d'algorithmes pour les équations de Navier-Stokes compressible". Thèse de Doctorat de l'Université de Nantes, juillet 1987.

COMPUTATION OF INVISCID VORTICAL FLOWS IN PISTON ENGINES

B. Binninger, M. Jeschke, H. Henke, D. Hänel
Aerodynamisches Institut, RWTH Aachen
5100 Aachen, West Germany

SUMMARY

Two and three-dimensional vortical flow in a cylinder of a piston engine is investigated by means of finite-difference solutions of the Euler equations. Since both, the physical understanding of piston flows is far from complete and adequate computational methods for such complex processes are missing, the restriction to inviscid flow is considered as a first step to achieve basic insight into the large-scale vortical motion of the flow.The discretization of the conservation equations is carried out in a time-dependent, node-centred grid. Central differences are used to approximate the spatial derivations. For the integration in time two methods are applied, an implicit factorization scheme for plane and axisymmetric flows and an explicit Runge-Kutta time-stepping scheme for the three-dimensional flow.The numerical results for plane flow are compared with experiments using Mach-Zehnder-interferometry. The comparison confirms that the results obtained with the Euler equations reflect essential features of the flow in the cylinder of piston engines. For three-dimensional flow two examples are chosen to discuss the influence of off-centred valves and of the shape of the piston crown on the onset of the swirling flow during the intake stroke and its subsequent development during the compression stroke.

INTRODUCTION

The flow in piston engines is governed by a variety of complex physical processes like turbulent interactions, inhomogeneity and mixture, as well as chemical reactions. Even though the investigation of this paper is restricted to the cold flow of homogeneous gases, various time and length scales are involved in this problem.Large time scales are prescribed by the speed of the engine. The smallest can be expressed in terms of the speed of sound. In general, numerical integration techniques cover this range of time scales, however, the problem becomes more challenging in space dimensions. Large vortex structures are limited by the size of the cylinder, the shape of the inlet or the valve and the piston. These structures are intermingled with turbulent fluid motion especially near shear layers, for example the jet like flow into the cylinder, or close to walls. Up to now it is not feasible to resolve the computational domain down to scales given by the Kolmogoroff-length. Especially in three-dimensional flow problems direct modeling is hindered by missing computer power. Instead turbulence modeling using the time averaged equations of motion with additional closure assumptions of algebraic or differential type is widely employed. A survey of such investigation is recently given by Heywood [1]. The applicability of such models is questionable, since they are derived for stationary boundary layers or jets. Therefore the present work intentionally focuses on the formation and development of large-scale vortices during the intake and compression stroke. Hence friction can be neglected and a description of the flow should be given by the Euler equations for time-dependent compressible flow.

GOVERNING EQUATIONS

The Euler equations are used to compute the time-dependent compressible flow in a piston engine. The domain of integration is defined by the fixed cylinder walls and the moving piston of which the crown may have special shapes. A suitable curvilinear time-dependent grid is described by the coordinates

$$x = x(\xi,\eta)\ ,\ y = y(\xi,\eta)\ ,\ z = z(\xi,\eta,\zeta,\tau)\ ,\ t = \tau\ , \tag{1}$$

where ζ maps the axial direction, ξ and η the cross-section of the cylinder onto the computational domain. The conservative form of the Euler equations then reads

$$\bar{U}_\tau + \bar{F}_\xi + \bar{G}_\eta + \bar{H}_\zeta = 0\ . \tag{2.1}$$

The vector

$$\bar{U} = J(\ \varrho,\ \varrho u,\ \varrho v,\ \varrho w,\ e\ )^T \tag{2.2}$$

is the vector of the conservative variables, $J = z_\zeta\,(x_\xi y_\eta - x_\eta y_\xi)$ the Jacobian of the transformation, and $\bar{F}$, $\bar{G}$, $\bar{H}$ are the corresponding fluxes with the contravariant velocities $\bar{u}$, $\bar{v}$, $\bar{w}$.

$$\bar{F} = \begin{pmatrix} \varrho\bar{u} \\ \varrho u\bar{u} + y_\eta z_\zeta p \\ \varrho v\bar{u} - x_\eta z_\zeta p \\ \varrho w\bar{u} \\ (e+p)\bar{u} \end{pmatrix},\quad \bar{G} = \begin{pmatrix} \varrho\bar{v} \\ \varrho u\bar{v} - y_\xi z_\zeta p \\ \varrho v\bar{v} + x_\xi z_\zeta p \\ \varrho w\bar{v} \\ (e+p)\bar{v} \end{pmatrix},\quad \bar{H} = \begin{pmatrix} \varrho\bar{w} \\ \varrho u\bar{w} + (y_\xi z_\eta - y_\eta z_\xi)\,p \\ \varrho v\bar{w} - (x_\xi z_\eta - x_\eta z_\xi)\,p \\ \varrho w\bar{w} + (x_\xi y_\eta - x_\eta y_\xi)\,p \\ (e+p)\bar{w} + (x_\xi y_\eta - x_\eta y_\xi)\,z_\tau p \end{pmatrix} \tag{2.3}$$

with

$$\bar{u} = (y_\eta u - x_\eta v)\,z_\zeta$$
$$\bar{v} = (-y_\xi u + x_\xi v)\,z_\zeta$$
$$\bar{w} = (y_\xi z_\eta - y_\eta z_\xi)u - (x_\xi z_\eta - x_\eta z_\xi)v + (-z_\tau + w)(x_\xi y_\eta - x_\eta y_\xi)\ .$$

The system of equations is closed with the equation of state for ideal gas.

$$p = (\kappa-1)\left(e - \frac{1}{2}\varrho(u^2 + v^2 + w^2)\right)\ . \tag{2.4}$$

For plane flow the term $\bar{F}_\xi$ vanishes, whereas for axisymmetric flow ξ and η are chosen as the circumferential and the radial direction respectively, so that $\bar{F}_\xi$ contributes together with some separated parts of $\bar{G}_\eta$ to a source term $\bar{Q}$ containing the curvature terms [3]

$$\bar{U}_\tau + \bar{G}_\eta + \bar{H}_\zeta = \bar{Q}$$

with

$$\bar{Q} = \frac{J}{\eta}\left(\varrho v,\ \varrho v^2 - \varrho w^2,\ 2\varrho vw,\ \varrho uv,\ (e+p)v\right)^T\ . \tag{3}$$

METHOD OF SOLUTION

For the discretization of the conservation equation a node-centred mesh is used with the variables and the geometry defined in one node point. Central differencing is used for the spatial derivatives, what is well suited for the considered low Mach number flow in piston engines. To avoid odd-even decoupling, fourth order damping terms are added. Thus for one direction the flux-balance reads e.g.

$$\left.\frac{\partial \bar{F}}{\partial \xi}\right|_j \longrightarrow \frac{\bar{F}_{j+1} - \bar{F}_{j-1}}{2\Delta\xi} + D_{E_\xi} \tag{4}$$

with

$$D_{E_\xi} = \varepsilon_E\, J\Delta\xi^4 \frac{\partial^4}{\partial\xi^4}(J^{-1}\bar{U})\ . \tag{5}$$

applied to the physical variable $J^{-1}\bar{U}$. For the resolution in time two methods are used, originally an implicit factorization method for the plane and axisymmetric flow and an explicit Runge-Kutta time-stepping scheme for three-dimensional calculations.

Implicit Factorization Method

The implicit factorization method used for the plane and axisymmetric calculations is based on the work of Beam and Warming [2] and was extended to axisymmetric flow in [3] by introducing source terms. For such flows the basic system of equations is given by

$$(I + \Delta\tau \frac{\partial \bar{A}^n}{\partial \eta} + D_{I_\eta})(I + \Delta\tau \frac{\partial \bar{B}^n}{\partial \zeta} + \Delta\tau \bar{C}^n + D_{I_\zeta})\Delta\bar{U}^n = = -\Delta\tau (\frac{\partial \bar{G}}{\partial \eta} + \frac{\partial \bar{H}}{\partial \zeta} + \bar{Q})^n - D_{E_\eta} - D_{E_\zeta} \quad , \quad \Delta\bar{U} = \bar{U}^{n+1} - \bar{U}^n . \tag{6}$$

with $\bar{A}$, $\bar{B}$, $\bar{C}$ the corresponding Jacobians of the fluxes $\bar{G}$ and $\bar{H}$, as well as the source term $\bar{Q}$ which is treated implicitly. Additional damping terms $D_{I\zeta}$ and $D_{I\eta}$ are included at the left hand side to compensate the stability restriction arising from the explicit damping terms. The tridiagonal structure of the scheme is retained by choosing the implicit damping as e.g.

$$D_{I_\eta} = -\varepsilon_I J \Delta\eta^2 \frac{\partial}{\partial \eta^2}(J^{-1}\bar{U}) . \tag{7}$$

Stability is ensured, if

$$\varepsilon_E \leq \frac{1}{16}(1 + \varepsilon_I)^2 . \tag{8}$$

Commonly $\varepsilon_I = 2\varepsilon_E$ was chosen with $\varepsilon_E = 0(\Delta\tau)$. A decoupling of that 5x5 block-tridiagonal system into 5 scalar equation is achieved by diagonalization of the matrices $\bar{A}$ and $\bar{B}$ using a similarity transformation [4].

$$\bar{A} = T_\eta \Lambda_\eta T_\eta^{-1} \quad , \quad \bar{B} = T_\zeta \Lambda_\zeta T_\zeta^{-1} \tag{9}$$

where Λ_η and Λ_ζ are diagonal matrices containing the eigenvalues of the Jacobian $\bar{A}$ and $\bar{B}$. Assuming the matrices T_η and T_ζ as locally constant the decoupled scheme reads

$$T_\eta^n (I + \Delta\tau \frac{\partial}{\partial \eta}(\Lambda_\eta^n) + D_{I_\eta})(T_\eta^{-1} T_\zeta)^n (I + \Delta\tau \frac{\partial}{\partial \zeta}(\Lambda_\zeta^n) + D_{I_\zeta})(T_\zeta^{-1})^n \Delta\bar{U}^n = = -\Delta\tau (\frac{\partial \bar{G}}{\partial \eta} + \frac{\partial \bar{H}}{\partial \zeta} + \bar{Q})^n - \Delta\tau \bar{H}^{n-1} . \tag{10}$$

Now the source term is included only in the explicit part of the equation, because its eigenvalues differ from those of the Jacobian $\bar{B}$ [3], so that an implicit treatment will hinder the decoupling of the system. The accuracy of first order in time is not affected as a comparison with the scheme (4.1) has shown.

Runge-Kutta Time-Stepping Scheme

In order to compute more realistic configurations of piston engines three-dimensional flows have to be considered. An extension of the proceeding factorization method, successfully used for the stiff axisymmetric equations, leads to stability restrictions in three dimensions due to the factorization error. Hence an explicit time-stepping scheme was chosen for three-dimensional computations. Since the restricted time steps of this method are of the same order as those required for the temporal accuracy of the computations, the efficiency is not impaired by using explicit instead of implicit schemes. Furthermore the simplicity and the suitability with regard to the architecture of vector computers gave preference to explicit methods.

Recently Runge-Kutta time-stepping schemes are brought up [5]. The method used in this paper belongs to a class of Runge-Kutta like schemes given by the following sequential algorithm.

$$\begin{aligned}
\bar{U}^{(0)} &= \bar{U}^{n} \\
&\vdots \\
\bar{U}^{(\nu)} &= \bar{U}^{(0)} - \alpha_\nu \Delta\tau \left(\bar{F}(\bar{U}^{(0)})_\xi + \bar{G}(\bar{U}^{(0)})_\eta + \bar{H}(\bar{U}^{(0)})_\zeta + D_E(\bar{U}^{(\mu)}) \right) \\
&\vdots \\
\bar{U}^{n+1} &= \bar{U}^{(N)} \qquad \mu = \min(\nu, M) \ , \ M \leq 5 \ .
\end{aligned} \tag{11.1}$$

In contrast to the classical four step Runge-Kutta Scheme this sequence requires less computer storage, but on the other hand the time accuracy for non-linear fluxes $\bar{F}$, $\bar{G}$ and $\bar{H}$ is limited to the second order. In this case the coefficients α_N and α_{N-1} are prescribed as

$$\alpha_N = 1 \qquad \alpha_{N-1} = \frac{1}{2} \ . \tag{11.2}$$

A stepping sequence with $N = 5$ has three free parameters α_3, α_2, α_1 left which serve to improve the stability properties of the scheme. Vichnevetsky [6] has shown that the maximum CFL-number for a stable algorithm of a scalar test problem is given by $CFL = N - 1$, if α_3, α_2 and α_1 are suitably chosen.

$$\alpha_3 = 3/8 \qquad \alpha_2 = 1/6 \qquad \alpha_1 = 1/4 \tag{11.3}$$

The stability property of a five step scheme are reflected by a von Neumann stability analysis for the scalar test equation

$$u_t + a u_x + \frac{\varepsilon}{\Delta t} \Delta x^4 u_{xxxx} = 0 \ , \quad a = \text{const} \ . \tag{12}$$

Fig. 1 shows the amount of the amplification factor G as a function of the wave angle $\theta = k\Delta x$ in the case of vanishing damping term $\varepsilon = 0$ for several CFL-numbers. Thereafter stability is gained up to $CFL = 4$.

The behaviour of the amplification factor with additional damping is plotted in Fig. 2 and Fig. 3. If the damping terms are evaluated at each step of sequence (11), $\mu = \nu - 1$, Fig. 2 reveals that complete damping of the highest frequences is not available. The function $G(\theta)$ however is smoothed and even reaches zero for the highest frequency, Fig. 3 , if

$$\mu = \min(\nu, 1) . \tag{13}$$

By freezing the damping term a considerable amount of computing time is saved, too.

BOUNDARY CONDITIONS

The boundary conditions are formulated explicitly. At rigid walls ξ, η or ζ = const vanishing normal velocities, $\bar{u}$, $\bar{v}$ or $\bar{w}$ Eq. (2.3.) respectively are required, whereas tangential velocities are linearly extrapolated. Likewise density and pressure are found by linear extrapolation except at moving walls where the value of the pressure is gained by evaluating the momentum equation in normal direction with respect to the wall.

The boundary conditions at open borders are put in order with the one-dimensional theory of characteristics. For incoming flow during the intake stroke density and velocity are prescribed, whereas the pressure is extrapolated from the inner region. The velocity normal to the cylinder head is chosen in a manner, so that the global density does not change during the intake stroke, the direction of the velocity is given by the angle of the valve seat. During the exhaust stroke only the mass flux is prescribed, all other flow quantities are extrapolated from the inner region.

RESULTS

Plane and Axisymmetric Flow

These preliminary investigations are used to study the influence of the piston shape and the effects arising from the inclination of the incoming jet and of the compression ratio on the onset and development of the vortical flow in the cylinder [3], [7]. The plane flow was computed to compare with the optical experiments done by Mach-Zehnder interferometry. The numerical simulation of the plane flow revealed that vortex patterns generated during the intake stroke are conserved until the end of the compression stroke, whereas in axisymmetric flows the flow patterns change during the compression stroke basicly due to vortex merging. Representative for the preceding investigations two examples for numerical and experimental results are discussed in the following section. The experimental set-up consists of a cylinder and a piston with rectangular cross-section for the plane flow simulation.

In the first case, the flow field in a cylinder with a step piston is investigated. The valve in the shape of a slit is located near the lower piston wall and aligned parallel to the axis of the piston which moves sinusoidally with an engine speed of 510 rpm. The stroke and the height of the piston is chosen to be equal, the compression ratio is $\varepsilon = 4.3$. The sequence of photographs in Fig.6 compares measured and computed flow flields during the intake stroke. In each case lines of constant density are presented. The experimental investigation shows that the incoming jet forms a fungoid structure at an early stage of the intake stroke (26° crank angle). Then the jet is deflected at the piston crown and rolls up to a counter-clockwise rotating vortex in the lower part of he flow field. At the same time a secondary vortex is generated at the corner of the piston step due to its accelerated movement. This secondary vortex is fed by the jet-like flow which is directed from the convex corner of the piston to the cylinder head. Finally the coarse structure of the flow field consists of two vortices with opposite rotation.The numerical prediction of the flow is in qualitative agreement with the experiments, although the numerical boundary conditions only roughly agree with the experimental conditions, as can be seen from the underestimation of the strength of the incoming jet. At the end of the intake stroke the lower vortex settles underneath the step of the piston with its strength increasing during the compression stroke. Several disturbing influences may have prevented to observe this vortex in the experiments up to now.

In the second example, Fig. 7, the investigation dealt with a flat piston in connection with an incoming jet inclined at 45°. The compression ratio now is $\varepsilon = 3.7$, the engine speed is 526 rpm. A large vortex is generated during the intake stroke which survives and even enlarges during the compression stroke. The numerical calculation predicts this vortex and its enlargement in close agreement with the experiment.

In the preceding computations a mesh with 83x53 grid points similar to those in Fig. 4 was used. To solve the Euler equations the diagonalized version of the factorized method Eq. (10) was employed. Its time accuracy was confirmed by comparison with the results obtained from the second order accurate Runge-Kutta scheme.

Three-dimensional Flow

Up to now several configurations have been investigated, e. g. flat and shaped pistons, and centred and off-centred valves. But the work is at an early stage, and the investigations have to be intensified in future. In all cases considered, symmetry about a plane containing the axis of the cylinder and the valve was assumed for saving computer time and storage. As indicated by Eq. (1) the computational grid is divided into a stationary part discretizing the cross-section of the cylinder and a time-dependent part mapping the axial direction onto the computational domain. The discretization of the cross-section is based on conformal mapping as given in [8] with additional smoothing and orthogonalization, Fig. 5. In axial direction the meshes are similar to those used for the two-dimensional flow, Fig.4. The number of points are 56x23x53.

The presented results concern with the onset of circumferential or swirl flow due to the off-centred intake valve. The case of a disc piston on the one side and a shaped piston on the other side shall be discussed. During the intake stroke (72° crank angle) the incoming flow forms an asymmetric primary toroidal vortex, Fig. 8 and 12, and a smaller secondary vortex in the corners at the cylinder head. The excentric location of the valve generates a significant swirl flow as can be seen in Fig. 9 and 10 in the case of a disc piston. This swirl flow strongly distorts the toroidal vortices which finally at a crank angle of 180° does not dominate the flow field any longer. The swirl flow is fully established during the compression stroke, Fig. 11.At the beginning the flow field with the shaped piston is similar to the one described above, Fig.12.The swirling flow, however, is now nonuniformly distributed over the cross-section of the clyinder and different at various axial locations, Fig. 13. Instead of a unique swirl flow which is formed during the intake stroke in the case of the disc piston, a complex system of vortices with their axis in circumferential direction occurs, Fig. 14. During the compression this system of toroidal vortices govern the flow, Fig. 15.

CONCLUSION

The time-dependent compressible inviscid flow in a cylinder of a piston engine is computed by finite difference schemes of the Euler equation. For plane flow the computed structure of large-scale vortices is compared with experimental results. The comparison confirms that some essential aspects of the flow can be described by the inviscid flow. First computations of three-dimensional flows representing more realistic configuration show the influence of off-centred valves and of the shape of the piston crown on the flow. Further questions related to stretching and tilting of vortex structures have not been answered yet. The symmetry condition assumed for the considered configurations have to be called in question. Calculations with a centred intake valve showed that the initial axisymmetric flow could not be maintained during the compression stroke. A possible influence on the solution may arise from the computational grid and has to be examined. Although special questions of practical interest can be answered through this investigation, efforts have to be initiated to include friction and turbulence in future investigations.

REFERENCES

[1] J. B. Heywood: Fluid Motion within the Cylinder of Internal Combustion Engines,The Freeman Scholar Lecture, Journal of Fluids Engineering, March 1987, Vol. 109/3.

[2] R. Beam, R. F. Warming: An Implicit Finite-Difference Algorithm for Hyperbolic Systems in Conservation-Law- Form, Journal of Comp. Physics, Sept. 1976, Vol. 22, pp. 87-110.

[3] H. Henke, D. Hänel: Numerical Solution of Gas Motion in Piston Engines. Ninth International Conference on Numerical Methods in Fluid Dynamics, 218, Springer-Verlag (1985), (Ed. Subbaramayer and J. P. Boujet).

[4] R. F. Warming, R. Beam, B. J. Hyett:Diagonalization and Simultaneous Symmetrization of the Gas-Dynamic Matrices. Mathematics of Computation, Vol. 29, Oct. 1975, pp.1037-1045.

[5] A. Jameson, W. Schmidt, E. Turkel: Numerical Solution of the Euler Equations by Finite Volume Methods Using Runge-Kutta Time-Stepping Schemes, AIAA-Paper 81-1259, 1981.

[6] R. Vichnevetsky: New Stability Theorems Concerning One-Step Numerical Methods for Ordinary Differential Equations. Math. Comp. Simulation, 25 (1983) pp. 199-205.

[7] H. Henke, D. Hänel: Numerical Simulation of Vortex Flow in Piston Engines. Proceedings of International Symposium on Diagnostics and Modelling of Combustion in Reciprocating Engines, Sept. 4-6, 1985, Tokyo, Japan.

[8] T. Berglind: Grid Generation around Car Configurations above a Flat Ground Plate Using Transfinite Interpolation. Flygtekniska Försöksanstalten, The Aeronautical Research Institute of Sweden, FFA Report 139, Stockholm (1985).

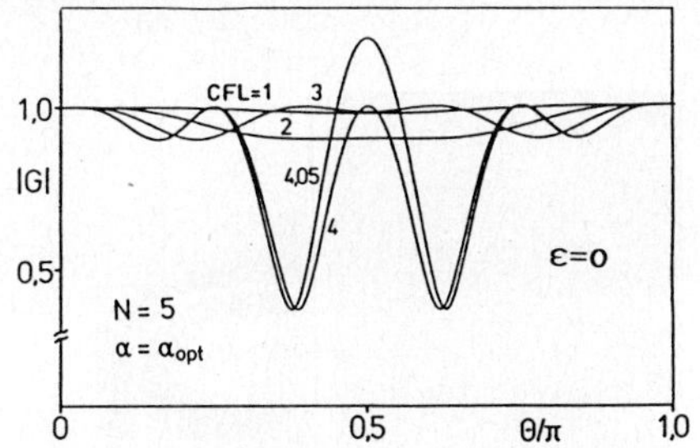

Fig. 1. Stability properties of the Runge-Kutta scheme Eq. (11.1) for linear equation (12) without damping.

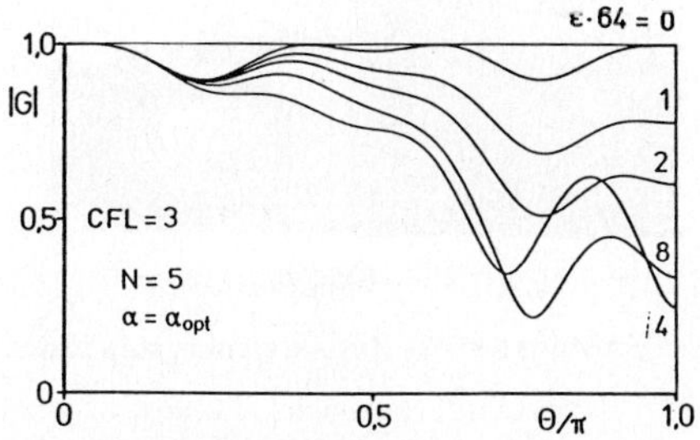

Fig. 2. Stability properties of system (11.1) with fourth order damping $\mu = v - 1$.

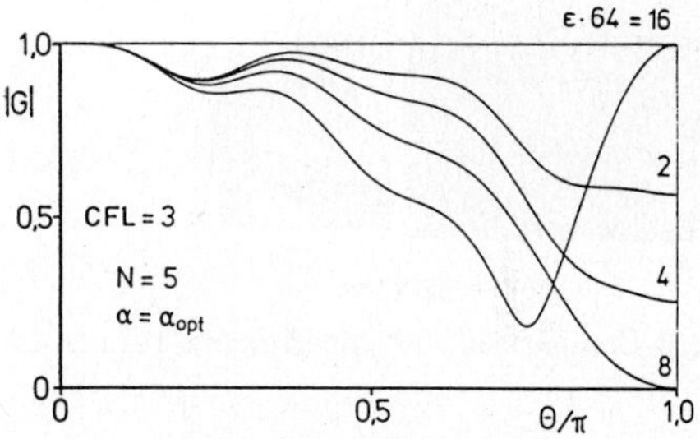

Fig. 3. Stability properties of system (11.1) with fourth order damping $\mu = \min(v, 1)$.

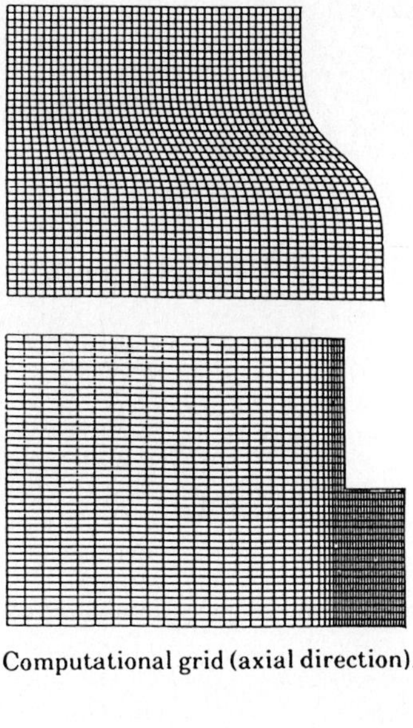

Fig. 4. Computational grid (axial direction).

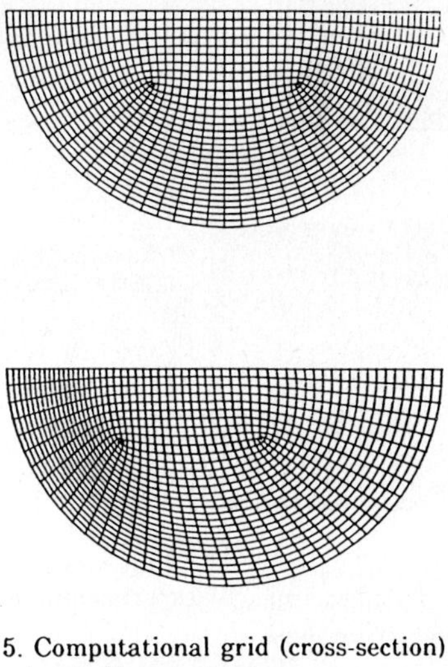

Fig. 5. Computational grid (cross-section) for centred and off-centred valves.

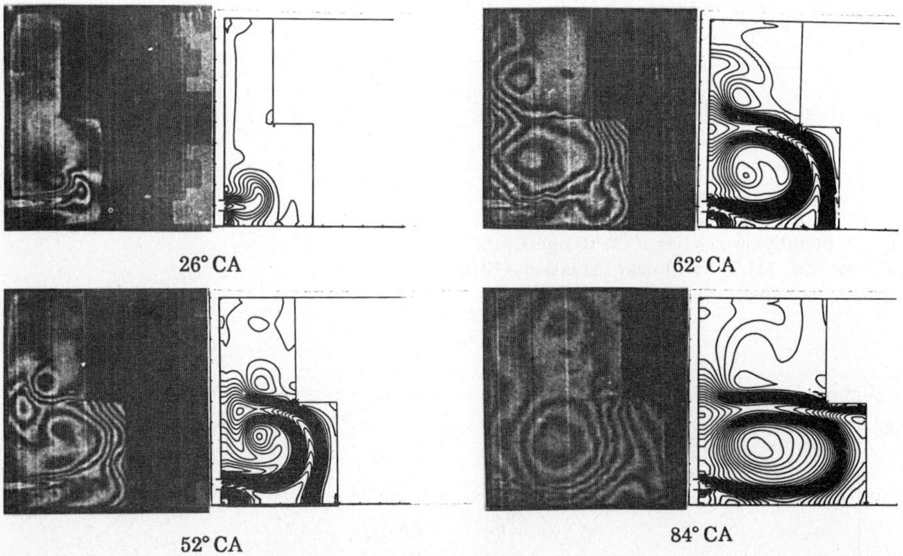

Fig. 6. Comparison with experimental results. Lines of constant density (Intake stroke, step piston).

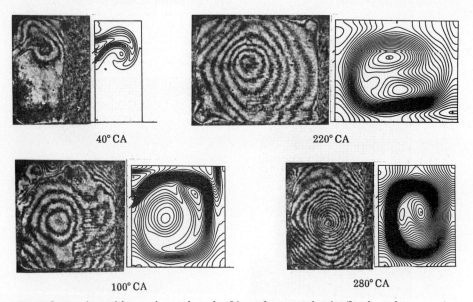

Fig. 7. Comparison with experimental results. Lines of constant density (Intake and compression stroke, flat piston).

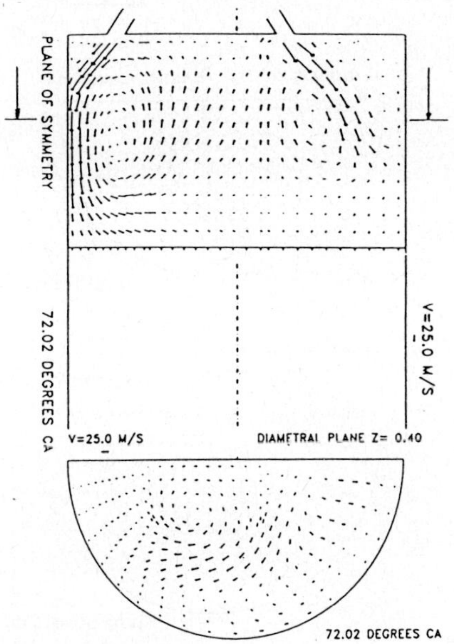

Fig.8

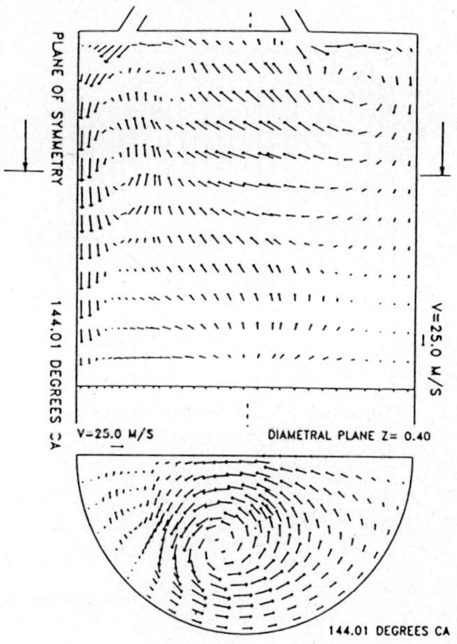

Fig. 9

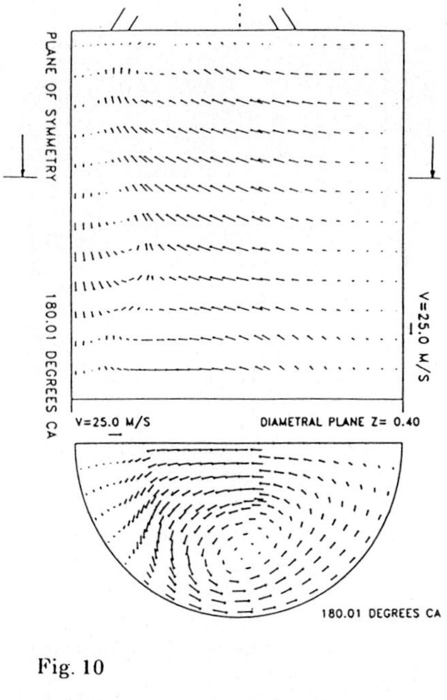

Fig. 10

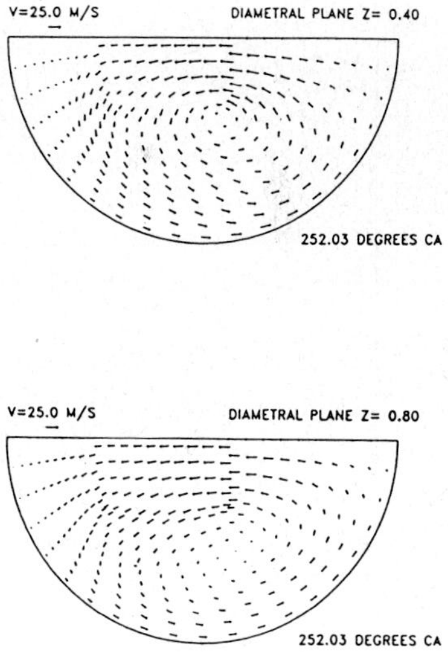

Fig. 11

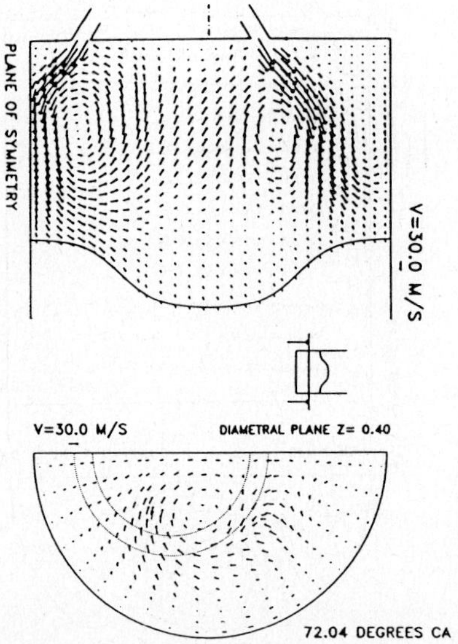

Fig. 12

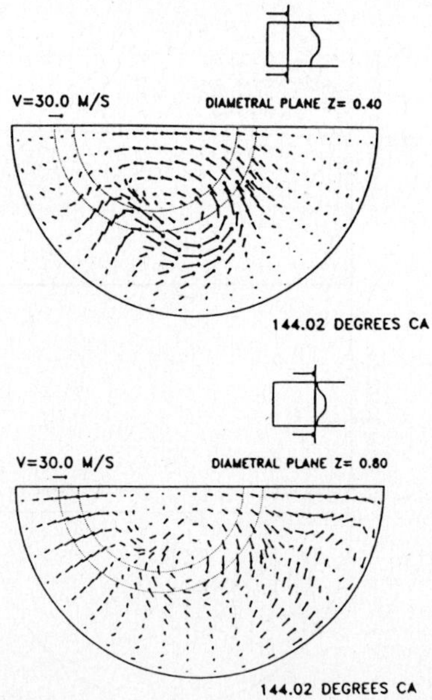

Fig. 13

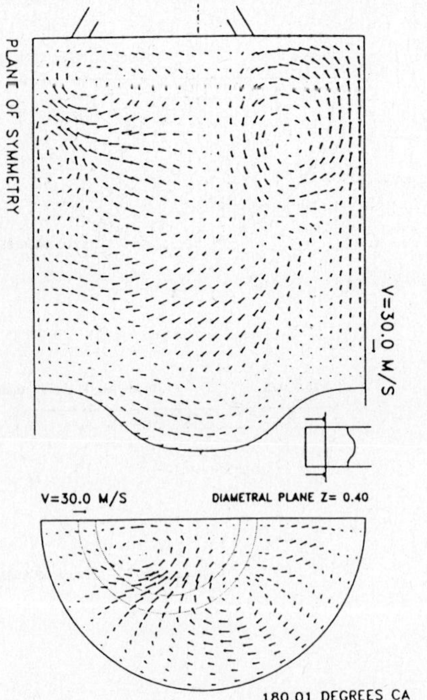

Fig. 14

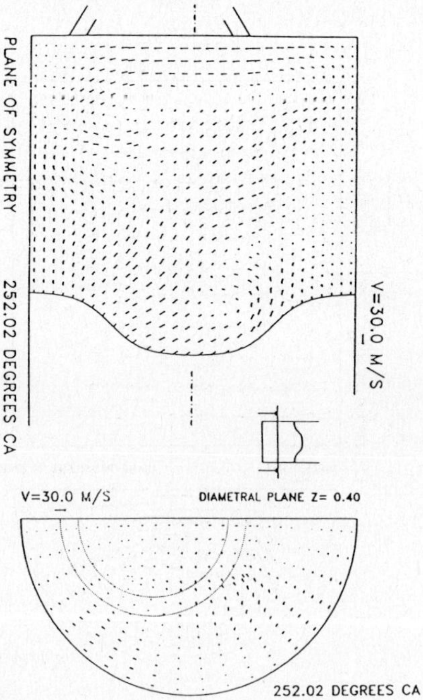

Fig. 15

CONVEXITY IN HYPERBOLIC PROBLEMS. APPLICATION TO A DISCONTINUOUS GALERKIN METHOD FOR THE RESOLUTION OF THE POLYDIMENSIONAL EULER EQUATIONS

F. BOURDEL*, PH. DELORME**, P.A. MAZET*
* ONERA-CERT, BP 4025, 31055 TOULOUSE CEDEX (FRANCE)
** ONERA, Division de l'Energétique, BP 72, 92322 CHATILLON CEDEX (FRANCE)

SUMMARY

For the physical hyperbolic problems one can exhibit a fundamental function of the entropic variables and of a space-time vector. In the case of Euler equations, this function can be expressed in a simple form and splitted into a convex function and a concave one, and it is possible to find a polydimensional scheme which generalizes the Courant scheme. Then we present some mono and bidimensional numerical results.

INTRODUCTION

When a hyperbolic system has a supplementary conservation law (on entropy) it is possible to define a function Σ^* of a space-time vector and the entropic variables. This function sounds having a very important place in those systems. Many of their properties can be interpreted by means of the convexity domain of Σ^* (as : Cauchy Kowalevska's characteristics, Rankine Hugoniot relations, inequation on the boundary, arrows of time, phenomenological relations, Courant scheme). We began the study of this function with a variational formulation where the convexity is essential (see Mazet [2]). We shall explain its properties in the first part. For the Euler equations, Σ^* can be written by using some ideas of the statistical mechanics (second Part). So it was possible to define a Galerkin discontinuous scheme, which is consistent with the weak equations and the entropy inequation, and to exhibit a global overestimation (Part III). This scheme looks like a polydimensional splitting method, but it is possible to imagine a space-time mesh refinement. In part IV numerical results will be presented.

I : DEFINITION AND PROPERTIES OF THE FUNDAMENTAL FUNCTION Σ^*

Notations : we shall use two manifolds :
- the space-time indexed with a latin index $i=0,n$ - 0 means time (dimension $n+1$)
- the state-space indexed with a greek index $\alpha=1,N$ (dimension N).

It is the space of the thermodynamical quantities which play a role in the considered physical problem (as : mass, momentum, energy in fluid mechanics). The system of partial differential equations (SPDE) obtained is written :

$$\partial_i f^{i,\alpha}(w) = g^{\alpha}(w) \quad \text{in } D' \qquad (1)$$
$$i=0,n \quad \alpha=1,N$$

(1) is a system of balance equations, the left-hand side is a space-time divergence.

We call phenomenological closure the relations

$f^{i,\alpha} = f^{i,\alpha}(w)$ or something equivalent.

In general we have $f^{o,\alpha}(w) = w^{\alpha}$

the SPDE is called hyperbolic if $\forall$ n, space vector $(\forall(n_i) i=1,3)$ the eigenvalues of $\frac{\partial f^{i,\alpha}}{\partial w^{\alpha}}$ are real.

* <u>Entropy, entropic variables, new phenomenological closure</u> :

(see Harten [8]) for the physical system the SPDE has a supplementary balance law :

$\exists\ \varphi_{\alpha}(w),\ S^{i}_{(w)}\ :\ \partial_i\ S^{i}(w) = \varphi_{\alpha}\ \partial_i\ f^{i,\alpha}$

$i=0,n \quad \alpha=1,N$, S^{o} convex in w

If such a law exists, the phenomenological closure verifies : $\frac{\partial S^{i}}{\partial f^{i,\alpha}} = \varphi_{\alpha}$,

φ_{α} are called entropic variables.

If we use the polar transform of $S^{o}(w)$, $S^{o*}(\varphi) = \sup_{w} \{w\ \varphi - S^{o}(w)\}$

$S^{o}(w)$ is convex in w so it is possible to define $w(\varphi)$ by $w^{\alpha} = \frac{\partial S^{o*}(\varphi)}{\partial \varphi}$

and the functions $S^{i*}(\varphi) = f^{i,\alpha}(w \circ \varphi)\varphi_{\alpha} - S^{i}(w \circ \varphi)$.

So the phenomenological closure can be defined with only n+1 functions

$S^{i*}(\varphi)$, S^{o*} convex in φ, $f^{i,\alpha} = \frac{\partial S^{i*}}{\partial \varphi_{\alpha}}$.

Let $\Sigma^{*}(\varphi,n)$ be : $\Sigma^{*}(\varphi,n) = n_i\ S^{i*}(\varphi)$

All the SPDE defined by Σ^{*} are hyperbolic because the eigenvalues of

$n_i \frac{\partial f^{i,\beta}}{\partial w^{\alpha}}$, $i=1,3$, $\alpha,\beta=1,N$

are the same as those of the matrix $\frac{\partial^2\ \Sigma^{*}}{\partial\varphi_{\alpha}\ \partial\varphi_{\beta}}(\varphi,w)$, where $n=(1,n_i)$, which are real because this matrix is symmetric.

* <u>Properties of the fundamental function $\Sigma^{*}(\varphi,n)$</u>

The arrows of time : if one studies the variance of the balance equations, i.e. how they change in another space-time frame, one finds another system of balance equations (cf. Delorme [4]), so : is the time only a geometric index ? In fact, not, because we have $S^{o*}(\varphi) = \Sigma^{*}(\varphi,n)$, where S^{o*} is the polar of the entropy in the new frame, and n the new time direction expressed in the old frame. So we can define a convex cone of the admissible "times" :

$$\mathcal{C}(\varphi) = \{n, \Sigma^{*}(\varphi,n) \text{ convex in } \varphi\} ;$$

the boundary of this cone is given by :

$\det\ [\frac{\partial^2\ \Sigma^{*}(\varphi,n)}{\partial\varphi_{\alpha}\ \partial\varphi_{\beta}}] = 0$, because if Σ^{*} is convex in φ the eigenvalues of the Hessian of Σ^{*} are positive.

n is the normal vector to a characteristic surface in the sense of Cauchy - Kowalevska (see [3])

$\mathcal{C}(\varphi)$ is the polar cone of the "future characteristic cone".

* Rankine-Hugoniot relations :

If $g(x^i) = 0 \quad i = 0,n$ is the equation of a discontinuity surface the R. & H. relations can be written. The gradient with respect to φ of $\Sigma^*(\varphi,\partial_i g)$ is constant through the jump (the gradient of g is a normal vector to the space-time discontinuity)

$$\frac{\partial \Sigma^*}{\partial \varphi_\alpha} = \frac{\partial S^{i*}}{\partial \varphi_\alpha} \partial_i g = f^{i,\alpha} \partial_i g.$$

* Boundary inequation

In order to describe the very difficult problem of the boundary conditions of the hyperbolic SPDE, AUDOUNET & MAZET and DUBOIS & LE FLOCH introduce the following boundary inequation :
If n is the outwards unit normal to a domain, φ_{in} the inside value, φ_{out} the outside one, we must have (cf. [1] & [7])

$$n_i \, [S^i_{out} - S^i_{in} - \varphi^{out}_\alpha \, (f^{i,\alpha}_{out} - f^{i,\alpha}_{in})] \leqslant 0 \qquad (2)$$

This inequation can be written

$$\boxed{\Sigma^*_{in} - \Sigma^*_{out} - \frac{\partial \Sigma^*}{\partial \varphi_\alpha}_{in} \; [\, \varphi^\alpha_{in} - \varphi^\alpha_{out} \,] \leqslant 0} \qquad (3)$$

(3) is the dual form of (2).
If the R. & H. relations are true : $n_i \, f^{i,\alpha}\,(out) = n_i \, f^{i,\alpha}\,(in)$, the boundary inequation gives the entropy inequation :

$$n_i \, S^i_{out} - n_i \, S^i_{in} \leqslant 0 \, .$$

For the linear SPDE it is very easy to see that the "good" boundary conditions are given by a convex-concave splitting of Σ^* (see below).

II : STUDY OF THE FUNDAMENTAL FUNCTION Σ^* FOR THE EULER SPDE

* the Σ^* function :

For the Euler equations in the 3 dimensional case we have n=3 and N=5. The w^α are mass(ϱ), momentum (q^i) and energy(e).

If v^i is the velocity $\frac{q_i}{\varrho}$ of the gas, the fluxes are :

$$f^i = \begin{pmatrix} \varrho \, v^i \\ \varrho \, v^i v^j + p \, \delta^{ij} \\ v^i \; (E + P) \end{pmatrix} \qquad i,j = 1,3$$

p is the pressure ; for polytropic gases, $p = (\gamma-1)\,[e - \varrho \frac{\Sigma v_i^2}{2}]$.
(γ is a constant which depends only from the gas).
The additional conservation law is given by :
$S^i = \varrho \, v^i \, G(P/_\varrho \gamma) \quad i = 0,3 \;, \; v^o = 1.$

S^o convex in w implies some conditions on the function G (see Harten [8]).
If we introduce the entropic variables :

$\varphi_\varrho, \varphi^i, \varphi_e$ we find the relations : $q^i \varphi_e + \varrho \varphi^i = 0$ (5)

$$\varrho\varphi_\varrho + \varphi_e . \varrho . \frac{V^2}{2} + \gamma \varphi_e [e - \varrho \frac{V^2}{2}] = S^o.$$

The general form of S^o is equivalent to :

S^o solution of the SPDE (5) ($\varphi^\alpha = \frac{\partial S^o}{\partial W^\alpha}$) and S^o convex in W.

In order to find the most general expression for Σ^*, we should calculate the entropic variables and eliminate w.

But it is easier to interpret (5) as a SPDE on S^{o*} ($W^\alpha = \frac{\partial S^{o*}}{\partial \varphi_\alpha}$), so

S^{o*} is a solution of (5) and is convex in φ.

By integrating (5) we find the general form of

$$S^{o*}(\varphi) = \frac{1}{(\varphi_e)^{1/\gamma-1}} \times H(\varphi_\varrho - \Sigma \frac{\varphi_i^2}{2\varphi_e})$$

with some conditions on H, which are necessary for the convexity of S^{o*}.

Σ (φ,n) is given by :

$$\Sigma^*(\varphi,n) = n_i S^{i*} = -\frac{\varphi^i n_i}{\varphi_e} \times \frac{1}{\varphi_e^{(\frac{\gamma}{\gamma-1})}} H(\varphi_\varrho - \sum_{j=1}^{3} \frac{(\varphi_j^2)}{2\varphi_e})$$

and $\varphi^o = -\varphi_e$ (i = 0,3).

The phenomenological closure is :

$$dS^{o*} = W^i d\varphi_i = \frac{H'}{H} [d\varphi_\varrho - \frac{\varphi_i}{\varphi_e} d\varphi^i + \frac{V^2}{2} d\varphi_e] - \frac{1}{\gamma-1} \frac{H}{\varphi_e^{(\frac{\gamma}{\gamma-1})}} d\varphi_e$$

$$\text{so } \varrho = \frac{H'}{H},\ E - \frac{1}{2}\varrho V^2 = -\frac{1}{\gamma-1} \frac{H}{\varphi_e^{(\frac{\gamma}{\gamma-1})}} \quad \text{(internal energy)}$$

$$\text{and} \quad V^i = \frac{\varphi_i}{\varphi_e}$$

$$dS^{i*} = -\frac{\varphi^i}{\varphi_e} [(\frac{H'}{H})(d\varphi_\varrho - \frac{\varphi_j}{\varphi_e} d\varphi_j + \frac{V^2}{2} d\varphi_e) - \frac{1}{(\gamma-1)} \frac{H}{\varphi_e^{(\frac{\gamma}{\gamma-1})}} d\varphi_e]$$

$$- \frac{H}{\varphi_e^{(\frac{\gamma}{\gamma-1})}} [d\varphi^i - \frac{\varphi^i}{\varphi_e} d\varphi_e]$$

$$\text{so we find } p = \frac{H}{\varphi_e^{(\frac{\gamma}{\gamma-1})}} = (\gamma-1)[E - \frac{1}{2}\varrho V^2].$$

* Statistical representation :
For the Euler equations, mass, momentum and energy can be defined through a statistical interpretation. Let us define the velocity distribution

function :

$$\theta\,(x^i,\ u^j,\ \eta),\ i\ j = 0,3$$

of $\begin{cases} x^i & : \text{space time coordinates} \\ u^j & : \text{velocity of the particles } (u^o = 1) \\ \eta & : \text{vibration (or rotation) velocity} \end{cases}$

as the density of number of particles whose velocities are u and η

so $W^\alpha = \int_{R^n} du \int_{R^+} d\eta \ \ K^\alpha(u,\eta)\ \theta$

$$\text{where } K^\alpha \text{ is : } \begin{bmatrix} 1 = u^o \\ u^i \\ \sum_{i=1}^{3} \frac{(u^i)^2}{2} + g(\eta) \end{bmatrix} \quad \text{(collision vector).}$$

We call $H = R^n \ X\ R^+$

The statistical definition of the entropy :

Let us consider the variational problem

$$\sup_{\varphi}\ \inf_{\theta} \int_H L(\theta) - \varphi_\alpha K^\alpha(u,\eta)\theta + \varphi_\alpha W^\alpha = S^o(W)$$

where

if L is a convex function, this problem is well-posed ;

$\hat\theta$ and $\hat\varphi$ are given by the stationarity equations :

$$\begin{cases} L'(\hat\theta) - \hat\varphi_\alpha K^\alpha = 0 & (6) \\ W^\alpha - \int_H K^\alpha\ \hat\theta = 0 & (7) \end{cases}$$

L is convex so (6) gives $\hat\theta = L^{*'}(\hat\varphi_\alpha K^\alpha)$

(L^* is the polar function of L)

and φ is given by the implicit equations

$W^\alpha = \int_H K^\alpha\ L^{*'}(\varphi_\beta K^\beta)$.

The Jacobian matrix :

$\frac{dW^\alpha}{d\varphi_\beta} = \int_H K^\alpha K^\beta L^{*''}(\varphi_\gamma . K^\gamma)$ is a positive matrix, as :

$$X_\alpha X_\beta \frac{\partial W^\alpha}{\partial \varphi_\beta} = \int (K^\beta X_\beta)^2 L^{*''} \geqslant 0 ,$$

and the equations (6) have a unique solution.

The gradient of $S^o(W)$ with respect to W is the vector φ :

$$dS = \int_H [L'(\hat\theta) - \varphi_\alpha k^\alpha]\, d\theta + [W^\alpha - \int_H k^\alpha\theta] + \hat\varphi_\alpha dW^\alpha$$

(6) & (7) give us the result : the entropic variables are $\hat\varphi$.

For physical systems we use the function $L = -b[\theta \mathrm{Log}\theta - \theta]$ where b is the Boltzmann constant. The Maxwell distribution is given by :

$\theta = \exp(-\varphi_\alpha k^\alpha / b)$.

The polar function of $S^o(W)$ is :

$$S^{o*}(\varphi) = \varphi_\alpha . W^\alpha - \int_H L(\hat{\theta}) = \int (\varphi_\alpha k^\alpha) . \hat{\theta} - L(\hat{\theta})$$

$$= \int L^*(\varphi_\alpha k^\alpha) \quad (\hat{\theta} = L^{*'}(\varphi_\alpha k^\alpha))$$

of course we have $\frac{\partial S^{o*}}{\partial \varphi} = \int L^{*'} k^\alpha = W^\alpha$.

Theorem : for the Euler equations for polytropic gases the fundamental function Σ^* is given by :

$$\Sigma^*(\varphi, n) = \int_H (u^i n_i) . L^*(\varphi_\alpha k^\alpha(u, \eta)) \, du \, d\eta \qquad (8)$$

$i = 0, n \qquad \alpha = 1, n+2$

with $u^o = 1$

$$K = \begin{bmatrix} 1 \\ u^i \\ \sum_{i=1}^{3} \frac{(u^i)}{2} + \frac{\eta^\delta}{2} \end{bmatrix}$$

and δ is a constant $\left(\frac{1}{\delta} + \frac{n}{2} = \frac{1}{\gamma - 1} \right)$

$H = R^n \times R^+$.

L^* is a convex function such that : $\forall \alpha, \quad \lim_{x \to +\infty} x^\alpha . L^*(x) = 0$

and this set of functions is not empty (e^{-x}, quickly decreasing functions).

<u>Proof</u> :

$$\varphi_\alpha k^\alpha = \varphi_\varrho + \varphi_i u^i + \varphi_e \left[\sum_i \frac{u_i^2}{2} + \frac{\eta^\delta}{2}\right]$$

we put $u'^i = u^i + \frac{\varphi^i}{\varphi_e}$

$\left(- \frac{\varphi^i}{\varphi_e}\right.$ is the mean velocity)

$$\text{so } \varphi_\alpha k'^\alpha = \varphi_\varrho - \Sigma \frac{\varphi_i^2}{2\varphi_e} + \varphi_e \left[\sum_i \frac{(u'^i)^2}{2} + \frac{\eta^\delta}{2}\right]$$

$$\Sigma^* = - \frac{\varphi^i}{\varphi_e} n_i \int_H L^* + \int u_i n^i L^* .$$

The right hand term is equal to zero by parity

Now we set $\begin{cases} \bar{u} = \sqrt{\varphi_e} \, u' \\ \bar{\eta} = (\varphi_e)^{1/\eta^\delta} \end{cases}$

$$\Sigma^*(\varphi,n) = -\frac{\varphi^i n_i}{\varphi_e} \times \frac{1}{\varphi_e\left(\frac{1}{\delta}+\frac{n}{2}\right)} \qquad \int_H L^*\left(\varphi_\varrho - \Sigma \frac{\varphi_i^2}{2} + \frac{\bar{u}^2}{2} + \frac{\eta^\delta}{2}\right)$$

Σ^* is a solution of the SPDE (5)
S^{o*} is convex in φ :

$$X^o_\alpha X_\beta \frac{\partial^2 S^{o*}}{\partial\varphi_\alpha \, \partial\varphi_\beta} = \int_H L^{*''} \; (\varphi_\alpha k^\alpha) \; (k^\beta X_\beta)^2 \geqslant 0 \; ,$$

because $L^{*''}$ is positive (L^* convex).

III : NUMERICAL APPROXIMATION

We shall begin by using the Σ^* function in a very simple case, the linear equation :

$$\partial_i \, [C^i\theta] = 0 \qquad i = 0,n \qquad C^o = 1$$

we can choose $S(\theta) = \frac{1}{2}\,\theta^2$, so $\varphi = \theta$

and $S^*(\varphi) = \frac{1}{2}\;\varphi^2$,

$\Sigma^*(\varphi,n) = \frac{1}{2}\; c^i n_i \; \varphi^2$.

Σ^* is convex in φ if $c^i n_i \geqslant 0$.

When $c^i n_i = 0$, (n_i) is a characteristic normal vector. We are able to split Σ^* into a convex function and a concave one :

$$\Sigma^*(\varphi,n) = \text{Min}(c^i n_i,0)\,\frac{\varphi^2}{2} + \text{Max}(c^i n_i,0)\,\frac{\varphi^2}{2}$$

Σ^{*-} is the concave part and Σ^{*+} the convex one.
Let us consider the classical Courant scheme in the monodimensional case :
$\partial_t\theta + c\,\partial_x\theta = 0$

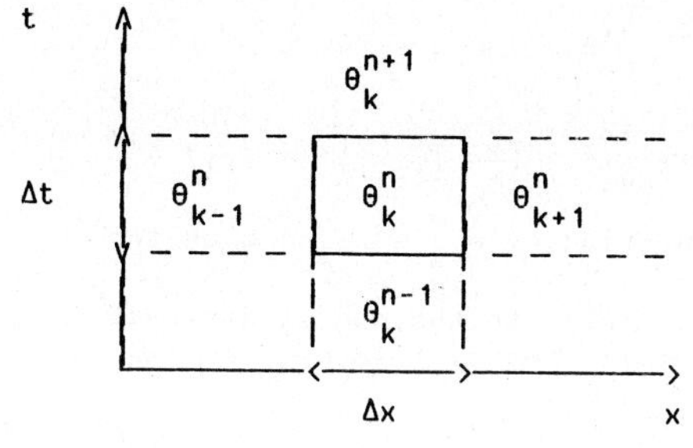

$$\theta_k^n = \theta_k^{n-1} + \frac{\Delta t}{\Delta x} \;\text{Min}\,(c,\,o)\;[\theta_{k+1}^n - \theta_k^n] + \text{Max}\,(c,o)\;[\theta_k^n - \theta_{k-1}^n] \qquad (9)$$

If we call ω the cell (n,k), $\partial\omega$ its space-time boundary, we can write (9) as :

$$\int_{\partial\omega} \min [c^i n_i(s), 0] [\theta_{out}(s) - \theta_{in}] ds = 0$$

where s is a point of $\partial\omega$, θ_{in} the interior value on the cell (θ_k^n), $\theta_{out}(s)$ the exterior value at the point s, $n_i(s)$ the outwards unit vector to $\partial\omega$ at the point s.

For the south branch :

$n = (-1,0) \quad - [\theta_k^n - \theta_k^{n-1}] \Delta x \ \mathrm{Min}(-1,0)$

north :

$n = (-1,0) \quad + [\theta_k^n - \theta_k^{n-1}] \Delta x \ \mathrm{Min}(1,0)$

west :

$n = (0,+1) \quad - \Delta t \ \mathrm{Min}(-c,0) [\theta_k^n - \theta_{k-1}^n]$

east :

$n = (0,-1) \quad + \Delta t \ \mathrm{Min}(c,0) [\theta_{k+1}^n - \theta_{k-1}^n]$.

So the classical Courant scheme can be written as :

$$\int_{\partial\omega} \frac{\partial\Sigma^{*-}}{\partial\varphi} (\varphi_{in}, n) - \frac{\partial\Sigma^{*-}}{\partial\varphi} (\varphi_{out}, n) = 0$$

One can notice that for the linear S.P.D.E. the "good" boundary condition is : $\Sigma^{*-}(\varphi_{in}, n)$ imposed, and the Courant scheme imposes the continuity of the convex part of Σ^* through $\partial\omega$.

We can also write a Galerkin discontinuous approximation (ref [9]) :
Let ω_h be a space time element, θ_h piecewise continuous (with discontinuities across the interelement boundaries only)
$\forall\varphi$ a test function with its support included in (ω_h)

$$0 = \int_{\omega_h} \Psi \, \partial_i (c^i \theta_h) + \int_{\partial\omega_h} \Psi \, [\mathrm{Min}(c^i n_i), 0] [\theta_{out} - \theta_{in}] \qquad (10)$$

where θ_{out} is the exterior trace of θ_h on $\partial\omega_h$
where θ_{in} is the interior trace of θ_h on $\partial\omega_h$
(10) can be written as :

$$0 = \int_{\omega_h} \Psi^\alpha \, \partial_i [\frac{\partial S^{*i}}{\partial\varphi_\alpha} (\varphi_h)] + \int_{\partial\omega_h} \Psi^\alpha \, [\frac{\partial\Sigma^-}{\partial\varphi_\alpha} (\varphi_{out}) - \frac{\partial\Sigma^-}{\partial\varphi_\alpha} (\varphi_{in})] \qquad (11)$$

if we choose for Ψ the characteristic function of ω_h we find back the Courant-scheme.
How to generalize those schemes from the linear to the non linear case ?
First it is necessary to have a natural splitting of $\Sigma^*(\varphi,n)$. For the Euler equations we have $\Sigma^* = \int_H u^i n_i \, L^*(\varphi_\alpha k^\alpha)$.

This natural splitting could be :

$$\int_H (u^i n_i) L^* = \int_{u^i n_i < 0} + \int_{u^i n_i > 0} ,$$

which is a concave/convex decomposition because :

$$\int_{u^i n_i > 0} (X_\alpha k^\alpha)^2 \quad L^{*''} (\varphi_\alpha k^\alpha) \geq 0.$$

If we choose $L^*(k\varphi) = e^{-k\varphi}$

Σ^{*+} and Σ^{*-} can be very easily computed numerically thanks to the function $\int_{-\infty}^{x} e^{-u^2} du$.

We are now able to present the scheme for the Euler equations with the simplest approximation φ_h piecewise constant on each ω_h space-time element on a domain Ω.
As we did for (11) we choose for Ψ the characteristic function of ω_h.

$$\forall h \quad \int_{\partial\omega_h} \frac{\partial \Sigma^{*-}}{\partial \varphi_\alpha} (\varphi_{in}, n) - \frac{\partial \Sigma^{*-}}{\partial \varphi_\alpha} (\varphi_{out}, n) = 0 \,. \qquad (12)$$

Remark :
If $\omega_h = [\Delta t \times \Delta x]$ we obtain the Kinetic flux vector splitting (KFVS) presented by DESHPANDE ([5] et [6]).

<u>Some results about the consistency</u> :

If we multiply (12) by φ_{in}^α we have :

$$(13) \quad \int_{\omega_h} \varphi_{in}^\alpha \, \partial_\alpha \, \Sigma^{*-}(\varphi_{in}, n) - \varphi_{in}^\alpha \, \partial_\alpha \, \Sigma^{*-}(\varphi_{out}, n) = 0$$

by definition of the polar function we have

$\varphi_{in} \nabla_\varphi \Sigma^{*-} = \Sigma_{in}^{*-} + \Sigma_{in}^{-}$ (in the left term of 13).

and we add $\varphi_{out} \nabla \Sigma_{out}^{*-} - \Sigma_{out}^{*-} - \Sigma_{out}^{-} = 0$ to (13) ;

we obtain :

$$(14) \quad \int_{\omega_h} \Sigma^{-}(\varphi_{in}, n) - \Sigma^{-}(\varphi_{out}, n) + \underbrace{\Sigma_{in}^{*-} - \Sigma_{out}^{*-} + \nabla_\varphi \Sigma^{*-}(\varphi_{out}) [\varphi_{out} - \varphi_{in}]}_{k}$$

but Σ^{*-} is concave so we have : $k \leq 0$
The left hand side of (14) is the scheme (12) applied to Σ^{-}.
So we have an inequality on each ω_h and k measures the dissipation.

Some algebraic relations close to consistency properties (both for conservative equations and entropy inequality), can also be derived from (12) :

Let Ω be an arbitrary reunion of adjacent elements ω_h, $\partial\Omega$ its boundary and Π the reunion of element boundaries belonging to the interior Ω of Ω.

By adding (12) over any ω_h in Ω we get :

$$\int_{\delta\Omega} [\nabla_\varphi \Sigma^{*-}(\varphi,n)] + \int_{\Pi} (\nabla_\varphi \Sigma^{*-}(\varphi_{out},n) - \nabla_\varphi \Sigma^{*-}(\varphi_{in},n)) - (\nabla_\varphi \Sigma^{*-}(\varphi_{in},n) - \nabla_\varphi \Sigma^{*-}(\varphi_{out},n)) = 0$$

but we have for any φ :

$\Sigma^*(\varphi,n) = \Sigma^{*-}(\varphi,n) - \Sigma^{*-}(\varphi,-n)$,

so we obtain

(15) $\langle \partial_i f^{i,\alpha}(\varphi), 1 \dot{\Omega} \rangle + \int_{\delta\Omega}[\nabla_\varphi \Sigma^{*-}(\varphi,n)] = 0$.

Thanks to (14) we get a similar result for the entropy inequality :

(16) $\langle \partial_i S^i(\varphi), 1 \dot{\Omega} \rangle + \int_{\delta\Omega} [\Sigma^-(\varphi,n)] \leqslant 0$.

Remark : We say that (15) (and (16)) are consistency properties because it is easy to see that if, by subdividing the elements ω_h of Ω in such a way that when $h \to 0$, $\varphi \to \Psi$ (so that for any piecewise regular bounded n-manifold V transversal to the discontinuities of Ψ we have :

$\int_V \nabla_\varphi \Sigma^{*-}(\varphi,n) \to \int_V \nabla_\varphi \Sigma^{*-}(\Psi,n)$ and

$\int_V \Sigma^-(\varphi,n) \to \int_V \Sigma^-(\Psi,n)$) we have both :

(if $\partial\Omega$ is a V-type manifold)

$\langle \partial_i f^{i,\alpha}(\Psi), 1 \dot{\Omega} \rangle = 0$ and $\langle \partial i\, S^i \Psi, 1 \dot{\Omega} \rangle \leqslant 0$.

Moreover, if we choose $\Omega = [0,T] \times \Omega x$ and if we assume that on $\partial\Omega x$, φ is constant (or, for Euler equations that the velocity is equal to zero), (16) leads to the overestimation :

$\int_{\Omega x} S(T,.) \leqslant \int_{\Omega x} S(0,.)$.

IV : FIRST NUMERICAL RESULTS

Numerically, we have implemented scheme (12) - Part III - where :

$\omega_h = [t_n, t_n + \Delta t] \times \omega_h^x$

and ω_h^x is : a segment in the 1-D case
a triangle in the 2-D case

This scheme has been explicited by putting :

φ equal to its value on the previous time step, on $[t_n, t_n + \Delta t] \times \partial\omega_h^x$.

This can be interpreted as the first step of a fixed point method used to solve the equations (12).

We can prove that there exists some Δt small enough to make this fixed point method convergent (in the linear case, it gives the C.F.L. condition).

Figures 1 and 2 show the results obtained in the monodimensional case of the Sod Shock tube.

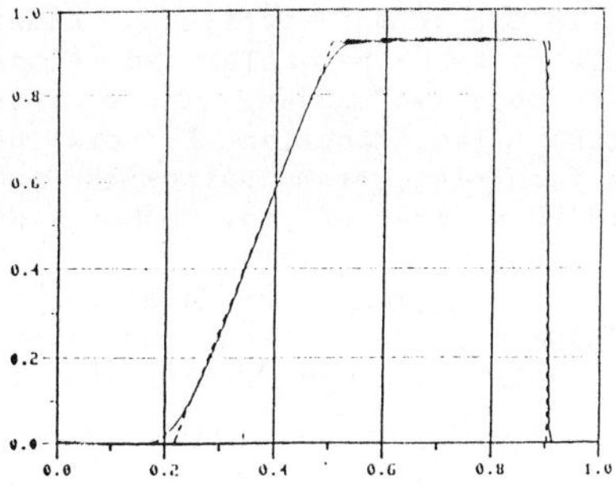

Figure 1 : Velocity

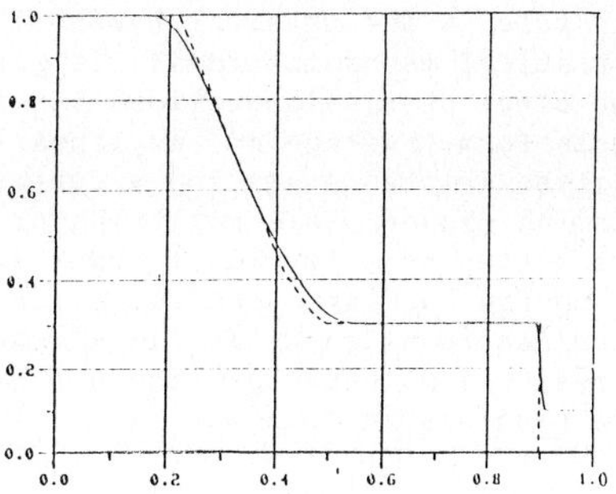

Figure 2 : Pressure

T = 0.288

Pictures 3 and 4 present a bidimensional case : an asymptotic stationary supersonic flow on a 20°-angled slope (the initial condition being a uniform flow at Mo = 2). The mesh has been auto-adaptatively refined.

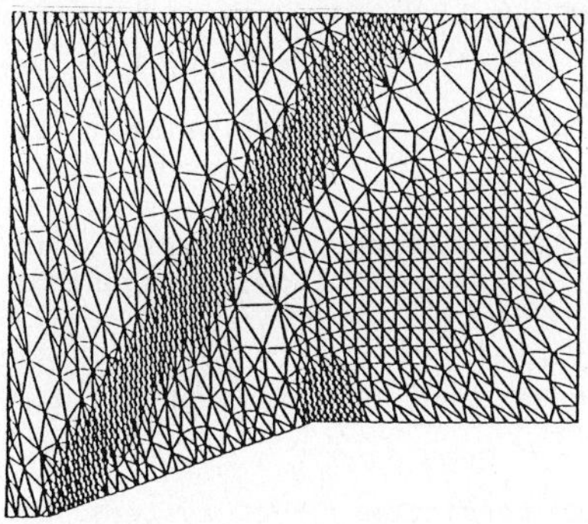

Figure 3 : Refined mesh

Number of
- elements : 2529
- nodes : 1346
- vertices : 3874
- boundary elements : 161

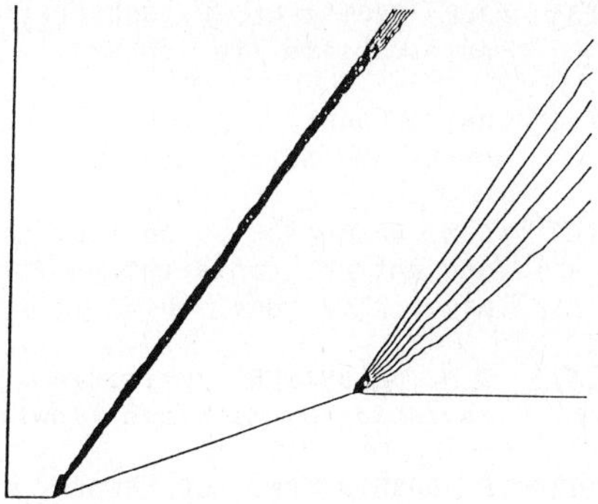

Figure 4 : Isopressure lines from 1.4 to 1.9 with $\Delta p = 0.1$

CONCLUSION

We have shown in the first part how the fundamental function Σ^* plays a main role in the physical hyperbolic SPDE. In the second part, by using statistical mechanics ideas, it was possible to set the Euler equations (and other physical equations as these of reactive gases) in a very simple form (average of a linear equation). The function Σ^* can be splitted into a convex and a concave part, resulting from outcoming and incoming fluxes. This splitting gives birth to schemes of any order. We have tested only the first order scheme, which gives satisfying results. Of course we are not able to demonstrate convergence results. In polydimensional cases we lack the knowledge of convenient functional spaces, but we hope that the use of the Σ^*-splitting may contribute to some progress in this way.

REFERENCES

[1] J. AUDOUNET : "Solutions de classe C1 par morceaux sous forme paramétrique des problèmes aux limites associés à un système de lois de conservation". Annales du séminaire d'Analyse Numérique 1984-1985 U.P.S. (Toulouse).

[2] F. BOURDEL, P.A. MAZET : "Multidimensional case of an entropie variational formulation of conservative hyperbolic systems". La Recherche Aérospatiale 1984-5, p. 67-76 (English edition).

[3] J.P. CROISILLE, F. BOURDEL, Ph. DELORME, P.A. MAZET : La Recherche Aérospatiale (to appear).

[4] Ph. DELORME : Ecriture intrinsèque des équations de bilan. La Recherche Aérospatiale 1985-6, p. 409-411.

[5] S.M. DESHPANDE : "On the Maxwellian Distribution, symmetric form and entropy conservation for the Euler equations". Nasa technical paper 2583, Nov. 86.

[6] S.M. DESHPANDE : Kinetic theory closed new upwind methods for inviscid compressible flows.

[7] F. DUBOIS, Ph. LE FLOCH : Boundary conditions for systems of hyperbolic conservation laws. CRAS, série I, Paris 1986.

[8] A. HARTEN : "On the symmetric form of systems of conservation laws" Journal of Computational Physics. Vol 49, 1983, p. 151-164.

[9] T.J.R. HUGHES : "Recent progress in the development and understanding of SUPG methods with special reference to the compressible Euler and Navier-Stockes Equations". Numerical methods in Fluids. Vol 7, n° 11, Nov. 87.

ADMISSIBILITY CONDITIONS FOR WEAK SOLUTIONS OF NONSTRICTLY HYPERBOLIC SYSTEMS.

M. Brio
Dept. of Mathematics, University of Arizona, Tucson, Az. 85721

SUMMARY

We discuss questions related to the choice of a proper class of waves and initial conditions for the well-posedness of the Riemann problem for nonstrictly hyperbolic systems of conservation laws. Since multiple eigenvalues represent strong or resonant wave interaction we propose to derive a relatively simple and universal set of model equations which describe qualitatively the underlying processes, like Burgers' equation does in a strictly hyperbolic case. Finally, we discuss numerical schemes utilizing various Riemann solvers to the above class of problems.

INTRODUCTION

Consider the following hyperbolic system of conservation laws

$$U_t + (F(U))_x = 0,$$

where U represents vector of dependent variables and $F(U)$ denotes the appropriate flux function. Riemann problem is an initial value problem with a piecewise constant initial data. It serves as an intermediate step in understanding of the general initial value problem and as an important model problem in various applications, such as piston and shock-tube problems in gas dynamics, Riemann solvers in numerical schemes, simpliest model of the field line reconnection, etc.

Solution is understood in a weak sense and consists of combinations of various waves propagating out from the initial discontinuity. There are two major problems:

a) to prove well-posedness (for example, existence, uniqueness, and stability) of the problem in some class of waves,

b) to choose the class of waves by studying physical equations on the next smaller length scale (for example, to introduce back into equations diffusive, dispersive, forcing or any other terms, which were dropped in the first place).

If the eigenvalues of the Jacobian matrix may coincide, then the appropriate system of conservation laws is called nonstrictly hyperbolic. Application of standard admissibility conditions worked out for a strictly hyperbolic case often leads to contradictions and various difficulties arising from the fact that those systems are fundamentally different. Physically, coinciding eigenvalues represent strong or resonant wave interaction compared to weak interactions represented by strictly hyperbolic systems. Mathematically, this manifests itself in the existence of free parameters for the solutions with discontinuous initial data, because interaction parameters are "hidden" in the discontinuity and are not introduced explicitly into the problem. Numerically, the discontinuous initial data produces different results depending on the numerical schemes used since the interaction parameters are introduced by truncation errors of the appropriate schemes (see article of B. Wendroff in this volume for additional illustration of this problem arising in applications of irregular grids).

For many systems nonstrict hyperbolicity implies nonconvexity, for example, the steepening rate of smooth waves may be zero at some points. For an example of such systems see a paper of H. Freistüler in this volume.

EXAMPLES OF THE EQUATIONS

One-dimensional equations of ideal magnetohydrodynamics (MHD) characterize the flow of conducting fluid in the presence of magnetic field and represent coupling of the fluid dynamical equations with Maxwell's equations of electrodynamics. Neglecting displacement current, electrostatic forces, effects of viscosity, resistivity, and heat conduction, one obtains the following ideal MHD equations [1]:

$$
\begin{aligned}
\rho_t + (\rho u)_x &= 0,\\
(\rho u)_t + (\rho u^2 + P^*)_x &= 0,\\
(\rho v)_t + (\rho u v - B_0 B)_x &= 0,\\
(\rho w)_t + (\rho u w - B_0 H)_x &= 0,\\
B_t + (Bu - B_0 v)_x &= 0,\\
H_t + (Hu - B_0 w)_x &= 0,\\
E_t + ((E + P^*)u - B_0(B_0 u + Bv + Hw))_x &= 0.
\end{aligned}
$$

In the above equations, the following notations are used: ρ for density, $\vec{u} = (u, v, w)$ for velocity, $\vec{B} = (B_0, B, H)$ for magnetic field, P for static pressure, P^* for full pressure, $P^* = P + \frac{1}{2}|\vec{B}|^2$, E for energy, $E = \frac{1}{2}\rho|\vec{u}|^2 + P/(\gamma - 1) + |\vec{B}|^2$, γ for ratio of specific heats, and $B_0 \equiv const.$

The eigenvalues of the Jacobian matrix can be written in nondecreasing order as

$$u - c_f, \quad u - c_a, \quad u - c_s, \quad u, \quad u + c_s, \quad u + c_a, \quad u + c_f,$$

where c_f, c_a, c_s are called fast, Alfvén, and slow characteristic speeds, respectively. They can be expressed as follows:

$$c_a^2 = b_x^2, \quad c_{f,s}^2 = \tfrac{1}{2}((a^*)^2 \pm \sqrt{(a^*)^4 - 4a^2 b_x^2}),$$

with the plus sign for c_f and minus sign for c_s. The following notations were used:

$$b_x^2 = B_0^2/\rho, \quad b^2 = (B_0^2 + B^2 + H^2)/\rho, \quad (a^*)^2 = a^2 + b^2,$$

where a is the sound speed.

There are two points where the eigenvalues may coincide:

(1) If $B_0 = 0$, then $c_s = c_a = 0$ and u is an eigenvalue of multiplicity 5.

(2) If $B^2 + H^2 = 0$, then $c_f^2 = max(a^2, b_x^2)$, and $c_s^2 = min(a^2, b_x^2)$.

Therefore, either $c_f^2 = c_a^2$ or $c_s^2 = c_a^2$, or both.

The usually used eigenvectors [1] are not well-defined near the above points and the matrix with these eigenvectors as its columns becomes singular. However, by proper renormalization a complete set can be obtained [2]. Using these eigenvectors we have shown that either slow or fast wave becomes linearly degenerate when the transverse component of the magnetic field passes through zero [2]. Therefore, ideal MHD equations form nonconvex nonstrictly hyperbolic system of conservation laws.

Recently, we have shown that isentropic (for example, dropping the energy equation) MHD equations in Lagrangean coordinates are equivalent to the equations describing isotropic hyperelastic materials. This can be shown as follows: denote the velocity variables as before and identify $\frac{1}{\rho}$, $\frac{B}{\rho}$ and $\frac{H}{\rho}$ with the strain components p_1, p_2, and p_3, respectively. Consider stress-strain relations for the isotropic hyperelastic materials:

$$\begin{aligned} s_1 &= f(p_1, p_2^2 + p_3^2), \\ s_2 &= p_2 g(p_1, p_2^2 + p_3^2), \\ s_3 &= p_3 g(p_1, p_2^2 + p_3^2), \end{aligned}$$

where f and g are arbitrary functions. If we choose f and g as

$$f = P(p_1) + \frac{p_2^2 + p_3^2}{2p_1^2}, \quad g = \frac{B_0}{p_1},$$

the resulting equations become identical to the elastic equations [3]:

$$\frac{\partial p_i}{\partial t} - \frac{\partial u_i}{\partial x} = 0, \quad \frac{\partial u_i}{\partial t} + \frac{\partial s_i}{\partial x} = 0, \quad i = 1, 2, 3.$$

Therefore, the above classification holds also in this case. This is contrary to the usual belief that the slow and the fast waves (in elasticity they are called longitudinal waves) are genuinely nonlinear [3], and that due to linear degeneracy of Alfvén waves (shear waves in elasticity), it is sufficient to consider a coplanar case of the Riemann problem [3], [4]. We would like to mention that both systems are a particular case of systems of conservation laws with rotationally symmetric flux function and nonstrict hyperbolicity and nonconvexity follow from this property (see an article of H. Freistüler in this proceedings).

As a result of a construction and numerical experiments with the second order upwind schemes [2], we have found that solutions to the Riemann problem may contain waves which are usually rejected by standard admissibility conditions (for example composite waves, consisting of a shock wave followed by a rarefaction wave of the same family). We also have observed the fact that various schemes produce different solutions from a one-parameter family of solutions.

In another work [5], we have shown by asymptotic analysis that some of the above solutions, like composite waves, exist for physically meaningful asymptotic limits of the original equations.

ADMISSIBILITY CONDITIONS

Hyperbolic systems of conservations laws can usually be obtained by assuming that the phenomena under consideration evolves on the advection time scale and that other effects, like viscosity, dispersion, capillarity, etc., can be neglected. This leads to discontinuities, non-uniqueness and "unphysical" solutions. To keep the discontinuities but to avoid the other two possibilities, solutions are considered in a weak sense together with some admissibility conditions. Following are the most common admissibility criteria for shock waves in case of strictly hyperbolic systems:

1) linearized stability analysis (also called Lax or evolutionary condition),

2) existence of a stable viscous profile (for example, Liu's condition),

3) physical entropy condition derived from the second law of thermodynamics,

4) requirement for hyperbolic equations to be a limit of the same equations perturbed by linear viscosity terms with the multiple of identity viscosity matrix (so-called entropy inequalities),

5) solutions should be admissible for the equations derived as a weakly nonlinear asymptotic limit of the full physical system of equations.

For example, in case of polytropic gas dynamics and other similar systems all of the above criteria reject the expansion shock waves.

Since the wave structure for nonstrictly hyperbolic systems is much more complicated then in strictly hyperbolic case, the above criteria admit different classes of solutions. The first approach, linearization around a constant state, is not applicable near the degenerate points since the wave interaction there is governed by the quadratic terms. The second condition is sufficient, but is not necessary, since for nonconvex systems a shock wave can expand near the degenerate points with the rate $\sqrt{t}$ and still be considered as a discontinuity on the advection time scale (see related discussion on the undercompressive shock waves in an article of D. Marchesin in this volume). The third condition is a necessary condition if one believes in the second law of thermodynamics. It is often not sufficient for the uniqueness. However, non-uniqueness in this case may reflect the fact that some relevant physical parameters are missing in the hyperbolic model. The forth condition is a necessary condition if the relevant physical equations contained only terms which could be approximated by linear viscosity with the multiple of the identity viscosity matrix. This is rarely a realistic assumption because very often viscosity coefficients are of various order of magnitude. The last criterion is a necessary condition and it deals with the equations describing the phenomena on the next time scale. The only assumption is that the wave under consideration is weakly nonlinear. For convex strictly hyperbolic systems this approach leads to Burgers' equation. Using similar technique, we have recently derived modified Burgers' equation [5]:

$$u_t + u^2 u_x = \epsilon u_{xx},$$

which governs the propagation of weakly nonlinear magnetoacoustic waves near degenerate points. It admits new class of solutions we have observed numerically in [2] for MHD equations. Recently, these solutions were observed in the data received from the space experiments [6].

We would like to note that usually used viscosity admissibility conditions for MHD shock waves, introduced by Germain [7] and followed by Conley and Smoller [8], require the existence of viscous profiles for all possible ratios of viscosity coefficients. In addition, they allow only coplanar shock structures by setting two variables to zero, and therefore, eliminating the possibility of three dimensional structures.

Finally, note that the above criteria were designed for a single wave. For the equations under consideration the waves corresponding to the coinciding eigenvalues may not exist separately in general. For example, Alfvén wave may be coupled to the fast or slow wave. Therefore, it seems natural to consider a two-dimensional manifold of solutions generated by these waves. This approach would explain observed one-parameter family of solutions for the coplanar MHD Riemann problem [2]. In contrast, only one-parameter per wave is required in case of strictly hyperbolic systems of conservation laws [9]. Physically, this distinction represents the difference between weak and strong or resonant wave interaction. Mathematically, it manifests itself as follows. The solution of a convex strictly hyperbolic system is not effected on the advection time scale by the small-scale processes. For a example, the width of the shock wave may vary, but it still represents the same discontinuity on the advection time scale.

In contrast, for nonstrictly hyperbolic systems small-scale processes, like various balances between dissipative terms, would have a global effect on the solution. This indicates that discontinuous initial data for such problems has to be supplemented by these small-scale interaction parameters in order to have a well-posed problem.

As a first step in investigating the above conjectures, we propose to derive and study a relatively simple and universal set of model equations.

MODEL PROBLEMS

Consider the following set of equations which admits only two waves corresponding to fast and slow waves discussed previously:

$$\begin{aligned} u_t + 0.5(u^2 + v^2)_x &= \epsilon_1 u_{xx}, \\ v_t + (v(u-1))_x &= \epsilon_2 v_{xx}. \end{aligned}$$

Initial data is assumed to be smooth. This system represents nonconvex strictly hyperbolic model of elastic or hydromagnetic coplanar case. It is in gradient form, the slow wave is nonconvex and the fast wave is genuinely nonlinear [9]. An interesting feature of this system is that the existence of travelling waves exhibits a global bifurcation [10]. Let $\epsilon = \epsilon_1/\epsilon_2$. In particular, there exists ϵ_0, such that for $\epsilon > \epsilon_0$ there are no travelling wave solutions corresponding to the slow shock waves violating Lax but satisfying Liu's condition. On the other hand, for $\epsilon < \epsilon_0$ there is a unique stable profile for such waves. Moreover, in this case the superfast-subslow shock waves have infinitely many viscous profiles, stability of which has to be investigated further. We would like to note, that the case considered by Germain, and by Conley and Smoller (see references) corresponds to $\epsilon \longrightarrow \infty$ for our model problem. The behavior of the system when $\epsilon > \epsilon_0$ is described by the above time dependent problem.

As a next step, we would like to introduce an intermediate wave into the system by generalizing the above model problem as follows :

$$u_t + 0.5(u^2 + v^2 + w^2)_x = \epsilon_1 u_{xx},$$
$$v_t + (v(u-1))_x = \epsilon_2 v_{xx},$$
$$w_t + (w(u-1))_x = \epsilon_2 w_{xx}.$$

This time dependent problem for smooth initial data allows to study the resonant wave interaction between slow and intermediate waves. For analytical description it seems promising to use the results for a bifurcation from a double eigenvalue and existence of secondary bifurcations [11].

Since some of the proposed studies are done using numerical experiments with various conservative numerical schemes, we would like to make the following remark.

At the moment, we are interested in resolving small-scales, so that the resulting solution is smooth. Also, the initial data is assumed to be smooth (in fact, the width of the initial discontinuity represents a natural length scale for the Riemann problem). The standard upwind schemes have several advantages in strictly hyperbolic case, like non-oscillatory behavior, high resolution of discontinuities, robustness (nonlinear stability?). But, they are not suitable for our purpose because they are not designed to model dicontinuities represented by smooth transitions with a particular ratio of viscosity coefficients. Eventually, one would like to avoid small-scales resolution (for an example of such technique for a single equation see a paper by E. Harabetian in this volume).

CONCLUSION

In this article we have discussed questions related to the choice of a proper class of waves and initial conditions for the well-posedness of the Riemann problem for nonstrictly hyperbolic systems of conservation laws. We briefly summarize them here.

1) Using weakly nonlinear asymptotics derive a universal set of equations describing propagation of two or more strongly or resonantly interacting waves for general nonstrictly hyperbolic systems of conservation laws. The equations may depend on special structure of the Jacobian matrix, like existence of invariant subspaces and dependencies between the blocks, see [2] for an example.

2) Describe the effect of small-scales parameters by studying smooth solutions of the model problems corresponding to shock waves satisfying physical entropy condition. Link the solutions of the model problems to the full systems and to the known properties of the solutions to the physical problems, like

mode coupling in elasticity, dependence of the solution to the reconnection problem on the initial and boundary conditions, existence of the free parameters for the shock waves in combustion and magnetohydrodynamics.

3) Using appropriate physical models work out principles for modifying upstream numerical schemes in order to avoid small-scales resolution in regions of strong or resonant wave interaction.

REFERENCES

[1] A. Jeffrey and T. Taniuti, Nonlinear Wave Propagation, Academic Press, New York (1964).

[2] M. Brio and C.C. Wu, "An upwind differencing scheme for the equations of ideal magnetohydrodynamics", J. Comput. Phys., vol. 75, April 1988.

[3] Z. Tang and T. C. T. Ting, "Wave curves for the Riemann problem of plane waves in simple isotropic elastic solids", University of Illinois, Chicago preprint, 1985.

[4] Gogosov, "Disintegration of an arbitrary discontinuity in magnetohydrodynamics", Journal of Appl. Math. and Mechanics", Vol. 25 (1961).

[5] M. Brio, "Propagation of weakly nonlinear magnetoacoastic waves", Wave Motion, Vol. 9, p. 455 (1987).

[6] C. C. Wu, Private communication, December 1987.

[7] P. Germain, "Shock waves and shock wave structure in magnetofluid dynamics", Rev. Mod. Phys. 32, pp. 951-958 (1960).

[8] C. Conley and J. Smoller, "On the structure of magnetohydrodynamic shock waves", Comm. Pure Appl. Math., 28, pp. 367-375 (1974).

[9] P. D. Lax, "Hyperbolic systems of conservation laws", Comm. Pure Appl. Math., 7, p. 159 (1954).

[10] J. Guckenheimer and P. Holmes, Nonlinear Oscillations, Dynamical Systems, and Bifurcations of Vector Fields. Springer-Verlag, 1983.

[11] D. Schaeffer and M. Shearer, "Riemann problems for nonstrictly hyperbolic 2×2 systems of conservation laws", to appear in Trans. AMS.

UNIFORMLY SECOND ORDER CONVERGENT SCHEMES FOR HYPERBOLIC CONSERVATION LAWS INCLUDING LEONARD'S APPROACH.

CAHOUET Jacques COQUEL Frederic
Research Branch , Laboratoire National d'Hydraulique
6 quai Watier 78401 CHATOU FRANCE

SUMMARY

We present a systematic procedure to correct 5-point linear schemes so that convergence towards the weak entropy solution of hyperbolic scalar conservation laws can be established while high order accuracy is achieved including at critical points. Our method can be described as a modification of TVD schemes which preserves the $BV \cap L^\infty$ stability ; entropy convergence is achieved by addition of an extra limiting mechanism which preserves accuracy.

INTRODUCTION

A very successful class of schemes for solving conservation laws problems is the class of TVD (Total Variation Diminishing) schemes. An important disadvantage of TVD schemes is that the TVD property makes the scheme necessarily degenerate to first order accuracy at local extrema, leading the overall accuracy to be at most first order in the L^∞-norm. We are lead to seek a weaker control over possible growth of the Total Variation of the numerical solution, to enable the design of full high order accurate schemes, still required to achieved $BV \cap L^\infty$ stability. The $BV \cap L^\infty$ stability is an important guide principle for scheme designing since it ensures the existence of a convergent subsequence in L^1_{loc} to a weak solution of the conservation laws problem, as the mesh size goes to 0. For this purpose, Harten and al [8],[9], have introduced the ENO schemes of uniformly high order accuracy. These schemes perform quite well, although it is not still proven that they achieved the required $BV \cap L^\infty$ stability. Quite recently, Shu [10] has proposed a Total Variation Bounded modification of some existing schemes involving the minmod function as a limiter, in order to combine BV stability and high order accuracy including at critical points (i.e.: sonic points or local extrema). Following Shu's approach, we present a simple systematic procedure to correct second order 5-point linear schemes to obtain $BV \cap L^\infty$ stable schemes of uniformly second order accuracy in smooth regions.

The format of this paper is as follows : In the first section, we present notations and recall basic features on numerical schemes. In section II, we develop a suited procedure to check TVD correction for 5-point linear schemes. It will allow us to show that, besides the usual TVD corrections of the Lax-Wendroff scheme, many other TVD schemes can be derived thanks to this procedure. As an example, we consider an original difference scheme, Modified-Exquisite, derived by Leonard in 1981 (see [5],[6]), based upon a monotonizing method of interpolations, and we prove that Modified-Exquisite is TVD. The third section is devoted to the description of the TVB modification procedure, applied to the TVD schemes derived in section II, making them uniformly second order accurate in smooth regions. Let us underline that the TVB modification we perform can be straightforward applied to well known TVD schemes, such as Roe's Superbee, Van Leer's scheme (see [4]). In section IV, we slightly modified the TVB (TVD) schemes previously designed, in order to enforce the convergence towards the entropy solution, as Vila pointed out [3]. In the last section, several numerical experiments are included to illustrate the efficiency of these schemes.

I. GENERALITIES

In this first part, we deal with numerical approximations of weak solutions to the initial value problem (I.V.P.) for one-dimensional scalar conservation laws :

$$\begin{cases} u_{,t} + f(u)_x = 0 & x \in R\,,\ t \in [0,1] \\ u(x,0) = u_0(x)\,. \end{cases} \tag{1.1}$$

We consider finite difference approximations of (1.1) in conservation form : Δx is the space step, Δt the time step and $\lambda = \Delta t / \Delta x$ the mesh ratio. Let $U^h(x,t)$ be an approximate solution :

$$U^h(x,t) = u_j^n \quad for \quad (x,t) \in\,](j-1/2)\,\Delta x\,,\,(j+1/2)\,\Delta x[\ x\ [n\,\Delta t\,,\,(n+1)\,\Delta t[\ .$$

$$u_j^{n+1} = u_j^n - \lambda\,(h_{j+1/2}^n - h_{j-1/2}^n)\,, \tag{1.2}$$

is a numerical (2p+1)-point scheme written in conservation form, where $h^n_{j+1/2}$ stands for the Lipschitz continuous, consistent numerical flux

$$h_{j+1/2}^n = h\,(u_{j-p+1}^n\,,\,.....\,,\,u_{j+p}^n)\,, \tag{1.3}$$

$$h\,(u\,,u\,,\,...\,,u) = f(u)\,.$$

We assume that scheme (1.2) can be written in incremental form :

$$u_j^{n+1} = u_j^n - C_{j-1/2}^n\,\Delta u_{j-1/2}^n + D_{j+1/2}^n\,\Delta u_{j+1/2}^n\,, \tag{1.4}$$

with the standard notation $\Delta u_{j+1/2}^n = u_{j+1}^n - u_j^n$.

The main interest of the incremental form for schemes (1.2) (1.3) is that sufficient conditions can be derived in order to achieve convergence of the family of approximate solutions (at least a subsequence) to a weak solution of the I.V.P. (1.1), as Δx vanishes. These conditions can be stated as follows, using the standard definition for the numerical BV-norm :

$$TV(u) = \sum_j |\,u_{j+1} - u_j\,|\,,$$

Lemma 1 (Harten [1] , Vila [2],[3]) : Let a scheme (1.2) (1.3) be given in its incremental form (1.4). We assume that C and D coefficients are positive. We denote by *(i)* and *(ii)* the following conditions on these coefficients :

$$(i) \qquad D_{j+1/2}^n + C_{j+1/2}^n \le 1 \quad , \forall j \in Z\,,$$

$$(ii) \qquad D_{j+1/2}^n + C_{j-1/2}^n \le 1 \quad , \forall j \in Z\,.$$

If condition *(i)* is satisfied, we have $TV\,(u^{n+1}) \le TV\,(u^n)$,

If condition *(ii)* is satisfied, we have $\|u^{n+1}\|_\infty \le \|u^n\|_\infty$.

Our aim is to design a systematic procedure to correct a set of 5-point linear schemes to obtain $BV \cap L^\infty$ stable schemes with respect to lemma 1. Our approach belongs to the flux limiters category previously analyzed by Sweby [4]. In the next section, we seek an attractive incremental decomposition well suited to our purpose. This investigation will lead us to a set of useful conditions allowing, for instance, the proof that the Modified-Exquisite scheme [5] is in fact TVD.

II. TVD CORRECTION OF 5-POINT LINEAR SCHEMES

We mean by 5-point linear schemes, schemes whose numerical flux are in a linear relationship with the physical flux values at grid points, described below, using the standard notation for local Courant numbers : $v^n_{j+1/2} = \lambda \, \Delta f^n_{j+1/2} \, / \, \Delta u^n_{j+1/2}$,

$$\text{if } v^n_{j+1/2} > 0 \ , \quad h^n_{j+1/2} = \alpha f(u^n_{j+1}) + \beta f(u^n_j) + \gamma f(u^n_{j-1}) ,$$

$$\text{if } v^n_{j+1/2} < 0 \ , \quad h^n_{j+1/2} = \alpha f(u^n_j) + \beta f(u^n_{j+1}) + \gamma f(u^n_{j+2}) .$$

It can be seen that these linear schemes are second order spatial accurate iff $\alpha-\gamma=1/2$. Let us recall that such schemes cannot be TVD and therefore need to be corrected. We observe that, before correction, this numerical flux can be broken up into an E-flux (at most first order accurate) [7], denoted g^E, and an added flux, considered as an antidiffusive flux :

$$h^n_{j+1/2} = g^{nE}_{j+1/2} + \frac{h^n_{j+1/2} - g^{nE}_{j+1/2}}{\Delta u^n_{j+1/2}} \Delta u^n_{j+1/2} , \tag{2.1}$$

which must be limited in order to achieve the TVD-property. If we restrict ourselves to E-schemes which behave like the first order upwind scheme away from sonic points, this antidiffusive flux admits a more attractive expression owing to the introduction of suitable notations. For clarity, we assume that we are away from a sonic point. For instance, for $v^n_{j+1/2}>0$, we introduce the following ratio of consecutive gradients :

$$r^n_{j+1/2} = \frac{\Delta f^n_{j+1/2}}{\Delta f^n_{j-1/2}} , \tag{2.2}$$

it can be easily seen that, owing to our restrictions, the so-called reduced numerical flux (2.3) only depends linearly on the ratio (2.2) :

$$\text{if } v^n_{j+1/2} > 0 \qquad \varphi(r^n_{j+1/2}) = \frac{h^n_{j+1/2} - g^{nE}_{j+1/2}}{\Delta f^n_{j+1/2}} \qquad \text{with} \quad g^{nE}_{j+1/2} = f(u^n_j) . \tag{2.3}$$

Let notice an obvious one to one correspondence between reduced flux and numerical flux associated to 5-point linear schemes. These notations allow the antidiffusive flux to be reformulated as follows :

$$a^n_{j+1/2} = \frac{1}{\lambda} v^n_{j+1/2} \frac{\varphi(r^n_{j+1/2})}{r^n_{j+1/2}} \Delta u^n_{j+1/2} . \tag{2.4}$$

This expression clearly indicates a way to limit the antidiffusive flux associated to a second order 5-point linear scheme by seeking a suited function φ, whose restriction to monotonic smooth regions is nothing else but the reduced flux associated to the second order 5-point linear scheme. In order to extend equality (2.4) to negative $v^n_{j+1/2}$, we define :

$$\text{if } v^n_{j+1/2} < 0 \qquad r^n_{j+1/2} = \frac{\Delta f^n_{j+1/2}}{\Delta f^n_{j+3/2}} \quad \text{and} \quad \varphi(r^n_{j+1/2}) = - \frac{(h^n_{j+1/2} - f(u^n_{j+1}))}{\Delta f^n_{j+3/2}} , \tag{2.5}$$

in such a way that φ is the same function as in (2.3), leading us to write the antidiffusive flux as

$$a_{j+1/2}^{n} = \frac{1}{\lambda}(-v_{j+1/2}^{n})\frac{\varphi(r_{j+1/2}^{n})}{r_{j+1/2}^{n}}\Delta u_{j+1/2}^{n} . \tag{2.6}$$

Owing to (2.4) and (2.6), we have actually derived the following useful equality for 5-point linear schemes :

$$h_{j+1/2}^{n} = g_{j+1/2}^{n\,E} + \frac{1}{\lambda}|v_{j+1/2}^{n}|\frac{\varphi(r_{j+1/2}^{n})}{r_{j+1/2}^{n}}\Delta u_{j+1/2}^{n} . \tag{2.7}$$

Thanks to this reformulation, TVD correction can be easily performed, seeking a suited function φ as previously pointed out. A way to design φ will be deduced from the following statement :

Theorem 2.1 : Assume that $\varphi(r)$ satisfies :

$$(\exists\,(M,\mu)\in R_{+}^{*}\times[-1,0]\,)\,/\,(\quad\forall r\in R,\ \mu\le\varphi(r)\le M \quad and \quad -M\le\frac{\varphi(r)}{r}\le(1+\mu)\,)\,,$$

then the corrected flux $h_{j+1/2}^{n\,C} = g_{j+1/2}^{n\,E} + a_{j+1/2}^{n\,C}$,

$$where \quad \begin{cases} a_{j+1/2}^{n\,C} = \dfrac{1}{\lambda}|v_{j+1/2}^{n}|\dfrac{\varphi(r_{j+1/2}^{n})}{r_{j+1/2}^{n}}\Delta u_{j+1/2}^{n} & \textit{away from a sonic point} \\ a_{j+1/2}^{n\,C} = 0 & \textit{otherwise} \end{cases} ,$$

defines a TVD scheme , which preserves the L^{∞}-norm.

Proof : see [15].

We deduce from Theorem 2.1 that the graph of $\varphi(r)$ must lie in the shaded area depicted on figure 1 for the corrected scheme to be TVD.

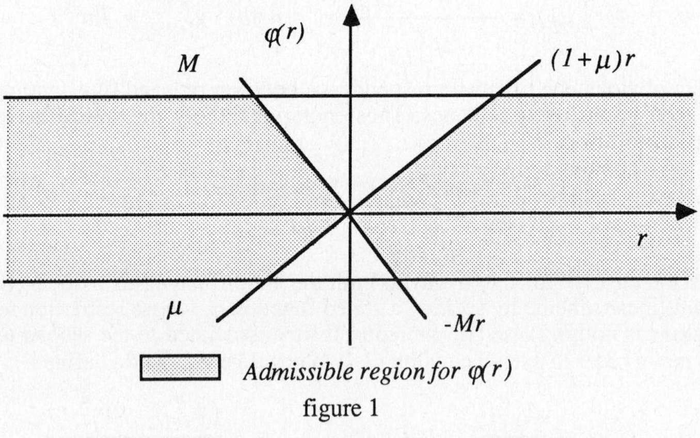

figure 1

Notice that the upper boundary M (and its associated one $-Mr$) is related to a C.F.L. like condition $max\,|\,v_{j+1/2}^{n}\,| \le 1/(2M+1) < 1$, that the resulting TVD scheme must satisfy. We now introduce a technical lemma of some importance in the following sections:

Lemma 2.2: Let be given an antidiffusive flux $a^n_{j+1/2}$ satisfying Theorem 2.1 assumptions, then $a^{n\,M}_{j+1/2} = \theta^n_{j+1/2}\, a^n_{j+1/2}$ *with* $\theta^n_{j+1/2} \in [0,1]$, $\forall\, (j,n) \in Z \times N$, still provides a TVD scheme under the same C.F.L. restriction.

Proof: Notice that $\varphi^M(r^n_{j+1/2}) = \theta^n_{j+1/2}\, \varphi(r^n_{j+1/2})$ can be associated to $a^{n\,M}_{j+1/2}$ so that $\varphi^M(r)$ fulfills Theorem 2.1 hypothesis. ∎

In order to achieve second order accuracy, we firstly require the function φ to fulfill the sufficient conditions stated below :

Lemma 2.3 :Assume that $\varphi(1) = 1/2$ and φ admits a derivative at $r = 1$, then second order accuracy is achieved with respect to space discretization (except at critical points).

Proof :See Theorem 2.4.

In order to improve time accuracy for transient computations (limited to first order since we have used the first order Euler backward scheme to approximate the time derivative), we slightly modify the function φ in respect with :

Theorem 2.4 : Assume that φ satisfies the conditions of Theorem 2.1 and Lemma 2.3, then the following function $\psi = (1 - |\nu|)\,\varphi$ provides a TVD scheme second order accurate in both time and space (except at critical points), under the same C.F.L. restriction as the φ - scheme.

Proof: Since $|\nu^n_{j+1/2}| \in [0,1]$; $\psi(r^n_{j+1/2}) = \theta^n_{j+1/2}\, \varphi(r^n_{j+1/2})$ *with* $\theta^n_{j+1/2} \in [0,1]$, and Lemma 2.2 ensures that the resulting scheme is still TVD under the same CFL condition. The Lax-Wendroff scheme can be defined thanks to $\psi(r) = 1/2\,(1 - |\nu|)\, r$, thus it can be seen that

$$h^{n\,M}_{j+1/2} - h^{n\,LW}_{j+1/2} = \frac{1}{\lambda} |\nu^n_{j+1/2}| (1 - |\nu^n_{j+1/2}|) \left(\frac{\varphi(r^n_{j+1/2})}{r^n_{j+1/2}} - \frac{1}{2} \right) \Delta u^n_{j+1/2} ,$$

Therefore for non critical points, since $r = 1 + (f_{,xx}/f_{,x})\,\Delta x + O(\Delta x^2)$, we have :

$$h^{n\,M}_{j+1/2} - h^{n\,LW}_{j+1/2} = \frac{1}{\lambda} |\nu^n_{j+1/2}| (1 - |\nu^n_{j+1/2}|) \left(\varphi(1) - \frac{1}{2} + B\,(\varphi_{,r}(1) - \varphi(1))\,\Delta x \right) u_{,x}\,\Delta x ,$$

with $B = f_{,xx}/f_{,x}$ such that $h^{n\,M}_{j+1/2} - h^{n\,LW}_{j+1/2} = O(\Delta x^2)$, under Lemma 2.3 assumptions ∎

According to this result and choosing a reduced flux $\varphi = 1/2\, r$ (obviously not TVD) which, after time improvement, leads to the Lax-Wendroff scheme, we see that the major part of the classical TVD corrections based upon the Lax-Wendroff scheme fall into the framework considered here. Besides these well-known corrections, a lot of TVD schemes can actually be derived from a TVD correction of any other 5-point linear scheme. We are able, for instance, to deduce Modified-Exquisite from a TVD correction of a five point linear scheme, nicknamed Quick [5], by choosing the appropriate pair $(M = 5/4 , \mu = 0)$ (see fig.2)

Schemes, derived under the latter guidelines, are second order accurate everywhere except at critical points where the TVD requirement makes them degenerate to first order accuracy. This systematic damping of local extrema leads the error to be at most $O(\Delta x)$ and $O(\Delta x^2)$ in the L^∞ and L^1-norm, respectively. To overcome this drawback, Harten and al [8], [9], have introduced the ENO schemes of high order accuracy in smooth regions. At this time, convergence results are unavailable but all numerical experiments enlighten their extreme stability. Quite recently, Shu [10] has proposed a Total Variation Bounded modification of some existing TVD schemes, involving the minmod function as a limiting process, in such a way that high order accuracy is achieved, including at critical points. Following Shu's approach, we present here a systematic procedure to convert TVD schemes into uniformly second order accurate schemes (in smooth regions), which provide bounded approximate solutions in $BV \cap L^\infty$. These schemes are less sophisticated than ENO schemes, and therefore less expensive ; on the other hand, they may produce more spurious oscillations at points of discontinuity. Anyhow, such oscillations are proved to be at most $O(\Delta x)$ in the L^∞-norm, numerical evidences ensures, in fact, $O(\Delta x^2)$.

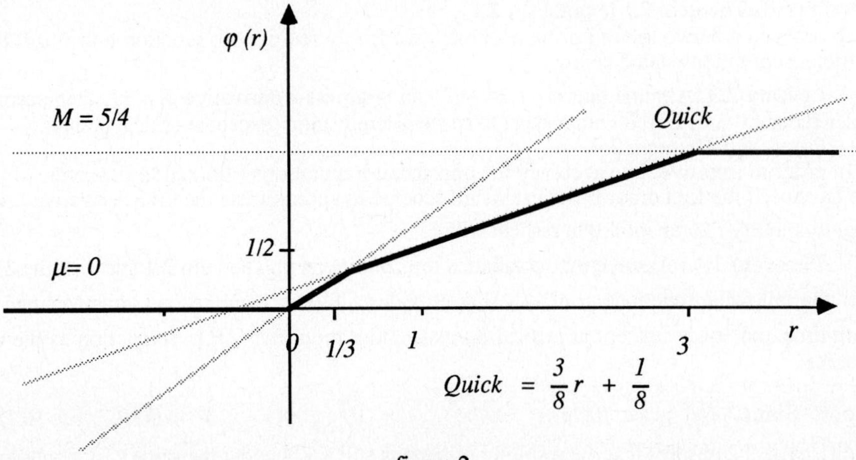

figure 2

III. TVB UNIFORMLY SECOND ORDER SCHEMES

A general requirement for the existence of a convergent subsequence in L^1_{loc} to a weak solution of (1.1), as Δt goes to 0, is that the family of approximate solutions satisfies to :

For all n and Δt such that $n\Delta t \leq T$

$$\begin{cases} TV(u^n) \leq A_{BV} \\ // u^n //_\infty \leq A_\infty \end{cases}, \tag{3.1}$$

for some fixed positive numbers A_{BV} (resp. A_∞) depending only on $TV(u_0)$, (resp. $// u_0 //_\infty$) and T. Let a numerical scheme be defined by :

$$u^n_{j+1} = u^n_j - \lambda (g^{nE}_{j+1/2} - g^{nE}_{j-1/2}) - \lambda (a^n_{j+1/2} - a^n_{j-1/2}), \tag{3.2}$$

$$\text{where} \quad a^n_{j+1/2} = \frac{1}{\lambda} | v^n_{j+1/2} | (1 - | v^n_{j+1/2} |) \frac{\varphi(r^n_{j+1/2})}{r^n_{j+1/2}} \Delta u^n_{j+1/2} . \tag{3.3}$$

Assume that φ fulfills Theorem 2.1 hypothesis then the scheme (3.2) (3.3) obviously satisfies inequalities (3.1) with $A_{BV} = TV(u_0)$ and $A_\infty = //u_0//_\infty$ since it preserves the BV and L^∞-norm. Let denote by φ^L the unlimited version of φ, in correspondence with the underlying 5-point linear scheme (see section II) and $a^n_{j+1/2}{}^L$ the associated (non corrected) antidiffusive flux. In order to provide the TVB modification procedure, we first modify the antidiffusive flux (3.3) thanks to :

$$\hat{a}^n_{j+1/2} = minmod(a^{n\,L}_{j+1/2} , a^n_{j+1/2}) , \tag{3.4}$$

where $\quad minmod(x , y) = \begin{cases} s\, min(|x| , |y|) & if\ s = sgn(x) = sgn(y) \\ 0 & otherwise \end{cases}$,

in such a way that we can state the :

Lemma 3.1 : Modification (3.4) preserves the $BV \cap L^\infty$ stability of the original scheme (3.2) (3.3) under the same C.F.L. like restriction.

Proof: Notice that : $\exists\ \theta^n_{j+1/2} \in [0,1]\ /\ \hat{a}^n_{j+1/2} = \theta^n_{j+1/2}\, a^n_{j+1/2} \qquad \forall j,n \in Z \times N$,

conclusion follows from Lemma 2.2.▮

Since TVD schemes usually require $a^n_{j+1/2}$ to vanish near a critical point, the minmod function will pick-up the second argument in (3.4), equals to 0 near such a point ; which is the source of local degeneracy of accuracy. In order to enforce the minmod function to pick-up the first argument in smooth regions, we introduce, following Shu's approach [10]

$$\bar{a}^n_{j+1/2} = minmod(a^{n\,L}_{j+1/2} , a^n_{j+1/2} + M\,\Delta x^2 sgn(a^{n\,L}_{j+1/2})) , \tag{3.5}$$

where M is a fixed positive constant. Therefore the resulting antidiffusive flux can be broken up into

$$\bar{a}^n_{j+1/2} = \hat{a}^n_{j+1/2} + d^n_{j+1/2} \qquad with \quad | d^n_{j+1/2} | \leq M\,\Delta x^2 . \tag{3.6}$$

The expected estimates on $BV \cap L^\infty$-norm can be performed thanks to (3.6) and give :

Theorem 3.2 : Assume that u_0 is compactly supported, then for any $M > 0$, the scheme (3.2) (3.5) is $BV \cap L^\infty$ stable in $0 \leq t \leq T$ with a fixed $T > 0$, under the same C.F.L. like restriction shared by the underlying TVD scheme (3.2) (3.3).

Proof: Let us consider $\bar{h}^n_{j+1/2} = (g^n_{j+1/2} + \hat{a}^n_{j+1/2}) + d^n_{j+1/2}$, and notice that $\hat{h}^n_{j+1/2} = g^{n\,E}_{j+1/2} + \hat{a}^n_{j+1/2}$, leads to a scheme (3.2) (3.4) satisfying both conditions *(i)* and *(ii)* of Lemma 1.1 . Therefore starting with

$$u^{n+1}_j = u^n_j - C^n_{j-1/2}\,\Delta u^n_{j-1/2} + D^n_{j+1/2}\,\Delta u^n_{j+1/2} - \lambda\,(d^n_{j+1/2} - d^n_{j-1/2}) , \tag{3.7}$$

we easily see that :

$$| u^n_{j+1} | \leq (1 - C^n_{j-1/2} + D^n_{j+1/2})\, | u^n_j | + C^n_{j-1/2}\, | u^n_{j-1} | + D^n_{j+1/2}\, | u^n_{j+1} | + 2\lambda M\,\Delta x^2 ,$$

from which we deduce: $\| u^{n+1} \|_\infty \leq \| u^n \|_\infty + 2M\,\Delta t\,\Delta x$, and indeed the expected estimate :

$$\| u^{n+1} \|_\infty \leq \| u_0 \|_\infty + (2MT)\,\Delta x \quad since \quad n\Delta t \leq T .$$

Moreover, in order to perform the BV estimate, let observe that thanks to (3.7) :

$$TV(u^{n+1}) \leq TV(u^n) + 2\lambda \sum_j | d^n_{j+1/2} - d^n_{j-1/2} | ,$$

$$\leq TV(u^n) + 4\lambda\, ML\,\Delta x ,$$

where $L = \Sigma\Delta x$ is the length of the x-support of u at time T (compact if u_0 is compactly supported) and therefore $TV(u^{n+1}) \leq TV(u_0) + 4MLT$. ▮

It seems to turn out from that proof that TVB modification is unable to cope with steady-state computations since T may become very large. In fact, inequality (3.6) is a bit too crude since, as Shu has pointed out [10], its upper bound is effective only near critical points or points of discontinuity, whose number does not depend on the mesh refinement. Numerical evidences lead us to think that the last estimates can be improved, in most cases, to :

$$\begin{cases} TV(u^n) \leq TV(u_0) + O(\Delta x) \\ \| u^n \|_\infty \leq \| u_0 \|_\infty + O(\Delta x^2) , \end{cases}$$

and actually allow steady-state computations. These estimates are also of some importance since they show that our TVB schemes do not suffer from a $O(1)$ Gibbs-like spurious oscillations at points of discontinuity. The price, we pay, in order to achieve second order accuracy at critical points, namely the loss of monotonicity, is not so expensive since spurious oscillations seem to be at most $O(\Delta x^2)$ in the L^∞-norm.

The following statement enlightens the role played by the constant M in the accuracy improvement of the TVB modification :

Theorem 3.3 : For any given $A > 0$, there exists a constant $M > 0$ such that the scheme (3.2) (3.5) is second order accurate in any region where the u-derivatives are bounded by A.

Proof : Let us consider $a^{n\,L}_{j+1/2} = \dfrac{1}{\lambda} | v^n_{j+1/2} | (1 - | v^n_{j+1/2} |) \dfrac{\varphi^L(r^n_{j+1/2})}{r^n_{j+1/2}} \Delta u^n_{j+1/2}$.

Notice that, near a critical point (sonic point or extremum) :

$$\frac{1}{\lambda} | v^n_{j+1/2} | \Delta u^n_{j+1/2} = sgn(v^n_{j+1/2}) \Delta f^n_{j+1/2} = sgn(v^n_{j+1/2}) f_{,xx} \frac{\Delta x^2}{2} + O(\Delta x^3) ,$$

we have also
$$\frac{1}{\lambda} | v^n_{j+1/2} | \frac{\Delta u^n_{j+1/2}}{r^n_{j+1/2}} = - f_{,xx} \frac{\Delta x^2}{2} + O(\Delta x^3) ,$$

and therefore
$$| a^{n\,L}_{j+1/2} | \leq Min \left(| \varphi^L(r^n_{j+1/2}) | , \left| \frac{\varphi^L(r^n_{j+1/2})}{r^n_{j+1/2}} \right| \right) B \, \Delta x^2 ,$$

$$\text{where} \qquad B = Max((1 - | v |) f_{,xx} / 2) .$$

Thanks to our assumptions on the u-derivatives, B is a bounded constant, depending on A. It can easily be seen that $Min \left(| \varphi^L(r) | , \left| \dfrac{\varphi^L(r)}{r} \right| \right)$, is also bounded by a fixed constant C for any $r \in R$, (since $\varphi^L(r) = ar + b$) and thus, if one chooses M greater than the product $B\,C$, the minmod function will pick-up the first argument $a^{n\,L}_{j+1/2}$ and leads to second order accuracy at critical points.∎

Our TVB modification procedure is also available for TVD schemes involving a more sophisticated limiter function such as Van Leer's limiter [11], for instance. Since the underlying linear scheme is, in this case, the Lax-Wendroff scheme whose reduced flux is

$\psi^{LW} = \frac{1}{2}(1-|\nu|)r$, we are lead to propose :

$$h_{j+1/2}^{n\,VL} = g_{j+1/2}^{n\,E} + minmod(\,a_{j+1/2}^{n\,LW}, a_{j+1/2}^{n\,VL} + M\,\Delta x^2\,sgn(\,a_{j+1/2}^{n\,LW}\,)\,)\,.$$

We underline the major advantage of the TVB schemes designed under the latter guidelines : They are second order accurate in smooth regions, including at critical points, while the approximate solutions they provide, remain bounded with respect to the $BV \cap L^\infty$-norm, they thus allow the existence of a convergent subsequence in $L^1{}_{loc}$ to a weak solution of (1.1), as Δx vanishes. If an additional entropy condition is satisfied, which implies, in the scalar case, that the limit solution is the unique physically relevant solution of (1.1), then the scheme is convergent. The next section is devoted to this additional requirement. The lack of consistency with an entropy inequality from which the TVD (or TVB) schemes suffer, is overcome by addition of a limiting mechanism to the antidiffusive flux which enforces the selection of the physical solution.(see Vila [3] and also Leroux and Quesseveur [12]).

IV. ENTROPY CORRECTION

For the theorical materials, the reader is referred to Lax [13]. Tadmor [14] has shown that E-schemes were converging towards the unique physically relevant solution, but such a result does not yield for a general TVD (or TVB) scheme. These schemes do not automatically select the physically relevant solution. In order to achieve convergence towards the entropy satisfying solution, we need to slightly modify the TVD (TVB) schemes so far considered, according to Vila's procedure [3]. In this paper, an extra-limiting correction is added to the antidiffusive flux $a^n{}_{j+1/2}$ in such a way that $a^n{}_{j+1/2}$ is enforced to vanish with the mesh size. We are unable to show any consistency with an entropy condition but thanks to such a correction, Vila has proved the following theorem :

Theorem 4.1 : Let be a finite difference scheme (1.2) defined by its numerical flux

(4.1) $h_{j+1/2}^{n} = g_{j+1/2}^{n\,E} + a_{j+1/2}^{n}$ such that the first order underlying E- scheme is entropy stable ([14]). Assume that the antidiffusive coefficients $a^n{}_{j+1/2}$ satisfy :

$$|\,a_{j+1/2}^{n}\,| \leq e(\,\Delta x\,) \qquad \forall j \in Z \qquad with \qquad \lim_{\Delta x \to 0} e(\,\Delta x\,) = 0$$

then if the approximate solution $U^h(x, t)$ computed by (1.2) (4.1) is bounded in BV and L^∞, $U^h(x, t)$ converges towards the unique physically relevant solution of the initial value problem (1.1).

Following Vila's approach, we correct the antidiffusive flux related to our previous TVD (or TVB) schemes by :

$$a_{j+1/2}^{n\,M} = sgn(\,a_{j+1/2}^{n}\,)\;Min\,(\,C\,\Delta x^{\beta}, |\,a_{j+1/2}^{n}\,|\,) \qquad \forall j \in Z \;\; where \;\; \beta \in \,]0,1[\quad , \quad (4.2)$$

in such a way that we can state :

Lemma 4.2 : The scheme (3.2) (4.2) is still $BV \cap L^\infty$ stable under the same C.F.L. like restriction.

Proof: $\exists\, \theta^n_{j+1/2} \in [0,1]\ /\ a^{n\,M}_{j+1/2} = \theta^n_{j+1/2}\, a^n_{j+1/2} \qquad \forall j,n \in Z \times N$

therefore $a^{n\,M}_{j+1/2} = \theta^n_{j+1/2}\, \hat{a}^n_{j+1/2} + \theta^n_{j+1/2}\, d^n_{j+1/2}$; notice that $\theta^n_{j+1/2}\, \hat{a}^n_{j+1/2}$ still defines a TVD schemes (Lemma 2.2) and that $|\, \theta^n_{j+1/2}\, d^n_{j+1/2} | \leq M\, \Delta x^2$.∎

Since $|\, a^{n\,M}_{j+1/2} | \leq C\, \Delta x^\beta$ *and* $\lim_{\Delta x \to 0} C\, \Delta x^\beta = 0$, thanks to lemma 4.2 and theorem 4.1, we claim that convergence is achieved towards the unique physically relevant solution. Notice that in any region where the u-derivatives are bounded by a fixed number *A*, there exists a constant *C* depending on *A* such that $|\, a^n_{j+1/2} | \leq C\, \Delta x^\beta$ and therefore entropy correction is not active This allows the accuracy to be preserved in smooth regions.

V. NUMERICAL EXPERIMENTS

In this section, we present several numerical results performed with the schemes designed in sections II and III. We use $M = 50$ for all computations and a fixed mesh ratio $\lambda = 0.1$. All tables are collected at the end.

Example 1: The Modified-Exquisite (denoted M.E.) TVD scheme and its TVB modification are used to solve a Riemann problem of Burger's equation :

$$u_{,t} + \left(\frac{u^2}{2}\right)_{,x} = 0 \qquad u_0(x) = \begin{cases} u_L = 2 & x < 0 \\ u_R = -1 & x > 0 \end{cases}$$

The exact solution is a moving shock at speed 1/2. We use $\Delta x = 1/80,\ T = 1.0$ and print out the numerical shock transition below (the starred postions are the positions of the shock) :

M.E. TVD scheme	: *2.0000*	*1.9999*	*1.9998*	*1.9058 **	*-0.7627*	*-1.0000*	*-1.0000* ,
M.E. TVB scheme	: *2.0005*	*1.9999*	*2.0041*	*1.9090 **	*-0.7642*	*-1.0000*	*-1.0000* ,
Unlimited M.E. (Quick):	*2.0536*	*1.8727*	*2.1744*	*1.9461 **	*-0.8866*	*-1.0000*	*-1.0000* ,

For $\Delta x = 1/40,\ T = 1.0$ the M.E. TVB scheme has computed:

*2.0024 1.9930 2.0158 1.9139 * -0.7759 -1.0000 -1.0000* .

We observe that the oscillations produced by the TVB scheme are small ($\leq 0.5\, M\, \Delta x^2$) while they grow larger (order $O(1)$) for the linear scheme Quick as expected.

Example 2 : TVD and TVB Modified-Exquisite schemes are used to solve two linear problems with smooth initial condition :

$$u_{,t} + u_{,x} = 0 \qquad u_0(x) = \sin(\pi x) \quad (5.1) \qquad \text{then} \quad u_0(x) = \exp(-x^2 / 0.04) \quad (5..2)$$

For I.V.P. (5.1), the solution is computed in $-1 \leq x < 1$ up to $t = 2$ (after one period in time); and for I.V.P (5.2), up to $t = 0.5$. The errors and numerical orders are listed in Table 1. Let observe that the TVD scheme is first order accurate in L^∞-norm and second order in L^1-norm while the TVB scheme has full order of accuracy both in the L^∞ and L^1-norm, as expected by the theory.

Example 3 : Same schemes are used to solve the nonlinear Burger's equation with periodic initial condition :

$$u_{,t} + \left(\frac{u^2}{2}\right)_{,x} = 0 \qquad -1 \leq x \leq 1 \qquad u_0(x) = \sin(\pi x + \pi).$$

The exact solution is smooth up to $t = 1/\pi$, when a stationnary shock develops at $x = 0$, which actually interacts with the expansion waves. The "exact" solution is computed by Newton-Raphson iterations (For details, see [9]). The errors of the approximate solutions are computed, at $t = 0.15$, in such a way that the exact solution is still smooth; then for $t = 1/\pi$, at which time the shock appears, in smooth region away from the shock ($|x\text{-}shock| \geq 0.05$). Errors and numerical orders are listed in Table 2. From this table, we see that the TVB scheme is globally second order accurate, in smooth regions, while the TVD one is only first order in the L^∞-norm.

Table 1

		I.V.P (5.1)				I.V.P (5.2)			
E	Δx	TVD	r	TVB	r	TVD	r	TVB	r
L^∞	1/10	9.26(-2)		3.96(-2)		2.18(-1)		1.43(-1)	
	1/20	3.26(-2)	1.51	8.89(-3)	2.16	9.07(-2)	1.27	3.87(-2)	1.89
	1/40	1.16(-2)	1.50	2.12(-3)	2.11	3.37(-2)	1.35	8.73(-3)	2.02
	1/80	4.18(-3)	1.49	5.49(-4)	2.06	1.21(-2)	1.39	2.06(-3)	2.04
L^1	1/10	5.69(-2)		3.28(-2)		6.24(-2)		6.70(-2)	
	1/20	1.55(-2)	1.88	6.65(-3)	2.30	2.15(-2)	1.54	1.50(-2)	2.16
	1/40	3.82(-3)	1.95	1.48(-3)	2.23	5.21(-3)	1.79	3.34(-3)	2.16
	1/80	9.25(-4)	1.98	3.58(-4)	2.17	1.37(-3)	1.84	7.94(-4)	2.13

Table 2

		$t = 0.15\ s$				$t = 1/\pi\ s \quad \lvert x - shock \rvert \geq 0.05$			
E	Δx	TVD	r	TVB	r	TVD	r	TVB	r
L^∞	1/10	1.72(-2)		1.49(-2)					
	1/20	9.16(-3)	0.91	3.71(-3)	2.01				
	1/40	3.43(-3)	1.16	8.78(-4)	2.04	5.76(-3)		6.02(-4)	
	1/80	1.26(-3)	1.26	2.14(-4)	2.04	1.85(-3)	1.64	1.44(-4)	2.06
L^1	1/10	1.23(-2)		8.12(-3)					
	1/20	3.40(-3)	1.85	1.84(-3)	2.14				
	1/40	9.69(-4)	1.83	4.30(-4)	2.12	1.48(-3)		4.56(-4)	
	1/80	2.34(-4)	1.90	1.04(-4)	2.10	3.36(-4)	2.14	1.04(-4)	2.13

REFERENCES

[1] A. Harten : "High resolution schemes for hyperbolic conservation laws.", J. Comp. Physics, vol 49, 1983, pp357-393.

[2] J.P. Vila : THESIS Université PARIS VI (1986).

[3] J.P. Vila : "High order schemes and entropy condition for nonlinear hyperbolic systems of conservation laws", C.M.A.P. report 111, Ecole Polytechnique, Palaiseau France (1986).

[4] P.K. Sweby : "High resolution schemes using flux limiters for hyperbolic conservation laws.", SIAM J. Numer. Anal., vol 21 ,1984, pp 995-1011.

[5] B.P. Leonard : "A survey of finite differences with upwinding for numerical modelling of the incompressible convective diffusion equation.", in Computational Techniques in Transient and Turbulent Flows, edited by Taylor & Morgan, 1981 , pp 1-23.

[6] I. Maekawa & T. Murumatsu : "High order differencing schemes in fluid flows analysis.", 5th IAHR Liquid Metal Working Group Meeting, Grenoble, France (June 23-27 1986).

[7] S. Osher & S. Chakravarthy : "High resolution schemes and the entropy condition.", SIAM J. Numer. Anal., vol 21, n° 5, 1984, pp 955-983.

[8] A. Harten & S. Osher : "Uniformly high order accurate nonoscillatory schemes, I.", SIAM J. Numer. Anal., vol 24, 1987, pp 279-309.

[9] A. Harten, B. Enquist, S. Osher & S. Chakravarthy : "Uniformly high order accurate essentially non-oscillatory schemes, III.", J. Comp. Physics, vol 71, 1987, pp 231-303.

[10] C. W. Shu : "T.V.B. uniformly high order schemes for conservation laws.", Math. Comp., vol 49, 1987, pp 105-121.

[11] B. Van Leer : "Towards the ultimate conservative difference scheme, IV. A new approach to numerical convection.", J. Comp. Physics , vol 23, 1977, pp 276-299.

[12] A. Y. Leroux & P. Quesseveur : "Convergence of an antidiffusive Lagrange-Euler scheme for quasi linear equations", SIAM J.Numer. Anal., vol 21, n° 5, 1984, pp.985-994.

[13] P. D. Lax : "Shock waves and entropy.", in "Contribution to nonlinear functionnal analysis.", E. A. Zerantonello (Ed.), Academic presss, 1971, pp 603-634.

[14] E. Tadmor : "Numerical viscosity and the entropy condition for conservative difference schemes.", Math. Comp., vol 43, 1984, pp 369-381.

[15] F.Coquel : EDF report, to be published (in French) .

HIGH RESOLUTION FINITE VOLUME SCHEMES AND COMPUTATIONAL AERODYNAMICS

D M Causon
Department of Mathematics & Physics
Manchester Polytechnic
Manchester M1 5GD
UK

SUMMARY

The underlying theory is reviewed for the construction of total variation diminishing (TVD) finite-difference schemes. Using these methods and artificial compression techniques, a high resolution version of the well-known MacCormack scheme is constructed. Applications of the method to high speed external aerodynamic flows are reported which use a finite volume formulation and operator-splitting. Existing production code implementations of MacCormack's method can be updated easily, as described, to reflect recent advances in one-dimensional schemes.

INTRODUCTION

Over the last fifteen years, substantial advances have been made in the numerical analysis of hyperbolic partial differential equations, particularly in those arising in computational aerodynamics. A popular approach is to solve the Euler equations numerically by marching forward through time using an appropriate finite-difference or finite-volume scheme, capturing any shock waves or contact discontinuities which may occur. A notable advance in the 1970's was the introduction of the method of operator-splitting by which a three-dimensional problem may be solved by applying a sequence of one-dimensional difference operators; each operator relating to a specific co-ordinate direction. This led to a focus on one-dimensional schemes. The early schemes of the 1970's were analysed using linear stability theory and those that were intended to capture shock waves were designed to be dissipative. However, the ability of these schemes accurately to resolve shock waves and contact discontinuities was impaired by the appearance of oscillations around the profile of the discontinuity. Whilst linear stability is useful for checking "local" stability of schemes applied to non-linear equations, it is often insufficient, particularly when strong discontinuities are present. The classical approach for damping oscillations and removing non-linear instabilities was to introduce extra artificial viscosity explicitly, in order to increase that which is present in the scheme implicitly. Unfortunately, this approach tended to cause a spurious entropy layer to form, emanating particularly from regions of high spatial gradient like shock waves and stagnation points. In cases of strong shock interactions this palliative often failed completely. The problem was that the added dissipative term invariably had a global and problem-dependent coefficient. In some parts of the flow field too much damping was being applied, and in others too little. What was needed was to apply the term more selectively.

In the 1980's various workers [1-9] began to put the procedures on a more systematic footing. There was renewed interest in Godunov methods and in Riemann solvers generally. Out of this work emerged the total variation diminishing (TVD) schemes, so-named by Harten [1], who gave a set of theorems, leading to conditions, which when satisfied by any three-point, conservative, second-order accurate finite-difference scheme are necessary and sufficient for the scheme to be TVD. As a sequel to this work Davis [2] showed that it is possible to formulate a classical Lax-Wendroff scheme in TVD form. This is accomplished by appending to the scheme a non-linear term which applies precisely the correct amount of artificial viscosity needed at each mesh point to limit overshoots and undershoots. This work opened the way to updating many existing production computer codes which employ the well-known MacCormack scheme, a variant of Lax-Wendroff and the subject of this paper. The theory of TVD schemes is developed for application to scalar conservation laws. The resulting schemes are then extended to non-linear equations and systems of conservation laws.

TVD FINITE DIFFERENCE SCHEMES

Consider the scalar conservation law

$$\frac{\partial u}{\partial t} + a\frac{\partial u}{\partial x} = 0 \tag{1}$$

with

$$u[x,0] = u_0(x),\ x \in R,$$

where a is a real constant. Let U_j^n be the numerical solution of (1) at $x = j\Delta x$, $t = n\Delta t$ with Δx the spatial mesh size and Δt the time step. Formally, if the total variation in the numerical solution is given by

$$\mathrm{TV}\left[U_j^n\right] = \sum_j \left|U_{j+1}^n - U_j^n\right| \tag{2}$$

then a scheme which is TVD satisfies the condition

$$\mathrm{TV}\left[U_j^{n+1}\right] \leqslant \mathrm{TV}\left[U_j^n\right]. \tag{3}$$

A general, three-point, conservative finite difference scheme for solving (1) can be written in the form

$$U_j^{n+1} = U_j^n - C_{j-\frac{1}{2}}\,\Delta U_{j-\frac{1}{2}}^n + D_{j+\frac{1}{2}}\,\Delta U_{j+\frac{1}{2}}^n \tag{4}$$

where

$$\Delta U_{j+\frac{1}{2}}^n = U_{j+1}^n - U_j^n\ ,\quad \Delta U_{j-\frac{1}{2}}^n = U_j^n - U_{j-1}^n.$$

Harten [1] showed that this scheme will be TVD iff the following conditions on coefficients C,D are satisfied

$$0 \leqslant C_{j-\frac{1}{2}}\ , \quad 0 \leqslant D_{j+\frac{1}{2}}\ , \quad 0 \leqslant C_{j-\frac{1}{2}} + D_{j+\frac{1}{2}} \leqslant 1. \tag{5}$$

Second-order accurate schemes which do not satisfy conditions (5) are not TVD and will exhibit oscillations around shock waves.

Sweby [3] used the concept of the "flux limiter" to construct a second-order accurate TVD scheme. His starting point was the well-known Lax-Wendoff (LW) scheme

$$U_j^{n+1} = U_j^n - \upsilon\Delta U_{j-\frac{1}{2}}^n - \nabla\left[\tfrac{1}{2}\ (1-\upsilon)\ \upsilon\ \Delta u_{j+\frac{1}{2}}^n\right] \tag{6}$$

$$\text{where } \upsilon = a\,\frac{\Delta t}{\Delta x} \tag{7}$$

and which can be written as the first-order scheme

$$U_j^{n+1} = U_j^n - \upsilon\Delta U_{j-\frac{1}{2}}^n \tag{8}$$

with the additional anti-diffusive flux term

$$-\ \nabla\left[\tfrac{1}{2}\ (1-\upsilon)\ \upsilon\ \nabla u_{j+\frac{1}{2}}^n\right]\ . \tag{9}$$

Since (8) is known to be non-oscillatory (because it is first-order) he proceeded by adding only a limited amount of the anti-diffusive flux (9) to give

$$U_j^{n+1} = U_j^n - \upsilon\Delta U_{j-\frac{1}{2}}^n - \nabla\left[\Phi\ \tfrac{1}{2}(1-\upsilon)\upsilon\ \Delta U_{j+\frac{1}{2}}^n\right] \tag{10}$$

where Φ is a flux limiter, to be defined.

More specifically, Sweby chose a particular form for scheme (10), given by

$$U_j^{n+1} = U_j^n - \upsilon\left\{\ 1 + \tfrac{1}{2}\ (1-\upsilon)\ \left[\ \Phi\left[r_j^+\right] \Big/\ r_j^+ - \Phi\left[r_{j-1}^+\right]\right]\right\}\ \Delta U_{j-\frac{1}{2}}^n \tag{11}$$

where

$$r_j^+ = \frac{\Delta U_{j-\frac{1}{2}}^n}{\Delta U_{j+\frac{1}{2}}^n} \tag{12}$$

and which, so constructed, reproduces the LW scheme if $\Phi(r) = 1$ and the explicit, second-order, upwind scheme (WB) of Warming and Beam [10] if $\Phi(r) = r$.

By comparing (11) with the general form (4) it is found that

$$C_{j-\frac{1}{2}} = \upsilon\left\{\ 1+\tfrac{1}{2}\ (1-\upsilon)\left[\frac{\Phi\left[r_j^+\right]}{r_j^+} - \Phi\left[r_{j-1}^+\right]\right]\right\} \tag{13}$$

$$D_{j+½} = 0 \tag{14}$$

and if we impose the constraints (5) on C,D we find that for scheme (11) to be TVD the flux limiter function $\Phi(r)$ must satisfy

$$0 \leqslant \left[\frac{\Phi(r)}{r}, \Phi(r) \right] \leqslant 2. \tag{15}$$

under the CFL condition $|\nu| \leqslant 1$. That is to say, the flux limiter must lie within the shaded region of Fig 1a. One can further show that the domain illustrated in Fig 1b corresponds to the second-order TVD region. Neither of the schemes LW or WB lies uniformly within the shaded region which explains why they are not TVD.

Following the work of Harten and Sweby, Davis [2] showed that it is possible to modify the LW scheme to be TVD. Since it is known that LW coefficients C,D do not satisfy constraints (5), his approach was to append to the scheme a term which changes the coefficients such that they do satisfy (5). A suitable term is

$$G^+\left[r_j^+\right]\Delta U_{j+½}^n - G^+\left[r_{j-1}^+\right]\Delta U_{j-½}^n \tag{16}$$

whereupon we can recover Sweby's TVD scheme (11) if we choose

$$G^+\left[r_j^+\right] = \frac{\nu}{2}(1-\nu)\left[1 - \Phi\left[r_j^+\right]\right], \tag{17}$$

subject to $\Phi(r)$ satisfying (15) and $|\nu| \leqslant 1$. Term (16) is effectively upwind weighted therefore the functional form (17) of $G^+(r_j^\pm)$ changes as the sign of a in (1) changes. This unwanted complication can be removed to produce a five-point, symmetric, TVD scheme by replacing (16) by

$$\begin{aligned} &\left[G^+\left[r_j^+\right] + G^-\left[r_{j+1}^-\right]\right]\Delta U_{j+½}^n \\ &- \left[G^+\left[r_{j-1}^+\right] + G^-\left[r_j^-\right]\right]\Delta U_{j-½}^n \end{aligned} \tag{18}$$

where

$$r_j^- = \frac{\Delta U_{j+½}^n}{\Delta U_{j-½}^n}, \tag{19}$$

$$G^\pm\left[r_j^\pm\right] = \frac{|\nu|}{2}(1-|\nu|)\left[1 - \Phi\left[r_j^\pm\right]\right]. \tag{20}$$

These results can be extended easily to scalar non-linear problems by defining a local Courant number to replace the global definition (7).

In order to extend the scheme to hyperbolic systems we first consider the constant coefficient system

$$\frac{\partial \underset{\sim}{U}}{\partial t} + A\frac{\partial \underset{\sim}{U}}{\partial x} = 0 \tag{21}$$

where U is a vector of m components and A is an m x m constant matrix. If P denotes the matrix whose columns are the right eigenvectors of A, then

$$P^{-1}AP = \Lambda = \mathrm{diag}\,(\lambda_\ell) \qquad (22)$$

when λ_ℓ are the eigenvalues of A.

We proceed by defining a new set of dependent variables

$$\underset{\sim}{V} = P^{-1}\underset{\sim}{U} \qquad (23)$$

then, multiplying (21) by P^{-1} we have

$$(P^{-1}\underset{\sim}{U})_t + P^{-1}A(\underset{\sim}{U})_x = 0 \qquad (24)$$

or $\underset{\sim}{V}_t + \Lambda\underset{\sim}{V}_x = 0$ (25)

which is an uncoupled set of scalar equations. We may solve (25) by applying the TVD LW scheme to each scalar equation in turn, that is

$$\begin{aligned} V_j^{n+1} &= V_j^n - \frac{\nu}{2}\left[V_{j+1}^n - V_{j-1}^n\right] + \frac{\nu^2}{2}\left[V_{j+1}^n - 2V_j^n + V_{j-1}^n\right] \\ &\quad + \left[G^+\left[r_j^+\right] + G^-\left[r_{j+1}^-\right]\right]\Delta V_{j+1/2}^n \\ &\quad - \left[G^+\left[r_{j-1}^+\right] + G^-\left[r_j^-\right]\right]\Delta V_{j-1/2}^n \,. \end{aligned} \qquad (26)$$

To obtain a scheme in terms of the original dependent variables we multiply by P^{-1}. The result is

$$\begin{aligned} \underset{\sim}{U}_j^{n+1} &= \underset{\sim}{U}_j^n - A\frac{\Delta t}{2\Delta x}\left[\underset{\sim}{U}_{j+1}^n - \underset{\sim}{U}_{j-1}^n\right] + A^2\frac{\Delta t^2}{2\Delta x^2}\left[\underset{\sim}{U}_{j+1}^n - 2\underset{\sim}{U}_j^n + \underset{\sim}{U}_{j-1}^n\right] \\ &\quad + P\left[G^+\left[r_j^+\right] + G^-\left[r_{j+1}^-\right]\right]P^{-1}\Delta\underset{\sim}{U}_{j+1/2}^n - P\left[G^+\left[r_{j-1}^+\right] + G^-\left[r_j^-\right]\right]P^{-1}\Delta\underset{\sim}{U}_{j-1/2}^n \end{aligned} \qquad (27)$$

However, scheme (27) requires that we compute matrices P and P^{-1}. This restriction can be removed by approximating the diagonal matrices $G^\pm$ by scalar matrices, ie

$$G^\pm(r^\pm) \simeq \bar{G}^\pm(r^\pm)I \qquad (28)$$

where I is the m x m unit matrix and $r^\pm$ are chosen to be scalar functions of $\underset{\sim}{U}$,

$$r_j^+ = \frac{\left[\Delta\underset{\sim}{U}_{j-1/2}^n\,,\ \Delta\underset{\sim}{U}_{j+1/2}^n\right]}{\left[\Delta\underset{\sim}{U}_{j+1/2}^n\,,\ \Delta\underset{\sim}{U}_{j+1/2}^n\right]} \qquad (29)$$

$$r_j^- = \frac{\left[\Delta\underset{\sim}{U}_{j-1/2}^n\,,\ \Delta\underset{\sim}{U}_{j+1/2}^n\right]}{\left[\Delta\underset{\sim}{U}_{j-1/2}^n\,,\ \Delta\underset{\sim}{U}_{j-1/2}^n\right]} \qquad (30)$$

where (.,.) denotes the inner product on R^m and m is the number of components of the solution vector $\underset{\sim}{U}_j^n$. If P does not vary significantly over adjacent mesh points, definitions (29),(30) can be interpreted as averages of the earlier scalar definitions. Other definitions of $r^{\pm}$ can be constructed but those indicated have been found to work well in numerical experiments.

With these approximations, the method takes the form:

$$U_j^{n+1} = U_j^n - A\frac{\Delta t}{2\Delta x}\left[U_{j+1}^n - U_{j-1}^n\right] + A^2\frac{\Delta t^2}{2\Delta x^2}\left[U_{j+1}^n - 2U_j^n + U_{j-1}^n\right]$$

$$+ \left[\bar{G}^-\left(r_j^+\right) + \bar{G}^-\left(r_{j+1}^-\right)\right]\Delta \underset{\sim}{U}_{j+\frac{1}{2}}^n$$

$$- \left[\bar{G}^+\left(r_{j-1}^+\right) + \bar{G}^-\left(r_{j+1}^-\right)\right]\Delta \underset{\sim}{U}_{j-\frac{1}{2}}^n \ . \tag{31}$$

The resulting scheme (31) does not depend explicitly on the transformation (22), so it can be used without modification for non-linear problems. In our applications we replace the first three terms of (31) by the finite volume MacCormack scheme. This scheme is equivalent to the LW scheme for linear problems.

TVD MACCORMACK FINITE-VOLUME SCHEME (TVDM)

Since we anticipate the use of a body-fitted mesh, we cast the equations of motion in generalised co-ordinates. The Euler equations, in strong conservation form, are

$$\frac{\partial U}{\partial t} + \frac{\partial}{\partial x^\ell}\left(\sqrt{g}^\ell \underline{\underline{F}}\right) = 0 \ , \ \ell = 1(1)3 \tag{32}$$

where $\underline{U} = \sqrt{g}\,[\rho, \rho w_1, \rho w_2, \rho w_3, e]^T$, $\sqrt{g}^\ell \underline{\underline{F}}(u) = \begin{bmatrix} \rho u^\ell \\ \rho w_1 u^\ell + p\dfrac{\partial x^\ell}{\partial Z_1} \\ \rho w_2 u^\ell + p\dfrac{\partial x^\ell}{\partial Z_2} \\ \rho w_3 u^\ell + p\dfrac{\partial x^\ell}{\partial Z_3} \\ (e+p)u^\ell \end{bmatrix}$,

$\sqrt{g}$ is the Jacobian and the flow velocity $\underline{q} = U^\ell \underline{g}_\ell = w_\ell \underline{a}_\ell$, where $\underline{a}_\ell$ are the Cartesian unit base vectors.

Equation (32) can also be cast in integral form, which is the basis of the finite volume method

$$\frac{\partial}{\partial t}\iiint_{vol}\underline{U}dvol + \iint_{S_1}{}^1\underline{\underline{F}}ds + \iint_{S_2}{}^2\underline{\underline{F}}ds + \iint_{S_3}{}^3\underline{\underline{F}}ds = 0 \ . \tag{33}$$

The appeal of the finite volume formulation becomes apparent by noting that $\partial x^\ell/\partial z_m = \underline{g}^\ell.\underline{a}_m$ and $\underline{q}.\underline{g}^\ell = u^\ell$ whereupon

$$\sqrt{g}^{\ell}\underline{\underline{F}}(U) = \sqrt{g}\begin{bmatrix} \rho\underline{g}.\underline{g}^{\ell} \\ \rho w_1\underline{g}.\underline{g}^{\ell} + p\underline{g}^{\ell}.\underline{a}_1 \\ \rho w_2\underline{g}.\underline{g}^{\ell} + p\underline{g}^{\ell}.\underline{a}_2 \\ \rho w_3\underline{g}.\underline{g}^{\ell} + p\underline{g}^{\ell}.\underline{a}_3 \\ (e+p)\underline{g}.\underline{g}^{\ell} \end{bmatrix} = \begin{bmatrix} \rho\underline{g} \\ \rho w_1\underline{g} + p\underline{a}_1 \\ \rho w_2\underline{g} + p\underline{a}_2 \\ \rho w_3\underline{g} + p\underline{a}_3 \\ (e+p)\underline{g} \end{bmatrix} . \sqrt{g}\, \underline{g}^{\ell} = \underline{\underline{H}}(U).\underline{S}^{\ell} \quad (34)$$

and we see that the computations can be performed with respect to the Cartesian flux tensor $\underline{\underline{H}}(U)$, rather than the curvilinear $\sqrt{g}^{\ell}\underline{\underline{F}}(U)$. Thus one need not become involved with the intricacies of the co-ordinate transformation.

We solve (33) using a factored sequence of one-dimensional finite volume operators where each component operator relates to its respective curvilinear co-ordinate direction. Further details may be found in [11].

The MacCormack finite volume operator $L_1(\Delta t)$ is

$$\overline{\underline{U}_{ijk}^{n+1}} = \underline{U}_{ijk}^{n} - \Delta t\left[\underline{\underline{H}}_{ijk}^{n}\,\underline{S}_{i+½} + \underline{\underline{H}}_{i-ijk}^{n}\,\underline{S}_{i-½}\right] \quad (35a)$$

$$U_{ijk}^{n+1} = ½\left[U_{ijk}^{n} + \overline{U_{ijk}^{n+1}} - \Delta t\left[\overline{\underline{\underline{H}}_{i+ijk}^{n+1}}\underline{S}_{i+½} + \overline{\underline{\underline{H}}_{ijk}^{n+1}}\,\underline{S}_{i-½}\right]\right] \quad (35b)$$

where $U_{ijk} = \text{vol}_{ijk}\left[\rho,\rho w_1,\rho w_2,\rho w_3,e\right]_{ijk}^{T}$ and $S_{i\pm½}$ are the area vectors on opposite faces of the cell, normal to the surface x^1 = constant.

Scheme (35) can be updated easily to TVD form by appending to the right-hand side of the "corrector" step (35b) the term:

$$\left[G_i^{+} + G_{i+1}^{-}\right]\Delta\underline{U}_{i+½}^{n} - \left[G_{i-1}^{+} + G_i^{-}\right]\Delta\underline{U}_{i-½}^{n} \quad (36)$$

where

$$G_i^{\pm} = G\left[r_i^{\pm}\right] = 0.5\ C(\nu)\left[1 - \Phi\left[r_i^{\pm}\right]\right] \quad (37)$$

and we have chosen

$$C(\nu) = \begin{cases} \nu(1-\nu) & , \nu \leqslant 0.5 \\ 0.25 & , \nu > 0.5 \end{cases}, \quad (38)$$

which is an upper bound to the Courant number dependent coefficient in (20). Also

$$\nu = \nu_i = \max_{\ell} |\lambda_\ell| \frac{\Delta t}{\Delta x^1} \quad (39)$$

where $\max_{\ell} |\lambda_\ell| = |u^1| + C$, with u^1 and C the local flow speed and sound speed respectively.

The flux limiter which we use is

$$\Phi(r) = \begin{cases} \min(2r,1) & , r > 0 \\ 0 & , r \leqslant 0 \end{cases} \tag{40}$$

which corresponds to the bold line illustrated in Fig 1b, and $r_i^{\pm}$ are as defined in (29),(30).

The M and TVDM schemes have been used to solve the one-dimensional Riemann problem. At time t = 0, the left and right states are:

$$\underline{U}_L = \begin{bmatrix} 0.445 \\ 0.311 \\ 8.928 \end{bmatrix} , \quad \underline{U}_R = \begin{bmatrix} 0.5 \\ 0.0 \\ 1.4275 \end{bmatrix} .$$

With increasing time, a mixing process takes place such that a rarefaction wave moves to the left and a contact discontinuity and a shock wave move to the right. The results shown in Fig 2a,b were obtained after 100 time steps with 140 cells. Other computer solutions are given in [1]. It can be seen that scheme TVDM exhibits none of the oscillations of the M scheme. However, the contact discontinuity is smeared over approximately thirteen points. The resolution of the contact discontinuity can be improved by applying artificial compression techniques.

TVD MACCORMACK SCHEME WITH ARTIFICIAL COMPRESSION (TVDMAC)

In order selectively to add artificial compression in cases where a contact discontinuity is anticipated, we construct a split operator C_Δ which compresses a "TVDM" solution at any given time level. Here, C_Δ is defined as

$$C_\Delta \underline{V}_i = \underline{V}_i - \frac{\lambda}{2}\left[\theta_{i+\frac{1}{2}}\, \underline{G}_{i+\frac{1}{2}} - \theta_{i-\frac{1}{2}}\, \underline{G}_{i-\frac{1}{2}}\right] \tag{41}$$

where

$$\underline{V}_i = L_1(\Delta t)\underline{U}_i^n$$

is the solution obtained by applying the TVDM scheme to the solution at the previous time level, and we have suppressed subscripts j,k. Also,

$$\theta_{i+\frac{1}{2}} = \max\left[\hat{\theta}_i \, , \, \hat{\theta}_{i+1}\right] \tag{42}$$

$$\hat{\theta}_i = \begin{cases} \left| \dfrac{\left|\Delta V^1_{i+\frac{1}{2}}\right| - \left|\Delta V^1_{i-\frac{1}{2}}\right|}{\left|\Delta V^1_{i+\frac{1}{2}}\right| + \left|\Delta V^1_{i-\frac{1}{2}}\right|} \right| & , \left|\Delta V^1_{i+\frac{1}{2}}\right| + \left|\Delta V^1_{i-\frac{1}{2}}\right| > \varepsilon \\ 0 & , \left|\Delta V^1_{i+\frac{1}{2}}\right| + \left|\Delta V^1_{i-\frac{1}{2}}\right| \leqslant \varepsilon \end{cases} \tag{43}$$

$$\varepsilon = 0.01 \max_i \left|V^1_{i+1} - V^1_i\right|$$

$$\underline{G}_{i+½} = G^m_{i+½} = g^m_i + g^m_{i+1} - \left|g^m_{i+1} - g^m_i\right| \operatorname{sgn}\left(V^m_{i+1} - V^m_i\right)$$

$$g^m_i = \alpha_i \left[V^m_{i+1} - V^m_{i-1}\right]$$

$$\alpha_i = \max\left\{0, \min_{1\leqslant m\leqslant m} \frac{\min\left[\left(V^m_{i+1} - V^m_i\right), \left[V^m_i - V^m_{i-1}\right] \operatorname{sgn}\left[V^m_{i+1} - V^m_i\right]\right]}{\left|V^m_{i+1} - V^m_i\right| + \left|V^m_i - V^m_{i-1}\right|}\right\}$$

when m = 1(1)M, M being the number of components of the solution vector $\underline{V}$.

Fig 2c shows the solution obtained by scheme TVDMAC for the Riemann problem. It can be seen that the resolution of the contact discontinuity is improved dramatically with artificial compression. Clearly, the operator C_Δ can be applied more than once, if desired, in order further to "compress" the data. In practice, we have rarely found this to be necessary. Further discussion of artificial compression techniques can be found in Harten [12].

APPLICATIONS

Fig 3a illustrates an aircraft forebody geometry typical of a combat aircraft. We wish to predict the disturbance field around the forebody at proposed intake locations, as well as the usual forces and moments, at Mach numbers ranging from high subsonic to supersonic and at various angles of attack. The forebody shown in Fig 3a has an elliptical nose dipped at 5°, sharp corners and flat sides to accommodate engine intakes. An analytical description of the body surface was formulated by fitting curves through the ordinates of particular cross sections. This provided a basis for the construction of a body-fitted H-O type grid in which the radial lines, emanating from the effective centre-line of the body, were segmented such that the radial spacing at the body surface was approximately the same as the streamwise spacing. The latter spacing was essentially uniform and corresponded to the stations at which the individual cross sections were defined. So constructed, the mesh had 57(x) x 16(θ) x 10(r) cells distributed around a "half-body" and a converged solution was obtained after 800 time steps, commencing the calculation with an impulsive start at Mach 1.40 and -5° angle of attack (Fig 3b). An operator-splitting three-dimensional implementation of the finite-volume TVD MacCormack scheme described earlier was used. Unlike similar calculations using the MacCormack method, there were no parameters to be adjusted to control the amount of added dissipation for shock capturing. Fig 4 shows a similar calculation for another forebody, typical of a civil aircraft. An analytical description of the geometry is given in the reference cited.

CONCLUSION

The theory underlying the construction of total variation diminishing (TVD) finite difference schemes has been reviewed. Using these methods, a high resolution version of the well-known MacCormack scheme has been constructed and some applications of the method reported. The required modifications to the standard MacCormack method are minor and will allow existing production code implementations of this scheme to be updated to reflect recent advances in one-dimensional schemes.

REFERENCES

1. Harten A. High Resolution Schemes for Hyperbolic Conservation Laws. *J Comp Phys*, vol 49, 357-392, 1983.

2. Davis SF. TVD Finite Difference Schemes and Artificial Viscosity. NASA CR 172373, 1984.

3. Sweby PK. High Resolution Schemes Using Flux Limiters for Hyperbolic Conservation Laws. *SIAM J Num Anal*, vol 21, no 5, 995-1011, 1984.

4. Roe PL. Approximate Riemann Solvers, Parameter Vectors and Difference Schemes. *J Comp Phys*, vol 43, 357-373, 1981.

5. Osher S. Riemann Solvers, the Entropy Condition and Difference Approximations. *SIAM J Num Anal*, vol 21, no 2, 217-235, 1984.

6. Van Leer B. On the Relation Between the Upwind Differencing Schemes of Godunov, Enquist-Osher and Roe. *SIAM J Sci Stat Comp*, no 5, 1-20, 1984.

7. Chakravarthy S. A New Class of High Accuracy TVD Schemes for Hyperbolic Conservation Laws. AIAA Paper No 85-0363, 1985.

8. Harten A. Uniformly High-Order Accurate Non-Oscillatory Schemes. *SIAM J Num Anal*, vol 24, no 2, 1987.

9. Yee HC. Upwind and Symmetric Shock Capturing Schemes. NASA TM 89464, 1984.

10. Warming RF and Beam RM. Upwind Second Order Difference Schemes and Applications in Aerodynamics. *JAIAA*, vol 14, 1241-1249, 1976.

11. Causon DM and Ford PJ. Numerical Solution of the Euler Equations Governing Axisymmetric and Three Dimensional Transonic Flow. *The Aeronautical Journal*, vol 89, no 886, 226-241, 1985.

12. Harten A. The Artificial Compression Method for Computation of Shocks and Contact Discontinuities: III. Self-Adjusting Hybrid Schemes. *Math of Comp*, vol 32, no 142, 363-389, 1978.

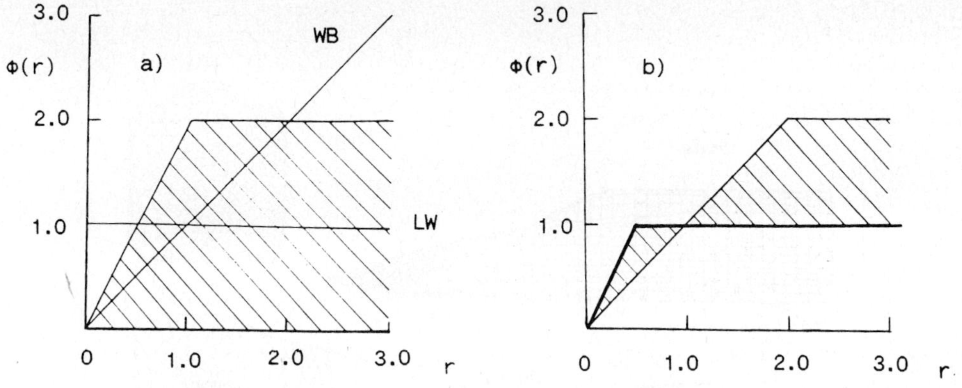

Fig. 1 Flux Limiter Functions

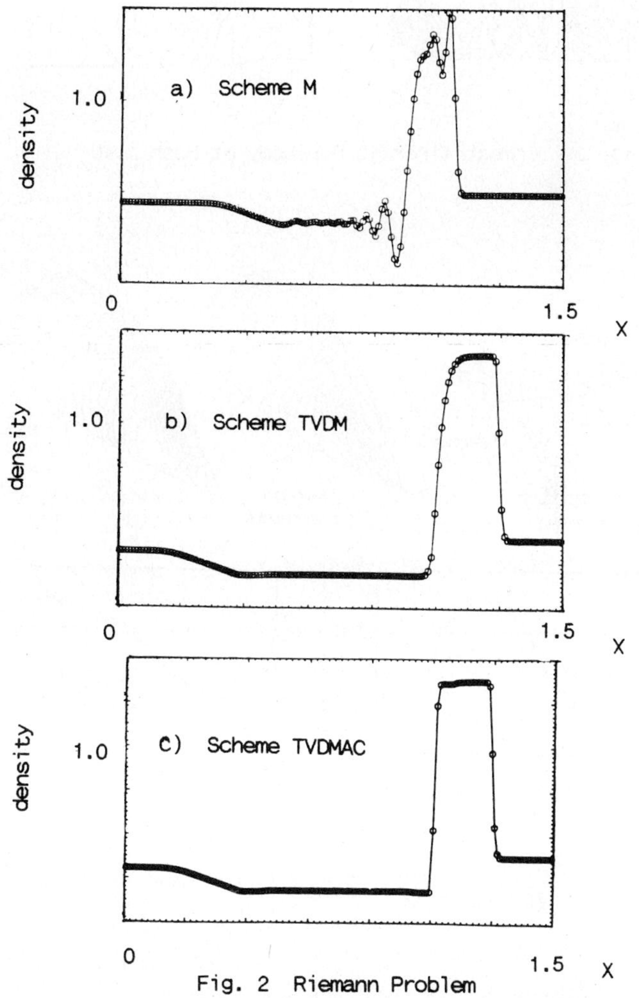

Fig. 2 Riemann Problem

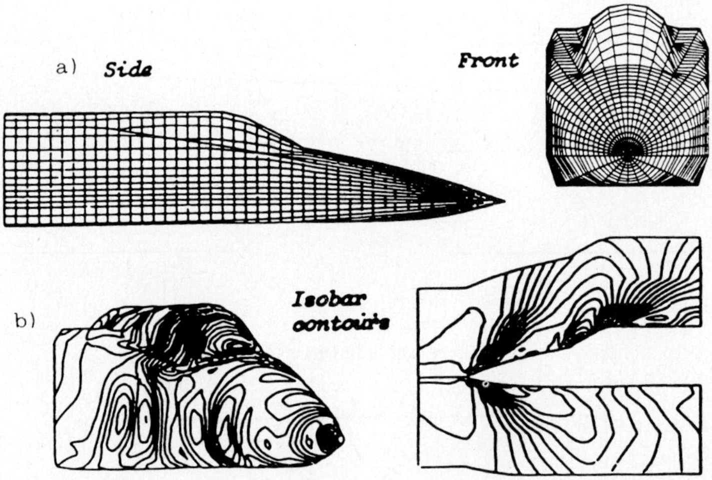

Fig. 3 Combat Aircraft Forebody at Mach 1.40

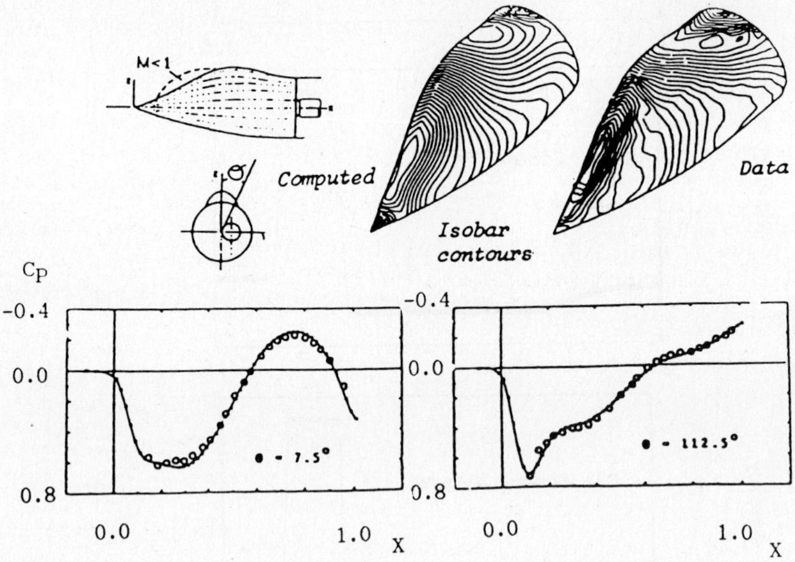

Fig. 4 Civil Aircraft Forebody from NASA TM 80062
at Mach 1.70

A STABILITY ANALYSIS OF A EULERIAN METHOD FOR SOME SURFACE GRAVITY WAVE PROBLEMS

Søren Christiansen
Laboratory of Applied Mathematical Physics
The Technical University of Denmark
DK-2800 Lyngby, Denmark

SUMMARY

For the solution of some surface gravity water wave problems a Eulerian method is given as a system of 2N ordinary differential equations where the moving boundary is described by means of N "Eulerian" points. Part of a stability analysis of the system has been carried out with respect to equilibrium solutions, periodic solutions, and numerical solutions. Numerical experiments with the system have not disclosed unprovoked saw-tooth instabilities, which have plagued some other methods.

1. INTRODUCTION

Surface water waves under gravity [55] constitute a moving boundary problem. (In special cases this problem can be formulated in terms of non-linear hyperbolic-type equations [55; Part E].) For the solution of the general problem various methods are at disposal [57]. Some of them can be characterized as "two-step methods" where (i) Laplace's equation (see § 2) is solved separately, and then (ii) the solution of Laplace's equation is used in a time-stepping procedure. Depending upon how Laplace's equation is solved the "two-step methods" are divided into two groups: (1) Solution using differential equation methods [3] [5] [18] [24] [41] [42] [57]. (2) Solution using integral equation methods [8] [18] [30] [31] [32] [34] [37] [38] [39] [43] [46] [54] [57]. In several cases it has been reported that in the course of the computation, when the equations are integrated forward in time, then a smooth surface curve may develop into a curve with a superimposed saw-tooth component. This instability has plagued several methods of computation [1] [3] [15] [16] [34] [39] [41] [45] [50]. Apparently, the first unmistakable example of this instability was presented by Longuet-Higgins & Cokelet [34, Fig. 4]. Traditionally, the instability has been overcome by an artificial smoothing. Also a method of Roberts [43] can be unstable, but a slight change of the algorithm eliminates the instability and results in a stable method [43]. In contrast to these difficulties a method of Vinje & Brevik [54] turns out to be remarkably stable without any instabilities. The integral equations used in [34] and [54] are of the first kind and of the second kind respectively. The analysis of [43] is very important when investigating the instabilities of [34], while the stability of [54] is still not fully understood [43; p. 25], apparently. Several methods, e.g. [34] [43] [54], employ on the surface curve N "Lagrangian" points for which the movement is described by a system of 3N ordinary differential equations (ODE's). Those methods are capable of treating difficult problems, e.g. overturning waves.

Initiated by the above methods we shall here consider a simpler method which employs on the surface curve N "Eulerian" points for which the movement is described by a system of 2N ODE's. (This method is not capable of treating overturning waves.) We derive the method (§ 2) and analyse the system of ODE's (§ 3) with respect to stability of equilibrium solutions (§ 3.1), stability of periodic solutions (§ 3.2), and stability of numerical solutions (§ 4). Finally, we discuss the applicability of some methods from nonlinear dynamics and give a summary of the results obtained (§ 5).

2. A EULERIAN METHOD

We consider a two-dimensional non-linear water problem, 2π-periodic in the (horizontal) x-direction, with horizontal bottom $y = 0$ with the y-axis pointing upwards (gravity $g = 1$) and with a moving surface curve described by $y = \eta(x,t)$, where t is time. Inside the region a 2π-periodic potential $\Phi = \Phi(x,y,t)$ satisfies Laplace's equation $\Delta\Phi = 0$, while $\Phi_y = 0$ on $y = 0$. On $y = \eta$ the potential Φ satisfies well-known boundary conditions [51; p. 16], which make the problem nonlinear. By introducing $\phi(x,t) \equiv \Phi(x,\eta(x,t),t)$, i.e., the potential Φ evaluated on the moving boundary, the conditions on the moving surface can also be written as the following partial differential equations (PDE's)

$$\eta_t = \Phi_y(x,\eta,t) - \eta_x(x,t)\,\Phi_x(x,\eta,t) \tag{2-1a}$$

$$\begin{aligned}\phi_t &= H - \eta(x,t) - \frac{1}{2}\,\Phi_x(x,\eta,t)^2 \\ &\quad + \frac{1}{2}\,\Phi_y(x,\eta,t)^2 - \eta_x(x,t)\,\Phi_x(x,\eta,t)\,\Phi_y(x,\eta,t)\ ,\end{aligned} \tag{2-1b}$$

where H is a suitably chosen constant.

The Eulerian method is derived as follows: For $x := x_i \equiv i\,2\pi/N$; $i = 1,2,\cdots,N$ (with N even) the N functions $\eta_i(t) := \eta(x_i,t)$ and the N functions $\phi_i(t) := \phi(x_i,t)$ are to be found. These functions are combined into one 2N-vector $\underset{\sim}{\psi} = [\psi_1, \cdots, \psi_N, \psi_{N+1}, \cdots, \psi_{2N}]^T \equiv [\eta_1, \cdots, \eta_N, \phi_1, \cdots, \phi_N]^T$, which depends on t. The potential Φ is expressed as (cf., e.g. [9])

$$\Phi = A_0 + A_n \cos nx \cosh ny + \sum_{j=1}^{n-1} (A_j \cos jx + B_j \sin jx) \cosh jy\ , \tag{2-2}$$

with $n = N/2$, each term 2π-periodic, satisfying Laplace's equation and the condition on $y = 0$; the coefficients $\{A_j\}_0^n$ and $\{B_j\}_1^{n-1}$ are time-dependent. By means of (2-2) and $\underset{\sim}{\psi}$ a system of linear algebraic equations is set up and solved [22], giving the value of the coefficients and thereby, through (2-2), various derivatives of Φ. The derivative η_x is approximated in terms of $\{\eta_i\}_1^N$, e.g. by means of periodic splines [21]. Hereby all the quantities on the right-hand side of (2-1), evaluated at $x := x_i$, are expressed in terms of $\underset{\sim}{\psi}$. We have therefore expressed $\underset{\sim}{\psi}'$ in terms of $\underset{\sim}{\psi}$ as a system of 2N ODE's

$$\underset{\sim}{\psi}' = \underset{\sim}{F}(\underset{\sim}{\psi})\ . \tag{2-3}$$

When $\underset{\sim}{\psi}$ is prescribed for $t = 0$ there is given an initial value problem where the evolution of $\underset{\sim}{\psi}(t)$ is found from (2-3).

3. ANALYSIS OF THE SYSTEM OF ODE's

The system (2-3) is to be analysed, in particular with respect to the stability of equilibrium solutions and of periodic solutions.

Only for very small values of N is it possible to write down the system in a tractable closed form. For general (even) values of N the system is given as a computer subroutine. Therefore part of the analysis of the system has to be carried out by investigating the subroutine by numerical methods. The computations referred to in this section are carried out using an IBM 3081, VM/CMS, FORTVS, double precision (i.e. using around 16 decimal digits).

The Jacobi matrix $\underline{\underline{J}}$, with elements $J_{ij} = \partial F_i / \partial \psi_j$, cannot, in general, be determined analytically, but it has to be computed numerically. The derivative is replaced by a two-sided difference approximation, viz. $(F_i(\psi_j+h) - F_i(\psi_j-h))/2h$, where h is chosen to be 10^{-5}, in accordance with a machine-epsilon around 10^{-16}, cf. [14; p. 286].

The description of the water waves, leading to (2-1), does not take any losses into account. Therefore the system (2-3) should be conservative [48; Ch. 6], which is the case when $\operatorname{div} \underset{\sim}{F} = 0$. Apparently, $\operatorname{div} \underset{\sim}{F} = 0$: This equality can simply be verified analytically in certain cases whereas numerical calculations corroborate the equality in several other cases.

3.1. Stability of equilibrium solutions.

The equilibrium points of the system (2-3), i.e. the points $\underset{\sim}{\psi}$ where $\underset{\sim}{F}(\underset{\sim}{\psi}) = \underset{\sim}{0}$, are not isolated [7; p. 122]. In fact $\underset{\sim}{\psi}$ determined by $\eta_1 = \eta_2 = \cdots = \eta_N =: \overline{\eta}$ and $\phi_1 = \phi_2 = \cdots = \phi_N =: \overline{\phi}$, with $\overline{\phi}$ arbitrary, is an equilibrium point. For these equilibrium points the eigenelements of the Jacobi matrix $\underline{\underline{J}}$ can be found analytically and given in closed form. The matrix $\underline{\underline{J}}$ has the following N+1 (imaginary) eigenvalues, with $\iota^2 = -1$, [6; Eq. 4]

$$\lambda^{(0)} = 0 \tag{3-1a}$$

$$\lambda^{(k)\pm} = \pm\, \iota\sigma^{(k)} \quad ; \quad k = 1,2,\cdots,\frac{N}{2}\ , \tag{3-1b}$$

with

$$\sigma^{(k)} = \sqrt{k \tanh(k\overline{\eta})}\ . \tag{3-1c}$$

Except for k = N/2 the eigenvalues have algebraic multiplicity two. The eigenvectors can similarly be given, and it turns out that there are only 2N-1 ordinary eigenvectors indicating that the matrix $\underline{\underline{J}}$ is defective [40; § 11.3]; the Jordan canonical form [10; Ch. 5] of the Jacobi matrix has one element outside the diagonal. The generalized eigenvector [40; § 11.6] pertains to the eigenvalue $\lambda^{(0)} = 0$ and has the form $[1,1,\cdots,1,0,0,\cdots,0]^T$. It can lead to a solution of the form $[0,0,\cdots,0,1,1,\cdots,1]^T$ growing proportional to t. This growth, however, does not influence the shape of the boundary curve described by $[\eta_1,\eta_2,\cdots,\eta_N]$. But strictly speaking the equilibrium solution is not stable [2; § 4.2, Th. 1].

3.2. Stability of periodic solutions.

Standing waves or progressive waves on the surface of the water can be considered as time-periodic solutions in a 2N-dimensional phase space. The stability of a 2N-dimensional periodic solution $\underset{\sim}{P}(t)$, satisfying (2-3), with period T_0, is investigated by adding to $\underset{\sim}{P}(t)$ a small 2N-dimensional component $\underset{\sim}{w}(t)$, which may represent the saw-tooth ripples superimposed on the wave characterized by $\underset{\sim}{P}(t)$, so that $\underset{\sim}{\psi} = \underset{\sim}{P} + \underset{\sim}{w}$ is also a solution satisfying (2-3). Here $\underset{\sim}{w}(t)$ satisfies the linear variational equation of (2-3) relative to the solution $\underset{\sim}{P}$, [7; pp. 176 ff]

$$\underset{\sim}{w}' = \underline{\underline{J}}(\underset{\sim}{P})\underset{\sim}{w} \ , \tag{3-2}$$

where $\underline{\underline{J}}(\underset{\sim}{P})$ is the Jacobi matrix evaluated around the solution $\underset{\sim}{P}$, having the period T_0, whereby the matrix has time-periodic coefficients, which leads to Floquet theory for the equation (3-2), [7; pp. 95 ff.] [23; § 8.3]. As mentioned above the matrix $\underline{\underline{J}}(\underset{\sim}{P})$ has to be found by numerical differentiation around the solution $\underset{\sim}{P}$, which satisfies (2-3).

As initial values for $\underset{\sim}{w}(t)$, i.e. $\underset{\sim}{w}(0)$, are chosen 2N different unit vectors, which are combined to a $2N \times 2N$ unit matrix $\underline{\underline{I}}$. By integrating $\underline{\underline{I}}$ through the period T_0 there results from (3-2) the so-called monodromy matrix $\underline{\underline{C}}$ [13; p. 119]. The simultaneous integration of (2-3) and (3-2) requires the solution of $2N + 4N^2$ coupled ODE's. The eigenvalues of $\underline{\underline{C}}$ are the so-called Floquet multipliers [7; p. 100], which give information about the stability of $\underset{\sim}{P}(t)$. The initial value for $\underset{\sim}{P}(t)$ depends upon the periodic solution in question.

<u>Example</u>: A periodic solution, viz. a standing wave, is obtained by the initial conditions [52]

$$\eta(x,0) = 1 - 0.0000425450 \cos 2x \tag{3-3a}$$

$$\phi(x,0) = 0.011459234 \cos x + 0.000000024 \cos 3x \tag{3-3b}$$

with the time period $T_0 = 7.19973938$. By means of a sampling af (3-3) for $x := x_i \equiv i\, 2\pi/N;\ i = 1,2,\cdots,N$ (with N even) the initial vector $\underset{\sim}{P}(0)$ is obtained. For N = 8 a system with 272 ODE's is integrated numerically over the period T_0, using [19] with TOL = 10^{-9}, in order to obtain the 16×16 monodromy matrix $\underline{\underline{C}}$, from which the (complex) eigenvalues, the Floquet multipliers, are determined numerically [20]. All the computed multipliers turn out to have absolute value very near one; in fact they are confined to the interval [0.999967, 1.000033]. With the multiplicity written in { } the multipliers are approximately: 1.00 {6}, $0.99 \pm \imath\, 0.13$ {2}, $-0.84 \pm \imath\, 0.54$ {2}, $-0.25 \pm \imath\, 0.97$ {1}.

The above Example, and other examples, indicate that all Floquet multipliers have absolute value equal to one, and that some of them appear with multiplicity larger than unity. This case is the most difficult one when it comes to the investigation of whether $\underset{\sim}{w}(t)$ has components growing with time t [7; p. 103, Th. 13]. It is here necessary to determine the Jordan canonical form of the monodromy matrix [12; § 2.7]; for example the eigenvalue one may have generalized eigenvectors which can give rise to components of the solution which grow proportional with t. However, the numerical determination of the Jordan canonical form of a given matrix is a very difficult numerical process [11] [25] [26]. For example the block structure of the Jordan form may be sensitive to perturbations of the matrix elements. In the present case the matrix elements, of the monodromy matrix, are found by means of a numerical integration which necessarily creates some computational errors.

Because of the numerical difficulties here mentioned it seems necessary to abandon the method here described for analysing the stability of periodic solutions. However, other methods are available (§ 5).

4. STABILITY OF NUMERICAL SOLUTIONS

Because the question about stability of the periodic solutions does not seem to be answered using Floquet theory, some numerical experiments have been carried out.

The system (2-3) has been integrated numerically [19]. A straightforward method to investigate whether saw-tooth instabilities are evolving is to carry out a harmonic analysis [58; § 6.13] of the equally spaced ordinates $\{\eta_i\}_1^N$ at various instances of time. A saw tooth can be obtained by sampling the function $\cos(N/2)x$ for $x := x_i \equiv i\,2\pi/N$; $i = 1,2,\cdots,N$, and therefore the Fourier coefficient corresponding to $\cos(N/2)x$, derived from $\{\eta_i\}_1^N$, gives information about the magnitude of the saw-tooth component.

With N = 8 and with various vectors $\underset{\sim}{\psi}(0)$ corresponding to periodic solutions (viz. standing waves derived from [52] and sampled as in § 3.2, Example) some integrations over 20 complete periods of time have been carried out. From these computations there has not been observed any systematic growth of the Fourier component responsible for the saw-tooth component. From this negative result one cannot, of course, infer that the periodic solutions of the system (2-3) are always stable.

It is, however, possible to conduct the numerical solution in such a manner that a saw-tooth surface curve will develop [6]: The system (2-3) can be integrated numerically using various numerical methods [28] [58]; each method is characterized by a region of absolute stability [47]. Let λ be an eigenvalue of the Jacobi matrix $\underline{\underline{J}}$, evaluated at a certain solution $\underset{\sim}{\psi}$, and let h be the time step, then if λh is outside the region, the solution $\underset{\sim}{\psi}$ will contain a growing component of the eigenvector of $\underline{\underline{J}}$ corresponding to λ.

In the case leading to (3-1), with a horizontal boundary curve, the two eigenvalues $\lambda = \lambda^{(N/2)\pm}$ have, correspondingly, the two linearly independent eigenvectors

$$[+\lambda,-\lambda,\cdots,-\lambda,-1,+1,\cdots,+1]^T\,, \tag{4-1}$$

which can be combined to a saw-tooth surface curve, where $\eta_1 = \eta_3 = \cdots = \eta_{N-1} \neq \eta_2 = \eta_4 = \cdots = \eta_N$. With λ being purely imaginary, cf. (3-1), the integration methods do not create saw-tooth instability, provided that the step length h satisfies $0 \leqq |\lambda| h \leqq S$ or

$$0 \leqq h \leqq h_{max} := \frac{S}{\sigma^{(N/2)}} \equiv \frac{S}{\sqrt{\frac{N}{2}\tanh\left(\frac{N}{2}\bar{\eta}\right)}}\,, \tag{4-2}$$

where S depends on the integration method used [6]. For a related problem [9] it is found that $\Delta t \approx (\Delta x)^{\frac{3}{2}}$ in analogy with (4-2), which, however, gives a more specific bound for $h = \Delta t$.

Actual numerical integrations have been carried out using various methods, in particular the method by Adams-Bashforth-Moulton, 4th order in PECE mode [58; p. 458] (which was used in [34]). For this method it is customary to put S = 0.92620161, cf. [47; Fig. 24] [53; Fig. 4.2], although it is not completely correct [6]. With N = 36 and $\bar{\eta} = 1$ then $h_{max} = 0.218308$ according to (4-2).

The initial functions

$$\eta(x,0) = 1 + \varepsilon \sin x \tag{4-3a}$$

$$\phi(x,0) = (\varepsilon/\sqrt{\tanh 1}) \cos x \tag{4-3b}$$

correspond to a wave of permanent form travelling to the right with phase speed $c = \sqrt{\tanh 1}$, provided that $|\varepsilon|$ is infinitesimal [27; Ch. IX]. From (4-3) an initial vector $\underset{\sim}{\psi}(0)$ can be derived by a sampling for $x := x_i$ as in § 3.2, Example.

The value $\varepsilon = 0.01$ is small enough so that the results of the numerical integration follow what could be expected from the linear theory around an equilibrium point corresponding to a horizontal surface curve: If $h = 0.24 > h_{max}$ a saw-tooth component will become visible with an amplitude which is independent of x, in accordance with (4-1). In principle, the Fourier component No. 18, which represents the saw-tooth component, will grow with the number of timesteps. If $h = 0.18 < h_{max}$ such a saw-tooth component apparently does not develop.

The value $\varepsilon = 0.1$ is so large that the above results are no longer applicable, because they are derived for an equilibrium point. If $h = 0.24$ the saw-tooth component again becomes visible, but now with a large amplitude near the crest and a small amplitude near the trough.

If the problem is to be investigated further, e.g. with respect to the development of the uneven saw-tooth component, it may be advantageous, at various stages of the evolution of $\underline{\psi}(t)$, to compute the Jacobi matrix $\underline{\underline{J}}(\underline{\psi})$ by numerical differentiation (as indicated in § 3) and subsequently determine numerically [20] the eigenvalues (and perhaps, some of, the eigenvectors) for $\underline{\underline{J}}(\underline{\psi})$. (For a somewhat related problem [43] a similar computation of the eigenvalues and eigenvectors has been carried out.) Actual numerical calculations have been carried out, and they can reproduce, with high accuracy, the eigenvalues (3-1), including the double eigenvalues. Because $\underline{\underline{J}}(\underline{\psi})$ in this case is defective (see § 3.1) the eigenvectors computed using [20] are not all reliable, and generalized eigenvectors ought to be determined [25] [26]. For other boundary elevations η and/or boundary potentials ϕ than those leading to (3-1) it is observed in some cases that two opposite, double, purely imaginary eigenvalues can split into a quadruple of simple eigenvalues, $\lambda = \pm\alpha \pm \iota\beta$, i.e. two with positive and two with negative real part. (A splitting of four eigenvalues zero into two pairs of real eigenvalues is reported in [43; Table 1].) If the corresponding eigenvectors/generalized eigenvectors of $\underline{\underline{J}}(\underline{\psi})$ were available it would then be possible to compute, from the characteristic polynomials of the actual integration method, how the various components will develop. However, such a calculation will give only a quasi-steady description of the evolution, while a full unsteady calculation may show that the eigenelements change as a function of time, so that growing components of $\underline{\psi}(t)$ at a later stage may turn into decaying components of $\underline{\psi}(t)$, and vice versa.

<u>Example</u>: A periodic solution, viz. a standing wave, is obtained by the initial conditions [52], cf. § 3.2, Example

$$\eta(x,0) = 1 - 0.00425449584 \cos 2x \tag{4-4a}$$

$$\phi(x,0) = 0.115047192 \cos x + 0.000023741 \cos 3x \tag{4-4b}$$

with the time period $T_0 = 7.197620947$. A vector $\underline{\psi}(0)$ is obtained by sampling. For $N = 16$ a system with 32 ODE's is integrated numerically over the period T_0, using [19] with $TOL = 10^{-9}$. For various values of t the matrix $\underline{\underline{J}}(\underline{\psi}(t))$ is determined by numerical differentiation (see § 3) and the eigenvalues are computed numerically [20]. For $t = 0$, $T_0/2$ and T_0 there is found a quadruple of eigenvalues: $\pm 0.08 \pm \iota\, 2.86$, and $\underline{\psi}(t)$ is not in accordance with the linear theory [27; Ch. IX], while for $t = T_0/4$ and $3T_0/4$ the eigenvalues do not have dominating real parts, and $\underline{\psi}(t)$ is in accordance with the linear theory.

5. CONCLUDING REMARKS

For the solution of some two-dimensional surface water wave problems under gravity a Eulerian method is given in the form of a system of 2N ODE's (2-3), where the moving boundary is described by means of N "Eulerian" points. On the moving boundary the problems in question satisfy some conditions written as a system of PDE's (2-1). It is known [33] [35] [44] [59] that two-dimensional water waves can travel with a permanent form with a certain phase speed, and that the waves are stable to superharmonic perturbations provided that the waves are not too high compared with the wave length.

In consequence of the stability of the problems under consideration described by the system of PDE's (2-1) the system of ODE's (2-3) has been investigated with respect to stability of equilibrium solutions and periodic solutions. Unfortunately, definitive results could not be obtained on the periodic solutions because of difficulties in determining the Jordan canonical form of the monodromy matrix. However, the monodromy matrix could be analysed further: Actual numerical calculations carried out on the monodromy matrix of § 3.2, Example, and other examples, indicate that the monodromy matrices are symplectic [29; pp. 181-183]. Such matrices which arise within Hamiltonian mechanics [35] can be analysed with respect to stability using more refined methods [17]. For the integration of symplectic problems specialized algorithms are available [4]. Furthermore, the stability of systems can also be analysed in terms of the Lyapunov exponents [29; § 5.2b, § 5.3], which can be computed numerically [56]. It would be of interest to continue the analysis of the system (2-3), but now applying the methods which are used in connection with nonlinear dynamics.

Numerical experiments based on the system (2-3) have not disclosed unprovoked saw-tooth instabilities, but it is possible to provoke such instabilities by carrying out an integration with a timestep which is too large. Some of the provoked instabilities have the same form as those shown in [34, Fig. 4]. But from this coincidence it is - of course - not possible to draw conclusions about the reason for the saw-tooth instability observed in [34], because the present method is Eulerian while the method of [34] has to be characterized as Lagrangian. With respect to stability there may be an essential difference between a Eulerian approach and a Lagrangian approach [49]. Therefore, it is not possible to conclude with certainty whether or not the instability of [34] and the stability of [54] are due to the application of first-kind integral equations in [34] and second-kind integral equations in [54].

ACKNOWLEDGEMENTS

Dr. Ulla Brinch-Nielsen and Jesper Mørk, Laboratory of Applied Mathematical Physics, are thanked for discussions on water waves and on nonlinear systems. Professor Per Grove Thomsen, Institute for Numerical Analysis, The Technical University of Denmark, is thanked for advice on numerical solution of ODE's. Professor Enok Palm, Mathematical Institute, University of Oslo, Oslo, Norway; Professor James Sethian, Department of Mathematics, University of California, Berkeley, CA, U.S.A., and Professor Jean-Marc Vanden-Broeck, Mathematical Research Center, Madison, VI, U.S.A., are thanked for discussions in relation to the conclusions of § 5, the last paragraph.

For bringing relevant literature to my knowledge I want to thank: Ulla Brinch-Nielsen [8] [36] [59], Virginia Muto [4], Jesper Mørk [12] [48], Mads Peter Sørensen [17], and Per Grove Thomsen [2] [47] [53].

Mrs. Kirsten Studnitz is thanked for efficient typing of the manuscript.

The numerical calculations were carried out at the Danish Computing Center for Research and Education, UNI·C, The Technical University of Denmark.

REFERENCES

[1] Baker, G. R., Meiron, D. I., Orzag, S. A.: "Generalized vortex methods for free-surface flow problems", Jour. Fluid Mech. 123 (1982) pp. 477-501.

[2] Braun, M.: "Differential equations and their applications", Springer-Verlag, New York et al. 2nd Edition (1978) 13 + 518 pp. (Applied Mathematical Sciences 15).

[3] Chan, R. K.-C., Street, R. L.: "A computer study of finite-amplitude water waves", Jour. Computational Phys. 6 (1970) pp. 68-94.
[4] Channell, P., Scovel, J. C.: "Symplectic integration algorithms", Preprint, Los Alamos National Laboratory, Los Alamos, NM, U.S.A.
[5] Cheng, S. I., Lu, Y.: "An Eulerian method for transient nonlinear free surface wave problems", Jour. Computational Phys. 62 (1986) pp. 429-440.
[6] Christiansen, S.: "An elementary analysis of saw-tooth instabilities on a moving boundary for surface gravity water waves", Zeit. Angew. Math. Mech. (ZAMM) (Sonderheft, GAMM-Tagung, Stuttgart 1987) 68 (1988) pp. T -T .
[7] Cronin, J.: "Differential equations, introduction and qualitative theory", Marcel Dekker, Inc., New York & Basel (1980) 8 + 372 pp.
[8] Dold, J. W., Peregrine, D. H.: "Steep unsteady water waves: An efficient computational scheme", Proceedings 19th International Conference on Coastal Engineering, Houston, 1 (1984) pp. 955-967. American Society of Civil Engineers, New York.
[9] Fenton, J. D., Rienecker, M. M.: "A Fourier method for solving nonlinear water-wave problems: application to solitary-wave interactions", Jour. Fluid Mech. 118 (1982) pp. 411-443.
[10] Franklin, J. N.: "Matrix theory", Prentice-Hall, Inc., Englewood Cliffs, New Jersey (1968) 12 + 292 pp.
[11] Golub, G. H., Wilkinson, J. H.: "Ill-conditioned eigensystems and the computation of the Jordan canonical form", SIAM Review 18 (1976) pp. 578-619.
[12] Haken, H.: "Advanced synergetics", Springer-Verlag, Berlin et al. (1983) 15 + 356 pp.
[13] Hale, J. K.: "Ordinary differential equations", Wiley-Interscience, New York et al. (1969) 16 + 332 pp.
[14] Hassard, B. D., Kazarinoff, N. D., Wan, Y.-H.: "Theory and applications of Hopf bifurcation", Cambridge University Press, Cambridge et al. (London Mathematical Society Lecture Note Series 41) (1981) 6 + 311 pp.
[15] Haussling, H. J.: "Two-dimensional linear and nonlinear stern waves", Jour. Fluid Mech. 97 (1980) pp. 759-769.
[16] Haussling, H. J., Coleman, R. M.: "Nonlinear water waves generated by an accelerated circular cylinder", Jour. Fluid Mech. 92 (1979) pp. 767-781.
[17] Howard, J. E., MacKay, R. S.: "Linear stability of symplectic maps", Jour. Mathematical Phys. 28 (1987) pp. 1036-1051.
[18] Hume III, E. C., Brown, R. A., Deen, W. M.: "Comparison of boundary and finite element methods for moving-boundary problems governed by a potential", Internat. Jour. Numerical Methods Engineering 21 (1985) pp. 1295-1314.
[19] IMSL Library (International Mathematical & Statistical Libraries, Inc., 7500 Bellaire Blvd., Houston, Texas 77036, U.S.A.), Edition 9.2, November 1984: "DGEAR (Differential equation solver - variable order Adams predictor corrector method or Gears method)".
[20] Idem: "EIGRF (Eigenvalues and (optionally) eigenvectors of a real general matrix in full storage mode)".
[21] Idem: "ICSPLN (Cubic spline interpolation with periodic end conditions)".
[22] Idem: "LEQT2F (Linear equation solution - full storage mode - high accuracy solution)".
[23] Jordan, D. W., Smith. P.: "Nonlinear ordinary differential equations", Clarendon Press, Oxford (1977) 8 + 360 pp.
[24] Kawahara, M., Miwa, T.: "Finite element analysis of wave motion", Internat. Jour. Numerical Methods Engineering 20 (1984) pp. 1193-1210.

[25] Kågström, B., Ruhe, A.: "An algorithm for numerical computation of the Jordan normal form of a complex matrix", ACM Trans. Math. Software 6 (1980) pp. 398-419.
[26] Kågström, B., Ruhe, A.: "Algorithm 560. JNF, an algorithm for numerical computation of the Jordan normal form of a complex matrix", ACM Trans. Math. Software 6 (1980) pp. 437-443.
[27] Lamb, H.: "Hydrodynamics", Dover Publications, New York, 6th Ed. (1945) 15 + 738 pp.
[28] Lambert, J. D.: "Computational methods in ordinary differential equations", John Wiley & Sons, Chichester et al. (1973) 15 + 278 pp.
[29] Lichtenberg, A. J., Lieberman, M. A.: "Regular and stochastic motion", Springer-Verlag, New York et al. (Applied Mathematical Sciences 38) (1983) 21 + 499 pp.
[30] Liggett, J. A.: "Hydrodynamics calculations using boundary elements", pp. 889-896 in: Kawai, T. (ed.): Finite element flow analysis, Proceedings of the Fourth International Symposium on Finite Element Methods in Flow Problems, Chuo University, Tokyo, 26-29 July, 1982; University of Tokyo Press, North-Holland Publishing Company, Amsterdam et al. (1982) 16 + 1096 pp.
[31] Liu, P. L.-F., Liggett, J. A.: "Applications of boundary element methods to problems of water waves", pp. 37-67 (Chapter 3) in: Banerjee, P. K., Shaw, R. P. (eds.): Developments in boundary element methods - 2, Applied Science Publishers, London, New Jersey (1982) 10 + 288 pp.
[32] Liu, P. L.-F., Liggett, J. A.: "Boundary element formulations and solutions for some non-linear water wave problems", pp. 171-190 (Chapter 7) in: Banerjee, P. K., Mukherjee, S. (eds.): Developments in boundary element methods - 3, Elsevier Applied Science Publishers, London, New York (1984) 11 + 313 pp.
[33] Longuet-Higgins, M. S.: "The instabilities of gravity waves of finite amplitude in deep water, I. Superharmonics", Proc. Roy. Soc. London A 360 (1978) pp. 471-488.
[34] Longuet-Higgins, M. S., Cokelet, E. D.: "The deformation of steep surface waves on water, I. A numerical method of computation", Proc. Roy. Soc. London A 350 (1976) pp. 1-26.
[35] MacKay, R. S.: "Stability of equilibria of Hamiltonian systems", pp. 254-270 in: Sarkar, S. (ed.): Nonlinear phenomena and chaos, Adam Hilger Ltd., Bristol & Boston (1986) 11 + 336 pp.
[36] MacKay, R. S., Saffman, P. G.: "Stability of water waves", Proc. Roy. Soc. London A 406 (1986) pp. 115-125.
[37] Nakayama, T.: "Boundary element analysis of nonlinear water wave problems", Internat. Jour. Numerical Methods Engineering 19 (1983) pp. 953-970.
[38] Nakayama, T., Washizu, K.: "The boundary element method applied to the analysis of two-dimensional nonlinear sloshing problems", Internat. Jour. Numerical Methods Engineering 17 (1981) pp. 1631-1646.
[39] New, A. L., McIver, P., Peregrine, D. H.: "Computations of overturning waves", Jour. Fluid Mech. 150 (1985) pp. 233-251.
[40] Noble, B.: "Applied linear algebra", Prentice-Hall, Inc., Englewood Cliffs, New Jersey (1969) 16 + 523 pp.
[41] Ohring, S.: "Nonlinear water wave generation using the method of lines", Jour. Computational Phys. 39 (1981) pp. 137-163.
[42] Prosperetti, A., Jacobs, J. W.: "Numerical method for potential flows with a free surface", Jour. Computational Phys. 51 (1983) pp. 365-386.
[43] Roberts, A. J.: "A stable and accurate numerical method to calculate the motion of a sharp interface between fluids", IMA Jour. App. Math. 31 (1983) pp. 13-35.
[44] Saffman, P. G.: "The superharmonic instability of finite-amplitude water waves", Jour. Fluid Mech. 159 (1985) pp. 169-174.

[45] Saffman, P. G., Yuen, H. C.: "A note on numerical computations of large amplitude standing waves", Jour. Fluid Mech. 95 (1979) pp. 707-715.
[46] Salmon, J. R., Liu, P. L-F., Liggett, J. A.: "Integral equation method for linear water waves", Jour. Hydraulics Division ASCE 106 HY12 (1980) pp. 1995-2010.
[47] Sand, J., Østerby, O.: "Regions of absolute stability", Computer Science Department, Aarhus University, Denmark, DAIMI PB-102 (September 1979) 2 + 57 pp.
[48] Schuster, H. G.: "Deterministic chaos", Physik-Verlag, Gmbh., Weinheim FRG (1984) 23 + 220 pp.
[49] Sethian, J.: "The design of algorithms for hypersurfaces moving with curvature-dependent speed", pp. 282-287 in: Book of abstracts, Hyperbolic Problems, Second International Conference, RWTH, Aachen, March 14-18, 1988. Aachen (1988) 12 + 357 pp.
[50] Soh, W. K.: "A numerical method for non-linear water waves", Computers and Fluids 12 (1984) pp. 133-143.
[51] Stoker, J. J.: "Water waves", Interscience Publishers, Inc., New York (1957) 28 + 567 pp.
[52] Tadjbakhsh, I., Keller, J. B.: "Standing surface waves of finite amplitude", Jour. Fluid Mech. 8 (1960) pp. 442-451.
[53] Thomsen, P. G., Zlatev, Z.: "Two-parameter families of predictor-corrector methods for the solution of ordinary differential equations", BIT, Nordisk Tidskrift. Inform. 19 (1979) pp. 503-517.
[54] Vinje, T., Brevig, P.: "Numerical simulation of breaking waves", Advances Water Res. 4 (1981) pp. 77-82.
[55] Wehausen, J. W., Laitone, E. V.: "Surface waves", pp. 446-778 in: Flügge, S. (ed.): Handbuch der Physik, IX. Strömungsmechanik III, Springer-Verlag, Berlin et al. (1960) 7 + 815 pp.
[56] Wolf, A., Swift, J. B., Swinney, H. L., Vastano, J. A.: "Determining Lyapunov exponents from a time series", Physica D 16 (1985) pp. 285-317.
[57] Yeung, R. W.: "Numerical methods in free-surface flows", Annual Rev. Fluid Mech. 14 (1982) pp. 395-442.
[58] Young, D. M., Gregory, R. T.: "A survey of numerical mathematics, I + II", Addison-Wesley Publishing Company, Reading, Massachusetts (1973) 51 + 63 + 1099 pp.
[59] Zufiria, J. A., Saffmann, P. G.: "The superharmonic instability of finite-amplitude surface waves on water of finite depth", Studies Appl. Math. 74 (1986) pp. 259-266.

A QUADRATURE APPROXIMATION OF THE BOLTZMANN COLLISION OPERATOR IN AXISYMMETRIC GEOMETRY AND ITS APPLICATION TO PARTICLE METHODS

P. DEGOND [1] F. J. MUSTIELES [1] B. NICLOT [2]

[1] Centre de Mathématiques Appliquées
Ecole Polytechnique, 91128 PALAISEAU cedex, France.
[2] Centre Technique CITROEN, DAT / CSI, Route de Gisy
78140 VELIZY-VILLACOUBLAY cedex, France.

SUMMARY

Abstract: This paper is devoted first, to the presentation of a new expression of the Boltzmann collision operator in axisymmetric geometry, and second, to its use for practical computations in connection with the deterministic particle method.

In numerous situations, the solution of the Boltzmann equation is invariant under the rotations of the velocities about a given axis. This invariance is seldom used to reduce the computational cost of the simulation. In this paper, we present an expression of the Boltzmann operator, which takes advantage of this invariance to reduce the dimension of the integrations involved in this operator.

We use this feature to propose a direct evaluation of the Boltzmann operator by quadrature formulae. This method is coupled with a particle method for the approximation of the differential part of the equation. The numerical scheme has been applied to different test cases, with an emphasis on the verification of the momentum and energy conservation by the discrete collision operator. The method is also applied to a real problem arising in semiconductor physics. The numerical results are presented and commented.

Acknowledgements: The authors wish to acknowledge the "Centre de Calcul Vectoriel pour la Recherche" for supporting the computer cost of the numerical simulations.

1. INTRODUCTION

The Boltzmann equation is not usually considered as part of the class of non-linear hyperbolic problems. However, there are numerous connections between these two areas, and particularly between the kinetic and fluid models of gas dynamics. From the viewpoint of numerical modelling, these two models have often been brought into conflicts together, since the kinetic models, although relying on safer and wider physical bases, are much more costly for a numerical computation. However, kinetic models provide interesting informations for instance on the internal structure of one dimensional shock waves [1]. Moreover, in hypersonic aerodynamics, the excitation and the relaxation of internal degrees of freedom (such as the rotation, vibration, and dissociation of the molecules) introduce additional features in the shock structure which must be taken into account by a reliable simulation program, and which are more accurately modelled by kinetic equations. To preserve the physical accuracy of kinetic models, with a moderate computational cost, a possible solution is to investigate mixed models, where the population of the thermal particles (which is expected to be large) is modelled by the fluid dynamics equations, and the suprathermal particles, by kinetic equations [2]. Thus, it seems important to develop accurate methods for the kinetic equations at the same pace as for fluid equations.

Most of the numerical computations make use of Monte-Carlo methods, which can be devided into Bird's method [6], and Nanbu's method [7,8]. A review on these methods can be found in [9]. Monte-Carlo methods are used in various areas of physics, such as semi-conductor physics [10]. Very few direct methods have been investigated (see Chorin [11]). This paper is a contribution to the development of direct simulation methods for the Boltzmann equation.

First, we investigate the Boltzmann collision operator in axisymmetric geometry. This situation arises in the study of shock or boundary layers, where the distribution function is invariant under rotations of the velocity about the normal axis of the shock layer. The distribution function can be expressed in terms of the reduced velocity coordinates $(v_1,v_2) \in \mathbb{R} \times \mathbb{R}_+$:

$$v_1 = (v,e_1) \quad ; \quad v_2 = | v - (v,e_1)\, e_1 | \tag{1}$$

where e_1 is the unit vector of the symmetry axis. The symmetry hypothesis reads $f(x,v) = \bar{f}(x,v_1,v_2)$. By the isotropy of the collision operator Q, there exists a reduced operator $\bar{Q}$, acting on functions $\bar{f}$, such that $Q(f,f) = \bar{Q}(\bar{f},\bar{f})$. In this paper, we give an explicit expression of $\bar{Q}$, which seems to be new. The details of the derivation of $\bar{Q}$ are given in [12]. Since $\bar{Q}$ involves integrations over a lower dimensional manifold than Q does, this expression can be helpful in numerical computations.

Second, we investigate the weighted particle method for solving the Boltzmann equation. This method has been proposed and analyzed by P. A. Raviart and S. Mas-Gallic [3,4,5], in the case of diffusion equations and linearized Boltzmann equations. This paper deals with its first application to the non-linear Boltzmann equation. Although this method can be used in any geometrical situation, we will restrict the presentation to the axisymmetric geometry, and to the reduced operator $\bar{Q}$. No error analysis is yet available in the non-linear case. In this paper, we present some numerical results for test problems.

2. THE BOLTZMANN OPERATOR IN AXISYMMETRIC GEOMETRY

Let $f(x,v,t)$, $x \in \mathbb{R}^3$, $v \in \mathbb{R}^3$, $t > 0$, be the distribution function, solution of the Boltzmann equation:

$$\frac{\partial f}{\partial t} + v.\nabla_x f = Q(f,f) \tag{2}$$

$$Q(f,f)(v_0) = \iint_{\mathbb{R}^3 \times S^2} \left(f' f'_0 - f f_0 \right) B\left(|v-v_0|, \cos\chi\right) d\omega\, dv \tag{3}$$

where

$$f_0 = f(x,v_0,t) \;;\; f = f(x,v,t) \;;\; f'_0 = f(x,v'_0,t) \;;\; f' = f(x,v',t)$$

$$v_0' = v_0 - \omega(\omega, v_0 - v) \quad ; \quad v' = v + \omega(\omega, v_0 - v) \tag{4}$$

$$\cos\chi = \frac{(v - v_0, v' - v_0')}{|v - v_0|\,|v' - v_0'|} .$$

We denote by S^2 the unit sphere of $\mathbb{R}^3$, and $\omega \in S^2$. B is a smooth function which is characteristic of the interaction potential.

From now on, we will concentrate on the collision operator, and will omit the x-dependence of the distribution functions. We suppose that f is invariant under the rotations of v about a given axis e_1, and introduce the reduced velocity coordinates according to formula (1). By the choice of a referential in the orthogonal plane to e_1, we can introduce the cylindrical coordinates of a vector v by:

$$v = \bar{v}(\theta) = (v_1, v_2\cos\theta, v_2\sin\theta) \ ; \ \theta \in [0,2\pi]$$

We let:

$$d\Omega(\bar{v}) = 2\pi\, v_2\, dv_1\, dv_2 \ ; \ \varepsilon(\bar{v}) = \frac{1}{2}\left(v_1^2 + v_2^2\right) .$$

For an axisymmetric function f, $f(v) = f(\bar{v}(\theta))$ is independent of θ in $[0,2\pi]$, and therefore, we may introduce $\bar{f}(\bar{v})$ such that:

$$\bar{f}(\bar{v}) = f(\bar{v}(\theta)) \ ; \ \forall\, \theta \in [0,2\pi] . \tag{5}$$

LEMMA 1: Let f be a continuous function from $\mathbb{R}^3$ into $\mathbb{R}$, symmetric about e_1. Let $\bar{f}$ be associated with f according to (5). Then, we have:

$$Q(f,f)(v_0) = \bar{Q}(\bar{f},\bar{f})(\bar{v}_0)$$

with

$$\bar{Q}(\bar{f},\bar{f})(\bar{v}_0) = \iint_{(\mathbb{R}\times\mathbb{R}_+)^2} \Big[J(\bar{v}_0', \bar{v}_0, \bar{v})\, \bar{f}(\bar{v})\, \bar{f}(\bar{v}_0)$$

$$- J(\bar{v}_0, \bar{v}_0', \bar{v})\, \bar{f}(\bar{v})\, \bar{f}(\bar{v}_0) \Big]\, d\Omega(\bar{v}_0')\, d\Omega(\bar{v}) \tag{6}$$

and

$$J(\bar{v}_0', \bar{v}_0, \bar{v}) = (2\pi)^{-2} \iint_{[0,2\pi]^2} H\left(\left| \bar{v}_0 - \bar{v}_0'(\theta_0') \right|, \left| \bar{v}(\theta) - \bar{v}_0'(\theta_0') \right| \right)$$

$$\delta\left[\varepsilon\left(\bar{v}(\theta) + \bar{v}_0'(\theta_0') - \bar{v}_0 \right) + \varepsilon_0 - \varepsilon_0' - \varepsilon \right] d\theta\, d\theta_0' \tag{7}$$

where δ stands for Dirac's measure, $\varepsilon = \varepsilon(v)$, $\varepsilon' = \varepsilon(v')$, and so on, and H is related to B according to:

$$B(|v - v_0|, \cos\chi) = H(|v_0 - v_0'|, |v_0 - v|)\, |v_0 - v_0'| \tag{8}$$

with

$$|v_0 - v_0'|^2 = \frac{1}{2} |v - v_0|^2 (1 - \cos\chi) .$$

PROOF: The details of the computation are performed in [12], and are omitted in this paper.

The outline is given below. The following expression of Q:

$$Q(f,f)(v_0) = \iiint H\left(|v_0 - v_0'|, |v - v_0|\right)\left(f_0' f' - f_0 f\right)$$

$$\delta\left(\varepsilon' + \varepsilon_0' - \varepsilon - \varepsilon_0\right)\delta\left(v' + v_0' - v - v_0\right) dv\, dv_0'\, dv'$$

is equivalent to (3) provided that H and B are related by (8). Indeed, the explicit integration of the Dirac measures leads to the conservation of momentum and energy, which is equivalent to (4). Such an expression is widely used in semiconductor physics [10]. Now by interchanging v' and v, we are led to:

$$Q(f,f)(v_0) = \tag{9}$$

$$\iiint H\left(|v_0 - v_0'|, |v - v_0'|\right) f_0' f\ \delta\left(\varepsilon + \varepsilon_0' - \varepsilon' - \varepsilon_0\right)\delta\left(v + v_0' - v' - v_0\right) dv\, dv_0'\, dv'$$

$$- \iiint H\left(|v_0 - v_0'|, |v - v_0|\right) f_0 f\ \delta\left(\varepsilon' + \varepsilon_0' - \varepsilon - \varepsilon_0\right)\delta\left(v' + v_0' - v - v_0\right) dv\, dv_0'\, dv'.$$

We only consider the first term G(f,f) of the right hand side of (9). The treatment of the other term is analogous. First, the integration over v' is achieved, which cancels the Dirac measure over momentum, and leads to:

$$G(f,f)(v_0) = \iint H\left(|v_0 - v_0'|, |v - v_0'|\right) f_0' f\ \delta\left(\varepsilon\left(v + v_0' - v_0\right) + \varepsilon_0 - \varepsilon_0' - \varepsilon\right) dv\, dv_0'.$$

Now, since f is axisymmetric, we use cylindrical coordinates to perform the v and v'_0 integrations. By a formal use of Fubini's theorem for Dirac measures, we get:

$$G(f,f)(v_0) = \iint \bar{f}(\bar{v}_0')\bar{f}(\bar{v})\ \frac{1}{(2\pi)^2}\iint_{[0,2\pi]^2} H\left(\left|\bar{v}_0 - \bar{v}_0'(\theta_0')\right|, \left|\bar{v}(\theta) - \bar{v}_0'(\theta_0')\right|\right)$$

$$\delta\left[\varepsilon\left(\bar{v}(\theta) + \bar{v}_0'(\theta_0') - \bar{v}_0\right) + \varepsilon_0 - \varepsilon_0' - \varepsilon\right] d\theta\, d\theta_0'\, d\Omega(\bar{v})\, d\Omega(\bar{v}_0')$$

which formally leads to (6), (7). These formal computations are proved in [12].

We can get a more explicit expression of J by introducing additional hypotheses and notations. We suppose that H only depends on $|v'_0 - v_0|$, and we let:

$$\bar{v}_0 = \left(v_1^0, v_2^0\right)\ ;\ \bar{v} = \left(v_1, v_2\right)\ ;\ \bar{v}_0' = \left(v_1'^0, v_2'^0\right).$$

We suppose that $v_2^0 v_2'^0 \neq 0$. We define:

$$b = \left((v_2)^2 - (v_2^0)^2 - (v_1 - v_1^0)(v_1'^0 - v_1^0)\right)\left(v_2^0 v_2'^0\right)^{-1}$$

$$c = \left[\left((v_2^0)^2 + (v_1 - v_1^0)(v_1'^0 - v_1^0)\right)^2 - \left(v_2 v_2'^0\right)^2 - \left(v_2 v_2^0\right)^2\right]\left(v_2^0 v_2'^0\right)^{-2}$$

$$\Delta = b^2 - c\ .$$

If $\Delta \geq 0$, we also introduce:

$$\gamma_1 = -b - \sqrt{\Delta} \quad ; \quad \gamma_2 = -b + \sqrt{\Delta}$$

$$a_1 = \mathrm{Max}\,(-1, \gamma_1) \quad ; \quad a_2 = \mathrm{Min}\,(1, \gamma_2)$$

and for u in [-1,1]:

$$P(u) = (1-u^2)(\gamma_2 - u)(u - \gamma_1)$$

$$h(u) = H\left(\left[\left(v_2^0\right)^2 + \left(v_1'^0 - v_1^0\right)^2 + \left(v_2'^0\right)^2 - 2 v_2^0 v_2'^0 u\right]^{1/2}\right).$$

LEMMA 2: (i) If $\Delta < 0$ or $|b| > \sqrt{\Delta} + 1$, then $J\left(\bar{v}'_0, \bar{v}_0, \bar{v}\right) = 0$.

(ii) If $\Delta \geq 0$ or $|b| \leq \sqrt{\Delta} + 1$, then:

$$J\left(\bar{v}'_0, \bar{v}_0, \bar{v}\right) = \frac{1}{\pi^2 v_2^0 v_2'^0} \int_{a_1}^{a_2} \frac{h(u)}{\sqrt{P(u)}}\, du . \tag{10}$$

REMARK 1: (i) The above expressions are only valid for $v_2^0 v_2'^0 \neq 0$. A complete expression of J for all the possible situations is given in [12].

(ii) The proof of Lemma 2 relies on the explicit integration of (7) with respect to θ and can be found in [12].

(iii) The integral (10) is undefined when the fourth degree polynomial P(u) has a multiple root in $[a_1, a_2]$, which occurs in the following cases:

$$\gamma_1 = \gamma_2 \text{ or } \gamma_1 = -1 \text{ or } \gamma_1 = 1 \text{ or } \gamma_2 = -1 \text{ or } \gamma_2 = 1$$

In these cases, the function J is singular. These singularities are reminiscent of the singularity of the Dirac measure $\delta\left(\varepsilon' + \varepsilon'_0 - \varepsilon - \varepsilon_0\right)$ which expresses the energy conservation. A numerical treatment of the integral (10) requires a smoothing of the singularities which introduces a discrepancy in the conservation of energy. This feature will be discussed in the next sections.

(iv) If the expression of H is sufficiently simple, which is the case for most intermolecular potentials, the integral J can be sampled and stored in the computer memory with respect to the three parameters $(\gamma_1, \gamma_2, \delta)$, where δ , is given by:

$$\delta = \left[\left(v_2^0\right)^2 + \left(v_1'^0 - v_1^0\right)^2 + \left(v_2'^0\right)^2\right]\left(2 v_2^0 v_2'^0\right)^{-1}.$$

EXAMPLE : For the Coulomb interaction between electrons in a semiconductor, under Thomas Fermi screening, the interparticle potential is given by $V(r) = C \exp(-\beta r)/r$ where β is the reciprocal Debye length. Then H and J are written [13]:

$$H\left(\left| v_0 - v'_0 \right|\right) = C\left(\left| v_0 - v'_0 \right|^2 + \beta^2\right)^{-2} \tag{11}$$

$$J\left(\bar{v}'_0, \bar{v}_0, \bar{v}\right) = \frac{C}{\pi^2 \left(v_2^0 v_2'^0\right)^3} \int_{a_1}^{a_2} \frac{du}{(S - 2u)^2 \sqrt{P(u)}} \tag{12}$$

where S is defined by

$$S = \left[\left(v_2^0\right)^2 + \left(v_1'^0 - v_1^0\right)^2 + \left(v_2'^0\right)^2 + \beta^2 \right] \left(v_2^0 v_2'^0\right)^{-1}.$$

PROPOSITION 1 : For any regular $\bar{f}$, we have:

(i) Conservation of mass: $\int \bar{Q}(\bar{f},\bar{f})(\bar{v}_0)\, d\Omega(\bar{v}_0) = 0$

(ii) Conservation of momentum: $\int \bar{Q}(\bar{f},\bar{f})(\bar{v}_0)\, v_1^0\, d\Omega(\bar{v}_0) = 0$

(iii) Conservation of energy: $\int \bar{Q}(\bar{f},\bar{f})(\bar{v}_0)\, \varepsilon(\bar{v}_0)\, d\Omega(\bar{v}_0) = 0$

(iv) For any triple of scalars $(\rho,u,T) \in \mathbb{R}_+ \times \mathbb{R} \times \mathbb{R}_+$, we denote by $\overline{M}_{\rho,u,T}(\bar{v})$, the Maxwellian:

$$\overline{M}_{\rho,u,T}(\bar{v}) = \frac{\rho}{(2\pi T)^{3/2}} e^{-\varepsilon(\bar{v}-\bar{u})/2T}$$

where $\bar{u} = (u,0)$. Then, we have: $\bar{Q}\left(\overline{M}_{\rho,u,T}, \overline{M}_{\rho,u,T}\right) = 0$.

PROOF: These properties are inherited from the usual properties of the three dimensional Boltzmann operators [14].

3. THE DETERMINISTIC PARTICLE METHOD

From now on, we will always deal with the reduced distribution $\bar{f}$ given by (5), and since no confusion will be possible, we drop the bars. We consider the following space homogeneous, axisymmetric Boltzmann equation:

$$\frac{\partial f}{\partial t} + E \frac{\partial f}{\partial v_1} = Q(f,f) \tag{13}$$

where Q (f,f) is the reduced operator (6), and E is a constant scalar. The $E\, \partial f/\partial v_1$ term describes the effect of the electric field on a population of charged particles (e.g. charge carriers in a plasma or a semi-conductor).

The deterministic particle method [3,4,5] relies on the approximation of f by a sum of weighted Dirac measures:

$$f(v_1, v_2, t) \approx \sum_i \omega_i\, f_i(t)\, \delta\left(v_1 - v_{1,i}(t)\right) \delta\left(v_2 - v_{2,i}\right)$$

where $\bar{v}_i(t) = \left(v_{1,i}(t), v_{2,i}\right)$ is the position of the i-th particle in the velocity space, ω_i its constant control volume, and $f_i(t)$ its variable weight. The convective part of equation (13) gives rise to the evolution of the position of the particles:

$$\frac{dv_{1,i}(t)}{dt} = E \,. \tag{14}$$

The collision term is accounted for by the variation of the weights:

$$\frac{df_i(t)}{dt} = Q_i \tag{15}$$

where Q_i is an approximation of $Q(f,f)\left(v_{1,i}(t), v_{2,i}\right)$. So far, any type of approximation of

Q_i can be used. Our method considers the particles as quadrature points for a numerical quadrature of the double integral involved in (6). This gives:

$$Q_i(t) = \sum_{j,k} \left[J^{\alpha}\left(\overline{v}_j(t), \overline{v}_i(t), \overline{v}_k(t)\right) f_k(t) f_j(t) - J^{\alpha}\left(\overline{v}_i(t), \overline{v}_j(t), \overline{v}_k(t)\right) f_k(t) f_i(t) \right] \omega_j \omega_k . \quad (16)$$

J^{α} is a smoothed approximation of J, which is needed because of the singularities of J (see remark 1 (iii)).

REMARK 2: (i) Equation (14) for the evolution of the particles is particularly simple due to the space homogeneity hypothesis. The extension of the method to space inhomogeneous problems does not lead to major difficulties [4]. Indeed, the main question remains the discretization of the Boltzmann operator, and equation (13) retains all the complexity of the problem.

(ii) Neither error estimates nor convergence proofs are available for this method. However, this method has been proved to be convergent for linearized Boltzmann operators [4]; most of the numerical tests are performed with linear operators [5,15], or weakly non-linear ones [16].

We now detail some points of the implementation of this method. We have investigated the screened Coulomb interaction, given by (11), (12). The computation of the integral (12) at each time step, for each triple of particles (i,j,k) would lead to a tremendous computer cost. Since J depends on $\left(\overline{v}_0, \overline{v}'_0, \overline{v}\right)$ through only three scalar parameters (γ_1, γ_2, S), we perform the sampling of J on a grid. The sampled values are computed by a numerical quadrature of the integral (12). The smoothed approximation J^{α} is obtained by a truncation of J near its singularities (see remark 1, (iii)), which is performed during the sampling procedure. The asymptotic behaviour of J for large values of γ_1, γ_2, and S is used to compute J outside the range of the sampling parameters, when needed in formula (16). Since J only depends on the intermolecular potential, the same sampling can be used for all the problems involving the same intermolecular potential. The algorithmic complexity of formula (16) is N^3 (N being the number of particles), which is very large, even for a moderate number of particles. To reduce the CPU time, formula (16) is computed using only a few number of "effective particles". The effective particles belong to a ball in the velocity space, centered at the maximum of f, and of specified radius, chosen large enough to retain more than 98 % of the total mass of the distribution function, but small enough to lead to a reasonable computer cost.

With this methodology, the sampling procedure (over 300,000 sample points) needs 10 mn of CPU time on CRAY-2, and the simulation itself over 50 time steps needs the same amount of time (with 1000 particles and 200 effective particles).

4. NUMERICAL RESULTS

Example 1: We consider equation (13) with a centered Maxwellian $\overline{M}_{\rho,0,T}(\overline{v})$ as initial data. Since the Maxwellians are in the kernel of the collision operator Q (proposition 1 (iv)), the exact solution is a translated Maxwellian $\overline{M}_{\rho,Et,T}(\overline{v})$. This test is intended to check whether the discrete collision operator acting on Maxwellians is vanishing, or at least small. The equations for the mean velocity u and the mean energy ε follow from the consevations properties (proposition 1 (ii) and (iii)):

$$\frac{dv}{dt} = E \ ; \ \frac{d\varepsilon}{dt} = E v \qquad (17)$$

and are also analytically solvable. The simulation has been performed in a context of semi-conductor physics, with $\rho = 10^{22}$ electrons per m^3 and T = 77 Kelvin. In this context, the strength of the Boltzmann operator is measured in terms of the electron-electron relaxation time τ_{ee}, which is estimated to 0.2 10^{-12} seconds. The simulation has been performed over a time equal to 5 10^{-12} seconds, corresponding to 50 time steps.

The error on the mean velocity u has been observed to be smaller than 1 %. Figure 1-a reproduces the time evolution of the mean energy. At the end of the simulation, a small discrepancy of about 3 % is visible. On the average, the mean velocity and the mean energy are quite accurately described. Figure 1-b shows snapshots of the v_1 dependence of the distribution function, at the beginning of the simulation and at time t = 2 ps. Although the peak has correctly moved towards the positive velocities, its magnitude has considerably decreased and its width has increased. This behaviour certainly comes from the non-conservativity of the energy by the discrete collision operator Q_i. This discrepancy is due to the smoothing of the singularities of J and to the approximation of the collision integral by a quadrature. These effects are particularly apparent in this test problem. Indeed, since the continuous collision operator vanishes while the discrete one does not, the numerical errors are not balanced by anything else, and the relative errors on the collision operator are infinite. Therefore this test problem can be considered as very severe.

Example 2: We now initialize equation (13) with a non equilibrium distribution function: the "half Maxwellian" :

$$f_0(\bar{v}) = \begin{cases} \overline{M}_{\rho,0,T}(\bar{v}) & \text{if } v_1 \geq 0 \\ 0 & \text{if } v_1 < 0 \end{cases} \quad .$$

The exact solution is not analytically known, but its qualitative behaviour is clear. As the time proceeds the solution must approach a Maxwellian with the same density and temperature, moving in the velocity speed at the speed E, just as in case 1. The "initial layer" during which the solution relaxes to a Maxwellian shape is or the order of a few τ_{ee}. Furthermore the evolutions of the mean velocity u and energy ε are still governed by equations (17), and are analytically known. As shown on figure 2-a, the relative error on ε is of the same order as in example 1. Figure 2-b again displays snapshots of the distribution function along the v_1 axis at time t = 0 , 0.2 and 2 picoseconds. It shows that the relaxation towards a Maxwellian shape (t = 0.2 ps) correctly happens, and that the degradation of the shape of the distribution function occurs afterwards. This confirms the conlusions of example 1, that the approximation of the collision operator is better when applied to distributions which do not cancel the continous operator.

Example 3: We now investigate a real problem of semi-conductor physics. We consider equation (13), where Q(f,f) is now the sum of the Boltzmann collision operator (denoted by Q_{ee} , representing electron-electron collisions) and of a linear operator (denoted by Q_1 , representing the collisions of electrons against optical phonons):

$$Q(f,f) = Q_{ee}(f,f) + Q_1(f) \ .$$

The operator Q_1 is written:

$$Q_1(f)(\bar{v}) = \int \left[s(\bar{v}',\bar{v}) f(\bar{v}') - s(\bar{v},\bar{v}') f(\bar{v}) \right] d\Omega(\bar{v}')$$

$$s(\bar{v},\bar{v}') = \beta(\bar{v},\bar{v}') \left[(N_0+1)\,\delta\left(\varepsilon' - \varepsilon + \hbar\omega_0\right) + N_0\,\delta\left(\varepsilon' - \varepsilon - \hbar\omega_0\right) \right]$$

where N_0 and $\hbar\omega_0$ are positive constants and β is a smooth function. The reader will find more details about these models in [13,15] . The phonon interaction exhibits a threshold phenomenon at the energy $\hbar\omega_0$: indeed, $Q_1(f)(\bar{v})$ is almost vanishing for energies $\varepsilon(\bar{v}) \leq \hbar\omega_0$. If the equation is initialized with a centered Maxwellian, the solution stabilizes at a stationary solution after some transient behaviour. The stationary solution results from the balance between the electric field and the collision operator Q_1 (which has a dissipative effect). The threshold behaviour of the phonon interaction induces a very particular shape of the transient solution: the mean velocity and the mean energy first reach larger values than those of the stationary state before decaying towards these values. This "overshoot" behaviour is due to the fraction of the electrons which are accelerated by the electric field up to the energy $\hbar\omega_0$ and suddenly undergo a phonon interaction which makes them lose all their energy. It is expected that including the electron-electron interaction should decrease the magnitude of the overshoot. Indeed, this interaction makes the distribution function more isotropic, and therefore diminishes the number of the electrons reaching the energy $\hbar\omega_0$. Figure 3-a and 3-b display the mean velocity and the mean energy versus time for the phonon interaction alone and for the phonon and the electron-electron interaction together. We actually observe the expected influence of the electron-electron interaction on the magnitude of the overshoot.

<u>Conclusion of the numerical tests</u>: The proposed discretization of the Boltzmann equation leads to a quite accurate description of the moments of the distribution function (mean velocity and mean energy). This is important since these are the quantities of interest for engineers. Moreover, it seems to (at least qualitatively) predict the correct behaviour in the physical case that has been tested. However, improvements should be made to reduce the discrepancy in the conservation of the energy by a better description of the singularities of J. More significant test problems should also be investigated.

5. CONCLUSION

We have presented a new formula for the Boltzmann operator in axisymmetric geometry, and its application to numerical simulations in connection with deterministic particle methods. This method is one of the very few direct methods (i.e. non Monte-Carlo) to be investigated for the solution of the Boltzmann equations. We have developped some computational aspects of the method and presented numerical results on test problems. Its relatively low cost (about 10 minutes CPU on CRAY-2 for one simulation) makes the method attractive, even though it displays some unpleasant features. Many improvements may be investigated, and this method may appear, in the near future, as an interesting alternative to Monte-Carlo methods.

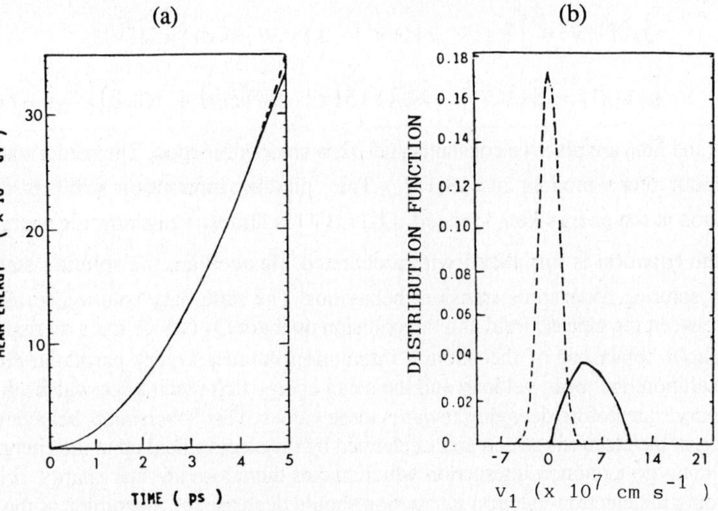

Figure 1: Equation (13) with a Maxwellian initial data

1-a: mean energy versus time for the exact solution (solid line) and for the computed solution (dashed line). The relative error reaches 3% at the end of the simulation.

1-b Distribution function versus v_1 at time t = 0 (dashed line) and t = 2 ps (solid line). The computed solution exhibits a non physical diffusion.

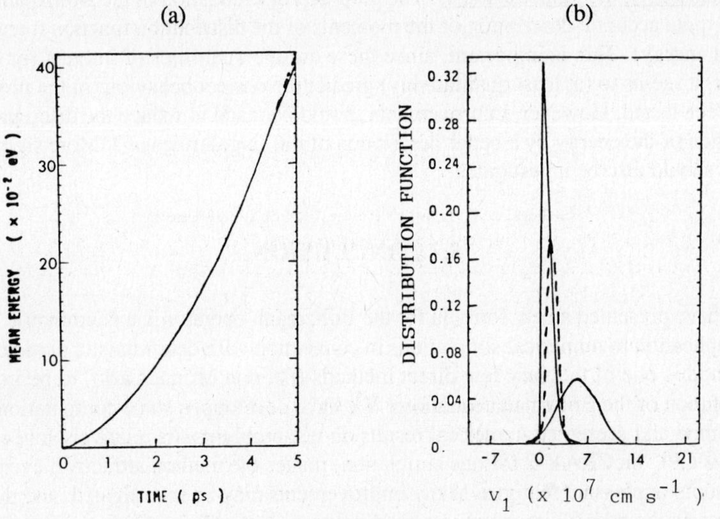

Figure 2: Equation (13) initialized with a half Maxwellian:

2-a: mean energy versus time for the exact solution (solid line) and for the computed solution (dashed line). Same observation as for figure 1-a.

2-b Distribution function versus v_1 at time t = 0 (solid line), t = 0.2 ps (dashed line), and t = 2 ps (solid line).Between 0 and 0.2 ps, the correct relaxation of the distribution function towards a Maxwellian shape is observed. Between 0.2 and 2 ps, a non physical diffusion occurs.

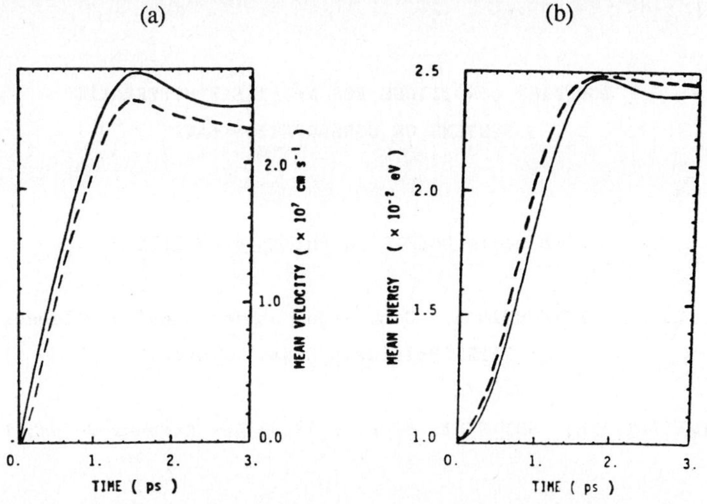

Figure 3: Equation (13) where Q(f,f) is the sum of an electron-electron collision term and of an electron-phonon collision term:

3-a mean velocity versus time for an electron-phonon collision term alone (solid line), and for the sum of an electron-electron and an electron-phonon collision term (dashed line).

3-b mean energy versus time; idem.

We observe a smoothing of the overshoot by the electron-electron collision term, in accordance to the qualitative prediction (see text).

REFERENCES

1 R. E. Caflish and B. Nicolaenko, Comm. Pure Appl. Math. 86 (1982), 161.

2 R.J. Mason, J. Comp. Phys. 51 (1983), 484.

3 S. Mas-Gallic and P. A. Raviart, Numer. Math. 51 (1987), 323.

4 S. Mas-Gallic, Transport Theory and Statistical Physics 16 (1987), 855.

5 S. Mas-Gallic and F. Poupaud, *Approximation of the Transport Equation by a Weighted Particle Method*, internal report, Ecole Polytechnique (1987), submitted to Transport Theory and Statistical Physics.

6 G. A. Bird, Phys. Fluids 6 (1963), 1518 ; Phys. Fluids 13 (1970), 2676 ; Molecular Gas Dynamics (Clarendon Press, Oxford, 1976). also S. M. Deshpande, Dept. Aero. Eng. Indian Inst. , Science Rep. 78, FM4 (1978).

7 K. Nanbu, J. Phys. Soc. Jpn. 49 (1982), 2042.

8 H. Babovsky, Math. Meth. Appl. Sci. 9 (1986).

9 R. Illner and H. Neunzert, Transport Theory and Statistical Physics, to appear.

10 P. Lugli and D. K. Ferry, Phys Rev. Lett. 46 (1985), 594 ; IEEE Electron Dev. Lett. EDL-6 (1985), 25.

11 A. J. Chorin, J. Comp. Phys. 8 (1971), 472 ; Comm. Pure Appl. Math. 25 (1972), 171.

12 B. Niclot, *"The two particle Boltzmann collision operator in axisymmetric geometry"*, internal report, Ecole Polytechnique (1987), sub. to Transport Theory and Statistical Physics.

13 L. Reggiani (ed), Hot Electron Transport in semi-conductors (Springer-Verlag, Berlin, 1985).

14 H. Grad, Principles of the Kinetic Theory of Gases, Handbuch der Physics 12 (1958), 205.

15 B. Niclot, P. Degond and F. Poupaud, *"Deterministic Particle Simulations of the Boltzmann Transport Equation of Semi-conductors"*, internal report, Ecole Polytechnique (1987), to appear in J. Comp. Phys.

16 P. Degond, F. Poupaud, B. Niclot and F. Guyot, *"Semiconductor Modelling via the Boltzmann equation"*, Proceedings of the 1987 AMS-SIAM-IMA Summer Seminar on Computational Aspects of VLSI Design.

BOUNDARY CONDITIONS FOR NONLINEAR HYPERBOLIC SYSTEMS OF CONSERVATION LAWS.

François DUBOIS* & Philippe LE FLOCH

ECOLE POLYTECHNIQUE, Centre de Mathématiques Appliquées, F-91128 Palaiseau Cedex, France.

* AEROSPATIALE, SDT-STMI, BP96, F-78133 Les Mureaux Cedex, France.

ABSTRACT

We propose two formulations of the boundary conditions for nonlinear hyperbolic systems of conservation laws. A first approach is based on the vanishing viscosity method and a second one is related to the Riemann problem. The equivalence between these two conditions is studied. The latter formulation is extended to treat numerically physically relevant boundary conditions. Monodimensional experiments are presented.

INTRODUCTION

We study initial-boundary value problems for nonlinear hyperbolic systems of conservation laws. Recall that with strong Dirichlet boundary conditions the associated problem is not well posed. Generally there is neither existence nor uniqueness. Thus weaker conditions are necessary ; in the linear case by example we know that data are given only on incoming characteristics.

In this paper we define the boundary condition in terms of **admissible values at the boundary**, related to the boundary datum. In our first formulation the set of admissible values is defined thanks to a **boundary entropy inequality** obtained by the vanishing viscosity method and the second set is related to the **resolution of a Riemann problem** at the boundary. The equivalence of these two formulations is established for nonconvex scalar conservation laws and strictly hyperbolic linear systems. The second formulation is naturally applied to Godunov-type numerical schemes : the numerical boundary condition reduces to the computation of a **boundary flux** thanks to some Riemann problem (or partial Riemann problem in physically relevant situations). As an application, outgoing waves from the Sod shock tube are presented.

BOUNDARY ENTROPY INEQUALITY (FIRST FORMULATION)

We consider a nonlinear hyperbolic system of conservation laws in one space dimension :

$$\frac{\partial u}{\partial t} + \frac{\partial}{\partial x} f(u) = 0 \qquad ; \qquad u(x,t) \in R^n \quad , \; x > 0 \; , \; t > 0 \qquad (1)$$

where $f : R \rightarrow R$ is a smooth flux-function. We suppose that there exists at least a pair (η,q) of entropy-flux in the sense of Lax [9]. The initial boundary value problem obtained by the viscosity method $(\epsilon>0)$:

$$\left\{\begin{array}{lll} \dfrac{\partial u^\epsilon}{\partial t} + \dfrac{\partial}{\partial x} f(u^\epsilon) = \epsilon \dfrac{\partial^2 u^\epsilon}{\partial x^2} & & x > 0 \; , \; t > 0 \\ u^\epsilon(x,0) = v_0(x) & & x > 0 \\ u^\epsilon(0,t) = u_0(t) & & t > 0 \end{array}\right. \qquad (2)$$

admits a unique solution u^ϵ and we study the behaviour of u^ϵ at the boundary as ϵ tends to zero. In fact a discontinuity appears, in general, at the boundary. The following result (essentially formal) yields an inequality at this discontinuity.

Theorem 1. Suppose that u^ϵ is bounded in $W^{1,1}_{loc}(R^+ x R^+, R^n)$ and converges in L^1_{loc} to u as $\epsilon \rightarrow 0$. Then for each admissible pair (η,q) of entropy-flux we have the following **boundary entropy inequality** :

$$q(u(0^+,t)) - q(u_0(t)) - d\eta(u_0(t)).(f(u(0^+,t))-f(u_0(t))) \le 0 \; , \; t > 0 \qquad (3)$$

between the taken value $u(0^+,t)$ and the prescribed value $u_0(t)$ at the boundary.

This result was first derived in [2] in the particular case of scalar conservation laws. The details concerning the derivation of the boundary entropy inequality (3) in the case of systems of conservation laws are presented in [6]. Remark that the latter inequality was independently obtained by other methods [1,12].

Given a state u_0 we define a (first) **set of admissible values at the boundary** :

$$E(u_0) = \{ v \in R^n ,\ q(v)-q(u_0)-d\eta(u_0).(f(v)-f(u_0)) \leq 0 ,\ \forall\ (\eta,q) \text{ pair of entropy-flux} \} .$$

Therefore let us extend the notion of Dirichlet boundary condition and define our (first) **formulation of the boundary condition** :

$$u(0^+,t) \in E(u_0(t)) \ , \ t > 0 \ . \tag{4}$$

The set $E(u_0(t))$ can be entirely explicited for both strictly hyperbolic linear systems and non-convex scalar conservation laws (see [6] for the proofs).

<u>Proposition 1. Strictly hyperbolic linear systems.</u>

Suppose that $f(u) \equiv A.u$, with a constant matrix A characterized by n eigenvalues λ_i (and n associated eigenvectors r_i) satisfying

$$\lambda_1 < \lambda_2 < \ldots < \lambda_p \leq 0 < \lambda_{p+1} < \ldots < \lambda_n \ . \tag{5}$$

Then the set $E(u_0)$ is the affine space containing u_0 and generated by the p first eigenvectors of A :

$$E(u_0) = \{ u_0 + \sum_{i=1}^{p} \alpha_i\ r_i \ ,\quad \alpha_1,\ldots,\alpha_p \in R \} .$$

The interpretation of the boundary condition (4) here is the following : the components of $u(0^+,t)$ on the (n-p) last eigenvectors (i.e. the incoming characteristics) are given by the boundary state $u_0(t)$. With the present approach we recover the classical one in this particular case.

Proposition 2. Scalar conservation laws.

Suppose that the flux $f(u)$ is a C^1 function from R to R. Then the set $E(u_0)$ of the admissible states u is entirely characterized by the **family of inequalities** :

$$\frac{f(u) - f(k)}{u - k} \leq 0 \qquad \forall\, k \in [u,u_0] \cup [u_0,u] \,. \tag{6}$$

This proposition was previously established in [10], and a geometrical interpretation is presented in [6]. In the particular case of **convex** scalar conservation laws the latter is simpler. Let us specify it for the Burgers equation.

Propositon 3. Burgers equation.

When $u \in R$ and $f(u) \equiv u^2/2$, the set $E(u_0)$ is given by :

(i) if $u_0 \geq 0$, $E(u_0) =]-\infty,-u_0] \cup \{u_0\}$

(ii) if $u_0 \leq 0$, $E(u_0) =]-\infty,0]$.

In the general case of an hyperbolic system of conservation laws, the lack of mathematical entropies does not allow a complete description of this boundary set $E(u_0)$.

APPROACH BY THE RIEMANN PROBLEM (SECOND FORMULATION)

For our second formulation of the boundary condition [5,6] we suppose that each Riemann problem $R(u_L,u_R)$ associated with (1) admits a unique entropy solution denoted by $w(x/t;u_L,u_R)$. Let us define a **second set of admissible values** by :

$$V(u_0) = \{\, w(0^+;u_0,u_R),\ u_R \text{ varying in } R^n \,\} \,.$$

Then we have the following result which generalizes [9] :

Theorem 2 Let v_0, u_0 be constant states of R^n. The problem

$$\begin{cases} \dfrac{\partial u}{\partial t} + \dfrac{\partial}{\partial x} f(u) = 0 & x > 0 \ , \ t > 0 \\ u(x,0) = v_0 & x > 0 \\ u(0,t) \in \mathbf{V}(u_0) & t > 0 \end{cases} \tag{7}$$

is **well posed** in the class of functions which consist of constant states separated by at most n elementary waves (rarefactions, shocks, contacts).

Proposition 4. Link between the two formulations.

In particular cases of strictly hyperbolic linear systems and (non necessarily convex) scalar conservation laws, **the two sets are identical**:

$$\mathbf{E}(u_0) = \mathbf{V}(u_0) \qquad \forall\, u_0 \in R^n .$$

The advantage of the second formulation is that $\mathbf{V}(u_0)$ can be easily computed. For the p-system, $\mathbf{V}(u_0)$ is exactly the 1-wave containing u_0. And, in [5,6] we have given details on the **V**-sets in the case of barotropic Euler-Saint Venant equations. For more precise relations concerning the **E** and **V** sets in the particular case of 2x2 systems of conservation laws, we refer to [3,6]. Refer also to [11] about a formulation of boundary conditions for weighted conservation laws.

APPLICATION TO THE EULER EQUATIONS OF GAS DYNAMICS

We apply now the ideas developed previously to Godunov-type finite volume numerical schemes [8]. We restrict ourselves to the first order accurate methods. The interval [0,1] is divided into N cells and the numerical approximation of the conservation law (1) at time $t_n = n\Delta t$ in the j° cell is given by :

$$\frac{1}{\Delta t}\left(u_j^{n+1} - u_j^{n} \right) + \frac{1}{\Delta x}\left(f_{j+1/2}^{n} - f_{j-1/2}^{n} \right) = 0 \ . \tag{8}$$

For the internal cells we have classically

$$f_{j+1/2}^{n} = \phi (u_j^{n} , u_{j+1}^{n}) , \quad j = 1,2,...,N-1 \tag{9}$$

for some numerical flux function ϕ that approaches the flux $f(w(0;u_j^n,u_{j+1}^n))$ of the Riemann problem $R(u_j^n,u_{j+1}^n)$ when $x/t=0$. We suppose that a boundary state u_L (resp u_R) is given for $x \le 0$ (resp $x \ge 1$) and we consider it intuitively as a limiting state for x tending towards $-\infty$ (resp $+\infty$). Thus the **numerical boundary condition** at time t_n results from the interaction of u_L (resp u_R) with the value u_1^n (resp u_N^n) of the field in the first (resp last) cell:

$$f_{1/2}^{n} = \phi (u_L , u_1^{n}) \; ; \; f_{N+1/2}^{n} = \phi (u_N^{n} , u_R) . \tag{10}$$

This kind of numerical boundary condition in terms of a **numerical flux** is natural with the approach of finite volumes. This fact was first recognized by Godunov (e.g. [7]).

The numerical scheme (8)(9)(10) has been applyed to the Sod shock tube [15] for the Euler equations of gas dynamics, i.e. with left and right states $u_L = (\rho_L, v_L, p_L) = (1,0,1)$, $u_R = (\rho_D, v_D, p_D) = (0.125,0,0.1)$, and N=100 cells. We used the Osher upwind scheme [13] and have performed the numerical computation for a time sufficiently long so that the different waves have been gone outside the computational domain [0,1] (see Figure 1). Some results are plotted on Figure 2. The boundary condition (10) appears numerically as transparent for all these nonlinear waves and the physical fields at x = 0 and x = 1 are correct (the difference with the exact solution is first due to the high level of numerical viscosity contained in the first order scheme). More details on this problem with the use of the exact linearized implicit Osher scheme are developed in [4].

We focus now on more realistic boundary conditions for the Euler equations. For most of the internal aerodynamics problems a state $u_0(t)$ is **not** physically given at the boundary. As usual, we distinguish between four cases : the fluid may be sub or super-sonic at the in or out-flow and physical parameters can be associated with each case [16] : **(i)** supersonic inflow : a state u_0 , **(ii)** subsonic inflow : total enthalpy H and physical entropy S , **(iii)** subsonic outflow : static pressure P , **(iv)** supersonic outflow : no numerical datum. We review briefly the main ideas of [4]. For each of the four above cases a manifold (eventually with boundary) **M** is defined by the boundary data; we have respectively

$$\text{(i)} \quad \mathbf{M} = \{ (\rho_0, v_0, p_0) \}$$

$$\text{(ii)} \quad \mathbf{M} = \left\{ (\rho, v, p) \ / \ \frac{1}{2} v^2 + \frac{\gamma \, p}{\gamma - 1} = H \ , \ p = S \, \rho^{\gamma} \right\}$$

$$\text{(iii)} \quad \mathbf{M} = \{ (\rho, v, p) \ / \ p = P \}$$

$$\text{(iv)} \quad \mathbf{M} = \left\{ (\rho, v, p) \ / \ v - c \geq 0 \ , \ c^2 = \frac{\gamma \, p}{\rho} \right\} .$$

Then the formula (10) relative to the computation of the boundary flux is adapted as follows (we consider only the case $x = 0$). A **partial Riemann problem** $P(\mathbf{M},z)$ is posed naturally by the boundary condition between the manifold **M** and the state z located in the (first) cell of the computational domain. We solve this problem in the same manner as Lax did [9] for the classical Riemann problem. A family of codim**M** (equal respectively to 3,2,1,1 in the previous cases) nonlinear waves issued from z intersects **M** at a state I. Interpreting those waves in the (x,t) plane, the solution of $P(\mathbf{M},z)$ joins the state I (of **M**) to the state z thanks to a fan of codim**M** waves (Figure 3). Then the boundary flux $f_{1/2}$ is given by

$$f_{1/2} = f(W) \tag{11}$$

where W is the state of this fan located at $x/t=0^+$. In [4] we have used the Riemann solver of Osher that contains only (eventually multivalued) rarefactions. Thus we have taken into account the (eventual) multiplicity of the states W. Furthermore in the particular case (iii) (given pressure P) and for a sufficiently weak nonlinearity (i.e. p(z) not too far from P) we recover previous results obtained by Osher-Chakravarthy [14]. We have also tested in [4] all thoses boundary conditions (i)-(iv) for one dimensional nozzles using both the explicit and linearized implicit versions of the scheme.

REFERENCES

[1] J. AUDOUNET, Solutions discontinues paramétriques des systèmes de lois de conservation et des problèmes aux limites associés, Seminar, Toulouse 3 University, 1983-1984.

[2] C. BARDOS, A.Y. LEROUX, J.C. NEDELEC, First Order Quasilinear Equations With Boundary Conditions, Comm. Part. Diff. Eqs. 4, n°9 (1979), 1017-1034.

[3] A. BENABDALLAH, D. SERRE, Problèmes aux limites pour les systèmes hyperboliques non-linéaires de deux équations à une dimension d'espace C. R. Acad. Sc. Paris **305**, Série 1 (1987), 677-680.

[4] F. DUBOIS, Boundary Conditions and the Osher Scheme for the Euler Equations of Gas Dynamics, Internal Report n° 170, Centre de Mathématiques Appliquées de l'Ecole Polytechnique , nov 1987.

[5] F. DUBOIS, P. LE FLOCH, Boundary Condition for Systems of Hyperbolic Conservation Laws, C. R. Acad. Sc. Paris **304**, Série 1 (1987), 75-78.

[6] F. DUBOIS, P. LE FLOCH, Boundary Conditions for Nonlinear Hyperbolic Systems of Conservation Laws, J. Diff. Eqs. **71**, n°1 (1988), 93-122.

[7] S.K. GODUNOV et al, **Résolution numérique des problèmes multidimensionnels de la dynamique des gaz**, Mir Edition, Moscou, 1979.

[8] A. HARTEN, P. LAX, B. VAN LEER, On Upstream Differencing and Godunov-Type Schemes for Hyperbolic Conservation Laws, SIAM Rev. **25** n°1 (1983), 35-61.

[9] P. D. LAX, **Hyperbolic Systems of Conservation Laws and the Mathematical Theory of Shock Waves**, SIAM, Philadelphia, 1973.

[10] P. LE FLOCH, Boundary Condition for Scalar Nonlinear Conservation Laws, Math. Meth. in Appl. Sciences (1988).

[11] P. LE FLOCH, J.C. NEDELEC, Weighted Scalar Conservation Laws with Boundary Conditions, Trans Amer. Math. Soc., in press; C. R. Acad. Sc. Paris **301**, Série 1 (1985), 1301-1304; and Internal Report n° 144, Centre de Mathématiques Appliquées de l'Ecole Polytechnique (1986).

[12] P.A. MAZET, F. BOURDEL, Analyse numérique des équations d'Euler pour l'étude des écoulements autour de corps élancés en incidence, CERT Report n° 1/3252, Toulouse (1986).

[13] S. OSHER, Numerical Solution of Singular Perturbation Problems and Hyperbolic Systems of Conservation Laws, in **Mathematical Studies n° 47**, North Holland (1981), 179-205.

[14] S. OSHER, S. CHAKRAVARTHY, Upwind Schemes and Boundary Conditions with Applications to Euler Equations in General Geometries, J. Comp. Phys. **50** (1983), 447-481.

[15] G. SOD, A Survey of Several Finite Difference Methods for Systems of Hyperbolic Conservation Laws, J. Comp. Phys. **27** (1978), 1-31.

[16] H. VIVIAND, J.P. VEUILLOT, Méthodes pseudoinstationnaires pour le calcul des écoulements transsoniques, ONERA T. P. n° 1978-4 (1978).

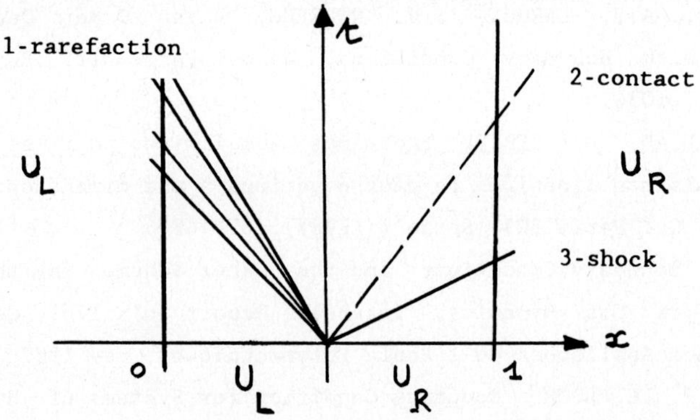

Figure 1. The Sod shock tube for time tending to infinity.

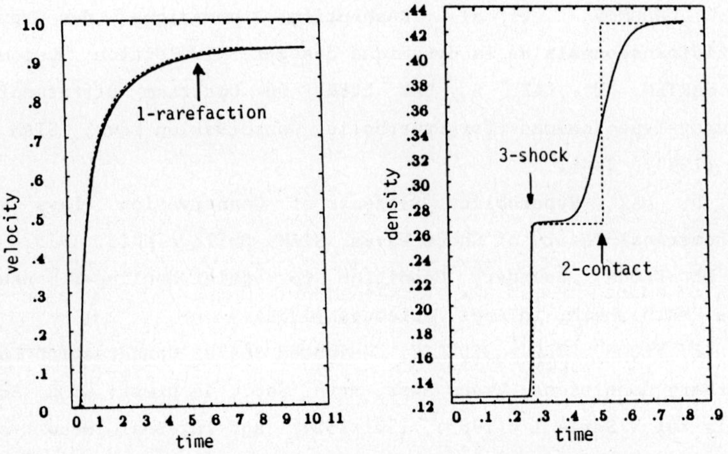

Figure 2. Evolution of the velocity at x=0 (left) and of the density at x=1 (right) for the Sod shock tube with N=100 cells. The dotted line is the exact solution.

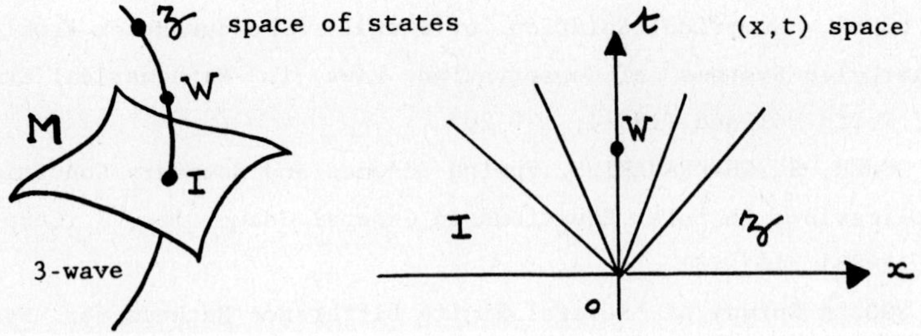

Figure 3. Resolution of the partial Riemann problem p(M,z) in the particular case (iii).

Time-Marching Method to Solve Steady Incompressible Navier-Stokes Equations for Laminar and Turbulent Flow

by

Peter Eliasson and Arthur Rizzi
FFA The Aeronautical Research Institute of Sweden
161 11 Bromma, Sweden
and
H.I. Andersson
Norwegian Institute of Technology
Trondheim, Norway

SUMMARY

For the solution of the steady incompressible Navier-Stokes equation, an explicit Runge-Kutta, finite-volume solver has been created using the artificial compressibility method. The standard $k - \epsilon$ turbulence model has also been included. A stability analysis was performed for the condition of the local time step for the Runge-Kutta scheme. Numerical results are presented for laminar and turbulent flow over two different backward facing steps.

INTRODUCTION

Recently Müller and Rizzi developed a Navier Stokes solver based on an explicit Runge-Kutta finite volume method to simulate laminar compressible flows over wings [1]. In this paper we are concerned with incompressible flow. If we were to simply apply the compressible code to this problem we would find that it would not converge well at all because with decreasing Mach number sound waves travel at a speed much larger than the speed of convection and they dominate the system making it stiff. This increasing disparity in wave speeds causes the governing system of equations to be poorly conditioned, and the stability of the computation is greatly impaired. If, however, the interest is only the steady flow, artificial compressibility is one way round the difficulty, because this approach removes the sound waves from the system by prescribing a pseudotemporal evolution for the pressure through the continuity equation which is hyperbolic and which converges to the true steady state value.

Our purpose here is to describe a rather general numerical method that takes the artificial compressibility approach for solving the steady incompressible Navier Stokes equations for laminar flow and also for turbulent flow with a $k - \epsilon$ turbulence model.

We show how it leads to a hyperbolic/parabolic system, we carry out a numerical study of its condition and set forth the CFL stability limit for the time integration.

Results are presented for internal laminar and turbulent flow over two backward facing steps. The numerical simulations are compared to the available experimental data.

MATHEMATICAL MODEL

Incompressible Navier-Stokes equations

Since the continuity equation for incompressible flow contains no time dependent term, an artificial time dependent term is added to the continuity equation. This is done by using the method proposed by Chorin [2]. The Navier-Stokes equations governing an incompressible flow, using the above method for the continuity equation, can then be stated in the following way:

$$\frac{1}{\rho_0}\frac{\partial p}{\partial t} + c^2\frac{\partial u_i}{\partial x_i} = 0 \qquad (1)$$

$$\frac{\partial u_i}{\partial t} + \frac{\partial u_i u_j}{\partial x_j} + \frac{1}{\rho_0}\frac{\partial p}{\partial x_i} - \frac{\mu}{\rho_0}\frac{\partial}{\partial x_j}\frac{\partial u_i}{\partial x_j} = 0 \qquad (2)$$

where ρ_0 is the constant density, u_i are the velocity components, μ is the viscosity coefficient and p the pressure. The viscosity coefficient μ is supposed to be constant, and c is an arbitrary parameter for optimal convergence. These equations have no physical meaning until steady state is obtained.

Introducing the integral formulation of (1, 2) the incompressible Navier-Stokes equations can be written:

$$\int_\Omega \frac{\partial \vec{q}}{\partial t} dV + \oint_{\partial\Omega} \mathbf{H}(\vec{q}) \cdot d\vec{S} = 0 \qquad (3)$$

where in two space dimensions:

$$\vec{q} = \begin{pmatrix} \frac{p}{\rho_0} \\ \vec{u} \end{pmatrix}, \quad \mathbf{H}(\vec{q}) = \begin{pmatrix} c^2\vec{u} \\ \vec{u}\vec{u} + \frac{p}{\rho_0}\mathbf{I} - \tau \end{pmatrix}, \quad \tau = \frac{\mu}{\rho_0} grad\vec{u}$$

I is the identity matrix and τ is the stress tensor.

The $k-\epsilon$ turbulence model

Assuming that all flow variables can be expanded in the form $f = \overline{f} + f'$ where $\overline{f}$ is a mean value and f' is a fluctuation around the mean, the transport equations governing an incompressible flow can be stated in the following way [3]:

$$\frac{\partial \overline{u}_i}{\partial t} + \frac{\partial \overline{u}_i \overline{u}_j}{\partial x_j} = -\frac{1}{\rho_0}\frac{\partial p}{\partial x_i} + \frac{\mu}{\rho_0}\frac{\partial}{\partial x_j}\frac{\partial \overline{u}_i}{\partial x_j} - \frac{\overline{\partial u_i' u_j'}}{\partial x_j} \quad . \qquad (4)$$

The turbulent Reynolds shear-stress $\overline{\partial u_i' u_j'}/\partial x_j$ is connected to the mean field by means of the generalized Boussinesq's hypothesis:

$$\frac{\overline{\partial u_i' u_j'}}{\partial x_j} = \frac{\mu_T}{\rho_0}(\frac{\partial u_i}{\partial x_j} + \frac{\partial u_j}{\partial x_i}) - \frac{2}{3}k\delta_{ij} \qquad (5)$$

where $\mu_T = \nu_T \rho_0$ is the eddy viscosity, δ_{ij} the Kronecker delta and $k = \frac{1}{2}\overline{u_i'^2}$ the turbulent kinetic energy. The bar denoting mean quantities will from now on be dropped.

Introducing a new variable ϵ for the dissipation rate of the turbulent kinetic energy, the standard $k-\epsilon$ turbulence model, consisting of two additional transport equations for k and ϵ, may be stated :

$$\frac{\partial k}{\partial t} + \frac{\partial u_i k}{\partial x_i} = \frac{\partial}{\partial x_i}(\frac{\nu_T}{\sigma_k}\frac{\partial k}{\partial x_i}) + \nu_T(\frac{\partial u_i}{\partial x_j} + \frac{\partial u_j}{\partial x_i})\frac{\partial u_i}{\partial x_j} - \epsilon \tag{6}$$

$$\frac{\partial \epsilon}{\partial t} + \frac{\partial u_i \epsilon}{\partial x_i} = \frac{\partial}{\partial x_i}(\frac{\nu_T}{\sigma_\epsilon}\frac{\partial \epsilon}{\partial x_i}) + \nu_T c_{1\epsilon}\frac{\epsilon}{k}(\frac{\partial u_i}{\partial x_j} + \frac{\partial u_j}{\partial x_i})\frac{\partial u_i}{\partial x_j} - c_{2\epsilon}\frac{\epsilon^2}{k} \tag{7}$$

where $\nu_T = c_\mu \frac{k^2}{\epsilon}$. c_μ, σ_k, σ_ϵ, $c_{1\epsilon}$ and $c_{2\epsilon}$ are empirically determined constants. The first terms on the right hand side in equations (6) and (7) are the diffusion terms, the second terms are the production terms and the third terms are the dissipation terms.

The integral formulation of the incompressible Navier-Stokes equations with the $k-\epsilon$ turbulence model can be introduced in accordance with (3). One additional term occurs though, a volume integral containing the production and the dissipation terms[3].

NUMERICAL METHOD

Spatial Discretization

The centered finite volume method is adopted here for solving the governing equations (3), the same method used for the incompressible Navier-Stokes equations [4] and the Euler equations [5]. The same method has been applied to the incompressible Navier-Stokes equations with the $k-\epsilon$ turbulence model. A short description of the method will be given below.

Let the computational domain Ω be divided into a number of quadrilateral subdomains $\Omega_{i,j}$ which form a structured grid with $m \times n$ cells. The solution to the volume integral in the governing equations is then approximated in the following way :

$$\int_{\Omega_{i,j}} \frac{\partial \vec{q}}{\partial t} dV \simeq \frac{d\vec{q}_{i,j}}{dt} vol_{i,j} \tag{8}$$

where $vol_{i,j}$ is the volume of cell (i,j).

The surface integral in equation (3) is approximated by the sum over all cell lattices in the quadrilateral of the averaged value of $\tilde{\mathbf{H}}$ at the lattice times the surface vector $\vec{S}$:

$$\oint_{\partial\Omega_{i,j}} \mathbf{H}(\vec{q}) \cdot d\vec{S} \simeq \left(\sum_{l=1}^{4} \mu_l \mathbf{H} \cdot \vec{S_l}\right)_{i,j} \tag{9}$$

where $\vec{S_l}$ is the outward pointing surface vector at cell lattice l and where μ_L is the averaging operator.

The flux tensor $\mathbf{H}$ is readily available in cell $\Omega_{i,j}$ except for the gradients of the velocity components. Following the definitions of the conservative variables as cell averages, the gradients in cell $\Omega_{i,j}$ are defined by:

$$(grad\phi)_{\Omega_{i,j}} = \frac{\int_{\Omega_{i,j}} grad\phi dV}{\int_{\Omega_{i,j}} dV} = \frac{\oint_{\partial\Omega_{i,j}} \phi d\vec{S}}{vol_{i,j}} \simeq \frac{1}{vol_{i,j}}\left(\sum_{l=1}^{4} \tilde{\phi}_l \cdot \vec{S_l}\right)_{i,j} \tag{10}$$

where $\phi = u, v$ (or w if three dimensions).

The approximations (8), (9) and (10) finally lead to the semi-discretized formulation

$$\frac{d\vec{q}_{i,j}}{dt} = -\frac{1}{vol_{i,j}} \left(\sum_{l=1}^{4} \tilde{\mathrm{H}}_l \cdot \vec{S}_l \right)_{i,j} \quad . \tag{11}$$

This equation will be referred to as the interior scheme. A corresponding interior scheme is obtained for the incompressible Navier-Stokes equations with the $k-\epsilon$ turbulence model, but we will restrict ourselves for the time being to talk about the laminar equations.

Boundary Conditions

At a solid wall the no-slip condition for the velocity is used. The wall pressure is obtained by assuming the boundary layer approximation to hold on the solid wall $\partial p/\partial n|_w = 0$.

It is physically meaningless to integrate the $k-\epsilon$ equations up to the wall since the standard high Reynolds number model is not valid in the vicinity of the wall. In order to avoid modifications to the model, a fictitious boundary is generally located inside the flow at a distance y_p from the wall. Then universal laws are used to describe the behaviour of the fluid at this boundary. This so-called wall-function approach can be obtained by assuming the near-wall region to be in local energy equilibrium, so that the velocity profile is logarithmic and the turbulent shear stress is constant. For a more detailed description of the wall-function approach, see the article by Chieng *et al.* [6] and also [3].

Numerical Damping

Using the interior scheme (11) and the boundary conditions described in the previous section, the physical flux over each cell has now been determined. The physical difference operator $\vec{F}_{ph}$ thus reads:

$$\vec{F}_{ph}(\vec{q}_{i,j}) = -\frac{1}{vol_{i,j}} \left(\sum_{l=1}^{4} \tilde{\mathrm{H}}_l \cdot \vec{S}_l \right)_{i,j} \quad . \tag{12}$$

The central differences in (12) give rise to oscillations and that is why some numerical damping have to be added to the scheme in order to damp the short wavelengths. The damping must then be of higher order than (12). Thus, the total difference operator $\vec{F}$ consists of the physical part $\vec{F}_{ph}$ and the numerical part $\vec{F}_n$. In interior cells the numerical damping is defined by a fourth order difference operator, and the semi-discrete approximation of the incompressible Navier-Stokes equations can be written:

$$\frac{\partial \vec{q}_{i,j}}{\partial t} = \vec{F}_{ph}(\vec{q}_{i,j}) + \vec{F}_n(\vec{q}_{i,j}) \quad , \quad \vec{F}_n(\vec{q}_{i,j}) = -\Gamma({\delta_i}^4 + {\delta_j}^4)\vec{q}_{i,j} \tag{13}$$

where $\delta_i \vec{q}_{i,j} = \vec{q}_{i+\frac{1}{2},j} - \vec{q}_{i-\frac{1}{2},j}$, equivalent for $\delta_j \vec{q}_{i,j}$. $\Gamma = \epsilon_4 IMAG/\Delta t$ with ϵ_4 a constant in the range 0.005 to 0.01, $IMAG$ being the maximum CFL number used and Δt the time step. Near boundaries $\vec{F}_n$ is defined by non-centered differences [7] [4] to ensure the dissipative property of damping.

Time Integration

The method to integrate ordinary differential equations like (12) is, in this paper an explicit one step, four stage, first order accurate Runge-Kutta algorithm. It has earlier been shown that this algorithm is superior to the standard three stage, second order Runge Kutta scheme [4]. For a general system

$$\frac{d\vec{q}}{dt} = \vec{F}(\vec{q}) \tag{14}$$

this scheme is defined as

$$\begin{aligned}
\vec{q}^{\,*}_{n+1} &= \vec{q}_n + \frac{1}{3}\Delta t\vec{F}(\vec{q}_n) \\
\vec{q}^{\,**}_{n+1} &= \vec{q}_n + \frac{4}{15}\Delta t\vec{F}(\vec{q}^{\,*}_{n+1}) \\
\vec{q}^{\,***}_{n+1} &= \vec{q}_n + \frac{5}{9}\Delta t\vec{F}(\vec{q}^{\,**}_{n+1}) \\
\vec{q}_{n+1} &= \vec{q}_n + \Delta t\vec{F}(\vec{q}^{\,***}_{n+1}))
\end{aligned} \tag{15}$$

and allows a CFL limit of 3 and the stability region is shown in Fig. 1.

Stability Analysis

A scalar model equation is used to study the stability of the semi-discretized incompressible Navier-Stokes equations (11)[4]:

$$\frac{\partial q}{\partial t} + \lambda_1\frac{\partial q}{\partial \xi} + \lambda_2\frac{\partial q}{\partial \eta} = \nu_1\frac{\partial^2 q}{\partial \xi^2} + \nu_2\frac{\partial^2 q}{\partial \xi\partial \eta} + \nu_3\frac{\partial^2 q}{\partial \eta^2} + \epsilon_1\frac{\partial^4 q}{\partial \xi^4} + \epsilon_2\frac{\partial^4 q}{\partial \eta^4} \tag{16}$$

Equation (16) is derived from the differential form of the incompressible Navier-Stokes equations written in curvilinear coordinates ξ and η by linearizing the momentum equation and from the differential form of the fourth order damping term [8]. The 1st derivatives are due to the convective terms, the 2nd derivatives to the viscous terms and the 4th derivatives to the numerical damping. They are discretized by second-order central differences, which are equivalent to the finite volume approximation on an equidistant Cartesian grid:

$$\left.\frac{\partial q}{\partial \xi}\right._{i,j} = \frac{1}{\Delta\xi}\mu_i\delta_i q_{i,j} + O(\Delta\xi^2) \quad , \quad \left.\frac{\partial q}{\partial \eta}\right._{i,j} = \frac{1}{\Delta\eta}\mu_j\delta_j q_{i,j} + O(\Delta\eta^2) \tag{17}$$

$$\left.\frac{\partial^2 q}{\partial \xi^2}\right._{i,j} = \frac{1}{\Delta\xi^2}(\mu_i\delta_i)^2 q_{i,j} + O(\Delta\xi^2) \quad , \quad \left.\frac{\partial^2 q}{\partial \eta^2}\right._{i,j} = \frac{1}{\Delta\eta^2}(\mu_j\delta_j)^2 q_{i,j} + O(\Delta\eta^2) \tag{18}$$

$$\left.\frac{\partial^2 q}{\partial \xi\partial \eta}\right._{i,j} = \frac{1}{\Delta\xi\Delta\eta}\mu_i\delta_i\mu_j\delta_j q_{i,j} + O(\Delta\xi^2, \Delta\eta^2) \ . \tag{19}$$

The fourth derivatives are discretized according to (13). The semi-discrete approximation of (16) is then obtained :

$$\begin{aligned}
\frac{\partial q}{\partial t} + [\frac{\lambda_1}{\Delta\xi}\mu_i\delta_i + \frac{\lambda_2}{\Delta\eta}\mu_j\delta_j - \frac{\nu_1}{\Delta\xi^2}(\mu_i\delta_i)^2 - \frac{\nu_2}{\Delta\xi\Delta\eta}\mu_i\delta_i\mu_j\delta_j - \\
- \frac{\nu_3}{\Delta\eta^2}(\mu_j\delta_j)^2 - \frac{\epsilon_1}{\Delta\xi^4}\delta_i{}^4 - \frac{\epsilon_2}{\Delta\eta^4}\delta_j{}^4]q_{i,j} = 0 \ .
\end{aligned} \tag{20}$$

The stablilty region of the Runge-Kutta scheme applied to the test equation

$$\frac{dq}{dt} = \Lambda q \tag{21}$$

is given by

$$|g(\Lambda \Delta t)| \leq 1 \tag{22}$$

(Fig. 1) where $g(z), z = \Lambda \Delta t$ is the growth factor.

A Fourier analysis is used to investigate the stability of the scheme (20) by setting

$$q = q(t, k_1, k_2)e^{i(k_1 \xi + k_2 \eta)} \tag{23}$$

where $k_1 = 2\pi/\lambda_1$ and $k_2 = 2\pi/\lambda_2$ are the wave numbers, λ_1 and λ_2 the wavelengths and $i = \sqrt{-1}$. The stability condition (22) is then satisfied if:

$$0 \geq Re(\Lambda \Delta t) \geq \Delta t[-\frac{\nu_1}{\Delta \xi^2} - \frac{\nu_2}{\Delta \xi \Delta \eta} - \frac{\nu_3}{\Delta \eta^2} + 16\frac{\epsilon_1}{\Delta \xi^4} + 16\frac{\epsilon_2}{\Delta \eta^4}] \geq REAL \tag{24}$$

and

$$|Im(\Lambda \Delta t)| \leq \Delta t[\frac{|\lambda_1|}{\Delta \xi} + \frac{|\lambda_2|}{\Delta \eta}] \leq IMAG \tag{25}$$

where Re and Im denote the real and imaginary part of a complex number. REAL and IMAG are choosen such that all complex numbers with $REAL \leq Re(z) \leq 0$ and $|Im(z)| \leq IMAG$ lie inside the stability region of scheme (14).

To be able to choose a stable time step for the present method, the coefficients in the model equation must be related to the incompressible Navier-Stokes equations. The coefficients $|\lambda_1|$ and $|\lambda_2|$ are choosen equal to the maximum eigenvalues for the Jacobian matrices for the Euler equations in ξ- and η-directions [4]. By using the relation between metric expressions in ξ- and η- coordinates and the geometric quantities in the finite-volume technique [9], the following relation is obtained:

$$\frac{|\lambda_1|}{\Delta \xi} + \frac{|\lambda_2|}{\Delta \eta} = [\tilde{U} + (\tilde{U}^2 + c^2 S^2)^{\frac{1}{2}}]/vol_{i,j} \tag{26}$$

where $S^2 = (|SIX| + |SJX|)^2 + (|SIY| + |SJY|)^2$, $\tilde{U} = |\vec{u} \cdot \vec{SI}| + |\vec{u} \cdot \vec{SJ}|$. $\vec{SI}$ and $\vec{SJ}$ are the arithmetic average surface vectors in I and J direction. $\vec{SI} = (SIX, SIY)$ etc.

The coefficients ν_1, ν_2 and ν_3 in (16) are derived by the viscous part of the linearized momentum equation. Again using the relation between metric expressions and the geometric quantities in the finite-volume technique leads to:

$$\frac{\nu_1}{\Delta \xi^2} = \nu \frac{\vec{SI} \cdot \vec{SI}}{{vol_{i,j}}^2} \quad , \quad \frac{\nu_2}{\Delta \xi \Delta \eta} = 2\nu \frac{|\vec{SI} \cdot \vec{SJ}|}{{vol_{i,j}}^2} \quad , \quad \frac{\nu_3}{\Delta \eta^2} = \nu \frac{\vec{SJ} \cdot \vec{SJ}}{{vol_{i,j}}^2} \tag{27}$$

where $\nu = \mu/\rho_0$. The coefficients ϵ_1 and ϵ_2 are choosen according to (13) :

$$\frac{\epsilon_1}{\Delta \xi^4} = -\Gamma \qquad \frac{\epsilon_2}{\Delta \eta^4} = -\Gamma \ , \tag{28}$$

This leads to the following relation for the local time step in two dimensions:

$$\begin{aligned} \Delta t \ = \ & min[IMAG \frac{vol_{i,j}}{\tilde{U} + (\tilde{U}^2 + c^2 S^2)^{\frac{1}{2}}}, \\ & |REAL - 32\epsilon_4 IMAG| \frac{vol_{i,j}^2}{\nu[\vec{SI} \cdot \vec{SI} + 2|\vec{SI} \cdot \vec{SJ}| + \vec{SJ} \cdot \vec{SJ}]}] \ . \end{aligned} \tag{29}$$

The condition for the local time step in three dimensions is derived analogously. Corresponding stability analysis can be made for the incompresible Navier-Stokes equations with the $k - \epsilon$ turbulence model in a similar manner[3].

RESULTS

Laminar flow

Results have been obtained for internal laminar flow, Re=50, 2:3 expansion, over a backward facing step, the problem of a 1984 GAMM workshop[10]. A parabolic shape of the velocity on the inflow boundary was given and the pressure is extrapolated upstream. The derivative of the flow variables in the streamwise direction is supposed to be zero at the outflow.

The point of reattachment can be seen in the streamline plot (Fig. 2) and the wall shear stress plot (Fig. 3). It was calculated to 2.83 step heights. The experiments state 3.0 for the point of reattachment, though most of the participants of the workshop managed to predict the reattachment point between 2.7 and 2.9. The agreement between numerical results and experimental data is quite satisfying in the wall shear stress plot. The evolution of the maximum velocity (the maximum velocity along the y-axis in x-direction, Fig. 4) also shows a good agreement between numerical and experimental data.

Turbulent flow

Flow over a two-dimensional backward facing step was modeled using the $k-\epsilon$ turbulence model. The calculations simulated the experiments by Westphal *et al.* [11] in which the Reynolds number *Re* based on the step height reaches 42.000, the expansion ratio is 3:5.

In the inflow cross-section the one-seventh power-law profile is assumed for the non-dimensional velocity u while k and ϵ are given fully turbulent profiles[3]. The v-velocity is set to zero and the pressure is extrapolated upstream. At the outflow cross-section it was assumed that the derivative of the flow variable is zero in the streamwise direction. In the wall region the standard wall function approach was used.

Numerical predictions were obtained for a 121×41. Results are presented at different locations at constant x (streamwise coordinate), $x = 4, 8, 12$ and 20 step heights downstream the step. The simulated profiles for the u-velocity and k are compared to experimental data [11] in Fig. 5 and Fig. 6 at $x = 4$ and 20. The velocity profiles are in good agreement close to the step but the discrepancies become more prominent as the outflow section is approached. A similar behaviour can be seen for the k-variable. The discrepancies also tend to become graeter near the upper wall and outflow where the simulations fail to the predict the second peak. Except for this the locations for local maxima and minima are well predicted.

The simulation lead to an underprediction of the reattachment length. In this case the reattachment length was calculated to 6.5 step heights while the experimental reattachment length was about 7.3. This underprediction has been reported also by others and in a paper by Autret *et al.* [12], the numerical estimate for the reattachment length was 5.22 for the same problem.

Autret *et al.* used a very coarse grid (28×44 nodes for a finite element solution) which might explain their bad estimate for the reattachment length. The mesh resolution is obviously important, thus a mesh of 121×41 was choosen which was the largest possible mesh in the y-direction in order to satisfy the boundary condition (11). However, (11) can never be satisfied in the vicinity of the reattachment point and that is why the wall function approach is bad in this area. In Fig. 7 two solutions for different mesh sizes are shown, 121×41 and 101×25. The u-profiles are compared at $x = 8, 20$. As indicated, the agreement between numerical results and experiments is better for the finer mesh. The reattachment length was calculated to 6.0 for the coarser mesh. On an even coarser mesh, 61×21, no converged solution was obtained.

CONCLUSIONS

An explicit central finite-volume Runge-Kutta method for the incompressible Navier-Stokes equation has been developed. The standard $k-\epsilon$ turbulence model has also been included. The code is simple and an extension to three dimensions is straight forward. The numerical simulations have proven to be in good agreement with other numerical results and with experimental data for laminar flow. For turbulent flow the agreement between numerical data and experimental data is acceptable for a separated flow, but the numerical solution seems to be mesh dependent.

REFERENCES

[1] Müller, B. and Rizzi A.: Navier Stokes Computation of Transonic Vortices Over a Round Leading Edge Delta Wing, AIAA Paper No. 87-1227, 1987.

[2] Chorin, A.J.: A Numerical Method for Solving Incompressible Viscous Flow Problems, J. Comp. Phys. 2, 12-26, 1967.

[3] Eliasson, P.: Solutions to the Navier-Stokes Equations using a $k-\epsilon$ Turbulence Model, FFA TN 1988-19, Apr. 1988.

[4] Eliasson, P.: Navier-Stokes Solutions for Laminar Flow over a NACA 0012 Profile and a Backward Facing Step, FFA TN 1987-50, Sep. 1987.

[5] Eliasson, P. and Krouthén, B.: The Construction of an Incompressible Inviscid Euler Solver for Rotating Systems and a Comparative Study of Two Euler Pump Flow Solutions, FFA TN 1987-03, 1987.

[6] Chieng, C.C. and Launder, B.E.: On the Calculation of Turbulent Transport Downstream from an Abrupt Pipe Expansion, Num. Heat Transfer 3, 189-207, 1980.

[7] Eriksson, L-E. and Rizzi, A.: Computer-Aided Analysis of the Convergence to Steady State of Discrete approximations to the Euler Equations, J. Comp. Physics, Vol. 57, pp 90-128, 1985.

[8] Peyret, R. and Taylor, T.: Computational Methods for Fluid Flow. Springer 1983.

[9] Müller, B.: Calculation of Separated Laminar Supersonic Flows over Blunt Bodies of Revolution at Zero Angle of Attack, ESA-TT-953, 1985.

[10] Morgan, K., Periaux, J. and Thomasset F.: Analysis of Laminar Flow over a Backward Facing Step, Notes on Numerical Mechanics, Vol. 9,1984.

[11] Westphal, R.V., Johnston, J.P. and Eaton, J.K.: Experimemtal Study of Flow Reattachment in a Single-sided Sudden Expansion, NASA Contract Report 3765-Report MD-41, Stanford University, 1984.

[12] Autret, A., Grandotto, M. and Dekeyser, I.: Finite Element Computation of a Turbulent Flow Over a Two-Dimensional Backward-Facing Step, International Journal for Numerical Methods in Fluids, vol. 7,89-102, 1987.

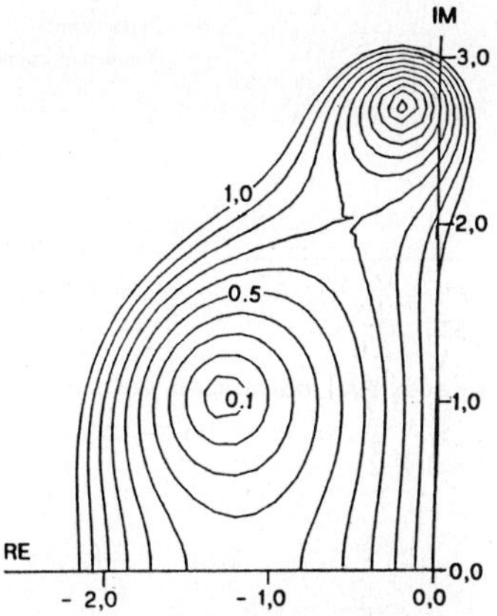

Figure 1: Stability region for the four stage, first order Runge-Kutta scheme, CFL = 3

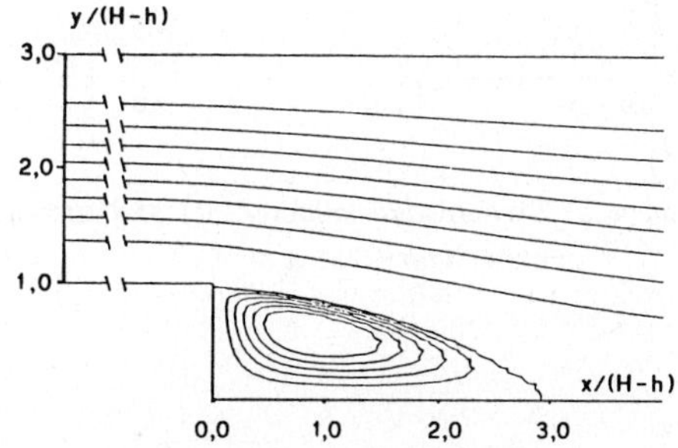

Figure 2: Streamline pattern, Re = 50

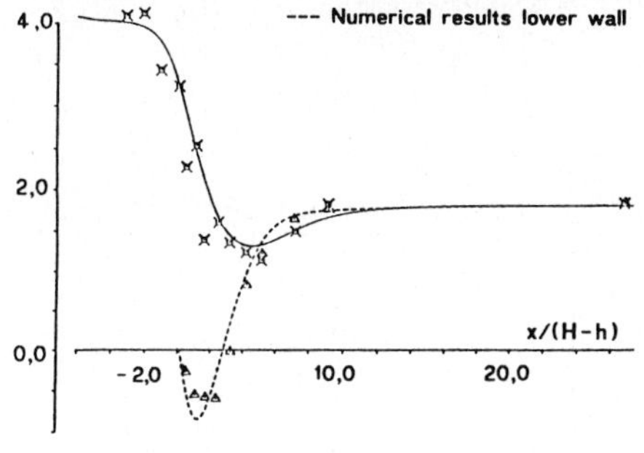

Figure 3: Wall shear stress distribution compared to experiments, Re = 50

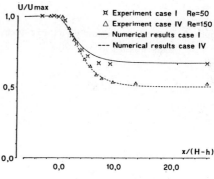

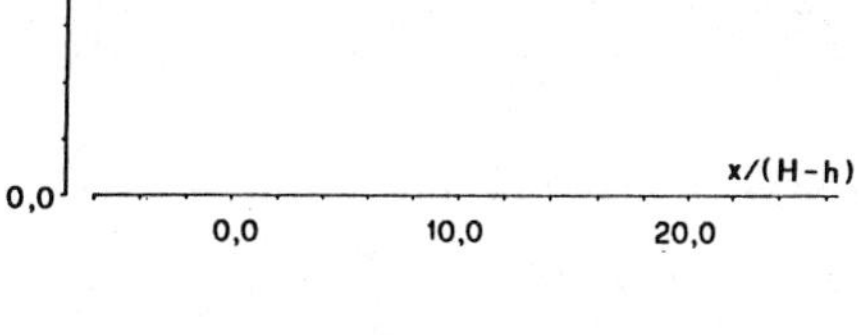

Figure 4: Evolution of maximum velocity compared to experiments, Re = 50 and Re = 150, expansion 2 : 3 and 1 : 2

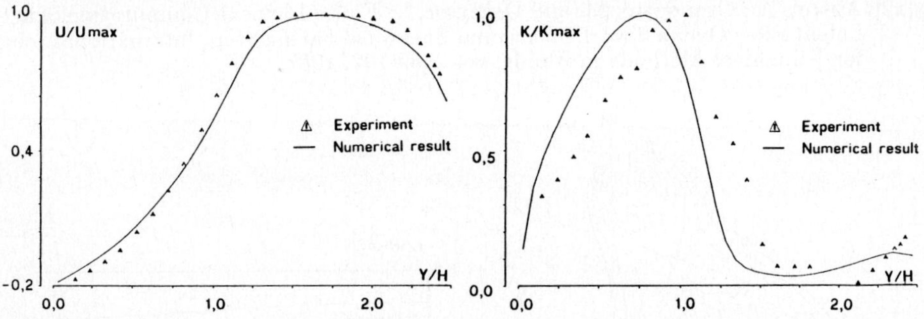

Figure 5: Streamwise velocity (u) and turbulent kinetic energy (k) compared with experiments at $x = 4$

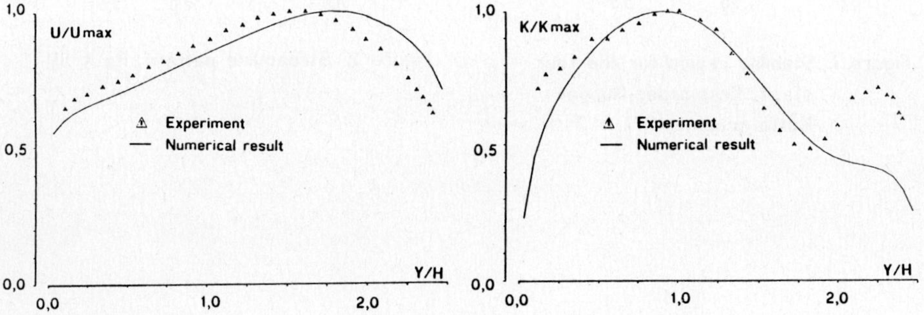

Figure 6: Streamwise velocity (u) and turbulent kinetic energy (k) compared with experiments at $x = 20$

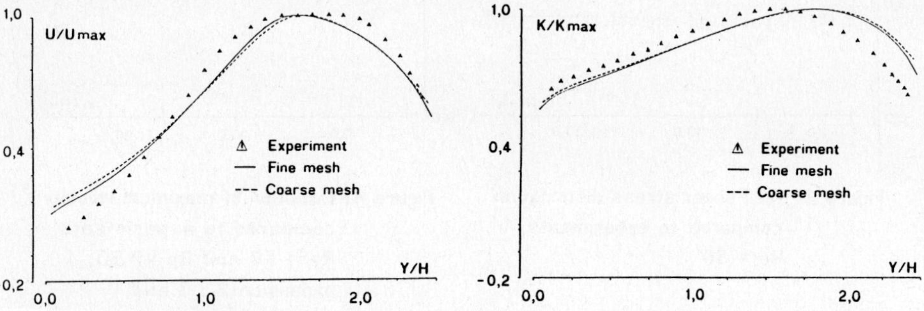

Figure 7: Streamwise velocity (u) compared with experiments at $x = 8$ and $x = 20$ for two mesh sizes, 121 x 41 (solid line) and 101 x 25 (dashed line)

ON THE FINITE VELOCITY OF WAVE MOTION MODELLED BY NONLINEAR EVOLUTION EQUATIONS

J. ENGELBRECHT
The Institute of Cybernetics
Estonian Academy of Sciences
Tallinn 200108
Estonia, USSR

SUMMARY

The problem of finite velocity in wave motion modelled by not strictly hyperbolic systems is discussed. The main idea is to establish associated systems by a certain perturbation scheme permitting to determine eigenvalues which serve as a basis for a moving frame. Such a procedure leads to the construction of evolution equations describing single waves with a certain accuracy. The final wave velocity is then determined from the evolution equation relative to the moving frame. This scheme is used for several nonlinear, weakly dispersive and/or dissipative systems. A model example of a solitary wave gives the explicit values of all the velocities under consideration.

1. INTRODUCTION

Hyperbolic equations and waves have been cornerstones of physics for a long time and they extend over most fields of contemporary physics. Waves are used as carriers not merely of energy but also of information. Every disturbance in the real physical world propagates with a finite velocity, which, generally speaking, should be easily related to the governing equations. The real physical model, however, is so complicated that waves are not necessarily governed by strictly hyperbolic equations since various asymptotic methods have been used for deriving the governing equations from conservation laws. Nevertheless, every mathematical model should be traced back to initial hyperbolic equations as complicated as they could be and every wave motion should be related to finite velocities. The main problem here is the following: how to determine these finite velocities if the initial system governing certain wave motion is not hyperbolic.

There are several characteristic features which should be taken into account. Variation of wave velocities with wave numbers and also with amplitudes is often of importance since most real processes are dispersive and nonlinear. The nonlinearity plays a crucial role in wave motion being responsible for discontinuities, solitary waves, interaction, etc., including chaotic motion which is intensively studied in contemporary physics. The complications in an analysis are also related

to the possible existence of many waves in a real process, which becomes even more essential in nonlinear systems. Here considerable sucess has been obtained by introducing the notion of evolution equations governing just one single wave. The celebrated Korteweg-de Vries, Schrödinger, Burgers, etc. equations have been the milestones in the contemporary mathematical physics. Very few evolution equations are hyperbolic except some single cases but as a rule they have been obtained in a moving frame, i. e. a certain velocity is already taken into account. The final velocity, however, may differ from this basic velocity and the question about the finite velocity remains with its utmost importance. Here one cannot also forget the dissipative systems with source terms which may govern certain progressive waves with finite velocities.
The need for an explicit theory explaining from one hand the correspondence of initial multi-wave systems to one-wave evolution equations and from the other hand the relationship between eigenvalues, phase and/or group velocities, and the final wave velocity (eigenvelocity) is obvious. This report does not pretend to present a full theory rather than to outline some examples which could be used as model problems in a general theory.
In Section 2 the main mathematical models used in describing wave motion are presented. Section 3 deals with a typical example of a perturbed system for which the eigenvalues, phase, group, and wave velocities are all explicitly determined. This example concerning the celebrated solitary wave serve as a model while the cases analysed in Section 4 do not permit exact determination of wave velocity. In Section 5 a system with a source is discussed. Finally, in Section 6 some conclusions are presented including the list of the possible stages in the analysis.

2. PRELIMINARY

In this Section several basic systems of equations and evolution equations used for describing wave motion are presented.

Hyperbolic systems. The system

$$\vec{u}_t + A_j \vec{u}_{x_j} + C\vec{u} = 0 , \tag{2.1a}$$

$$A_j = A_j(x_j, t, \vec{u}), \; C = C(x_j, t), \; x_j \in R^n. \tag{2.1b}$$

where $\vec{u}$ is an n-element vector, is strictly hyperbolic in R^n if the eigenvalues $\lambda^{(i)}$ of A_j, satisfying the determinant $(A_j - \lambda^{(i)} I) = 0$ are all real and the corresponding left eigenvectors $l^{(i)}$ satisfying $l^{(i)} A_j = A_j l^{(i)}$, $i = 1, 2, \dots , n$ are linearly independent [1, 2]. Here and further, the summation convention over repeated indices is used and the subscripts denote the differentiation.

Perturbed hyperbolic systems. In this case a system may also contain the higher derivatives modelling weak dispersive, dissipative e. a. effects [1, 3-5]

$$\vec{u} + A_j \vec{u}_{x_j} + \varepsilon \Big[\sum_{\beta=1}^{s} \prod_{\alpha=1}^{p} \Big(H_\alpha^\beta \frac{\partial}{\partial t} + K_{j\alpha}^\beta \frac{\partial}{\partial x_j} \Big) \Big] \vec{u} + C\vec{u} = 0 , \tag{2.2}$$

where ε is a small parameter. Existence of such a parameter enables us to use certain asymptotics in form of system (2.1) called an associated system [3]. The eigenvalues of this system determine the velocities as well as the number of waves used further for constructing higher asymptotics.
There is an interesting subcase of (2.2) which leads to wave hierarchies [1, 4]. In this case the operators in parenthesis describe also a wave motion. As an example for n= 1, j = 1, $x_1 = x$ it may read

$$u_t + c_1 u_x + \varepsilon (u_{tt} - c_2^2 u_{xx}) = 0 , \tag{2.3}$$

where c_1 and c_2 are constants.

Dispersive systems. These systems are characterized by the form of the solution

$$u_i (x_j, t) = a \exp (ik_j x_j - i\omega t) , \tag{2.4}$$

where k_j are the wave numbers and ω is the frequency. The system itself may be of the form (2.2) or also of the form [4]

$$\vec{u}_t + \Big[\sum_{\beta=1}^{s} \prod_{\alpha=1}^{p} (H_\alpha^\beta \frac{\partial}{\partial t} + K_{j\alpha}^\beta \frac{\partial}{\partial x_j}) \Big] \vec{u} + C \vec{u} = 0 . \tag{2.5}$$

Perturbed dispersive systems. In this case the elementary solutions (2.4) with k_j = const., ω = const. do not exist but the periodicity in $\Theta = k_j x_j - \omega t$ holds with $k_j \neq$ const., $\omega \neq$ const. [4] and the dispersion relation depends upon the amplitude. The system may be either of form (2.2) or of form (2.5) usually having a small parameter which emphasizes weak nonlinearity and weak dispersion.

Evolution equation. As mentioned in the Introduction, the evolution equations are single—wave equations written in the most cases in the moving frame [1, 3, 5]. For a certain $u_j = u \in \vec{U}$ a typical evolution equation reads

$$[u_\tau + F (u, u_\xi, u_{\xi\xi}, u_{\xi\xi\xi}, ...)]_\xi = G (u) . \tag{2.6}$$

where $\xi = \xi (c_i t - x_j)$ is the phase variable (moving frame) and $\tau = \tau (t)$ (or $\tau = \tau (x_j)$) is the variable characterizing propagation. The operator $F(u, u_\xi, ...)$ does not contain the derivatives with respect to τ. The velocity c_i used for the moving frame may be: (i) the eigenvalue (characteristic speed [6]) from a corresponding associated hyperbolic system [3, 5]; (ii) phase velocity for dispersive systems [4, 5] ; (iii) group velocity for strongly dispersive systems [5]. If G (u) = 0 then under certain conditions at infinity (2.6) yields

$$u_\tau + F(u_\xi, u_{\xi\xi}, u_{\xi\xi\xi}, ...) = 0 , \tag{2.7}$$

which is the most common form of evolutions equations [1, 3-5, 7]. The formal procedures of derivation of evolution equations are given elsewhere [1, 3, 5, 7].

3. A PERTURBED DISPERSIVE SYSTEM - A MODEL EXAMPLE

The moving frame used for deriving evolution equations needs a finite velocity to start with but the final wave velocity (signal velocity) may be different. Further a specific example is considered which is solved explicitly in order to demonstrate the possible differences in velocities.
Let us consider the one-dimensional ($x_1 = x$) wave motion in a nonlinear medium with microstructure. The governing equation is usually written in terms of the longitudinal displacement U_1

$$c_0^2 (1 + m U_{1,x}) U_{1,xx} + c_0^2 l_0^2 U_{1,4x} - U_{1,tt} = 0 , \tag{3.1}$$

where $c_0^2 = (\lambda + 2\mu)\rho_0^{-1}$; λ, μ are Lamé parameters, ρ_0 is the density, m and l_0 are the nonlinear and scale parameters, respectively, the latter characterizing the microstructure [3]. The deformation is small but finite and there are two small parameters in this problem: (i) the maximum deformation and (ii) the scale parameter. Using the matrix notation as in Section 2, (3.1) yields

$$\vec{U}_t + A\vec{U}_x + l_0^2 B\vec{U}_{3x} = 0 , \tag{3.2 a}$$

$$\vec{U} = \begin{vmatrix} U_{1,t} \\ U_{1,x} \end{vmatrix} = \begin{vmatrix} u_1 \\ u_2 \end{vmatrix} , \quad A = \begin{vmatrix} 0 & -c_0^2(1 + m u_2) \\ -1 & 0 \end{vmatrix} , \quad B = \begin{vmatrix} 0 & -c_0^2 \\ 0 & 0 \end{vmatrix} . \tag{3.2 b}$$

It is easily seen that according to Section 2 system (3.2) belongs to perturbed dispersive systems with weak nonlinearity. The corresponding linear associated system

$$\vec{U}_t + A_0 \vec{U}_x = 0 , \qquad A_0 = \begin{vmatrix} 0 & -c_0^2 \\ -1 & 0 \end{vmatrix} \tag{3.3}$$

gives the eigenvalues $\lambda^{(1)} = c_0$, $\lambda^{(2)} = -c_0$ as characteristic speeds. According to a well-established procedure [3, 5] the ray coordinates (moving frame)

$$\xi = c_0 t - x , \quad \tau = \varepsilon x \tag{3.4}$$

are introduced. Then together with a series expansion the following evolution equation describing the wave motion to the right is obtained

$$\frac{\partial u_1}{\partial \tau} + \frac{m}{2\varepsilon c_0} u_1 \frac{\partial u_1}{\partial \xi} + \frac{1}{2} \frac{l_0}{\varepsilon} \frac{\partial u_1}{\partial \xi^3} = 0 . \tag{3.5}$$

Here u_1 actually denotes u_{10} , i. e. the first term in the respective expansion of $\vec{U}$ (see [3]). In the dimensionless form (3.5) yields

$$\frac{\partial v}{\partial \sigma} + \operatorname{sign}|m|\, v \frac{\partial v}{\partial \zeta} + \nu \frac{\partial^3 v}{\partial \xi^3} = 0 , \tag{3.6 a}$$

$$v = u_1 a_0^{-1} \quad , \quad \sigma = \frac{1}{2}\tau\tau_c^{-1} \quad , \quad \zeta = \xi\tau_c^{-1} \quad , \qquad (3.6b)$$

and the small parameter ε is related to the initial amplitude a_0 through the relation

$$\varepsilon = |m|\, a_0 c_0^{-1} \quad . \qquad (3.6c)$$

Here τ_c denotes the characteristic wave length.
Assuming $m > 0$ without any loss of generality the model Korteweg-de Vries (KdV) equation

$$\frac{\partial v}{\partial \sigma} + v\frac{\partial v}{\partial \xi} + \nu\frac{\partial^3 v}{\partial \zeta^3} = 0 \qquad (3.7)$$

is obtained. The KdV equation possesses the soliton-type solutions which are of the form [8, 9]

$$v = a\,\mathrm{sech}^2 \left[\left(\frac{a}{12\nu} \right)^{\frac{1}{2}} \left(\zeta - \frac{a}{3}\sigma \right) \right] , \qquad (3.8)$$

i. e. this solution propagates relative to the moving frame with a velocity $\frac{a}{3}$ which according to Bhatnagar is called eigenvelocity [10] and according to Drazin - wave velocity [9]. The phase and group velocities, determined from the corresponding linearized system

$$\vec{u}_t + A_0\vec{u}_x + l_0^2 B\vec{u}_{3x} = 0 \qquad (3.9)$$

are following

$$c_{ph} \cong c_0 - \frac{1}{2}c_0 l_0^2 k^2 \quad , \qquad (3.10a)$$

$$c_{gr} \cong c_0 - \frac{3}{2}c_0 l_0^2 k^2 \quad , \qquad (3.10b)$$

corresponding to the solution with a phase $\Theta = kx - \omega t$. It is clear that this result cannot be directly derived from the linearized form of the evolution equation (3.5). Here the phase

$$\Theta_* = -k_*\xi + \omega_*\tau \qquad (3.11)$$

must be used resulting in

$$c_{ph}^* = \frac{1}{2}k_* l_0 \varepsilon \quad , \qquad (3.12a)$$

$$c_{gr}^* = \frac{3}{2}k_*^2 l_0^2 \varepsilon^{-1} \quad . \qquad (3.12b)$$

It is easily verified that the expressions (3.10) and (3.12) are identical provided (3.9) and (3.11) are taken into account [11]. Now it is clear that the real finite velocity of a solitary wave in a nonlinear dispersive system (3.2) is neither the characteristic speed $c_0 = \lambda^{(1)}$ nor the phase (group) velocity but completely different depending on the balance of nonlinear and dispersive effects. This is certainly a limit case and the solution to (3.7) may also be obtained in the form of cnoidal waves with another limit case being the usal sinusoidal wavetrain for which (3.10) holds [12].

If the wave motion is perturbed in such a way that the evolution equation (3.7) has also a r.h.s., then the wave speed is influenced again. In order to demonstrate this effect let us present (3.7) in its normalized form

$$V_T - 6\,V V_X + V_{XXX} = 0\,, \tag{3.13a}$$

$$V = -\frac{1}{6}\,\nu^{-\frac{1}{3}}\,v\,, \quad X = \nu^{-\frac{1}{3}}\,\zeta\,, \quad T = \sigma\,. \tag{3.13b}$$

Then the solution (3.8) takes the form

$$V = -2\varkappa^2\,\mathrm{sech}^2[\varkappa(X - 4\varkappa^2 T)]\,, \tag{3.14}$$

where $\varkappa$ is the eigenvalue from the corresponding Schrödinger equation for the potential [9, 10]. The wave velocity is equal to $4\varkappa^2$. If a source-like term is included to (3.13) describing the energy influx [13, 14] then the governing equation reads

$$V_T - 6\,V V_X + V_{XXX} = \eta R(V)\,. \tag{3.15}$$

where η is a certain new small parameter and $R(V)$ — a smooth function. According to Lamb [14], a soliton-type excitation is perturbed and the eigenvalue $\varkappa$ to (3.15) is governed by the equation

$$\varkappa_T = -\frac{\eta}{4\varkappa}\int_{-\infty}^{\infty} dz\, R(V_s)\,\mathrm{sech}^2 z\,, \tag{3.16}$$

where z is the phase function and V_s — the initial soliton—type excitation. Suppose $R(V) = -bV^3$, b = const. Then

$$\varkappa = \varkappa_0\left(1 + \frac{8}{3}\,\eta a \varkappa_0^4 T\right)^{-\frac{1}{4}}\,, \quad a \doteq \frac{156}{35}\,b\,. \tag{3.17}$$

It is obvious that in this case the eigenvalue $\varkappa$ is decreasing in the course of T growing and the corresponding wave speed is also decreasing.
Finally, let us remark that this is a model example - on the basis of the characteristic speed $c_0 = \lambda^{(1)}$ the wave velocity is explicitly determined being dependent on the amplitude. However, this result is correct only for a solitary wave, formed as a result of balance between nonlinear and dispersive effects. With nonlinear effects decreasing, the corresponding cnoidal waves are in effect with a limiting case of a sinusoidal wavetrain. For this limit the usual notions of phase and group velocities hold but k, ω, and k_*, ω_*, obtained from the initial system and the evolution equation, respectively, should be clearly distinguished.

4. PERTURBED HYPERBOLIC SYSTEMS

4.1. Thermoelastic medium.

The classical theory of thermoelasticity is based on Fourier's law of heat conduction which leads to the paradox of infinite thermal wave velocities in the dynamic theory of heat conducting continuous media [15]. There has been a long discussion about the proper form of governing equations for modified thermoelasticity in order to remove this paradox [16]. This problem is not yet solved and in this paper only the mathematical questions are discussed while the physical considerations about the validity of one or another model need to be analysed separately.
The rate-type law of heat conduction in one-dimensional approximation ($x_1 = x$) leads to the linear system (2.1a) with

$$\vec{u} = \begin{vmatrix} u_{1,t} \\ u_{1,x} \\ \Theta \\ Q \end{vmatrix}, \quad A = \begin{vmatrix} 0 & -c_0^2 & \varkappa\rho_0 & 0 \\ -1 & 0 & 0 & 0 \\ \frac{\varkappa T_0}{\rho_0 c_E} & 0 & 0 & -\frac{1}{\rho_0 c_E} \\ 0 & 0 & -\frac{k_0}{\tau_0} & 0 \end{vmatrix}, \quad C\vec{u} = \begin{vmatrix} 0 \\ 0 \\ 0 \\ \frac{Q}{\tau_0} \end{vmatrix}, \tag{4.1}$$

where Θ is the dimensionless temperature, Q is the heat flux, $c_0^2 = (\lambda + 2\mu)\rho_0^{-1}$, $\varkappa = (3\lambda + 2\mu)\alpha_T$; ρ_0, T_0 are the initial density and temperature, respectively, c_E is the specific heat, k_0 is the thermal conductivity and τ_0 is the relaxation time of the heat flux [3].
The dimensionless relative to c_0 velocities - the eigenvalues of the matrix A are determined by the formulae

$$\lambda^{(1),(2)} = \pm(M - N)^{\frac{1}{2}}, \quad \lambda^{(3),(4)} = \pm(M + N)^{\frac{1}{2}}, \tag{4.2a}$$

$$M = \tfrac{1}{2}[1 + (\omega_d \beta_r)^{-1} + e]^{\frac{1}{2}}, \tag{4.2b}$$

$$N = \tfrac{1}{2}[1 + (\omega_d \beta_r)^{-2} + e^2 - 2(\omega_d\beta_r)^{-1} + 2e + 2e(\omega_d\beta_r)^{-1}]^{\frac{1}{2}}, \tag{4.2c}$$

where the dimensionless parameters are determined by the expressions

$$e = \varkappa^2 T_0[(\lambda + 2\mu)\rho_0 c_E]^{-1}, \tag{4.3a}$$

$$\beta_r = c_0 \tau_0 L^{-1}, \tag{4.3b}$$

$$\omega_d = L\rho_0 c_E c_0 k_0^{-1}. \tag{4.3c}$$

where L is a scale parameter. The parameter e is the standard coupling parameter, the relaxation parameter β_r was introduced by Lord and Shulman [17], and the diffusion parameter ω_d by Johnson [18]. The phase velocity c_1 for sinusoidal waves is easily determined in terms of a dimensionless frequency $\chi = \omega\omega_*^{-1}$ where ω_* is the characteristic frequency [15]

$$\omega_* = c_0^2 \rho_0 c_E k_0^{-1}. \tag{4.4}$$

The phase velocity depends upon the relaxation time τ_o through a parameter $n_* = \tau_o\, \omega_*$. The plot of $c_1\, c_o^{-1}$ over χ is shown in Fig. 1 [3] where the curve for $n_* = 0$ is found by Chadwick [15].

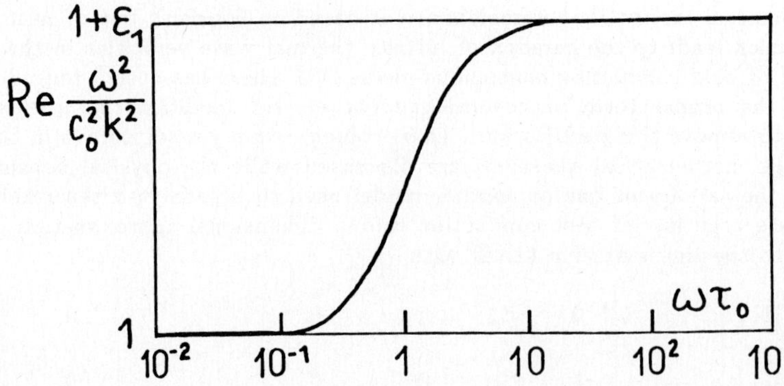

Figure 1

The evolution equation for $v = u_1\, a_o^{-1}$ (see Section 3) using again the dimensionless variables takes the form of the Burgers' equation [3]

$$\frac{\partial v}{\partial \sigma} + \text{sign}\, |m+e|\, v\, \frac{\partial v}{\partial \zeta} = \Gamma^{-1}\, \frac{\partial^2 v}{\partial \zeta^2}\ , \qquad (4.5)$$

where the nonlinearity is also taken into account. The parameter Γ is known as the acoustic Reynolds number. As it is known, the Burgers' equation allows the Taylor shock profile [19], the velocity of which depends upon the amplitude

$$V = \frac{1}{2}\, (\, v_1 + v_2)\ , \qquad (4.6)$$

where v_1, v_2 are the amplitudes before and after the shock, respectively.

Consequently, in the case of thermoelastic waves the eigenvalues (4.2 a) determine the velocities which differ from the phase velocities (see Fig. 1). The moving frame for the evolution equation (4.5) is given by $\zeta = \zeta[(\,1 + e)^{1/2}\, t\, -\, x]$ and the possible shock wave velocity (4.6) is determined relative to this velocity. A stable solitary wave as shown in Section 3 is not possible.

4.2. Relaxing medium.
According to the standard viscoelastic body [20] the governing equation belongs to the type (2.3) [3]

$$c_o^2\, (\,1 + m\, U_{1,x})\, U_{1,xx} - U_{1,tt} + \varepsilon_1 c_o^2\, \tau_o U_{1,xxt}\, +$$

$$+\, \tau_o [\, c_o^2\, (1 + m U_{1,x}\,)\, U_{1,xx}\, -\, U_{1,tt}]_{,t} = 0\ , \qquad (4.7)$$

where ε_1 , τ_o are material parameters. The linearized dispersion relation is

$$\omega = \pm\, c_e\, k_0 \left(1 + \frac{\varepsilon_1\, \omega^2 \tau_0^1}{1 + \omega^2 \tau_0^2} + i \frac{\varepsilon_1\, \omega \tau_0}{1 + \omega^2 \tau_0^2} \right)^{\frac{1}{2}} . \tag{4.8}$$

where c_e is the equilibrium velocity determined by

$$c_e^2 = (\lambda + 2\mu)\, \rho_0^{-1} . \tag{4.9}$$

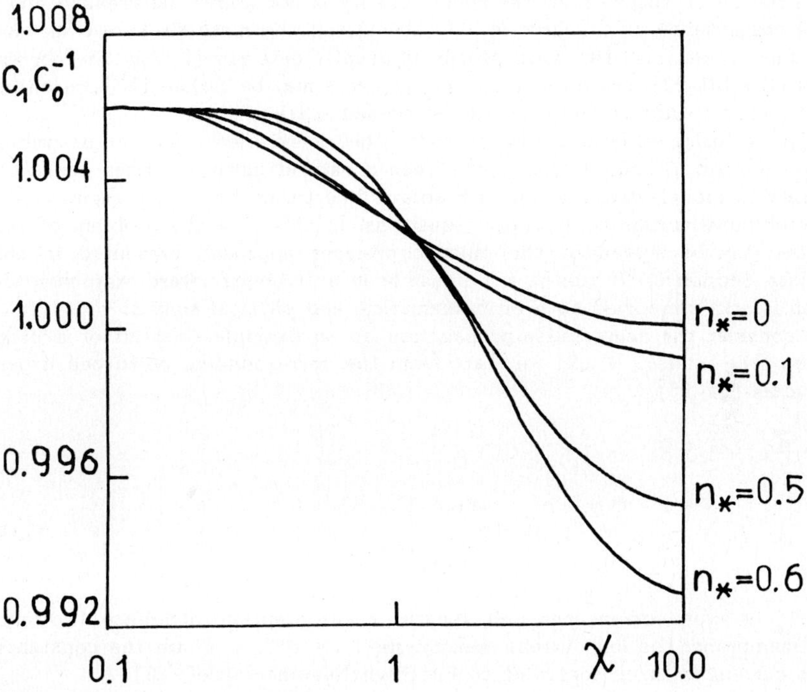

Figure 2

The real part of the expression (4.8) versus $\omega \tau_0$ is plotted in Fig. 2 and it is easily seen that for high-frequency processes the equilibrium velocity c_e does not correspond to the real velocity. In this case the instantaneous velocity c_i must be introduced

$$c_i^2 = (1 + \varepsilon_1)(\lambda + 2\mu)\rho_0^{-1} > c_e^2 . \tag{4.10}$$

The evolution equations may be constructed on the basis of both velocities [3] with suitable approximations for low- and high-frequency processes. However, there is no distinct estimation for a real velocity c between the equilibrium (minimum) and instantaneous (maximum) velocities.

5. THE SYSTEM WITH SOURCE TERMS

There has been a considerable interest to dissipative systems with source terms which also may lead to wave-type solutions [21, 22]. Here the governing system is of diffusion type and the progressive wave corresponds to a certain closed orbit in the phase space. Since the finite velocity is not known beforehand and a wave at the absence of a source is also absent, all the methods known in wave theory fail in analysis. The wave profile is usually determined from the corresponding ODE [21, 22]. The problems of convergence may be serious [23] preventing to calculate the wave profile over all the needed space.

Since the evolution equations have proved to be an excellent tool for describing the wave motion in complicated media, then a natural question arises about the possibility to model wave motion with source like terms by such systems which permit the construction of evolution equations. In this case the problem of real velocities may be solved by the most convenient approach presented in this paper (see Section 3). If this is not possible in a straightforward way then the problem is still simplified both in mathematical and physical senses.

Let us consider the nerve pulse propagation as an example. Instead of a usual diffusion-type model [21, 22] we start from the corresponding perturbed hyperbolic model [24, 25]

$$\vec{u}_t + A_1 \vec{u}_{x_1} + \vec{H} = 0, \tag{5.1a}$$

$$\vec{u} = \begin{vmatrix} v \\ i_a \end{vmatrix}, \quad A_1 = \begin{vmatrix} 0 & m_1 \\ m_3 & 0 \end{vmatrix}, \quad \vec{H} = \begin{vmatrix} m_2 I \\ m_4 i_a \end{vmatrix}, \tag{5.1b}$$

where i_a is axon current per unit length, v is potential difference (voltage) across membrane; I is ion current density, m_i, $i = 1, 2, 3, 4$ are the constants. The ion current is taken according to FitzHugh-Nagumo model [26]

$$I = K_1 v + K_3 v^3 + R, \quad K_1 > 0, \quad K_3 > 0. \tag{5.2}$$

and the recovery variable R is governed by its own equation

$$R_t = q_0 (v + q_1), \quad q_1, \quad q_0 - \text{const.} \tag{5.3}$$

The telegraph equations (5.1) give the eigenvelocity $\lambda^{(1),(2)} = \pm (m_1 m_3)^{\frac{1}{2}} = \pm c_0$ but every wave with $I = 0$ will be heavily damped. The source term $I \neq 0$ leads to a progressive wave with the velocity which differs from c_0. The evolution equation, written in the moving frame $\xi = c_0 t - x_1$ in terms of $z = v + q_1$ takes the form

$$\frac{\partial^2 z}{\partial \xi \partial x_1} + \mu (b_0 - b_1 z + b_2 z^2) \frac{\partial z}{\partial \xi} + b_3 z = 0, \tag{5.4}$$

where μ, b_0, b_1, b_2, b_3 are positive constants [27]. The full analysis of the corresponding ODE

$$z'' + \mu(b_0 - b_1 z + b_2 z^2) z' + b_3 \Theta^{-1} z = 0, \tag{5.5}$$

where $(\ldots)' = \frac{d}{d\eta}$, $\eta = x_1 + \Theta\xi$ is given elsewhere [27]. The final velocity of a constant profile is given by the expression

$$c = \Theta c_0 (\Theta - 1)^{-1} , \quad \Theta > 1 , \tag{5.6}$$

but the procedure for determining Θ is not known. Nevertheless, the convergence of a solution is not a problem here while the numerical calculations are stable for every Θ giving $c > c_0$. In some sense, the problem is similar to that of a relaxing medium (see Section 4.2).

DISCUSSION

The examples given above reflect several faces of the problem of wave velocities. A possible way to deal with not strictly hyperbolic nonlinear systems some of which are listed in Section 2, is the transformation to single-wave equations, i. e. to evolution equations. Such an approach goes through the following steps:

(i) the analysis of the associated systems in order to establish the eigenvalues, phase and/or group velocities;

(ii) the construction of nonlinear evolution equations using the velocities determined from the corresponding associated system for determining the moving frame;

(iii)the analysis of evolution equations in order to establish the final wave velocity. Here the corresponding ordinary differential equations may also be needed for determination of stable trajectories in the phase space.

A model example of a solitary wave is presented in Section 3 where all the velocities are explicitly determined. Other examples demonstrate in one way or another the possible shortcomings of the theory which does not permit to establish the exact velocities except the limit estimations. The evolution equations, however, seem to be the best tools in handling of finite velocities especially for typical nonlinear systems in contemporary mathematical physics.

REFERENCES

[1] JEFFREY, A., KAWAHARA, T.: "Asymptotic Methods in Nonlinear Wave Theory", Pitman, Boston e. a. 1982.

[2] OLIGER, J.: "Numerical Methods for Hyperbolic Equations ". In: " Numerical Methods for Partial Differential Equations ", ed. by S. I. Hariharan and T. H. Moulden, Longman, Essex, 1986, p. 59-100 .

[3] ENGELBRECHT, J.: " Nonlinear Wave Processes of Deformation in Solids ". Pitman, Boston e. a. , 1983.

[4] WHITHAM, G.: " Linear and Nonlinear Waves ", Wiley, New York, 1974.

[5] TANIUTI, T., NISHIHARA, K.: " Nonlinear Waves ", Pitman, Boston e. a., 1983.

[6] DAFERMOS, C. M.: " Quasilinear Hyperbolic Systems that Result from Conservation Laws ". In: " Nonlinear Waves ", ed. by S. Leibovich and A. R. Seebass, Cornell University Press, 1974, p. 82-102.

[7] PELINOVSKI, E., FRIDMAN, V., ENGELBRECHT, J.: " Nonlinear Evolution Equation ", Tallinn, Valgus, 1984 (in Russian, English edition to be published by Longman).

[8] EILENBERGER, G.: " Solitons. Mathematical Methods for Physicists ", Springer, Berlin e. a., 1983.

[9] DRAZIN, P.: " Solitons ". Cambridge University Press, 1985.

[10] BHATNAGAR, P. L. : " Nonlinear Waves in One-dimensional Dispersive Systems ", Clarendon, Oxford, 1979.

[11] ENGELBRECHT, J.: " The Evolution Equations in the Theory of Nonlinear Waves ". In : " Proceedings IUTAM Symp. Nonlinear Deformation Waves, Tallinn, 1982", ed. by U. Nigul and J. Engelbrecht. Springer, Berlin e. a., 1983, p. 44-62.

[12] JEFFREY, A., KAKUTANI, T. : " Weak Nonlinear Dispersive Waves : A Discussion Centered Around the Korteweg-de Vries Equation ". SIAM Review, 14 (1972) 4, p. 582-643.

[13] OSTROVSKY, L. A.: " Solitons in Active Media ". In: " Proceedings IUTAM Symp. Nonlinear Deformation Waves, Tallinn, 1982 ", ed. by U.Nigul and J. Engelbrecht. Springer, Berlin e. a., 1983, p. 30-43.

[14] LAMB. G. L. Jr.: " Elements of Soliton Theory ", Wiley, New York, 1980.

[15] CHADWICK, P.: " Thermoelasticity. The Dynamical Theory ". In: " Progress in Solid Mechanics ", vol. 1 , North - Holland, Amsterdam, 1960, p. 265-328.

[16] CHANDRASEKHARAIAN, D. S.: " Thermoelasticity with a Second Sound : A Review ", AMR, 39 (1986) No. 3, p. 355-376.

[17] LORD, H. W., SHULMAN, Y.: "A Generalized Dynamical Theory of Thermoelasticity ", J. Mech. Phys. Solids, 15 (1967) p. 299-309.

[18] JOHNSON, A. F.: " Pulse Propagation in Heat-conducting Elastic Materials", J. Mech. Phys. Solids, 23 (1975) p. 55-75.

[19] SACHDEV, P. L.: " Nonlinear Diffusive Waves ", Cambridge University Press, 1987.

[20] CHRISTENSEN, R. M.: " Theory of Viscoelasticity ", Academic Press, New York e. a., 1971.

[21] ZYKOV, V. S.: " Modelling of Wave Processes in Active Media ", Nauka, Moscow, 1984 (in Russian).

[22] SCOTT, A. C.: " Neurophysics ", Wiley, New York, 1977.

[23] MIURA, R. M.: "Accurate Computation of the Stable Solitary Wave for the FitzHugh-Nagumo Equations ", J. Math. Biol., 13 (1982) p. 247-269

[24] LIEBERSTEIN, H. M.: " On the Hodgkin - Huxley Partial Differential Equation ", Math. Biosci., 1 (1967) p. 45-69.

[25] ENGELBRECHT, J.: " On Theory of Pulse Transmission in a Nerve Fibre ", Proc. Royal Soc. London, A375 (1981) p. 195-209.

[26] NAGUMO, J., ARIMOTO, S., YOSHIZAWA , S. : " An Active Pulse Transmission Line Simulating Nerve Axon ", Proc. IRE, 50 (1962) p. 2061-2070.

[27] ENGELBRECHT, J., TOBIAS, T.: " On a Model Stationary Nonlinear Wave in an Active Medium ", Proc. Royal Soc. London, A411 (1987) p. 139-154.

HYPERBOLIC SCHEMES FOR MULTI-COMPONENT EULER EQUATIONS

G. Fernandez , B. Larrouturou
INRIA, Sophia Antipolis, 06560 Valbonne, FRANCE

INTRODUCTION

Our purpose is to build efficient conservative schemes for the computation of multi-species (possibly reactive) flows. On this way we consider simplified models in which the governing equations include Euler's hyperbolic terms for several species. Even in the case where the different species have different molecular weights and specific heat ratios, the model is shown to remain hyperbolic. We present several numerical results obtained by extending to this multi-component Euler system the classical flux-splitting scheme of Roe [5].

MULTI-COMPONENT EULER EQUATIONS

For the sake of simplicity, we consider in a first step the one-dimensional flow of a mixture of only two species Σ_1 and Σ_2, and we neglect any reactive or diffusive effect. We therefore consider the following "multi-component Euler equations":

$$W_t + F_x = 0, \tag{1}$$

with:

$$W = \begin{pmatrix} W_1 \\ W_2 \\ W_3 \\ W_4 \end{pmatrix} = \begin{pmatrix} \rho \\ \rho u \\ E \\ \rho Y \end{pmatrix}, \qquad F = \begin{pmatrix} \rho u \\ \rho u^2 + p \\ u(E+p) \\ \rho u Y \end{pmatrix}. \tag{2}$$

The notations for the density ρ, velocity u, pressure p and total energy per unit volume E are classical; moreover, Y is the mass fraction of the first component Σ_1 [that is, ρY (resp: $\rho(1-Y)$) is the separate density of Σ_1 (resp: Σ_2)]. To close the system (1), we need to express the pressure as a function of the dependent variables W_i. We classically assume that the mixture is locally at thermal equilibrium, which means that the temperature field is the same for both species, and that the species Σ_1 and Σ_2 behave as perfect gases. Using Mayer's relation we can write:

$$p_i = \rho Y_i \frac{\mathcal{R}}{M_i} T = (\gamma_i - 1)\rho Y_i C_{vi} T \ , \quad i = 1, 2 \ .$$

In these relations, p_i, M_i, γ_i and C_{vi} are respectively the partial pressure, the molecular weight, the specific heat ratio and the specific heat at constant volume of species Σ_i; moreover, $\mathcal{R}$ is the universal gas constant, and $Y_1 = Y$, $Y_2 = 1 - Y$. On the other hand, the total specific energy E is given by:

$$E = \sum_i (\rho Y_i C_{vi} T + \frac{1}{2}\rho Y_i u^2) \ .$$

Using now Dalton's law $p = \sum_i p_i$, we obtain:

$$p = (\gamma - 1)(E - \frac{1}{2}\rho u^2) , \tag{3}$$

where γ, the local specific heat ratio of the mixture, is given by:

$$\gamma = \frac{Y_1 C_{v1}\gamma_1 + Y_2 C_{v2}\gamma_2}{Y_1 C_{v1} + Y_2 C_{v2}} = \frac{W_4 C_{v1}\gamma_1 + (W_1 - W_4)C_{v2}\gamma_2}{W_4 C_{v1} + (W_1 - W_4)C_{v2}} . \tag{4}$$

Thus $\gamma = \gamma(W)$ is an homogeneous function of degree 0, and the flux vector $F = F(W)$ is homogeneous of degree 1, as in the single component case:

$$\gamma(rW) = \gamma(W) , \quad F(rW) = rF(W) \text{ for } r > 0 . \tag{5}$$

Setting $F(W) = \mathcal{F}(W, \gamma(W))$, we can write the Jacobian matrix as:

$$A(W) = \frac{dF}{dW} = \frac{\partial \mathcal{F}}{\partial W} + \frac{\partial \mathcal{F}}{\partial \gamma}\frac{d\gamma}{dW} ,$$

that is (only the terms appearing in bold are new compared to the single-component case):

$$A(W) = \begin{pmatrix} 0 & 1 & 0 & \mathbf{0} \\ (\gamma - 3)\frac{u^2}{2} + \mathbf{X} & (3-\gamma)u & \gamma - 1 & \mathbf{Z} \\ -uH + (\gamma - 1)\frac{u^3}{2} + \mathbf{uX} & H - (\gamma - 1)u^2 & \gamma u & \mathbf{uZ} \\ -\mathbf{uY} & \mathbf{Y} & \mathbf{0} & \mathbf{u} \end{pmatrix} ,$$

where $H = \frac{E + p}{\rho}$ is the enthalpy per unit mass, and where we have set:

$$X = \frac{p}{\gamma - 1}\frac{\partial \gamma}{\partial W_1} , Z = \frac{p}{\gamma - 1}\frac{\partial \gamma}{\partial W_4} .$$

The developed expression of Z will be useful in the sequel; we have:

$$Z = \frac{p}{\gamma - 1}\frac{C_{v1}C_{v2}(\gamma_1 - \gamma_2)}{\rho[Y C_{v1} + (1 - Y)C_{v2}]^2} = \frac{C_{v1}C_{v2}(\gamma_1 - \gamma_2)T}{Y C_{v1} + (1 - Y)C_{v2}} .$$

A remarkable result is that the matrices A and $\frac{\partial \mathcal{F}}{\partial W}$ have the same real eigenvalues. The eigenvalues of $A(W)$ are indeed the roots of the polynomial:

$$\mathcal{P}(\lambda) = (\lambda - u)^2[(\lambda - u)^2 - \frac{\gamma p}{\rho} - (X + YZ)] ;$$

but the homogeneity (5) of γ implies:

$$X + YZ = \frac{p}{\rho(\gamma - 1)} \left(W_1 \frac{\partial \gamma}{\partial W_1} + W_4 \frac{\partial \gamma}{\partial W_4} \right) = 0 \ ,$$

and the eigenvalues of A are:

$$\lambda_1 = u \ , \quad \lambda_2 = u + c \ , \quad \lambda_3 = u - c \ , \quad \lambda_4 = u \ ,$$

where the sound speed c still has the classical expression:

$$c = \sqrt{\frac{\gamma p}{\rho}} \ ,$$

with the local value (4) of γ. The corresponding eigenvectors

$$\vec{e}_1 = \begin{pmatrix} 1 \\ u \\ \dfrac{u^2}{2} - X \\ 0 \end{pmatrix} , \qquad \vec{e}_2 = \begin{pmatrix} 1 \\ u + c \\ H + uc \\ Y \end{pmatrix} , \qquad \vec{e}_3 = \begin{pmatrix} 1 \\ u - c \\ H - uc \\ Y \end{pmatrix} , \qquad \vec{e}_4 = \begin{pmatrix} 0 \\ 0 \\ X \\ Y \end{pmatrix} ,$$

are linearly independant, which shows that the system (1)-(4) is hyperbolic (although non strictly hyperbolic since $\lambda_1 = \lambda_4$).

There is no difficulty in checking that these results also hold in several space dimensions or if the number of species is greater than two.

NUMERICAL APPROACH

Following previous studies on the numerical simulation of perfect gas flow or reactive gas flow, we use for the approximation of system (1) a finite-volume approach, which, in multidimensional situations, may operate on an unstructured finite-element mesh (see [2], [3]). Our goal is therefore to investigate how the classical flux-splitting hyperbolic schemes perform when extended to the full system (1) with the species equation added.

To present the numerical method, we describe its explicit first-order accurate form, which we write as:

$$\frac{W_j^{n+1} - W_j^n}{\Delta t} + \frac{F_{j+1/2}^n - F_{j-1/2}^n}{\Delta x} = 0 \ , \tag{6}$$

with:

$$F_{j+1/2}^n = \Phi(W_j^n, W_{j+1}^n) \ . \tag{7}$$

All numerical results presented below have been obtained using the following numerical flux function Φ based on Roe's average [5]:

$$\Phi(W_L, W_R) = \frac{F(W_L) + F(W_R)}{2} + \frac{1}{2} \mid A(\tilde{W}) \mid (W_L - W_R) \ , \tag{8}$$

where $\tilde{W} = \tilde{W}(W_L, W_R)$ is defined by the density $\tilde{\rho}$, the velocity $\tilde{u}$, the enthalpy $\tilde{H}$ and the mass fraction $\tilde{Y}$ given by the relations:

$$\tilde{\rho} = \frac{\rho_L \sqrt{\rho_L} + \rho_R \sqrt{\rho_R}}{\sqrt{\rho_L} + \sqrt{\rho_R}} \ , \quad \tilde{u} = \frac{u_L \sqrt{\rho_L} + u_R \sqrt{\rho_R}}{\sqrt{\rho_L} + \sqrt{\rho_R}} \ , \tag{9}$$

$$\tilde{H} = \frac{H_L\sqrt{\rho_L} + H_R\sqrt{\rho_R}}{\sqrt{\rho_L} + \sqrt{\rho_R}} \; , \quad \tilde{Y} = \frac{Y_L\sqrt{\rho_L} + Y_R\sqrt{\rho_R}}{\sqrt{\rho_L} + \sqrt{\rho_R}} \; . \tag{10}$$

An interesting question now is to know whether the fundamental property of Roe's scheme still holds for the mixture, that is if:

$$F(W_L) - F(W_R) = A(\tilde{W})(W_L - W_R) \; . \tag{11}$$

The answer is yes in the case where both species have the same specific heat ratio, that is if $\gamma_1 = \gamma 2 = \gamma(W)$ for all W. In fact, checking the property (11) in this particular case almost amounts to checking it in the single component case, since we then have $X = Z = 0$; this is easily done using the following arithmetic rules, for any variable U:

$$\Delta UV = \underline{U}\Delta V + \overline{V}\Delta U \; , \quad \overline{\rho U} = \underline{\rho}\overline{U} \; ,$$

where:

$$\Delta U = U_R - U_L \; , \quad \overline{U} = \frac{U_L\sqrt{\rho_L} + U_R\sqrt{\rho_R}}{\sqrt{\rho_L} + \sqrt{\rho_R}} \; , \quad \underline{U} = \frac{U_L\sqrt{\rho_R} + U_R\sqrt{\rho_L}}{\sqrt{\rho_L} + \sqrt{\rho_R}} \; .$$

In the case where $\gamma_1 \neq \gamma_2$, the property (11) no longer holds just as it is. To recover this property, we have to slightly modify the flux function (8) and use instead of $A(\tilde{W})$ a modified matrix $\tilde{A} \approx A(\tilde{W})$; more precisely, the property:

$$F(W_L) - F(W_R) = \tilde{A}(W_L - W_R)$$

holds for all W_L, W_R if we define $\tilde{A}$ as:

$$\tilde{A} = \begin{pmatrix} 0 & 1 & 0 & 0 \\ (\tilde{\gamma} - 3)\dfrac{\tilde{u}^2}{2} + \tilde{X} & (3 - \tilde{\gamma})\tilde{u} & \tilde{\gamma} - 1 & \tilde{Z} \\ -\tilde{u}\tilde{H} + (\tilde{\gamma} - 1)\dfrac{\tilde{u}^3}{2} + \tilde{u}\tilde{X} & \tilde{H} - (\tilde{\gamma} - 1)\tilde{u}^2 & \tilde{\gamma}\tilde{u} & \tilde{u}\tilde{Z} \\ -\tilde{u}\tilde{Y} & \tilde{Y} & 0 & \tilde{u} \end{pmatrix} , \tag{12}$$

where $\tilde{u}$, $\tilde{H}$, $\tilde{Y}$ are still given by (9)-(10), and where:

$$\tilde{\gamma} = \gamma(\tilde{W}) = \frac{\tilde{Y}\gamma_1 C_{v1} + (1 - \tilde{Y})\gamma_2 C_{v2}}{\tilde{Y} C_{v1} + (1 - \tilde{Y})C_{v2}} \; , \tag{13}$$

$$\tilde{Z} = \frac{C_{v1}C_{v2}(\gamma_1 - \gamma_2)\tilde{T}}{\tilde{Y}C_{v1} + (1 - \tilde{Y})C_{v2}} \; , \quad \tilde{X} = -\tilde{Y}\tilde{Z} \; , \tag{14}$$

with:

$$\tilde{T} = \frac{T_L\sqrt{\rho_L} + T_R\sqrt{\rho_R}}{\sqrt{\rho_L} + \sqrt{\rho_R}} \neq T(\tilde{W}) \; . \tag{15}$$

We refer the reader to [1] for the details of the derivation of (12)-(15).

NUMERICAL RESULTS

We first consider a Riemann problem for system (1), namely Sod's shock tube problem [6] with two different components on both sides of the discontinuity at $t = 0$, with $\gamma_1 = \gamma_2 = 1.4$. The exact solution of this problem consists of the classical known solution of Sod's problem for the hydrodynamical variables ρ, u and p, with a jump of the mass fraction located at each time at the contact discontinuity.

In Figure 1, we compare the results obtained using the scheme (6)-(8) with those obtained using the usual Roe's scheme for the classical Euler equations $\rho_t + (\rho u)_x = 0$, $(\rho u)_t + (\rho u^2 + p)_x = 0$, $E_t + [u(E+p)]_x = 0$, combined with a donor-cell approximation of the species equation $(\rho Y)_t + (\rho u Y)_x = 0$ (in the sequel, this second scheme is referred to as the "donor-cell scheme"). Incidentally, it is easy to check that, in this particular case where $\gamma_1 = \gamma_2$, the numerical values of the hydrodynamical variables obtained with both schemes are exactly identical; they are shown in Figure 1.a/. In Figure 1.b/, we show the species profiles $Y(x)$ obtained with both schemes in an experiment where the species Σ_1 is initially on the left side of the discontinuity (i.e. $Y(x, t = 0) = 1$ for $x < 0.5$, 0 for $x > 0.5$); on the opposite, the species profile of Figure 1.c/ correspond to the initial condition $Y(x, t = 0) = 0$ for $x < 0.5$, 1 for $x > 0.5$. For both experiments, the results obtained with the scheme (6)-(8) are much better than those obtained with the "donor-cell scheme". In case b/ the differences between the two schemes disappear when time increases. Such is not the case for the second experiment c/ where the observed kink for the "donor-cell scheme" remains as long as we pursue the computation. Moreover, when the grid point number is increased, this kink becomes thinner and thinner but keeps the same amplitude.

Another remark on these results concerns the comparisons of both experiments b/ and c/. For sake of clarity, let $Y_b(x,t)$ and $Y_c(x,t)$ denote the exact solutions of the Riemann problems b/ and c/. It is clear that these exact solutions satisfy the relation:

$$Y_b(x,t) + Y_c(x,t) = 1 \ . \tag{16}$$

But the analogous relation does not hold for the numerical solution; in other words, $(Y_b)_j^n + (Y_c)_j^n \neq 1$, which means that calling Σ_1 or Σ_2 the gaseous species which is initially in the left compartment of the shock tube has an influence on the numerical results ! This surprising fact comes from the conservative formulation in which the mass fraction is obtained as a nonlinear function of the dependent variables: $Y = \dfrac{W_4}{W_1}$. In practice, this difference between experiments b/ and c/ is very small when the scheme (6)-(8) is used, but it appears to be important for the "donor-cell scheme".

This nonlinear character of the scheme also has the drawback that the inequalities $0 \leq Y \leq 1$, which are required from a physical point of view, do not necessarily hold for the numerical solution (again, this drawback is more important for the "donor-cell scheme"). In fact, the only way of guaranteeing the discrete maximum principle for the mass fraction would be to use a non-conservative formulation (i.e. to solve the equation $\rho Y_t + \rho u Y_x = 0$), which would have several other disadvantages for the shock problem considered here.

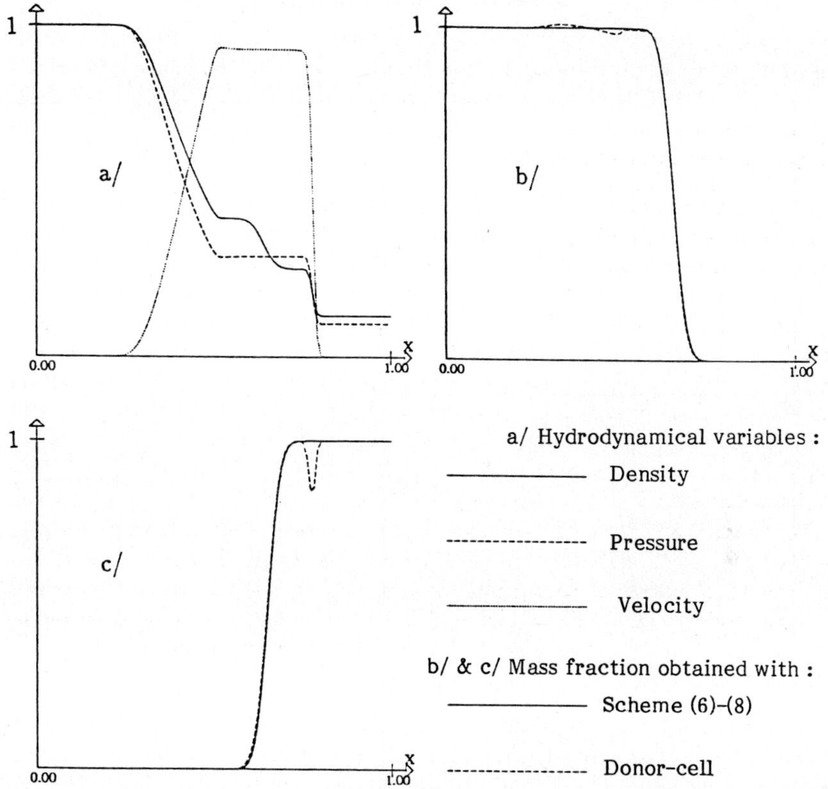

Figure 1: Density, pressure, velocity and mass fraction profiles for a two-component shock tube.

Next, we address a case where the species Σ_1 and Σ_2 have not the same specific heat ratio: $\gamma_1 = 1.2$, $\gamma_2 = 1.4$. The results are less satisfactory. When we use the scheme (6)-(7) with the flux function:

$$\Phi(W_L, W_R) = \frac{F(W_L) + F(W_R)}{2} + \frac{1}{2} \mid \tilde{A} \mid (W_L - W_R) \; ,$$

[with $\tilde{A}$ given by (12)-(15)], we obtain the results shown on Figure 2.a/, where one can observe a small but non physical pressure jump at the contact discontinuity. Almost identical results are obtained with the flux function (8). Lastly, when we use the flux function (8) while neglecting the terms X and Z in the Jacobian $A(\tilde{W})$, we obtain the results presented on Figure 2.b/, which are worse than the preceding ones. This shows that neglecting the derivatives of γ in the dissipative term $\frac{1}{2} \mid A(\tilde{W}) \mid (W_L - W_R)$ negatively affects the numerical results.

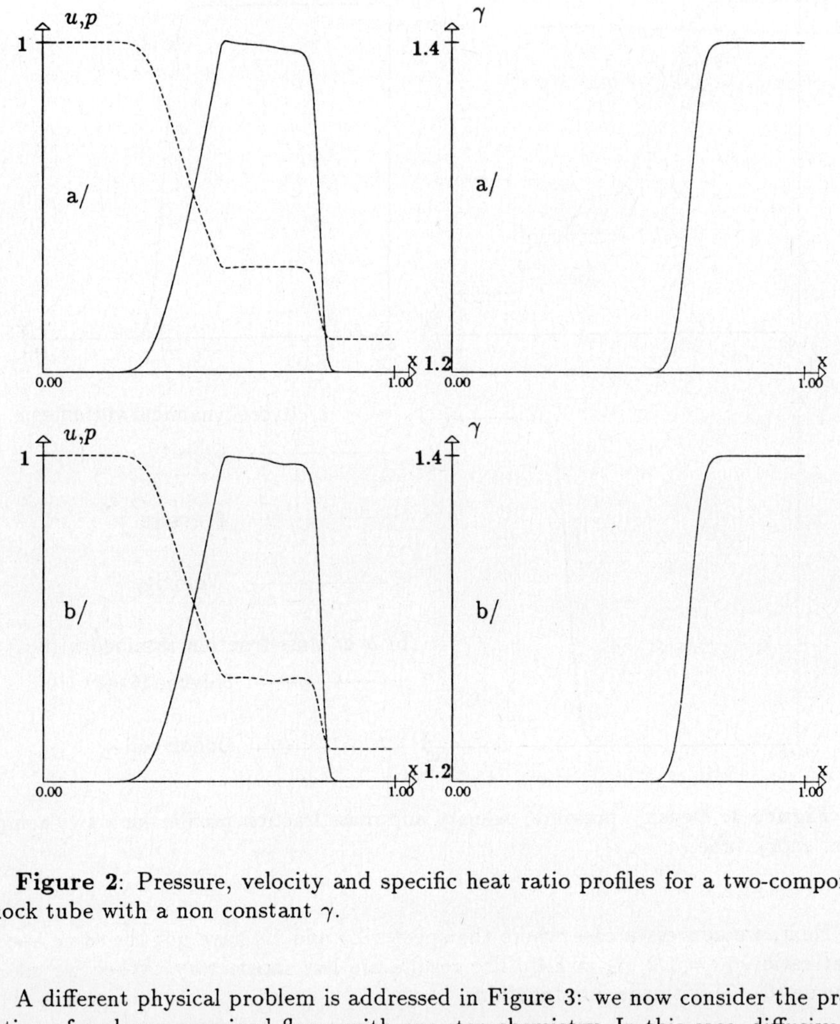

Figure 2: Pressure, velocity and specific heat ratio profiles for a two-component shock tube with a non constant γ.

A different physical problem is addressed in Figure 3: we now consider the propagation of a planar premixed flame with one-step chemistry. In this case, diffusive and reactive terms are added to the energy and species equations, and system (1) becomes:

$$W_t + F_x = \begin{pmatrix} 0 \\ 0 \\ D_T \dfrac{\partial^2 T}{\partial x^2} + Q\mathcal{A}\,\rho Y\ exp(-\dfrac{\mathcal{E}}{\mathcal{R}T}) \\ D_Y \dfrac{\partial^2 Y}{\partial x^2} - \mathcal{A}\,\rho Y\ exp(-\dfrac{\mathcal{E}}{\mathcal{R}T}) \end{pmatrix} .$$

Here, D_T and D_Y are the thermal and molecular diffusion coefficients, Q is the heat released by the chemical reaction, $\mathcal{E}$ is the activation energy of the reaction, and

$\rho Y\ exp(-\frac{\mathcal{E}}{\mathcal{R}T})$ is the reaction rate (see e.g. [4]).

The results of several numerical experiments are shown on Figure 3, where we compare the results obtained with three different schemes: the scheme (6)-(8), the "donor-cell scheme", and a "centered scheme" in which the usual Roe's scheme for the Euler equations is combined with a centered approximation of the species equation. We also use two different computational grids, a coarse mesh of 61 equally spaced nodes in the interval [0,1], and a fine uniform mesh of 401 nodes. The error in the flame location which is observed when the "centered" and "donor-cell" schemes are used on the coarse grid is considerably reduced when the "global flux-splitting" (6)-(8) is employed.

Lastly, we show in Figure 4 the two-dimensional interaction of two supersonic gaseous jets; the impinging jets are made up of two different species, with different molecular weights and different specific heat ratios. The system of governing equations is simply the two-dimensional analogue of (1)-(4), with no diffusive and reactive terms. Although the diffusive effect of the scheme (6)-(8) clearly appears when observing the mass fraction contours (obtained on a uniform non adaptive mesh), the scheme behaves in a very promising fashion. In particular, we want to emphasize here that no acceptable result can be obtained for this experiment with the "donor-cell scheme" (that is, as above, with a donor-cell approximation of the mass fraction equation coupled to the usual Roe's scheme for the Euler equations): indeed, the maximum value of the mass fraction Y rapidly increases above 1 and even exceeds 1.5 when the "donor-cell" calculation proceeds, while the mass fraction remains in the interval [0,1] (with an error of the order of 10^{-2}) when the scheme (6)-(8) is used.

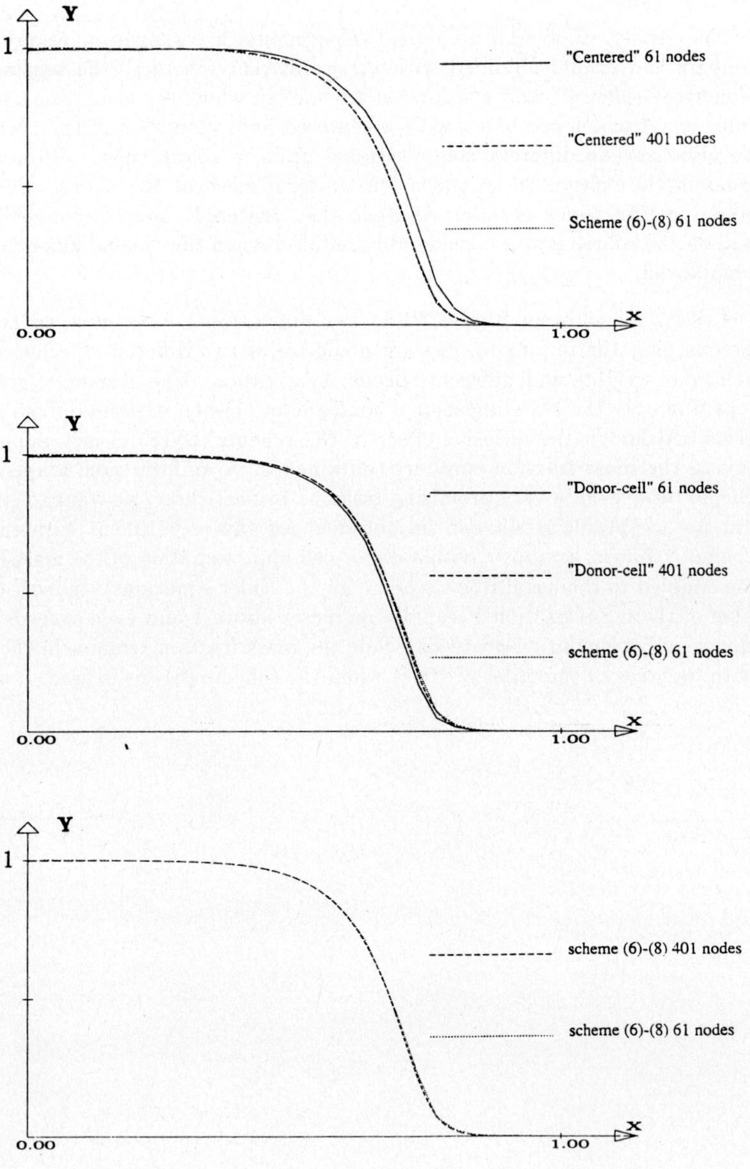

Figure 3: Mass fraction profiles across a planar premixed flame.

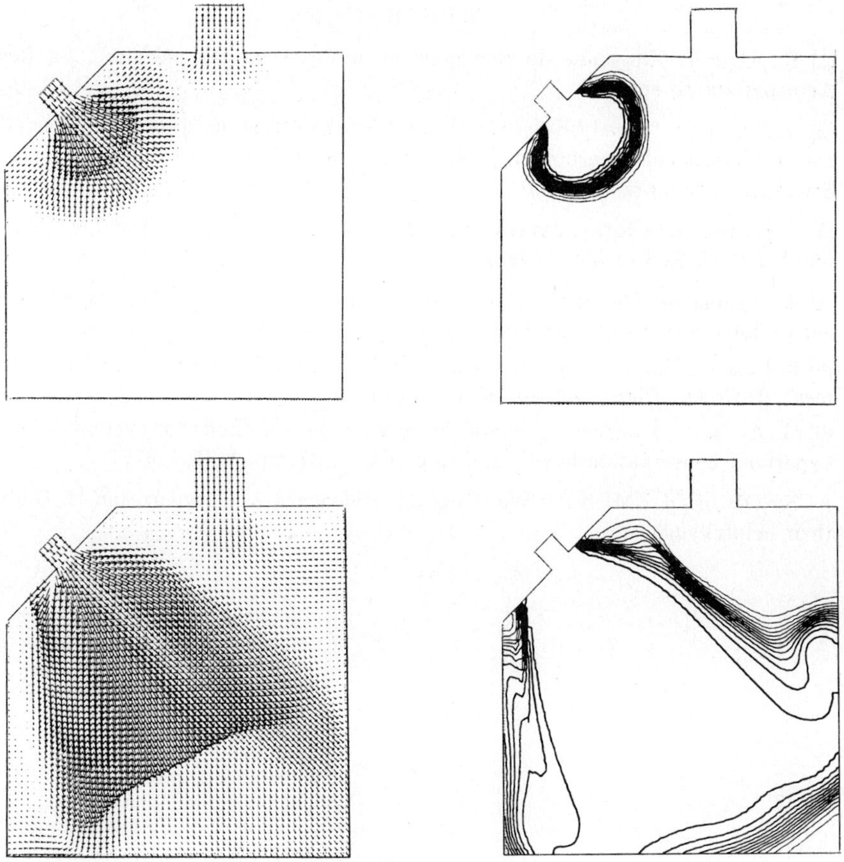

Figure 4: Gaseous jets interaction: velocity field and mass fraction contours.

CONCLUSION

We have presented several numerical experiments which show that, for the simulation of (possibly reactive) multi-component inviscid flows, the results of calculations in which the usual Euler terms and the added continuity equations are treated separately can be substantially improved by using a global flux-splitting approach, in which one of the classical upwind schemes developed in these last years for the Euler equations is globally applied to the whole system of conservation laws, as in the scheme (6)-(8).

REFERENCES

[1] R. Abgrall, "Moyenne de Roe pour un mélange de gaz parfaits", La Recherche Aérospatiale, to appear.

[2] A. Dervieux, "Steady Euler simulations using unstructured meshes", Partial differential equations of hyperbolic type and applications, Geymonat ed., pp. 33-111, World Scientific, Singapore, (1987).

[3] F. Fezoui, "Résolution des équations d'Euler par un schéma de Van Leer en éléments finis", INRIA Report 358, (1985).

[4] A. Habbal, A. Dervieux, H. Guillard, B. Larrouturou, "Explicit calculation of reactive flows with an upwind hydrodynamical code", INRIA Report 690, (1987).

[5] P. L. Roe, "Approximate Riemann solvers, parameters vectors and difference schemes", J. Comp. Phys., **43**, pp. 357-372, (1981).

[6] G. A. Sod, "A survey of several finite-difference methods for systems of nonlinear hyperbolic conservation laws", J. Comp. Phys., **27**, pp. 1-31, (1977).

ACKNOWLEDGEMENTS: We thank our colleagues A. Dervieux and H. Guillard for their helpful suggestions.

Multigrid Methods for the Solution of Porous Media Multiphase Flow Equations

T. W. Fogwell   F. Brakhagen
Gesellschaft für Mathematik und Datenverarbeitung
Fed. Rep. Germany*

ABSTRACT

Two solution methods for the solution of the equations

$$\phi\frac{\partial S_w}{\partial t} = \nabla\cdot[\frac{k_{rw}K}{\mu_w}(\nabla P_w + \rho_w g)] - Q_w$$

$$-\phi\frac{\partial S_w}{\partial t} = \nabla\cdot[\frac{k_{ro}K}{\mu_o}(\nabla P_o + \rho_o g)] - Q_o$$

for incompressible, two phase flow in a porous medium were employed together with multigrid solvers. The IMPES method is implicit in pressure and explicit in saturation. The simultaneous solution (SS) method is a linearized fully implicit method. Both methods relied on special interpolation operators for multigrid transfers based on the discretized differential equation. This provided a method of overcoming the difficulties associated with the discontinuous coefficients. The application to the SS method was completely new and showed the usual increase of efficiency associated with multigrid methods.

1. INTRODUCTION

For flow in a porous medium Darcy discovered that a good method to use to calculate the superficial velocity was to regard the pressure gradient as the driving force and take the properties of the medium as transmission factors. The velocity, of course, was inversely proportional to the viscosity and the pressure could be due to gravity effects. For each phase ℓ, then, we have the velocity

$$v_\ell = -\frac{K_\ell}{\mu_\ell}(\nabla P_\ell + \rho_\ell g) \tag{1}$$

where μ is the viscosity, K is the permeability, P is the pressure, ρ is the density, and g is the gravity vector.

The volumetric fraction (saturation) of phase ℓ is denoted by S_ℓ, so that $\sum S_\ell = 1$. It was discovered that for multiphase flow the permeability tensor could be more

---

Part of the Brazilian–German Cooperation in the Field of Informatics in the Area of Computational Mathematics; P. J. Paes-Leme, Brazilian Project Leader of ORESIM in Rio de Janeiro, Brazil.

accurately split into two parts, so that $K_\ell = k_{r\ell}K$, where $k_{r\ell}$ is a scalar depending on the saturations and K is a tensor depending on the spatially varying structure of the medium. The mass conservation is represented by the continuity equation

$$\frac{\partial(\phi\rho_\ell S_\ell)}{\partial t} = -\nabla\cdot(\rho_\ell v_\ell) - q_\ell \tag{2}$$

where q_ℓ is a production (sink) term depending on time and location and ϕ is porosity (fraction void space).

In order to solve the equations boundary and initial conditions must be specified together with the parameters in the equation. Usually the boundary conditions are no-flow Neumann type conditions. For constant temperature, the following are specified: $\phi(P,X)$, $\rho_\ell(P_\ell)$, $\mu_\ell(P_\ell)$, and $k_{r\ell}(S_1,\ldots,S_n)$. Another relationship is then needed in order to close the set of equations. This is usually taken to be the capillary pressure function P_c defined to be the difference in pressure between a wetting and a non-wetting phase, and is usually considered an empirical function of the saturations.

The Darcy's law equation can be combined with the continuity equation to eliminate the velocity and gives

$$\frac{\partial(\phi\rho_\ell S_\ell)}{\partial t} = -\nabla\cdot[\frac{\rho_\ell k_{r\ell}K}{\mu_\ell}(\nabla P_\ell + \rho_\ell g)] - q_\ell. \tag{3}$$

The unknowns are then the P_ℓ and S_ℓ. For incompressible flow ρ_ℓ and ϕ are constant and this becomes

$$\phi\frac{\partial S_\ell}{\partial t} = \nabla\cdot[\frac{k_{r\ell}K}{\mu_\ell}(\nabla P_\ell + \rho_\ell g)] - Q_\ell \tag{4}$$

where $Q_\ell = q_\ell/\rho_\ell$. In order to eliminate the saturations we can sum the above equations to give

$$\nabla\cdot\sum_\ell[\frac{k_{r\ell}K}{\mu_\ell}(\nabla P_\ell + \rho_\ell g)] - Q_t = 0 \tag{5}$$

where Q_t is the total production. This is an elliptic equation and demonstrates why one might expect multigrid methods to work.

Unfortunately, there is a strong hyperbolic part to this set of equations resulting from the mass conservation requirement. In order to demonstrate this suppose there are just two phases, o and w, where w is the wetting phase. Then the capillary pressure $P_c = P_o - P_w$. If the capillary pressure is taken to be zero, then $P_o = P_w = P$. If the flow is horizontal, then $\rho_\ell g = 0$. We now have

$$v_\ell = -\frac{k_{r\ell}K}{\mu_\ell}\nabla P$$

for each phase ℓ, and

$$v_w = \frac{k_{rw}\mu_o}{k_{ro}\mu_w}v_o.$$

Let $v_t = v_o + v_w$ and

$$f_\ell = \frac{\frac{k_{r\ell}}{\mu_\ell}}{\frac{k_{rw}}{\mu_w} + \frac{k_{ro}}{\mu_o}},$$

for $\ell = o$ or w. Then $v_\ell = f_\ell v_t$ and the equation for phase ℓ becomes

$$\phi \frac{\partial S_\ell}{\partial t} = -\nabla \cdot (f_\ell v_t) - Q_\ell.$$

If we look at the places where there is no source or sink term, then $Q_\ell = 0$. Expanding the right side, we get

$$\phi \frac{\partial S_\ell}{\partial t} = -f_\ell \nabla \cdot v_t - v_t \cdot \nabla f_\ell.$$

For incompressible flow this becomes

$$\phi \frac{\partial S_\ell}{\partial t} = -v_t \frac{df_\ell}{dS_\ell} \cdot \nabla S_\ell$$

which has the form of a first order hyperbolic equation. The discontinuities in the initial conditions are then propagated. This is why one might expect multigrid methods not to work so well. T. F. Russel, J. Douglas, and R. E. Ewing, among others, have used properties of characteristics in solution methods designed to take advantage of the hyperbolic nature of the equations describing some kinds oil reservoir simulation problems.

The discontinuities in the solution are not the only problems. There are large discontinuities in the coefficients as well. K can have discontinuous jumps of several orders of magnitude due to changes in the medium's geological features. These are aggravated by large discontinuous changes in $k_{r\ell}$ as a result of the sudden changes in the saturations. These difficulties give rise to special discretization problems as well as difficulties with the grid transfer operators required with multigrid methods.

2. EQUATIONS SOLVED

The equations we solve are those for incompressible two phase flow in two dimensions. We use two general techniques, both using multigrid, one called IMPES (implicit pressure, explicit saturation) and the other called the simultaneous solution (SS) method. The general equations for both methods are

$$\phi \frac{\partial S_w}{\partial t} = \nabla \cdot [\frac{k_{rw} K}{\mu_w} (\nabla P_w + \rho_w g)] - Q_w \tag{6}$$

$$-\phi \frac{\partial S_w}{\partial t} = \nabla \cdot [\frac{k_{ro} K}{\mu_o} (\nabla P_o + \rho_o g)] - Q_o. \tag{7}$$

For the IMPES method we add the two above equations to obtain a "pressure equation," as follows,

$$\nabla \cdot [(\lambda_o + \lambda_w) \nabla P_o - \lambda_w \nabla P_c - (\lambda_o \rho_o + \lambda_w \rho_w) g] = Q_o + Q_w \tag{8}$$

where $\lambda_\ell = k_{rl} K / \mu_\ell$ for $\ell = o$ or w. This equation is elliptic. The solution method proceeds as follows: All saturation related terms are taken at the old time level and equation (8) is solved for P_o at the new time. Once P_o is known, equation (7) is then solved explicitly for the new saturation S_w. From this P_c is known and the process is started again.

The problem is discretized by the standard finite difference method with reflection boundary conditions. The space discretization is carried out as follows: For

$$\frac{\partial}{\partial x}(\lambda \frac{\partial P}{\partial x})$$

we approximate by

$$\frac{1}{\Delta x_i}[\lambda_{i+\frac{1}{2}} \frac{P_{i+1} - P_i}{\Delta x_{i+\frac{1}{2}}} - \lambda_{i-\frac{1}{2}} \frac{P_i - P_{i-1}}{\Delta x_{i-\frac{1}{2}}}],$$

defined as $\Delta(\lambda \Delta P)$. This produces a five point discretization pattern which we use on grids which are uniform in each direction. The λ_i changes with P_i, which changes with time, where

$$\lambda_{i+\frac{1}{2}} = (\frac{k_r K}{\mu})_{i+\frac{1}{2}}.$$

k_r is taken upstream, while K is the harmonic mean. μ is taken to be constant. The standard multigrid method will fail because of the large jump discontinuities in the coefficients λ_ℓ. The difference operator itself is used for the interpolation operator as originally proposed by Alcouffe, Brandt, Dendy, and Painter [1]. This then more closely follows the continuity of the $\lambda_\ell \nabla P_\ell$ terms rather than attempting to interpolate the discontinuous ∇P_ℓ terms. The IMPES method has the advantage for multigrid methods that the implicit part of the procedure is an elliptic equation. This is particularly well suited to multigrid solution techniques. Because of the explicit treatment of saturation, however, the following restriction on the size of the time step, as was shown by Aziz and Settari [2], is imposed for stability:

$$\Delta t < \min_i \frac{\Delta x \Delta y \phi}{K_x \frac{\Delta y}{\Delta x} + K_y \frac{\Delta x}{\Delta y}} \min_{i,S_w} \frac{\frac{\mu_w}{k_{rw}} + \frac{\mu_o}{k_{ro}}}{|P_c'|}.$$

If the mesh sizes are very small or P_c' is very large, then the sizes of the time steps required for stability are unacceptably small. In this case the simultaneous solution method preferable.

The system we use in the simultaneous solution method is symmetric. The equations (6) and (7) are transformed into

$$\nabla \cdot [\lambda_o(\nabla P_o + \rho_o g)] = -\phi S_w'(\frac{\partial P_o}{\partial t} - \frac{\partial P_w}{\partial t}) + Q_o \tag{9}$$

$$\nabla \cdot [\lambda_w(\nabla P_w + \rho_w g)] = \phi S_w'(\frac{\partial P_o}{\partial t} - \frac{\partial P_w}{\partial t}) + Q_w. \tag{10}$$

by taking $S_w' = \partial S_w / \partial P_c$, on the assumption that $P_c(S_w)$ is invertible and S_w' exists. The discretization of the spatial derivatives is carried out the same as for the IMPES method to produce a five point pattern. The discretization of the time derivative is produced as follows:

$$\frac{P_{ci,n+1} - P_{ci,n}}{\Delta t} = \Delta_t P_c.$$

Weighting by θ in time taken for space derivatives gives

$$\phi S' \Delta_t P_c = \theta[\Delta(\lambda \Delta P)]_{n+1} + (1-\theta)[\Delta(\lambda \Delta P)]_n$$

According to the stability analysis we performed, this is stable for values of θ greater than a half and unstable for values of θ less than a half, with conditional stability at the value of one half. Although the possibility of using this time weighting parameter, θ, was programmed as an option, it has so far not been used.

The equations are linearized by taking λ and S_w' at the old time, except for nonlinear $P_c(S_w)$ where an iterative approximation to S_w' at the new time is made (see [2,9]). The fully implicit case takes λ and S_w' at new times, but then this is nonlinear. There is also an unsymmetric form given by

$$\nabla \cdot [\lambda_o(\nabla P_o + \rho_o g)] = -\phi \frac{\partial S_w}{\partial t} + Q_o$$

$$\nabla \cdot [\lambda_w(\nabla P_o - P_c' \nabla S_w + \rho_w g)] = \phi \frac{\partial S_w}{\partial t} + Q_w,$$

but, generally, unsymmetric sets of equations are not as well handled by multigrid methods. For this reason, we chose to tackle the symmetric form first.

The equations are solved simultaneously for P_o and P_w. The new saturations are then found by $S_w(P_c)$. The finite difference discretization of the system leads to a symmetric, block–pentadiagonal system where the blocks are 2x2 submatrices. The off–diagonal blocks are diagonal matrices. The discretization is backward in time with explicit mobilities λ_o and λ_w. Again the difficulties in a multigrid scheme are the discontinuities in the coefficients λ_ℓ. We use a generalization of the interpolation procedure we use for the IMPES method applied to systems, which was originally proposed by Dendy [8].

3. SOLUTION BY A MULTIGRID METHOD

Simulation of large reservoirs by the methods described above requires the solution of very large sets of linear equations. The number of unknowns may run up to several thousands. If direct methods are used for the solution of such systems, the amount of work increases as the square of the number of unknowns. In terms of computer work and storage, it is generally much more economical to use a fast iterative method.

Very rapid convergence is provided by multigrid methods. These are 'asymptotically optimal' iterative methods, i. e., the computational work required for achieving a fixed accuracy is proportional to the number of discrete unknowns. The multigrid methods work in the following way: It is well known that relaxation methods are very efficient at the elimination of high frequency errors which have a wavelength of the order of a grid spacing but are very inefficient at the elimination of long wavelength errors. This fact is exploited in multigrid methods. By the application of a suitable relaxation method approximations with smooth errors are obtained very efficiently. Such smooth errors can be accurately represented on a coarser grid. Thus, corrections of approximations with smooth errors can be calculated efficiently on this coarser grid. This basic idea is employed on coarser and coarser grids. Finally we arrive at a very coarse grid, on which a linear system of equations can easily be solved. Having found a sufficiently good correction on a coarser grid, we return to the next finer grid by interpolating this correction and adding it to the fine grid approximation. This is continued back to the finest grid.

Let a sequence of grids G^k $(k = 1(1)M)$ be defined with $h_1 > h_2 > \ldots > h_M$, where h_k is the mesh size of G^k. Let L^M be a finite difference approximation to the differential operator L on G^M and let L^k be an approximation to L^M on G^k for $k < M$. By X^k we denote the linear space of real valued grid functions defined on G^k:

$$X^k : \quad G^k \rightarrow \mathbf{R}.$$

Obviously $X^k \cong \mathbf{R}^{N_k}$, where N_k is the number of grid points of G^k. By

$$I_k^{k-1} : \quad X^k \rightarrow X^{k-1}$$

we denote a restriction operator and by

$$I_{k-1}^k : \quad X^{k-1} \rightarrow X^k$$

an interpolation operator. For convenience we introduce the notation

$$\bar{u}^k := S_k^\sigma (u^k, L^k, f^k)$$

to denote that $\bar{u}^k$ is the result of σ relaxation steps applied to $L^k u^k = f^k$ starting with u^k as first approximation, where $u^k. \bar{u}^k, f^k \in X^k$. Similarly we denote by

$$\bar{w}^k := MG_k^\gamma (u^k, L^k, f^k)$$

so that $\bar{w}^k$ is the result of γ successive applications of the following algorithm MG_k to $L^k u^k = f^k$ starting with u^k as first approximation:

Algorithm MG_k:

Step 1: If $k > 1$ go to step 2, else solve exactly

$$L^1 \bar{u}^1 = f^1. \tag{11}$$

$$MG_1 := \bar{u}^1. \tag{12}$$

Step 2: (Smoothing)

$$\bar{u}^k := S_k^{\sigma_1} (u^k, L^k, f^k) \tag{13}$$

Step 3: (Transfer to the next coarser grid)

$$\bar{f}^{k-1} := I_k^{k-1} [f^k - L^k \bar{u}^k], \tag{14}$$

$$v^{k-1} := 0. \tag{15}$$

Step 4:

$$\bar{v}^{k-1} := MG_{k-1}^\gamma (v^{k-1}, L^{k-1}, \bar{f}^{k-1}), \tag{16}$$

Step 5: (Transfer to the next finer grid)

$$w^k := \bar{u}^k + I_{k-1}^k \bar{v}^{k-1}, \tag{17}$$

Step 6: (Smoothing)

$$\bar{w}^k := S_k^{\sigma_2}(w^k, L^k, f^k), \tag{18}$$

$$MG_k := \bar{w}^k. \tag{19}$$

The multigrid method consists of the following iteration:

$$\text{Initialize: } u^{(0)} \in X^M, \tag{20}$$

$$u^{(j+1)} := MG_M(u^{(j)}, L^M, f^M). \tag{21}$$

γ in step 4 is either equal to 1 or 2. With $\gamma = 1$ we obtain a V-Cycle and with $\gamma = 2$ a W-Cycle. If the coefficients of the differential operator L are continuous, it is usually very efficient to use bilinear interpolation for I_{k-1}^k and the transpose of this interpolation operator for the restriction I_k^{k-1}. Moreover, in this case very good results are generally obtained if L^k for $k < M$ is the same finite difference approximation of L on G^k as L^M on G^M.

If we solve the oil reservoir differential equations by the multigrid method, we have the difficulty that the transmissibilities may have large jump discontinuities of several orders in magnitude. Discontinuities in permeablity occur between different layers having different geological structures. Furthermore the relative permeabilities are dependent on the saturations, which have discontinuities at the oil-water-interface. As noted by Alcouffe et al. (see [1]) the multigrid method exhibits poor convergence for problems with large discontinuities if the bilinear interpolation is used. If the coefficients of the differential equations jump by orders of magnitude, the use of a more appropriate interpolation is necessary to achieve the usual multigrid efficiency. The interpolation operator should mimic the properties of the difference operator. Consequently, the difference operator itself was used for the interpolation operator I_{k-1}^k (see [1,3,4,6]). Moreover, it must be guaranteed that the difference equations on the next coarser grid approximate those on the given grid. To achieve this we defined the coarse grid difference operator L^{k-1} by the Galerkin approach

$$L^{k-1} = I_k^{k-1} L^k I_{k-1}^k \qquad (k = M(-1)2). \tag{22}$$

L^{k-1} $(k = M(-1)2)$ are nine point operators, although L^M is a five point operator.

The interpolation operator used for the solution of the IMPES pressure equation is defined in the following way. Let L^k be defined at point $(x_i, y_j) \in G^k$ by

$$\begin{aligned} L_{i,j}^k u_{i,j}^k &= SW_{i,j}^k\, u_{i-1,j-1}^k + W_{i,j}^k\, u_{i-1,j}^k + NW_{i,j}^k\, u_{i-1,j+1}^k \\ &+ S_{i,j}^k\, u_{i,j-1}^k + C_{i,j}^k\, u_{i,j}^k + N_{i,j}^k\, u_{i,j+1}^k \\ &+ SE_{i,j}^k\, u_{i+1,j-1}^k + E_{i,j}^k\, u_{i+1,j}^k + NE_{i,j}^k\, u_{i+1,j+1}^k. \end{aligned} \tag{23}$$

Suppose $(x_{i+1}, y_j) \in G^k \setminus G^{k-1}$ and $(x_i, y_j), (x_{i+2}, y_j) \in G^k \cap G^{k-1}$, where $k = M(-1)2$. On G^{k-1} the latter two grid points are denoted by $(x_I, y_J), (x_{I+1}, y_J)$ respectively. Form $\tilde{W}_{i+1,j}^k = SW_{i+1,j}^k + W_{i+1,j}^k + NW_{i+1,j}^k$, $\tilde{C}_{i+1,j}^k = S_{i+1,j}^k + C_{i+1,j}^k + N_{i+1,j}^k$, $\tilde{E}_{i+1,j}^k = SE_{i+1,j}^k + E_{i+1,j}^k + NE_{i+1,j}^k$. Then for horizontal lines embedded in the coarse grid the interpolation I_{k-1}^k is given by

$$(I_{k-1}^k u^{k-1})_{i+1,j} = -(\tilde{W}_{i+1,j}^k u_{I,J}^{k-1} + \tilde{E}_{i+1,j}^k u_{I+1,J}^{k-1}) / \tilde{C}_{i+1,j}^k. \tag{24}$$

A similar formula is used for the interpolation on vertical lines embedded in the coarse grid. At fine grid points which do not lie on coarse grid lines [for example (x_{i+1}, y_{i+1})] the interpolation formula is found by solving the equation

$$L^k_{i+1,j+1}\, u^k_{i+1,j+1} \;=\; 0 \tag{25}$$

for $u^k_{i+1,j+1}$. This interpolation approximates the continuity of $(\lambda_o + \lambda_w)\nabla p_o$ over the entire domain (see [1]). The restriction operator I_k^{k-1} is defined by the transpose of the interpolation operator

$$I_k^{k-1} \;=\; (I^k_{k-1})^T. \tag{26}$$

For relaxation we used the point, the line, or the alternating line Gauss Seidel method.

In the simultaneous solution method a system of two differential equations is to be solved numerically. Here again we have the difficulty of jump discontinuities in the coefficients of the differential equations. Therefore, we implement a multigrid method which is a generalization of the method used for IMPES (see [8]), in which the differential operator is used as the basis for the interpolation. This tends to follow the continuity of $\lambda_l \nabla p_l$ rather than trying to follow ∇p_l (which can be discontinuous here) as in the usual schemes. Interpolation and coarse grid matrices are calculated in the same manner as for the IMPES-method. In doing this we replace scalar operations by matrix operations, so that, for example, the division by $\tilde{C}^K_{i+1,j}$ in equation (24) is replaced by the matrix inversion $[\tilde{C}^K_{i+1,j}]^{-1}$. Relaxation is done by the collective point, line, or alternating line Gauss Seidel method.

4. RESULTS

We tested the multigrid method described above for the following situation. The reservoir was represented by a horizontal square. Initially the entire reservoir was (almost) saturated with oil. Water was injected in one corner and oil was produced in the opposite corner of the square. We assumed no flow Neumann boundary conditions and considered the following cases:

1) Isotropic case, i. e. $K_x = K_y$, where K_x, K_y are the permeabilities in x- and y-direction respectively.

2) Anisotropic case with $K_x = 10^3\, K_y$.

3) Anisotropic case with jump discontinuities. There were three subregions with different permeabilities K_x, K_y. In the first subregion we assumed $K_x = 10^2\, K_y$, in the second $K_x = K_y$ and in the third $K_y = 10^2\, K_x$.

In case 1) the point Gauss Seidel method was a very good smoother. In the second case it was necessary to use the line Gauss Seidel relaxation, whereas in the third case only the alternating line Gauss Seidel was a good smoothing procedure. The capillary pressure was assumed to be a linear function of saturation. The values of the capillary pressure were assumed to be rather small, they ranged only between 0.0 and 0.1. The relative permeabilities were approximated by piecewise linear functions.

In all these cases the multigrid method described above exhibited the usual multigrid efficiency. If the IMPES-method was used, for a grid with $65 \times 65 = 4225$ grid points for example execution times between 3.5 and 7.0 seconds were necessary on the IBM 3090 computer, to make the residuals smaller than 10^{-10}. For the simultaneous solution method we found frequently that the same number of cycles and about the same number of relaxations were necessary to achieve a certain accuracy as for IMPES. The execution times of the simultaneous solution method were of course larger than for IMPES because we had to solve a system of two differential equations instead of a single equation. On the average they were between 1.4 and 1.5 times as large as for IMPES.

If the derivative of the capillary pressure function is very small, then the discrete system of the simultaneous solution method becomes nearly singular (see [2]). In this case we sometimes observed divergence of the multigrid method after it had converged very well for a large number of time steps. When this happened, it was always possible to obtain convergence by choosing the coarsest grid sufficiently fine. So far we haven't observed any loss of efficiency in this case. The execution times were about the same.

5. CONCLUSIONS

We have programmed the IMPES method of solution for two phase incompressible flow using a multigrid solver originally proposed by Alcouffe, et al [1]. This method avoids problems which arise in usual multigrid techniques due to discontinuities of the coefficients. The discretized operator equation itself is taken as the basis for the interpolation operator, and, thus, the coarse grid operator, by means of the Galerkin approximation. The efficiencies we have achieved are typical for those found when the multigrid method is used.

We then programmed the simultaneous solution method (linearized fully implicit method), which included the time discretized terms. We used a generalization of the multigrid method used in the IMPES case for systems of equations [8]. The inclusion, however, of the mass conservation terms in the system made the system much more hyperbolic. As a result, although the method worked well, one had to take care not to choose the coarsest grid too coarse. If it was chosen too coarse, then the equations were dominated by the mass conservation part of the equations, rather than the diffusion transport part. This restriction caused, however, no practical limitations to the use of the method. Generally, it was sufficient to have at least several interior point on the coarsest grid. With this slight caution, this form of the equations was solved as efficiently as is usually the case for multigrid methods.

REFERENCES

[1] R. E. Alcouffe, A. Brandt, J. E. Dendy, J. W. Painter. *The Multi-Grid Method for the Diffusion Equation with Strongly Discontinuous Coefficients.* SIAM J. Sci. Stat. Comput., Vol. 2, 430 - 454 (1981).

[2] K. Aziz, A. Settari. *Petroleum Reservoir Simulation.* Elsevier Applied Science Publishers, London (1979).

[3] A. Behie, P. A. Forsyth. *Multi-Grid Solution of the Pressure Equation in Reservoir Simulation.* Sixth SPE Symposium on Reservoir Simulation, New Orleans, Louisiana (1982).

[4] A. Behie, P. A. Forsyth. *Comparison of Fast Iterative Methods for Symmetric Systems.* IMA Journal of Numerical Analysis 3, 41 - 63 (1983).

[5] A. Brandt. *Guide to Multigrid Development.* in Proceedings, Köln-Porz, Springer-Verlag, Berlin, Heidelberg, New York, pp. 220 - 312 (1981).

[6] J. E. Dendy. *Black Box Multigrid.* Journal of Computational Physics 48, 366 - 386 (1982).

[7] J. E. Dendy. *Black Box Multigrid for Nonsymmetric Problems.* Applied Mathematics and Computation 13, 261 - 283 (1983).

[8] J. E. Dendy. *Black Box Multigrid for Systems.* Applied Mathematics and Computation 19, 57 - 74 (1986).

[9] D. W. Peaceman. *Fundamentals of Numerical Reservoir Simulation.* Elsevier Scientific Publishing Company, Amsterdam (1977).

A standard model of generic rotational degeneracy

Heinrich Freistühler
Institut für Mathematik, RWTH Aachen
Templergraben 55, D-5100 Aachen, Fed. Rep. of Germany

Summary

Riemann's initial value problem has been studied in our paper [9] for a rather broad class of such hyperbolic systems with which rotational symmetry creates a specific kind of degeneracy: under natural genericity assumptions it is found to have a unique stable centered solution. This result, which holds locally near degenerate points in state space, applies to important examples from continuum mechanics. In sec.3 of the present report we will see that in all these cases the pattern of the degenerate transverse waves is isomorphic to that of a rather simple standard model. In sec.4 important qualitative features of the wave pattern in the general case are observed by means of simple explicit calculations on the model. Sec.1 introduces to the situation by reviewing previous results for reference and sec.2 discusses the concept of stability that we use.

1. Rotationally degenerate systems

We consider hyperbolic systems

$$\mathbf{u}_t(\xi,t) + (f(\mathbf{u}(\xi,t)))_\xi = 0, \quad (\xi,t) \in \mathbb{R} \times \mathbb{R}_+, \tag{1.1}$$

of conservation laws: the flux function $f : U \to \mathbb{R}^n$ is a smooth map defined on an open state space $U \subset \mathbb{R}^n$, and its Jacobian $Df(u)$ is $\mathbb{R}$-diagonalizable at any $u \in U$. We restrict our search for solutions to those of the Riemann problem, i.e. (1.1) together with initial data of the special form

$$\mathbf{u}(\xi,0) = \begin{cases} u_l & ,\ \xi < 0 \\ u_r & ,\ \xi > 0 \end{cases} \qquad u_l, u_r \in U. \tag{1.2}$$

Def. 1.1. For $m, k, n = m + k \in \mathbb{N}$ we decompose $u \in \mathbb{R}^n$ as $u = (x, y)$ with $x \in \mathbb{R}^m, y \in \mathbb{R}^k$ and define for any (proper or improper) rotation $O \in \mathbf{O}(m)$,the orthogonal group on $\mathbb{R}^m$, $\overline{O} \in \mathbf{O}(n)$ by $\overline{O}(x,y) = (Ox, y)$.A hyperbolic system given by $f : U \to \mathbb{R}^n$ is *rotationally symmetric (to degree m ; with respect to x)* if

$$f \circ \overline{O} = \overline{O} \circ f \quad \text{for any } O \in \mathbf{O}(m). \tag{1.3}$$

We call the set $C = \{(x,y) \in U; x = 0\}$ the *center* of f.

Clearly, (1.3) presupposes $\overline{\mathbf{O}(m)}U = U$, i.e. rotational symmetry of U, and implies that the notion of solution is invariant: With $\mathbf{u}$ any solution of (1.1), also $\overline{O} \circ \mathbf{u}$ is a solution.

For $m \geq 2$ the hyperbolicity of any such system is always non-strict at its center: (1.3) implies

$$Df \circ \overline{O}\, D\overline{O} = D\overline{O}\, Df \ , \tag{1.4}$$

especially at the center

$$Df|C = D\overline{O}\, Df|C\, D\overline{O}^{\top} \ , \tag{1.5}$$

and with respect to the modes (=(eigenvalue,eigenspace)-pairs (λ, R) of Df) this yields the rotational invariance

$$R|C = D\overline{O}\, R|C \ , \tag{1.6}$$

so that there exists at least one eigenvalue λ of multiplicity m at C. Generically (in a geometric sense; gasdynamics with its additional Galilean invariance is a prominent counterexample), λ splits into two different eigenvalues (branches over U): a simple one and another one that is $(m-1)$-fold and linearly degenerate. The latter corresponds to rotations in state space: its eigenspace bundle R has $(m-1)$-spheres as integral manifolds. We are especially interested in the degenerate situation near C.

Obviously, any rotationally symmetric hyperbolic system f is of the form

$$f(x,y) = (X(x,y), Y(x,y))\text{with} \tag{1.7}$$

$$X(x,y) = \tilde{X}(|x|,y)x, Y(x,y) = \tilde{Y}(|x|,y), \tag{1.8}$$

where $\tilde{X} : I\!R \times I\!R^k \to I\!R, \tilde{Y} : I\!R \times I\!R^k \to I\!R^k$ are smooth maps and even in the first argument.

Def 1.2. To f we define the corresponding *radial system* $\hat{f}$ and the corresponding *central system* $\check{f}$ by

$$\hat{f}(\hat{x},y) = (\tilde{X}(\hat{x},y)\hat{x}, \tilde{Y}(\hat{x},y)) \text{ on } \hat{U} = \{(\hat{x},y) \in I\!R \times I\!R^k; \hat{x}S^{m-1} \times \{y\} \subset U\}, \tag{1.9}$$

$$\check{f}(y) = Y(0,y) \text{ on } \check{U} = \{y \in I\!R^k; (0,y) \in U\}. \tag{1.10}$$

We fix $u_0 \in C$ and denote by $\hat{u}_0, \check{u}_0$ the corresponding points in $\hat{U}, \check{U}$. If $\hat{f}$ is strictly hyperbolic, then $D\hat{f}$ has exactly k eigenvalues $\tilde{\lambda}$ whose restrictions $\tilde{\lambda}|\hat{C}$ to the center $\hat{C} = \{(\hat{x},y) \in \hat{U}; \hat{x} = 0\}$ of $\hat{f}$ are also eigenvalues of $D\check{f} \cong D(\hat{f}|\hat{C})$, and there is exactly one further eigenvalue $\hat{\lambda}$ whose corresponding eigenspace $\hat{R}$ is transverse to $\hat{C}$ at $\hat{C}$. For reasons of symmetry, $\hat{R}|\hat{C} = I\!R \times \{0\}$ and $\hat{\lambda}$ is even with respect to $\hat{x}$; so $(\partial/\partial\hat{x})\hat{\lambda}(\hat{u}_0) = 0$, i.e. $\hat{\lambda}$ cannot be genuinely nonlinear at $\hat{C}$. Nothing, however, generally prevents $\hat{\lambda}$ from being genuinely nonlinear outside $\hat{C}$; e.g. it is locally if $(\partial^2/\partial\hat{x}^2)\hat{\lambda}(\hat{u}_0) \neq 0$. This motivates the following

Genericity assumptions. We consider a rotationally symmetric system which fulfils

$$\hat{f} \text{ is strictly hyperbolic at } \hat{u}_0, \tag{1.11}$$

$$\hat{\lambda} \text{ fulfils } (\partial^2/\partial\hat{x}^2)\hat{\lambda}(\hat{u}_0) \neq 0, \tag{1.12}$$

$$\text{any other eigenvalue of } D\hat{f} \text{ is either g. nl. or l. dg.} \tag{1.13}$$

The following is the main result of [9].

Theorem 1.1. Let f be a hyperbolic system that is rotationally symmetric and u_0 a state in its center C. Assume (1.11),(1.12),(1.13) hold. Then locally near u_0 the Riemann problem has a unique stable solution, which depends continuously on the data. (See below for the meaning of "stable solution of a Riemann problem".)

A major motivation of our study of rotationally symmetric systems is that important systems of continuum mechanics are rotationally symmetric or have rotationally symmetric systems associated to them. We mention a general result, which has been proved in [9] as a consequence of theorem 1.1.

Theorem 1.2. Assume $f : U \to I\!R^n$ is the flux function of a hyperbolic system, and the eigenvalues of Df are all positive. Then the system

$$\begin{aligned} \mathbf{u}_t(\xi,t) - \mathbf{v}_\xi(\xi,t) = 0 \\ \mathbf{v}_t(\xi,t) - (f(\mathbf{u}(\xi,t)))_\xi = 0 \end{aligned} \tag{1.14}$$

of p.d.e. is hyperbolic.
If f is rotationally symmetric (see(1.3)) and fulfils the genericity assumptions ((1.11) to (1.13)) at a point $u_0 \in C$, then locally near (u_0, v_0) (,where $v_0 \in I\!R^n$ is arbitrary,) the Riemann problem of (1.14) has a unique stable solution, which depends continuously on the data.

Magnetohydrodynamic plane waves and elastic plane waves in isotropic bodies are governed by systems of the form (1.14). It is isotropy which induces rotational symmetry of the corresponding f in either case; so this symmetry is present in magnetohydrodynamics, and it is present in elasticity if the material is isotropic. In realistic cases assumptions (1.11) to (1.13) are satisfied for these physical systems, and in contrast to gasdynamics, the geometrically generic form of rotational degeneracy is realized. Theorem 1.2 seems to give the first rigorous results on existence and uniqueness of solutions to the Riemann problem of these systems near degenerate states. Note also, for comparison, that another well-known system of the form (1.14) with rotationally symmetric f, that of the elastic string (see [4], where its Riemann problem began to be studied) has $C = \emptyset$ and so lacks a central degeneracy.
For further details and proofs we refer the reader to [9].

2. Stable solutions of Riemann problems

In this section we define and interpret the concept of *stable solutions*. From the outset we assume centeredness and a minimum amount of regularity:

Def. 2.1. A solution (of the Riemann problem (1.1),(1.2)) is a piecewise smooth function $\mathbf{u} \in \mathcal{L}^1_{\text{loc}}(I\!R, U)$ that solves weakly the equation

$$-s\mathbf{u}'(s) + (f \circ \mathbf{u})'(s) = 0, \quad s \in I\!R \tag{2.1}$$

and has

$$\lim_{s\to-\infty} \mathbf{u}(s) = u_l, \; \lim_{s\to\infty} \mathbf{u}(s) = u_r. \tag{2.2}$$

Equip the set of all solutions with the $\mathcal{L}^1_{\text{loc}}$-topology.

A solution may contain discontinuities where $u^- = u(s-)$ and $u^+ = u(s+)$ disagree.

Def. 2.2. A discontinuity is said to be *linearly stable* if it fulfils the Rankine-Hugoniot conditions

$$f(u^+) - f(u^-) = s(u^+ - u^-) \tag{2.3}$$

and, with

$$R^-(u,s) = \sum_{\lambda<s} ker(Df(u) - \lambda), R^+(u,s) = \sum_{\lambda>s} ker(Df(u) - \lambda), \tag{2.4}$$

also

$$R^+(u^-,s) + R^-(u^+,s) \subset R^-(u^-,s) \oplus R^+(u^+,s) \oplus \mathbb{R}(u^+ - u^-). \tag{LS}$$

Def. 2.3. A solution of a Riemann problem is called *linearly stable* if all its discontinuities are; it is called a *stable solution* if it is the $\mathcal{L}^1_{\text{loc}}$-limit of linearly stable solutions.

In order to motivate condition (LS) we look at the solution

$$\mathbf{u}_0(\xi,t) = \begin{cases} u^- \ , & \xi < st \\ u^+ \ , & \xi > st \ , \end{cases} \tag{2.5}$$

of (1.1.),(1.2) corresponding to u^-, u^+ which fulfil the RH conditions. The next arguments follow [5], pp. 25-27. We assume that to slightly perturbed initial data ${}^0\mathbf{u}$ there is a solution $\mathbf{u}$ of similar structure as $\mathbf{u}_0$: smooth outside a smooth curve, along which the RH conditions are satisfied:

$$\begin{aligned} \mathbf{u}_t(\xi,t) + (f(\mathbf{u}(\xi,t)))_\xi = 0 \ &, \ \xi \neq \zeta(t) \ , \\ -\zeta'(t)(\mathbf{u}^r(t) - \mathbf{u}^l(t)) + (f(\mathbf{u}^r(t)) - f(\mathbf{u}^l(t))) = 0 \ & \qquad (\xi,t) \in \mathbb{R} \times \mathbb{R}_+ \\ \mathbf{u}(\xi,0) = {}^0\mathbf{u}(\xi) \ &, \ \zeta(0) = 0 \end{aligned} \tag{2.6}$$

where $\mathbf{u}^{r/l}(t) = \mathbf{u}(\zeta(t) \pm 0, t)$; we further assume that $\mathbf{u}$ depends smoothly on ${}^0\mathbf{u}$ in the way that there are families $\mathbf{u}_\varepsilon$ depending smoothly on a real parameter ε pointing in appropriately arbitrary "directions" ${}^0\mathbf{w} = (\frac{\partial}{\partial\varepsilon}({}^0\mathbf{u}_\varepsilon))_{\varepsilon=0}$. If ζ_ε parametrize the corresponding discontinuity curves, the transformation

$$\bar{\mathbf{u}}_\varepsilon(\xi,t) = \mathbf{u}_\varepsilon(\xi - \zeta_\varepsilon(t), t), \mathbf{w} = (\frac{\partial}{\partial\varepsilon}\bar{\mathbf{u}}_\varepsilon)_{\varepsilon=0}, \sigma = (\frac{\partial}{\partial\varepsilon}\zeta_\varepsilon)_{\varepsilon=0} \tag{2.7}$$

yields the linearized problem

$$\mathbf{w}_t(\xi,t) + (A(u^\pm) - sI)\mathbf{w}_\xi(\xi,t) = 0 \ , \ \pm\xi > 0 \tag{2.8.1}$$

$$-\sigma'(t)(u^+ - u^-) + (A(u^+) - sI)\mathbf{w}(0+,t) - (A(u^-) - sI)\mathbf{w}(0-,t) = 0 \tag{2.8.2}$$

$$\mathbf{w}(\xi,0) = {}^0\mathbf{w}(\xi) \ , \ \sigma(0) = 0. \tag{2.8.3}$$

Let C^∞_{00} consist of all functions $\in C^\infty(\mathbb{R}, \mathbb{R}^n)$ whose support is compact and does not contain 0.

Lemma 2.1. The linearized problem has a unique solution for any ${}^0\mathbf{w} \in C^\infty_{00}$ if and only if (LS) holds.

Proof. Decompose any solution

$$\mathbf{w}(\xi,t) = \sum_{k=1}^{n} \mathbf{w}_k^{\pm}(\xi,t) r_k^{\pm} \;,\; \pm\xi > 0, \tag{2.9}$$

where $\{r_k^-\}, \{r_k^+\}$ are complete sets of eigenvectors at the left and at the right hand state, respectively:

$$Df(u^{\pm}) r_k^{\pm} = \lambda_k^{\pm} r_k^{\pm} \;; \tag{2.10}$$

(2.8.1) decomposes into the characteristic equations

$$\left(\frac{\partial}{\partial t} + (\lambda_k^{\pm} - s)\frac{\partial}{\partial \xi}\right)\mathbf{w}_k^{\pm}(\xi,t) = 0 \;,\; \pm\xi > 0. \tag{2.11}$$

Now (2.8.2) is the linear algebraic system

$$-\sigma'(t)(u^+ - u^-) + \sum_{k=1}^{n} \mathbf{w}_k^+(0,t)(\lambda_k^+ - s) r_k^+ - \sum_{k=1}^{n} \mathbf{w}_k^-(0,t)(\lambda_k^- - s) r_k^- = 0 \tag{2.12}$$

of equations between the quantities $-\sigma'(t), \mathbf{w}_k^{\pm}(0,t)$. Depending on whether the corresponding characteristics impinge or do not impinge on the boundary $\xi = 0$, the $\mathbf{w}_k^{\pm}(0,t)$ are determined by the initial values (2.8.3) or should be, as should be $-\sigma'(t)$, by (2.12). The wellposedness of (2.12) in this sense is however equivalent to (LS).

3. The standard model

For any $m \in \mathbb{N}$ the system

$$\mathbf{u}_t(\xi,t) + (f_m(\mathbf{u}(\xi,t)))_\xi = 0 \;,\; (\xi,t) \in \mathbb{R} \times \mathbb{R}_+, \tag{3.1}$$

of p.d.e. with the flux function

$$f_m : \mathbb{R}^m \to \mathbb{R}^m \;,\; f_m(u) = |u|^2 u \tag{3.2}$$

is hyperbolic: the Jacobian

$$Df_m(u) = |u|^2 I_m + u u^\top \tag{3.3}$$

has a *radial* mode (λ^r, R^r):

$$\lambda^r(u) = 3|u|^2 \;,\; R^r(u) = \mathbb{R}u \tag{3.4}$$

and,for $m \geq 2$, an *azimuthal* mode (λ^a, R^a):

$$\lambda^a(u) = |u|^2 \;,\; R^a(u) = u^\perp. \tag{3.5}$$

It is rotationally symmetric to (maximal) degree m: def. 1.1 applies with $k = 0$ and

$$f \circ O = O \circ f \quad , \quad O \in \mathbf{O}(m). \tag{3.6}$$

It fulfils the genericity assumptions (1.11) to (1.13): since the corresponding radial system has the flux function

$$\hat{f} = f_1 : I\!R \to I\!R \quad , \quad \hat{f}(\hat{x}) = \hat{x}^3 \; , \tag{3.7}$$

(1.11) and (1.13) are trivial; (1.12) follows from

$$\hat{\lambda}(\hat{x}) = \hat{f}'(\hat{x}) = 3\hat{x}^2 \; . \tag{3.8}$$

The center of this system in the sense of def. 1.1 is the origin $\{0\} \subset I\!R^m$.

Certainly we do not need theorem 1.1 and the rather complicated arguments used in its proof in order to treat the Riemann problem of this simple system. The aim of this paper is to point out important qualitative features of the degenerate part of the wave pattern in the general case, that is the pattern of *transverse* waves: In the general situation defined by the genericity assumptions (1.11) to (1.13), there is an interval I_0, a neighborhood of $\hat{\lambda}(\hat{u}_0)$, such that elementary waves (RH discontinuities as well as simple waves) moving at speeds $s \in I_0$ oscillate nearly orthogonal to C ("transverse"), whereas waves of speeds $s \notin I_0$ oscillate nearly parallel to C ("almost central"). The almost central waves can be considered as perturbations of waves of $\check{f}$, and so their pattern is the usual one of a system with eigenvalues either g.nl. or l.dg. (see [1]). Different phenomena appear with the transverse waves, and it is to observe these that we look at the above simple system. It is a fairly true model of the general case in the following precise sense:

Lemma 3.1. For an arbitrary hyperbolic system with a generic rotational degeneracy, project (locally near C) all pairs of states (u^-, u^+) that can be joined by transverse waves on their x-component (the component with respect to which the flux function is symmetric) to get pairs (x^-, x^+) and interpret x^-, x^+ as points in the state space of the standard model f_m of same degree m of symmetry. This yields a true picture in the sense that these (x^-, x^+) also correspond to (locally all) elementary waves to f_m, which moreover are "of the same kind": the speeds of these waves are in the same order with respect to the characteristic speeds of the model as their preimages are with respect to the transverse characteristic speeds of the original system. Also (in)stability of discontinuities is carried over properly.

On purpose we avoid formulae here and will not give a technical proof of this lemma. A look at the parametrization of transverse waves introduced in the proof of theorem 1.1 in [9] brings the above statement immediately to the (inner mind's) eye, since this projection is the key idea of that proof.
Also by the isomorphisms $J^{\pm}$ (constructed in [9]) between solutions (of positive or negative speed) of (1.14) and solutions of the original system with flux function f, it is clear at once that the wave pattern of the standard model produces a true picture also of that of any system of form (1.14) with rotationally symmetric f; note that this applies to the said physical examples !

We begin our study of the standard model by surveying its elementary waves:

Property 3.1. To any point $u^- \in I\!R^m \backslash \{0\}$ the following (stable or unstable) elementary waves are possible:
(i) u^- can be joined by a simple wave to any state $u^+ = \mu u^-$ with $\mu > 1$;
(ii) u^- can be joined by a RH discontinuity to any state u^+ with $|u^+| = |u^-|$;
(iii) u^- can be joined by a RH discontinuity to any state $u^+ = \mu u^-$ with $\mu \in I\!R$.

The simple waves (i) belong to the fast mode λ^r, the discontinuities (ii) are contact discontinuities of the slow mode λ^a (for $m \geq 2$): there will be no doubt that these waves should be admitted. The discontinuities (iii), just all other solutions of the RH conditions

$$(|u^-|^2 - s)u^- = (|u^+|^2 - s)u^+ \ , \tag{3.9}$$

cannot be generally assigned to any of the both modes and must be checked for admissibility, which clearly depends on the criterion one applies. Before we compare different criteria, we distinguish a special solution.

Def 3.1. The *standard solution* to the Riemann problem of the standard model is constructed as follows:
If $m = 1$ (,which yields the cubic standard example of a nonconvex scalar law), take the usual solution by one simple wave, one shock or one mixed wave (:the type introduced by Liu in [3] consisting of a shock and an adjacent simple wave).
If $m \geq 2$, proceed like this: To given $u_l, u_r \in I\!R^m \setminus \{0\}$ find a unique intermediate state u_m as that point in state space where the sphere $|u_l|S^{m-1}$ and the ray $I\!R_+ u_r$ intersect. u_l can be joined to u_m by a slow contact discontinuity, and unless already $u_m = u_r$, u_m can be joined to u_r by a shock or a fast simple wave, depending on whether $0 < \mu < 1$ or $1 < \mu$ in $u_m = \mu u_r$. If $u_l = 0$ or $u_r = 0$ there is a unique fast simple wave or shock connecting the both initial states.

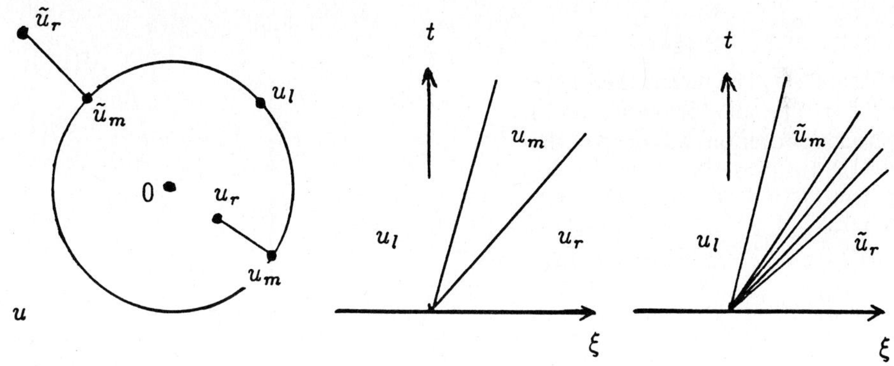

It is easy to see that this procedure leads to a unique solution and that admitting more discontinuities than those used in it leads to nonuniqueness. Except for a certain geometric intuition, however, up to now the procedure lacks any justification.

4. The wave pattern

We discuss the wave pattern of the standard model; this can be done by very simple calculations. By lemma 3.1 we are sure that interesting phenomena we observe in this way also appear with any other system exhibiting a generic rotational degeneracy. We compare different *admissibility criteria*, discuss the question of *embedding* systems into each other and that of the so-called *anomalous shocks*, and point out a *continuous version of lability* of intermediate states in solutions.

Property 4.1. A discontinuity with left hand state $u^- \neq 0$ and right hand state $u^+ = \mu u^-$ is admissible
(a) in the sense of Lax's shock inequalities (:"SI"):
if $m = 1$: exactly for $\mu \in (-\frac{1}{2}, 1)$,
if $m = 2$: exactly for $\mu \in (-1, -\frac{1}{2}) \cup (0, 1)$,
if $m > 2$: exactly for $\mu \in (0, 1)$;
(b) in the sense of Liu's condition (E): (for any m:) exactly for $\mu \in [-\frac{1}{2}, 1)$;
(c) in the sense of criterion (LS) of linear stability:
if $m = 1$: exactly for $\mu \in (-\frac{1}{2}, 1)$,
if $m \geq 2$: exactly for $\mu \in (0, 1)$.

Proof. The shock speed

$$s(\mu) = |u^-|^2(1 + \mu + \mu^2) \tag{4.1}$$

must be compared with the characteristic speeds $\lambda^r(u^-) = 3|u^-|^2, \lambda^r(u^+) = 3\mu^2|u^-|^2$, and, for $m \geq 2$, $\lambda^a(u^-) = |u^-|^2, \lambda^a(u^+) = \mu^2|u^-|^2$.

If $m = 1$, the (SI) require $\lambda^r(u^-) > s > \lambda^r(u^+)$; if $m \geq 2$, they mean

$$\begin{array}{ll} \text{either} & \lambda^a(u^-) < s \ \text{ and } \ \lambda^r(u^-) > s > \lambda^r(u^+) \\ \text{or} & \qquad\qquad \lambda^a(u^-) > s > \lambda^a(u^+) \ \text{ and } \ s < \lambda^r(u^+). \end{array} \tag{4.2}$$

Liu's condition (E) means (here) that $s(\mu) \leq s(\tilde{\mu})$ for all $\tilde{\mu}$ between μ and 1. In checking criterion (LS) observe that $R^\pm = R^\pm(u, s)$ are for

$s < \lambda^a(u)$: $R^- = \emptyset, R^+ = I\!R^m$,
$s = \lambda^a(u)$: $R^- = \emptyset, R^+ = I\!Ru$,
$\lambda^a(u) < s < \lambda^r(u)$: $R^- = u^\perp, R^+ = I\!Ru$,
$s = \lambda^r(u)$: $R^- = u^\perp, R^+ = \emptyset$,
$s > \lambda^r(u)$: $R^- = I\!R^m, R^+ = \emptyset$.

In the following comparison of (SI), (E), (LS) with regard to the question of uniqueness of solutions, we consider also limiting cases as admissible, since certainly the set of admissible solutions should be $\mathcal{L}^1_{\text{loc}}$-closed.

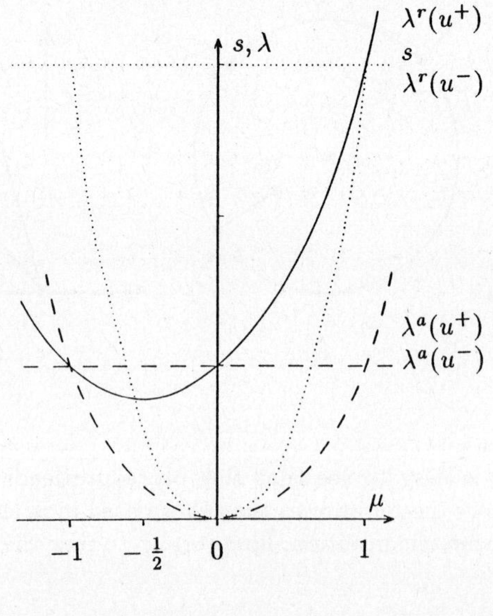

Property 4.2. If $m = 1$, (SI),(E),(LS) effect the same selection of admissible discontinuities; it leads to the unique standard solution (see def. 3.1).
If $m = 2$, (LS) is the most restrictive criterion and leads to the unique standard solution; (E) as well as (SI) admit more discontinuities, each in a different way, but both leading to nonuniqueness.
If $m > 2$, (LS) and (SI) lead to the same selection of discontinuities and to the unique standard solution; (E) admits more discontinuities and thus "causes" nonuniqueness.

This follows from property 4.1. The difference between (LS) and (SI) in the case $m = 2$ is worth an extra notice:

Property 4.3. With degree $m = 2$ of symmetry there are shocks that correspond to the linearly degenerate (!) slow mode in the sense that (SI) are fulfilled with the characteristics of this mode impinging on the shock on both sides. These shocks are, however, not linearly stable. (In the above notation, these shocks are given by values $\mu \in (-1, -\frac{1}{2})$.)

An old and famous example are the magnetohydrodynamic so-called intermediate shocks (more precisely: one species of them), about whose stability there was an interesting discussion about thirty years ago (, see [2]). In 1980, Keyfitz and Kranzer encountered such "anomalous entropy shocks" in their well-known treatment [4] of the Riemann problem of the elastic string.

A comment might be in order about what we do not intend to say here. Lax and Liu designed their conditions for situations with separate eigenvalues, in which they are well-known to be very fruitful and have good justifications. In the situation that Lax originally considered in his famous paper [1], the local Riemann problem for a strictly hyperbolic system with each eigenvalue either g.nl. or l.dg., the (SI) are equivalent to linear stability e.g. in the sense of (LS). On the other hand linear stability as an admissibility criterion does not lead to uniqueness even in the simple case of a nonconvex scalar law with several inflection points, which is treated perfectly well by condition (E). Finally, in our situation, (E) and (SI) together are equivalent to (LS) for any m.

We now turn to the question of embedding systems into each other. For the standard model of degree m any linear subspace $L \subset I\!R^m$ of dimension $\tilde{m} < m$ is an invariant submanifold in the sense of [6]. The restriction $f_m|L$ has just $f_{\tilde{m}}$ as a coordinate representation, so the standard model of degree $\tilde{m}$ might be called a subsystem of that of degree m. This must, however, be handled with care:

Property 4.4. The criterion (LS) may make a different choice when applied to a subsystem as when used with the system itself. Here this is the case for the subsystem given by $\hat{f} = f_1$ of the system given by $f_m, m \geq 2$: whereas (LS) allows discontinuities across the origin and, as a consequence, Liu type mixed waves for f_1, the same discontinuities and waves are forbidden as such of $f_m, m \geq 2$.

This is an analogue on the level of dependent variables of the known fact that stability of a multidimensional planar shock front with respect to perturbations that are one-dimensional (in the independent space variable) is rather different from stability with respect to perturbations that are themselves multidimensional.

Finally we look at a phenomenon of structural lability.

Property 4.5. The standard solution depends continuously on the data. This is, however, not true for the intermediate state u_m: there are initial value pairs (u_l, u_r) such that arbitrary small changes of them draw u_m near any point of a whole $(m-1)$-sphere.

Continuous dependence of the solution as well as of the intermediate state on the initial data is of course no question as long as $u_r \neq 0$, since the speeds of the elementary waves and also the state u_m are continuous functions of (u_l, u_r), then. In the neighborhood of the origin, small changes of u_r can go hand in hand with large changes in its polar angle, which determines the polar angle of u_m; so u_m may jump anywhere near the sphere of radius $|u_l|$ around the origin. In the limit case $u_r = 0$ the intermediate state is undetermined: any u_m on the said sphere can be reached from u_l by a contact discontinuity; since, however, in this case the "shock" joining u_m to u_r has speed equal to that of the contact discontinuity, u_m will not appear in the solution. If $u_r \to 0$, the width of the sector in (ξ, t)-space (the length of the s-interval) on which $\mathbf{u} = u_m$ shrinks to zero, so that continuous dependence of the solution is maintained.
Similar things were found in [4],[7]: with systems considered in these papers intermediate states may jump near neighborhoods of usually two discrete points in state space. In distinguishing by the type of the manifold near whose arbitrary points the state may jump, one might call this a *discrete version* and the phenomenon described above a *continuous version* of lability of intermediate states.

References

[1] P. Lax: *Hyperbolic systems of conservation laws II*, Comm. Pure Appl. Math. 10 (1957), 537-566

[2] P. Germain: *Contribution à la théorie des ondes de choc en magnétodynamique des fluides*, O.N.E.R.A. Publ. No 97 (1959)

[3] T.-P. Liu: *The Riemann problem for general systems of conservation laws*, J. Diff. Eqs. 18 (1975), 218-234

[4] B. Keyfitz, H. Kranzer: *A system of non-strictly hyperbolic conservation laws arising in elasticity theory*, Arch. Rat. Mech. Anal. 72 (1980), 219-241

[5] A. Majda: *Stability of multidimensional shockfronts*, AMS Mem. No.275 (1983)

[6] B. Temple: *Systems of conservation laws with invariant submanifolds*, Trans. AMS 280 (1983), 781-795

[7] E. Isaacson, D. Marchesin, B. Plohr, B. Temple: *The classification of quadratic Riemann problems (I)*, Madison Research Center Rep. No. 2891 (1986)

[8] M. Brio, C.C. Wu: *Characteristic fields for the equations of magnetohydrodynamics*, in: Non-strictly hyperbolic conservation laws, Proc. of an AMS special session, held January 9-10,1985, Contemporary Mathematics, vol. 60 (1987), 19-23

[9] H. Freistühler: *Rotational degeneracy of hyperbolic systems of conservation laws* (1987), to appear

A NUMERICAL METHOD FOR A SYSTEM OF EQUATIONS MODELLING ONE-DIMENSIONAL THREE-PHASE FLOW IN A POROUS MEDIUM

Tore Gimse
Dept. of Mathematics,
University of Oslo,
P.O.Box 1053,
N - 0316 Oslo, Norway

SUMMARY

A numerical method for Riemann problems for a class of equations (system of conservation laws) is presented. Stability and convergence in a case of three-phase flow in porous media is shown and the application to general Cauchy problems is discussed.

INTRODUCTION

Based upon ideas presented by Dafermos [1], a numerical method for one-dimensional, scalar conservation laws :

$$u_t + f(u)_x = 0 \qquad \text{with} \quad u(x,0) = u_0(x) \tag{1}$$

was developed by H.Holden, L.Holden and Høegh-Krohn [2]. The algorithm is tracing envelopes of the flow-function f. By approximating f by a piecewise linear function and $u_0(x)$ by a piecewise constant function one obtains a solution consisting of shocks only, finitely many at any time and a finite number of shock collisions as $t \to \infty$. Hence, this method is different from the usual methods of finite differences, first presented by Lax [3]. If f actually is piecewise linear and u_0 is piecewise constant, the solution is exact, else one has good error estimates (e.g. Lucier [4]). Existence and uniqueness for (1) are well known (e.g. Oleinik [5]). For a 2x2 system with some restrictions of f, Isaacson and Temple [6] have shown uniqueness of a weak solution for a global problem. As an approach to a system of conservation laws L.Holden and Høegh-Krohn have studied Riemann problems for a specific class of equations :

$$u_{i,t} + f_i(u_1,..u_i)_x = 0 \qquad i = 1,...,N$$

$$u_0(x) = u(x,0) = \begin{matrix} u_- & \text{if} \quad x < 0 \\ u_+ & \text{if} \quad x > 0. \end{matrix} \tag{2}$$

The results are presented in a preprint [7] in which their proofs suggest a numerical method for the problem (2).

THE NUMERICAL METHOD

The algorithm works inductively, so assume that

$$u_t + f(u)_x = 0 \quad , \quad u(x,0) = \begin{matrix} u_- & \text{if} \ x < 0 \\ u_+ & \text{if} \ x > 0 \end{matrix} \tag{3}$$

is solved. u may be either a vector or a scalar. Different kinds of assumptions on f may be made, see [7] for more details, here we assume that f is continuous and piecewise linear. Let $u_1 = u_-$, u_2, $u_3, \dots, u_n = u_+$ be the constant states that make up the solution and $s_1, s_2, \dots, s_{n-1}$ the corresponding shock speeds.

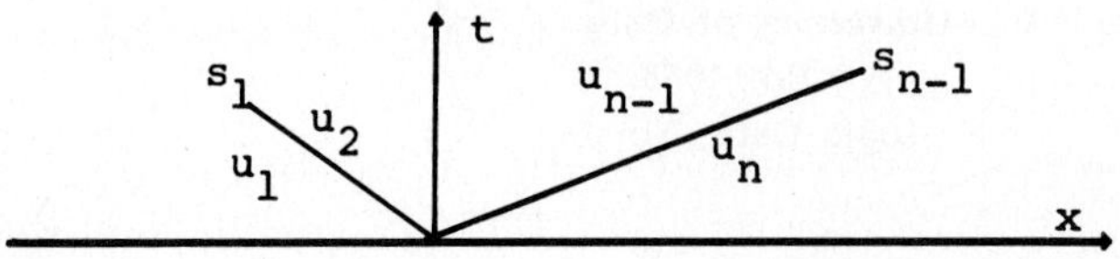

Fig. 1 The solution of : $u_t + f(u)_x = 0$.

Consider the next equation :

$$v_t + g(u,v)_x = 0 \quad , \quad v(x,0) = \begin{matrix} v_- & \text{if } x < 0 \\ v_+ & \text{if } x > 0 \end{matrix} \tag{4}$$

v is a scalar variable. We know that passing from $g(u_i, \ )$ to $g(u_{i+1}, \ )$, that is, passing from an area in the x-t plane of $u = u_i$ to an area of $u = u_{i+1}$, we have a shock of speed s_i. On the other hand, within each area u is constant, so there we have a scalar problem with a restriction of the permitted shock speeds : $s_{i-1} < s < s_i$. Hence, the sequence of u-values induces a sequence of g-functions to be considered; let g_i denote $g(u_i, \ )$. To help us explicitely constructing the solution we define two kinds of sets: $H_{i,in}$ is the set of v-values where we may land after having made a jump from g_{i+1} to g_i, and $H_{i+1,out}$ is the set of points from where this jump may originate. Hence, $H_{i,in}$ and $H_{i+1,out}$ are the permitted values to the left and to the right (respectively) of the u_i/u_{i+1} shock. We start out on the function g_1 at the point $v=v_-$. We then find the set of points on g_1 from where we may jump to $g_1(v_-)$ with speed less than or equal to s_1. By a jump we mean to find a path along the upper/lower convex envelope [2]. These are the points that we may invoke where $u=u_1$, and so make up $H_{1,in}$. Next we consider g_2 and find the points upon it from where we may jump with speed s_1 and land on g_1 at a point of $H_{1,in}$. This points make up $H_{2,out}$. $H_{2,out}$ is contained in $H_{2,in}$ (if we may jump from a point, we may of course come there first). In addition we know that we may move along g_2 with speed between s_1 and s_2. Therefore we have to include in $H_{2,in}$ the points of g_2 from where we may jump to $H_{2,out}$ with such speeds. From $H_{2,in}$ we now repeat the process for g_3 as we did with g_2 from $H_{1,in}$. The process is repeated until $H_{n,out}$ is constructed. We are now prepared to trace the solution, and start out in the point $g_n(v_+)$. If this point is in $H_{n,out}$ we jump across to g_{n-1}. Otherwise we first have to jump along g_n with decreasing speeds larger than s_{n-1} until we reach a point of $H_{n,out}$. Then pass to g_{n-1} (where we know we land in $H_{n-1,in}$), from where we may have to jump into $H_{n-1,out}$ before passing to g_{n-2} etc. In this way we construct our solution path all way down to $g_1(v_-)$. As shown by L.Holden and Høegh-Krohn [7], the solution exists, but is not generally unique. However, the set of initial values where we do not have uniqueness is finite, and in the case of two equations there is uniqueness. That will be the kind of system we will examine closer, a system of equations modelling a case of three-phase flow in a porous medium.

FLOW EQUATIONS

We write the equations :

$$\begin{aligned} u_t + f(u)_x &= 0 \\ v_t + g(u,v)_x &= 0 \end{aligned} \qquad (5)$$

and the initial states (to the left and right of x=0 respectively) (u_-,v_-) and (u_+,v_+). Interpretated physically u denotes gas- and v is oil-saturation. The saturation of water $w = 1 - u - v$. The equations describe a system where gas-flow is independent of whether it takes place in oil or water environment, whereas the oil-flow is sensible to the amount of both water and gas present. We have approximated f with a piecewise linear function. (Usually f is determined experimentally, and so it is piecewise linear in most applications.) We will denote the approximation f. Furthermore we assume that both f( ) and g(u,) are strictly increasing, continuous functions with at most one point of inflection. We also assume $f(0) = 0$ and $g(u,0) = 0$ (no substance gives no flow) and that $g_u < 0$ (the more gas present, the less relative amount of the flow is oil flowing). The physical situation implies that g is not defined for negative arguments nor for arguments so that $u+v > 1$.

EXISTENCE

We first state that the solution exists, that is, we can always find the H-sets and the solution always remains within the phase-space $0 \le u + v \le 1$.

Lemma 1.
The slope of the line connecting two g-functions at their endpoints $g_i(1-u_i)$ and $g_{i+1}(1-u_{i+1})$ equals s_i, the u_i/u_{i+1} shock-speed.
Proof:
The slope of the line is

$$s = \frac{g(u_{i+1},\ 1-u_{i+1}) - g(u_i,\ 1-u_i)}{(1-u_{i+1}) - (1-u_i)} .$$

Now, if h is the fractional flow function of water, we always have $f(u) + g(u,v) + h(u,v,w) = 1$ (all that flows is u, v and w). If $v = 1 - u$, $w = 0$, and so $h = 0$. Hence $g(u, 1-u) = 1 - f(u)$, and :

$$s = \frac{(1 - f(u_{i+1})) - (1 - f(u_i))}{(1- u_{i+1}) - (1 - u_i)} = \frac{f(u_{i+1}) - f(u_i)}{u_{i+1} - u_i} = s_i .$$

This lemma guarantees we do not pass out of the phase-space at $u+v=1$. It also implies that if $v=1-u_i \in H_i$ (in or out), then $1-u_j \in$

H_j for all j. The same is true for v=0, since all g-functions coincide here. To show that the H-sets are non-empty, observe that $v_- \in H_{1,in}$ (by construction). If $H_{1,in} = \{v_-\}$ then (by Lemma 1) the line through v_- with slope s_1 will cut g_2. This cutting point will be in $H_{2,out}$. Then, by induction no H is empty. On the other hand, if $H_{1,in}$ consists of more than one point, at least one of the two points 0 and $1-u_1$ is in it, and again no H is empty. Hence :

Theorem 1.
The solution of the Riemann problem (5) exists and is well-defined inside the phase-space $0 \leq u + v \leq 1$.

THE H–SETS

For the S-shaped functions that we will be interested in, there are basically three kinds of H-sets.

1) A cutting H-set. A H-set including one point of g where g_v is greater than s (fig.3a).
2) An upper-touching H . A H-set including all v > v' where $g(u,v')_v = s$ (fig.3b).
3) A lower-touching H . As for 2), but for v smaller than some v' (fig.3c).

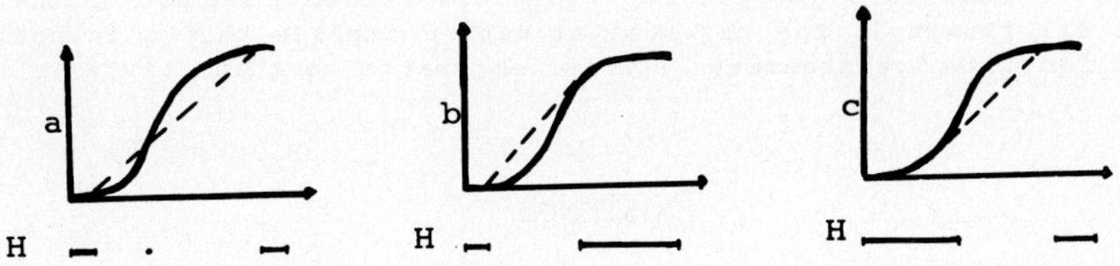

Fig. 2 The three kinds of H-sets.

If $H_i = (0,1-u_i)$ we may name it both upper- and lower-touching. (These names are motivated by the properties of the so called h-functions of [7], our H-sets are the intervals where h = g.) We define the points that determine the H-s in the following way:
A H-set consists of at most three parts, two intervals and possibly one single point, for each index "i,xx" we call the right point of the left part $v_{i,xx,l}$, the middle point (cutting point) $v_{i,xx,c}$ and the lefthand point of the right part $v_{i,xx,r}$. The v_l or v_r ("i,xx" is omitted when no confusion is possible) is called the touching point if H is touching and $g_v(v_l)$ or $g_v(v_r)$ equals the corresponding s. Denoting the u-solution sequence $u_1 = u_-$, u_2, u_3, . . , $u_n = u_+$, we order the H-sets :

$$H_{1,in}\ ,\ H_{2,out}\ ,\ H_{2,in}\ ,\ H_{3,out}\ ,\ \ldots\ ,\ H_{n,out}.$$

With respect to this order we have the following useful property :

Lemma 2.
Except for the first H-set we may divide the sequence into two parts (possibly one is empty). The first part consists of cutting, the latter part consists of only upper or lower-touching sets.

Proof :

Assume that $u_1 < u_2 < . . . < u_n$. Then $g_i > g_{i+1}$. Assume that H_i is upper-toucing. We will prove that the next H is upper touching. If H_i is an "out" set, we construct $H_{i,in}$ by adding the interval (v', $v_{i,out,r}$) where $g_i(v')_v = s_i$. (Or $v' = 0$ if the slope of g_i is always smaller.) In addition we include the interval ($v_{i,out,l}$, v'') where v'' is the point where the line through v' with slope s_i cuts g_i. Then, $H_{i,in}$ is upper touching. If H_i is an "in" set, $H_{i+1,out}$ is constructed by tracing g_{i+1} from the right until $g_{i+1,v} = s_i$ (touching point), then including the part to the left of the point where the line through this point with slope s_i cuts g_{i+1}. So $H_{i+1,out}$ will be upper touching also in this case. A lower touching or a cutting H may induce a cutting or an upper touching H in the next step. The case of a decreasing u-sequence is treated symmetrically, upper should be substituted with lower, right with left and vice versa. #

By simple use of Lemma 2 we find the following monotonity property of the solution (see Gimse [8]).

Theorem 2 :

There is a value s_0, so that both : $v(s)$, for $s < s_0$ and $v(s)$, for $s > s_0$ are monotone functions. #

The following lemma determines how H_{in} is related to H_{out} :

Lemma 3:

If the points are defined :

$v_{i,in,l} > v_{i,out,l}$ and $v_{i,in,r} < v_{i,out,r}$.

Proof:

Since we have a finite number of u-shocks, there is only a finite number of shock speeds to consider. Hence, there is some minimum difference $\delta s_{min} = \min(s_i - s_{i-1}) > 0$. Consider the following figure

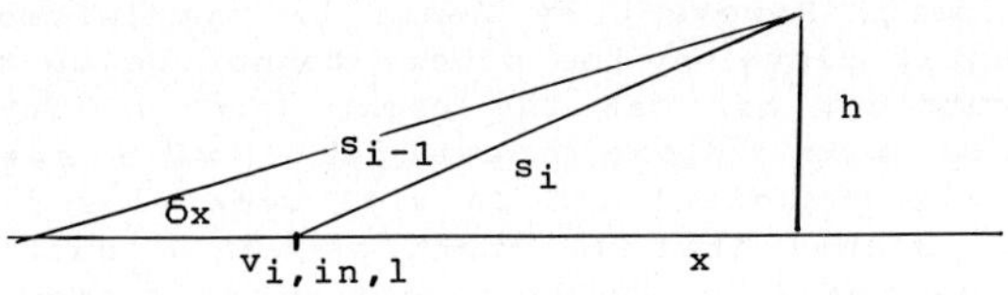

Here $s_i = h/_x$ and $s_{i-1} = h/_{(x+\delta x)}$ (the H-sets are assumed cutting, if one (or both) are touching , the upper line is higher above, and so δx will be even greater). Thereby $\delta s_{min} \le$ $(h\, \delta x / x(x+\delta x)) \le (h\, \delta x / x^2)$, which gives: $\delta x \ge \delta s_{min}\, x^2/h$. Since g_i is increasing, $\delta x \le v_{i,in,l} - v_{i,out,l}$. The argument for the upper part is similar, if H_i is upper touching, $vi_{,in,r}$ is the point where $g_v = s_i$ and where $g_v = s_{i-1}$, $v_{i,out,r}$ is greater . #

Before investigating continuity and stability properties we make the following observations concerning jumps between the H-sets. (We assume that the u-sequence is increasing, the case of a decreasing sequence is treated symmetrically.)

1) The points of the right part of $H_{i,out}$ is mapped continuously onto the right part of $H_{i-1,in}$ (say (v'', $1-u_{i-1}$)).
2) Either : The middle point of $H_{i,out}$ is mapped into the middle point of $H_{i-1,in}$.

Or : There is an interval $(v', v_{i,out,l})$ that is mapped continously onto the interval $(v_{i-1,in,r}\ ,\ v'')$

3) The rest of $H_{i,out}$ is mapped continuously onto $(0, v_{i,in,l})$.
4) The rightmost part of the left and the leftmost part of the right part of $H_{i,in}$ is mapped into the middle or touching point of $H_{i,out}$.

CONTINUITY AND STABILITY

With respect to v_+ and v_-.

Consider two values v_+ and v_+'. We will demonstrate that if v_+' is close to v_+, the solution paths are close (L_1-close). Assume the last H_{out} is upper-touching. (The case of H_{out} lower-touching (decreasing u-sequence) is treated symmetrically.) If both v_+ and v_+' are outside H_{out}, we have to jump into the touching point, and so their paths coincide from there. Next, if v_+ is on H_{out}, but v_+' is not, they move closer if v_+ is in the upper part, while they are separated if v_+ is in the lower part. In the latter case, v_+' passes to the touching point. This situation is however equivialent to the case of v_+' in the rightmost point of the lower part of H_{out}. The only difference in the solution path is the jump up to the touching point and down. Observe that the speed of these two jumps are close, and that the difference tends to zero as $v_+' - v_+$. It remains to consider the case when both v_+ and v_+' is on H_{out} (or have come there by jumping as above). If the two points pass to the same part of the next H_{in}, (and if g_{vv} is not zero in some interval), it is obvious that the mapping is continuous with respect to the distance between the v-values (the observations above). Then assume the two points do not pass to the same part of H_{in}. The leftmost point then end in the lower part of H_{in}, while the right point goes to the upper part. However, by Lemma 2, the leftmost point, if it was sufficiently close to the other, cannot be on the next H_{out}. (If H_{out} is cutting, nor can the right point.) Thus, the leftmost point have to pass to the upper part (with speed between the incoming and the outgoing) and so will come closer to the right point. Finally, assume that we start out on a cutting H_{out}, If both v_+' and v_+ are on H_{out}, we pass continuously over as above. If none of them are, both jump into the cutting point, from where the paths are identical. If one is and the other is not part of H_{out}, the latter jumps into the cutting point, from where it continues back to v_-, while the other passes over, but (by Lemma 3) it will not land in a point of the next H_{out}. Therefore, in the next step this point also passes into the secuence of cutting points. Hence, the solutions, as curves in phase-space, are close in the two cases ; Single shocks are not necessarily stable, but rarefaction waves are.

Remark : If g is approximated by piecewise linear functions, one may have an interval of slope equal to some s_i. Thereby the jumps between the g's do not map the distance from v' continuously in a small neighbourhood of v'. However, as the approximation is done finer, the discontinuities tend to zero.

In fig.3 we have illustrated the construction of the solution in a simple case of three different u-values ($u_1 < u_2 < u_3$).

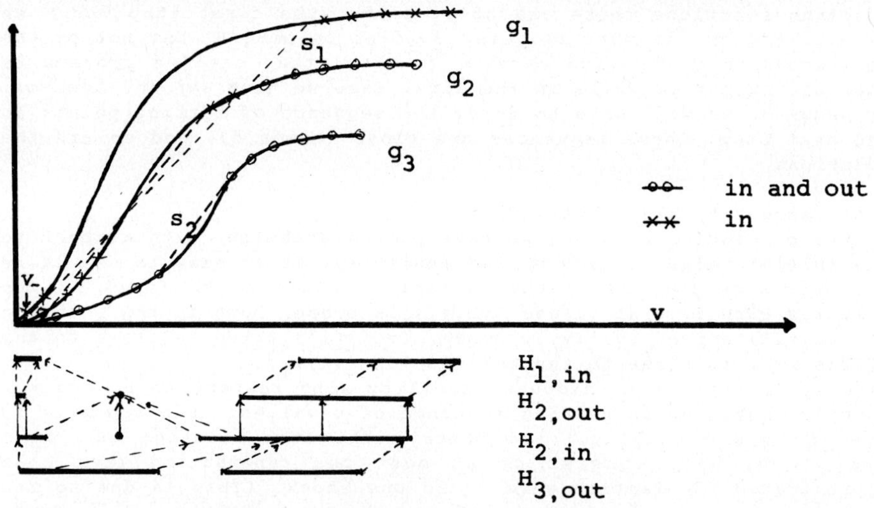

Fig. 3 An example of H-sets and solution paths.

We now turn our attention to the problem of variation of v_-. For any value v_-' in a small neighbourhood of v_-, let H' be the H-sets constructed from v_-'.

Lemma 4 :
If H_j' and H_j are both touching, then $H_i = H_i'$ for all $i \geq j$.
Proof:
The slope at the touching point is the same (independent of v_-), hence the touching points are equal, and so the entire sets.#

We turn to the case of two cutting H-sets. We have :

Lemma 5 :
The perpendicular distance between the two parallell lines through the cutting points is less than the distance $|v_- - v_-'|$.
Proof:
The lemma trivially holds for the first pair of H-sets. Assume it is valid for index j. Since g(u,) is increasing, the upper line intersects with $g(u_j,)$ above and to the right of the intersection of the lower line. The algorithm tells us to tilt the lines a bit more ($s_{j-1} < s_j$), and so the perpendicular distance shrinks. #
Observe that this lemma is also valid if one H_j is touching, while the other is still cutting. Then assume H_j' is touching while H_j is cutting. Since $g_u < 0$, we know that the perpendicular distance between the lines from the point of g_j with slope s_j to the similar point of g_{j+1}, δ, is greater than 0. If $|v_- - v_-'| < \delta$ then the perpendicular distance between the touching line of H' and the cutting line of H is also < δ, (by Lemma 5), and so the next H cannot be cutting. Hence,

Lemma 6 :
If H_j' is touching and H_j is cutting, then $H_i' = H_i$ for $i > j$ provided v_-' is sufficiently close to v_-. #

We are now prepared to trace the solutions. As long as we have touching functions there are no problems, the first step where we have to differ, is when reaching a point on some H' but not on the corresponding H (or vice versa). In the latter case we proceed to the cutting point, while in the first case we jump across. However, by Lemma 3, we will have to enter the sequence of cutting points in the next step. These sequences are close (Lemma 5), and so are the solutions.

With respect to all initial data.
In the proceeding sections we have proved stability with respect to the initial values v_- and v_+ independently. It is easy to see it is not necessary for one of the initial v-values to be fixed : The solution with initial values (v_-,v_+) is close both to the solution of (v_-,v_+') and of (v_-',v_+). Hence, (v_-',v_+') which is close to any of the two, is close to the solution of (v_-,v_+).
Finally it remains to discuss stability with respect to u_- and u_+. Consider again an increasing sequence of u-values, (the opposite is treaded symmetrically,) and observe that the solution of $u_t + f(u)_x = 0$, will consist of at most one rarefaction wave (an approximated rarefaction wave) and one shock. (This is due to the shape of f.) Assume it starts out with a rarefaction wave. Then, by taking some u_-' close to u_-, we introduce or lose one (or a few) u-value(s). The remaining u-values of the approximated wave are the same. (Assume $u_-' > u_-$, else, rename.) If $H_{1,in}$ is upper touching, so is all H'. If $H_{1,in}$ is lower-touching or cutting we know, by the continuity of g, that the touching/cutting point of the first H' is close to the corresponding point of H. Hence, the situation will be similar to the problem of variation of v_-. On the other hand, if u_+ is varied slightly, the last g-function will be slightly different (again by continuity of g), and so will H_{out}. (If there is no distinct shock at u_+, one (or a few) g(s) may be added or subtracted at the end of the sequence.) Thus the jump from the last function will be slightly altered only. We have:

<u>Theorem 3 :</u>
The essential structures of the solution of the initial value problem is stable with respect to variation of the initial values. (By essential structures we mean rarefaction waves, approximated rarefaction waves or major discontinuities. The structure of single peaks are not necessarily stable.) #
<u>Corollary .</u>
The solution (as a curve in phase-space) depends L_1-continuously upon the initial data. #
This Corollary is weaker, since approximated rarefaction waves consist of single points in phase-space.

THE APPROXIMATION

Finally we investigate the correspondance between the exact solution, where f is not approximated, and the numerical solution where it is. If the not-approximated f gives a discontinuity in u, so does the approximated, so assume that u is continuous. If u is constant, we solve a scalar problem exactly (g is not approximated), so it remains to consider a not-constant u. Then, by carrying out the differentiation of $v_t + g(u,v)_x = 0$ (e.g. [7]) :

$$s = g_u \frac{u_s}{v_s} + g_v .$$

The algorithm for the approximated case gives us a sequence of g-functions ($g_{ij} = g(u_i,v_j)$) :

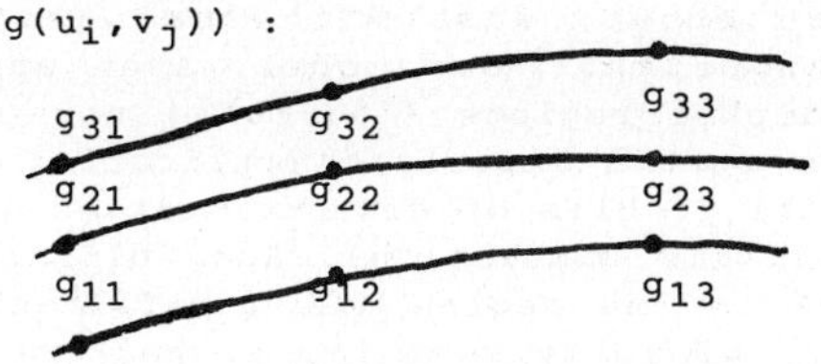

We make a difference approximation to s at g_{22} by setting :

$$g_u = \frac{g_{32} - g_{12}}{u_3 - u_1} \quad \text{and} \quad g_v = \frac{g_{23} - g_{21}}{v_3 - v_1}$$

and $$u_s = \frac{u_3 - u_1}{\delta s} \qquad v_s = \frac{v_3 - v_1}{\delta s} .$$

Then put these approximations into the expression for s:

$$s \approx (g_{32} - g_{12} + g_{23} - g_{21})/(v_3 - v_1) .$$

Now, by using the same approximation for g_u and g_v in a first order Taylor's formula (expanded for the point g_{22}) we find :

$$g_{ij} \approx g_{22}+[(g_{32}-g_{12})/(u_3-u_1)]\cdot(u_i-u_2)+[(g_{23}-g_{21})/(v_3-v_1)]\cdot(v_j-v_2).$$

We solve for $s_1 = (g_{22}-g_{11})/(v_2-v_1)$ and $s_2 = (g_{33}-g_{22})/(v_3-v_2)$:

$$s_1 \approx [(g_{32} - g_{12})\cdot(v_3 - v_1)/(2(v_2 - v_1)) + g_{23} - g_{21}]$$

$$s_2 \approx [(g_{32} - g_{12})\cdot(v_3 - v_1)/(2(v_3 - v_2)) + g_{23} - g_{21}] .$$

Where we have assumed : $u_{i+1} - u_i = \delta u$ for all i (uniform approximation).

Hence : $s_1 \le s \le s_2$, or $s_1 \ge s \ge s_2$, and $s_1 \to s_2$ when $\delta u \to 0$. Furthermore, we know that a difference approximation will converge, so our approximated s will tend to the exact s value. Thus,

Theorem 4 :

The solution when f is approximated by a piecewise linear function will converge to the exact solution as the approximation converges.

APPLICATIONS

We close this paper with some remarks on applications. In general Cauchy problems we may approximate the initial value function by a piecewise constant function to obtain a finite number of Riemann problems. These may be solved as above. We may apply the results of [2] to the (scalar) gas-flow equation. They proved that when a finite number of Riemann problems are considered, there is only a finite number of shock collisions, and so the method solves the

problem in a finite number of steps. In our case, in a bounded spatial area, we will find a single, constant u-value after some time (since all speeds are greater that zero, all shocks will move out of our area of interest). Then our second equation is scaler, and we may apply [2]'s argument once more. In [8] some examples of such problems are shown, also with some comparison to upwind schemes. Also note that our model, when applicable, gives no problems with elliptic regions ([9],[10]) nor unbounded variation (e.g. [11]). The second important application of the ideas presented here, is the problem of discontinuities (e.g. in geological datas in oil reservoir simulation). Such differences give rise to one flow function in one region and a different flow function in the neighbouring region. By assuming a shock of zero speed at the discontinuity we may solve the problem by making a jump from the one to the other as we do between different g-functions. Alternatively we may add an equation with a zero flow function :
$u_{0,t} = 0$, with appropriate initial conditions.

ACKNOWLEDGEMENTS

The author would like to thank Helge Holden, Lars Holden, Nils-Henrik Risebro and the late Prof. Raphael Høegh-Krohn for fruitful discussions and suggestions concerning the work presented in this paper. Also thanks to VISTA and the Norwegian Science Research Council NAVF for financial support.

REFERENCES

[1] Dafermos C.M. Polygonal approximation of the initial value problem for a conservation law. J.Math.Anal.Applic.**38**(1972) 33-41.
[2] Holden H., Holden L. and Høegh-Krohn R. A numerical method for first order nonlineal scalar hyperbolic conservation laws in one dimension. Preprint **12** (1986), Univ. Oslo.
[3] Lax P.D. Weak solutions of hyperbolic equations and their numerical computation. Comm. Pure Appl. Math.**7** (1954) 159-193.
[4] Lucier B.J. A moving mesh numerical method for hyperbolic conservation laws. Math. Comput. **46** (1986) 59-69.
[5] Oleinik O.A. Uniqueness and stability of the generalized solution of the Cauchy problem for a quasilinear equation. Math. Soc. Transl. Ser. 2 **33** (1964) 285-290.
[6] Isaacson J. and Temple B. The Riemann problem for general 2x2 conservation laws. Trans. Amer. Soc. **199** (1974) 89-112.
[7] Holden L. and Høegh-Krohn R. A class of N nonlinear hyperbolic conservation laws. Preprint **6** (1987), Univ. Oslo.
[8] Gimse T. Thesis, (To appear). (1988), Univ. Oslo.
[9] Bell J.B., Trangenstein J.A. and Shubin G.R. Conservation laws of mixed type describing threephase flow in porous media. SIAM J. Appl. Math. **46** (1986) 1000-1017.
[10] Holden L. On the strict hyperbolicity of the Buckley-Leverett equation for three phase flow in a porous medium. Preprint, Norwegian Computing Centre (1988).
[11] Temple B. Global solution of the cauchy problem for a class of 2x2 nonstrictly hyperbolic conservation laws. Adv. Appl. Math. 3 (1982) 335-375.

Nonuniqueness of Solutions for Riemann Problems

J. Glimm [1, 2, 3, 4]

New York University
Courant Institute of Mathematical Sciences
251 Mercer Street
New York, N.Y. 10012

ABSTRACT

Arguments are advanced that physically meaningful nonunique solutions to Riemann problems can occur. The implications of this point of view for both theory and computation are developed, as part of a review of recent progress concerning the interaction of nonlinear waves and the front tracking method for computation.

I. Computations

We emphasize the possibility of nonuniqueness for solutions of Riemann problems. The implications of nonuniqueness for computational science are clear: there is a decisive advantage for computational methods which allow explicit choice under user control among possible nonunique solutions. The case of flames and reactive fluid flow is a well explored test case. Either it is necessary to use exceedingly fine computational scales, to resolve the chemistry and internal fluid layers fully (which would normally be prohibitive in a large scale computation) so that the flame speed is determined correctly as a consequence of the computation, or it is necessary to add the flame speed or some equivalent information explicitly to the computational algorithm. Although front tracking is not the only way to do this, it has been recognized as a promising vehicle to achieve the second of these possibilities. The second route, because it does not require the full resolution of internal layers, allows

1. Supported in part by the National Science Foundation, grant DMS - 8619856
2. Supported in part by the Applied Mathematical Sciences subprogram of the Office of Energy Research, U. S. Department of Energy, under contract DE-AC02-76ER03077
3. Supported in part by the Army Research Office, grant DAAG29-85-0188
4. Supported in part by the Air Force Office Office of Scientific Research AFSOR-88-0025.

enhanced resolution in the computation. To the author's knowledge, the second route is the dominant one for the large scale computation of reactive fluids, especially in the case of flames and detonation waves.

Enhanced resolution in fluid computations is a problem of major importance. Front tracking is an adaptive computational method which is especially oriented to fluid discontinuities. For problems with significant discontinuities, such as shock waves and fluid interfaces, it has yielded enhanced resolution by factors typically in the range of 3 to 5 per linear dimension (27 to 125 per space time grid block in two spatial dimensions) and occasionally up to 50 per linear dimension (1.25×10^5 per space time grid block) [10]. One could expect further advantages in the more complex cases involving nonunique Riemann solutions, such as reactive flow, where the full resolution of a complex internal structure can be avoided by the use of front tracking.

Front tracking is based on marker particles, which locate a discontinuity surface sharply, without numerical smearing. The particles are propagated by a fluid velocity, characteristic velocity or shock speed velocity, and thus represent Lagrangian or characteristic particles embedded in an Eulerian computation.

Front tracking is also based on mathematical theory, and necessity being the mother of invention, it has motivated some of the recent developments in the theory of hyperbolic wave interactions (Riemann problems), as is discussed in the next section.

II. Mathematical Theory

II.1. The Isolated Umbilic Point. We study the hyperbolic conservation law

$$U_t + \nabla \cdot F(U) = 0 . \tag{2.1}$$

Let

$$A = \frac{\partial F}{\partial U} \tag{2.2}$$

be the Jacobian matrix and

$$\lambda_1 \leq \cdots \leq \lambda_n \tag{2.3}$$

its eigenvalues, assumed to be real but not necessarily distinct. Let $r_1, \cdots, r_n$ be the corresponding right eigenvectors. Points U_0 with non distinct λ's, i.e. $\lambda_i(U_0) = \lambda_{i+1}(U_0)$, are called umbilic points.

Let the solution values $U(x,t)$ lie in a state space $\mathcal{S}$:

$$U(x,t) \in \mathcal{S} \subset R^n . \tag{2.4}$$

Then the hyperbolic wave structure of (2.1) defines a geometry on $\mathcal{S}$ and umbilic points are generically singular points in this geometry as we now explain. Let $\xi = x/t$. Then $U = U(\xi) = U(x/t)$ is a solution of (2.1) provided

$$(\xi I - A)U_\xi = 0 , \tag{2.5}$$

or in other words $\xi = \lambda_i(U)$ and up to a scalar factor $U_\xi = r_i(U)$. These solutions, called rarefaction waves, define coordinate lines on $\mathcal{S}$, one for each i, $1 \leq i \leq n$. For distinct eigenvalues λ_i, standard perturbation theory for matrices is available, and yields a regular geometry defined by the rarefaction wave curves. However for degenerate eigenvalues, matrix perturbation theory involves fractional powers and the singular wave geometry at umbilic points reflects this fact. The singular geometry of the wave curves associated with umbilic points $U \in \mathcal{S}$ gives rise to striking and novel phenomena for the interaction of nonlinear hyperbolic waves. A survey of this work, due to E. Gomes, H. Holden, Eli Isaacson, D. Marchesin, P. Paes-Leme, F. Palmeira, B. Plohr, D. Schaeffer, M. Shearer and B. Temple, is contained in [12]. An essential tool in the development of this theory was the development of a computational code for the numerical solution of such Riemann problems by Marchesin, Plohr and co-workers.

II.2. Entropy Conditions. Shock waves are jump discontinuous weak solutions of (2.1), characterized by the relations

$$s[U] = [F] , \tag{2.6}$$

where

$$[U] = U^+ - U^- , \quad [F] = F^+ - F^- \tag{2.7}$$

are the jumps in U and F and s is the shock speed. These weak solutions are nonunique, and supplementary conditions, known as entropy conditions, are imposed to reject undesired solutions and hence to yield uniqueness for solutions of the Riemann problem.

The Riemann problem is the Cauchy problem with scale invariant data, so that in one space dimension, $U(x,0) = U_L$ for $x < 0$, $U(x,0) = U_R$ for $x > 0$. Gomes [14] has observed that a fundamental entropy condition for solutions of the Riemann problem fails. She obtained this result in the process of completing the solution of Riemann problems for quadratic flux F models.

The Lax entropy condition counts the number of characteristics which enter the shock wave; it is required that for one family (i), the characteristics enter the shock wave from both sides while for all other families, the characteristics cross. Thus they enter from one side while leaving from the other. Such shocks are clearly associated with a single (i) characteristic family. Gomes has given what seems to be the first example of a system (2.1) with a Lax shock which fails to have a viscous profile. A shock has a viscous profile if it is the limit as $\epsilon \to 0$ of solutions U^ϵ of the associated parabolic equation,

$$U_t^\epsilon + \nabla \cdot F(U^\epsilon) = \epsilon \Delta U^\epsilon \ , \tag{2.8}$$

see also [3]. A second fundamental entropy condition is shock profilability, as defined above; namely $U = \lim_{\epsilon \to 0} U^\epsilon$. Gomes finds profilable shocks which are not Lax shocks in her examples. Thus these two notions are properly disjoint. She shows that the Riemann problem has a satisfactory existence theory, when solved in the class of profilable shocks, but it may fail to have solutions in the class of Lax shocks. Gomes' examples are simple mathematically and are motivated by three phase flow in oil reservoirs, so they cannot be rejected on either aesthetic or pragmatic grounds.

II.3 A New Paradigm For Differential Equations of Mathematical Physics. The ideas of Hadamard hold that mathematical equations modeling physics should be well posed, in the sense that the solution should exist, and be uniquely and continuously determined by the data. For more than forty years, examples of an equation of state with nonunique Riemann solutions were known. See [21] for example. The shocks are profilable Lax shocks in these examples. The equation of state is thermodynamically consistent but appears to be pathological and perhaps does not correspond to any real material. For whatever reason, these examples have been somewhat overlooked.

The Riemann problem associated with polymer flood of oil reservoirs has a line of umbilic points. An entropy condition gives a unique solution of this

Riemann problem. Eli Isaacson and B. Temple have analyzed the large time asymptotics for solutions with fixed left and right states, U_L and U_R [17]. The asymptotics is not uniquely determined by U_L and U_R. There is a one parameter family of asymptotics, each a solution of the same $U_L - U_R$ Riemann problem. In other words these particular non unique Riemann solutions are physically meaningful and should not be rejected. In forthcoming work, Eli Isaacson, D. Marchesin and B. Plohr [18] have shown that the allowed solutions of a Riemann problem may depend on the explicit form of the viscosity matrix (taken as the identity in (2.8)), in the neighborhood of an isolated umbilic point.

Brio [1], in studying Riemann problems for MHD, has found umbilic points with two and three coinciding eigenvalues. Admissibility of shocks depends on the form of the viscosity matrix and numerical computation shows that the nature of the computationally resolved wave patterns depends sensitively on the choice of numerical method. Freistuhler [5] has also studied Riemann problems associated with MHD. He has a simple model with an isolated umbilic point and a cubic flux function. The solution has a very different character from the previously studied isolated umbilic point with a quadratic flux function. Again most commonly used entropy conditions give inconsistent and unsatisfactory results. He also has an example of nonuniqueness of solutions of the Riemann problem for shocks with viscous profiles. Freistuhler proposes dynamical linearized stability as an acceptable entropy condition in his model.

Combustion waves also depend on a ratio of length scales to set the speed and hence the wave structure between a given left and right state. Computations with combustion are also very sensitive to numerical methods when these internal length scales are not fully resolved computationally.

Physically meaningful nonunique solutions are known from other examples. Combustion equations with unburnt $U_L = U_R$ have two solutions, one for unburnt $U = U_L$ for all t and one for ignition of combustion at $x = 0, t = 0$. Both solutions are physically meaningful.

A paradox is associated with shock reflection problems, in that simple arguments cannot distinguish between two possible solution geometries (regular reflection and Mach reflection). Non uniqueness of weak solutions for the

Euler equation is a possible resolution of this paradox.

The Riemann problem for steady two dimensional flow with supersonic left and right states has a high degree on nonuniqueness. In this problem there is a body of lore for selecting the "correct" solution, some of which seems to imply a further specification of the problem physically, such as the existence of boundary layers.

We propose a new paradigm for the study of uniqueness in Riemann problems. It is convenient to express our ideas in the language of dynamical systems. Nonuniqueness corresponds to a bifurcation. A mathematical characterization of nonuniqueness is given by a complete unfolding of the Riemann problem, which would identify the multiplicity and or dimensionality of nonuniqueness. Some of this nonuniqueness should be rejected as "universally unphysical", and then canonical forms for the remaining "possibly physical" solutions could be found.

Next standard unfoldings can be constructed, and mapped onto the canonical forms above. The viscosity profile criterium has been described above. Limits of smooth Cauchy data approaching jump data is a regularization related to both the combustion and the polymer nonuniqueness examples discussed above. Most systems are subsystems or asymptotic limits of larger systems. Enlarging the system is or might be a regularization in agreement with fundamentally correct physics. This is most commonly the correct regularization for equations with embedded elliptic regions. Heterogeneous, noisy or stochastic perturbations to the data or the equations may also resolve some examples of nonuniqueness.

The conservation law (2.1) is invariant under the scale transformation group

$$x \to ax , \quad t \to at , \quad a > 0 . \tag{2.9}$$

It is a general feature of scale invariant problems that they may be underspecified, and hence sitting on bifurcation points. Scale invariant problems are characterized by the setting to zero of all length scales. There is no reason to require that the unfolding of this bifurcation has a unique physically meaningful branch. For example if the physics of well posed problems in a neighborhood of a conservation law depends in an essential manner on two length scales, then their ratio is a dimensionless number, which must be specified as an addition to the conservation law, in order to give it a unique physical content.

II.4. Wave Structures for Real Systems. A comprehensive treatment of the wave structures implied by real (as opposed to ideal) equations of state has been written [21]. In this work, the older theory is reworked and unified and new results have been added. A presentation of the equations of elastic flow in Eulerian conservation form has been given [22] using the language of modern differential geometry [20]. Modeling related to three phase Buckley-Leverett equations for oil reservoirs was mentioned above [15,16]. (See Section 2.1.)

II.5. Two Dimensional Waves. The elementary waves are the building blocks out of which a Riemann solution is constructed. The Riemann solution is characterized by invariance under scale transformations (2.9), while the elementary wave is invariant under an additional symmetry: it moves at a constant velocity with fixed form as a traveling wave. The elementary waves for a polytropic fluid equation of state were classified in [8]. In two dimensions the Riemann solutions will in general have an infinite number of pieces, and cannot be described explicitly. However this objection does not apply to certain restricted classes of two dimensional Riemann problems [19,27], nor does it apply to the approximate solution of general Riemann problems. A very successful line of work has been the study of two dimensional (wave front curvature) corrections to one dimensional wave motion for reacting fluids. Let $D(\kappa)$ denote the detonation velocity for an unsupported wave, as a function of curvature κ, and let $\delta D(\kappa) = D(0) - D(\kappa)$. For expanding waves, Bukiet, Jones, Bdzil and Stewart [2,26] have determined δD as a function of κ and the reaction rate δ, to leading order in κ. They find

$$\delta D = c_1 \kappa\,, \qquad \delta < 1\,,$$

$$\delta D = c_2 \kappa \ln(\kappa) + c_3 \kappa\,, \qquad \delta = 1\,,$$

$$\delta D = c_4 \kappa^{\frac{1}{\delta}}\,, \qquad 1 < \delta\,.$$

The coefficients $c_1, \cdots, c_4$ depend on the equation of state, the reaction rate and the kinetics. They can be determined analytically in simple cases and numerically in any case.

III. Applications

Our primary applications to technology have been to the modeling of oil reservoirs [4, 7, 11]. Those to science have concerned chaotic mixing associated with unstable fluid interfaces. We have given what appears to be the first correct computation of a single mode Rayleigh-Taylor finger and bubble, both for the incompressible [9] and compressible [6] cases, for general values of the Atwood number A. These computations were compared with a large body of experiments for a range of values of A, and to theory and incompressible computations (for $A = 1$ only, because the theory and prior validation of computation was available in this case only).

However, in a discovery which should be a warning to chaos workers in other, related, problems, we found [13] by analysis of experiments of Read [23] that the single mode theory is irrelevant to chaotic flow, and gives a terminal bubble velocity which is incorrect by a factor of two or more. The cause of this discrepancy has been traced to bubble-bubble nearest neighbor correlations.

A statistical model for bubble interactions due to Sharp and Wheeler [24, 25] was tested and found [13] to give agreement with experiment.

References

1. M. Brio, "Admissibility Conditions for Weak Conditions of Non-strictly Hyperbolic Systems", paper in this volume.
2. B. Bukiet and J. Jones, "The Competition Between Curvature and Chemistry in a Spherically Expanding Detonation", *Appl. Phys. Leteers*, in press.
3. C. Conley and J. Smoller, "Viscosity Matrices for Two–Dimensional Nonlinear Hyperbolic Systems", *Comm. Pure Appl. Math.*, vol. 23, pp. 867–884, 1970.
4. P. Daripa, J. Glimm, B. Lindquist, and O. McBryan, "Polymer floods: a case study of nonlinear wave analysis and of instability control in tertiary oil recovery", *SIAM J Appl Math*, vol. 48, pp. 353–373, 1988.
5. H. Freistühler, "A Standard Model of Generic Rotational Degeneracy", paper in this volume.

Conference on Hyperbolic Problems, To appear.

6. C. Gardner, J. Glimm, O. McBryan, R. Menikoff, D. H. Sharp, and Q. Zhang, "The Dynamics of Bubble Growth for Rayleigh-Taylor Unstable Interfaces," *Phys. of Fluids*, vol. 31, pp. 447-465, 1988.
7. J. Glimm, B. Lindquist, O. McBryan, and L. Padmanhaban, "A Front Tracking Reservoir Simulator: 5-spot Validation Studies and the Water Coning Problem," *Frontiers in Applied Mathematics*, vol. 1, SIAM, Philadelphia, 1983.
8. J. Glimm, C. Klingenberg, O. McBryan, B. Plohr, D. Sharp, and S. Yaniv, "Front Tracking and Two Dimensional Riemann Problems," *Advances in Appl. Math.*, vol. 6, pp. 259-290, 1985.
9. J. Glimm, O. McBryan, D. Sharp, and R. Menikoff, "Front Tracking Applied to Rayleigh Taylor Instability," *SIAM J. Sci. Stat. Comput.*, vol. 7, pp. 230-251, 1986.
10. J. Glimm, J. Grove, and X.L. Li, "Three Remarks on the Front Tracking Method," in *Proceedings of Taormina Conference*, Oct 1987.
11. J. Glimm, B. Lindquist, O. McBryan, and G. Tryggvason, "Sharp and Diffuse Fronts in Oil Reservoirs: Front Tracking and Capillarity," in *Mathematical and Computational Methods in Seismic Exploration and Reservoir Modeling*, ed. William E. Fitzgibbon, pp. 68-84, SIAM, Philadelphia, 1987.
12. J. Glimm, "The Interactions of Nonlinear Hyperbolic Waves," *Comm. Pure Appl. Math.*, vol. 41, pp. 569-590, 1988.
13. J. Glimm and X.L. Li, "On the Validation of the Sharp-Wheeler Bubble Merger Model from Experimental and Computational Data," *Phys. of Fluids*, vol. 31, pp. 2077-2085, 1988.
14. Maria Elasir Gomes, "Singular Riemann Problem for a Fourth Order Model for Multiphase Flow," Thesis (Portuguese) Departamento de Matematica, Pontificia Universidade Catolica do Rio Janeiro, 1987.
15. Eli Isaacson, D. Marchesin, B. Plohr, and B. Temple, "The Classification of Solutions of Quadratic Riemann Problems I," MRC Report, 1985.
16. Eli Isaacson and B. Temple, "The Classification of Solutions of Quadratic Riemann Problems II, III," *SIAM J Appl. Math.*, To appear.

17. Eli Isaacson and B. Temple, *The Structure of Asymptotic States in a Singular System of Conservation Laws*, To Appear.

18. Eli Isaacson, D. Marchesin, and B. Plohr, *Viscous Profiles for Transitional Shock Waves*, To Appear.

19. B. Lindquist, "The Scalar Riemann Problem in Two Spatial Dimensions: Piecewise Smoothness of Solutions and its Breakdown," *SIAM J Anal*, vol. 17, pp. 1178-1197, 1986.

20. J. Marsden and T. Hughes, *Mathematical Foundations of Elasticity*, Prentice-Hall, Englewood Cliffs, 1983.

21. R. Menikoff and B. Plohr, "Riemann Problem for Fluid Flow of Real Materials," *Rev. Mod. Phys.*, To Appear.

22. B. Plohr and D. H. Sharp, "A Conservative Eulerian Formulation for the Equation of Elastic Flow," *Adv. Appl. Math.*, To appear.

23. K. I. Read, "Experimental Investigation of Turbulent Mixing by Rayleigh-Taylor Instability," *Physica 12D*, pp. 45-48, 1984.

24. D. H. Sharp and J. A. Wheeler, "Late Stage of Rayleigh-Taylor Instability," Institute for Defense Analyses, 1961.

25. D. H. Sharp, "Overview of Rayleigh-Taylor Instability," *Physica 12D*, pp. 3-17, 1984.

26. S. Stewart and J. B. Bdzil, "The Shock Dynamics of Stable Multidimensional Detonation," *J. Fluid Mech.*, To appear.

27. D. Wagner, "The Riemann Problem in Two Space Dimensions for a Single Conservation Law," *Math. Ann.*, vol. 14, pp. 534-559, 1983.

SIMPLE STABILITY CRITERIA FOR DIFFERENCE APPROXIMATIONS OF HYPERBOLIC INITIAL-BOUNDARY VALUE PROBLEMS

Moshe Goldberg*
Department of Mathematics
Technion
Haifa 32000, Israel

Eitan Tadmor**
School of Mathematical Sciences
Tel Aviv University
Tel Aviv 69928, Israel

SUMMARY

In this note we discuss new, simple stability criteria for a wide class of finite difference approximations for initial-boundary value problems associated with the hyperbolic system $\partial u/\partial t = A\partial u/\partial x + Bu + f$ in the quarter plane $x \geqslant 0$, $t \geqslant 0$. With these criteria, stability is easily achieved for a multitude of examples that incorporate and generalize most of the cases studied in recent literature.

Consider the first order system of hyperbolic partial differential equations

$$\partial u(x,t)/\partial t = A\partial u(x,t)/\partial x + Bu(x,t) + \mathbf{f}(x,t), \quad x \geqslant 0,\ t \geqslant 0, \tag{1a}$$

where $\mathbf{u}(x,t) = \left(u^{(1)}(x,t),\dots,u^{(n)}(x,t)\right)'$ is the unknown vector (prime denoting the transpose), $\mathbf{f}(x,t) = \left(f^{(1)}(x,t),\dots,f^{(n)}(x,t)\right)'$ is a given n-vector, and A and B are fixed $n \times n$ matrices such that A is diagonal of the form

$$A = \operatorname{diag}\left(A^{I},A^{II}\right), \quad A^{I} > 0, \quad A^{II} < 0, \tag{2}$$

with A^{I} and A^{II} of orders $k \times k$ and $(n-k) \times (n-k)$, respectively.

The solution of (1a) is uniquely determined if we prescribe intial values

$$u(x,0), \quad x \geqslant 0 \tag{1b}$$

and boundary conditions

$$u^{II}(0,t) = Su^{I}(0,t) + \mathbf{g}(t), \quad t \geqslant 0, \tag{1c}$$

where S is a fixed $(n-k) \times k$ coupling matrix, $\mathbf{g}(t)$ a given $(n-k)$-vector, and

$$u^{I} = \left(u^{(1)},\dots,u^{(k)}\right)', \quad u^{II} = \left(u^{(k+1)},\dots,u^{(n)}\right)' \tag{3}$$

a partition of $\mathbf{u}$ into its outflow and inflow components, respectively, corresponding to the partition of A in (2).

---

* Research sponsored in part by U.S. Air Force Grants AFOSR-83-0150 and AFOSR-88-0175.

** Research sponsored in part by NASA Contract NAS1-17070 and U.S.-Israel BSF Grant 85-00346.

Introducing a mesh size $\Delta x > 0$, $\Delta t > 0$, such that $\lambda \equiv \Delta t/\Delta x$ is constant, and using the notation $v_\nu(t) = v(\nu\Delta x, t)$, we approximate (1a) by a general, basic difference scheme — explicit or implicit, dissipative or not, two-level or multilevel — of the form

$$Q_{-1}v_\nu(t+\Delta t) = \sum_{\sigma=0}^{s} Q_\sigma v_\nu(t-\sigma\Delta t) + \Delta t b_\nu(t), \quad \nu = r, r+1, \ldots,$$

$$Q_\sigma = \sum_{j=-r}^{p} A_{j\sigma}E^j, \; Ev_\nu = v_{\nu+1}, \quad \sigma = -1, \ldots, s, \tag{4}$$

where the $n \times n$ coefficient matrices $A_{j\sigma}$ are polynomials in λA and ΔtB, and the n-vectors $b_\nu(t)$ depend on $f(x,t)$ and its derivatives.

The difference equations in (4) have a unique solution $v_\nu(t+\Delta t)$ if we provide initial values

$$v_\nu(\mu\Delta t), \quad \mu = 0, \ldots, s, \quad \nu = 0, 1, 2, \ldots, \tag{5}$$

and specify, at each time level $t = \mu\Delta t$, $\mu = s, s+1, \ldots$, boundary values $v_\nu(t+\Delta t)$, $\nu = 0, \ldots, r-1$. Such boundary values are determined by conditions of the form

$$T_{-1}^{(\nu)}v_\nu(t+\Delta t) = \sum_{\sigma=0}^{q} T_\sigma^{(\nu)}v_\nu(t-\sigma\Delta t) + \Delta t d_\nu(t), \quad \nu = 0, \ldots, r-1,$$

$$T_\sigma^{(\nu)} = \sum_{j=0}^{m} C_{j\sigma}^{(\nu)}E^j, \quad \sigma = -1, \ldots, q, \tag{6a}$$

where the $n \times n$ matrices $C_{j\sigma}^{(\nu)}$ depend on A, ΔtB and S, and the n-vectors $d_\nu(t)$ are functions of $f(x,t)$, $g(t)$ and their derivatives.

Our intention is to interpret the difficult and often stubborn Gustafsson-Kreiss-Sundström (GKS) stability criterion in [4] in order to obtain simple and convenient stability criteria for approximation (4)-(6a). While we were unable to meet this goal for general boundary conditions of type (6a), we managed to achieve rather satisfactory results under the further assumption that, in accordance with the partition of A in (2), the $C_{j\sigma}^{(\nu)}$ are of the form

$$C_{j\sigma}^{(\nu)} = \begin{bmatrix} C_{j\sigma}^{II} & C_{j\sigma}^{I\,II(\nu)} \\ C_{j\sigma}^{II\,I(\nu)} & C_{j\sigma}^{II\,II(\nu)} \end{bmatrix}, \tag{6b}$$

where

$$\text{the } C_{j\sigma}^{II} \text{ are independent of } \nu, \tag{6c}$$

$$\text{the } C_{j\sigma}^{II} \text{ are diagonal when } B = 0, \tag{6d}$$

$$\text{the } C_{j\sigma}^{I\,II(\nu)} = 0 \text{ when } B = 0, \tag{6e}$$

$$C_{j\sigma}^{II\,II(\nu)} = 0 \text{ for } j > 0 \text{ and } \sigma > -1 \text{ when } B = 0. \tag{6f}$$

The essence of (6c)-(6e) is that for $B = 0$, the outflow boundary conditions are *translatory* (i.e., determined at all boundary points by the same coefficients), *separable* (i.e., split into independent scalar conditions for the different outflow unknowns), and independent of inflow values. Assumption (6f) implies that for $B = 0$ the inflow values at the boundary depend essentially on the outflow.

It should be pointed out that our outflow boundary conditions are quite general, despite the apparent restrictions in (6c)-(6e). Indeed, (6c) is not much of a restriction, since in practice the outflow boundary conditions are translatory. In particular, if the numerical boundary consists of a single point, then the boundary conditions are translatory by definition, so (6c) holds automatically. The restrictions in (6d),(6e) pose no great difficulties either, since they are satisfied by all reasonable boundary conditions, where for $B = 0$ the $C_{j\sigma}^{II}$ usually reduce to polynomials in the block A^{I}, and the $C_{j\sigma}^{I\,II(\nu)}$ vanish.

We realize that in view of the restriction in (6f) our inflow boundary conditions are not quite as general as the outflow ones. They can, however, be constructed to any degree of accuracy (see [1]); and if the boundary consists of a single point, then such conditions can be achieved in a trivial manner, simply by duplicating the analytic condition (1c), i.e.,

$$v_0^{II}(t+\Delta t) = Sv_0^{I}(t+\Delta t) + g(t+\Delta t).$$

Throughout our work we assume, of course, that the basic scheme (4) is stable for the pure Cauchy problem, and that the other assumptions which guarantee the validity of the GKS theory in [4] hold.

The first step in our analysis was to reduce the above stability question to that of a scalar, homogeneous problem. This is obtained by considering the outflow scalar equation

$$\partial u/\partial t = a\partial u/\partial x, \quad x \geqslant 0, \quad t \geqslant 0, \quad a = \text{constant} > 0, \tag{7}$$

for which the basic scheme (1.4) reduces to the homogeneous scheme

$$Q_{-1}v_{\nu}(t+\Delta t) = \sum_{\sigma=0}^{s} Q_{\sigma}v_{\nu}(t-\sigma\Delta t), \quad \nu = r, r+1, \ldots$$
$$Q_{\sigma} = \sum_{j=-r}^{p} a_{j\sigma}E^{j}, \quad \sigma = -1, \ldots, s, \tag{8a}$$

and the boundary conditions (1.6) reduce to translatory conditions of the form

$$T_{-1}v_{\nu}(t+\Delta t) = \sum_{\sigma=0}^{q} T_{\sigma}v_{\nu}(t-\sigma\Delta t), \quad \nu = 0, \ldots, r-1,$$
$$T_{\sigma} = \sum_{j=0}^{m} c_{j\sigma}E^{j}, \quad \sigma = -1, \ldots, q, \tag{8b}$$

where $a_{j\sigma}$ and $c_{j\sigma}$ are scalar coefficients.

Referring to (8) as the *basic approximation*, we proved:

THEOREM 1 [3, Theorem 1.1]. *Approximation* (4)-(6) *is stable if and only if the reduced outflow scalar approximation* (8) *is stable for every eigenvalue* $a > 0$ *of* A^I. *That is, approximation* (4)-(6) *is stable if and only if the scalar outflow components of its principal part are all stable.*

This reduction theorem implies that from now on we may restrict our stability study to the basic approximation (8).

In order to introduce our stability criteria for the basic approximation, we use the coefficients of the basic scheme (8a) to define the *basic characteristic function*

$$P(z,\kappa) = \sum_{j=-r}^{p} \left[a_{j,-1} - \sum_{\sigma=0}^{s} a_{j\sigma} z^{-\sigma-1} \right] \kappa^j .$$

Similarly, using the coefficients of the boundary conditions in (8b) we define the *boundary characteristic function*

$$R(z,\kappa) = \sum_{j=0}^{m} \left[c_{j,-1} - \sum_{\sigma=0}^{q} c_{j\sigma} z^{-\sigma-1} \right] \kappa^j .$$

Now putting

$$\Omega(z,\kappa) \equiv |P(z,\kappa)| + |R(z,\kappa)| ,$$

it is not difficult to combine Theorems 3.1' and 3.2' of [3] in order to obtain:

THEOREM 2. *The basic approximation* (8) *is stable if:*

(*i*) *either*

$$\left. \frac{\partial P(z,\kappa)}{\partial z} \cdot \frac{\partial P(z,\kappa)}{\partial \kappa} \right|_{z=\kappa=-1} < 0 \tag{10a}$$

or

$$\Omega(z=-1,\kappa=-1) > 0. \tag{10b}$$

(ii) $$\Omega(z,\kappa) > 0 \text{ for all } |z| = |\kappa| = 1,\ \kappa \neq 1,\ (z,\kappa) \neq (-1,-1), \tag{10c}$$

$$\Omega(z,\kappa=1) > 0 \text{ for all } |z| = 1,\ z \neq 1, \tag{10d}$$

$$\Omega(z,\kappa) > 0 \text{ for all } |z| \geqslant 1,\ 0 < |\kappa| < 1. \tag{10e}$$

The advantage of this setting of Theorem 2 is clarified by the following lemma, in which we provide helpful sufficient conditions for each of the four inequalities in (10b-e) to hold:

LEMMA 1 [3, Theorem 2.2].

(i) *Inequalities (10b,c) hold if either the basic scheme (8a) or the boundary conditions (8b) are dissipative.*

(ii) *Inequality (10d) holds if any of the following is satisfied:*
(a) The basic scheme is two-level.
(b) The basic scheme is three-level and

$$\Omega(z=-1,\kappa=1) > 0. \tag{11}$$

(c) The boundary conditions are two-level and at least zero-order accurate as an approximation of equation (7).

(d) The boundary conditions are three-level, at least zero-order accurate, and (11) is satisfied.

(iii) *Inequality (10e) holds if the boundary conditions fulfill the von Neumann condition, and are either explicit or satisfy*

$$T_{-1}(\kappa) \equiv \sum_{j=0}^{m} c_{j,-1}\kappa^{j} \neq 0 \text{ for } 0 < |\kappa| \leqslant 1.$$

As mentioned earlier, we always assume that the basic scheme is *stable* for the pure Cauchy problem, i.e.,

(i) The basic scheme fulfills the von Neumann condition; that is, the roots $z(\kappa)$ of the equation

$$P(z,\kappa) = 0$$

satisfy

$$|z(\kappa)| \leqslant 1 \quad \text{for all } \kappa \text{ with } |\kappa| = 1.$$

(ii) If $|\kappa| = 1$ and $z(\kappa)$ is a root of $P(z,\kappa)$ with $|z(\kappa)| = 1$, then $z(\kappa)$ is a simple root of $P(z,\kappa)$.

As usual, we say that the basic scheme is *dissipative* if the roots of $P(z,\kappa)$ satisfy

$$|z(\kappa)| < 1 \text{ for all } \kappa \text{ with } |\kappa| = 1,\ \kappa \neq 1.$$

Analogous definitions hold for the boundary conditions with $P(z,\kappa)$ replaced by $R(z,\kappa)$. Clearly, both for the basic scheme and the boundary conditions, dissipativity implies the von Neumann condition.

The stability criteria obtained in Theorem 2 depend both on the basic scheme and the boundary conditions, but not on the intricate and often complicated interaction between the two. Consequently, Theorem 2, aided by Lemma 1, provides in many cases a convenient alternative to the celebrated GKS stability criterion in [4].

Having the new criteria, one can now easily establish stability for a host of examples that incorporate and generalize most of the cases studied in recent literature (e.g., [3]). We conclude this note with three of these examples:

EXAMPLE 1. Consider an arbitrary basic scheme, and let the boundary conditions be generated by either the explicit, first-order accurate, right-sided Euler scheme:

$$v_\nu(t+\Delta t) = v_\nu(t) + \lambda a\bigl[v_{\nu+1}(t) - v_\nu(t)\bigr], \quad 0 < \lambda a < 1, \quad \nu = 0,\dots,r-1, \tag{12}$$

or by its implicit analogue:

$$v_\nu(t+\Delta t) = v_\nu(t) + \lambda a\bigl[v_{\nu+1}(t+\Delta t) - v_\nu(t+\Delta t)\bigr], \quad \lambda a > 0, \quad \nu = 0,\dots,r-1. \tag{13}$$

These two-level boundary conditions are dissipative (see [1], Examples 3.5 and 3.6), hence fulfill the von Neumann condition. Further, for (13) we have

$$\mathrm{Re}\bigl[T_{-1}(\kappa)\bigr] = 1 + \lambda a\bigl[1 - \mathrm{Re}(\kappa)\bigr] \neq 0, \quad |\kappa| \leqslant 1.$$

By Lemma 1, therefore, inequalities (10b–e) hold, and Theorem 2 implies stability.

EXAMPLE 2. Take an arbitrary two-level basic scheme, and define the boundary conditions by horizontal extrapolation of order $\ell-1$:

$$v_\nu(t+\Delta t) = \sum_{j=1}^{\ell} \binom{\ell}{j}(-1)^{j+1} v_{\nu+j}(t+\Delta t), \quad \nu = 0,\dots,r-1.$$

Here,

$$R(z,\kappa) = \sum_{j=1}^{\ell} \binom{\ell}{j}(-1)^{j+1}\kappa^j = (1-\kappa)^\ell,$$

so $R(z,\kappa) \neq 0$ for $\kappa \neq 1$, which directly gives (10b,c,e). Moreover, since the basic scheme is two-level, Lemma *1(ii)(a)* implies (10d), and Theorem 2 again proves stability.

It is interesting to note (e.g. [2]) that this result may fail, both for dissipative and nondissipative basic schemes, if the basic scheme consists of more than two time levels.

EXAMPLE 3. Consider the Leap-Frog scheme

$$v_\nu(t+\Delta t) = v_\nu(t-\Delta t) + \lambda a\bigl[v_{\nu+1}(t) - v_{\nu-1}(t)\bigr], \quad 0 < \lambda a < 1, \quad \nu = 1,2,3,\dots,$$

with oblique extrapolation of order $\ell-1$ at the boundary:

$$v_0(t+\Delta t) = \sum_{j=1}^{\ell} \binom{\ell}{j}(-1)^{j+1} v_j\bigl[t - (j-1)\Delta t\bigr].$$

We have

$$P(z,\kappa) = 1 - z^{-2} - \lambda a z^{-1}(\kappa - \kappa^{-1}),$$

so

$$\frac{\partial P}{\partial z} \cdot \frac{\partial P}{\partial \kappa}\bigg|_{z=\kappa=-1} = \frac{-1}{\lambda a} < 0.$$

Also,

$$\Omega(z,\kappa) \geqslant |P(z,\kappa)| > 0 \quad \text{for } z = \kappa \neq \pm 1,$$

and

$$\Omega(z,\kappa) \geqslant |R(z,\kappa)| = |1 - z^{-1}\kappa|^{\ell} > 0 \quad \text{for } z \neq \kappa.$$

Hence, (10a,c-e) hold, and by Theorem 2 stability follows.

REFERENCES

[1] M. Goldberg and E. Tadmor, Scheme-independent stability criteria for difference approximations of hyperbolic initial-boundary value problems. II, *Math. Comp.* 36 (1981), 605-626.

[2] M. Goldberg and E. Tadmor, Convenient stability criteria for difference approximations of hyperbolic initial-boundary value problems, *Math. Comp.* 44 (1985), 361-377.

[3] M. Goldberg and E. Tadmor, Convenient stability criteria for difference approximations of hyperbolic initial-boundary value problems. II, *Math. Comp.* 48 (1987), 503-520.

[4] B. Gustafsson, H.-O. Kreiss and A. Sundström, Stability theory of difference approximations for mixed initial boundary value problems. II, *Math. Comp.* 26 (1972), 649-686.

Hyperbolic Heat Transfer Problems with Phase Transitions

J.M. Greenberg*
Department of Mathematics and Statistics
UMBC
Catonsville, Maryland 21228

*This research was partially supported by the Air Force Office of Scientific Research.

1. Problem Formulation

In this note we consider melt problems for solid heat conductors. The basic conservation law is the balance of energy

$$\rho_1 \hat{e}_{\hat{t}} + \hat{q}_{\hat{x}} = 0, \tag{1.1}$$

where $\hat{e}$ is the internal energy density, $\hat{q}$ the heat flux, and ρ_1 the constant mass density of the material in its reference configuration. We shall replace the Fourier law: $\hat{q} = -k_1 \hat{T}_{\hat{x}}$ by the first order relaxation process

$$\delta_1 \hat{q}_{\hat{t}} + \hat{q} + k_1 \hat{T}_{\hat{x}} = 0, \quad \delta_1 > 0 \text{ and } k_1 > 0, \tag{1.2}$$

where $\hat{T}$ is the local temperature. The internal energy is given by

$$\hat{e} = c_1 \hat{T} + l\phi, \tag{1.3}$$

where

$$c_1 = \frac{\partial \hat{e}}{\partial \hat{T}} > 0 \quad \text{and} \quad l = \frac{\partial \hat{e}}{\partial \phi} > 0 \tag{1.4}$$

are the specific and latent heats respectively. ϕ is an order parameter or "phase field" variable. Here ϕ can be identified with the strain or specific volume at a material point $\hat{x}$ at time $\hat{t}$; that is the strain $\hat{\gamma}$ can be recaptured from ϕ via

$$\hat{\gamma} = f(\phi), \tag{1.5}$$

where f is a nonnegative increasing function. The order parameter ϕ is assumed to satisfy a Landau-Ginzburg type equation

$$\delta_2 \phi_{\hat{t}} - k_2 \phi_{\hat{x}\hat{x}} = -\phi(\phi^2 - 1) + \mu_2 \hat{T}, \tag{1.6}$$

where δ_2, k_2, and μ_2 are positive constants.[1]

It is convenient to scale (1.1)–(1.6) so that the coefficients in (1.1)–(1.3) transform to unity. We let

$$\frac{\partial}{\partial \hat{t}} = \frac{1}{\delta_1}\frac{\partial}{\partial t}, \quad \frac{\partial}{\partial \hat{x}} = \sqrt{\frac{c_1 \rho_1}{\delta_1 k_1}}\frac{\partial}{\partial x}, \quad \hat{e} = le,$$

$$\hat{T} = \frac{l}{c_1}T, \quad \text{and} \quad \hat{q} = l\sqrt{\frac{k_1 \rho_1}{c_1 \delta_1}}q. \tag{1.7}$$

[1]Similar models have been considered by Caginalp et. al. when the Fourier law is applicable; see [1–5].

Then, (1.1)–(1.3) become

$$e = T + \phi, \tag{1.8}$$

$$e_t + q_x = 0, \tag{1.9}$$

and

$$q_t + q + T_x = 0. \tag{1.10}$$

The Landau-Ginzburg equation transforms to

$$\lambda\phi_t - \lambda^2 a^2 \phi_{xx} = -\phi(\phi^2 - 1) + \frac{2T}{3^{\frac{3}{2}}T_*}, \tag{1.11}$$

where

$$\lambda = \frac{\delta_2}{\delta_1}, \quad a^2 = \frac{c_1\rho_1\delta_1 k_2}{k_1\delta_2^2}, \quad \text{and} \quad T_* = \frac{2}{3^{\frac{3}{2}}}\frac{c_1}{l\mu}. \tag{1.12}$$

Our interest is in this system when

$$0 < \lambda << 1, 0 < a^2, \quad \text{and} \quad 0 < \frac{c_1}{l\mu} \leq 1. \tag{1.13}$$

One might surmise that in this parameter range the equation (1.11) rapidly equilibrates and that the heat flow process is adequately described by (1.9) and (1.10) where e and T are given parametrically in of ϕ by

$$e = \phi + \frac{3^{\frac{3}{2}}T_*\phi(\phi^2 - 1)}{2} \quad \text{and} \quad T = \frac{3^{\frac{3}{2}}T_*\phi(\phi^2 - 1)}{2}.\text{[2]} \tag{1.14}$$

In fact, the reduced system does yield a satisfactory description of the heat-flow process so long as ϕ is in the solid phase ($\phi < -\frac{1}{\sqrt{3}}$) or the liquid phase ($\phi > \frac{1}{\sqrt{3}}$). In both regions the reduced equations (1.9), (1.10), and (1.14) represent a genuinely nonlinear hyperbolic system and the standard theory for such systems yields a physically correct description of the underlying thermal process. Difficulties arise when we wish to consider situations where both the solid and liquid phases are present.

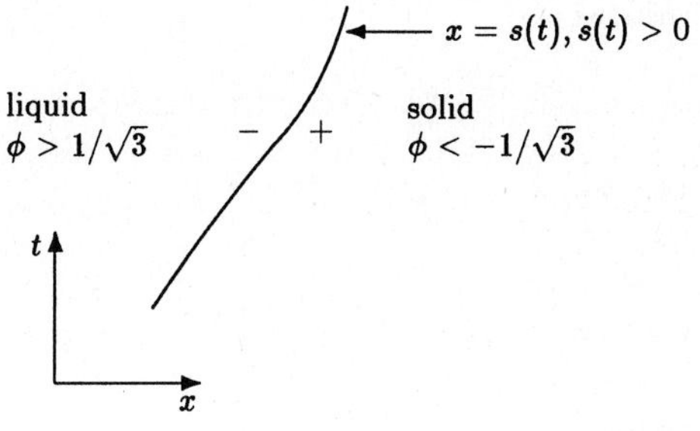

Figure 1

To see this, consider the configuration shown in Figure 1. We assume liquid to the left of a melt interface (shock) $x = s(t), \dot{s}(t) > 0$, and solid to the right. In each region the reduced system

[2]The constraint $0 < T_* \leq \frac{2}{3^{\frac{3}{2}}}$ guarantees that the map $\phi \rightarrow e(\phi)$ is 1-1 and thus that T can be written as a function of e.

holds and we insist that (1.9) and (1.10) hold as conservation laws in the whole x-t plane. Then, across the interface we obtain the Rankine-Hugoniot equations:

$$\dot{s}(e(\phi_-) - e(\phi_+)) = (q_- - q_+), \tag{1.15}$$

and

$$\dot{s}(q_- - q_+) = (T(\phi_-) - T(\phi_+)) \tag{1.16}$$

where $e(\cdot)$ and $T(\cdot)$ are given by (1.14). These, of course, are equivalent to

$$0 \leq \dot{s} = \sqrt{\frac{T(\phi_-) - T(\phi_+)}{e(\phi_-) - e(\phi_+)}}, \tag{1.17}$$

and

$$q_- - q_+ = \sqrt{(T(\phi_-) - T(\phi_+))(e(\phi_-) - e(\phi_+))}. \tag{1.18}$$

Equation (1.17) does imply that the states $\phi_- > \frac{1}{\sqrt{3}}$ and $\phi_+ < -\frac{1}{\sqrt{3}}$ must satisfy $T(\phi_-) - T(\phi_+) \geq 0$, but places no other constraints upon the phase field ϕ. This calculation naturally leads us to question the physical relevancy of states (ϕ_-, q_-) and (ϕ_+, q_+) satisfying (1.17) and (1.18) and the constraints $\phi_- > \frac{1}{\sqrt{3}}$, $\phi_+ < \frac{-1}{\sqrt{3}}$, and $T(\phi_-) \geq T(\phi_+)$.

To see which solutions of (1.17) and (1.18) are meaningful we go back to the full system (1.8)–(1.12) when (1.13) holds. We introduce inner variables

$$y = \frac{x}{\lambda} \quad \text{and} \quad \tau = \frac{t}{\lambda}, \tag{1.19}$$

and rewrite the equation in the stretched coordinates. The result is

$$e = T + \phi, \tag{1.20}$$

$$e_\tau + q_y = 0, \tag{1.21}$$

$$q_\tau + T_y = -\lambda q, \tag{1.22}$$

and

$$\phi_\tau - \phi_{yy} = -\phi(\phi^2 - 1) + \frac{2T}{3^{\frac{3}{2}}T_*}. \tag{1.23}$$

We now look for right facing travelling wave solutions of (1.20)–(1.23) in the $\lambda = 0^+$ limit. These solutions are functions of

$$\xi = y - c\tau = \frac{x - ct}{\lambda}, c > 0, \tag{1.24}$$

and satisfy $e = e_- + \dfrac{(\phi - \phi_-)}{1 - c^2}$, $\quad T = T_- + \dfrac{c^2(\phi - \phi_-)}{1 - c^2}$, $\quad q = q_- + \dfrac{c(\phi - \phi_-)}{1 - c^2}$ and

$$a^2\phi_{\xi\xi} + c\phi_\xi = \phi(\phi^2 - 1) - \frac{2}{3^{\frac{3}{2}}T_*}\left(T_- + \frac{c^2}{1 - c^2}(\phi - \phi_-)\right), \tag{1.25}$$

and the boundary condition

$$\lim_{\xi \to -\infty} \phi(\xi) = \phi_- > \frac{1}{\sqrt{3}}. \tag{1.26}$$

At $\xi = -\infty$ we assume the equilibrium constitutive equations (1.14) hold. This guarantees that

$$e_- = \phi_- + \frac{3^{\frac{3}{2}}T_*\phi_-(\phi_-^2 - 1)}{2} \quad \text{and} \quad T_- = \frac{3^{\frac{3}{2}}T_*\phi_-(\phi_-^2 - 1)}{2} \tag{1.27}$$

where again $0 < T_* \leq \frac{2}{3^{\frac{3}{2}}}$. Equation (1.27) reduces the problem to finding $c > 0$ and $\phi(\cdot)$ such that

$$a^2\phi_{\xi\xi} + c\phi_\xi = \phi(\phi^2 - 1) - \phi_-(\phi_-^2 - 1) - \frac{2c^2(\phi - \phi_-)}{3^{\frac{3}{2}}T_*(1 - c^2)} \tag{1.28}$$

and

$$\lim_{\xi \to -\infty} \phi(\xi) = \phi_- > \frac{1}{\sqrt{3}}. \tag{1.29}$$

The solutions representing a phase-change have the further property that

$$\lim_{\xi \to \infty} \phi(\xi) = \phi_+ < -\frac{1}{\sqrt{3}}. \tag{1.30}$$

2. Travelling Waves

In this section we shall investigate the travelling wave problem formulated in the last section. We seek $c > 0$ and $\phi(\cdot)$ such that

$$a^2\phi_{\xi\xi} + c\phi_\xi = \phi(\phi^2 - 1) - \phi_-(\phi_-^2 - 1) - \frac{2c^2(\phi - \phi_-)}{3^{\frac{3}{2}}T_*(1 - c^2)}, \tag{2.1}$$

and

$$\lim_{\xi \to -\infty} \phi(\xi) = \phi_- > \frac{1}{\sqrt{3}}. \tag{2.2}$$

For $\frac{1}{\sqrt{3}} < \phi_- < 1$ we have

$$\phi(\phi^2 - 1) - \phi_-(\phi_-^2 - 1) = (\phi - \phi_L)(\phi - \phi_I)(\phi - \phi_-) \tag{2.3}$$

where

$$\phi_L = \frac{-\phi_-}{2} - \frac{1}{2}\sqrt{4 - 3\phi_-^2} < 0 < \phi_I = \frac{-\phi_-}{2} + \frac{1}{2}\sqrt{4 - 3\phi_-^2} < \phi_- \tag{2.4}$$

and the inequality $\phi_I > 0$ guarantees there is no solution of (2.1) and (2.2) satisfying $\phi_+ = \lim_{\xi\to\infty} \phi(\xi) < -\frac{1}{\sqrt{3}}$.

For $1 < \phi_- \leq \frac{2}{\sqrt{3}}$ the factorization (2.3) and (2.4) holds except now $-\frac{1}{\sqrt{3}} < \phi_I(\phi_-) < 0$. Additionally, for $0 < c^2 < \frac{3^{\frac{3}{2}}T_*(3\phi_-^2 - 1)}{2 + 3^{\frac{3}{2}}T_*(3\phi_-^2 - 1)}$, we have

$$\phi(\phi^2 - 1) - \phi_-(\phi_-^2 - 1) - \frac{2c^2(\phi - \phi_-)}{3^{\frac{3}{2}}T_*(1 - c^2)} = (\phi - \Psi_L)(\phi - \Psi_I)(\phi - \phi_-) \tag{2.5}$$

where

$$\Psi_L = \frac{-\phi_-}{2} - \frac{1}{2}\sqrt{4 - 3\phi_-^2 + \frac{8c^2}{3^{\frac{3}{2}}T_*(1 - c^2)}} < \Psi_I = \frac{-\phi_-}{2} + \frac{1}{2}\sqrt{4 - 3\phi_-^2 + \frac{8c^2}{3^{\frac{3}{2}}T_*(1 - c^2)}} < \phi_-. \tag{2.6}$$

In this parameter regime the travelling waves are given by

$$\phi(\xi) = \frac{\phi_- + \Psi_L}{2} - \frac{(\phi_- - \Psi_L)}{2} \tanh\left(\left(\frac{\phi_- - \Psi_L}{2^{\frac{1}{2}}a}\right)(\xi - \alpha_0)\right) \tag{2.7}$$

where $c > 0$ satisfies

$$c = \frac{3a}{2^{\frac{3}{2}}}\left(\phi_- - \sqrt{4 - 3\phi_-^2 + \frac{8c^2}{3^{\frac{3}{2}}T_*(1-c^2)}}\right) \tag{2.8}$$

and α_0 is an arbitrary phase shift. The state $\phi_+ < -\frac{1}{\sqrt{3}}$ is given by

$$\phi_+ = \Psi_L > -1.^3 \tag{2.9}$$

When $\phi_- = 1, c = 0$ and $\Psi_L = -1$ and the travelling wave is stationary. This solution corresponds to the equal area Maxwell line on a $T - e$ diagram (see (1.14)). It should be noted that when $1 < \phi_- \leq \frac{2}{\sqrt{3}}, 4 - 3\phi_-^2 \geq 0$ and $\phi_- - \sqrt{4 - 3\phi_-^2} > 0$ and that (2.8) has a unique positive solution for any $a > 0$.

Similar results obtain when $\frac{2}{\sqrt{3}} < \phi_-$. Here, the issue is whether (2.8) has a solution, c, in the interval $\sqrt{\frac{3^{\frac{3}{2}}T_*(3\phi_-^2 - 4)}{8 + 3^{\frac{3}{2}}T_*(3\phi_-^2 - 4)}} \leq c \leq \sqrt{\frac{3^{\frac{3}{2}}T_*(3\phi_-^2 - 1)}{8 + 3^{\frac{3}{2}}T_*(3\phi_-^2 - 1)}}$. In general, (2.8) will not have such a solution for all $a > 0$.

Three points are worthy of note about these travelling waves. The first is that their speed, c, satisfies

$$c < \sqrt{\frac{T'(\phi_-)}{e'(\phi_-)}} \quad \text{and} \quad c < \sqrt{\frac{T'(\phi_+)}{e'(\phi_+)}} \tag{2.10}$$

where $e(\cdot)$ and $T(\cdot)$ are the constitutive functions defined in (1.14) and prime denotes differentiation. *Thus, the melt interface moves subsonically relative to the states at* $\xi = \pm\infty$. The second is that both the speed of these interfaces and the downstream state ϕ_+ are completely determined by the upstream state ϕ_-. The third point is that speed c depends upon the higher order physics of the problem through the parameter a and thus the jump relations for the reduced description (equations (1.15) and (1.16)) do not suffice to determine the melt interface. We will have more to say about this in section 3 when we discuss the Stefan problem (see (3.4) and (3.5)).

We conclude this section by noting there are other travelling wave solutions to (2.1) and (2.2) when $\phi_- > \frac{1}{\sqrt{3}}$. These solutions do not represent change of phase waves but rather shocks in the liquid phase; that is $\frac{1}{\sqrt{3}} < \phi_+ < \phi_-$. If $0 < a << 1$, these solutions exist for any $\frac{1}{\sqrt{3}} < \phi_+ < \phi_-$ with speed c satisfying

$$\sqrt{\frac{T'(\phi_+)}{e'(\phi_+)}} < c = \sqrt{\frac{T(\phi_-) - T(\phi_+)}{e(\phi_-) - e(\phi_+)}} < \sqrt{\frac{T'(\phi_-)}{e'(\phi_-)}}. \tag{2.11}$$

3. Stefan Problem

In the Stefan Problem one seeks a solution to (1.8)–(1.11) in the quarter plane $x > 0$ and $t > 0$ which satisfies the following initial and boundary conditions:

$$\phi(x,0) = -1, \quad T(x,0) = 0, \quad e(x,0) = -1, \quad \text{and} \quad q(x,0) = 0, x > 0, \tag{3.1}$$

and

[3] That the last inequality is true in a neighborhood of $\phi_- = 1$ follows from $\frac{d\Psi_L}{d\phi_-}|_{\phi_-=1} = \frac{1}{4}$.

$$\phi(0,t) = \phi_0 \in (1, \frac{2}{\sqrt{3}}) \quad \text{and} \quad T(0,t) = \frac{3^{\frac{3}{2}} T_*}{2} \phi_0(\phi_0^2 - 1) > 0, t > 0. \tag{3.2}$$

The initial condition guarantees the material is in the solid phase at the nominal melt temperature at $t = 0$, and the boundary condition implies that the end $x = 0$ is in the liquid phase. Our interest is in the reduced description of this problem. Here we let

$$e = \phi + \frac{3^{\frac{3}{2}} T_* \phi(\phi^2 - 1)}{2} \quad \text{and} \quad T = \frac{3^{\frac{3}{2}} T_* \phi(\phi^2 - 1)}{2} \tag{3.3}$$

and seek a weak solution of (1.8)–(1.10) and (3.3) satisfying the initial and boundary conditions (3.1) and (3.2). The structure of the solution is shown in Figure 2.

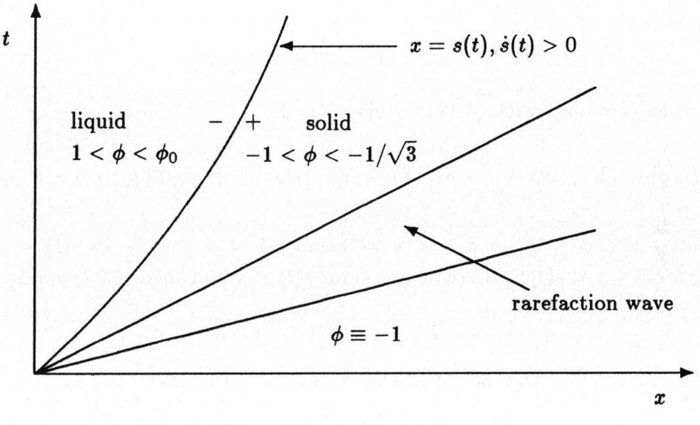

Figure 2

At the melt interface we impose the boundary conditions:

$$\dot{s} = \hat{c}(\phi_-, a) \tag{3.4}$$

where $\hat{c}(\phi_-, a)$ is the unique solution of (2.8) and

$$\phi_+ = \Psi_L(\phi_-, \hat{c}(\phi_-, a)) = \frac{-\phi_-}{2} - \frac{1}{2}\sqrt{4 - 3\phi_-^2 + \frac{8\hat{c}^2(\phi_-, a)}{3^{\frac{3}{2}} T_*(1 - \hat{c}^2(\phi_-, a))}}.^4 \tag{3.5}$$

The calculations leading to (1.25) imply that across the melt interface

$$e(\phi_+) - e(\phi_-) = \frac{(\phi_+ - \phi_-)}{1 - \hat{c}^2(\phi_-, a)}, \tag{3.6}$$

$$T(\phi_+) - T(\phi_-) = \frac{\hat{c}^2(\phi_-, a)(\phi_+ - \phi_-)}{1 - \hat{c}^2(\phi_-, a))}, \tag{3.7}$$

and

$$q_+ - q_- = \frac{\hat{c}(\phi_-, a)(\phi_+ - \phi_-)}{(1 - \hat{c}^2(\phi_-, a))}, \tag{3.8}$$

[4]These are just the admissibility conditions derived in (2.8) and (2.9).

and thus that the Rankine-Hugoniot equations for (1.8)–(1.10) hold

$$\hat{c}(\phi_-, a)(e(\phi_+) - e(\phi_-)) = (q_+ - q_-) \tag{3.9}$$

$$\hat{c}(\phi_-, a)(q_+ - q_-) = (T(\phi_+) - T(\phi_-)). \tag{3.10}$$

We conclude with a set of identities which give a growth estimate for the melt interface. A similar estimate was obtained by Greenberg in [6]. We observe that solutions of the reduced system (1.8)–(1.10) and (3.4) and (3.5) satisfy

$$\frac{d}{dt}\left(\int_0^{s(t)} x(e(\phi) - 1)(x,t)\,dx + \int_{s(t)}^{\infty} x(e(\phi) + 1)(x,t)\,dx + s^2(t)\right) = \int_0^{\infty} q(x,t)\,dx \tag{3.11}$$

and

$$\frac{d}{dt}\int_0^{\infty} q(x,t)\,dx + \int_0^{\infty} q(x,t)\,dx = T(\phi_0) > 0. \tag{3.12}$$

(3.12) implies that

$$\int_0^{\infty} q(x,t)dx = T(\phi_0)(1 - e^{-t}) \tag{3.13}$$

and (3.13), when combined with (3.11), yields

$$\int_0^{s(t)} x(e(\phi) - 1)(x,t)dx + \int_{s(t)}^{\infty} x(e(\phi) + 1)(x,t)dx + s^2(t) = T(\phi_0)(t + e^{-1} - 1). \tag{3.14}$$

The inequalities $1 < \phi \leq \phi_0$ in $0 \leq x < s(t)$ and $-1 < \phi < -\frac{1}{\sqrt{3}}$ in $s(t) < x$ imply that $e(\phi) - 1 > 0$ for $0 \leq x < s(t)$ and $e(\phi) + 1 > 0$ for $s(t) < x$ and then (3.14) yields

$$s^2(t) \leq T(\phi_0)(t + e^{-t} - 1). \tag{3.15}$$

The interesting fact is that this estimate is independent of the parameter a.

REFERENCES

[1] Caginalp, G., An analysis of a phase field model of a free boundary, *Archive for Rational Mechanics and Analysis* **92**. 205–245 (1986).

[2] Caginalp, G., Phase field models of solidification: Free boundary problems as systems of nonlinear parabolic differential equations, *Free Boundary Problems: Applications and Theory*, A. Bossavit, et.al. 107–121 (1985).

[3] Caginalp, G., Solidification problems as systems of nonlinear differential equations, *Lectures in Applied Mathematics* **23**, 347–369 (1986).

[4] Caginalp, G., and Fife, P.C., Phase-field methods for interfacial boundaries, *Physical Review B* **33**, 7792–7794 (1986).

[5] Caginalp, G, and Fife, P.C., Dynamics of layered interfaces arising from phase boundaries, *SIAM Journal of Applied Mathematics* (to appear).

[6] Greenberg, J.M., A Hyperbolic Heat Transfer Problem with Phase Changes, *I.M.A. Jour. of Appl. Math.* **38**, 1–21 (1987).

UNSYMMETRIC HYPERBOLIC SYSTEMS AND ALMOST INCOMPRESSIBLE FLOW

Bertil Gustafsson
Dept. of Scientific Computing,
Uppsala University, Sturegatan 4B 2tr,
S-752 23 Uppsala, Sweden

1. INTRODUCTION

When solving flow problems at low Mach-numbers ε, the coefficient matrices for the inviscid part becomes very unsymmetric. This leads to severe difficulties when constructing numerical methods, since the problem is ill conditioned. We shall discuss the symmetrization procedure for the Euler equations. For smooth solutions it is possible to express the solution as an asymptotic expansion in ε. By defining a new dependent variable closely related to the speed of sound and subtracting the first term in the expansion, we arrive at a symmetric system which is convenient for computation. Even with the symmetric system, a fundamental difficulty remains. The low Mach-numbers give rise to time-scales of different magnitude, since the sound waves are much faster than the advection waves. If the fast waves are present in the solution and are of interest, then an explicit method is natural to use for the time dependent problem.The stability limit on the time step is no real restriction, since the fast waves must be resolved anyway. In the case where the fast waves are negligible, the situation is different. There is no need to take small time steps of accuracy reasons, and implicit or semi-implicit methods should be used. As we shall see, the form of the equations which is the result of our symmetrization is very convenient for constructing semi-implicit approximations.

Of special interest is the limit solutions as $\varepsilon \to 0$ corresponding to incompressible flow. The limit system is

$$
\begin{aligned}
&(a) \quad u_t + uu_x + vu_y + p_x = 0 \\
&(b) \quad v_t + uv_x + vv_x + p_y = 0 \qquad (1.1) \\
&(c) \quad u_x + v_y = 0
\end{aligned}
$$

which is time incomplete. Consequently regular methods for hyperbolic systems cannot be applied. Many ways of overcoming this difficulty have been proposed, and we shall not try to list all such methods. However, most methods for the time dependent problem are based on a different form of the system, even when the primitive variables u, v, p are kept. One such form is obtained by differentiating and adding the first two equations, which substitute the third:

$$
\begin{aligned}
u_t + uu_x + uv_y + p_x &= 0 \\
v_t + uv_x + vv_y + p_y &= 0 \\
p_{xx} + p_{yy} &= f \quad .
\end{aligned}
$$

Here f depends on the velocity components and its space derivatives. For each time step the Poisson equation is solved for p. One difficulty with this approach is to keep the divergence zero or small.

Another approach is to introduce p_t as an extra term in (1.1c). This is called the artificial compressibility method, and was originally introduced by Chorin [1] for the Navier-Stokes equations. In this way the sound waves are brought back into the solution, but now with an artificial speed of propagation. The true solution is obtained only at steady state.

Chorin [2] also introduced another way of solving the true time-dependent equations directly with difference methods. It is applied to the Navier-Stokes equations, and is based on a decomposition of the solution into one part with zero divergence and another part with zero curl.

Finally it should be mentioned that finite element methods can be directly applied to (1.1). However, there are some practical difficulties to construct basis functions which satisfy the divergence condition, in particular near the boundaries. We propose solution methods which are based on the true compressible equations. (After all, there is always some compressibility in any fluid.) We transform these equations into an appropriate form for computation. For the linearized problem we prove that the solutions converge to solutions of the corresponding incompressible problem, both for open boundaries and solid wall boundaries.

Some of the material in Section 2 and 3 is found in [4], [5], [6]. References of immediate interest are also found in these papers.

2. THE EULER EQUATIONS

The Euler equations are

$$\bar{U}_t + \bar{A}(\bar{U})\bar{U}_x + \bar{B}(\bar{U})\bar{U}_y = 0$$

where

$$\bar{U} = \begin{pmatrix} \bar{p} \\ \bar{u} \\ \bar{v} \end{pmatrix}, \quad \bar{A}(\bar{U}) = \begin{pmatrix} u & \rho a^2 & 0 \\ 1/\rho & u & 0 \\ 0 & 0 & u \end{pmatrix}, \quad \bar{B}(\bar{U}) = \begin{pmatrix} v & 0 & \rho a^2 \\ 0 & v & 0 \\ 1/\rho & 0 & v \end{pmatrix}$$

$$\rho = \kappa\rho^{\gamma}, \quad a^2 = dp/d\rho.$$

We use a γ-law for the pressure here but we could as well use any equation of state of the form $p = g(\rho)$ where g is a smooth function of ρ. Non-dimensional varables are introduced by

$$\frac{x}{x_*} \to x, \quad \frac{y}{x_*} \to y, \quad \frac{tu_*}{x_*} \to t, \quad \frac{u}{u_*} \to u, \quad \frac{v}{u_*} \to v,$$

$$\frac{p}{\rho_* u_*^2} \to p, \quad \frac{\rho}{\rho_*} \to \rho$$

where * indicates a typical value of the corresponding variable. Let a_* be the typical speed of sound, and define the Mach-number ε by $\varepsilon = u_*/a_*$. Then the Euler equations take the non-dimensional form

$$U_t + A(U)U_x + B(U)U_y = 0$$

where

$$U = \begin{pmatrix} p \\ u \\ v \end{pmatrix}, \quad A(U) = \begin{pmatrix} u & \rho^\gamma/\varepsilon^2 & 0 \\ 1/\rho & u & 0 \\ 0 & 0 & u \end{pmatrix}, \quad B(U) = \begin{pmatrix} v & 0 & \rho^\gamma/\varepsilon^2 \\ 0 & v & 0 \\ 1/\rho & 0 & v \end{pmatrix}$$

$$p = \frac{\rho^\gamma}{\gamma\varepsilon^2}.$$

The non-dimensional speed of sound is

$$a = \frac{1}{\varepsilon}\rho^{\frac{\gamma-1}{2}}. \tag{2.1}$$

We shall consider small Mach-numbers ε. Then there are two drawbacks with this system. The pressure variable takes very large values and the coefficient matrices are very unsymmetric. In order to illustrate the effect of the unsymmetric coefficients we consider the simple model problem

$$U_t + AU_x = 0, \quad A = \begin{pmatrix} 0 & 1/\varepsilon^2 \\ 1 & 0 \end{pmatrix}, \quad U = \begin{pmatrix} p \\ u \end{pmatrix}$$

with periodic boundary conditions. With

$$T = \begin{pmatrix} 1/\varepsilon & 0 \\ 0 & 1 \end{pmatrix}, \quad V = T^{-1}U \tag{2.2}$$

the system becomes

$$V_t + \tilde{A}V_x = 0, \quad \tilde{A} = \begin{pmatrix} 0 & 1/\varepsilon \\ 1/\varepsilon & 0 \end{pmatrix} \tag{2.3}$$

which is energy-conserving, i.e. for the L_2-norm we have

$$\|V(t)\| = \|V(0)\|.$$

This leads to the inequality

$$\|U(t)\| \le \mathrm{cond}(T)\|U(0)\|, \quad \mathrm{cond}(T) = |T|\cdot|T^{-1}|. \tag{2.4}$$

In our case cond(T) = $1/\varepsilon$, and the problem is obviously ill conditioned even if it is formarly well posed for any fixed value of ε. The estimate (2.4) is sharp. For example, the initial condition

$$p(x,0) = 0$$
$$u(x,0) = f(x)$$

gives the solution

$$p = \frac{1}{2\varepsilon}[f(x-t/\varepsilon)-f(x+t/\varepsilon)]$$
$$u = \frac{1}{2}\,[f(x-t/\varepsilon)+f(x+t/\varepsilon)]\,.$$

The unsymmetry of the coefficient matrix causes the variable p to become large even when f is a bounded function.

When applying numerical methods we must expect the same type of ill-conditioning. Consider the Crank-Nicholson method with centered difference operators in space

$$\left(I+\frac{k}{2}AD_0\right)U^{n+1} = \left(I-\frac{k}{2}AD_0\right)U^n$$

where k is the time-step. The Fourier transformed system is

$$\left(I+\frac{\lambda}{2}Ai\sin\xi\right)U^{n+1} = \left(I-\frac{\lambda}{2}Ai\sin\xi\right)U^n, \quad \lambda = \frac{k}{h}$$

(h is the step-size in space).
Let T be defined by (2.2) and let $\hat{V} = T^{-1}\hat{U}$. We get

$$\left(I+\frac{\lambda}{2}\hat{A}i\sin\xi\right)\hat{V}^{n+1} = \left(I-\frac{\lambda}{2}\hat{A}i\sin\xi\right)\hat{V}^{n},$$

where $\tilde{A}$ is defined in (2.3). This system is a direct approximation of the Fourier transformed version of (2.3). Since $\tilde{A}$ is symmetric, there is a unitary matrix S such that with W = S*V we obtain a diagonal system

$$\left(I+\frac{\lambda}{2}Di\sin\xi\right)\hat{W}^{n+1} = \left(I-\frac{\lambda}{2}Di\sin\xi\right)\hat{W}^{n},$$

$$D = \frac{1}{\varepsilon}\begin{pmatrix}1 & 0\\ 0 & -1\end{pmatrix}, \quad |\hat{W}| = |\hat{V}|.$$

Each component obviously fulfills

$$|\hat{W}^{(\nu)n+1}| = |\hat{W}^{(\nu)n}|, \quad \nu=1,2$$

and we get for the original vector

$$|\hat{U}^{n+1}| \le |T|\cdot|\hat{V}^{n+1}| = |T|\cdot|\hat{W}^{n+1}| = |T|\cdot|\hat{W}^{n}| = \ldots = |T|\cdot|\hat{W}^{0}|$$
$$= |T|\cdot|\hat{V}^{0}| \le |T|\cdot|T^{-1}|\cdot|\hat{U}^{0}| = \mathrm{cond}(T)|\hat{U}^{0}|.$$

Parseval's relation then gives

$$\|U^n\| \le \mathrm{cond}(T)\|U^0\| \tag{2.5}$$

which is a direct analogue of (2.4). However, in practice we must expect even more severe difficulties than the ill-conditioning represented by (2.5). For each time-step we must solve a linear system of equations. In Fourier space it has the form

$$\hat{C}\hat{U}^{n+1} = \hat{F},$$

$$\hat{C} = \begin{pmatrix} 1 & \dfrac{\lambda}{2\varepsilon^2}\mathrm{i}\sin\xi \\ \dfrac{\lambda}{2}\mathrm{i}\sin\xi & 1 \end{pmatrix}.$$

If, for example, there is a perturbation $\delta\hat{F}$ in the right hand side, a well known result from linear algebra shows that the perturbation δU^{n+1} in the solution satisfies

$$\frac{|\delta\hat{U}^{n+1}|}{|\hat{U}^{n+1}|} \le \mathrm{cond}(\hat{C}) \frac{|\delta\hat{F}|}{|\hat{F}|} .$$

In our case

$$|\hat{C}| = \mathcal{O}(1/\varepsilon^2)$$
$$|\hat{C}^{-1}| = \mathcal{O}(1),$$

i.e.

$$\mathrm{cond}(\hat{C}) = \mathcal{O}(1/\varepsilon^2).$$

This shows that even rounding errors will have a serious effect if ε is very small. The conclusion is that the unsymmetric form of the system is inappropriate for computation.

3. SYMMETRIC FORM OF THE EULER EQUATIONS

The most straightforward way of eliminating the major part of the unsymmetry is to make a simple scaling of the pressure

$$\tilde{p} = \varepsilon p$$

i.e. the nondimensionalization in Section 2 is done differently. This gives the coefficient matrices

$$A(U) = \begin{pmatrix} u & \rho^\gamma/\varepsilon & 0 \\ 1/(\rho\varepsilon) & u & 0 \\ 0 & 0 & u \end{pmatrix}, \quad B(U) = \begin{pmatrix} v & 0 & \rho^\gamma/\varepsilon \\ 0 & v & 0 \\ 1/(\rho\varepsilon) & 0 & v \end{pmatrix}$$

and since ρ is of the order 1, these matrices are almost symmetric. However, the new variable p is still of the order $1/\varepsilon$ which is inconvenient. It is well known that the error in the solution of a stable method is bounded by δ/k where δ is the absolute rounding error and k is the time step. This is normally accepted. If the dependent variable are all of the order 1, the large coefficients in the matrices introduces rounding errors $\delta = \mathcal{O}(1/\varepsilon)$ and the error in the solution is $(k^{-1}\varepsilon^{-1})$ which must be accepted. However, in our case, when computing $p_x/(\rho\varepsilon)$ we get $\delta = (1/\varepsilon^2)$, and the accumulated error may be unacceptable.

The natural scaling of p seems to be $\varepsilon^2 p \to \tilde{p}$. However, this gives the coefficient matrices

$$A(U) = \begin{pmatrix} u & \rho^\gamma & 0 \\ 1/(\rho\varepsilon^2) & u & 0 \\ 0 & 0 & u \end{pmatrix}, \quad B(U) = \begin{pmatrix} v & 0 & \rho^\gamma \\ 0 & v & 0 \\ 1/(\rho\varepsilon^2) & 0 & v \end{pmatrix}$$

and we are back to an ill conditioned system.

When considering smooth solutions corresponding to almost incompressible flow, the momentum equations show that p_x and p_y are bounded independent of ε. This means that the large part of p is almost constant, and it is natural to define a new variable by subtracting this part out. In order to have the matrices exactly symmetric, we use the transformation

$$\phi = \frac{2}{(\gamma-1)\varepsilon}(\rho^{\frac{\gamma-1}{2}} - 1) .$$

When taking the definition (2.1) into account, it is seen that the new variable is closely related to the speed of sound:

$$\phi = \frac{2}{\gamma-1}\left(a - \frac{1}{\varepsilon}\right) .$$

The new system is in component form

$$(a)\ \phi_t + u\phi_x + \left(\frac{1}{\varepsilon}+c\right)u_x + v\phi_y + \left(\frac{1}{\varepsilon}+c\right)v_y = 0$$

$$(b)\ u_t + \left(\frac{1}{\varepsilon}+c\right)\phi_x + uu_x + vu_y = 0 \qquad (3.1)$$

$$(c)\ v_t + uv_x + \left(\frac{1}{\varepsilon}+c\right)\phi_y + vv_y = 0$$

$$c = \frac{\gamma-1}{2}\phi .$$

Next assume that all the variables and its first derivatives are bounded independent of ε. The equations (3.1b,c) imply that $\phi_x = \mathcal{O}(\varepsilon)$, $\phi_y = \mathcal{O}(\varepsilon)$, and with proper boundary conditions we therefore have $\phi = \mathcal{O}(\varepsilon)$. With $\tilde{\phi} = \phi/\varepsilon$ we let $\varepsilon \rightarrow 0$ and obtain from (3.1)

$$u_t + uu_x + vu_y + \tilde{\phi}_x = 0$$

$$v_t + uv_x + vv_y + \tilde{\phi}_y = 0 \qquad (3.2)$$

$$u_x + v_y = 0 .$$

This is the well known system for incompressible flow. The variable

$$\tilde{\phi} = \frac{2}{(\gamma-1)\varepsilon}\left(a-\frac{1}{\varepsilon}\right)$$

is usually denoted by p and called pressure. Apparently, the solutions to (3.1) converge to solutions to (3.2) if they are bounded together with its derivatives.

The system (3.1) is well suited for numerical methods. It has the form

$$U_t + \left(\frac{1}{\varepsilon}P_0+P_1\right)U = 0$$

where P_0 has constant coefficients. Hence, a semi-implicit method is efficient. In [4] we have proposed the leap-frog/backwards Euler difference scheme

$$\frac{U^{n+1}-U^{n-1}}{2k} + \frac{1}{\varepsilon} Q_0U^{n+1} + Q_1U^n = 0$$

where Q_0, Q_1 are centered difference approximations of P_0, P_1. The solution U^{n+1} is obtained efficiently since Q_0 is linear, and with a regular grid it has constant coefficients. The LU-decomposition of the coefficient matrix can be made once and for all, and the solution is obtained by two resubstitutions.
The scheme has proved to be very robust. This depends on the fact that the backwards Euler part has a damping property which keeps the divergence low when ε is small. We refer to [4] and [5] for details about the scheme.

4. ANALYSIS OF THE LINEARIZED EQUATIONS

In this section we consider the linearized system with constant coefficients corresponding to the state $\bar{U} = (\bar{\phi},\bar{u},\bar{v})$:

$$\text{(a)} \quad \phi_t = -\bar{u}\phi_x - \left(\frac{1}{\varepsilon}+\bar{c}\right)u_x - \bar{v}\phi_y - \left(\frac{1}{\varepsilon}+\bar{c}\right)v_y$$

$$\text{(b)} \quad u_t = -\bar{u}u_x - \left(\frac{1}{\varepsilon}+\bar{c}\right)\phi_x - \bar{v}u_y \qquad (4.1)$$

$$\text{(c)} \quad v_t = -\bar{u}v_x - \left(\frac{1}{\varepsilon}+\bar{c}\right)\phi_y - \bar{v}v_y$$

$$\bar{c} = \frac{\gamma-1}{2}\bar{\phi} .$$

We take the domain in space as the unit square, and it is assumed that the solutions are periodic in the y-direction. Since we shall derive estimates also for the derivatives, it is assumed that the initial and boundary conditions are compatible such that no discontinuities are introduced initially.
We consider the case $u > 0$ such that $x = 0$ is an inflow

boundary and $x = 1$ is an outflow boundary. Since we treat the linearized system, we prescribe homogeneous boundary conditions:

Inflow, $x = 0$

$$\phi_0 + \varepsilon\alpha u_0 = 0 \qquad (4.2)$$
$$v_0 = 0 .$$

Outflow, $x = 1$

$$u_1 = 0 \qquad (4.3)$$

(The notation U_0, U_1 is used for $U(0,y,t), U(1,y,t)$ respectively.)

Here $\alpha \neq 0$ is a constant.
The scalar product and the norm are defined by

$$(U,V) = \int_0^1\int_0^1 U^*V dxdy$$
$$\|U\|^2 = (u,U) .$$

We shall also use the boundary norm

$$|U(t)|_B = \left[\int_0^1 (|U(0,y,t)|^2 + |U(1,y,t)|^2)dy\right]^{1/2} .$$

We shall now prove

Theorem 4.1 Assume that the conditions (4.2), (4.3) hold with

$$\alpha > \frac{\bar{u}}{2} > 0 . \qquad (4.4)$$

Then there is a constant $\delta > 0$ such that the solutions to (4.1) satisfy

$$\|\frac{\partial^{i+j}U}{\partial y^i \partial t^j}(t)\|^2 + \delta\int_0^t |\frac{\partial^{i+j}U}{\partial y^i \partial t^j}(\tau)|_B^2 \, d\tau \le \|\frac{\partial^{i+j}U}{\partial y^i \partial t^j}(0)\| \tag{4.5}$$

$$\int_0^t\int_0^1 |\frac{\partial^{i+j}\phi_0}{\partial y^i \partial t^j}(\tau)|^2 dy d\tau \le \frac{\alpha^2\varepsilon^2}{\delta} \|\frac{\partial^{i+j}U}{\partial y^i \partial t^j}(0)\|^2 \tag{4.6}$$

for ε sufficiently small and for all non-negative integers i,j.

<u>Proof</u>. When using integration by parts and the boundary conditions we obtain

$$\frac{1}{2}\frac{d}{dt}\|\phi\|^2 = \frac{\bar{u}}{2}\varepsilon^2\alpha^2\int_0^1 |u_0|^2 dy - \left(\frac{1}{\varepsilon}+\bar{c}\right)\alpha\varepsilon\int_0^1 |u_0|^2 dy$$

$$- \frac{\bar{u}}{2}\int_0^1 |\phi_1|^2 dy + \left(\frac{1}{\varepsilon}+\bar{c}\right)(u,\phi_x) + \left(\frac{1}{\varepsilon}+\bar{c}\right)(v,\phi_y)$$

$$\frac{1}{2}\frac{d}{dt}\|u\|^2 = \frac{\bar{u}}{2}\int_0^1 |u_0|^2 dy - \left(\frac{1}{\varepsilon}+\bar{c}\right)(\phi_x,u)$$

$$\frac{1}{2}\frac{d}{dt}\|v\|^2 = -\frac{\bar{u}}{2}\int_0^1 |v_1|^2 dy - \left(\frac{1}{\varepsilon}+\bar{c}\right)(\phi_y,v) .$$

By adding these three equalities and integrating, we obtain (4.5) for i = j = 0. The boundary condition then implies (4.6) for i = j = 0.

By differentiating the system (4.1) and the boundary conditions (4.2), (4.3) with respect to t we obtain exactly the same system for U_t, U_{tt} e t c. The same procedure works also for U_y, U_{yy}, U_{yt}...; hence (4.5), (4.6) follows for all positive integers i,j.

Since the boundary conditions cannot be differentiated with respect to x, we must use a different technique in order to obtain estimates for the x-derivatives. We shall prove

<u>Theorem 4.2</u>. Assume that the assumptions given in Theorem 4.1 hold. Then there are constants δ_1, C_1, C_2 such that the solutions to (4.1) satisfy

$$\|U_x(t)\|^2 + \delta_1\int_0^t |U_x(\tau)|_B^2 \, d\tau \le C_1(\|U_x(0)\|^2+\|U_y(0)\|2+\|U_t(0)\|^2) \tag{4.7}$$

$$\int_0^t\int_0^1 [(u_x)_0^2+(\phi_x)_0^2]dyd\tau \le C_2\ \varepsilon^2(\|U_y(0)\|^2+\|U_t(0)\|^2)\,. \tag{4.8}$$

Proof. We first use the differential equations (4.1a,b) at the boundary $x = 0$, and solve for ϕ_x, u_x (v_y vanishes at $x = 0$). We get

$$(\phi_x)_0^2 + (u_x)_0^2 \le \text{const}\ \varepsilon^2[(\phi_y)_0^2+(u_y)_0^2 + (\phi_t)_0^2 + (u_t)_0^2] \tag{4.9}$$

and (4.8) follows from (4.5).
By differentiating the system (4.1) with respect to x and using integration by parts we obtain

$$\text{(a)}\quad \frac{1}{2}\frac{d}{dt}\|\phi_x\|^2 = \frac{\bar{u}}{2}\int_0^1 [(\phi_x)_0^2-(\phi_x)_1^2]dy + \left(\frac{1}{\varepsilon}+\bar{c}\right)\int_0^1 [(\phi_x)_0(u_x)_0- (\phi_x)_1(u_x)_1]dy + \left(\frac{1}{\varepsilon}+\bar{c}\right)(u_x,\phi_{xx})+\left(\frac{1}{\varepsilon}+\bar{c}\right)(v_x,\phi_{xy})$$

$$\text{(b)}\quad \frac{1}{2}\frac{d}{dt}\|u_x\|^2 = \frac{\bar{u}}{2}\int_0^1 [(u_x)_0^2-(u_x)_1^2]dy - \left(\frac{1}{\varepsilon}+\bar{c}\right)(\phi_{xx},u_x) \tag{4.10}$$

$$\text{(c)}\quad \frac{1}{2}\frac{d}{dt}\|v_x\|^2 = \frac{\bar{u}}{2}\int_0^1 [(v_x)_0^2-(v_x)_1^2]dy - \left(\frac{1}{\varepsilon}+\bar{c}\right)(\phi_{xy},v_x)\,.$$

At $x = 0$ we use

$$\left|\left(\frac{1}{\varepsilon}+\bar{c}\right)(\phi_x)_0(u_x)_0\right| \le \frac{1}{2}\left(\frac{1}{\varepsilon}+\bar{c}\right)^2(\phi_x)_0^2 + \frac{1}{2}(u_x)_0^2\,. \tag{4.11}$$

Furthermore, we observe that

$$v_0 = (v_y)_0 = (v_t)_0 = 0$$

hence, it follows from (4.1c) that

$$(v_x)_0 = -\frac{1}{\bar{u}}\left(\frac{1}{\varepsilon}+\bar{c}\right)(\phi_y)_0 \,. \tag{4.12}$$

At x = 1 we get from (4.1b)

$$\left(\frac{1}{\varepsilon}+\bar{c}\right)(\phi_x)_1 = -\bar{u}(u_x)_1$$

i.e.,

$$-\left(\frac{1}{\varepsilon}+\bar{c}\right)(\phi_x)_1(u_x)_1 = \bar{u}(u_x)_1^2 \,. \tag{4.13}$$

By adding the equations (4.10) and using (4.11)-(4.13) we obtain

$$\frac{d}{dt}\,\|U\|^2 \le \left[\bar{u}+\left(\frac{1}{\varepsilon}+\bar{c}\right)^2\right]\int_0^1 (\phi_x)_0^2 dy+(\bar{u}+1)\int_0^1 (u_x)_0^2 dy$$

$$+\frac{1}{\bar{u}}\left(\frac{1}{\varepsilon}+\bar{c}\right)^2 \int_0^1 (\phi_y)_0^2 dy - \bar{u}\int_0^1 (v_x)_1^2 dy + \bar{u}\int_0^1 [(u_x)_1^2-(\phi_x)_1^2]dy \,. \tag{4.14}$$

Since (4.1a) applied at x = 1 implies

$$(u_x)_1 = -(v_y)_1 - \frac{\varepsilon}{1+\varepsilon\bar{c}}(\phi_t+\bar{v}\phi_y)_1 - \frac{\varepsilon}{1+\varepsilon\bar{c}}\,\bar{u}(\phi_x)_1$$

we get for the last term in (4.14)

$$\bar{u}\int_0^1 [(u_x)_1^2 - (\phi_x)_1^2]dy \le C_1 \int_0^1 [(v_y)_1^2 + (\phi_y)_1^2 + (\phi_t)_1^2]dy$$

$$- C_2 \int_0^1 (\phi_x)_1^2 dy$$

where C_1, C_2 are positive constants. When integrating (4.14) with respect to t and using (4.5), (4.6) and (4.9)

we obtain (4.7).
The linearized incompressible equations are

$$\begin{aligned} u_t + \bar{u}u_x + \bar{v}u_y + \tilde{\phi}_x &= 0 \\ v_t + \bar{u}v_x + \bar{v}v_y + \tilde{\phi}_y &= 0 \\ u_x + v_y &= 0 \,. \end{aligned} \tag{4.15}$$

This system requires three boundary conditions, and in accordance with (4.2), (4.3) we prescribe

$$\begin{aligned} \phi_0 + \alpha u_0 &= 0 \\ v_0 &= 0 \\ u_1 &= 0 \,. \end{aligned} \tag{4.16}$$

The theorems above show that all first derivatives of the solutions to the compressible equations (4.1) are bounded if the system is properly initialized. Furthermore (4.6) and (4.7) show that $\phi = \mathcal{O}(\varepsilon)$, and consequently $\tilde{\phi} = \phi/\varepsilon$ is well defined in the limit $\varepsilon = 0$. We have proved

<u>Theorem 4.3</u>. Assume that $\|U(0)\|$, $\|U_x(0)\|$, $\|U_y(0)\|$, $\|U_t(0)\|$ are bounded independent of ε and that the condition (4.4) is fulfilled. Then the solutions to the compressible problem (4.1), (4.2), (4.3) converge to the solutions to the incompressible problem (4.15), (4.16) as $\varepsilon \to 0$, with $\bar{\phi} = \phi/\varepsilon$.
<u>Remark</u>. The assumption that $\|U_t(0)\|$ is bounded is equivalent to

$$\|u_x+v_y\| = \mathcal{O}(\varepsilon), \ \|\phi_x\| = \mathcal{O}(\varepsilon), \ \|\phi_y\| = \mathcal{O}(\varepsilon) \text{ at } t = 0 \,. \tag{4.17}$$

We next turn to the case where the boundaries are solid walls. In this case we linearize around $\bar{U} = (0,\bar{v},\bar{\phi})^T$. We have
<u>Theorem 4.4</u>. Assume that $\bar{u} = 0$ in (4.1) and that the boundary conditions

$$u_0 = u_1 = 0 \tag{4.18}$$

hold. Then the solutions to (4.1) satisfy

$$\|\frac{\partial^{i+j}U}{\partial y^i \partial t^j}(t)\| \leq \|\frac{\partial^{i+j}U}{\partial y^i \partial t^j}(0)\| , \qquad i,j \geq 0$$

$$\|U_x(t)\|^2 \leq \|U_x(0)\| .$$

Proof. The proof is analogous to the ones for Theorem 4.1 and Theorem 4.2, and we don´t give the details. Since u = 0, all boundary terms disappear when applying the energy mehtod to U. This is true also for all y- and t-derivatives. For U_x we need a new condition. This is obtained from (4.1b), which implies $\phi_x = 0$ at the boundaries. This is sufficient to cancel the only remaining boundary terms, and the theorem follows.

Convergence to the incompressible problem follows as in Theorem 4.3.

Estimates of essentially the same type as the ones derived in this section can be obtained also for the system (4.1) with variable coefficients $\bar{U} = \bar{U}(x,y,t)$.

REFERENCES

1. Chorin J.: A numerical method for solving incompressible viscous flow problems, J. Comp. Phys, Vol. 2, pp 12-26 (1967).
2. Chorin J.: Numerical solution of the Navier-Stokes equations. Math. Comp. Vol. 22, pp 745-762 (1968).
3. Gustafsson B. and Kreiss H.-O.: Difference approximations of hyperbolic problems with different time scales. The reduced problem. SIAM J. Numer. Anal., Vol. 20, No. 1, pp 46-58 (1983).
4. Guerra J. and Gustafsson B.: A semi-implicit method for hyperbolic problems with different time-scales. SIAM J. Numer. Anal., Vol. 23, No. 4, pp 734-749 (1986).
5. Guerra J. and Gustafsson B.: A numerical method for incompressible and compressible flow problems with smooth solutions. J. Comp. Phys., Vol. 63, No. 2, pp 377-397 (1986).
6. Gustafsson, B.: Unsymmetric hyperbolic systems and the Euler equations at low Mach-numbers. J. Scient. Comp., Vol. 2, 123-136 (1987).

FREQUENCY DECOMPOSITION MULTI-GRID METHODS FOR HYPERBOLIC PROBLEMS

Wolfgang Hackbusch
Sigrid Hagemann

Institut für Informatik und Praktische Mathematik
Christian-Albrechts-Universität zu Kiel
Olshausenstraße 40
2300 Kiel 1

SUMMARY

Constructing solvers for discretised partial differential equations one can follow two different aims,

- adapting a method to a given class of problems which leads to very efficient solvers for these particular equations or
- devising a robust method in the sense that the class of problems the algorithm works for will be as large as possible.

Multi-grid methods are a well known tool for the treatment of partial differential equations. Attempts toward robustness have dealt until now mostly with inventing new elaborate smoothing iterations. The results are, especially considering the treatment of singularly perturbed problems like anisotropic equations or the convection diffusion equation, still not completely satisfactory, at least in case of 3D.
A new approach is the so-called Frequency Decomposition Multi-Grid Method, being content with a very simple smoothing procedure and laying emphasis on a multiple coarse grid correction, every correction dealing with a different part of the frequency spectrum.
The application of this new method to the anisotropic equation is described in [4] and [5] whilst we are concerned here with hyperbolic and parabolic problems.

1. SOLVING SINGULARLY PERTURBED PROBLEMS WITH STANDARD MULTIGRID TECHNIQUES

1.1. EQUATIONS UNDER CONSIDERATION

Let $a, a_x, b_x, b_y, c \in \mathbb{R}$, $a, a_x \geq 0$ and $\Omega=(0,1)\times(0,1)$ the unit square. The model problem of the two dimensional convection diffusion equation

$$-a(\partial^2/\partial x^2 u(x,y)+\partial^2/\partial y^2 u(x,y)) +$$
$$b_x\partial/\partial x u(x,y) + b_y\partial/\partial y u(x,y) + cu(x,y) = f(x,y) \text{ in } \Omega \qquad (1.1a)$$

with periodic boundary conditions

represents for $a\to 0$ a singularly perturbed problem since its type changes from elliptic to hyperbolic.

In contrast to this the problem

$$-a_x(\partial^2/\partial x^2 u(x,y)) - 1*(\partial^2/\partial y^2 u(x,y)) +$$
$$b_x\partial/\partial x u(x,y) + b_y\partial/\partial y u(x,y) + cu(x,y) = f(x,y) \text{ in } \Omega \quad (1.1b)$$

with periodic boundary conditions

turns into parabolic type with $a_x \to 0$.
To ensure stability we choose for the discretisation the "upwind differencing scheme" ([1], p. 219), yielding the difference stars

$$L^a = h_l^{-2}\begin{bmatrix} & -a+b_y^-h_l & \\ -a-b_x^+h_l & 4a+|b_x|h_l+|b_y|h_l+ch_l^2 & -a+b_x^-h_l \\ & -a-b_y^-h_l & \end{bmatrix} \quad (1.2a)$$

and

$$L^b = h_l^{-2}\begin{bmatrix} & -1+b_y^-h_l & \\ -a_x-b_x^+h_l & 2+2a_x+|b_x|h_l+|b_y|h_l+ch_l^2 & -a_x+b_x^-h_l \\ & -1-b_y^-h_l & \end{bmatrix} \quad (1.2b)$$

respectively.
$h_l=1/N_l$, $N_l \in \mathbb{N}$, is the step size on grid Ω_l,
$b_z^+=\max(0,b_z)$, $b_z^-=\min(0,b_z)$ for z=x or z=y.

1.2. STANDARD MULTI-GRID METHODS

The application of the standard multi-grid method, consisting of

(1.3a) coarsening by doubling the meshsize in both directions,
(1.3b) a matrix dependent prolongation p ([1], pp. 212),
(1.3c) the adjoint restriction $r=p^*$,
(1.3d) the Galerkin coarse grid matrix $L_{l-1}=rL_lp$ and
(1.3e) $(\nu_1+\nu_2)$ smoothing steps performed by a relaxation operator S

onto the discretized problem $L_lu_l=f_l$ on the finest grid Ω_l with meshsize h_l can clearly be represented by the flow chart (fig. 1; cf. [3], p. 45).

1.3. APPLICATION OF STANDARD MULTI-GRID METHODS

To analyse the convergence of an iteration treating (1.1a) and (1.1b) note that

$$e_l^{\nu,\mu}(x,y)=1/2\ e^{\pi i(\nu x+\mu y)},\ (x,y)\in\Omega_l, 1-N_l\le\nu,\mu\le N_l, N_l=1/h_l, \quad (1.4)$$

are the complex eigenfunctions of the unsymmetric grid operators L^a and L^b of (1.2a) and (1.2b) on the extended domain

$$\Omega=(-1,1)\times(-1,1),\ \Omega_l=\{(x,y)\in\Omega: x=\nu h_l,\ y=\mu h_l\},$$

considering periodic boundary conditions.

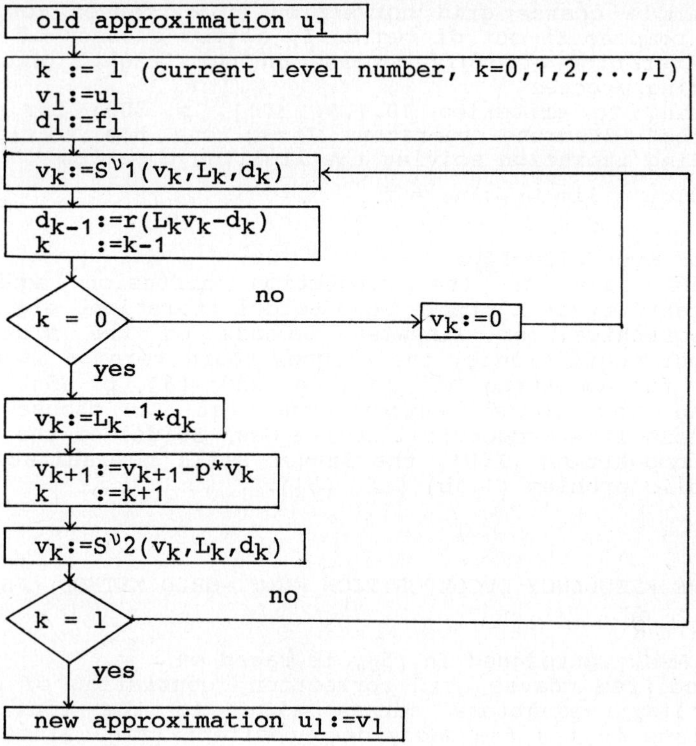

Fig. 1 Flow chart of the standard multi-grid method

Classifying the frequencies σ into high and low, where low means $1-N_l/2 \leq \sigma \leq N_l/2$, one can decomposite the spectrum into four different regions.

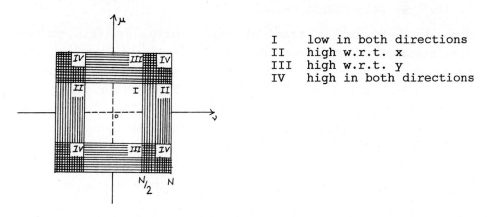

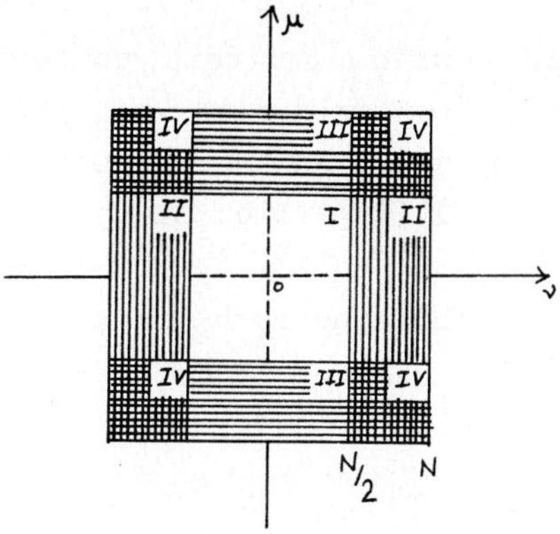

I low in both directions
II high w.r.t. x
III high w.r.t. y
IV high in both directions

Fig. 2 Spectrum decomposition

The simple coarse grid correction only reduces low frequency error components out of region I, so the success of the multigrid iteration strongly depends on the "right" choice of the smoothing process.
According to criterion 10.1.1. ([1], p. 302) for singularly perturbed discrete operators $L_l(\varepsilon)$ one has to look for a smoothing iteration solving the limiting equation

$$L_l(0)u_l = \lim_{\varepsilon\to 0} L_l(\varepsilon)u_l = f_l$$

fast or even directly.
In the case of the convection diffusion equation the applicability of a simple Gauß-Seidel iteration, e.g. pointwise lexicographical or linewise depends on the signs of the gradient coefficients, the chequer-board version is not at all suited for smoothing (cf. [1], pp. 220; [6], p. 35).
Looking for robust methods one has to choose elaborate smoothers like symmetric linewise Gauß-Seidel or the incomplete LU-decomposition (ILU), the latter still not suitable for the parabolic problem (1.1b) (cf. [5]).

2. THE FREQUENCY DECOMPOSITION MULTI-GRID METHOD (FDMGM)

The FDMGM, introduced in [5], is based on
- a modified coarse grid correction, consisting of up to four auxiliary equations corresponding to the four different regions in the frequency decomposition (fig. 2) to be solved on each level instead of the standard one, and
- a very simple smoothing method, e.g.
  - damped Jacobi or
  - (damped) chequer-board Gauß-Seidel.

2.1. THE MODIFIED COARSE GRID CORRECTION

2.1.1. THE FOUR COARSE GRIDS

Through shifting the standard coarse grid, in this context distinguished by the index (0,0), $\Omega_{l-1}^{(0,0)}$ of grid size $h_{l-1}=2h_l$ by h_l in x-, in y- or in both x- and y-direction, we obtain three further grids $\Omega_{l-1}^{(1,0)}$, $\Omega_{l-1}^{(0,1)}$ and $\Omega_{l-1}^{(1,1)}$ (fig. 3a-e).
Let $J=\{(0,0),(1,0),(0,1),(1,1)\}$ denote the set of the four coarse grid indices.
The double index $\alpha=(\alpha_x,\alpha_y)$ indicates a shift by $\alpha_x * h_l$ in x- and $\alpha_y * h_l$ in y-direction, where $\alpha_x,\alpha_y \in \{0,1\}$.

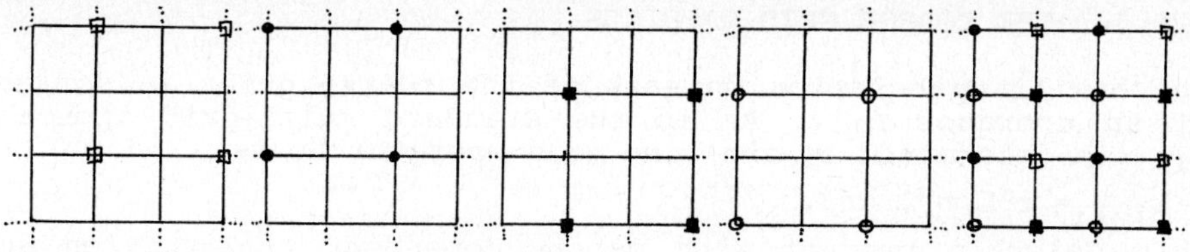

Fig. 3a $\Omega_{l-1}^{(0,0)}$ Fig. 3b $\Omega_{l-1}^{(1,0)}$ Fig. 3c $\Omega_{l-1}^{(0,1)}$ Fig. 3d $\Omega_{l-1}^{(1,1)}$ Fig. 3e $\Omega_l = \bigcup_{\alpha\in J} \Omega_{l-1}^{\alpha}$

2.1.2. THE ASSOCIATED PROLONGATIONS AND RESTRICTIONS

The matrix dependent prolongation p^{00}, symbolised by the difference star

$$p^{00} = \begin{bmatrix} q_{-1,1} & q_{0,1} & q_{1,1} \\ q_{-1,0} & q_{0,0} & q_{0,1} \\ q_{-1,-1} & q_{0,-1} & q_{1,-1} \end{bmatrix}$$

with entries $q_{i,j} \geq 0$, $i,j \in \{-1,0,1\}$ corresponds to the standard coarse grid Ω_{l-1}^{00} (cf. fig. 3a); for the three additional coarse grids we choose

$$p^{10} = \begin{bmatrix} -q_{1,1} & q_{0,1} & -q_{-1,1} \\ -q_{1,0} & q_{0,0} & -q_{-1,0} \\ -q_{1,-1} & q_{0,-1} & -q_{-1,-1} \end{bmatrix}$$ with alternating sign in x-direction,

$$p^{01} = \begin{bmatrix} -q_{-1,-1} & -q_{0,-1} & -q_{1,-1} \\ q_{-1,0} & q_{0,0} & q_{1,0} \\ -q_{-1,1} & -q_{0,1} & -q_{1,1} \end{bmatrix}$$ with alternating sign in y-direction,

$$p^{11} = \begin{bmatrix} q_{1,-1} & -q_{0,-1} & q_{-1,-1} \\ -q_{1,0} & q_{0,0} & -q_{-1,0} \\ q_{1,1} & -q_{0,1} & q_{-1,1} \end{bmatrix}.$$ with alternating sign in both directions

The latter three prolongations are no interpolations.

Remark (2.1): Grid functions which are highly oscillating in x-direction (connected with region II in fig. 2) belong to the range of prolongation p^{10}, in the same way are correlated the prolongation p^{01} with region III and p^{11} with region IV.

The restrictions are defined in the usual way as adjoints of the prolongations, i.e. $r^{\alpha}=(p^{\alpha})^*$, $\alpha\in J$.

Remark (2.2): This choice implies $r^{\alpha}p^{\beta}=0$ if $\alpha\neq\beta$.

2.1.3. THE COARSE GRID MATRICES

The auxiliary equation on each of the coarse grids is defined by an operator L_{l-1}. As in the standard multi-grid approach L_{l-1} is determined by the fine grid operator L_l as

$$L_{l-1} = r^{\alpha} L_l p^{\alpha}, \ \alpha \in J.$$

This Galerkin product with matrix dependent prolongation and adjoint restriction preserves stable one-sided differences (cf. [1], p. 222).

2.1.4. THE MULTIPLE COARSE GRID CORRECTION

Evaluation of the defect equations on up to all four coarse grids produces the new multiple coarse grid correction

$$u_l \leftarrow u_l - \sum_{\alpha \in K} p^{\alpha} (L_{l-1}^{\alpha})^{-1} r^{\alpha} (L_l u_l - f_l), \ K \subseteq J.$$

If $K=\{(0,0)\}$ we get back the standard correction. The extra terms are added to provide a reduction of the high frequency error components of regions II - IV, which have not been sufficiently diminished during the smoothing process.

2.2. THE SMOOTHING PROCEDURE

We use in the following examples the pointwise damped Jacobi iteration and the pointwise Gauß-Seidel iteration with chequer-board ordering of the grid points as smoothers to ensure the robustness of the FDMGM originating from the multiple coarse grid correction and not a suitable choice of a sophisticated smoothing operator.

2.3. THE FREQUENCY DECOMPOSITION TWO-GRID METHOD

The FD-two grid method consisting of ν pre-smoothing steps and the performance of new multiple coarse grid correction can be analysed by local mode analysis (cf. [3]) to obtain exact two grid rates of convergence.
In the following tables we contrast the spectral radii of the standard multi-grid method (S) with the new results (FD).

Table 1: Two-grid rates for (1.1a), two Gauß-Seidel steps as smoother, h=1/8.

b_x	b_y	log(a) =	2.0	1.5	1.0	0.5	0.0	-0.5	-1.0	-1.5	-2.0	-2.5	-3.0	-10
0	0	S	0.062	0.062	0.062	0.061	0.059	0.052	0.035	0.032	0.011	0.001	0.000	0.000
		FD	0.084	0.083	0.083	0.080	0.074	0.053	0.030	0.007	0.001	0.000	0.000	0.000
1	0	S	0.063	0.063	0.064	0.066	0.074	0.098	0.159	0.259	0.520	0.711	0.789	0.829
		FD	0.084	0.084	0.083	0.081	0.077	0.075	0.089	0.131	0.257	0.350	0.389	0.410
1	1	S	0.062	0.062	0.062	0.061	0.054	0.088	0.188	0.340	0.511	0.586	0.612	0.624
		FD	0.084	0.084	0.083	0.081	0.078	0.081	0.110	0.277	0.474	0.571	0.607	0.624

Table 2: Two-grid rates for (1.1b), two Gauß-Seidel steps as smoother, h=1/8.

b_x	b_y	$\log(a_x)$ =	2.0	1.5	1.0	0.5	0.0	-0.5	-1.0	-1.5	-2.0	-2.5	-3.0	-10
0	0	S	0.960	0.880	0.675	0.323	0.059	0.303	0.611	0.784	0.851	0.873	0.881	0.884
		FD	0.292	0.277	0.235	0.149	0.074	0.125	0.148	0.104	0.051	0.042	0.042	0.042
1	0	S	0.955	0.865	0.641	0.280	0.074	0.341	0.643	0.805	0.866	0.886	0.893	0.896
		FD	0.300	0.283	0.236	0.144	0.077	0.135	0.153	0.107	0.051	0.053	0.053	0.054
0	1	S	0.960	0.880	0.678	0.336	0.074	0.168	0.264	0.293	0.317	0.325	0.327	0.327
		FD	0.292	0.277	0.236	0.153	0.077	0.128	0.230	0.377	0.454	0.483	0.493	0.493
1	1	S	0.955	0.865	0.645	0.292	0.054	0.201	0.309	0.338	0.363	0.372	0.374	0.375
		FD	0.300	0.283	0.237	0.148	0.078	0.139	0.246	0.399	0.477	0.506	0.516	0.521

2.4. THE FREQUENCY DECOMPOSITION MULTI-GRID METHOD

Replacing the exact solution of the defect equations on the coarser grids by recursive application of the method itself generates the frequency decomposition multi-grid method (FDMGM).

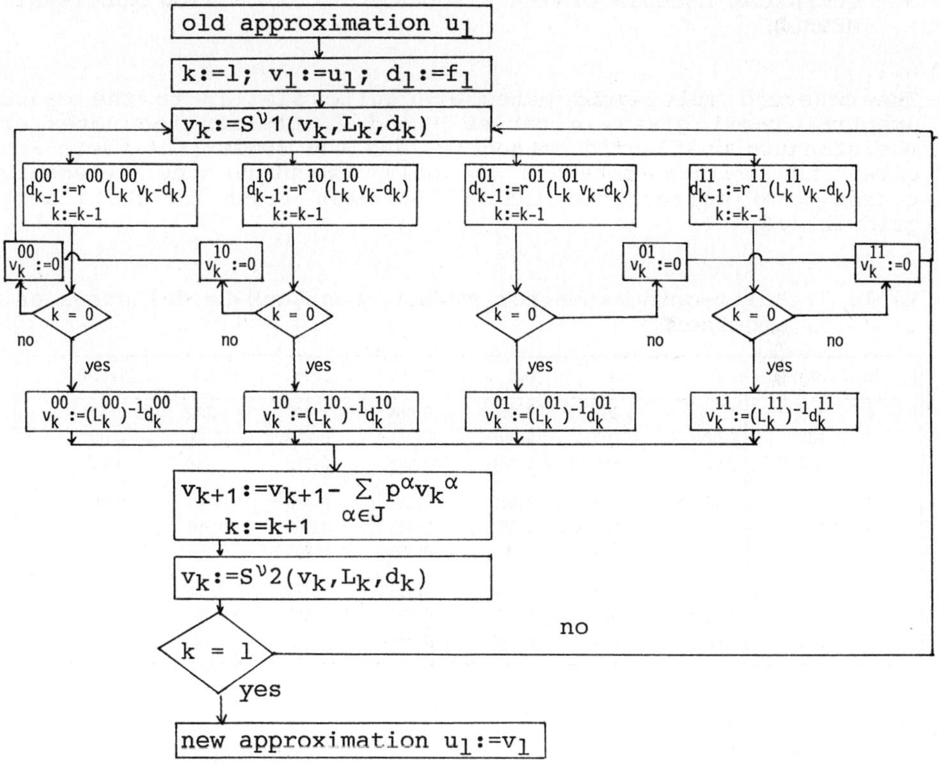

Fig. 4: Flow chart of FDMGM

To reduce the cost of the algorithm it is sufficient to restrict oneself to the execution of the so-called necesssary coarse grid corrections (cf. [4, 5]) without considerable loss of accuracy. Instead of quadrupling the number of equations to be solved on each coarser level we get the calling tree

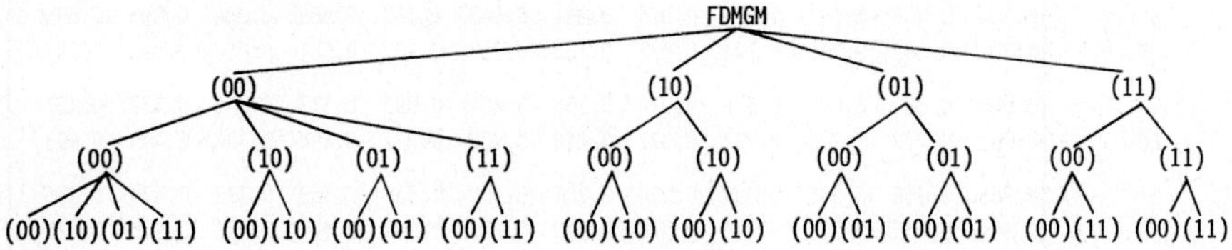

Fig. 5: Calling tree of necessary coarse grid corrections

where (α) denotes the performance of the coarse grid correction of index α∈J.

3. NUMERICAL RESULTS OF THE FREQUENCE DECOMPOSITION MULTI-GRID METHOD

The observed multi-grid rates are quite similar to the exact spectral radii given in tables 1 and 2. Besides the rates of the standard multi-grid method (S) and the FDMGM (FD) there are given the values obtained by only executing the necessary coarse grid corrections (NFD). The mesh width of the finest grid is 1/8.

Table 3: Multi-grid rates for (1.1a), two Gauß-Seidel steps as smoother.

b_x	b_y	log(a) =	2.0	1.0	0.0	-1.0	-2.0	-3.0	-10.0
0	0	S	0.058	0.057	0.056	0.054	0.040	0.002	0.000
		FD	0.084	0.083	0.081	0.061	0.010	0.000	0.000
		NFD	0.084	0.084	0.081	0.061	0.010	0.000	0.000
1	0	S	0.058	0.058	0.067	0.129	0.425	0.722	0.650
		FD	0.084	0.083	0.082	0.098	0.309	0.701	0.776
		NFD	0.084	0.084	0.083	0.099	0.297	0.657	0.713
1	1	S	0.058	0.058	0.056	0.098	0.473	0.725	0.784
		FD	0.084	0.083	0.082	0.094	0.613	0.743	0.750
		NFD	0.084	0.083	0.083	0.094	0.612	0.739	0.747

Table 4: Multi-grid rates for (1.1b), two Gauß-Seidel steps as smoother.

b_x	b_y	$\log(a_x)$ =	2.0	1.0	0.0	-1.0	-2.0	-3.0	-10.0
0	0	S	0.960	0.672	0.056	0.661	0.924	0.967	0.967
		FD	0.303	0.247	0.081	0.210	0.112	0.048	0.048
		NFD	0.303	0.247	0.081	0.210	0.113	0.051	0.051
1	0	S	0.957	0.655	0.067	0.676	0.928	0.964	0.969
		FD	0.300	0.249	0.082	0.213	0.112	0.051	0.051
		NFD	0.308	0.249	0.083	0.217	0.114	0.050	0.050
0	1	S	0.960	0.673	0.063	0.462	0.561	0.583	0.586
		FD	0.303	0.246	0.082	0.253	0.589	0.683	0.695
		NFD	0.303	0.246	0.082	0.252	0.588	0.682	0.694
1	1	S	0.957	0.657	0.056	0.482	0.581	0.602	0.604
		FD	0.300	0.250	0.082	0.261	0.600	0.692	0.695
		NFD	0.308	0.250	0.083	0.261	0.600	0.691	0.702

REFERENCES

[1] Hackbusch, W., Multi-Grid Methods and Applications. Springer, Heidelberg 1985.

[2] Hackbusch, W., and Trottenberg, U. (eds.), Multi-Grid Methods. Proceedings. Lecture Notes in Mathematics 960. Springer, Berlin, Heidelberg 1982.

[3] Stüben, K., and Trottenberg, U., Multi-Grid Methods: Fundamental Algorithms, Model Problem Analysis and Applications. In [2], pp. 1 - 176.

[4] Hackbusch, W., The Frequency Decomposition Multi-Grid Algorithm (to appear).

[5] Hackbusch, W., A New Approach to Robust Multi-Grid Solvers. Proceedings. ICIAM Paris, June 1987.

[6] Hagemann, S., Mehrgitterverfahren zur Lösung anisotroper dreidimensionaler Randwertprobleme. Diplomarbeit, Kiel 1986.

EXISTENCE AND UNIQUENESS FOR LINEAR HYPERBOLIC SYSTEMS WITH UNBOUNDED COEFFICIENTS

Andrzej Hanyga
Institute of Geophysics
Polish Academy of Sciences
Warszawa, Pasteura 3
POLAND

Mauro Fabrizio
Istituto di Matematica
Università di Bologna
Bologna, piazza Porta San Donato
ITALY

We consider formally hyperbolic linear systems with unbounded propagation speeds .
Existence, uniqueness and continuous dependence on the data **g**, **h**, **e** are proved for the Cauchy problem

$$\mathbf{u}_{tt} = \mathbf{A}^{ij}(\mathbf{x},t)\,\mathbf{u}_j + \mathbf{B}(\mathbf{x},t)\,\mathbf{u}_t + \mathbf{C}^i(\mathbf{x},t)\,\mathbf{u}_i + \mathbf{D}(\mathbf{x},t)\,\mathbf{u} + \mathbf{e}(\mathbf{x},t)$$

for $(\mathbf{x},t) \in [0,T] \times \Omega$,
$\mathbf{u}(\mathbf{x},0) = \mathbf{h}(\mathbf{x})$, $\mathbf{u}_t(\mathbf{x},0) = \mathbf{g}(\mathbf{x})$, $\mathbf{x} \in \Omega$
and either $\Omega = R^n$ or Ω bounded and homogeneous Neumann conditions :

$$\mathbf{A}^{ij}(\mathbf{x},t)\, n_i\, \mathbf{u}_j = 0 .$$

It is assumed that the m×m coefficient matrices $\mathbf{A}^{ij}$ are integrable :

$$\mathbf{A}^{ij} \in L^1(D) \ \text{ if } \Omega \text{ is bounded }, \ \mathbf{A}^{ij} \in L^2(D,\rho) \text{ if } \Omega = R^n$$

(ρ is a weight possibly decaying at ∞),
and satisfy a pseudo-Lipschitz condition with respect to t:

$$\beta\Delta v_i \cdot A^{ij}(x,t)v_j \leqslant v_i \cdot [A^{ij}(x,t+\Delta)-A^{ij}(x,t)]v_j \leqslant \alpha\Delta v_i \cdot A^{ij}(x,t)v_j$$

for arbitrary sets of vectors $v_i \in R^m$, $i = 1,\dots,n$ and a sufficiently small Δ. The m×m matrix $B(x,t)$ is assumed to be bounded by two continuous functions :

$$b_1(t)v^2 \leqslant v \cdot B(x,t)\, v \leqslant b_2(t)v^2 \text{ for every } v \in R^m \text{ and a.a. } (x,t).$$

References:

[1] M.Fabrizio & A.Hanyga (1983), Existence and uniqueness for linear hyperbolic systems with unbounded coefficients , Ann. Univ. Ferrara , sez. VI, Sc. matematiche, **29** , 137-151.

[2] M.Fabrizio & A.Hanyga (1987), Solutions in the variational sense to hyperbolic systems with unbounded coefficients, Archives of Mechanics , No 1.

A NUMERICAL METHOD FOR COMPUTING VISCOUS SHOCK LAYERS

Eduard Harabetian
Department of Mathematics
The University of Michigan

ABSTRACT

We introduce a high resolution numerical scheme for the computation of viscous shock layers. The novel feature consists in the use of travelling wave solutions as approximating tools. This is a departure from previous methods which are based on splitting the viscous part from the hyperbolic part of the equation. We present stability results as well as numerical tests for one dimensional models.

1. INTRODUCTION

In this paper we consider the numerical approximation of viscous perturbations of nonlinear hyperbolic conservation laws.

Let us consider the scalar equation:

$$u_t + f(u)_x = \varepsilon(a(u)u_x)_x, \quad a(u)\geq a_o>0, \quad f \text{ convex} \quad (1.1)$$

and define the (nondimensional) quantities:

$$R = \frac{|f'|\cdot h}{\varepsilon}, \text{ (Cell Reynolds Number)},$$

$$\lambda = \frac{|f'|\cdot \Delta t}{h}, \text{ (CFL Number)},$$

where $|f'| = \sup|f'(u)|$, and h and Δt are the spatial and temporal mesh widths of some discrete approximation.

Equations such as (1.1) are considered as one dimensional scalar models for the equations governing the flow of fluids with small viscosity.

In order to understand the significance of R, consider a travelling wave solution of (1.1), that is a solution of the form $u(x,t)=w((x-st)/\varepsilon)$, and for simplicity assume $|f'|=1$. The existence of a monotone decreasing profile w, with w tending to $w_\pm$ as x tends to $\pm\infty$ is discussed for example in [4]. Projecting on a grid of size h, $\Delta t=\lambda h$, we obtain the discrete profile:

$$w_j = w(jh/\varepsilon - s\lambda h/\varepsilon) = w(R\cdot(j-s\lambda)).$$

We now point out that the larger the R the fewer the points w_j in the transition layer between w_- and w_+ (see Fig.4).

Our goal is a stable numerical method that resolves such viscous profiles very accurately even when they have but two or three points in the transition layer. The interest in such methods arises, for example, when computing reactive flows where the layer itself plays a major role.

The standard approach for approximating equations like (1.1) has been a centered difference approximation to the viscous term,

$$\varepsilon(a(u)u_x)_x \simeq (\varepsilon/h^2)\Delta_{j+1/2}(a(u_j)\Delta u_{j-1/2}),$$
$$\Delta u_{j-1/2} = u_j - u_{j-1}$$

and either a centered difference or an upwind difference approximation to the hyperbolic term [3], [6]. The characteristic feature of this approach is to separate the viscous part from the hyperbolic part in the equation. We performed numerical experiments on the viscous Burger equation, using both Richtmeyer's second order centered difference scheme and then the second order upwind (MUSCL [7]) scheme for the hyperbolic part. We computed a travelling wave moving with speed one for which we had the exact solution. The results are shown in Figs.1 and 2 below.

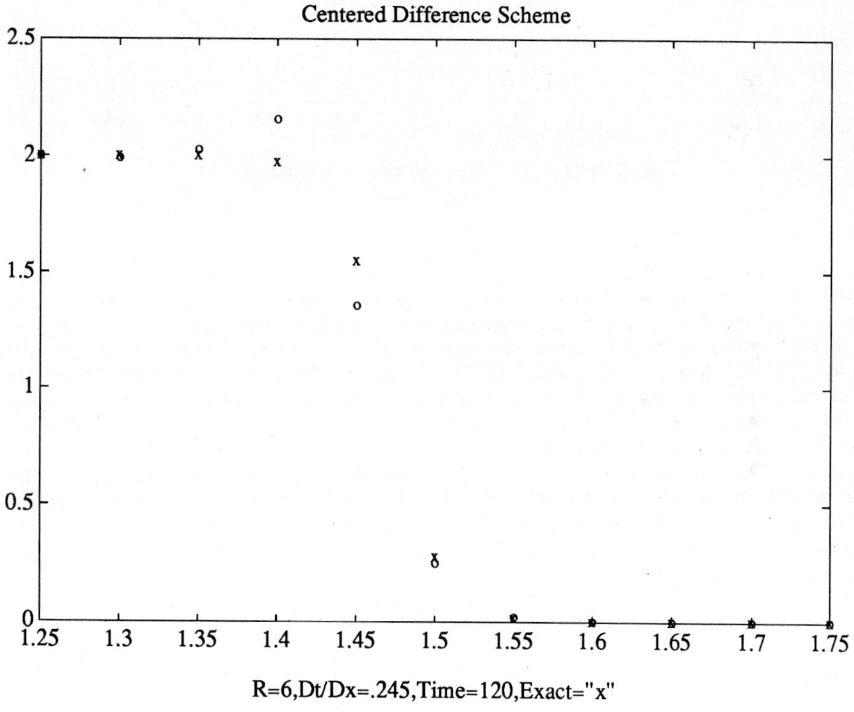

Fig.1

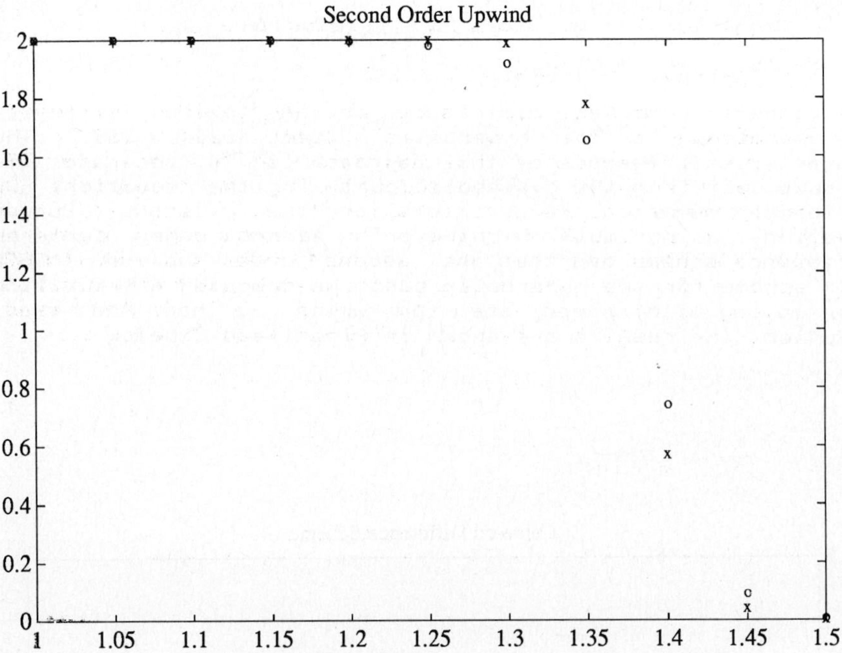

R=6,Dt/Dx=.28,Time=120,Error=.172,Exact="x"

Fig. 2

We point out that the centered difference scheme is oscillatory since, when R is not small there is not sufficient viscosity to stabilize it. The upwind scheme, on the other hand, is total variation stable for all values of R, but suffers from loss of accuracy inside the transition layer. The numerical results indicate roughly a 10% error (pointwise). This can be explained by the fact that, near shocks, upwind schemes create a significant amount of artificial viscosity of their own and this overshadows the effect of the real viscosity in the equation. We believe that any algorithm which is simply based on splitting the equation into its viscous and hyperbolic parts will have such problems. In this paper we will construct an unsplit scheme which uses the travelling wave solutions of (1.1) as approximating tools between grid points. The key idea is to interpolate a piece of a travelling wave between grid points and then let the exact solution operator to (1.1) act on it. Then, we evaluate the flux $\varepsilon a(u)u_x - f(u)$ at cell boundaries. As we will see this scheme borrows from both the centered difference and upwind methodologies. As a consequence it has both the high resolution property of centered differencing and the total variation diminishing property of the upwind approach.

In what follows we will describe the travelling wave scheme in detail and prove it is TVD stable for a fairly large set of values of (λ,R), that includes, for example, cases where the other schemes fail to perform. We also present numerical evidence that supports the theoretical results. A more complete account, including accuracy and optimal stability results, can be found in [2].

2. THE TRAVELLING WAVE SCHEME

We will only consider the case $u_{j+1}<u_j$ to derive the numerical flux $h_{j+1/2}$. This is the case of a shock for the hyperbolic part where the dissipation plays the most important role. In case of rarefactions, $u_j<u_{j+1}$, we advocate the use of any of the two schemes mentioned above.

As we mentioned in the introduction we wish to interpolate a travelling wave solution between u_j and u_{j+1}, that is find w such that w(x-st) is a solution to (1.1), and $w(x_j)=u_j$, $w(x_{j+1})=u_{j+1}$. For simplicity we let $x_j=0$, $x_{j+1}=h$, and rescale time and space so that t goes to t/h, x goes to x/h. Then, (1.1) becomes

$$u_t + f(u)_x = \mu(a(u)u_x)_x, \qquad \mu=|f'|/R. \tag{2.0}$$

The travelling wave solution will satisfy the ODE

$$\mu(a(w)')' = -sw' + f(w)' \tag{2.1}$$

We consider this with boundary conditions

$$w(0)=u_j,\ w(1)=u_{j+1}. \tag{2.2}$$

Since the speed s remains yet undetermined, we have the freedom of prescribing one more boundary condition. For instance we let $w'(0)=w'_j$.

Using the convexity of f, one easily proves:

PROPOSITION 2.1

Suppose $u_j > u_{j+1}$, and $w'(0) = w'_j <0$. Then there exists a unique s and w(x) that solves (2.1) and (2.2).

Moreover, one has the following estimates:

$$\text{a)}\quad \frac{\Delta f_{j+1/2}}{\Delta u_{j+1/2}} - \mu\,\frac{w'_j}{\Delta u_{j+1/2}}\,a(u_j)\,g^{-1}\left[\frac{a(u_j)w'_j}{a_1\Delta u_{j+1/2}}\right] \le s \tag{2.3}$$

$$s \le f'(u_j) - \mu\,\frac{w'_j}{\Delta u_{j+1/2}}\,a(u_j)\,g^{-1}\left[\frac{a(u_j)w'_j}{a_0\Delta u_{j+1/2}}\right]$$

$$\text{b)}\quad \frac{\Delta f_{j+1/2}}{\Delta u_{j+1/2}} + \mu\,\frac{w'_{j+1}}{\Delta u_{j+1/2}}\,a(u_{j+1})\,g^{-1}\left[\frac{a(u_{j+1})w'_{j+1}}{a_1\Delta u_{j+1/2}}\right] \le s$$

$$s \le f'(u_j) + \mu\,\frac{w'_{j+1}}{\Delta u_{j+1/2}}\,a(u_{j+1})\,g^{-1}\left[\frac{a(u_{j+1})\,w'_{j+1}}{a_0\Delta u_{j+1/2}}\right]$$

where $w'_{j+1} = w'(1)$, $g(z) = (1/z)\log(1+z)$, and $a_0 = \inf a(u)$, $a_1 = \sup a(u)$.

REMARKS. The function $zg^{-1}(z)$ is defined on $(0,\infty)$, decreasing and tending to $+\infty$ at 0 and to $-\infty$ at $+\infty$. Therefore, s becomes large when $w_j'/\Delta u_{j+1/2}$ is near 0 or near ∞. Also, since $g^{-1}(1) = 0$, one obtains maximum control on s when $w_j'/\Delta u_{j+1/2}$ is chosen to be 1 (assuming $a_0 = a_1 = 1$).

Given u_j, u_{j+1} and w_j' we define the numerical flux for the travelling wave scheme by

$$h^{TW}_{j+1/2} = \mu a(w(1/2 - s\Delta t/2h))w'(1/2 - s\Delta t/2h) - \qquad (2.6)$$
$$- f(w(1/2 - s\Delta t/2h)), \quad \mu = \varepsilon/h .$$

REMARK. The numerical flux $h^{TW}_{j+1/2}$ is obtained by first using the exact solution to (1.1) with initial data given by the travelling wave $w(x)$, that is $w(x-st)$, and then evaluating the flux $\varepsilon a(w)w_x - f(w)$ at $x = x_{j+1/2}$, $t = \Delta t/2$.

3. TVD STABILITY

In this section we analyze the travelling wave scheme when $w_j'/\Delta u_{j+1/2} = 1$. This choice, even though makes the scheme second order accurate only, yields optimal stability results.

For simplicity, we will assume $a_0=a_1=1$, so that from (2.3a)

$$\Delta f_{j+1/2}/\Delta u_{j+1/2} \le s \le f'(u_j), \text{ since } g^{-1}(1)=0. \qquad (3.0)$$

For TVD stability we use the framework introduced by Harten [1], in which one writes the scheme in incremental form

$$u^{n+1}_j = u^n_j + C_{j+1/2}\Delta u_{j+1/2} - D_{j-1/2}\Delta u_{j-1/2}, \quad \text{where}$$
$$C_{j+1/2} = -(h_{j+1/2}-f(u_j))/\Delta u_{j+1/2},$$
$$D_{j-1/2} = (h_{j-1/2}-f(u_j))/\Delta u_{j-1/2}$$

and then shows that

$$C_{j+1/2} \ge 0, \quad D_{j+1/2} \ge 0, \quad 1-C_{j+1/2}+D_{j+1/2} \ge 0. \qquad (3.1)$$

The inequalities (3.1) imply the scheme is TVD. In our case one easily verifies that

$$C^{TW}_{j+1/2} = (\Delta t/h)(\mu - s\,\frac{w_* - u_j}{\Delta u_{j+1/2}}), \qquad (3.2)$$

$$D^{TW}_{j+1/2} = (\Delta t/h)(\,\frac{\mu w'_{j+1}}{\Delta u_{j+1/2}} + s\,\frac{u_{j+1} - w_*}{\Delta u_{j+1/2}}\,),$$

where $w_* = w(x_*)$, $x_* = 1/2 - s\Delta t/2h$.

THEOREM 3.1. *If*

$$\text{a)}\ R \le \sup_{0\le z\le 1} \frac{2}{(1 - \lambda z)(1 - z)}\log(1/z), \qquad (3.3)$$

$$\text{b)}\ \lambda(2/R + 1) \le 1$$

then the travelling wave scheme is TVD, i.e. the coefficients (3.2) satisfy (3.1).

REMARKS. The inequality (3.3b) is a CFL-like condition. The values of the quantity which appears on the right of (3.3a) were computed as a function of λ below

λ	0	.25	.5	.6	.65	.75	.8	1
$R_*(\lambda)$	2	2.6	3.6	3.9	4.1	4.3	4.5	4.9

From (3.3) $2\lambda/(1-\lambda) \le R \le R_*(\lambda)$ so, for example, we have TVD stability if $\lambda=.65$ and $R=4.1$. Let us also mention that the results of Theorem 3.1 are not optimal, meaning that we were able to compute at higher values of R and λ (see Fig.4 for numerical results). One could explain the good stability properties of this scheme as follows. To have $C_{j+1/2} \ge 0$ in (3.2) for example, we need $s(w_* - u_j)/\Delta u_{j+1/2} \le \mu$ when $s\ge 0$. The quantity on the left of the inequality never gets too large since whenever s is getting large positive, $w_* = w(1/2 - s\Delta t/2h)$ is getting closer to u_j. This property appears to be specific to this scheme.

Proof of Theorem 3.1. From (3.1) and (3.2) we need to show

$$\text{a) } s \le \mu\Delta u_{j+1/2}/(w_* - u_j) = \mu w_j'/(w_* - u_j) \qquad (3.4)$$

$$\text{b) } -s \le \mu w_{j+1}'/(u_{j+1} - w_*)$$

$$\text{c) } (\Delta t/h)\left[2\mu + \Delta f_{j+1/2}/\Delta u_{j+1/2} + \frac{2s}{\Delta u}(u_j - w_*)\right] \le 1,$$

since $\mu w_{j+1}' = \mu w_j' - s\Delta u_{j+1/2} + \Delta f_{j+1/2}$.

We proceed by showing (3.4a) and (3.4b).

CASE 1: $s\ge 0$.

Here (3.4b) is immediate, since the right hand side is always positive. To show (3.4a) we use the estimate

$$s \le f'(u_j) - \mu \frac{w_j'}{w_* - u_j} g^{-1}\left[\frac{w_j'}{w_* - u_j} x_*\right], \text{ where} \qquad (3.5)$$

$$x_* = 1/2 - s\Delta t/2h \ge 0 \text{ if } s\Delta t/h \le 1.$$

The estimate above is obtained easily from (2.5) with $x = x_*$, the same way (2.3a) was obtained from (2.4) in Prop.2.1.

From (3.5) one gets

$$x_* \frac{-s + |f'|}{\mu} \ge \frac{w_j'}{w_* - u_j} g^{-1}\left[\frac{w_j'}{w_* - u_j} x_*\right]$$

As we remarked before, the function $zg^{-1}(z)$ is decreasing on $(0,\infty)$. Let $K(z)$ be its inverse. Then,

$$K(x_* \frac{-s + |f'|}{\mu}) \le \frac{x_* w_j'}{w_* - u_j} .$$

Now, (3.4a) follows from

$$s \le (\mu/x_*)K(x_*(-s+|f'|)/\mu),$$

which, after some algebrais equivalent to

$$g((-s+|f'|)/s) \ge sx_*/\mu, \quad \text{or}$$

$$1/\mu \le \frac{1}{x_*(-s + |f'|)} \log(|f'|/s).$$

Let $z = s/|f'|$. Then $0 \le z \le 1$ (from (3.0)) and

$$R = |f'|/\mu \le \frac{1}{(-z+1)x_*/|f'|} \log(|f'|/s).$$

Substituting for x_* from (3.2) we see that (3.3a) implies (3.4a).

CASE 2: $s \le 0$.

Here (3.4a) is trivial and (3.4b) follows as in Case 1, only, now, we use the estimate

$$0 \ge -|f'| + \mu \frac{w'_{j+1}}{u_{j+1} - w_*} g^{-1}\left[(1-x_*)\frac{w'_{j+1}}{u_{j+1} - w_*}\right]$$

instead. This estimate, again, can be obtained from (2.5) the same way (2.3b) was obtained from (2.4).

We now turn to (3.4c), the CFL condition. Since , from (3.4a) and (3.4b) the quantity on the left of (3.4c) is positive, it suffices to show

$$(\Delta t/h)(2\mu + \Delta f_{j+1/2}/\Delta u_{j+1/2} - 2s) \le 1 \quad \text{whenever } s \le 0,$$

since $-1 \le (u_j - w_*)/\Delta u_{j+1/2} \le 0$.

From (3.0) we get

$$\Delta f_{j+1/2}/\Delta u_{j+1/2} - 2s \le -\Delta f_{j+1/2}/\Delta u_{j+1/2},$$

so it suffices to have

$$(\Delta t/h)(2\mu + |f'|) \le 1,$$

which follows from (3.3b).

4. NUMERICAL EXPERIMENTS

We now turn to the numerical experiments. We have computed a moving profile for Burger's equation

$$u_t + (u^2/2)_x = \varepsilon u_{xx}$$

$$u \to 2 \text{ at } x=-\infty, \quad u \to 0 \text{ at } x=+\infty$$

The exact solution is given by

$$u(t,x) = \frac{1}{1 + \exp((x-t)/\varepsilon)}.$$

We started with initial data given by the exact solution and cmputed with different values of R ranging from 4 to 6.6. At $R=4$, $\lambda=.65$, all schemes were numerically stable. We know this to be theoretically true for the upwind scheme and for the travelling wave scheme. However, the travelling wave scheme was about 10 times more accurate then the other two (Figs.3 and 4). At higher values of R the centered scheme starts to oscillate (Fig.1) and the travelling wave scheme remains superior in accuracy to the upwind scheme (Figs.2 and 4).

An interesting feature of the travelling wave scheme, which is not shared by the other two schemes is, that to be stable for larger values of R it needs a larger CFL number (Thm. 3.1). One can therefore compute in the large R regime by taking relatively large time steps (Compare Figs.2 and 4).

Centered Difference Scheme

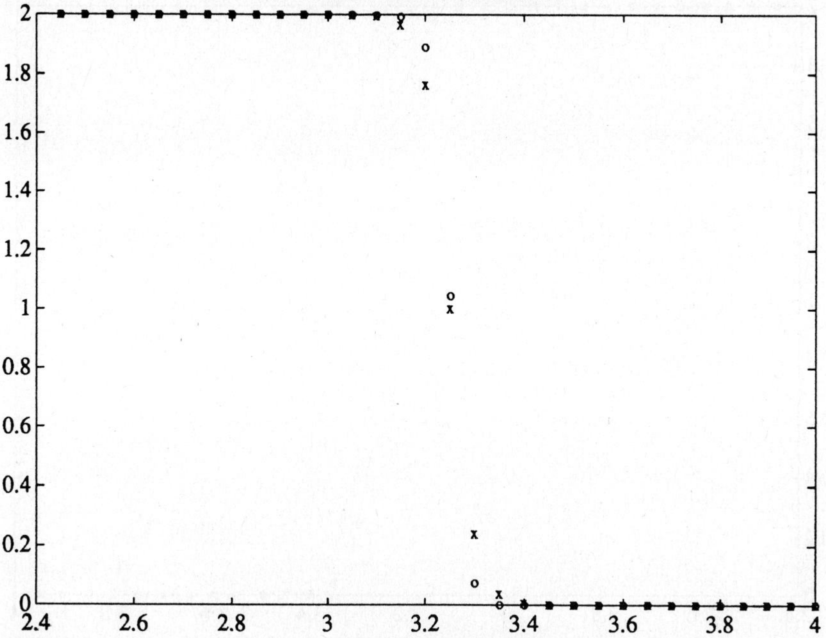

R=4,Dt/Dx=.325,Time=200,Error=0.165

Second Order Upwind

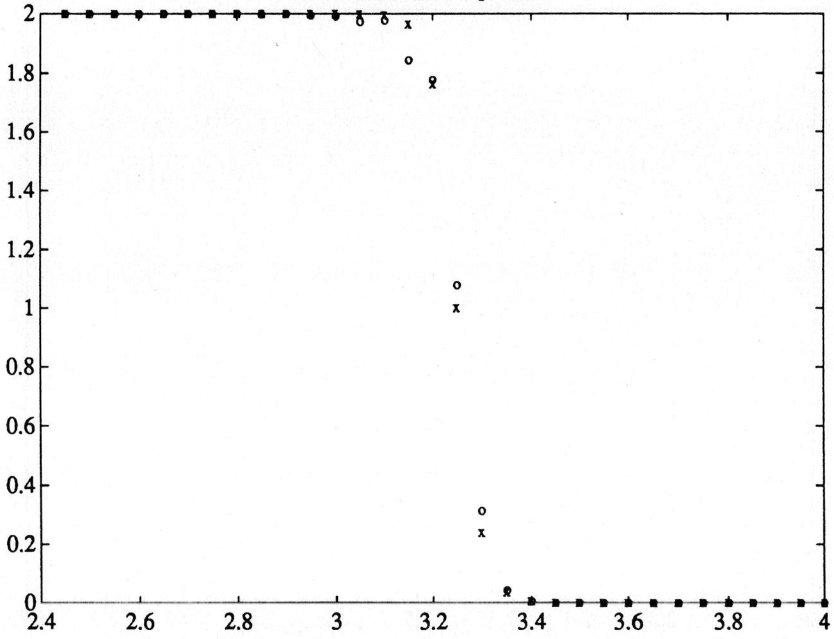

R=4,Dt/Dx=.325,Time=200,Error=.120,Exact="x"

Travelling Wave Scheme

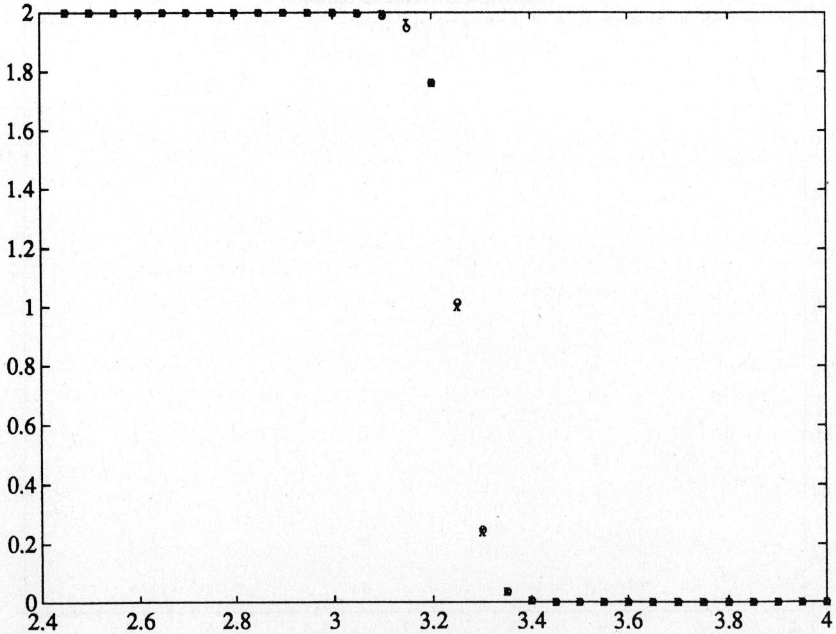

R=4,Dt/Dx=.325,Time=200,Error=.016,Exact="x"

Travelling Wave Method

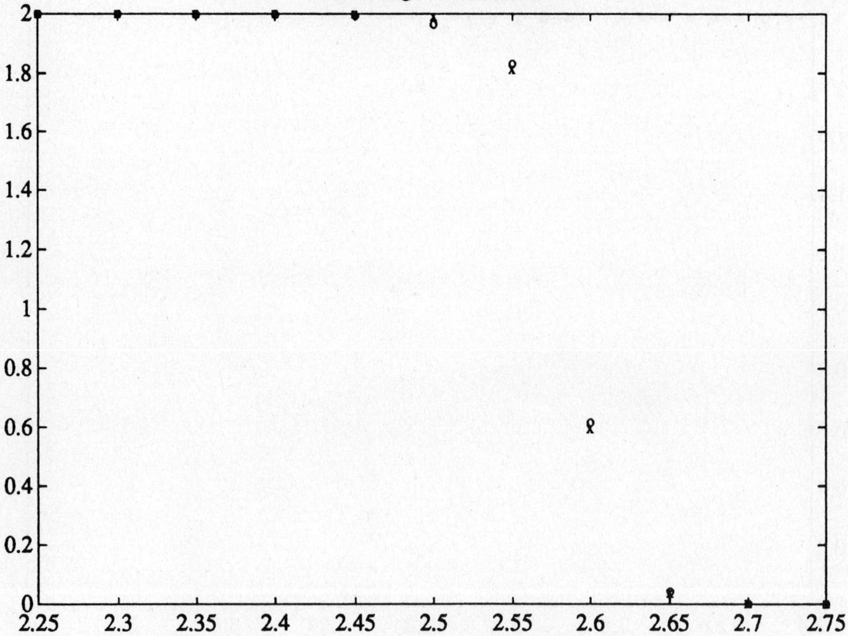

R=6.66,Dt/Dx=.43,Time=120,Error=.026,Exact="x"

REFERENCES

[1] HARTEN, A, "High Resolution Schemes for Hyperbolic Conservation Laws", J. Compet. Phys., v.49, (1983) pp. 357-393.

[2] HARABETIAN, E., "A Numerical Method for Viscous Perturbations of Hyperbolic Conservation Laws", SIAM J. Num. Ana., submitted, (1988).

[3] MACCORMACK, R.W., "A Numerical Method for Solving the Equations of Compressible Viscous Flow, AIAA Journal, v.20, No.9, pp. 1275-1281.

[4] OSHER, S., RALSTON, J., "L^1 Stability of Travelling Waves with Applications to Convextive Porous Media Flow", Comm. Pure Appl. Math., 35 (1982), pp. 737-751.

[5] OSHER, S., CHAKRAVARTHY, S., "High Resolution Schemes and the Entropy Condition", SIAM J. Numer. Anal., v.21, (1984), pp. 955-984.

[6] THOMAS, J.L., WALTERS, R.W., "Upwind Relaxation Algorithms for the Navier-Stokes Equation", AIAA Conf., Cincinnati, Ohio, (1985).

[7] VAN LEER, B., "Towards the Ultimate Conservation Difference Scheme, a Second Order Sequel to Godunov's", J. Comp. Phys., v.32, (1979), pp. 101-136.

SOLUTION OF THE EULER EQUATIONS FOR UNSTEADY, TWO-DIMENSIONAL, TRANSONIC FLOW

H. H. Henke
Messerschmitt-Bölkow-Blohm GmbH
2800 Bremen, FRG
Hünefeldstraße 1-5

INTRODUCTION

The numerical simulation of the aerodynamic forces acting on an oscillating wing-section in transonic flow by solving the Euler equations is described. The method of solution is the approximate-factorization method of Beam and Warming /1/. Since time-dependent calculations are the primary concern, an implicit algorithm was developed because it allows considerably larger time-steps than for explicit schemes, and calculation must be carried out over several periods of oscillation on wing-sections. In this paper several calculations for steady and unsteady transonic flow cases were carried out.

GOVERNING EQUATIONS

For the present investigation the Euler equations are written in curvilinear coordinates (ξ,η,τ):

$$\bar{U}_\tau + \bar{F}_\xi + \bar{G}_\eta = 0 \tag{1}$$

where

$$\bar{U} = \frac{1}{J}\begin{bmatrix}\rho\\ \rho u\\ \rho v\\ e\end{bmatrix} \quad \bar{F} = \frac{1}{J}\begin{bmatrix}\rho\tilde{U}\\ \rho u\tilde{U}+\xi_x p\\ \rho v\tilde{U}+\xi_y p\\ (e+p)\tilde{U}-\xi_t p\end{bmatrix} \quad \bar{G} = \frac{1}{J}\begin{bmatrix}\rho\tilde{V}\\ \rho u\tilde{V}+\eta_x p\\ \rho v\tilde{V}+\eta_y p\\ (e+p)\tilde{V}-\eta_t p\end{bmatrix}.$$

In the above relations $\tilde{U}$ and $\tilde{V}$ are the contravariant velocities

$$\tilde{U} = \xi_t+\xi_x u+\xi_y v \ ; \quad \tilde{V} = \eta_t+\eta_x u+\eta_y v \tag{2}$$

and J is the Jacobian of the coordinate transformation

$$J^{-1} = x_\xi y_\eta - x_\eta y_\xi = 1/(\xi_x\eta_y-\xi_y\eta_x). \tag{3}$$

METHOD OF SOLUTION

The method of approximate factorization of Beam and Warming /1/ was used for solving the Euler equations:

$$\left(I+\Delta\tau\frac{\partial}{\partial\xi}\bar{A}^n+D_{I\xi}\right)\left(I+\Delta\tau\frac{\partial}{\partial\eta}\bar{B}^n+D_{I\eta}\right)\Delta\bar{U}^n=-\Delta t\left(\frac{\partial\bar{F}}{\partial\xi}+\frac{\partial\bar{G}}{\partial\eta}\right)^n+D_E=RHS^n \quad (4)$$

$$\Delta\bar{U}^n = \bar{U}^{n+1} - \bar{U}^n .$$

The flux-vectors are locally linearized in time

$$\bar{F}^{n+1} = \bar{F}^n + \bar{A}^n(\bar{U}^{n+1} - \bar{U}^n) + O(\Delta\tau^2)$$

$$\bar{G}^{n+1} = \bar{G}^n + \bar{B}^n(\bar{U}^{n+1} - \bar{U}^n) + O(\Delta\tau^2) \quad (5)$$

with the Jacobian matrices $\bar{A} = \dfrac{\partial\bar{F}}{\partial\bar{U}}$, $\bar{B} = \dfrac{\partial\bar{G}}{\partial\bar{U}}$.

The terms D_I and D_E are implicit and explicit nonlinear damping terms defined in /2/.

The spatial derivatives were approximated by central differences of second order accuracy, so that 4x4 blocktridiagonal systems results. The solution of these systems of equations requires relatively long computational time.

A substantial reduction of the computational time is obtained by diagonalizing the matrices $\bar{A}$ and $\bar{B}$ with the similarity transformation of the form /3,4/:

$$\bar{A} = T_\xi\Lambda_\xi T_\xi^{-1} \quad ; \quad \bar{B} = T_\eta\Lambda_\eta T_\eta^{-1} \quad (6)$$

where

$$\Lambda_\xi = \begin{bmatrix} \tilde{U} & & & 0 \\ & \tilde{U} & & \\ & & \tilde{U}+k_\xi c & \\ 0 & & & \tilde{U}-k_\xi c \end{bmatrix}, \quad \Lambda_\eta = \begin{bmatrix} \tilde{V} & & & 0 \\ & \tilde{V} & & \\ & & \tilde{V}+k_\eta c & \\ 0 & & & \tilde{V}-k_\eta c \end{bmatrix}$$

with $k_\xi=(\xi_x{}^2+\xi_y{}^2)^{1/2}$, $k_\eta=(\eta_x{}^2+\eta_y{}^2)^{1/2}$ and c is the speed of sound, yielding systems of 4 scalar equations /4,5/

$$T_\xi{}^n\left(I+\Delta\tau\frac{\partial}{\partial\xi}\Lambda_\xi+D_{I\xi}\right)N^n\left(I+\Delta\tau\frac{\partial}{\partial\eta}\Lambda_\eta+D_{I\eta}\right)(T_\eta{}^{-1})^n\Delta U^n=RHS^n \quad (7)$$

with $N^n = (T_\xi{}^{-1})^n\, T_\eta{}^n$.

The matrices $T_\xi{}^n$, $(T_\xi{}^{-1})^n$, ... are taken outside of the spatial derivatives and outside of the brackets, so that the block matrices are diagonalized, whereby a decoupling is achieved.

The latter method is nonconservative in unsteady transonic flow and was therefore only used for the calculation of steady flow with embedded shocks. For unsteady flow calculations scheme (4) was used.

The boundary conditions for the correction variable (in the implicit part of the difference equations) is assumed to be $\Delta\bar{U}^n=0$ and the physical values on the boundary are formulated explicitly. This formulation being first order in time is easy to implement for all types of boundary conditions.

The boundary condition at the body surface is given for $\eta=0$ by the condition of impermeability

$$\tilde{V} = 0 . \tag{8}$$

The tangential component of the velocity is obtained by linear extrapolation. Then the Cartesian velocity components on the body (for an orthogonal grid) are given by

$$\begin{bmatrix} u \\ v \end{bmatrix} = J^{-1} \begin{bmatrix} \eta_y & -\xi_y \\ -\eta_x & \xi_x \end{bmatrix} \begin{bmatrix} \tilde{U}-\xi_t \\ -\eta_t \end{bmatrix} . \tag{9}$$

The value of the density on the profile is found likewise by linear extrapolation. A relation for the pressure along the surface is obtained from the normal momentum relation by combining the two transformed momentum equations in nonconservative form:

$$\begin{aligned} p_n(\eta_x{}^2+\eta_y{}^2) = (\xi_x\eta_x+\xi_y\eta_y)p_\xi + (\eta_x{}^2+\eta_y{}^2)p_\eta + \\ + \rho(\eta_{tt}+\eta_{xt}u+\eta_{yt}) - \rho\tilde{U}(\eta_x u_\xi+\eta_y v_\xi) \end{aligned} \tag{10}$$

where n is the normal direction to the contour η=const. The derivatives tangential to the body are approximated by central differences and the normal derivatives by one-sided differences.

In the far-field characteristic compatibility relations based on the one-dimensional characteristics similar to that given in /6/ are employed. The local linearized characteristic variables are

$$\begin{aligned} w_{\eta,1} &= \frac{J}{|\nabla\eta|}\left[\eta_x\left(\rho - \frac{p}{c_0{}^2}\right)\right] , \quad w_{\eta,2} = \frac{J}{|\nabla\eta|}\left[\eta_y u-\eta_y v\right] \\ w_{\eta,3,4} &= \frac{J}{|\nabla\eta|}\frac{1}{\sqrt{2}}\left[\frac{p|\nabla\eta|}{\rho_0 c_0} \pm(\eta_x u+\eta_y v)\right] \end{aligned} \tag{11}$$

with the corresponding eigenvalues

$$\lambda_{\eta,1} = \lambda_{\eta,2} = \tilde{V} \; ; \quad \lambda_{\eta,3,4} = \tilde{V} \pm k_\eta c$$

Additionally a vortex-correction formulation /7,8/ must be taken into account, so that there is no or little change in lift due to the extend of the computational domain.

The linear stability analysis shows unconditional stability for the approximate-factorization method, but the amplification factor approaches unity for large Courant numbers, as a result of the factorization error. The consequence is a decreasing rate of convergence with increasing time step, and in practice the Courant number is restricted to O(10). This restriction is not so weighty for unsteady flow computation because physical aspects (wave motion) must be taken into account.

RESULTS

For all calculations carried out a C-type grid, given for transonic test problems in /9/ with 141x21 or 141x31 points was used. The response characteristics of the airfoil surface pressure to the airfoil motions can be depicted using Fourier representation. If the unsteady angle of attack is expressed as $\alpha(t)=\alpha_m+\mathrm{Im}(\alpha_0 e^{i\omega t})$, the Fourier series representation of the pressure coefficient can be written as $c_p(x,t)=c_{pm}(x) +\Sigma\mathrm{Im}(c_{p,\alpha_0}^{n}(x)\alpha_0 e^{in\omega t})$.

In Fig. 1 the pressure distribution for several time points for a harmonically oscillating profile at a Mach-number of 0.80 is given. For the same profile the steady pressure distribution for $M_\infty=0.85$ is given in Fig. 2a, and in Fig. 2b the first mode harmonic components of the pressure on the lower and upper side of the profile are shown. The unsteady pressure distributions, as can be seen here, are symmetric. For the same case the lift coeeficient as function of the angle of attack for the unsteady motion is given in Fig. 3, and in Fig. 4 the L_2-Norm in the change of the density $(\Delta\rho/\Delta t)$ as a function of the time steps for steady and unsteady flow is presented.

For the steady flow computation the influence of the vortex-correction formulation on the pressure distribution is shown in Fig. 5. As can be seen here, this formulation leads to nearly 10% more lift in comparison to the formulation without a correction. The extend of the computational domain is nearly 12 chord lengths in each direction.

In Fig. 6 the steady pressure distribution and the lines of constant Mach-number for a MBB A3 profile in transonic flow is shown. In Fig. 7a the corresponding unsteady pressure distribution for the oscillating wing section, and in Fig. 7b the number of supersonic points as function of the time-steps is used as a crude indication of convergence (for the steady flow case), for the time-step n>400 the change of supersonic points for the oscillating airfoil can be seen. For all unsteady flow calculations 200 time-steps for a period of oscillation were used.

CONCLUSION

In this paper a method has been developed for calculating aerodynamic forces on a wing-section oscillating harmonically in transonic inviscid flow. The Euler equations are taken as governing flow equations. They are solved by the approximate-factorization method of Beam and Warming. Results have been presented for steady flow cases and for unsteady flow around an oscillating airfoil.

REFERENCES

/1/ Beam, R., Warming, R.F.: An Implicit Finite-Difference Algorithm for Hyperbolic Systems in Conservation-Law-Form
Journal of Comp. Physics, Vol. 22, Sept. 1976, pp 87-110.

/2/ Henke, H., Hänel, D.: Artificial Damping in Approximate Factorization Methods
Proceedings of the Sixth GAMM-Conference on Numerical Methods in Fluid Mechanics, D. Rues and W. Kordulla (Eds.), Notes on Numerical Fluid Mechanics, Vol. 13, Vieweg Verlag, Braunschweig.

/3/ Warming, R.F., Beam, R., Hyett, B.J.: Diagonalization and Simultaneous Symmetrization of the Gas Dynamic Matrices
Mathematics of Computation, Vol. 29, Oct. 1975, pp 1037-1045.

/4/ Pulliam, T.H., Chaussee, D.S.: A Diagonal Form of an Implicit Approximate-Factorization Algorithm
Journal of Comp. Physics, Vol. 39, 1981, pp 347-363.

/5/ Henke, H., Hänel, D.: Numerical Simulation of Gas Motion in Piston Engines
Ninth International Conference on Numerical Methods in Fluid Dynamics, Paris 1984
Lecture Notes in Physics, Vol. 218, Springer Verlag 1985, Ed. Soubbaramayer and J.P. Boujet.

/6/ Whitfield, D.L., Janus, J.M.: Three-Dimensional Unsteady Euler Equations Solution Using Flux Vector Splitting
AIAA Paper 84-1552, AIAA 17th Fluid Dynamics, Plasma Dynamics, and Lasers Conference, June 1984, Snowmass, Colorado.

/7/ Thomas, J.L., Salas, M.D.: Far-Field Boundary Conditions for Transonic Lifting Solutions to the Euler Equations
AIAA Journal, Vol. 24, July 1986, pp 1074-1080.

/8/ Pulliam, T.H., Steger, J.L.: Recent Improvements in Efficiency, Accuracy, and Convergence for Implicit Approximate-Factorization Algorithms
AIAA Paper 85-0360.

/9/ Rizzi, A., Viviand, H., (Eds.): Numerical Methods for the Computation of Inviscid Transonic Flows with Shock Waves.
GAMM Workshop, Notes on Numerical Fluid Mechanics, Vol. 3, Vieweg Verlag, Braunschweig, 1981.

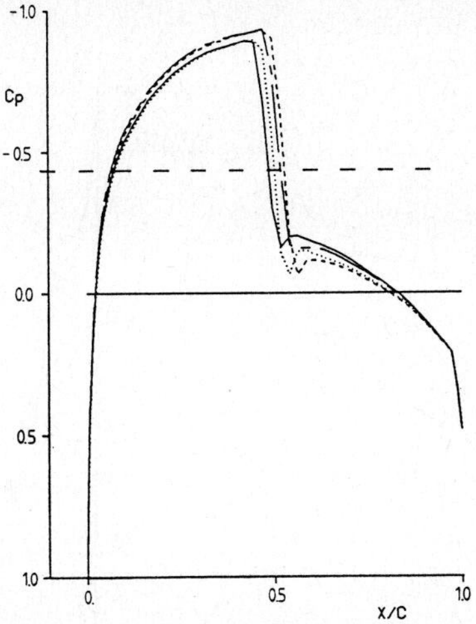

Fig. 1: Pressure distribution for several time points for a pitching airfoil-section
NACA 0012, $M_\infty=0.80$, $\alpha_m=0^0$, $\alpha_0=0.5^0$

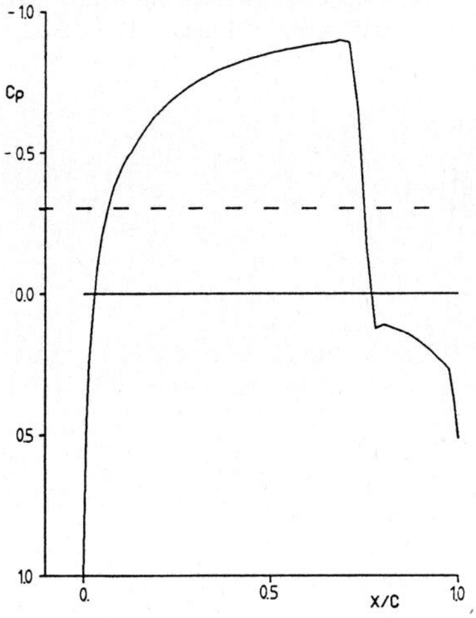

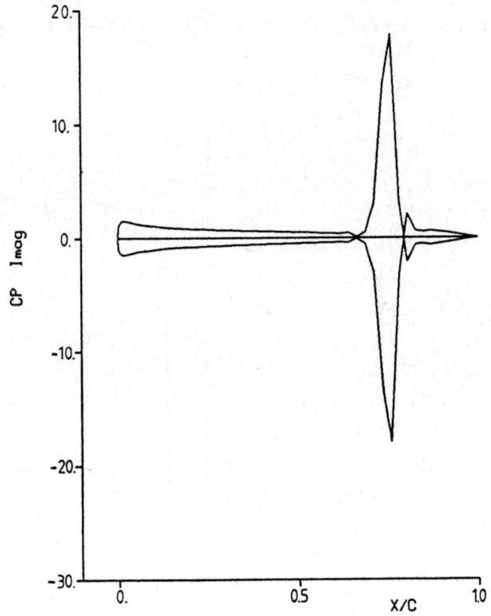

Fig. 2: Pressure distributions for NACA 0012 profile
$M_\infty=0.85$, $\alpha_m=0^0$

a) steady pressure

b) unsteady pressure: $c_{p\,Imag}$ for upper and lower side

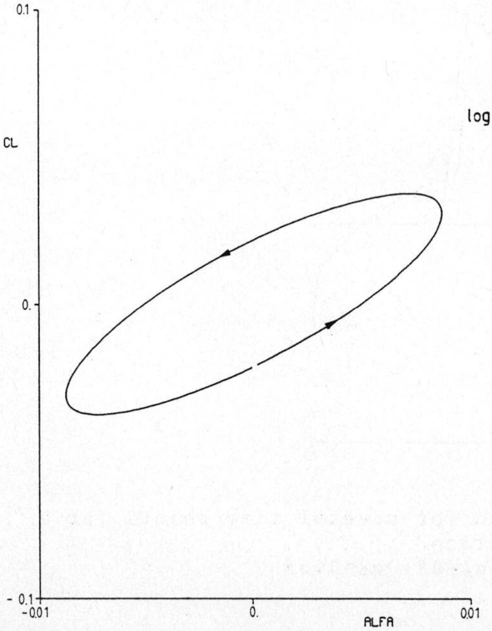

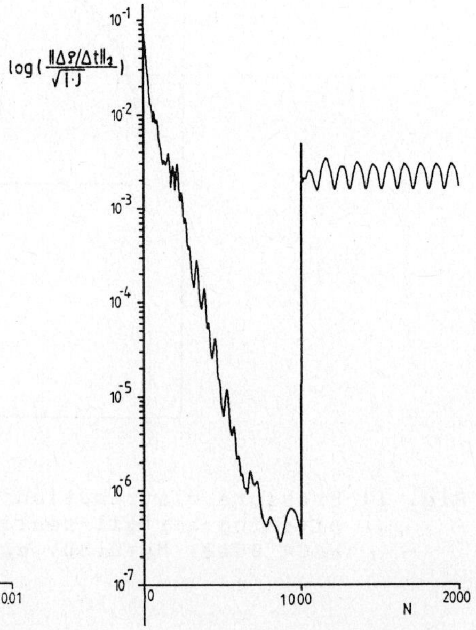

Fig. 3: Lift coefficient as function of the angle of attack for one period of pitch-motion $M_\infty=0.85$, $\alpha_m=0.^0$, $\alpha_0=0.5^0 a_m=0^0$,

Fig. 4: L_2-Norm in the change of the density $(\Delta\rho/\Delta t)$ as function of the number of time steps. Unsteady flow: N>1000

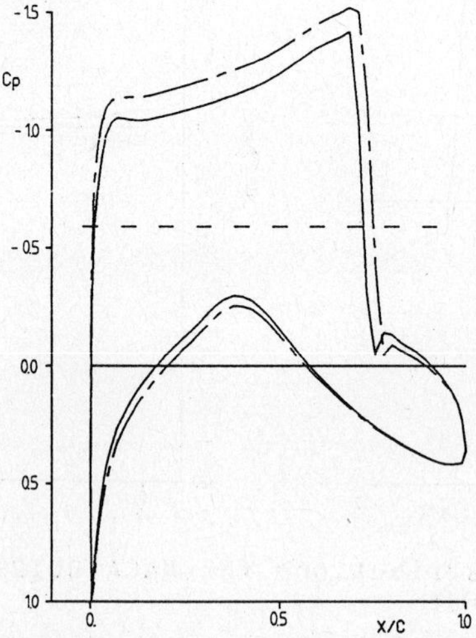

Fig. 5 Steady pressure distribution for a RAE 2822 profile $M_\infty=0.75$, $\alpha_m=3.^0$

—— without vortex-correction in the far-field

—- vortex-correction in the far-field

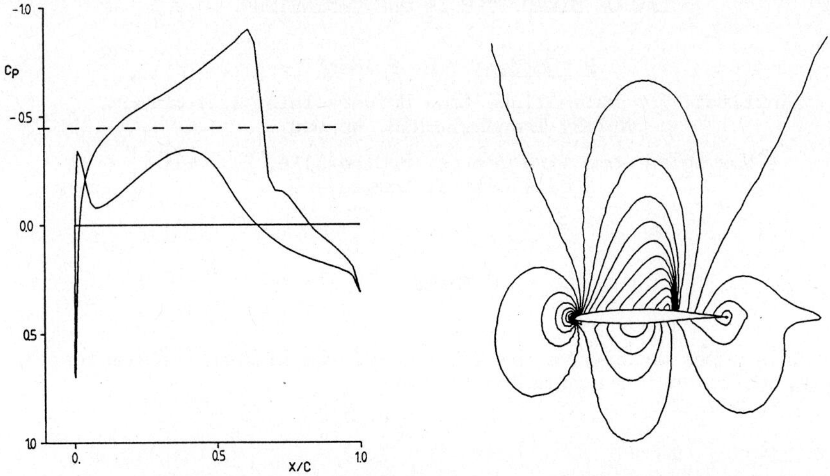

Fig. 6: Steady pressure distribution for a MBB A3 profile and lines of constant Mach-numbers $M_\infty=0.80$, $\alpha_m=-0.20^0$

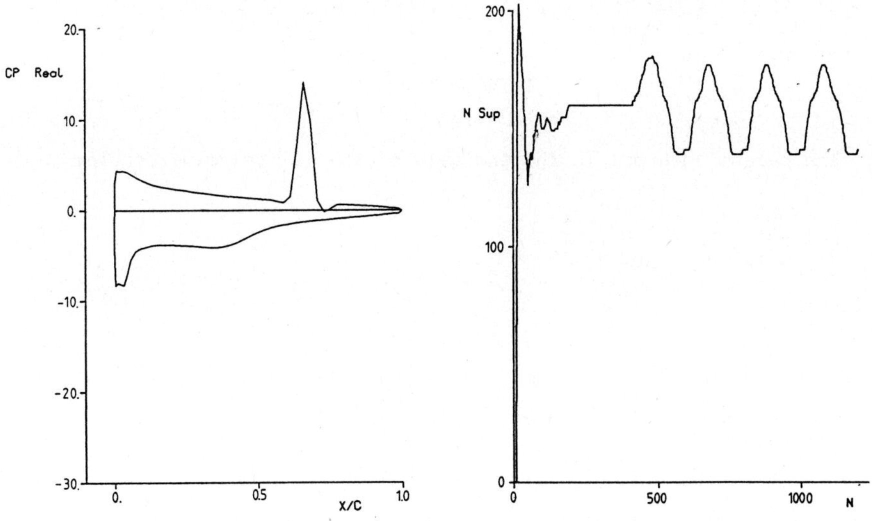

Fig. 7: Unsteady pressure distribution and number of supersonic points as function of the time steps

a) c_{pReal} for upper and lower side

b) NSUP: N<400 steady flow, N>400 unsteady flow

ON SOME RECENT RESULTS FOR AN EXPLICIT CONSERVATION LAW OF MIXED TYPE IN ONE DIMENSION

H. Holden[1], L. Holden[2]

[1] Institutt for matematiske fag, Universitetet i Trondheim, N-7034 Trondheim-NTH, Norway.

[2] Norwegian Computing Centre, P.O.Box 114, Blindern, N-0314 Oslo 3, Norway.

SUMMARY

In this paper we analyze the solution of the Riemann problem for the following one-dimensional conservation law

$$\frac{\partial}{\partial t}\begin{bmatrix} u \\ v \end{bmatrix} + \begin{bmatrix} \frac{1}{2}(\frac{u^2}{2} + v^2)+\rho v \\ u(v-\rho) \end{bmatrix} = 0 \, .$$

This differential equation is hyperbolic when $u^2+16v^2 \geq 16\rho^2$ and elliptic when $u^2+16v^2 < 16\rho^2$, and has been studied for $\rho = 0$ by Isaacson and Temple [1]. It corresponds to a symmetric case II in the classification of Schaeffer and Shearer [2].

INTRODUCTION

The recent interest in the analysis of the initialvalue problem

$$\frac{\partial}{\partial t}\begin{bmatrix} u \\ v \end{bmatrix} + \frac{\partial}{\partial x}\begin{bmatrix} f(u,v) \\ g(u,v) \end{bmatrix} = 0, \tag{1}$$

with

$$u(x,0) = \begin{cases} u_L \, , & x < 0 \\ u_R \, , & x > 0 \end{cases} , \qquad v(x,0) = \begin{cases} v_L \, , & x < 0 \\ v_R \, , & x > 0 \end{cases} , \tag{2}$$

the so-called Riemann problem, has revealed many interesting new phenomena.

Part of the motivation to study the problem (1), (2) has been the need to understand the behavior of three-phase flow in a porous medium, see [2 - 5]. In particular a numerical investigation [6] with realistic data in the context of three-phase flow showed that one should expect regions in phase-space (i.e. (u,v)-space) where (1) is not hyperbolic, but elliptic (hyperbolic/elliptic meaning that $\begin{bmatrix} f_u(u,v) & f_v(u,v) \\ g_u(u,v) & g_v(u,v) \end{bmatrix}$ has two real/non-real eigenvalues). Indeed more recent theoretical work has proved that one cannot in general rule out the existence of elliptic regions for three-phase flow in porous media [7], [8]. The solution of the Riemann problem was shown numerically to be quite stable even with initial data quite close to the elliptic region [6]. The starting point for the renewed interest in (1), (2) was the surprising complexity which showed up in the solution of a prototype problem with a umbilic point, i.e. an isolated point where the (real) eigenvalues are degenerate. This resulted in a very detailed analysis of the case where f and g both are homogeneous 2nd degree polynomials in u and v with one umbilic point, and it was shown [2] that the problem could be classified in four distinct classes in this case. A perturbation of a umbilic point will in general lead to the existence of an elliptic region.

In order to obtain rigorous results for a model problem in the mixed type case, one of us analyzed a Riemann problem in detail in the symmetric case I [9]. The solution of the Riemann problem was found to be quite wellbehaved except for a loss of uniqueness close to the elliptic region.

Recently we have extended the analysis to the symmetric case II which we want to report on here [10]. Cases I and II are believed to be most relevant for three-phase flow in porous media [7].

The solution in the case at hand is more complex. There are regions where there is no solution when one essentially only allows Lax-shocks and so-called compressive shocks. The analytical results are supplemented by numerical simulations using a Lax-Friedrichs difference scheme. The detailed results of this analysis will appear in [11].

RAREFACTIONS AND HUGONIOT LOCI

Consider the conservation law

$$\frac{\partial}{\partial t}\begin{bmatrix} u \\ v \end{bmatrix} + \frac{\partial}{\partial x}\begin{bmatrix} \frac{1}{2}(\frac{u^2}{2} + v^2)+\rho v \\ u(v-\rho) \end{bmatrix} = 0 \tag{3}$$

with $\rho > 0$. The fundamental property of this equation is contained in

<u>Proposition 1.</u> The system (3) is hyperbolic when $u^2+16v^2 \geq 16\rho^2$, strictly hyperbolic when $u^2+16v^2 > 16\rho^2$ and elliptic when $u^2+16v^2 < 16\rho^2$. ■

The two basic "atoms" which build up the general solution of the Riemann problem are the shock solutions and the rarefaction waves. A shock solution of (2) and (3) is a solution of the form

$$z(x,t) = \begin{cases} z_L \ , & x < st \\ z_R \ . & x > st \end{cases} \tag{4}$$

with $z = \begin{bmatrix} u \\ v \end{bmatrix}$. The shock speed s satisfies the Rankine-Hugoniot relation

$$s(z_R - z_L) = F(z_R) - F(z_L) \tag{5}$$

with $F = \begin{bmatrix} f \\ g \end{bmatrix}$. The Hugoniot locus relative to a point $z_L \in \mathbb{R}^2$ is by definition

$$H(z_L) = \{ z \in \mathbb{R}^2 | \exists s \in \mathbb{R} : s(z - z_L) = F(z) - F(z_L) \}. \tag{6}$$

The Hugoniot locus may have quite complicated behavior.

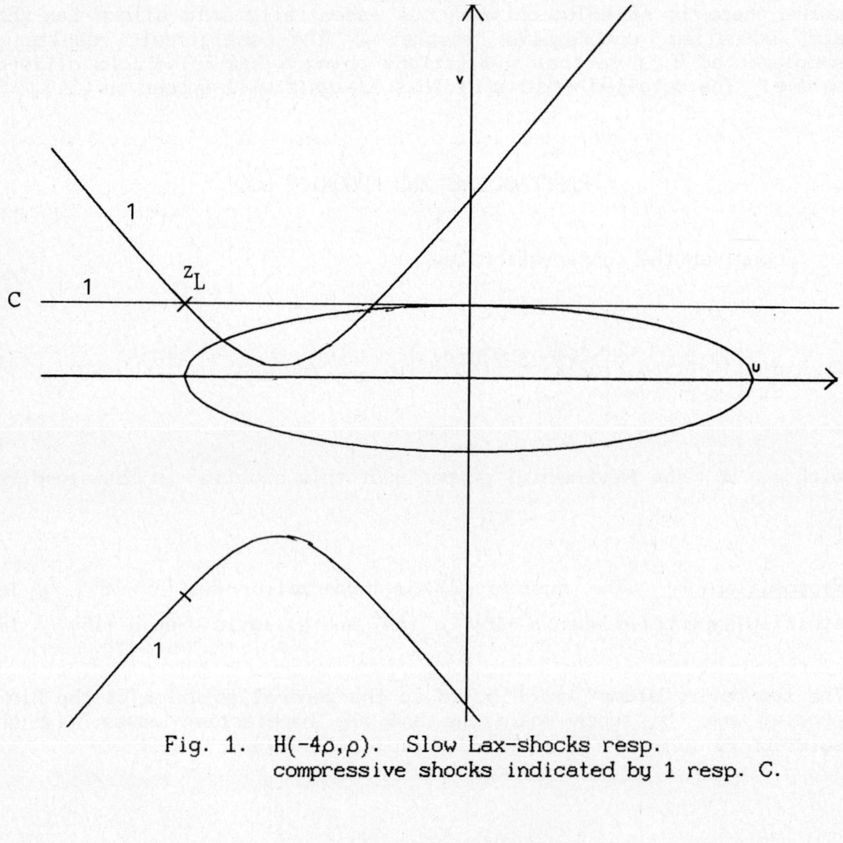

Fig. 1. $H(-4\rho,\rho)$. Slow Lax-shocks resp. compressive shocks indicated by 1 resp. C.

A rarefaction wave is a solution of (2) and (3) on the form

$$z(x,t) = \begin{cases} z_L , & x < \lambda(z_L)t \\ \eta(\frac{x}{t}) , & \lambda(z_L)t < x < \lambda(z_R)t \\ z_R , & x > \lambda(z_R)t \end{cases} \quad (7)$$

where $\lambda(z)$ is an eigenvalue of the matrix $dF(z)$ and $\eta(\xi)$ satisfies

$$\frac{d\eta}{d\xi}(\xi) = r(\eta(\xi)) \quad , \quad \eta(\lambda(z_k)) = z_k \quad , \quad k \in \{L,R\} \quad (8)$$

where $r(z)$ is the eigenvector corresponding to the eigenvalue $\lambda(z)$ normalized such that

$$\nabla\lambda(z)\cdot r(z) = 1 \ . \quad (9)$$

The where (9) is not possible, i.e. where

$$\nabla\lambda(z)\cdot r(z) = 0 \quad (10)$$

is called the inflection locus and plays an important role in the solution of the Riemann problem.

Allowing for weak solutions of the conservation law one has to face uniqueness problems. The so-called entropy condition is supposed to single out the correct solution. We will here allow essentially for Lax-shocks satisfying either (12) or (13) and compressive shocks satisfying

$$\lambda_2(z_R) < s < \lambda_1(z_L) \ . \quad (11)$$

trying to follow the Liu-construction [12] using the Liu-Oleinik entropy condition.

THE SOLUTION OF THE RIEMANN PROBLEM

The general solution of the equations (2) and (3) is very complicated indeed. Therefore we will here concentrate on an explicit characterization of the solution for one particular left state, which however will be representative of the solution for a set of left states.

In addition we will give examples of simulations using a Lax-Friedrichs difference scheme.

The solution will be given in $z_R = \begin{bmatrix} u_R \\ v_R \end{bmatrix}$ space.

Let

$$z_L = \begin{bmatrix} -4\rho \\ \rho \end{bmatrix} . \tag{12}$$

The slow waves consist of slow shocks and slow rarefactions. The slow rarefactions curve terminates at the boundary of the ellipse. For points z_R on this curve the solution is a simple rarefaction connecting z_L and z_R, denoted R. The slow Lax-shocks, satisfying

$$s < \lambda_1(z_L) \quad , \quad \lambda_1(z_R) < s < \lambda_2(z_R) \, , \tag{13}$$

constitute two branches, one connected to z_L and one branch in 3rd quadrant. From each of these slow waves we can continue with fast shocks and fast rarefactions as shown in Fig. 2.

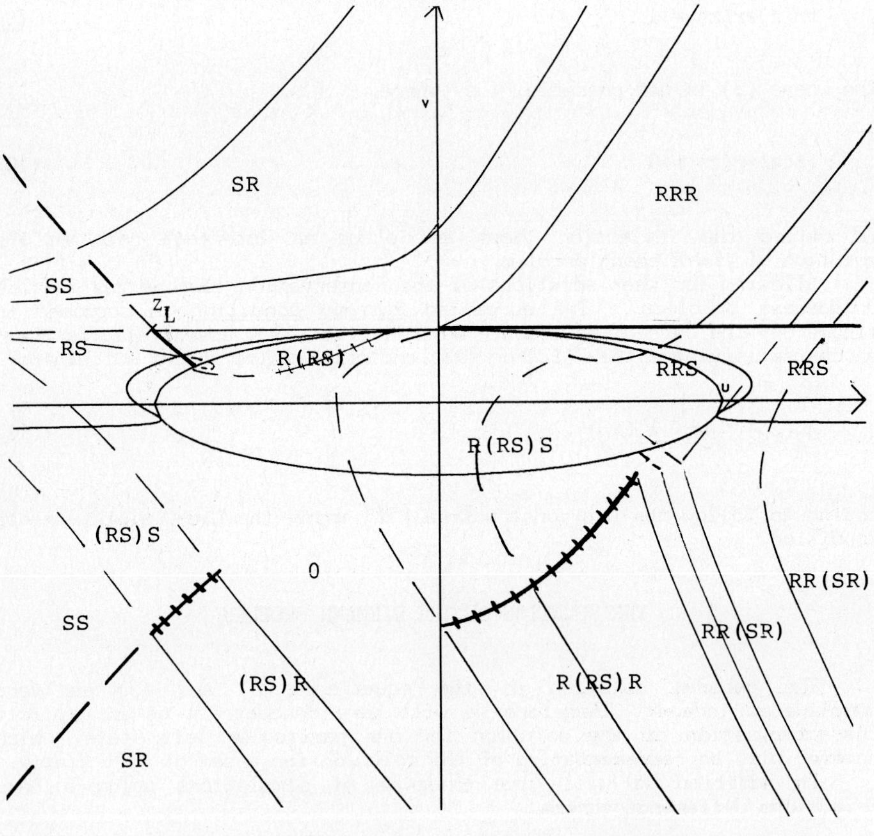

Fig. 2. The solution of Riemann problem in the z_R-plane for $z_L = \begin{bmatrix} -4\rho \\ \rho \end{bmatrix}$.

The fast rarefaction wave follows the line $v = \rho$ to $u = 0$ where it bifurcates into one fast branch leaving $v = \rho$ and a slow branch along $v = \rho$. We will write the solution as RR for $z_R = (u_R, \rho)$, $u_R > 0$. The fast Lax-shocks, satisfying

$$s > \lambda_2(z_R) , \quad \lambda_1(z_L) < s < \lambda_2(z_L) \tag{14}$$

lie on the line $v = \rho$, $u < u_L = -4\rho$.

From each point on the slow rarefaction we can jump with a shock with speed equal to the fastest speed in the rarefaction wave. Such a wave is called a composite and is denoted by (RS). From the composite one can continue with fast waves. The solution close to the slow rarefaction is complicated. To the left of this curve we use fast shocks. However, as the slow rarefaction crosses the inflection locus, we have to use a fast rarefaction followed by a fast composite and eventually a fast shock. To the right of the slow rarefaction we use fast rarefactions which however terminate on the inflection locus. From here we use a fast composite which eventually turn into a shock looping back to the slow rarefaction. In the area marked 0 there is no solution if we only allow Lax and compressive shocks.

We now turn to the line $v = \rho$, $u > 0$. For each such point on this line with $u \in (0, 2\sqrt{6}\rho)$ we can jump with a speed equal to the fastest speed in the rarefaction, i.e. a slow composite. With this curve as a starting point we continue with fast waves. As $u \in (2\sqrt{6}\rho, 4\sqrt{2}\rho)$ there are detached fast shocks, see Fig. 2.

In this way we have constructed the solution of the Riemann problem where it exists for a particular left state. For the general statement we refer to [10]. Finally we would like to report on some preliminary numerical result for this model. Using the Lax-Friedrichs difference scheme

$$z_i^{n+1} = \frac{1}{2}(z_{i+1}^n + z_{i-1}^n) - \frac{\Delta t}{\Delta x}(F(z_{i+1}^n) - F(z_{i-1}^n)) \tag{15}$$

with $z_i^0 = z_L$ when $i < 0$ and $z_i^0 = z_R$ when $i \geq 0$ and z_i^n approximates $z(n\Delta t, i\Delta x)$. Where (3) and (4) has one solution, the difference scheme is found to converge numerically to that solution, see Fig. 3, even with initial states close to or inside the elliptic region.

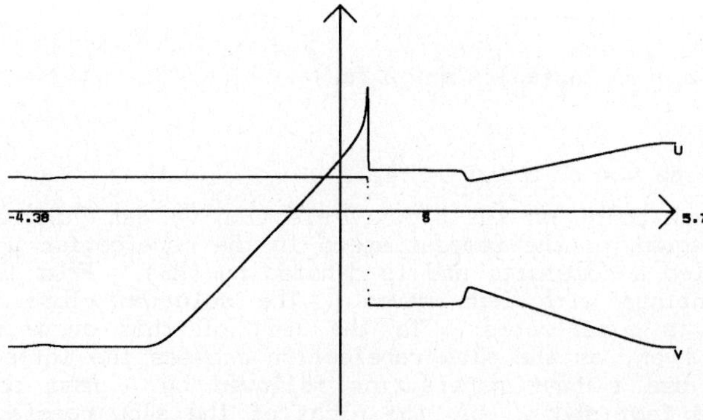

Fig. 3. The solution with $z_R = \begin{bmatrix} 2\rho \\ -4\rho \end{bmatrix}$, described by R(RS)R.

In the cases where we have stated that there is no solution, one finds numerically solutions with crossing shocks. However, by allowing for all crossing shocks, there is a continuum of solutions, and it is not possible to select a unique solution by comparing only shock speeds and the eigenvalue of the left and right state. For further details we refer to [11].

Acknowledgements. We are grateful for support from Norwegian Research Council for Science and the Humanities (H.H.) and the Royal Norwegian Council for Scientific and Industrial Research (L.H.).

REFERENCES.

[1] ISAACSON, E., TEMPLE, B.: "The classification of solutions of quadratic Riemann problems (II), (III)", SIAM J. Math. Anal.

[2] SHAEFFER, D.G., SHEARER, M.: "The classification of 2×2 systems of non-strictly hyperbolic conservation laws, with applications to oil recovery". Comm. Pure Appl. Math. 40 (1987) pp. 141-178.

[3] ISAACSON, E., MARCHESIN, D., PLOHR, B., TEMPLE, B.: "The classification of solutions of quadratic Riemann problems (I)", SIAM J. Math. Anal.

[4] SCHAEFFER, D.G., SHEARER, M.: "Riemann problems for nonstrictly hyperbolic 2×2 systems of conservations laws", Trans. Amer. Math. Soc. 340 (1987) pp. 267-306.

[5] SCHAEFFER, D.G., SHEARER, M., MARCHESIN, D., PAES-LEME, P.L.: "Solution of the Riemann problem for a prototype 2×2 system of non-strictly hyperbolic conservation laws", Arch. Rat. Mech. Anal. 97 (1987) pp. 299-320.

[6] BELL, J.B., TRANGENSTEIN, J.A., SHUBIN, G.R.: "Conservation laws of mixed type describing three-phase flow in porous media", SIAM J. Appl. Math. 46 (1986) pp. 1000-1017.

[7] SHEARER, M.: "Loss of strict hyperbolicity of the Buckley-Leverett Equations for thre-phase flow in porous media", Proc. IMA Workshop on Oil Reservoir Simulation (December 1986).

[8] HOLDEN, L.: "On strict hyperbolicity for the Buckley-Leverett equations for three-phase flow in a porous medium, in preparation.

[9] HOLDEN, H.: "On the Riemann problem for a prototype of a mixed type conservation", Comm. Pure Appl. Math. 40 (1987) pp. 229-264.

[10] HOLDEN, H., HOLDEN, L.: "On the Riemann problem for a prototype of a mixed type conservation law II", Preprint, University of Trondheim 1988.

[11] HOLDEN, H., HOLDEN, L.: In preparation.

[12] LIU, T.P.: "The Riemann problem for general 2×2 conservation laws", Trans. Amer. Math. Soc. 199 (1974) pp. 89-112.

Qualitative behavior of solutions for Riemann problems of conservation laws of mixed type

L. Hsiao

Department of Mathematics, University of Washington
Seattle, Washington, U.S.A.

Institute of Mathematics, Academia Sinica
Beijing, China

1. Introduction.

It is known that the system of conservation equations of describing the time evolution of unsteady flows is typically of hyperbolic type when dissipative effects are neglected. However, for complex materials there may be open regions where the linearized system is elliptic. This occurs in various models in applications which has caused a lot of attention and there have been some mathematical analysis performed on it to extend the theory of hyperbolic conservation laws, especially Riemann problem and admissibility condition to include equations of mixed type, such as [BN] on traffic flow; [J] [M] [Sh] [Sl] [Hs], [Ha] on vibrations of elastic bars and Van der Waal fluids; and recent results in [K] [Ho] and [HM].

In the first case the elliptic region is ellipsoidal, however the Riemann problem is only discussed in special cases when the states are not in the elliptic region. In the second case the elliptic region is a strip and the Riemann problem is solved with different considerations about the admissible discontinuity by different authors. In [Ho] the system is elliptic inside a circle and it is claimed that the Riemann problem always has a weak solution which however is not necessarily unique.

The following system arising in modelling certain nonlinear advection process in ecology is discussed in [HM] for which the elliptic region is unbounded with boundary curve $(v-u+a-1)^2+4(a-1)u=0$

$$\begin{cases} u_t + [u(1-v)]_x = 0 \\ v_t + [v(a+u)]_x = 0 \end{cases} \tag{1.1}$$

where $a > 1$.

The Hugoniot locus of (1.1) can be at most three disconnected branches for each family and can be parametrized by u for the first family and by v for the second family.

Combining and generalizing Lax-admissibility criterion and Liu-Oleinick criterion for admissible discontinuity used to strictly hyperbolic system of conservation laws, the authors in [HM] introduced the generalized entropy condition by which it was shown that the Riemann problem always has a weak solution, however, certain regions exhibit multiple solutions.

In order to have uniqueness, a minimum principle was introduced in the definition of an admissible weak solution in [HM]. The existence and uniqueness of the admissible weak solution is proved then for any given Riemann data.

The purpose of this paper is to investigate the structure and the qualitative behavior of the admissible weak solution defined in [HM]. Referring to the Lax-criterion, a shock satisfying the generalized entropy condition may agree with or violate it at different level.

Research supported by the Science Fund of the Chinese Academy of Sciences and the National Science Foundation under Grant.No. DMS-8657319

We give a complete classification concerning the relation between the speed of the admissible shock and the characteristics associated in Secion 2. Continuous dependence is discussed in Section 3 then. It is to be expected in general in problem of mixed type that the solution is not continuously dependent on the initial data. However, some kind of stable behavior can still be expected for the admissible weak solution. It is shown in Theorm 3.2 that the entire U_--plane ($U = (u, v)$) is divided into ten regions such that the topological structure of admissible wave curves keep the same as $U_- = (u_-, v_-)$ varies in each one of the regions. The topology may vary only when U_- across the boundaries in the U_--plane. In other words, for fixed $U_- = (u_-, v_-)$, the admissible weak solution of the Riemann problem may divide the U_+-plane into different regions representing different combinations of shocks, rarefaction waves and composite waves, the qualitative structure of the U_+-regions changes only as U_- across the U_--boundary curves in the U_--plane.

2. Classification of admissible shocks.

The i-characteristic speed of (1.1) is defined as

$$\begin{aligned} \lambda_1 &= \frac{1}{2}\{u - v + a + 1 - [(v - u + a - 1)^2 + 4(a - 1)u]^{1/2}\} \\ \lambda_2 &= \frac{1}{2}\{u - v + a + 1 + [(v - u + a - 1)^2 + 4(a - 1)u]^{1/2}\} \end{aligned} \tag{2.1}$$

at any point (u, v) where $\Delta(u, v) = (v - u + a - 1)^2 + 4(a - 1)u \geq 0$. The elliptic region where $\Delta(u, v) < 0$ is shown in Fig.2.1. λ_1 is equal to λ_2 on the curve $\Delta(u, v) = 0$. The system is strictly hyperbolic at any (u, v) where $\Delta(u, v) > 0$. Moreover, the system is genuinely nonlinear when it is hyperbolic except the set of the fognals, i.e. the curves where genuine nonlinearity fails, which is made up of four rays:

$$\begin{cases} v = 0 \\ u \geq -(a-1) \end{cases}, \quad \begin{cases} u = 0 \\ v \leq -(a-1) \end{cases}; \quad \begin{cases} v = 0 \\ u \leq -(a-1) \end{cases}, \quad \begin{cases} u = 0 \\ v \geq -(a-1) \end{cases}.$$

On the first group of the rays, the first family of characteristics is linearly degenerate and $\lambda_1 = 1$ and $\lambda_1 = a$ respectively; on the second group of the rays, the second family of characteristics is linearly degenerate and $\lambda_2 = 1$ and $\lambda_2 = a$ respectively. The distribution of rarefaction wave curves is shown in Fig.2.1 where the curves R_1 ending at the point $\{u = -(a-1), v = 0\}$; R_1 ending at the point $\{u = 0, v = -(a-1)\}$ and rays $v = 0$ with $u \geq -(a-1)$; $u = 0$ with $v \leq -(a-1)$ will play important roles as the boundary curves in the dividing of the U_--plane.

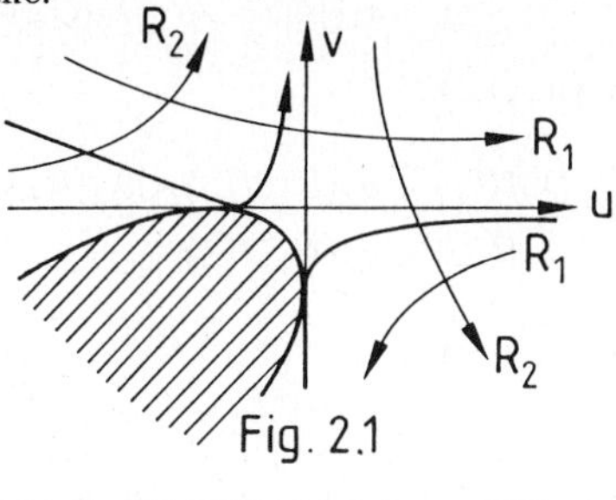

Fig. 2.1

A discontinuity is determined by Rankine-Hugoniot con di tion, namely

$$\begin{cases} \sigma[u] = [(1 - v)u] \\ \sigma[u] = [(a + u)v] \end{cases} \tag{2.2}$$

where $[w] = w_\gamma - w_\ell$ denotes the jump of the quantity w across the discontinuity with speed σ.

For any given (u_0, v_0), (2.2) defines the Hugoniot locus $H_1(u_0, v_0)$ and $H_2(u_0, v_0)$. It is shown in [HM] that $H_1(H_2)$ is a single-valued function of $u(v)$ along which it holds

$$\begin{cases} \sigma_1 = \frac{1}{2}\{u - v_0 + a + 1 - [(v_0 - u + a - 1)^2 + 4(a-1)u]^{1/2}\} \\ (\sigma_1 - a - u)(v - v_0) - v_0(u - u_0) = 0 \end{cases} \tag{2.3$_1$}$$

$$\left(\begin{cases} \sigma_2 = \frac{1}{2}\{u_0 - v + a + 1 + [(v - u_0 + a - 1)^2 + 4(a-1)u_0]^{1/2}\} \\ (\sigma_2 - 1 + v)(u - u_0) + u_0(v - v_0) = 0 \;. \end{cases}\right) \tag{2.3$_2$}$$

For handling the elliptic region in the system (1.1), a generalized entropy condition was introduced in [HM]. For any given $(u_-, v_-) = U_-$, a discontinuity (σ, U_+, U_-) is called admissible according to the generalized entropy condition if $U_+ \in H_1(U_-)$ (similar to $U_+ \in H_2(U_-)$, taking v as the variable) such that either

I. For any u between u_- and u_+ where σ_1 is defined, it holds

$$\sigma_1(u; u_-, v_-) \geq \sigma_1(u_+; u_-, v_-)$$

or

II. $\sigma_1(u; u_-, v_-)$ is non-increasing with respect to $|u - u_-|$ for all $u \in \mathcal{J}_{+,-}$ for which it is defined, where

$$\mathcal{J}_{+,-} = \begin{cases} [u_-, u_+) & \text{if } u_- < u_+ \\ (u_+, u_-] & \text{if } u_+ < u_- \;. \end{cases}$$

Denote the set of states, belonging to $H_i(U_-)$ and satisfying the above generalized entropy condition (G.E.C.) by $S_i(U_-)$ which has different distributions corresponding to different location of $(u_-, v_-) = U_-$ as shown in [HM]. We discuss the relation between the speed of the admissible shock and the characteristics associated next respectively.

Case 1. $u_- > 0, v_- > 0$

It can be easily shown by using the formulae (2.1), (2.3) that σ_1 is decreasing along $S_1(u_-, v_-)$ as u decreasing and σ_2 is decreasing along $S_2(u_-, v_-)$ as v increasing (see Fig. 2.2), satisfying Lax-condition as follows

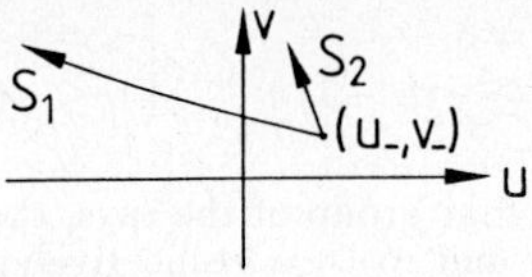

Fig. 2.2

$$\begin{aligned} \lambda_1(U_+) < \sigma_1(U_+; U_-) &< \lambda_1(U_-) \\ \sigma_1(U_+; U_-) &< \lambda_2(U_+) \end{aligned} \tag{2.4$_1$}$$

for any $U_+ = (u_+, v_+) \in S_1(u_-, v_-)$;

$$\begin{aligned} \lambda_2(U_+) < \sigma_2(U_+; U_-) &< \lambda_2(U_-) \\ \lambda_1(U_-) &< \sigma_2(U_+; U_-). \end{aligned} \tag{2.4$_2$}$$

for any $U_+ = (u_+, v_+) \in S_2(u_-, v_-)$.

Case 2. $u_- > 0, v_- < 0$

The distribution of the set $S_i(u_-, v_-)$ is shown in Fig. 2.3 and 2.4 corresponding to $v_- \geq -(a-1)$ or $v_- < -(a-1)$ respectively where the arrows indicate the direction of decrease of the corresponding shock speeds, the same as in figure 2.2.

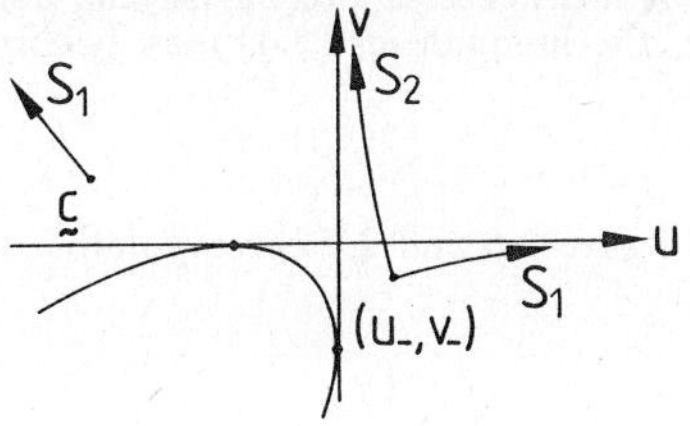

Fig. 2.3 $v_- \geq -(a-1)$

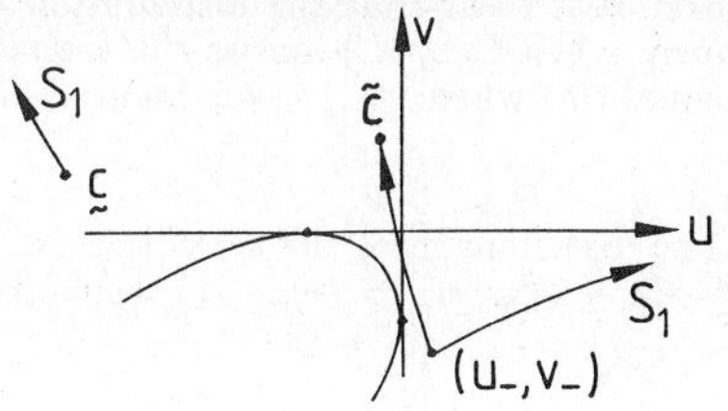

Fig. 2.4 $v_- < -(a-1)$

For any $U_+ = (u_+, v_+) \in S_i(u_-, v_-)$ which is connected branch to (u_-, v_-), (2.4) can be obtained in the same way as in case 1. As far as the branch S_1 disconnected to (u_-, v_-) is concerned, we consider $\sigma_1(\tilde{c}, U_-)$ first, where $\tilde{c}$ is defined by (see [HM])

$$\begin{cases} u_{\tilde{c}} = -(\sqrt{a-1} + \sqrt{-v_-})^2 \\ v_{\tilde{c}} = \frac{\sqrt{-v_-}}{\sqrt{a-1}+\sqrt{-v_-}} u_- + \sqrt{(a-1)(-v_-)} \end{cases}$$

and $\sigma_1(\tilde{c}; U_-) = 1 - \sqrt{(a-1)(-v_-)}$ by $(2.3)_1$,

Due to $u_- - u_- - (a-1) < 0$, it is easy to see that $\sigma_1(\tilde{c}; U_-) < \lambda_1(U_-)$ which implies that for any U_+ belonging to the same branch S_1 as $\tilde{c}$ does, it holds

$$\sigma_1(U_+; U_-) < \lambda_1(U_-) < \lambda_2(U_-) \tag{2.5}$$

we show $\sigma_1(\tilde{c}; U_-) < \lambda_2(\tilde{c})$ next which is equivalent to

$$v_{\tilde{c}} - u_{\tilde{c}} - (a-1) < 2\sqrt{(a-1)(-v_-)} + \sqrt{(v_{\tilde{c}} - u_{\tilde{c}} + a - 1)^2 + 4(a-1)u_{\tilde{c}}} \ .$$

This is equivalent, since $v_{\tilde{c}} - u_{\tilde{c}} - (a-1) > 0$, to

$$(a-1)(v_{\tilde{c}} - v_-) + \sqrt{(a-1)(-v_-)} \cdot \sqrt{(v_{\tilde{c}} - u_{\tilde{c}} + a - 1)^2 + 4(a-1)u_{\tilde{c}}} > 0$$

which is true because $v_{\tilde{c}} - v_- = \sqrt{-v_-}\{\frac{u_-}{\sqrt{a-1}+\sqrt{-v_-}} + (\sqrt{a-1} + \sqrt{-v_-})\} > 0$.
We prove that for any U_+ belonging to the same branch S_1 as $\tilde{c}$ does, it holds

$$\sigma_1(U_+; U_-) < \lambda_2(U_+) \ . \tag{2.6}$$

It is shown that (2.6) holds when $U_+ = \tilde{c}$. Suppose $U_+ = U^*$ is the first point on the same branch where $u^* < u_{\tilde{c}}$, (so that $v^* > v_{\tilde{c}} > 0$), $\sigma_1(U^*; U_-) = \lambda_2(U^*)$. It is easy to check that $\lambda_2(U^*) > 1$ since $v^* > 0$, namely $\sigma_1(U^*; U_-) > 1$. However, $\sigma_1(U^*, U_-) < \sigma_1(\tilde{c}; U_-) < 1$. This contradiction implies (2.6).

(2.7) can be verified by a straightforward calculation, we omit it.

In summary, for any $U_+ \in S_1(u_-, v_-)$, disconnected branch starting from $\tilde{c}$, $(2.4)_1$ still holds.

Remark. It is clear that the distribution of the set S_i is not necessary to be varying continuously when (u_-, v_-) across the fognals. Moreover, the inequalities (2.4) may become into equalities when (u_-, v_-) is located on the fognals.

Case 3. $u_- < 0, v_- > 0$.

The distribution of the set $S_i(u_-, v_-)$ is shown in figure 2.6 and 2.7 corresponding to $u_- \leq -(a-1)$ or $u_- > -(a-1)$ respectively.

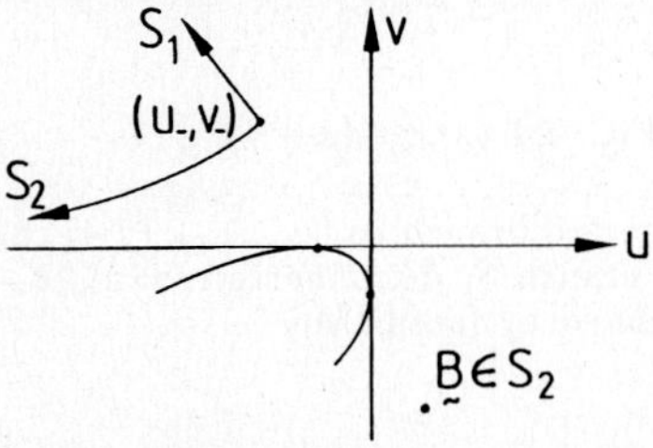

Fig. 2.5 $u_- \leq -(a-1)$

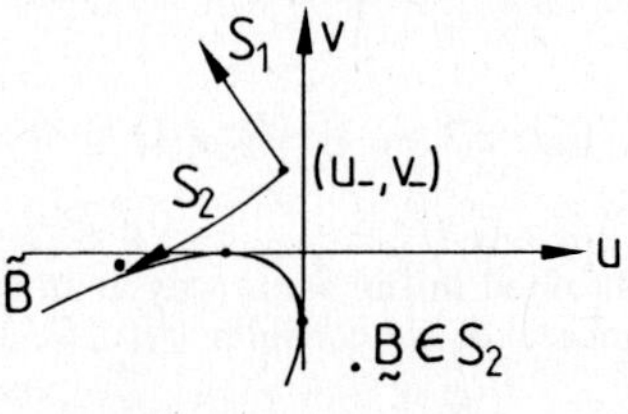

Fig 2.6 $u_- > -(a-1)$

For any $U_+ = (u_+, v_+) \in S_i(u_-, v_-)$ which is connected branch to (u_-, v_-), (2.4) can be obtained in the same way as before. Furthermore it is not difficult to show that

$$\sigma_2(\underset{\sim}{B}; U_-) > \lambda_2(U_-) \tag{2.8}$$

$$\sigma_2(\underset{\sim}{B}; U_-) > \lambda_1(\underset{\sim}{B}) \tag{2.9}$$

and

$$\sigma_2(\underset{\sim}{B}; U_-) < \lambda_2(\underset{\sim}{B}) \tag{2.10}$$

where $\underset{\sim}{B}$ is defined by (see [HM])

$$v_{\underset{\sim}{B}} = -(\sqrt{a-1} + \sqrt{-u_-})^2 \tag{2.11}$$

$$u_{\underset{\sim}{B}} = \frac{\sqrt{-u_-}}{\sqrt{-u_-} + \sqrt{a-1}} \cdot v_- + \sqrt{(a-1)(-u_-)} \tag{2.12}$$

Divide the region $\{u < 0, v < 0\}$ into subregions I–IV, as shown in Fig. 2.7. The discussion in case 4: $(u_-, v_-) \in I$ or case 5: $(u_-, v_-) \in II$ is the same as before, we omit it.

Case 6. $(u_-, v_-) \in III$.

The distribution of the set $S_i(u_-, v_-)$ is shown in Figure 2.8 and 2.9 corresponding to $u_- < -(a-1)$ or $u_- > -(a-1)$ respectively.

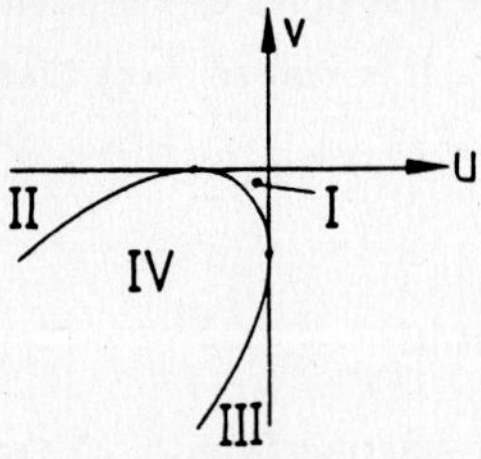

Fig. 2.7

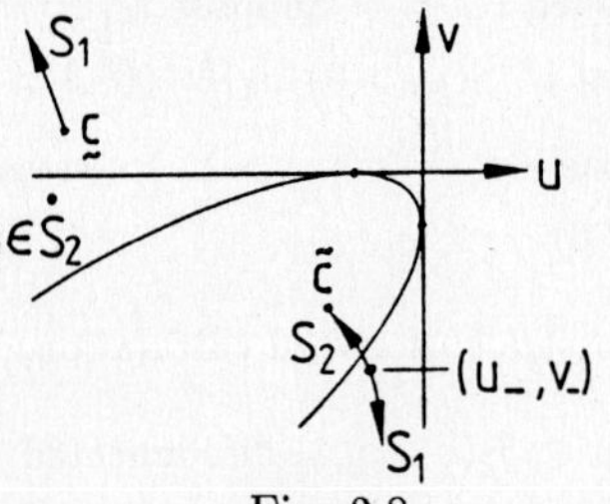

Fig. 2.8

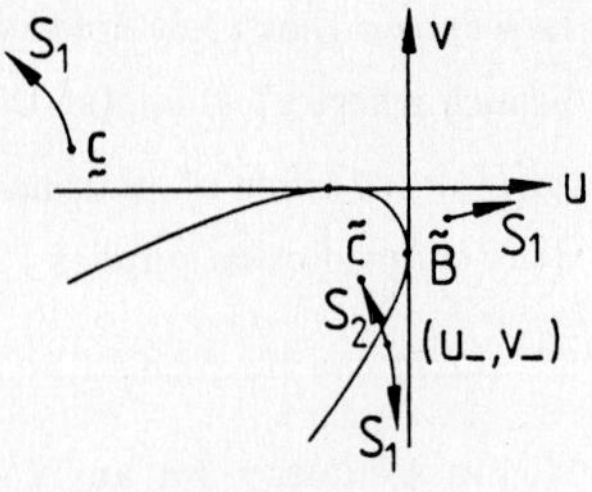

Fig. 2.9

For any $U_+ = (u_+, v_+) \in S_i(u_-, v_-)$ which is connected branch to (u_-, v_-), (2.4) can be obtained in the same way as before.

For any $U_+ \in S_1(u_-, v_-)$, disconnected branch starting from $\underset{\sim}{c}$, the similar argument shows (2.5) and (2.7). As far as (2.6) is concerned, the difference is the case when $v^* < 0$. Suppose $U_+ = U^*$ is the first point on the same branch where $v^* < 0$, $\sigma_1(U^*; U_-) = \lambda_2(U^*)$, then $v_{\underset{\sim}{c}} < 0$ and $\underset{\sim}{c}$ is located on a R_2 curve along which λ_2 is decreasing untill the point U' where this R_2 curve intersects to the curve $\Delta(u,v) = 0$. Clearly, $v' > v_-$. On the other hand, $\sigma_1(\underset{\sim}{c}; U_-) = 1 - \sqrt{(a-1)(-v_-)} > \sigma_1(U^*; U_-) = \lambda_2(U^*) > \lambda_2(U') = 1 - \sqrt{(a-1)(-v')}$ which implies $v_- > v'$. This contradiction implies $\sigma_1(U_+; U_-) < \lambda_2(U_+)$.

For any $U_+ \in S_1(u_-, v_-)$, disconnected branch starting from $\tilde{B}$, we show that

$$\sigma_1(U_+; U_-) < \lambda_1(U_-) \tag{2.14}$$

$$\sigma_1(U_+; U_-) > \lambda_1(U_+) \tag{2.15}$$

$$\sigma_1(U_+; U_-) < \lambda_2(U_+) \tag{2.16}$$

where

$$v_{\underset{\sim}{B}} = -(\sqrt{a-1} - \sqrt{-u_-})^2 \tag{2.17}$$

$$u_{\tilde{B}} = \frac{\sqrt{-u_-}}{\sqrt{-u_-} - \sqrt{a-1}} v_- - \sqrt{(a-1)(-u_-)} \tag{2.18}$$

$$\sigma_1(\tilde{B}; U_-) = a - \sqrt{(a-1)(-u_-)}\ . \tag{2.19}$$

(2.14) and (2.15) can be verified by a argument similar to case 3 for (2.8) and (2.10) except we have to distinguish now the different cases corresponding to different sign of $v_{\underset{\sim}{B}} - u_{\underset{\sim}{B}} + a - 1$.

As far as (2.16) is concerned, we show $\sigma_1(\tilde{B}; U_-) < \lambda_2(\tilde{B})$ first which is equivalent to

$$v_{\underset{\sim}{B}} - u_{\underset{\sim}{B}} + a - 1 < 2\sqrt{(a-1)(-u_-)} + \sqrt{(v_{\underset{\sim}{B}} - u_{\underset{\sim}{B}} + a - 1)^2 + 4(a-1)u_{\underset{\sim}{B}}}$$

which is true when $v_{\underset{\sim}{B}} - u_{\underset{\sim}{B}} + a - 1 < 0$; otherwise which is equivalent to $G(u_-, v_-) < 0$ where

$$\begin{aligned} G(u_-, v_-) = v_-(\sqrt{-u_-} + \sqrt{a-1}) + \{\sqrt{a-1}^3 + 2\sqrt{-u_-}(\sqrt{a-1} - \sqrt{-u_-})^2 \\ -\sqrt{-u_-}\sqrt{(a-1)(-u_-)} + \sqrt{-u_-}(\sqrt{a-1}^2 - \sqrt{-u_-}^2)\}\ . \end{aligned}$$

Investigate the distribution of the set $G(u_-, v_-) = 0$, we can verify that (u_-, v_-) is located in the region $G(u_-, v_-) < 0$ as shown in Fig. 2.10 in the case 6.

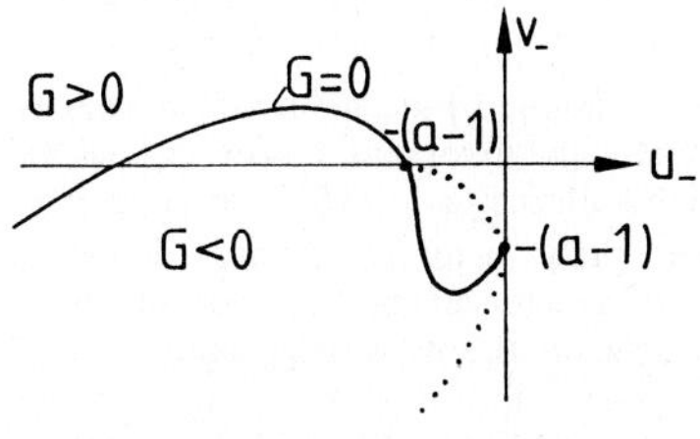

Fig. 2.10

Therefore, $\sigma_1(\widetilde{B};U_-) < \lambda_2(\widetilde{B})$, which means (2.16) holds when $U_+ = \widetilde{B}$. We show next (2.16) holds for any U_+ on the same branch. Suppose not, U^* is the first point where $\sigma_1(U^*;U_-) = \lambda_2(U^*)$, $u^* > u_{\underset{\sim}{B}}$, then U^* must be on a R_2-curve along which λ_2 is decreasing to a as u increasing. Namely, $\sigma_1(U^*;U_-) = \lambda_2(U^*) > a$. On the other hand, $\sigma_1(U^*;U_-) < \lambda_1(\widetilde{B};U_-) = a - \sqrt{(a-1)(-u_-)} < a$. This contradiction implies (2.16).

For the isolated point $\widetilde{B} \in S_2(u_-, v_-)$, it holds that

$$\sigma_2(\widetilde{B};U_-) < \lambda_1(U_-) \tag{2.21}$$

$$\sigma_2(\widetilde{B};U_-) > \lambda_1(\widetilde{B}) \tag{2.22}$$

$$\sigma_2(\widetilde{B};U_-) < \lambda_2(\widetilde{B}) \tag{2.23}$$

where $\sigma_2(\widetilde{B};U_-) = a - \sqrt{(a-1)(-u_-)}$.

The proof is similar, we omit it.

In summary, when both U_+ and U_- are in the hyperbolic region.

1. Any state U_+ located on $S_i(U_-)$ connected to the state U_- supplies an admissible shock satisfying the Lax condition $(2.4)_1$.
2. Any state U_+ located on $S_i(U_-)$ with ending point $\underset{\sim}{c}$ or $\widetilde{B}$, disconnected to U_-, supplies an admissible shock satisfying the Lax condition $(2.4)_1$
3. When the state U_+ takes the isolated state $\widetilde{B} \in S_2(U_-)$, the admissible shock agrees with the Lax condition on the number of characteristics which enter or leave the shock but violates the Lax condition on the style, namely
$$\sigma_2(\widetilde{B};U_-) < \lambda_1(U_-)$$
$$\lambda_1(\widetilde{B}) < \sigma_2(\widetilde{B};U_-) < \lambda_2(\widetilde{B}) \,.$$
4. When the state U_+ takes the isolated state $\underset{\sim}{B} \in S_2(U_-)$, the admissible shock violates the Lax condition on both the numbers and the style. Namely
$$\sigma_2(\underset{\sim}{B};U_-) > \lambda_2(U_-)$$
$$\lambda_1(\underset{\sim}{B}) < \sigma_2(\underset{\sim}{B};U_-) < \lambda_2(\underset{\sim}{B}) \,.$$

3. Continuous dependence of admissible weak solution

Consider (1.1) with initial data

$$(u,v)|_{t=0} = \begin{cases} (u_-, v_-), x < 0 \\ (u_+, v_+), x > 0 \end{cases} \tag{3.1}$$

where $U_- = (u_-, v_-)$ and $U_+ = (u_+, v_+)$ are arbitrary states in the (u,v)-plane. Since both the system and the initial data are invariant under the transformation $s \to \alpha x$, $t \to \alpha t$, $\alpha > 0$, we look for self-similar solution $u = u(\xi)$, $v = v(\xi)$, $\xi = \frac{x}{t}$.

It is shown in [HM] that there is no uniqueness if we use the same definition as for a purely hyperbolic system of conservation laws for an admissible weak solution and a minimum principle is introduced in the definition then.

Definition 3.1. An single-valued function is called an admissible weak solution of the Riemann problem (1.1) (3.1) if

I. It satisfies the boundary conditions $(u,v) \to (u_{\mp}, v_{\mp})$ as $\xi \to \mp\infty$.

II. It is either a rarefaction wave or a constant state whenever it is smooth.

III. Any discontinuity satisfies the Rankine-Hugoniot condition (2.2) and the above generalized entropy condition (G.E.C.)

IV. The sum of the strength of all of the jumps takes the minimum value among all possible single-valued function $(u(\xi), v(\xi))$ satisfying *I-III*.

It is to be expected in general in problem of mixed type that the solution is not continuously dependent on the initial data. However, some kind of stable behavior can be expected for our admissible weak solution which will be described in theorem 3.2.

It is shown in [HM] that for any fixed $U_- = (u_-, v_-)$, the admissible weak solution of the Riemann problem may divide the U_+-plane into as many as 12 regions, representing different combinations of shocks, rarefaction waves and composite waves.

Now we show that the U_--plane is divided by certain boundary curves into different regions such that the qualitative structure of the U_+-regions changes only as U_- across the U_--boundary curves in the U_--plane. Denote the points $\{u = -(a-1), v = 0\}$ and $\{u = 0, v = -(a-1)\}$ by P and Q respectively.

The R_1 curve starting from P; the ray $v = 0, u \geq -(a-1)$; the R_1 curve starting from Q; the ray $u = 0, v \leq -(a-1)$ and the rays $v \leq 0$ $u = -(a-1)$; $u \leq 0$, $v = -(a-1)$ play essential roles which divide the U_--plane into different regions, as shown in Figure 3.1.

Where the hyperbolic region is divided into six subdomains, numbered by I-VI. Now we discuss each of them.

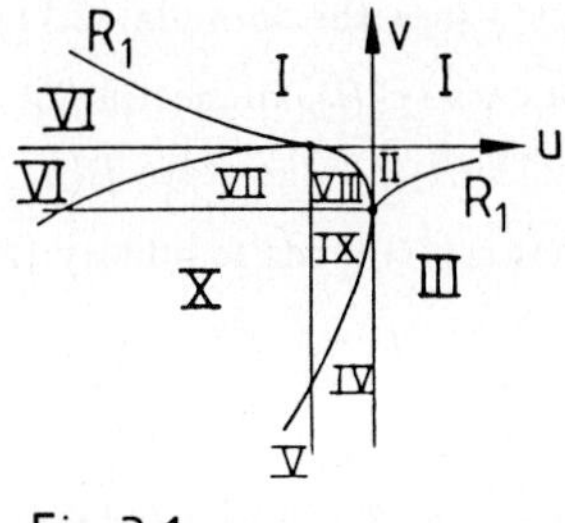

Fig. 3.1

Case 1. $U_- \in I$ in the U_--plane.

It is shown in [HM] that the structure of the U_+-regions in shown in Fig. 3.2 when $-(a-1) < u_- < 0$. The wave pattern for the solution of Riemann problem is $S_1 - S_2$ when $U_+ \in I$; $S_1 - R_2$ when $U_+ \in II$; $R_1 - R_2$ when $U_+ \in III$ or III^*; $R_1 - S_2$ when $U_+ \in IV$ or IV^*; $S_1 - S_2 - S_2 - S_2$ when $U_+ \in V$;

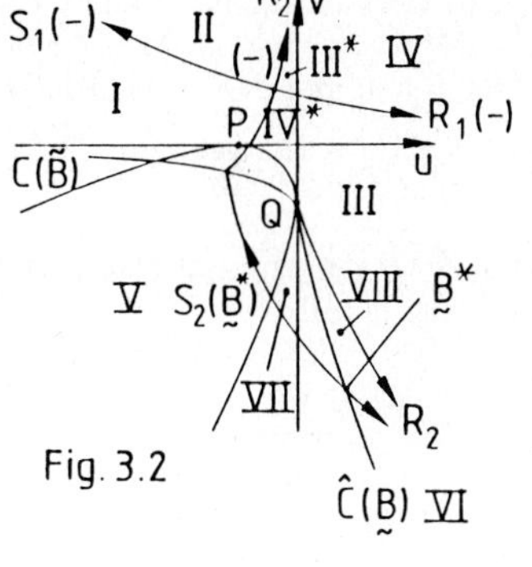

Fig. 3.2

$S_1 - S_2 - S_2 - R_2$ when $U_+ \in VI$; $R_1 - S_2 - S_2 - S_2$ when $U_+ \in VII$; $R_1 - S_2 - S_2 - R_2$ when $U_+ \in VIII$.

When $u_- \to -(a-1)$ ($v_- > 0$ is fixed), $v_{\tilde{B}} \to 0$ and $u_{\tilde{B}} \to -\infty$ which implies $v_{\underset{\sim}{B}^*} \to -\infty$ and $u_{\underset{\sim}{B}^*} \to +\infty$, therefore, the regions V and VI disappear. When $u_- \to 0$ ($v_- > 0$ is fixed), $\tilde{B} \to Q$ and $\underset{\sim}{B}^* \to Q$, the regions II^*, IV^*, VII and VIII disappear. When u_- varies from $u_- < 0$ to $u_- > 0$, the qualitative structure of the U_+-regions keeps the same.

Case 2. $u_- \in II$ in the U_--plane.

It can be proved that the structure of the U_+-regions is shown in Fig. 3.3, when $u_- > 0$. Where $R_1(u_-, v_-)$ intersects to the curve $\Delta(u,v) = 0$ at N. When (u_1, v_1) varies on $R_1(u_-, v_-)$ from $u_1 = 0$ to $u_1 = u_N$, the corresponding point $\widetilde{B}$ (see the formula (2.17) and (2.18), replacing (u_1, v_1) for (u_-, v_-) there) forms

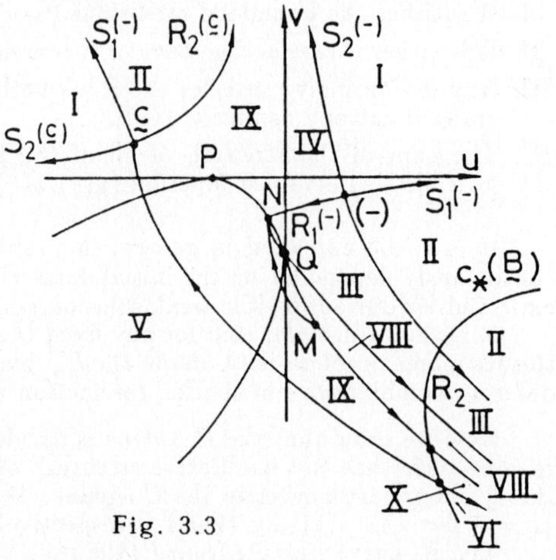

Fig. 3.3

the curve $c(\widetilde{B})$, connecting N and Q. Corresponding to each point (u^*, v^*) on this curve $c(\widetilde{B})$, there is the point $\underset{\sim}{B}(u^*, v^*)$ (see the formula (2.11), (2.12), replacing (u^*, v^*) for (u_-, v_-) there) which forms the curve $\widehat{c}(\underset{\sim}{B})$, connecting Q and M. Corresponding to each point (u_*, v_*) on the curve $S_2(\underset{\sim}{c}) \cup R_2(\underset{\sim}{c})$, the state $\underset{\sim}{B}(u_*, v_*)$ is defined which forms the curve $c_*(\underset{\sim}{B})$. It can be shown that $c_*(\underset{\sim}{B})$ tends to infinity $(u \to \infty, v \to -\infty)$ as $u_* \to -\infty$, $v_* \to 0$; tends to

$$\begin{cases} v \to -(a-1) \\ u \to +\infty \end{cases} \quad \text{as} \quad \begin{cases} u_* \to 0 \\ v_* \to +\infty \end{cases}.$$

The wave pattern for the solution of Riemann problem is the same as in case 1 corresponding to the subregions I–VIII respectively. The wave pattern is $S_1 - R_2 - S_2 - S_2$ when $U_+ \in IX$; $S_1 - R_2 - S_2 - R_2$ when $U_+ \in X$. Where the region confined by the curves $c(\widetilde{B}), u = 0$ and $R_1(u_-, v_-)$ belongs to IV, the region confined by the curves $\widehat{c}(\underset{\sim}{B})$, $c(\widetilde{B})$ and $S_2(M)$ is the subregion VII. It is easy to show that the structure of the U_+-regions is similar for $u_- \leq 0$.

Case 3. $U_- \in III$ and Case. 4 $U_- \in IV$ can be discussed in a similar way to Case 5, we omit the detail.

Case 5. $u_- \in V$.

The structure of the U_+-regions is shown in Fig. 3.4 where the curves $c(\widehat{c})$ is made up of the point $\widehat{c}(u_1, v_1)$ when (u_1, v_1) varies along the curve $S_1(u_-, v_-) \cup R_1(u_-, v_-)$ and $c_*(\underset{\sim}{B})$ is defined in the same way as in Case 2. $\widetilde{B} \in S_2(u_-, v_-)$. The curve $S_2(\widetilde{B})$ will end up at (u_-, v_-) which intersect to the curve $\Delta(u,v) = 0$ at the point T. Corresponding to each point on

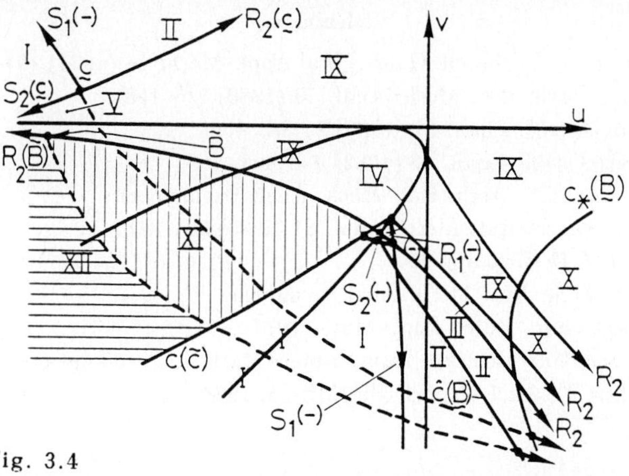

Fig. 3.4

$R_2(\widetilde{B}) \cup S_2(\widetilde{B})$ with $u \le u_T$, there is a point $\underset{\sim}{B}$ which forms the curve $\widehat{c}(\underset{\sim}{B})$ starting from the point T.

The wave pattern is the same as before when U_+ belongs to the corresponding subregions. $I - V$, IX and X respectively. The wave pattern is $S_2 - S_2 - S_2 - S_2$ when $u_+ \in XI$; $S_2 - R_2 - S_2 - S_2$ when $U_+ \in XII$.

Case 6. $u_- \in VI$.

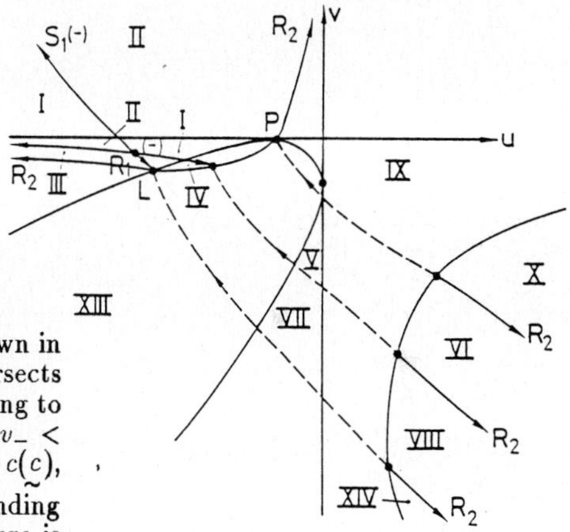

Fig. 3.5

The structure of the U_+-regions is shown in Fig. 3.5. Where the curve $R_1(u_-, v_-)$ intersects to the curve $\Delta(u, v) = 0$ at L. Corresponding to each point on $R_1(u_-, v_-) \cup S_1(u_-, v_-)$ with $v_- < 0$, there is a point $\underset{\sim}{c}$ which forms the curve $c(\underset{\sim}{c})$, connecting the points L and P. Corresponding to each point on $R_2(L) \cup c(\underset{\sim}{c}) \cup R_2(P)$, there is a point $\underset{\sim}{B}$ which forms the curve $c^*(\underset{\sim}{B})$.

The wave pattern is the same as before when U_+ belongs to the corresponding subregions I-VI, IX and X respectively. The wave pattern is $R_1 - R_2 - S_2 - S_2$ when $u_+ \in XIII$; $R_1 - R_2 - S_2 - R_2$ when $u_+ \in XIV$. When the state (u_-, v_-) varies from $v_- < 0$ to $v_- > 0$, the qualitative structure of the U_+-regions keeps the same.

We end up with the theorem.

Theorem 3.2. The U_--plane is divided by certain boundary curves into ten regions (see Fig. 3.1) such that the topological structure of admissible wave curves keep the same as $U_- = (u_-, v_-)$ varies in each one of the regions. The topology may vary only when U_- across the boundaries in the U_--plane.

References

[BN] J.H. Bick and G.F. Newell, Quart. J. of Appl. Math. 18 (1961) 191–204.
[J] R.D. James, Arch. Rat. Mech. Anal. 13 (1980) 125–158.
[M] M.S. Mock, J. Diff. Eqan. 37 (1980) 70–88.
[Sh] M. Sheaver, J. Diff. Eqan. 46 (1982) 426–445.
________________, Arch. Rat. Mech. Anal. (to appear)
[Sl] M. Slemrod, Arch. Rat. Mech. Anal. 81 (1983) 301–315.
[Hs] L. Hsiao, J. *P.D.E.* to appear.
[Ha] H. Hattori, Arch. Rat. Mech. Anal. to appear.
[Ho] H. Holden, Comm. Pure. Appl. Math. 1987.
[HM] L. Hsiao and P. de Mottoni, Trans. Amer. Math. Soc. to appear.
[K] B.L. Keyfitz, Research Report, UH/MD-25, 1988.

STRONGLY NONLINEAR HYPERBOLIC WAVES

John K. Hunter
Colorado State University
Fort Collins, Colorado 80523 USA

SUMMARY

We describe geometrical optics theories for nonlinear waves and derive a theory for hyperbolic waves with large–amplitude, rapidly varying initial data. We consider initial data which is either compactly supported or periodic in a phase variable. We also analyze the decay of periodic solutions of hyperbolic conservation laws and the resonant interaction of weakly nonlinear sawtooth waves.

1. GEOMETRICAL OPTICS

Geometrical optics is an asymptotic theory for short waves. Let λ be a typical lengthscale of the wave e.g. the width of a pulse, or the wavelength of a periodic wave. Let L be a lengthscale of modulations in the wave. That is, significant changes in the wave occur as it propagates over distances of the order L. Geometrical optics provides an asymptotic approximation for the wave in the limit

$$\epsilon = \frac{\lambda}{L} \to 0+.$$

Let us give two examples. Suppose that a finite source of sound waves generates a spherical pulse of waves whose width is of the order λ. The wave will decay as it spreads out from the source and a characteristic lengthscale of this effect is the distance $|x|$ of the wave from the source. Geometrical optics describes the far field, where $|x| \gg \lambda$. Second, suppose that a periodic soundwave of wavelength λ propagates through a stratified fluid. The scale height, L, is the distance over which the fluid density changes by a factor of e. Geometrical optics applies when the wavelength is much less than the scale height. For example, this condition is usually satisfied in the atmosphere, where $L \sim 7$ km.

The behaviour of a wave in the geometrical optics limit depends on whether the wave is hyperbolic or dispersive, and on the strength of the wave. Linear wave equations are usually obtained by linearizing the original, nonlinear equations. Linearization fails if: (a) the wave amplitude is too large; or (b) the propagation distance is too long. In case (a), we say that the wave is strongly nonlinear, and in case (b) we say that the wave is weakly nonlinear. The basic equations of nonlinear geometrical optics are summarized in Table 1.

Table 1. Nonlinear geometrical optics theories

	HYPERBOLIC	DISPERSIVE
weakly nonlinear	Inviscid Burgers' eq. $a_t + aa_x = 0$	Nonlinear Schrodinger eq. $ia_t + a_{xx} \pm a\|a\|^2 = 0$
strongly nonlinear	?	Averaged Lagrangian eqs. $\omega = \omega(k, a)$, $k_t + \omega_x = 0$ $\frac{\partial}{\partial t}L_\omega(\omega,k,a) - \frac{\partial}{\partial x}L_k(\omega,k,a)=0$

It is not known what theory, if any, is appropriate for strongly nonlinear hyperbolic waves. We shall describe one such theory in Sections 2 and 3, developed through joint work with J. B. Keller [1].

The most successful approximate theory for large–amplitude, hyperbolic waves is Whitham's shock dynamics [2], which describes the propagation of strong shocks in gas dynamics. Although shock dynamics involves rays and wavefronts, which are often associated with geometrical optics, it is not a geometrical optics theory in the sense defined above. This is because shock dynamics does not use a short wave assumption.

2. NONLINEAR HYPERBOLIC WAVES

We shall derive a formal asymptotic approximation for the solution $u(x, t; \epsilon) \in \mathbb{R}^m$ of the initial value problem

$$u_t + \sum_{i=1}^{n} \partial_{x_i} f_i(x, u) = 0 \tag{2.1}$$

$$0 < \epsilon \ll 1.$$

$$u(x, 0; \epsilon) = u_0\left[x, \frac{\psi(x)}{\epsilon}\right] \tag{2.2}$$

We assume that (2.1) is strictly hyperbolic and genuinely nonlinear. In this section we suppose that $u_0(x, \eta)$ is compactly supported in η for each $x \in \mathbb{R}^n$. The case when u_0 is periodic in η is discussed in Section 3.

We shall show that for short times, of the order ϵ, the solution is described asymptotically by a constant coefficient system of conservation laws in one space dimension. Shocks form and the solution decays rapidly. For large times, of the order one, the solution is described by weakly nonlinear theory. Initial data for the weakly nonlinear solution is obtained by matching with the large time behaviour of the solution of the one–dimension system. In the matching region, $\epsilon \ll t \ll 1$, the solution is both approximately one dimensional and of small amplitude.

It is interesting to note that no coupling between the ray geometry and the wave amplitude is necessary in this theory. When the wave is strong, one–dimensional theory suffices. When the wave is weak, the rays are the same as in the linearized theory. This differs from a number of heuristic theories (e.g. [3], [4]) for strongly nonlinear hyperbolic waves, which do propose such a coupling.

An example is illustrated in Figure 1. The initial data is supported in a region of width order ϵ about the unit sphere. The geometrical optics assumption is that this width is much less than the radius of the sphere i.e. $\epsilon \ll 1$. We show schematically the decay of the initial data to N–waves. This decay is described by a one–dimensional equation, with radial distance as the space variable. For larger times, the weak N–waves focus or defocus, because they are nonplanar. This is described by the usual weakly nonlinear inviscid Burger's equation.

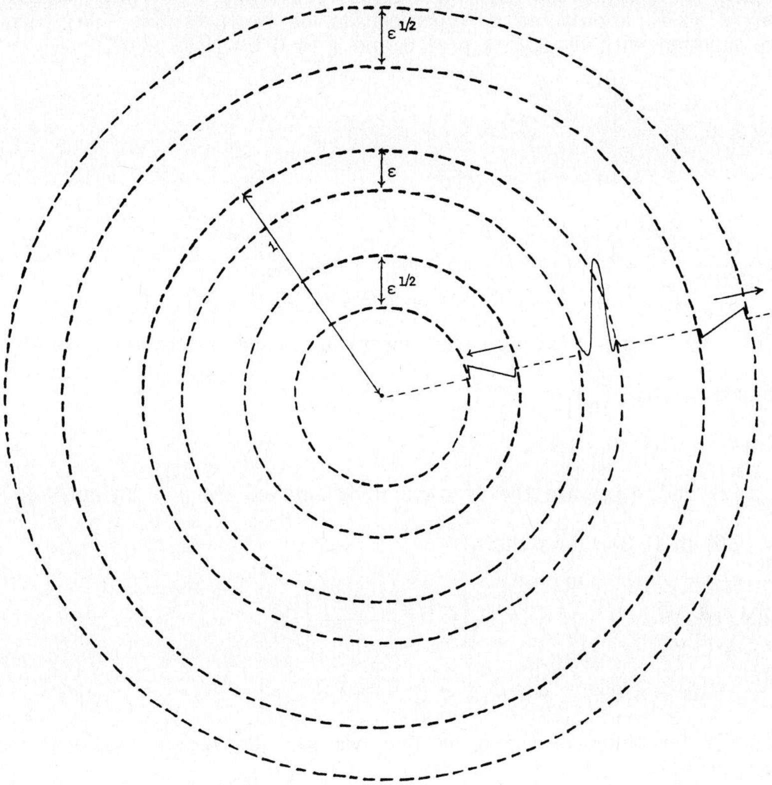

Figure 1. Schematic illustration of the evolution of a large-amplitude, rapidly varying hyperbolic wave.

First, we derive the asymptotic equations in the initial layer, when $t = O(\epsilon)$. The leading order approximation for a solution of (2.1) and (2.2) is

$$u(x,t;\epsilon) = v\left[x,\frac{x}{\epsilon},\frac{t}{\epsilon}\right]+O(\epsilon),\ \epsilon \to 0+ \text{ with } \frac{t}{\epsilon} =O(1). \tag{2.3}$$

Use of (2.3) in (2.1) and (2.2) implies that $v(x, \eta, \tau)$ satisfies

$$v_\tau + g_\eta = 0 \tag{2.4}$$

$$v(x, \eta, 0) = u_0(x, \eta) \tag{2.5}$$

where

$$g(x, v) = \sum_{i=1}^{n} \psi_{x_i}(x) f_i(x, v).$$

Equation (2.4) is a system in one space variable η. The slow space variable x occurs in (2.4) as a parameter.

Next, we consider the matching region, $\epsilon \ll t \ll 1$. For large values of τ, a solution of (2.4) approaches a superposition of N–waves [5]. We denote the N–wave function with invariants $p < 0$ and $q < 0$ by $N(\theta; p, q)$,

$$N(\theta; p, q) = \begin{cases} 0 & \theta > \sqrt{2p} \\ \theta & -\sqrt{-2q} < \theta < \sqrt{2p} \\ 0 & \theta < -\sqrt{2q}. \end{cases}$$

Then,

$$v(x,\eta;\tau) \sim \sum_{j=1}^{m} \tau^{-1/2} N_j[\tau^{-1/2}(\eta - \lambda_j\tau)] r_j, \qquad \tau \to +\infty. \tag{2.6}$$

In (2.6), $\lambda_j(x)$ and $r_j(x)$ are the eigenvalues and eigenvectors of $\nabla_u g(x, 0)$ (normalized so that $\nabla_u \lambda_j \cdot r_j = 1$). Also,

$$N_j(\theta) = N(\theta; p_j, q_j),$$

where $p_j(x)$ and $q_j(x)$ are the N–wave invariants of the jth asymptotic N–wave. Use of (2.6) in (2.3) shows that

$$u(x, t; \epsilon) \sim \epsilon^{1/2} \sum_{j=1}^{m} t^{-1/2} N_j\left[t^{-1/2}\left[\frac{x - x_j t}{\epsilon^{1/2}}\right]\right] r_j \tag{2.7}$$

as $\epsilon \to 0+$, $t \to 0+$, $\epsilon \ll t$.

Finally for times of the order one, we use the weakly nonlinear expansion described in [6]:

$$u(x,t;\epsilon) = \epsilon^{1/2} \sum_{j=1}^{m} a_j\left[x, t, \frac{\phi_j(x, t)}{\epsilon^{1/2}}\right] r_j + O(\epsilon) \tag{2.8}$$

as $\epsilon \to 0+$, $t = O(1)$.

The phase ϕ_j in (2.8) is a solution of the eikonal equation

$$\det[\phi_t I + \sum_{i=1}^{n} \phi_{jx_i} \nabla_u f_i(x, 0)] = 0, \tag{2.9}$$

and R_j is a normalized null vector of the matrix in (2.9). The amplitudes $\{a_j(x, t, \theta): j = 1, ..., m\}$ satisfy decoupled inviscid Burgers' equations,

$$\partial_{s_j} a_j + \partial_\theta\left[\frac{1}{2} a_j^2\right] + \frac{\partial_{s_j} \gamma_j}{2\gamma_j} a_j = 0. \tag{2.10}$$

In (2.10), ∂_{s_j} is a derivative along the rays associated with ϕ_j,

$$\partial_{s_j} = \partial_t + \sum_{i=1}^{n} L_j \cdot \nabla_u f_i(x, 0) R_j \partial_{x_i},$$

and $\gamma_j(x, t)$ is a known coefficient [6], with $\gamma_j(x, 0) = 1$.

Use of (2.8) in the matching condition (2.7) shows that

$$\begin{aligned}
&\phi_j(x, 0) = \psi(x),\\
&\phi_{jt}(x, 0) = \lambda_j(x),\\
&a_j(x, t, \theta) \sim t^{-1/2} N_j(t^{-1/2}\theta), \qquad t \to 0+.
\end{aligned} \tag{2.11}$$

The solution of (2.10) and (2.11) is

$$a_j(x, t, \theta) = \gamma_j^{-1/2} \sigma_j^{-1/2} N_j(\sigma_j^{-1/2}\theta)$$

where $\sigma_j(x, t)$ is the solution of

$$\partial_{s_j} \sigma_j = \gamma_j^{-1/2}, \qquad \sigma_j(x, 0) = 0.$$

Thus, each wave remains an N–wave. The amplitude and widths of each N–wave are affected by focusing and nonuniformities in the wave medium, which are described by the function $\gamma_j(x, t)$.

3. PERIODIC INITIAL DATA

In this section we consider (2.1) and (2.2) when $u_0(x, \eta)$ is 2π–periodic in η. For short times, $t = O(\epsilon)$, we obtain (2.4) and (2.5) just as before. However, two possible difficulties arise in extending the theory to longer times. First, to determine the solution in the matching region, $\epsilon \ll t \ll 1$, we need to know the large time behaviour of periodic solutions of (2.4). No general result is known at present and we consider this question further in Section 4. Matching with a weakly nonlinear solution is only possible if the solution of (2.4) and (2.5) decays like τ^{-1} as $\tau \to \infty$. The weakly nonlinear solution then has the form

$$u(x, t; \epsilon) = \epsilon \sum_{j=1}^{\tilde{m}} a_j\left[x, t, \frac{\phi_j(x, t)}{\epsilon}\right] R_j + O(\epsilon^2)$$

$$\text{as } \epsilon \to 0+, \qquad t = O(1).$$

Here, $a_j(x, t, \theta)$ is 2π–periodic in θ and satisfies matching conditions of the form

$$a_j(x, t, \theta) \sim t^{-1} \sigma_j(x, \theta), \qquad t \to 0+.$$

The second possible difficulty comes from the fact that the periodic waves do not decouple, unlike the N–waves. The available theory for resonantly interacting, weakly nonlinear periodic waves [7] requires certain assumptions on the phases $\{\phi_j\}$, which may not be satisfied here.

For scalar equations and systems in two independent variables, these difficulties do not arise. Glimm and Lax [8] proved that solutions of 2×2 systems decay like τ^{-1}. Also, weakly nonlinear periodic waves decouple (at leading order in the wave amplitude) for systems in two independent variables. Thus, for times t of the order one, the wave amplitudes are 2π–periodic solutions of (2.10). A simple scalar example is worked out in detail in [1].

Finally, we remark that if (2.1) has linearly degenerate characteristics, then only minor modifications are required when u_0 has compact support. The genuinely nonlinear waves still decay to N–waves (of course, the linearly degenerate waves do not). However, when u_0 is periodic in η, the analysis is different, and much harder, if there are linearly degenerate characteristics.

4. LARGE TIME BEHAVIOUR OF PERIODIC SOLUTIONS

We shall use formal asymptotics to determine possible large time behaviours of periodic solutions of hyperbolic conservation laws which decay like t^{-1} as $t \to +\infty$. This question arises in Section 3 and is also of independent interest. To fix notation, we consider the initial value problem for u(x, t),

$$u_t + f(u)_x = 0, \tag{4.1}$$

$$u(x, 0) = u_0(x). \tag{4.2}$$

Here, $u: \mathbb{R}^2 \to \mathbb{R}^m$ and $f: \mathbb{R}^m \to \mathbb{R}^m$. We assume that (4.1) is strictly hyperbolic and genuinely nonlinear and that u_0 is a 2π–periodic function of x with zero mean. We let $A = \nabla f(0)$ and denote the jth eigenvalue, left eigenvector and right eigenvector of A by λ_j, ℓ_j and r_j. The eigenvectors are normalized so that

$$\nabla\lambda_j(u)\cdot r_j(u)\big|_{u=0} = 1, \quad \ell_j\cdot r_j = 1.$$

Suppose that u(x, t) is a solution of (4.1) which decays like t^{-1} as $t \to +\infty$ (in the L^∞–norm). Then, for times of the order ϵ^{-1}, $0 \ll \epsilon \ll 1$, the solution has amplitude of the order ϵ. The evolution of such a solution may be described using the asymptotic equations for weakly nonlinear waves which are derived by Majda and Rosales [9]. The result is that

$$u \sim t^{-1} \sum_{j=1}^{m} \sigma_j(x - \lambda_j t) r_j + O(t^{-2}) \quad \text{as } t \to +\infty, \tag{4.3}$$

where $\{\sigma_j(\theta): j = 1, ..., m\}$ are 2π–periodic functions satisfying

$$\frac{d}{d\theta}\left\{\frac{1}{2}\sigma_j^2(\theta) + \sum_{p<q}^{j} I_{jpq}[\sigma_p, \sigma_q](\theta)\right\} = \sigma_j(\theta), \quad j=1,...,m. \tag{4.4}$$

In (4.4), $\sum_{p<q}^{j}$ stands for the sum over all $1 \leq p < q \leq m$, with p and q distinct from j. The equations are coupled through the integral averages

$$I_{jpq}[\sigma_p,\sigma_q](\theta)=\Gamma_{jpq}\lim_{T\to\infty}\frac{1}{T}\int_0^T\sigma_p(s)\sigma_q(-\rho_{pjq}\theta-\rho_{jpq}s)ds, \tag{4.5}$$

where

$$\begin{aligned}\Gamma_{jpq} &= \ell_j\cdot\nabla^2 f(0)\cdot(r_p, r_q),\\ \rho_{jpq} &= \frac{\lambda_q-\lambda_j}{\lambda_j-\lambda_p}.\end{aligned} \tag{4.6}$$

If only one wave is present,

$$u = t^{-1}\sigma(x-\lambda t)r + O(t^2)$$

where $\sigma(\theta)$ is a solution of

$$\frac{d}{d\theta}\left[\frac{1}{2}\sigma^2\right] = \sigma, \quad \sigma(\theta+2\pi) = \sigma(\theta). \tag{4.7}$$

We shall call such a solution a ***generalized sawtooth wave***. If σ is piecewise smooth, then it is equal either to zero or to $\theta-\theta_0$ on intervals where it is smooth. At a jump discontinuity, the jump and entropy conditions imply that $\sigma(\theta-) = -\sigma(\theta+) > 0$ (see Figure 2). More generally, one can consider weak solutions of (4.7) which are of bounded variation.

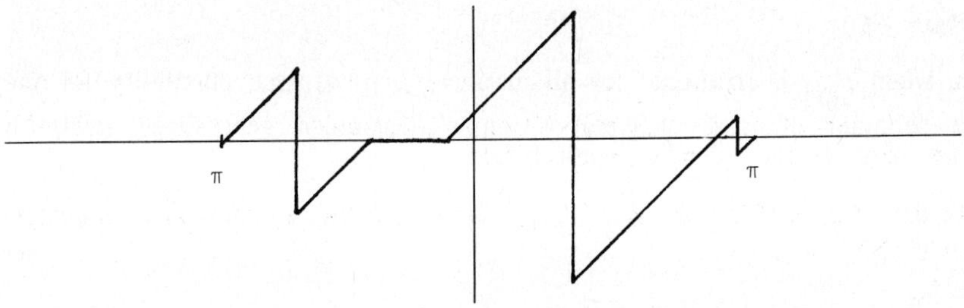

Figure 2. A generalized sawtooth wave.

The next proposition gives a condition for (4.4) to have decoupled sawtooth wave solutions. An analogous proposition is proved in [9] for smooth, periodic waves.

PROPOSITION 4.1. *Suppose that ρ_{jpq} is irrational and σ_p, σ_q are piecewise smooth, generalized sawtooth waves. Then*

$$\partial_\theta I_{jpq}[\sigma_p, \sigma_q](\theta) = 0.$$

We omit the details of the proof. To see the main idea, suppose that σ_q is a sawtooth wave

$$\sigma_q(\theta) = S(\theta),$$

where

$$S(\theta) = \theta, \qquad |\theta| < \pi,$$
$$S(\theta + 2\pi) = S(\theta). \tag{4.8}$$

Then,

$$\frac{d}{d\theta}\sigma_q(\theta) = 1 - 2\pi \sum_{n=-\infty}^{\infty} \delta[\theta - (2n + 1)\pi].$$

Using this equation and (4.5), together with the fact that σ_p has zero mean, implies that $\partial_\theta I_{jpq}$ is proportional to

$$\lim_{N\to\infty} \frac{1}{N} \sum_{n=1}^{N} \sigma_p\left[-\frac{(2n + 1)\pi}{\rho_{jpq}} - \frac{\nu_{jpq}\theta}{\rho_{jpq}}\right].$$

Since the points $\left\{\frac{(2n + 1)\pi}{\rho_{jpq}}: n = 1, 2, 3, ...\right\}$ are distributed uniformly over a period of σ_p when ρ_{jpq} is irrational, this limit equals the mean of σ_p and and therefore it is zero.

Thus, when ρ_{jpq} is irrational for all distinct (j, p, q), one possibility for the large time behavior of u is a superposition of decoupled generalized sawtooth waves. This suggests the following conjecture.

CONJECTURE 4.2. *Suppose that (4.1) is strictly hyperbolic and genuinely nonlinear and that*

$$\frac{\lambda_j - \lambda_p}{\lambda_j - \lambda_q} \textit{ is irrational for all distinct } (j,p,q). \tag{4.9}$$

Then a periodic solution of (4.1) of bounded variation and zero mean decays like t^{-1} *(in the sup and total variation norms) as* $t \to +\infty$ and *it approaches a superposition of generalized sawtooth waves.*

In the next section we show what can happen if (4.9) does not hold.

5. RESONANTLY INTERACTING SAWTOOTH WAVES

Let us assume that (4.1) is a 3 × 3 system of strictly hyperbolic conservation laws. Then (4.1) has the following weakly nonlinear asymptotic solution [9]:

$$u = \epsilon \sum_{j=1}^{3} a_j(k_j x - \omega_j t, \epsilon t) r_j + O(\epsilon^2), \quad \epsilon \to 0+, \quad \epsilon t = O(1). \tag{5.1}$$

Here, $\omega_j = \lambda_j k_j$ and $\{a_j(\theta, T)\}$ satisfy

$$\partial_T a_j + \partial_\theta \left\{ \frac{1}{2} M_j a_j^2 + I_j[a_p, a_q](\theta) \right\} = 0, \tag{5.2}$$

where (j, p, q) is a cyclic permutation of (1, 2, 3) and

$$I_j[a_p, a_q](\theta) = \Gamma_j \lim_{T\to+\infty} \frac{1}{T}\int_0^T a_p(s) a_q(-\rho_{pjq}\theta - \rho_{jpq}s) ds.$$

The coefficients are

$$\begin{aligned} M_j &= k_j \nabla_u \lambda_j \cdot r_j \\ \Gamma_j &= k_j \ell_j \cdot \nabla^2 f \cdot (r_p, r_q) \\ \rho_{jpq} &= \frac{k_q(\lambda_q - \lambda_j)}{k_p(\lambda_j - \lambda_p)}. \end{aligned} \tag{5.3}$$

Solutions of (5.2) of the form

$$a_j = T^{-1}\sigma_j(\theta)$$

lead to equations (4.4) considered in the last section.

Instead, let us consider sawtooth wave solutions of (5.2),

$$a_j(\theta, T) = \alpha_j(T) S(\theta - \xi_j). \tag{5.4}$$

Here, ξ_j is a constant and $S(\theta)$ is defined in (4.8). When $\{\rho_1, \rho_2, \rho_3\}$ are irrational, the sawtooth waves decouple and one obtains

$$\mathring{\alpha}_j = M_j \alpha_j^2, \qquad \circ = \frac{d}{dT},$$

with the solution ($M_j \neq 0$) $\alpha_j = M_j^{-1}(T - T_0)^{-1}$. The other extreme is $\rho_1 = \rho_2 = \rho_3 = 1$. This is a resonance condition because, from (5.3), if $\{\mu_j\}$ are such that

$$\begin{aligned} &\mu_1\omega_1 + \mu_2\omega_2 + \mu_3\omega_3 = 0, \\ &\mu_1 k_1 + \mu_2 k_2 + \mu_3 k_3 = 0, \\ &\omega_j = \lambda_j k_j, \end{aligned}$$

then $\mu_p/\mu_q = \rho_{jpq}$. Thus, when $\rho_1 = \rho_2 = \rho_3 = 1$, the frequencies and wavenumbers of the sawtooth waves sum to zero. (We shall not discuss rational ρ_j's here, when the interaction is more complicated.)

PROPOSITION 5.1. *Suppose that* $\rho_1 = \rho_2 = \rho_3 = 1$. *Then (5.2) has solutions of the form (5.4) iff the phase shifts satisfy*

$$\xi_1 + \xi_2 + \xi_3 \equiv \pi \ (\mathrm{mod}\ 2\pi),$$

and the wave amplitudes satisfy the following system of ODE's:

$$\begin{aligned} \mathring{\alpha}_1 &= \Gamma_1\alpha_2\alpha_3 - M_1\alpha_1^2, \\ \mathring{\alpha}_2 &= \Gamma_2\alpha_3\alpha_1 - M_2\alpha_2^2, \\ \mathring{\alpha}_3 &= \Gamma_3\alpha_1\alpha_2 - M_3\alpha_3^2. \end{aligned} \tag{5.5}$$

The proof is a straightforward calculation, using (5.4) in (5.2).

Majda, Rosales and Schonbek [10] derive (5.5) in the special case of the gas dynamics equations, when it reduces to a second-order system of ODE's (because $M_2 = \Gamma_2 = 0$).

For simplicity we now assume that (4.1) is genuinely nonlinear and we let

$$M_1 = M_2 = M_3 = 1 \tag{5.6}$$

without loss of generality. The shocks in the sawtooth waves are admissible if $\alpha_j > 0$. One can prove that if $\Gamma_j > 0$ and $\alpha_j(0) > 0$, for $j = 1, 2, 3$, then $\alpha_j(T) > 0$ for all $T > 0$, and the solution remains admissible. However, if one or more of the Γ_j's is negative, then numerical integration of (5.5) shows that one of the α_j's typically changes sign after a finite time and the sawtooth wave solution becomes inadmissible. It seems likely that a "cusped rarefaction wave" appears [10].

A particular solution of (5.5) is

$$\alpha_j(T) = \frac{K_j}{T - T_0}, \tag{5.7}$$

where T_0 is an arbitrary constant and $\{K_1, K_2, K_3\}$ satisfy the algebraic equations

$$\begin{aligned} K_1^2 - K_1 &= \Gamma_1 K_2 K_3, \\ K_2^2 - K_2 &= \Gamma_2 K_3 K_1, \\ K_3^2 - K_3 &= \Gamma_3 K_1 K_2. \end{aligned} \tag{5.8}$$

There are three possibilities, depending on the signs of the K_j's:

(a) If $K_1, K_2, K_3 > 0$, then (5.7) decays as $T \to +\infty$;

(b) If $K_1, K_2, K_3 < 0$, then (5.7) blows up as $T \to T_0-$;

(c) If K_1, K_2, K_3 have mixed signs, then (5.7) is inadmissible.

A special case occurs when $\Gamma_1\Gamma_2\Gamma_3 = 1$ and $\Gamma_j > 0$. Then (5.5) has constant admissible solutions

$$\alpha_j = \Gamma_j^{1/3}K.$$

Perturbing off this solution, we find that for $\Gamma_1\Gamma_2\Gamma_3$ close to one, there are solutions of the form (5.7) with

$$K_j \sim \Gamma_j^{1/3}K,$$

$$K = \frac{\Gamma_1^{1/3} + \Gamma_2^{1/3} + \Gamma_3^{1/3}}{1 - \Gamma_1\Gamma_2\Gamma_3}.$$

Thus, if $1 \gg 1 - \Gamma_1\Gamma_2\Gamma_3 > 0$, the solution decays to zero as $T \to +\infty$, but for $1 \gg \Gamma_1\Gamma_2\Gamma_3 - 1 > 0$ it blows up in finite time.

This simultaneous blow–up of all the sawtooth waves is rather peculiar. We note one possible consequence. When blow–up occurs, the weakly nonlinear solution becomes inconsistent. However, taking $T_0 = 1$, suppose that the solution remains valid in a matching region, $1 - T = O(\delta(\epsilon))$, where $\epsilon \ll \delta \ll 1$ as $\epsilon \to 0+$. (Asymptotic solutions usually have this kind of property.) Then, the sup and total variation (per period) of the asymptotic solution (5.1), (5.4), (5.7) at $T = 0$ are of the order ϵ. However, when $1-T = O(\delta)$, they are of the order $\delta^{-1}\epsilon$. Since $\delta \to 0$ as $\epsilon \to 0$, this would imply that the sup or total variation at time T of periodic solutions of general conservation laws is not bounded by the sup and total variation of the initial data.

REFERENCES

1. J. K. Hunter and J. B. Keller, Nonlinear hyperbolic waves, to appear in ***Proc. Roy. Soc. A.***

2. G. B. Whitham, ***Linear and Nonlinear Waves***, Wiley, New York, 1974.

3. E. Cumberbatch and E. Varley, Large amplitude waves in stratified media: acoustic pulses, ***J. Fluid Mech.***, 43 (1970), 513–537.

4. V. E. Fridman, Self refraction of small amplitude shock waves, ***Wave Motion*** 4 (1982), 151–161.

5. T. P. Liu, Decay of N–waves of solutions of general systems of nonlinear hyperbolic conservation laws, ***Comm. Pure Appl. Math.*** 30 (1977), 585–610.

6. J. K. Hunter and J. B. Keller, Weakly nonlinear high frequency waves, ***Comm. Pure Appl. Math.*** 36 (1983), 547–569.

7. J. K. Hunter, A. Majda and R. R. Rosales, Resonantly interacting, weakly nonlinear hyperbolic waves II. Several space variables, ***Stud. Appl. Math.*** 75 (1986), 187–226.

8. J. Glimm and P. Lax, Decay of solutions of systems of nonlinear hyperbolic waves I. A single space variable, ***Stud. Appl. Math.*** 101 (1970), 1–111.

9. A. Majda and R. R. Rosales, Resonantly interacting weakly nonlinear hyperbolic waves I. A single space variable, ***Stud. Appl. Math.*** 71 (1984), 149–179.

10. A. Majda, R. R. Rosales and M. Schonbek, A canonical system of integrodifferential equations in resonant nonlinear acoustics, to appear in ***Stud. Appl. Math.***

THE STRUCTURE OF THE RIEMANN SOLUTION FOR NON-STRICTLY HYPERBOLIC CONSERVATION LAWS

Eli L. Isaacson[1] Dan Marchesin[2] Bradley J. Plohr[3]

[1]Department of Mathematics, University of Wyoming, Laramie, WY 82071, USA

[2]Instituto de Matemática Pura e Aplicada and PUC-RJ, Rio de Janeiro, 22453, Brazil

[3]Computer Sciences Department, University of Wisconsin, Madison WI 53706, USA

SUMMARY

We describe the full Riemann solution for a system of two equations which possesses an umbilic point where the characteristic speeds coincide. The solution contains many nontrivial topological features to be expected in non-strictly hyperbolic problems. The model we solve describes the flow of three immiscible fluids in porous media. Despite simplifying assumptions on physical properties of the fluids, the model captures the essential global features of the flow.

The solution is too complicated to be obtained analytically. Rather, we used a program designed for the numerical solution of 2×2 Riemann problems. The program has modules which (1) construct local wave curves by a continuation algorithm, (2) construct non-local wave curves using a global search algorithm, (3) construct boundaries across which the nature of the wave curves change, and (4) verify whether shocks are limits of parabolic viscous profiles.

The difficulties in the solution arise because the Hugoniot curves contain disconnected branches of non-contractible shocks, because the classical coordinate system of Lax for the construction of Riemann solutions does not exist near the umbilic point and because there are shocks with viscous profiles which do not obey Lax's entropy condition. These non classical shocks have to be considered to insure the existence of the solution. The solution depends continuously on the initial data; numerical evidence indicates that it is unique.

1. THE MODEL

For several years we have been studying the Riemann problem for a system of two conservation laws that models the flow of oil, water, and gas in porous media. The importance of such problems was emphasized in 1941 by Leverett and Lewis [10]. The two-phase scalar problem was solved in 1942 in the classical work of Buckley and Leverett [2] who established the formation of saturation shocks, or oil banks, as the mechanism responsible for oil recovery in petroleum reservoirs. The simplified model represents the conservation of mass of oil, water, and gas combined with Darcy's force law. Compressibility, capillarity, and gravity effects are neglected. In one spatial dimension with appropriate boundary conditions, this model is represented by the system

$$u_t + f(u,v)_x = 0$$
$$v_t + g(u,v)_x = 0$$

with $f = U/D$, $g = V/D$, and $D = U+V+W$. Here U, V, and W are the permeabilities of oil, water, and gas, respectively; they depend on the saturations u, v, w of these three fluids. The saturations are non-negative and sum to 1; hence, we take $w = 1-u-v$. We denote the viscosities of the fluids by a, b, and c. Laboratory measurements are consistent with $V = V(v) = v^2/b$, $W = W(w) = w^2/c$, and $U = U(u,v)$ where U depends only weakly on v [4]. Thus, in our simplified model we take $U = U(u) = u^2/a$. In this work we present the Riemann solution for the fully symmetric case where $a = b = c = 1$.

The novel feature of this model is that interior to the domain of physical interest, there is a unique point at which the characteristic speeds coincide. This is called an umbilic point. For our model, this point is determined by the equality of the derivatives $U' = V' = W'$. The Jacobian matrix of the system becomes a multiple of the identity at the umbilic, and the characteristic directions are undetermined. As a result, the usual construction of the solution of the Riemann problem for stricly hyperbolic problems, which relies on a local coordinate system of characteristic directions, cannot be used at this point.

The failure of strict hyperbolicity occurs in Stone's model, which possesses a region with complex characteristic speeds (or elliptic region) [1]. This is a model commonly used in Petroleum Engineering, with a special functional dependence of U on u and v [4]. Surprisingly, even though general perturbations of isolated umbilic points produce compact elliptic regions, gravity effects in our model preserve the isolated umbilic point [12]. Therefore we believe that solving this problem is an important step towards obtaining the general solution of realistic multiphase flow problems.

Our purpose is to solve the Riemann problem for the model, i.e. to find the solution of the Cauchy problem for the system above with initial data at $t=0$ (u_L,v_L) for $x < 0$ and (u_R,v_R) for $x > 0$.

2. RAREFACTIONS AND SHOCKS

Except at the umbilic point, the Jacobian derivative matrix of the flux $\underline{f} = (f,g)$ has distinct real eigenvalues $\lambda_1 < \lambda_2$ corresponding to right eigenvectors $\underline{r}_1$ and $\underline{r}_2$. The integral curves of the fields of eigenvectors give rise to rarefaction waves, which are smooth solutions $\underline{u} = (u,v)$ that depend only on the ratio x/t and satisfy $\lambda(\underline{u}(x,t)) = x/t$. A shock wave consists of two constant states $\underline{u}_-$ and $\underline{u}_+$ separated by a discontinuity traveling with speed s; the states are related by the Rankine-Hugoniot jump condition $s(\underline{u}_+-\underline{u}_-) = \underline{f}(\underline{u}_+) - \underline{f}(\underline{u}_-)$. For a given state $\underline{u}_-$, the set of states $\underline{u}_+$ satisfying the jump conditions forms the Hugoniot curve $\mathcal{H}(\underline{u}_-)$, which parametrizes shock waves. The Hugoniot curve consists of two branches that emanate from $\underline{u}_-$ in the directions of the eigenvectors.

For the model described in §1, the rarefaction curves are depicted in Fig. 2.1. Rarefactions corresponding to the smaller characteristic speed are drawn as double lines, while those for the larger speed are single lines. The arrows on the curves indicate the direction of increasing characteristic speed. It is clear that there is no longer a coordinate system in a neighborhood of the umbilic point, although there is one in a neighborhood of any other point.

Another new feature is that the Hugoniot curve has detached branches, as indicated in Fig. 2.3. The origin of the Hugoniot curve is the point where two branches cross. Points on the curve where the shock speed coincides with a characteristic speed for either side of the discontinuity are marked; these marks are important for constructing wave curves and viscous profiles. For our model it is easy to express the Hugoniot curve in polar coordinates as $\underline{u}_+ = \underline{u}_- + R(\underline{u}_-,\theta)$, where R is a quotient which vanishes for certain angles. Using this expression it is easy to find the bifurcation loci - in our case coinciding with the straight lines through the umbilic shown in Fig. 2.2. In general, a Hugoniot curve changes topology as its origin $\underline{u}^-$ crosses certain curves, called the bifurcation loci [7,8]. This behavior, which has no analogue in scalar

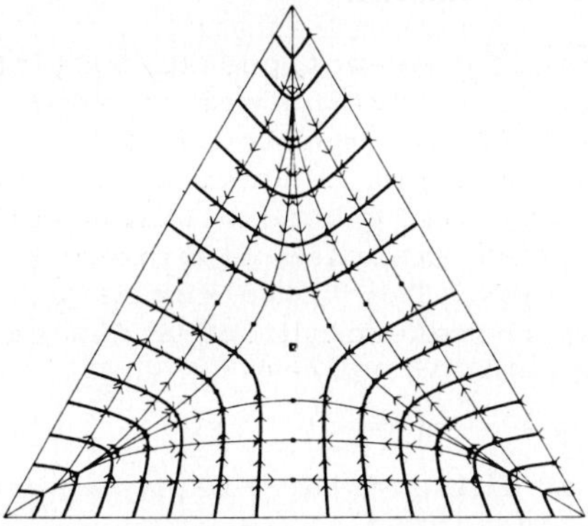

Fig. 2.1 Rarefaction curves

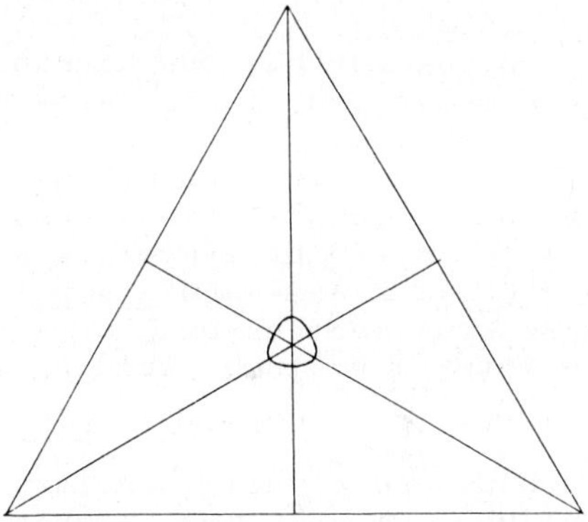

Fig. 2.2 Inflection locus

conservation laws, makes it even more difficult to find a coordinate system in which to determine the Riemann problem solution. The Hugoniot curve for a point $\underline{u}_-$ on the bifurcation loci is shown in Fig. 2.4. It consists of the dashed curves plus the bifurcation segment through $\underline{u}_-$.

On the rarefaction curves the speed $\lambda_k(\underline{u}) = x/t$ must be monotone increasing as x increases from left $(\underline{u} = \underline{u}_L)$ to right $(\underline{u} = \underline{u}_R)$. Therefore these curves have to stop at points where the speed has an extreme. These points constitute the inflection loci; their nomenclature arises from analogy with scalar conservation laws. These loci are generically given by $\nabla\lambda_k(\underline{u})\cdot r_k(\underline{u}) = 0$, k=1,2 [8]. For our model, the 1-inflection locus consists of three segments from the vertices to the umbilic point in Fig. 2.2. Remarkably, the 2 inflection locus consists not only of the opposite segments but also of the small closed curve near the umbilic. This indicates that the behavior of the solution near and far from the umbilic point is substantially different.

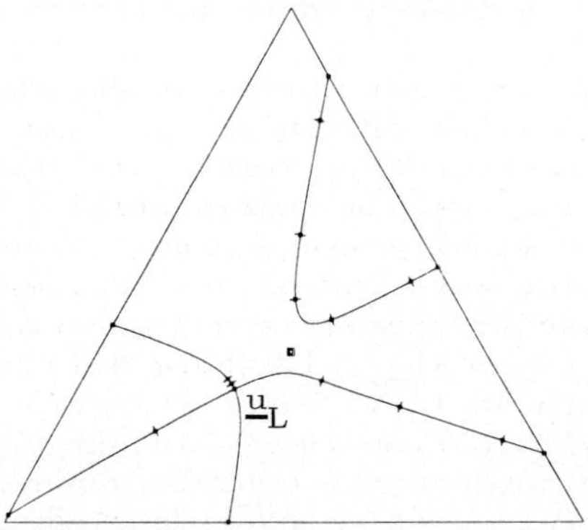

Fig. 2.3 A Hugoniot curve

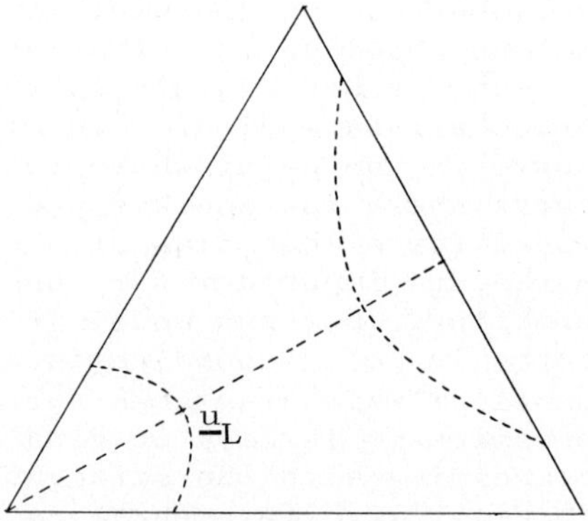

Fig. 2.4 Bifurcating Hugoniot curve

3. SHOCKS WITH VISCOUS PROFILES

To insure that the Riemann problem does not possess multiple weak solutions, it is necessary to restrict the set of admissible shocks. This is done for multiphase models by adding a small diffusive term $\varepsilon(D(\underline{u})_x)_x$ to the right hand side of the system of conservation laws. We admit only shocks which are limits as $\varepsilon\searrow 0$ of traveling waves of a parabolic equation with $\underline{u} \to \underline{u}_-$ as $x \to -\infty$ and $\underline{u} \to \underline{u}_+$ as $x \to +\infty$. The traveling wave is a smooth function of $\xi = (x-st)/\varepsilon$; therefore the equation can be integrated once yielding the dynamical system

$$D(\underline{u})\underline{u}_\xi = \underline{f}(\underline{u}) - \underline{f}(\underline{u}_-) - s(\underline{u}-\underline{u}_-).$$

Both $\underline{u}_-$ and $\underline{u}_+$ are singularities of this vector field; a traveling wave solution is an orbit connecting these states.

For strictly hyperbolic, genuinely nonlinear systems there are connecting orbits for weak Lax shocks [3]: 1-Lax shocks are repeller-saddle connections, 2-Lax shocks saddle-attractor connections. In our model, we verified numerically that even strong Lax shocks have the appropriate connections. However, even including limiting cases, Lax shocks are not sufficient to complete the solution of the Riemann problems. We have to introduce a new kind of discontinuity whose profile is the orbit between two saddle points of the vector field [13,6]. Since for this new discontinuity we have $\lambda_2(\underline{u}_-) > s > \lambda_1(\underline{u}_-)$ and $\lambda_1(\underline{u}_+) < s < \lambda_2(\underline{u}_+)$ it can be preceded by a 2-wave and succeeded by a 1-wave. Therefore, it is not associated with any family: we call it a transitional shock.

Using symmetry considerations it is easy to prove that there are crossing shocks on the bifurcation loci such as in [13] for the viscosity matrix D = I that we employ. For this viscosity matrix, one can prove that for certain models which approximate ours near the umbilic, all pairs $\underline{u}_-$, $\underline{u}_+$ connected by saddle-to-saddle orbits lie on opposite sides of the bifurcation loci relative to the umbilic point [6]. This fact was verified numerically for our model.

Using profilable shocks (Lax's shocks and transitional shocks) we were able to find the complete solution of the Riemann problem.

4. ADMISSIBLE WAVE CURVE BOUNDARIES

In physical space, the solution of the Riemann problem consists of a sequence of rarefaction fans, discontinuities and constant states; these elementary curves are grouped into waves that belong to the first family (1-waves), to the second family (2-waves), or constitute transitional waves. The solutions obey the geometrical constraint that wave speeds in physical space increase from left to right.

Wave curves are represented in state space by sequences of three types of elementary segments: shocks, rarefactions and composite waves. These are shock waves adjacent to rarefaction waves; the shock speed coincides with the characteristic speed at the adjacency state. They appear when the problem is genuinely nonlinear [11]. Each elementary segment must stop wherever its speed attains an extreme, and the type of elementary segment that follows is determined by certain rules [8,7] based on the Bethe-Wendroff theorem [14]. Finally, since the Hugoniot curves possess disconnected branches, wave curves also have detached branches.

Using Bethe-Wendroff theorem it can be shown that certain loci play a crucial role in determining the nature of wave curves: they are the bifurcation locus, the inflection locus, the hysteresis locus and the double contact locus [8,7]. Rarefaction curves stop at the inflection locus, shock curves change topology when its base point crosses the bifurcation locus. The double contact locus consists of states $\underline{u}$ for which there is $\underline{u}'$ such that

$$\underline{u}' \in \mathcal{H}(\underline{u}), \quad \lambda_i(\underline{u}) = s(\underline{u},\underline{u}') = \lambda_j(\underline{u}').$$

Composite segments end at points $\underline{u}'$. In our model, only the 2-family double contact locus in Fig. 4.1 plays a role ($i=j=2$). The correspondence between $\underline{u}$ and $\underline{u}'$ points is established by A,B,C → A′,B′,C′ and symmetry considerations. We remark that the curves in Fig. 4.1 have been considerably blown up; the trefoil and the curvilinear triangle are actually tangent to the small closed curve in Fig. 2.2.

There are other less basic loci; the only one that plays a role here is the interior boundary contact, which satisfies

$$\lambda_k(\underline{u}) = s(\underline{u},\underline{u}'), \quad \underline{u} \in \mathcal{H}(\underline{u}'), \quad \underline{u}' \text{ lies on the boundary.}$$

The relevant part of these loci are shown in Fig. 4.2; the part A, B, with corresponding points A′, B′, belongs to family 1 while the part F, G, with corresponding points F′, G′, belongs to family 2. In this section we only show the

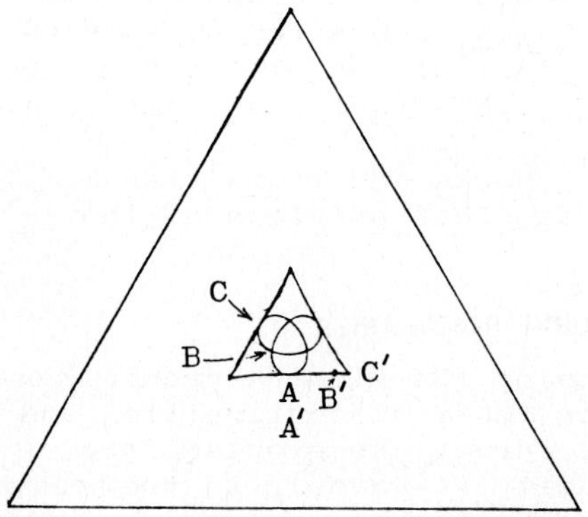

Fig. 4.1 The 2 double contact

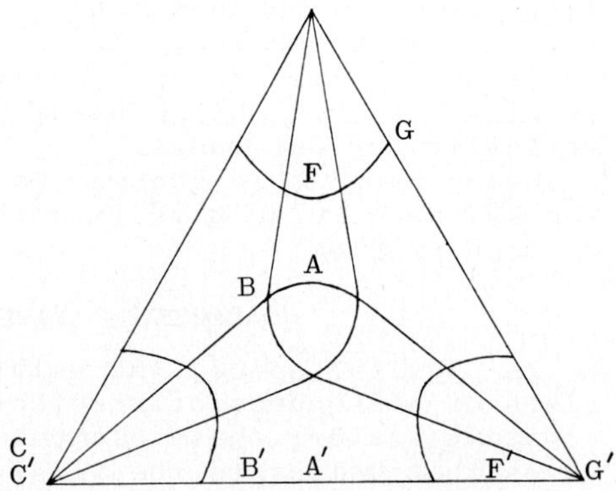

Fig. 4.2 One and two interior boundary contact

parts of any locus which correspond to shocks with viscous profiles (§3); the admissibility was verified numerically. Because of lack of admissibility, the hysteresis is irrelevant in our model, and we omit it. Parts of other loci were eliminated for the same reason.

We draw the $\underline{u}_L$ boundaries for the 1-wave curves and 2-wave curves in Fig. 4.3 and 4.4 respectively, in one-sixth of the domain triangle. The wave curves with $\underline{u}_L$ in each of the regions shown in the figures have the some topology and consist of the same sequence of segments. In Fig. 4.3 the curves AB and CD are sections of the 1-boundary contact from Fig. 4.2. The curve CD is a rarefaction ending at C; it plays a role because 1-wave curves for $\underline{u}_L$ to the left of CD have only a local branch while for $\underline{u}_L$ to the right of CD have also a nonlocal branch (Fig. 5.1). In Fig. 4.4 for the $\underline{u}_L$ boundaries for family 2, EF is a 2-boundary contact, BC is a 2-double contact, BD is an inflection locus and AB is a 2-rarefaction ending at B. Therefore, there are 5 different types of 1-wave curves for $\underline{u}_L$ in each region of Fig. 4.3 and 6 types of 2-wave curves for $\underline{u}_L$ in each region of Fig. 4.4.

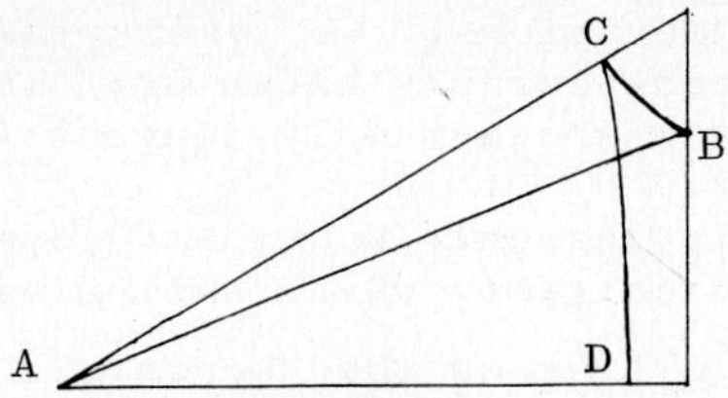

Fig. 4.3 Boundaries for 1-wave curves

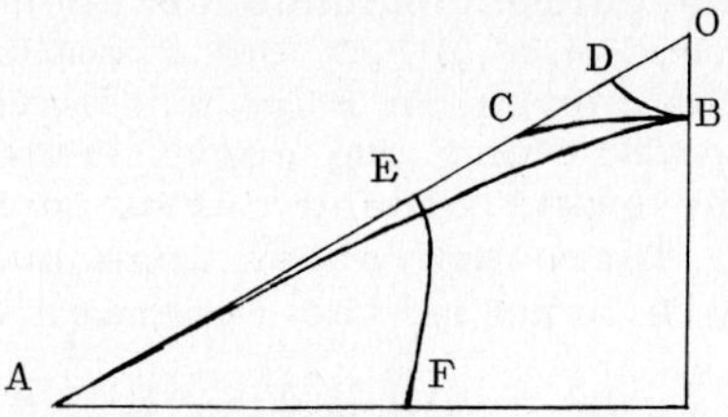

Fig.4.4 Boundaries for 2-wave curves

5. WAVE CURVES

Wave curves of the first family for $\underline{u}_L$ and $\underline{u}_{L'}$ in two of the regions of Fig. 4.3 are shown in Fig. 5.1. One wave curves are the set of right states which can be constructed with a succession of 1-waves for the fixed state $\underline{u}_L$. Only wave curves whose shocks have viscous profiles are shown. Solid lines represent rarefactions, with arrows indicating the direction of increasing speed. Shocks and composites are represented by dashed lines or crossed lines, respectively. The wave curve for $\underline{u}_{L'}$ consists of a rarefaction (L′a), a shock (L′c) and a composite (ab). Each point of (ab) is a shock starting at a point in (L′a) where it is characteristic. If we move L′ above (AB), a new shock appears such as (gh) for $\underline{u}_L$ (g lies on the boundary if L is on (AB)). If we move L′ to the left of (CD), a nonlocal branch shows up. Thus for $\underline{u}_L$ the 1-wave curve is (dLfgh) and (ijk). The curve (ij) is a composite based on (Lf) and (jk) is a shock based on L. If we lower L below (CB) the shock segment (jk): disappears for L on (CB) j lies on the boundary.

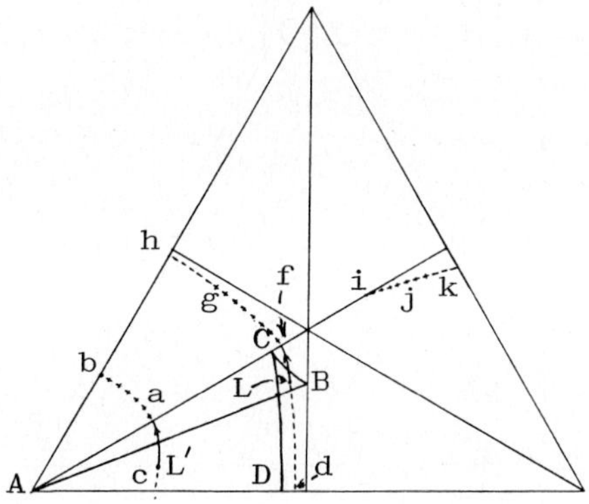

Fig. 5.1 One wave curves

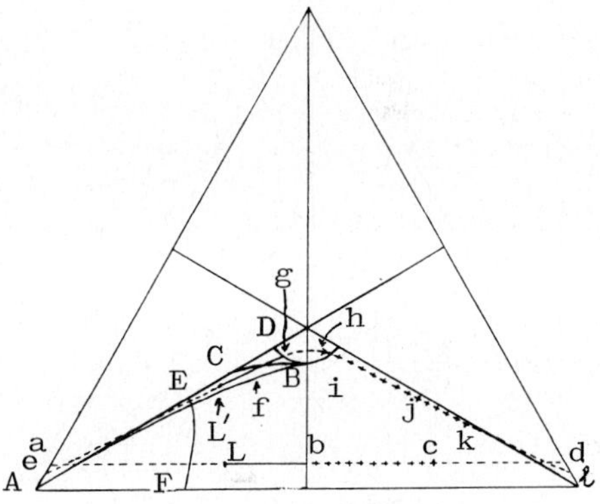

Fig. 5.2 Two wave curves

In Fig. 5.2 we show 2-wave curves for $\underline{u}_L$ and $\underline{u}_{L'}$ in two of the regions of Fig. 4.4. The two wave curves displayed are the set of right states which can be constructed with a succession of 2-waves for the fixed state $\underline{u}_L$, using only admissible shocks. In the wave curve for $\underline{u}_L$ (aLbcd), if L is moved to the left of (EF), the point c reaches the boundary and the shock segment (cd) disappears. The 2-wave curve for $\underline{u}_{L'}$ (eL′fghijkℓ) for $\underline{u}_{L'}$ above (AB) has the same general shape but other segments arise near the umbilic region above (DBG). The curves (eL′) and (kℓ) are shocks; (L′fg) is a rarefaction, (ij) is a composite based on (fi) and (jk) is a composite based on (L′f). The point f lies on the double contact locus (EB); it corresponds to the point h on the trefoil (Fig. 4.1).

Because of the existence of crossing shocks for $\underline{u}_L$ points on the bifurcation locus, it is necessary to take into account the 2-wave curves for points on each of the segments (AE), (EC), (CC′), (C′D), (DO) in Fig. 5.3. The boundaries shown in this figure are those of Fig. 4.4, except for the addition of (C′B′) which corresponds to (CB) of Fig. 4.3, a 1-interior boundary contact. In Fig. 5.3 we show the 2-wave curve for $\underline{u}_L$ on (EC); the (eLfghijkℓ) parts of this wave curve may be obtained as limits of the corresponding parts of the wave curve in Fig. 5.2, when we allow $\underline{u}_L$ to lie on (EC). There are two new segments; (mn) consists of crossing shocks based on L and (km) consists of crossing composites based on (Lf).

If the point L crosses C′ so it lies on (C′O), the segment of crossing shocks (mn) ceases to reach the boundary. Rather, n becomes an interior point on the bifurcation locus, coinciding with point i in Fig. 5.1, where a non local 1-wave curve branch starts.

In the next section, the 1-, 2- and transitional wave curves will be used to construct the Riemann solution. Each of these wave curves uses only admissible shocks and satisfies the geometric constraint of increasing speeds from left to right.

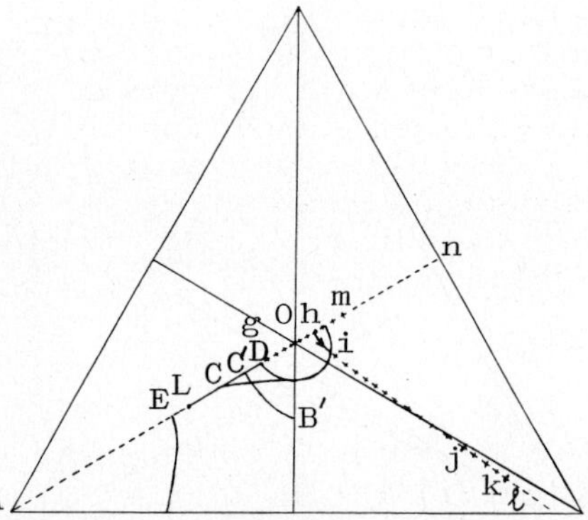

Fig. 5.3 Transitional wave curve

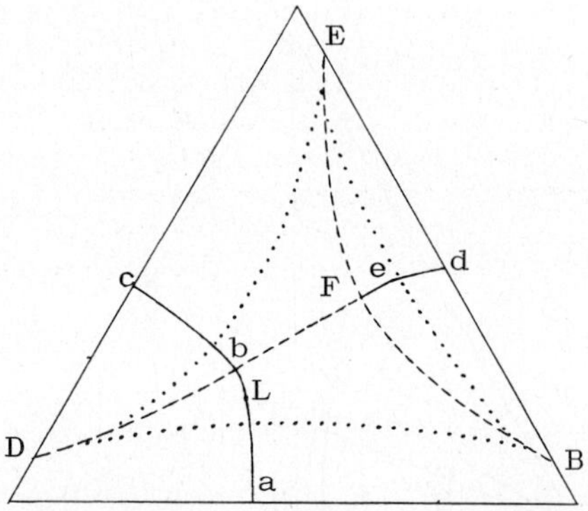

Fig. 6.1 Solution of Riemann problem

6. THE SOLUTION OF THE RIEMANN PROBLEM AND CONCLUSIONS

We will follow as much as possible the classical construction of the Riemann solution-a 1-wave curve from state L to M and a 2-wave curve from state M to R; as we will see, transitional waves will play a role to complete the solution for all possible pairs of states (L,R). We show an example of the construction in Fig. 6.1, for L at the right of (CD) in Fig. 4.3. (If L lies at the left of (CD), the segment (ed) in Fig. 6.1 disappears while (fe) extends all the way to the boundary.

Given L, we construct the local 1-wave curve (abc) through L. For any state M on (abc) we construct the 2-wave curve through M, defining points R which can be reached by a local 1-wave curve from L to M followed by a 2-wave curve from M to R. It is necessary to verify that the wave speed from M to R is larger than the wave speed from L to M. This restriction eliminates the ending portion of the 2-wave curves. Thus the allowed 2-wave curves end on the dashed lines (Db), (EF) and (FB). In this way we find the solution for the given L to any R to the left of curve (EFB). To obtain the solution for states R to the right of (EFB), we have to use the nonlocal part (ed) of the 1-wave curve based on L as well as the transitional wave (Fe) based on b. We then construct the 2-wave curves through points on (Fed) as shown in Fig. 6.1. Numerical experiments and the triple shock theorem [6] show that the geometric consistence requirements force these 2-wave curves to stop precisely on (EFB).

The solution for the given $\underline{u}_L$ is now complete. In general, it consists in physical space of one of the following sequences: (i) the left state, 1-waves, a constant state, 2-waves, the right state or (ii) the left state, 1-waves, a constant state, possibly 2-waves, a transitional shock, a constant state, 2-waves and the right state. For other states $\underline{u}_L$, the solution may be obtained by symmetry considerations. This solution is L^1_{loc} continuous in the Cauchy data $\underline{u}_L$, $\underline{u}_R$; the continuity may be verified by inspection.

The largest shocks in the solution seem to be the nonlocal 1-wave shocks and the non classical crossing shocks. Since oil recovery is optimized by large shocks, it is important that numerical methods used in oil reservoir simulation be accurate on these shocks. However, the theory of standard numerical methods indicates that they are accurate only for small or contractible shocks. This is an indication that improved numerical methods are needed.

REFERENCES

1. J. Bell, J. Trangenstein, and G. Shubin, "Conservation Laws of Mixed Type Describing Three-Phase Flow in Porous Media," SIAM J. Appl. Math. 46 (1986), pp. 1000-1017.

2. S. Buckley and M. Leverett, "Mechanisms of Fluid Displacement in Sands," Trans. AIME 146 (1942), pp. 107-116.

3. C. Conley and J. Smoller, "Viscosity Matrices for Two-Dimensional Nonlinear Hyperbolic Systems," Comm. Pure Appl. Math. XXIII (1970), pp. 867-884.

4. F. Fayers and J. Matthews, "Evaluation of Normalized Stone's Methods for Estimating Three-Phase Relative Permeabilities," Soc. Pet. Eng. J. 24 (1984), pp. 225-232.

5. M.E. Gomes, "Singular Riemann Problem for a 4th Order Model for Multi-Phase Flow," Thesis (in Portuguese), Departamento de Matemática, Pontifícia Universidade Católica do Rio de Janeiro, Sept. 1987.

6. E. Isaacson, D. Marchesin, B. Plohr, and J.B. Temple, "The Classification of Solutions of Quadratic Riemann Problems (I)," SIAM J. Math. Anal., 1988, To appear. MRC Technical Summary Report #2891, 1985.

7. E. Isaacson, D. Marchesin, and B. Plohr, "Construction of Nonlinear Waves for Conservation Laws," in preparation, 1988.

8. E. Isaacson, D. Marchesin, B. Plohr and J.B. Temple, "Multiphase flow model with singular Riemann problems,", in preparation, 1988.

9. P. Lax, "Hyperbolic Systems of Conservation Laws II," Comm. Pure Appl. Math. X (1957), pp. 537-556.

10. M. Leverett and W. Lewis, "Steady Flow of Gas-Oil-Water Mixtures through Unconsolidated Sands," Trans. AIME 142 (1941), pp. 107-116.

11. T.-P. Liu, "The Riemann Problem for General Systems of Conservation Laws," J. Diff. Eqs. 18 (1975), pp. 218-234.

12. D. Marchesin and H. Medeiros, "A Note on Gravitational Effects in Multiphase Flow," in preparation, 1988.

13. M. Shearer, D. Schaeffer, D. Marchesin, and P. Paes-Leme, "Solution of the Riemann Problem for a Prototype 2X2 System of Non-Strictly Hyperbolic Conservation Laws," Arch. Rat. Mech. Anal. 97 (1987), pp. 299-320.

14. B. Wendroff, "The Riemann Problem for Materials with Non-Convex Equations of State: I Isentropic Flow; II General Flow," J. Math. Anal. and Appl. 38 (1972), pp. 454-466; 640-658.

DETONATION INITIATION DUE TO SHOCK WAVE-BOUNDARY INTERACTIONS

Rupert Klein
Institut f. Allgem. Mechanik, RWTH Aachen
Templergraben 55, 5100 Aachen, West-Germany

Abstract

Shock induced ignition and the subsequent development of reactive-gasdynamic waves in two-dimensional confined ducts are investigated by means of numerical simulations. The *inhomogeneous* Euler equations are employed to describe the gasdynamic-chemical interactions. A second order accurate two-step Godunov-type scheme, which directly accounts for the source terms, is proposed. Its performance is demonstrated by solving a test problem, whose exact solution is available. Two examples of flows within L-shaped configurations reveal interesting mechanisms, which support the formation of reactive Mach-stems, and thus trigger the onset of multidimensional detonation waves. A relation of the present idealized model problems to knock damage in internal combustion engines is pointed out.

1. Introduction and discussion of the model problem

The present studies are aimed to explain a mechanism, which can lead to the typical patterns of knock damage in internal combustion engines. Erosive surface destructions appear within the narrow gap between piston and cylinder wall as is sketched in **Fig. 1a**. We believe that shock waves, which may be driven by local autoignition within the unburnt endgas, can penetrate into the gap and give rise to a flow field, which is dominated by gasdynamic wave propagation. Sudden temperature increases, induced by shock reflections at the boundary walls, should be fast enough to overcome any heat loss and to ignite the mixture within the gap. Subsequent violent shock-reaction waves may be responsible for the observed destructions. In order to support this point of view, the Reynolds number $\mathrm{Re} = \rho_1\, u_1\, h/\eta_1$ is estimated by $\mathrm{Re} \approx 5000, \ldots, 10000$. Here we used a width $h \sim 0.05$mm of the gap, and evaluated density ρ_1, velocity u_1 and viscosity η_1 behind an inert shock with strength $\Delta p/p_0 \approx 0.5, \ldots, 1.0$ running over precompressed gases of $p_0 \approx 40$bar and $T_0 \approx 1200$K. The precompression of the unburnt gases is due to the piston motion as well as due to thermal expansions of the gas, that is already burnt. The shock strength is estimated from measurements of Pischinger et al. [1], who experimentally investigate several aspects of knocking combustion. The above range of Reynolds' number justifies the assumption that wave phenomena play an important role for the onset of combustion even within the narrow slabs considered above. In order to reveal the basic gasdynamic-chemical interactions and to show the crucial influence of the geometry, the following simplified model is employed. An L-shaped two-dimensional duct, as depicted in **Fig. 1b**, replaces the according part of the combustion chamber in Fig. 1a. An initially plane inert shock enters the configuration from the open side, thereby passing precompressed unburnt gases. Molecular transport and real gas effects are neglected and the gasdynamic properties of the system are described by means of Euler's equations for an ideal gas with constant specific heats. The chemical heat release is taken into account by supplying a source term to the conservation equation for thermal and kinetic energy (see also the remarks following (2)), and finally two additional quasi-conservation laws model the progress of chemical reactions. Thus we are left with the following set of governing equations

$$\underline{u}_t + \underline{f}^1(\underline{u})_x + \underline{f}^2(\underline{u})_y = \underline{q}\,. \tag{1}$$

Here $\underline{u} = (\rho, m, n, e, \rho\alpha, \rho\beta)$ is the vector of (quasi-)conserved variables, which are the densities of mass (ρ), momentum (m,n), thermal and kinetic energy (e), and of two density weighted reaction progress variables (α,β). $\underline{f}^1 = (m, m^2/\rho + p, mn/\rho, m(e+p)/\rho, \alpha m, \beta m)$ and $\underline{f}^2 = (n, mn/\rho, n^2/\rho + p, n(e+p)/\rho, \alpha n, \beta n)$ are the flux densities of the conserved quantities in x- and y-direction, respectively. The source vector is $\underline{q} = (0, 0, 0, Q\rho r_\beta, -\rho r_\alpha, -\rho r_\beta)$, with Q the chemical heat per unit mass of the unburnt gas and r_α, r_β the reaction rates.

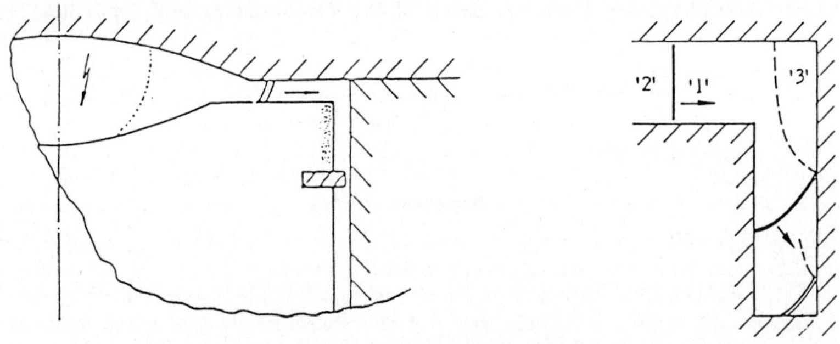

Figure 1: A critical region of knock damage in internal combustion engines (a), and its idealized two-dimensional counterpart (b).

The equations of state, relating pressure and temperature to the quasi-conserved quantities are

$$p = (\gamma - 1)(e - (m^2 + n^2)/2\rho)\,, \qquad T = p/\rho\,, \tag{2}$$

respectively. It should be mentioned that by replacing e with the total energy $\tilde{e} = e + \rho\beta Q$ one obtains a homogenous conservation equation for $\tilde{e}$. In principle this seems to be desirable, since a lot of the recent theoretical results regarding numerical algorithms for hyperbolic systems rely on the conservation property. However, a lengthy but straight-forward calculation shows that the equivalence of the energy balances in both formulations is mirrored exactly by the related difference equations of our numerical scheme, provided that $\tilde{e}$ is linear in β as it is assumed here. The advantage of using e instead of $\tilde{e}$ is that the eigenvectors of the Jacobian matrices $\partial \underline{f}^{(i)}/\partial \underline{u}$, which are used extensively in the scheme, simplify considerably, and thus lead to a more efficient formulation.

The reaction rates obey the Arrhenius-type laws

$$r_\alpha = B_\alpha T^2 \exp(-E_\alpha/T)H(\alpha) \stackrel{\text{def}}{=} \frac{1}{\tau}\,, \qquad r_\beta = B\beta \exp(-E/T)(1 - H(\alpha))\,. \tag{3}$$

Here the frequency factors B_α, B and activation energies E_α, E are constants, H is the heavyside step function and T denotes the temperature. These rate laws, together with the definition of the source term $Q\rho r_\beta$ of the energy balance equation imply that the first, energetically neutral reaction has to be completed before the exothermal second reaction can start. The approach is similar to that of Korobieinikov et al. [2] and to the *induction time model* of Oran et al. [3]. It allows to model independently the typical two phases of explosive gaseous reactions. These are a rather temperature sensitive but energetically almost neutral induction period, and a subsequent highly exothermal phase, in which all the chemical heat is released. Note, that with a zero initial condition for α one obtains a simple one-step irreversible Arrhenius reaction. It is well known (see e.g. Williams [4]), that for sufficiently large activation energy E and a correspondingly large frequency factor $B \sim \exp(E)/E$ this one-step model also shows the above mentioned two-phase behaviour. However, in this case the temperature dependences of the two stages of combustion are strongly related, since they are both determined by the parameters B and E of only one Arrhenius rate law. Therefore, we prefer the two-step model (3) but construct the rate law r_α such that τ exactly mimics the temperature behaviour of the ignition delay time of the one-step, large activation energy model. In this way we could in [5] directly compare the results obtained with the one- and two-step reaction schemes and show the deficiencies of an oversimplified kinetic modelling. In turn the simple two-step reaction assumed here also seems to be incapable of representing all important features of explosive hydrocarbon reactions. Thus for *quantitative* predictions one should employ more sophisticated reduced kinetic mechanisms, derived from detailed chemistry. Such systematic reductions, valid for flame combustion are already devised by Peters [6], [7], whereas reduced schemes, which reliably describe the properties of ignition processes are work in progress and have so far been obtained only for the hydrogen-oxygen system [7]. As regards the particular choice of the model parameters in (3), the data of Schmidt [8] as well as curve fits for ignition delay times of CH_4-air-mixtures of Oran et al. [3] provide estimates of the overall activation energy

$E_{ind} \approx 7, \ldots, 20$, which are valid in the ranges of pressure and temperature considered here. (We let $E_{ind} = E_\alpha$ for the two-step mechanism and $E_{ind} = E$ for the one-step Arrhenius-model.) In order to obtain appropriate values of the frequency factors B, B_α it is useful to consider first the characteristic time scales imposed by the several physical effects, that are included in our model. The induction phase of the chemical reaction is characterized by the ignition delay time $t_{I,3}$, referred to state "3", which occurs at a fixed wall after the incident shock wave has suffered a head on reflection (see Fig. 1b for the related position in the model configuration). The rate of chemical heat release during the stage of exothermal reaction provides the time scale $t_Q = 1/r_{\beta,ZND}$, where $r_{\beta,ZND}$ is the maximum rate of the second reaction in a Chapman-Jouguet (CJ)-detonation-wave described by means of the Zeldovich-von Neumann-Döring (ZND)-model (see e.g. Fickett and Davis [9] for details of the theory). Finally, one has to account for pressure wave propagation with its characteristic time $t_s = h/c$, where h is the width of the gap, and c is a representative speed of sound propagation. In explosive gases at ambient conditions and in systems with spatial extensions of a centimeter or more one usually has the estimates $t_I >> t_Q$, $t_s >> t_Q$ and often $t_I >> t_s$. But for the present application the estimates of time scales have to be revisited, because we are dealing with precompressed gases at high temperatures and with systems of very small extensions. Generally the induction time t_I goes down rapidly with increasing temperature and also t_s becomes small with decreasing size of the system and increasing temperatures. A rough estimate of ignition delay times at higher temperatures and pressures, derived from data of Schmidt [9] shows, that ignition delay times of the order of $t_{I,3} \approx 0.5\mu s$ may occure. This is just the range of the acoustic timescale, which is $t_{s,3} \approx 0.1\mu s$, when an initial temperature of $T_1 = 1200K$ and a strength of the incident shock of $\Delta p/p_1 \approx 1.0$ is assumed. Thus it is reasonable to consider a regime, in which at least two of the physical effects act on the same time scale and therefore can be expected to show interesting mutual interactions. Since in addition the exothermal reactions in general are much less sensitive to temperature variations than the induction phase, a crossover with $t_Q = O(t_I)$ will occur under sufficient precompression. For the results presented below we actually assumed the regime

$$t^*_{I,3} = t_{I,3}/t_{s,3} = O(1) \qquad \text{and} \qquad t^*_Q = t_Q/t_{s,ZND} = O(1)\,. \tag{4}$$

Now, instead of choosing the preexponentials B, B_α directly, we rather prescribe $t^*_{I,3}$ and t^*_Q, which can uniquely be related to the former, once E, E_α are fixed (cf. [5]). The a priori estimates (4) are also crucial for numerical simulations, since in the *usual* regime, stated previously, the chemical reactions give rise to very stiff source terms on the r.h.s. of (1). In this case one can hardly obtain reasonable numerical approximations using shock capturing schemes on equally spaced grids. Instead sophisticated automatic adaptive gridding or front tracking methods would have to be employed in order to resolve the thin shock-reaction structures, which in general emerge under such conditions. Fortunately, these problems are less dominating, when the estimates (4) hold.

For the present application the nondimensional chemical heat $Q = (-\Delta h)/RT_1$ is of the order $O(10)$. Although Q is thus not a very large quantity, we follow an advise by Oran and Boris [10], who propose to restrict the time step in an explicite numerical algorithm such that the heat released per time step in a cell cannot exceed a fixed percentage (e.g. 30%) of its internal energy.

2. The numerical scheme for Euler's equations with source terms

In order to perform numerical integrations of the system (1) on domains with piecewise straight, right angled boundaries (see Fig. 1b), we use a cartesian numerical grid, which is equally spaced in each direction. The directional operator splitting technique of Strang [11] is employed to extend a discrete second order accurate one-dimensional solution operator to two dimensions. Thereby half the source density is assigned to each of the one-dimensional splitted equations

$$\underline{v}_t + \underline{f}^1(\underline{v})_x = \underline{q}(\underline{v})/2\,, \qquad \underline{w}_t + \underline{f}^2(\underline{w})_y = \underline{q}(\underline{w})/2\,. \tag{5}$$

The one-dimensional scheme used to solve (5) reads as

$$\underline{u}_i^{n+1} = \underline{u}_i^n - \frac{\Delta t}{\Delta x}\left[\underline{f}_{h,i+1/2}^{n+1/2} - \underline{f}_{h,i-1/2}^{n+1/2} + \frac{\Delta t}{2}\left[(\underline{\underline{A}}\,\underline{q}/2)_{i+1/2}^{n+1/2} - (\underline{\underline{A}}\,\underline{q}/2)_{i-1/2}^{n+1/2}\right]\right] + \Delta t\,(\frac{1}{2}\,\underline{q}_i^{n+1/2})\,. \tag{6}$$

Here $\underline{u}_i^n$ approximates the average of the quasi-conserved quantities $\underline{u}$ at time level t^n within the i-th cell $[x_{i-1/2} \leq x \leq x_{i+1/2}]$ and

$$\underline{f}_{h,i+1/2}^{n+1/2} = \underline{f}^{HLLE}(\underline{u}_{i+}^{n+1/2}, \underline{u}_{(i+1)-}^{n+1/2}) . \tag{7}$$

is the numerical flux of a second order MUSCL-type scheme for the *homogeneous* version of (5). It is obtained by means of the approximate Riemann solver of Einfeldt [12], which is based on a proposition of Harten, Lax and vanLeer [13]. In the version used here for a γ-law gas it is equivalent to Roe's [14] linearized solver, except that it is much more efficient and can be extended to more general equations of state in a straight forward way (see Einfeldt [15]). The preliminary states $\underline{u}_{i\pm}^{n+1/2}$ at the cell interfaces within the i'th cell are obtained by means of a first order characteristic method. E.g. for $\underline{u}_{i+}^{n+1/2}$ we have

$$\underline{u}_{i+}^{n+1/2} = \underline{u}_i^n + \sum_{\nu=1}^{6} \Delta\eta_{i+}^{(\nu),n} \underline{R}_i^{(\nu),n} , \tag{8}$$

where

$$\Delta\eta_{i+}^{(\nu),n} = \begin{cases} \frac{1}{2}(\Delta x - a^{(\nu)}\Delta t)\,\delta\eta_i^{(\nu),n}, & \text{if } a^{(\nu)} \geq 0 \\ \frac{1}{2}(\Delta x \delta\eta_i^{(\nu),n} - a^{(\nu)}\Delta t \,\text{minmod}\left[\delta\eta_{i+1}^{(\nu),n}, \delta\eta_{i+}^{(\nu),n}\right]), & \text{if } a^{(\nu)} < 0 \end{cases} \tag{9}$$

and, dropping the superscripts (ν), n for the moment

$$\delta\eta_{i+} = 2\delta\eta_{i+1/2} - \delta\eta_i\,, \qquad \delta\eta_{i+1/2} = \underline{L}_i \cdot (\underline{u}_{i+1} - \underline{u}_i)/\Delta x\,, \qquad \delta\eta_i = S_k(\delta\eta_{i-1/2}, \delta\eta_{i+1/2}) . \tag{10}$$

$\underline{u}_{i-}^{n+1/2}$ is obtained by replacing i+ with i- and reversing the ordering signs in (8) to (10). The limiters used in (9) and $(10)_3$ are

$$\begin{aligned} \text{minmod}(a,b) &= \text{sgn}(a)\max(0, \min(\text{sgn}(a)a, \text{sgn}(a)b)) \\ S_k(a,b) &= \text{sgn}(a)\max(|\,\text{minmod}(ka,b)|, |\,\text{minmod}(a,kb)|) . \end{aligned} \tag{11}$$

The slope limiter S_k corresponds to a class of flux limiters given by Sweby [16], as is shown by Munz [17]. The corresponding flux correction schemes are second order accurate and TVD for $1 \leq k \leq 2$. Following Munz [17] we use $k = 1.4, \ldots, 1.6$ on the genuinely nonlinear and $k = 1.8, \ldots, 2.0$ on the linearly degenerate characteristic fields. In eqs. (8) to (10) $\underline{R}_i^{(\nu),n}$, $\underline{L}_i^{(\nu),n}$ are the right and left eigenvectors of the Jacobian matrix $(\partial \underline{f}/\partial \underline{u})_i^n$, $a^{(\nu)}$ are the corresponding eigenvalues and the several $\delta\eta$'s represent wave amplitudes on the different characteristic fields. They are projections of corresponding differences of the conserved quantities $\underline{u}$ onto the right eigenvectors $\underline{R}^{(\nu)}$. Especially $\delta\eta_i^{(\nu),n}$ is the limited slope of the ν-th local characteristic variable within cell i. While the relation for $a^{(\nu)} \geq 0$ is a straight forward evaluation of the characteristic equations, based on the linear distribution $\eta^{(\nu)} = \eta_i^{(\nu),n} + (x - x_i)\delta\eta_i^{(\nu),n}$ for $x_{i-1/2} \leq x \leq x_{i+1/2}$, the case $a^{(\nu)} < 0$ requires an explanation. Characteristics with $(\partial x/\partial t)^{(\nu)} = a^{(\nu)} < 0$ reach the i-th cell from outside at the right cell interface $x_{i+1/2}$. Since we are dealing with weak solutions of (quasi-)conservation laws, there may be a discontinuity within cells i, i+1 in the exact solution. Accross it the characteristic equations may not be applied. Colella and Glaz [18] circumvent this difficulty by replacing the characteristics coming from outside by a *pseudo-characteristic* with the speed $a^* = \max_\nu(0, a^{(\nu)})$. Thus their updating procedure $t^n \rightarrow t^{n+1/2}$ only uses information from inside the cell under consideration. Obviously in that way an unphysical transport of information is buildt into the halftime-step. However, Colella and Glaz point out, that this does not restrict the performance of the scheme, since the numerical fluxes, used in the final timestep (6) (with $\underline{q} \equiv 0$), are obtained from the exact (or approximate) solutions of a Riemann problem and the Riemann solver automatically drops out all information that approaches from the wrong side. In several tests for the homogeneous equations we could verify these considerations. But in (6) with $\underline{q} \neq \underline{0}$, evaluations of the source terms at the cell interfaces $\underline{q}_{i+1/2}^{n+1/2}$ are required (see (12)). For this purpose it is a self-suggesting choice to employ the Roe-average $\underline{u}_{LR}(\underline{u}_{i+}^{n+1/2}, \underline{u}_{(i+1)-}^{n+1/2})$, which is obtained anyway when the fluxes $\underline{f}_{h,i+1/2}^{n+1/2}$ are calculated. Since $\underline{u}_{LR}$ is symmetric in both arguments (see Roe [14]), it does not select the right information with respect to the characteristic directions. This explains, why it was necessary to construct more elaborate approximations of the interfacial states in order to obtain satisfactory results, e.g. for the test problem of section 3.1. The essential idea, leading to $(9)_2$ for $a^{(\nu)} < 0$ is to choose reasonable distributions of the local characteristic variables outside a cell, such that an evaluation of the compatibility relations does not amount to an application of a characteristic scheme across a discontinuity. **Fig. 2** shows the several situations, which were taken into account in the construction of $(9)_2$. Let us consider the case of nondecreasing η as in Fig. 2a. In order to avoid uncontrolled extrapolations of the linear distribution within cell i, we use as a basis the outer interpolation between $(x_{i+1/2}, \eta^n_{i+})$ and (x_{i+1}, η^n_{i+1}) with slope $\delta\eta^n_{i+}$ outside the cell. Here $\eta^n_{i+} = \eta^n_i + \Delta x \delta\eta^n_i/2$. Since $|a^{(\mu)}|\Delta t/2 \leq \Delta x/2$ for all μ, due to the CFL-condition, this Ansatz ensures, that no value of η outside $[\eta^n_{i+}, \eta^n_{i+1}]$ can be selected. Furthermore, by means of the limitation $(10)_3$ η^n_{i+} obeys $\eta^n_i \leq \eta^n_{i+} \leq \eta^n_{i+1}$ and therefore stays within the range of the cell averages. Nevertheless, if there is a large gradient between cells i and i+1, then one has to be careful in employing the outer interpolation. As is shown in Figs. 2b,c there may be a discontinuity of either the local characteristic variable η itself (Fig. 2b) or of its first derivative (Fig. 2c). In the first case the slope of η outside the cell must be limited in order to avoid application of a characteristic scheme across a shock, whereas in the second case the use of the outer interpolation is allowed, since discontinuities of slopes move with characteristic speed and do not imply special jump conditions. We account for this requirement by means of the minmod-function in (9), which employs the limited slope of cell i+1 as soon as the slope $\delta\eta_{i+}$ of the outer interpolation becomes large. Comparing Figs. 2b and 2c we note further, that $\delta\eta_{i+} >> \delta\eta_i \approx \delta\eta_{i+1}$ in 2b, while $\delta\eta_{i+1} \approx \delta\eta_{i+} >> \delta\eta_i$ in 2c. Therefore the use of $\delta\eta_{i+1}$ in (9) instead of the more apparent choice $\delta\eta_i$ allows to distinguish between the two cases of Figs. 2b, 2c.

The terms $(\Delta t/2)(\underline{\underline{A}}\,\underline{q})^{n+1/2}_{i+1/2}$ in (6) are second order corrections to the numerical fluxes, which account for the influence of the sources on the wave interactions between adjacent cells. This aspect is discussed in some detail in [19], where a formulation of these expressions in terms of eigenvectors of $\underline{\underline{A}}^{n+1/2}_{i+1/2}$ is given. To evaluate the source wave corrections we employ the Roe-averaged state $\underline{u}_{LR}$, which is obtained, when the *homogeneous* fluxes $\underline{f}^{n+1/2}_{h,i+1/2}$ are calculated by means of the $HLLE$-Riemann solver. Thus we have

$$(\underline{\underline{A}}\,\underline{q})^{n+1/2}_{i+1/2} = \left[\left(\frac{\partial \underline{f}}{\partial \underline{u}}\right)(\underline{u}_{LR})\,\underline{q}(\underline{u}_{LR})\right]^{n+1/2}_{i+1/2}, \qquad \underline{u}_{LR} = \underline{u}_{LR}(\underline{u}^{n+1/2}_{i+}, \underline{u}^{n+1/2}_{(i+1)-}). \tag{12}$$

It should be noted that we do not apply the slope limiting procedure to the sources directly as proposed by Roe [20]. Instead the limiting enters the evaluations of sources only via the update (8) to (11).

The cell centered source terms $\underline{q}^{n+1/2}_i$ are calculated by means of the following iterative method. For $0 \leq j \leq J-1$ we let

$$\underline{u}^{n+1,j+1}_i = \underline{u}^{n+1}_{h,i} + \frac{1}{2}\Delta t\,[\underline{q}^{n+1,j}_i + \underline{q}^n_i], \tag{13}$$

where

$$\underline{u}^{n+1}_{h,i} = \underline{u}^{n+1/2}_{i+} + \underline{u}^{n+1/2}_{i-} - \underline{u}^n_i, \qquad \underline{u}^{n+1,0}_i = \underline{u}^{n+1}_{h,i} + \Delta t\,\underline{q}^n_i,$$

and finally

$$\underline{q}^{n+1/2}_i = \frac{1}{2}[\underline{q}^{n+1,J}_i + \underline{q}^n_i]. \tag{14}$$

As usual $\underline{q}^\beta_\alpha$ denotes $\underline{q}(\underline{u}^\beta_\alpha)$. The results given in section 3 are obtained with $J = 2$. This completes the description of scheme (6).

Additional complications are imposed by the sharp convex corner of the duct. In [19] special boundary conditions are proposed, which are based on asymptotic considerations of Euler's equations in the vicinity of the corner. The procedure reduces numerical dissipation next to the edge and also the undesired sensitive dependence of the results on details of the numerical scheme, which was observed by Woodward and Colella [21] in a comparison of several shock capturing algorithms.

3. Results

3.1 Clarke's Testproblem

The performance of scheme (6) will be demonstrated in this subsection by means of a one-dimensional test problem, given by Clarke and Toro [22]. The sources of mass, x-momentum and energy are set to $\underline{q} = H(\frac{1}{2}-x)(G/c)(\rho, m, e+p)^T$. Here H is again the heavyside step-function. This choice amounts to an

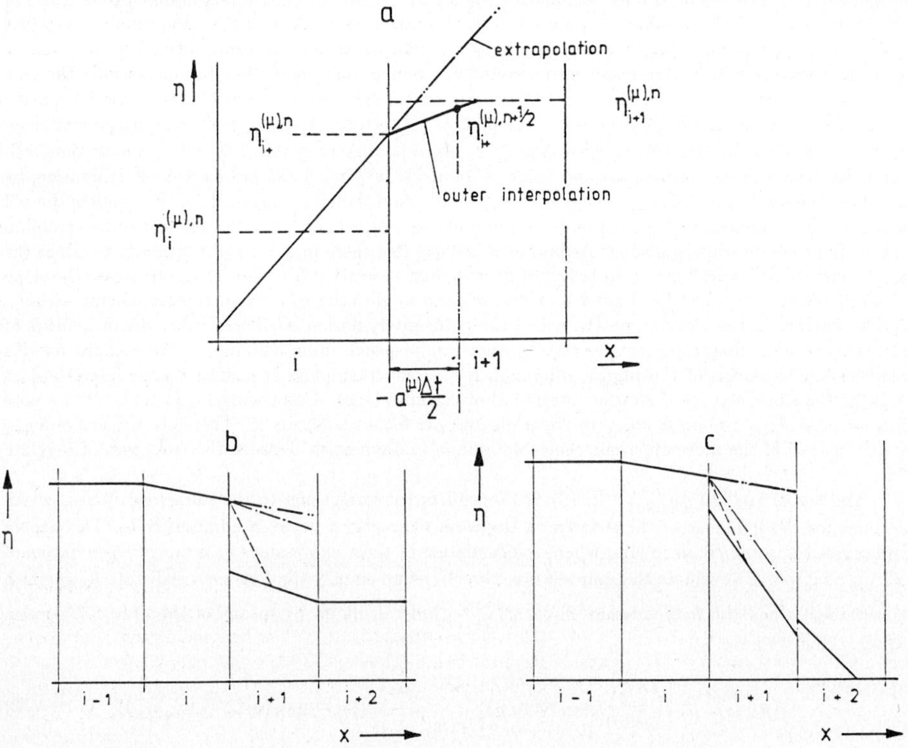

Figure 2: The distributions of local characteristic variables η, whose characteristics meet the cell from outside. (a) Extrapolation may lead to the introduction of new extrema; (b) discontinuous distribution of η; (c) discontinuous slope $\partial\eta/\partial x$; different outer slopes:
$\delta\eta_i$ —·· —, $\delta\eta_{i+}$ — — —, $\delta\eta_{i+1}$ —·—.

isentropic addition of mass in the region $x < \frac{1}{2}$ at a rate G/c. Using homogeneous quiescent gas initial conditions, the flow is also homentropic and one can introduce characteristic coordinates, corresponding to the paths of pressure waves in the x-t-plane. With the present choice of $\underline{q}$ the compatibility relations along the characteristic curves have a constant r.h.s., which allows an analytic integration.

Up to the time when shocks or centered expansions form, the exact solution can be constructed from these characteristic results. **Fig. 3** shows results based on the data ($p = 1$atm, $c = 330.4$ms, $m \equiv 0$) at $t = 0$ and $G = 1294301 m/s^2$ (see Clarke and Toro [22]). Exact and approximate distributions of the density at times 0.1, 0.4, 0.7ms are given together with the related relative errors $\Delta\rho = (\rho - \rho_{exact})/\rho_{exact}$ on the last time level. In these calculations we employed the limiters $S_{1.6}$ and $S_{1.8}$ on the soundwave and particle path characteristics, respectively. The results obviously are well within the range of second order accuracy except at $x = 0.5$ and $x \approx 0.75$, where singularities in the characteristic solutions develop. There no higher accuracy can be expected from a shock capturing scheme, applied on an equally spaced grid. (Extended calculations can be found in [5], where different updating formulae $t^n \to t^{n+1/2}$ and different limiters are tested and where a comparison with *time operator splitting* for the source terms is performed.)

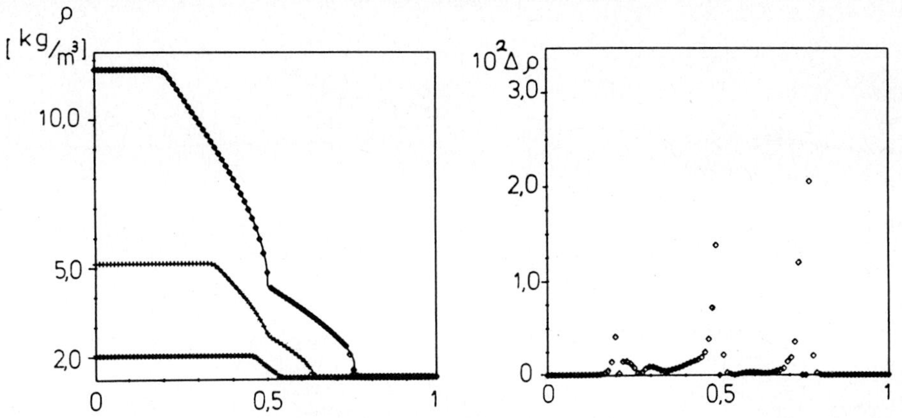

Figure 3: Distribution of the density ρ and its relative error $\Delta\rho$ for Clarke's test problem.

3.2 Shock induced ignition in an L-shaped duct

Here we present two examples of the development of reacting waves after shock diffraction and reflection within the idealized geometry of Fig. 1b. The evolution of the flow is discussed mainly in terms of sequences of density distributions represented by plots of contour lines. When interpreting the diagrams it should be kept in mind, that steep density gradients may occur at shock fronts, at contact discontinuities and across thin zones of chemical heat release. The latter gradients are due to thermal expansions of the burning gas and can be identified easily by comparison with related distributions of the reaction progress variables. Due to lack of space we will not display these distributions throughout.

Consider now the sequence of **Figs. 4**. In this first example the induction and reaction time scales are $t^*_{I,3} = 0.42$ and $t^*_Q = 0.38$, respectively. We further impose a high temperature sensitivity letting $E_\alpha = 20.0$. **Fig. 4a** shows contour lines of the density at a time, when the reflected shock front has already removed from the reflecting wall a considerable distance. It interacts with the vortex, which has formed downstream of the edge when the incident shock diffracted there. In the present case, the shock is not yet supported and enhanced by chemical reactions and the vortex is able to break up the front. On the right of the vortex core, the shock is accelerated towards the wall, whereas on the left it is decelerated. Consequently the shock is stretched and weakened substantially in the center of the vortical region. At the opposite concave corner, the highly sensitive induction reaction selects those mass particles, which rested at high temperatures for the longest time. The induction period is finished locally and chemical heat is released. Approaching the concave corner from inside of the flow field, the contour lines first show an increase of the density in a pressure wave which is driven by the local explosion, and then a decrease of ρ towards the corner, which is due to corresponding thermal expansions. In **Fig. 4b** the reaction driven pressure wave has steepened and is catching up the initially reflected shock on its way back to the entrance of the duct. Ahead of the convex corner the part of the shock front, that was previously accelerated by the vortex, is now reflected and begins to traverse the duct again in opposite direction. A short time later in **Fig. 4c$_1$** the wave system facing the oncoming flow from the entrance has developed into a slightly curved detonation wave. It now merely consists of a strong precursor shock and a subsequent expansion due to combustion. The small front enclosing the tip of the convex corner indicates a previous reflection of this precursor shock. The transverse wave, which has emerged in Fig. 4b, now interacts with the reactive pressure wave, that spreads out from the concave corner. It can be read from **Fig. 4c$_2$**, which depicts the corresponding distribution of the reaction progress variable β, that a wedge of unburnt material has formed in front of the wall. It should be emphasized, that the transverse wave here results from multiple reflections of an essentially inert shock front and that it is not driven by chemical heat release. Nevertheless it leads to a substantial temperature increase in front of the wall and accelerates the induction reactions within the wedge of unburnt. **Fig. 4d** shows the result of a subsequent explosion of this amount of end gas. The inert

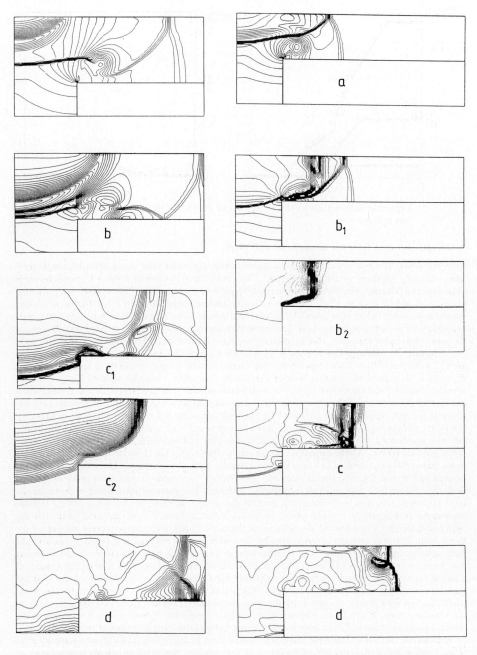

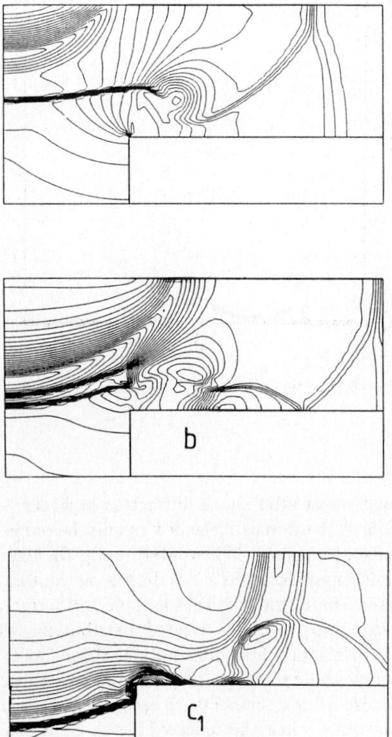

Figure 4: Two-step chemical model
$E_\alpha = 20.0$, $t^\star_{I,3} = 0.43$, $M_{sh} = 1.6$,
$E = 5.0$, $t^\star_Q = 0.38$.

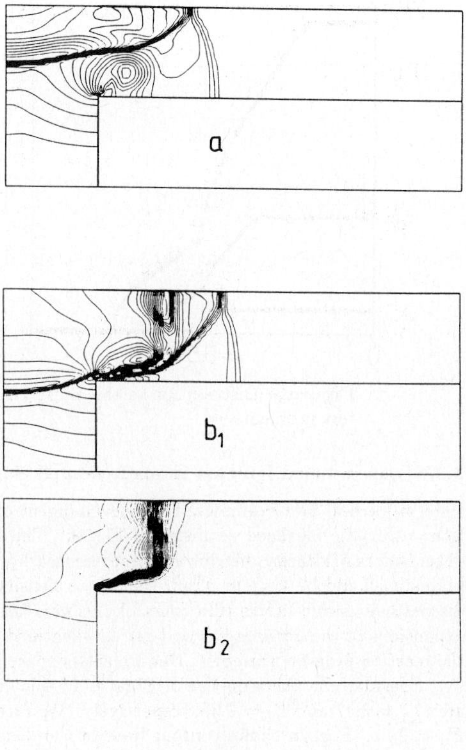

Figure 5: One-step chemical model
$E = 10.0$, $t^\star_{I,3} = 0.23$, $M_{sh} = 1.6$,
$t^\star_Q = 0.21$.

transverse wave has already passed the duct as is indicated by the weak, nearly horizontal gradient at the top right corner of the diagram. At the same time the explosion in front of the corner wall has led to high peaks of pressure and density connected with a new secondary transverse shock. This wave is now directly coupled to the chemical reactions and may be considered as part of a weak detonative triple shock, although in the present case the width of the duct would most likely be too small to allow the establishment of a (quasi-)steadily propagating reactive Mach-stem if the whole structure would be allowed to propagate down a continued straight channel. However, the following example will show, how some changes of the model parameters, which do not touch the orders of magnitude of the characteristic time scales, can substantially support the onset of typical patterns of multidimensional detonations.

We decrease the temperature sensitivity of the induction time to $E_{ind} = 10$ and consider slightly faster chemical reactions according to $t^*_{I,3} = 0.23$ and $t^*_Q = 0.21$. In contrary to the constellation of the first example the present regime can be and actually is simulated by means of the one-step model. **Fig. 5a** gives the situation just before the reflected shock begins to interact with the vortex. The density gradient behind the shock indicates that there is already a substantial progress of the chemical reaction. No well bounded, localized explosion takes place here due to the decreased temperature sensitivity of the induction phase and due to the fact that there is no sharp threshold between the induction period and the phase of heat release. In **Fig. 5b$_1$** the shock-reaction front just passes the vortical separation flow ahead of the corner. It is seen that the vortex is no longer able to break the front, but only turns it around slightly such that a part of it is now nearly parallel to the wall. **Fig. 5b$_2$** shows the distribution of the reaction progress variable. Obviously there is again a wedge of unburnt material. But it should be noted that the time of the present situation corresponds to that of Fig. 4a of the previous example, where the reflected shock is seen just before it hits the corner wall, while Fig. 4c$_2$, which also shows an enclosure of unburnt, corresponds to a later time. Also in the present case an already developed shock reaction wave reflects at the wall instead of an inert shock as in Fig. 4b. In consequence the resulting transverse wave is much stronger than that of the first example. In **Fig. 5c** the subsequent complicated wave interactions in front of the corner wall are depicted. As before a sharp peak of pressure and density builds up at a point, where the reflected transverse wave meets the leading front and the wall. Later on this peak, together with the transverse wave departs from the wall and passes the duct, thereby interacting with the lead shock. The result is a typical (Mach-stem) configuration, which in the present case seems to be strong enough to survive multiple reflections between the opposite walls of the duct. **Fig. 5d** shows the triple point after it has developed a very clear and presumably stable structure.

4. Conclusions

The shock induced reactive-gasdynamic flow within a two-dimensional L-shaped configuration is examined by means of numerical simulations. A regime is considered, where ignition delay, chemical heat release and also pressure wave propagation all occur on comparable time scales and thus show interesting mutual effects. Some order of magnitude estimates suggest that this regime may be particularly relevant for the problem of knock damage in internal combustion engines. The calculations are based on the balance equations of an inviscid perfect gas, supplied by a two-stage chemical model. The results reveal different inherently multidimensional mechanisms, which provoke the formation of reacting Mach-stems. Violent reflections of these triple shocks at the bounding walls may well be responsible for erosive damages, which are typically observed within engines running under knocking conditions.

Discrete approximations are obtained using a MUSCL-type shock capturing scheme, designed for *direct* integration of the *inhomogeneous* Euler equations. Although the construction of the scheme was stimulated by considerations of Roe [20], *no upwind weighting* of the sourceterms in the sense of his suggestions is performed. Instead second order source corrections to the numerical fluxes are obtained by straight forward expansions about the fluxes of a second order scheme for the homogeneous equations. Special attention is paid to the determination of cell interface states, which are employed in the according source evaluations. Results for Clarke's [22] test problem are in satisfactory agreement with the exact solutions and the approximations do not show any source induced oscillations.

References

[1] Pischinger, F., Kollmeier, H.P., Spicher, U., *Das Klopfen im Ottomotor—Ein altes Problem aus neuer Sicht*, Mitteilungen des Instituts f. Verbrennungskraftmaschinen u. Thermodynamik, TU Graz **49**, (1987).

[2] Korobieinikov, V.P., Levin, V.A., Markov, V.V., Citenyi, G.G., *Propagation of Blast Waves in a Combustible Gas*, Astronautica Acta **17**, 529 (1972).
[3] Oran, E.S., Boris, J.P., Young, T., Flanigan, M., Burks, T., Picone, M., *Numerical Simulations of Detonations in Methane-Air Mixtures*, Proc. of the 18th Symp. on Combustion, The Combustion Institute (1982).
[4] Williams, F.A., *Combustion Theory*, 2nd Edition, Benjamin/Cummings (1985).
[5] Klein, R., *Stoßinduzierte Zündung und der Übergang zur Detonation in engen Spalten*, Dissertation, RWTH Aachen (1988).
[6] Peters, N., *Numerical and Asymptotic Analysis of Systematically Reduced Reaction Schemes for Hydrocarbon Flames*, in: *Numerical Simulations of Combustion Phenomena*, Lecture Notes in Physics, **241**, 90-109, Springer (1985).
[7] Peters, N., *Systematic Reduction of Flame Kinetics—Principals and Details*, 11th ICODERS, Warsaw, August 1987.
[8] Schmidt, F.A.F., *Verbrennungskraftmaschinen*, 4. Aufl., Springer (1967).
[9] Fickett, W., Davis, W.C., *Detonation*, University of California Press (1979).
[10] Oran, E.S., Boris, J.P., *Detailed Modelling of Combustion Systems*Progr. Energy Comb. Sci. **7**, 1-72 (1981).
[11] Strang, G., *On the Construction and Comparison of Difference Schemes*, SIAM, J. Num. Anal. **5**, 506-517 (1968).
[12] Einfeldt, B., *On Godunov-Type Methods for Gasdynamics*, SIAM, J. Num. Anal. **25**, No. 2 (1988).
[13] Harten, A., Lax, P.D., vanLeer, B., *On Upstream-Diferencing and Godunov-Type Schemes for Hyperbolic Conservation Laws*, SIAM Review **25**, 35-61 (1983).
[14] Roe, P.L., *Approximate Riemann-solvers, Parameter Vectors and Difference Schemes*, J. Comp. Phys. **43**, 357-372 (1981).
[15] Einfeldt, B., *On Godunov-Type Methods for the Euler Equations with a General Equation of State*, Proc. of the 16th Int. Conf. on Shock Tubes and Waves, Ed. H. Grönig, VCH-Verlagsgesellschaft, Weinheim, West-Germany (1988).
[16] Sweby, P.K., *High-Resolution Schemes Using Flux Limiters for Hyperbolic Conservation Laws*, Siam, J. NUm. Anal. **21**, 995-1011 (1984).
[17] Munz, C.D., *Näherungsverfahren höherer Ordnung zur Approximation von Stoßwellen*, Berichte **26**, **28**, Fak. f. Mathem., Universität Karlsruhe (1985).
[18] Colella, P., Glaz, H.M., *Efficient Solution Algorithms for the Riemann-Problem for Real Gases*, J. Comp. Phys. **59**, 264-289 (1985).
[19] Klein, R., *Shock Initiated Ignition in a L-Shaped Duct: Two Aspects of its Numerical Simulation*, to appear in Proc. of the 7th GAMM-Conference on Numerical Methods in Fluid Mechanis, Louvain La Neuve, Belgium (1987).
[20] Roe, P.L., *Upwind Differencing Schemes for Hyperbolic Conservation Laws with Source Terms*, Eds.: A. Dold, B. Eckmann, Lecture Notes in Mathematics, 1270, Springer (1987).
[21] Woodward, P., Colella, P., *The Numerical Simulation of Two-Dimensional Fluid Flow with Strong Shocks*, J. Comp. Phys. **54**, 115-173 (1984).
[22] Clarke, J.F., Toro, E.F., *Gas Flows Generated by Solid Propellant Burning*, Lecture Notes in Physics **241**, Numerical Simulation of Combustion Phenomena (Eds.: R. Glowinsky, B. Larroutourou, R. Temam), Springer (1985).

NONCONVEX SCALAR CONSERVATION LAWS IN ONE AND TWO SPACE DIMENSIONS

Christian Klingenberg
Dept. of Applied Mathematics, University of Heidelberg,
Im Neuenheimer Feld 294, 6900 Heidelberg, W.-Germany

Stanley Osher
Dept. of Mathematics, University of California,
Los Angeles, CA 90024, USA

1. Introduction

Consider the initial value problem for a scalar conservation law

$$\begin{aligned} &u_t + \nabla . f(u) = 0 , \qquad t \geq 0, \qquad \bar{x} \in \mathbb{R}^n \\ &u(x,0) = u_0 (x) . \end{aligned} \tag{1.1}$$

For the nonlinear flux function even smooth initial data in general may not prevent the development of jumps in the solution to (1.1). Thus we consider (1.1) in the sense of distributions.

When trying to understand the qualitative behaviour of solutions to conservation laws in more than one space dimension, as a first step one may consider selfsimilar solutions. This way the problem becomes more tractable, because the number of independent variables is reduced by one. In particular one has considered Riemann problems.

Definition: If (1.1) is invariant under the transformation

$$(c\bar{x}, ct) \rightarrow (\bar{x}, t), \; c > 0,$$

then it is called a Riemann problem.

Thus in two space dimensions initial data in (1.1) which is piecewise constant in sectors meeting at the origin is an example of a two dimensional Riemann problem (2-d R.P.).

These arise naturally in front tracking, a numerical scheme for conservation laws, see e.g. [GK]. The essential feature of this method is that a lower dimensional grid is fitted to and follows the jump surfaces. At the intersection points of these discontinuities 2-d R.P. occur. We mention in passing that one can give a short list of these generic types of such intersection points for two dimensional gas dynamics, which presumably constitutes the pieces that the solution to a 2-d R.P. for the Euler equations is made up of, [GK].

Recently some progress was made in understanding the 2-d R.P. for the scalar conservation law:

$$u_t + f(u)_x + g(u)_y = 0 . \tag{1.2}$$

One knows existence [CS] and uniqueness [K] of the weak solution satisfying the entropy condition. It was natural to ask next what these solutions for the case of a Riemann problem look like. Wagner [W] constructed the solution for a convex f very close to a convex g. In [HK] this was extended to a generic case with f = g, where f is a quadratic and g a cubic polynomial, see section 2. For general flux functions, for the case f = g, the solution was constructed in [CK], see section 3. There something reminiscent of a large time Godunov method for the scalar equation in one space dimension was used. This method inspired two results for the one dimensional scalar equation, which are reported in section 4. We close with some examples in section 5.

2. The Riemann problem for the scalar conservation law in two space dimensions with unequal flux functions

The selfsimilar solution may be described completely by giving the solution in the plane, say t = 1. Far away from the origin, the solution is given by solving a 1-d R.P. across the jumps given in the initial data. [HK] proceeded to describe the interaction of these waves for the two flux functions being a cubic and a quadratic polynomial.

We shall give an illustrative example on what may happen. Suppose we consider the equation

$$u_t + (u^2)_x + (u^3)_y = 0 .$$

Say across the positive x-axis there was an initial jump that gave rise to a jump followed by a rarefaction wave, see Fig. 2.1.

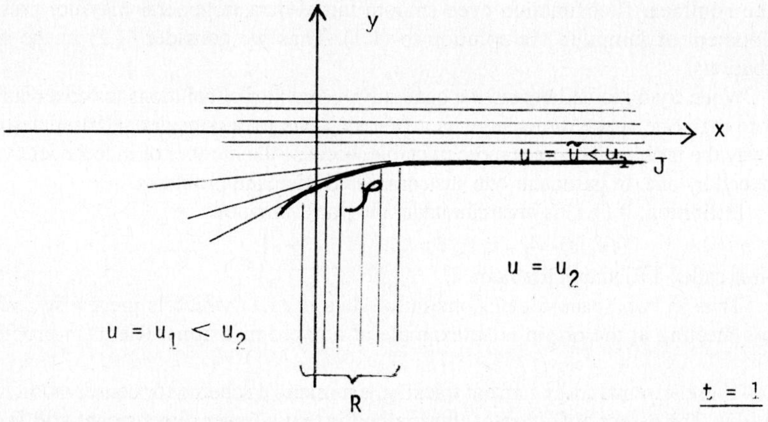

Fig. 2.1 Part of the solution of a particular two dimensional Riemann problem with initial discontinuities across the x-axis and the y-axis.

The jump J interacts with the rarefaction wave R to form a surface $\mathcal{S}$ which is a smooth continuation of J that bends into R. To the left of $\mathcal{S}$ a new rarefaction appears which is tangential to $\mathcal{S}$, see Fig. 2.1. We may now define an ordinary differential equation for $\mathcal{S}$, show that it is well defined on the interval $[(u_2 + \tilde{u})/2, u_2]$ and satisfies the entropy condition.

Depending on the location of u_1 relative to u_2 and $\tilde{u}$ we have two possibilities:

a) $(u_2 + \tilde{u})/2 < u_1$.

This gives rise to a onesided contact discintinuity (c.d.) with the constant state u_1 on one side and $u_2 + \tilde{u} - u_1$ on the other side.

b) $u_1 < (u_2 + \tilde{u})/2$.

At $Z = \{f'((\tilde{u} + u_2)/2),\ g'((\tilde{u} + u_2)/2)\}$ the shock strength of $\mathcal{S}$ has decayed to zero. Then $\mathcal{S}$ continues on smoothly into a curve Γ given by $(f'(u), g'(u))$, $u_1 < u < (\tilde{u} + u_2)/2$, where the two rarefaction waves meet.

In this way we could construct the solution to our 2-d R.P., see Fig. 2.2 for an example of a solution.

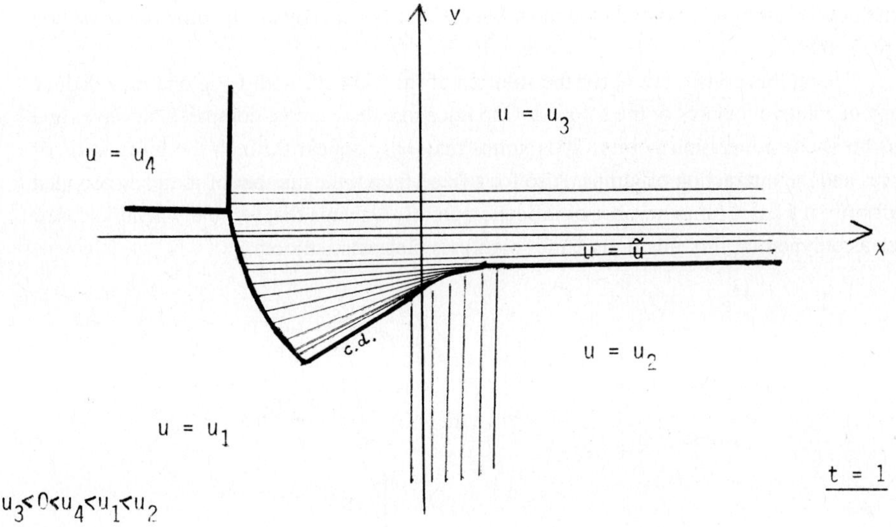

Fig. 2.2 An example of a solution to

$$u_t + (u^2)_x + (u^3)_y = 0$$

with particular initial data constant in each of the four quadrants. Notice how this example contains the piece shown above and mentioned in a) of section 2.

3. The Riemann problem for the scalar conservation law in two space dimensions with equal flux functions

The solution for the 2-d R.P. to (1.2) with $f = g$ is constructed by considering an equivalent problem in one space dimension. Under the coordinate transformation

$\xi = (x+y)/2$ and $\eta = (x-y)/2$ the equation (2.1) for f = g becomes

$$u_t + f(u)_\xi = 0 \tag{3.1}$$

with η a parameter. The initial data for η = const. is piecewise constant with a finite number of jumps. It is easy to see that the 2-d R.P. is now reduced to constructing the solution to (3.1) for $\eta > 0$ and for $\eta < 0$.

For small t, near each jump we may construct the solution to the 1-d R.P. by the method of convex hull (see Fig. 3.1). After some finite time an interaction between two adjacent waves is possible. This interaction may be described qualitatively by using the a timedependent version of the convex hull, see Fig. 3.2. For details see [CK]. One finds that the union of convex hulls given initially gets deformed in a unique way towards the final convex hull, which consists of the solution to 1-d R.P. with the two constant states being the left most state in the initial data for η = const. and the right most state there. This construction is reminiscent of a Godunov scheme which is taken past the time of interaction of the waves.

Using this construction, for the solution of the 2-d R.P. with f = g one may deduce many qualitative features of the solution. One finds that there are no compression waves and thus no shock generation points. Thus jumps may only appear through the bifurcation of jumps and the interaction of jumps. Also for a fixed time t, the number of jumps is bounded uniformly if f has a finite number of inflection points. It seems that for many generic cases, such as polynomial flux functions f, the solution is piecewise smooth.

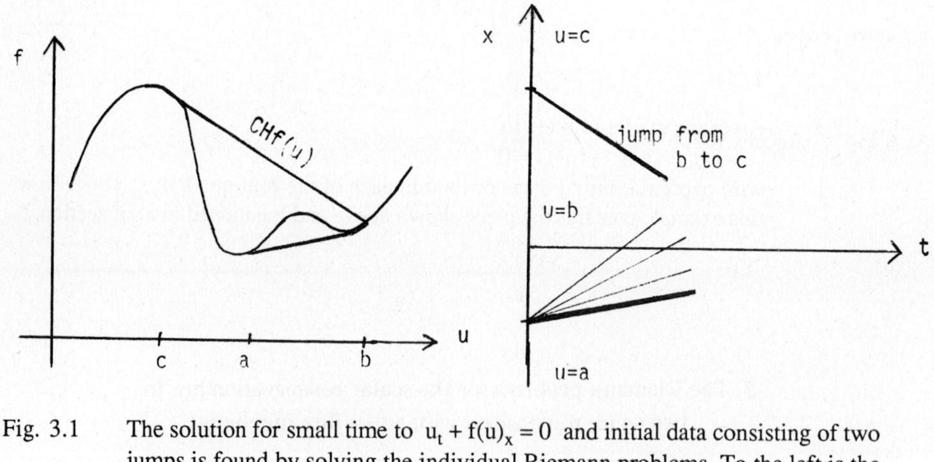

Fig. 3.1 The solution for small time to $u_t + f(u)_x = 0$ and initial data consisting of two jumps is found by solving the individual Riemann problems. To the left is the graph of f together with the convex hull between the jumps. To the right is the solution plane tilted, so that the slopes of the jumps in the convex hull CHf(u) and in the solution plane are parallel.

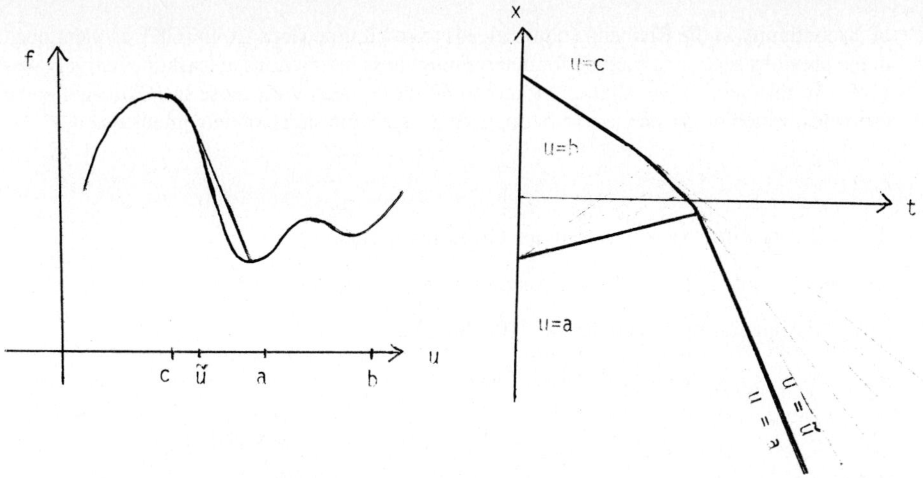

Fig. 3.2 The complete solution to the problem in Fig. 3.1 after the interaction of the waves. This is constructed using a time dependent version of the method of convex hull. The convex hull drawn to the left gives the solution for large time.

4. The scalar conservation law in one space dimension without convexity

Consider

$$u_t + f(u)_x = 0\,, \quad f \in C^2 \tag{4.1}$$

with the initial condition

$$u(x,0) = u_0\,(x)\;.$$

A classical method for approximating the solution is due to Godunov [G]. There the initial data is approximated by piecewise constant data on intervals of length Δx. At every jump discontinuity in the new initial data this leads to a Riemann problem. The Riemann problem resolves into a well known fan solution, as mentioned in the previous section. For Godunov's method one solves all the Riemann problems found by the constant states exactly. One takes this exact solution of the piecewise constant initial data up to time Δt, such that two neighbouring fans solutions don't interact. At this time level Δt the solution u is again approximated by a piecewise constant function, obtained by averaging over each cell x:

$$\bar{u} = \frac{1}{\Delta x}\int_{\Delta_x} u(x,\Delta t)\,dx\;.$$

Now we may proceed as before.

The inability to determine the interaction between waves propagating from the points of discontinuity in the Riemann solution leads to small time steps Δt. In [CK], as mentioned in the previous section, a method for determining these interactions at least qualitatively was given. In this section we shall try to combine these ideas with those in [O] to get some estimates, which might play a role in improving large time step Godunov methods.

4.1 A large time flux written as Godunov's flux

By integrating (4.1) over Δx X $[0,t]$ we obtain

$$0 = \int_0^t \int_{\Delta x} u_t + f(u)_x \, dx \, dt$$

$$= \int_{\Delta x} u(x,t) - u(x,0) \, dx + \int_0^t f(u(x_1,s)) - f(u(x_0,s)) \, ds .$$

Take the mean value of u at time t in Δx to be $\bar{u}$ and at time $t = 0$ to u_0. Then we obtain

$$\bar{u} = u_0 - \frac{1}{\Delta x} \int_0^t f(u(x_1,s)) - f(u(x_0,s)) \, ds . \tag{4.2}$$

Now suppose the initial data is piecewise constant with jumps at the points x_i, $i \in \mathbb{Z}$, and $x_i < x_{i+1}$, and values

$$u(x,0) = u_i , \quad x \in [x_i , x_{i+1}] , \quad i = ..., -k, ..., -1,0,1, ..., k, ...$$

For t small enough, the fans emanating from two neighbouring Riemann problems in this initial value problem don't intersect. The solution of a Riemann problem is a function of $(x - x_i)/t$ alone. Thus for t small u is constant along $x = x_i$, say $\tilde{u}_i$, and $\tilde{u}_i$ is only a function of u_{i-1} and u_i, the initial constant states to the left and right of x_i. Thus we may define h^G as (now for convenience we set $i = 0$)

$$h^G(u_{-1},u_0) = f(\tilde{u}_0) = \frac{1}{t} \int_0^t f(u(x_0,t)) \, dt , \quad t \text{ small}$$

and (4.2) becomes

$$\bar{u} = u_0 - \frac{t}{\Delta x} (h^G(u_{-1},u_0) - h^G(u_0, u_1)) , \qquad t \text{ small.} \tag{4.3}$$

Now let t become larger. Then u along $x = x_i$ becomes a function of several initial cells neighbouring x_i and we may define

$$h(u_{-k}, \dots, u_{k-1}) = \frac{1}{t} \int_0^t f(u(x_0,t)) \, dt , \tag{4.4}$$

with the property that if the initial constant values in all the cells are equal we obtain

$$h(u, ..., u) = f(u) .$$

Theorem 1 There exist a u_r in the convex hull of $\{u_{-k}, ..., u_{-1}\}$ and a u_L in the convex hull of $\{u_0, ..., u_{k-1}\}$ such that

$$h(u_{-k}, ..., u_{k-1}) = h^G (u_r,u_L) . \tag{4.5}$$

Proof: Using the notaion in (4.4) we may write (4.2) as

$$\bar{u} = u_0 - \frac{t}{\Delta x} (h(u_{-k+1}, \dots, u_k) - h(u_{-k}, \dots, u_{k-1})) .$$

Note that $\bar{u}$ is a nondecreasing function of all its variables since it is the average of the exact solution. Thus

$$\frac{\partial \bar{u}}{\partial u_k} = -h_k \geq 0 \Rightarrow h_k \leq 0$$

$$\frac{\partial \bar{u}}{\partial u_{k-1}} = -h_{k-1} + h_k \geq 0 \Rightarrow h_{k-1} \leq h_k$$

in general

$$h_1 \leq h_2 \leq \ldots \leq h_k \leq 0 \leq h_0 \leq \ldots \leq h_{-k+2} \leq h_{-k+1} .$$

We prove the theorem by induction. The case k = 1 is immediate. Suppose the claim is true for k. Then add a value of u_{k+1} on the right. Let $u \in [u_k, u_{k+1}]$. Now consider

$$\begin{aligned} g(u) &= h(u_{k+1}, u_k, \ldots, u_{-k+1}) - h(u, u, u_{k-1}, \ldots, u_{-k+1}) \\ &= (u_{k+1} - u)\, h_{k+1} + (u_k - u)\, h_k , \quad \text{by M.V.Th.} . \end{aligned}$$

We have

$$\begin{aligned} g(u_{k+1}) &= (u_k - u_{k+1})\, h_k \\ g(u_k) &= (u_{k+1} - u_k)\, h_{k+1} . \end{aligned}$$

Thus g(u) changes sign in (u_k, u_{k+1}) or vanishes at the endpoints. Thus there exist $\tilde{u}_k$ such that

$$h(u_{k+1}, u_k, \ldots, u_{-k+1}) = h(\tilde{u}_k, u_{k-1}, \ldots, u_{-k+1})$$

and by induction hypothesis the right hand side

$$= h^G(u_r, u_L)$$

for some u_r and u_L .

Next add a value u_{-k} on the left. For $u \in [u_{-k+1}, u_{-k}]$ we consider

$$\tilde{g}(u) = h(u_{k+1}, u_k, \ldots, u_{-k+1}, u_{-k}) - h(u_{k+1}, \ldots, u, u)$$

by the above result

$$= h(u_{k+1}, u_k, \ldots, u_{-k+1}, u_{-k}) - h(\tilde{u}_k, u_{k-1}, \ldots, u, u) .$$

By repeating the above argument we see that $\tilde{g}(u)$ vanishes in $[u_{-k}, u_{-k+1}]$. Thus

$$\begin{aligned} h(u_{k+1}, u_k, \ldots, u_{-k+1}, u_{-k}) &= h(\tilde{u}_k, u_{k-1}, \ldots, \tilde{u}_{-k+1}) \\ &= h^G(u_r, u_L) \end{aligned}$$

some u_r and some u_L, which proves the induction step, and finishes the proof.

To recapitulate, we have shown the following:

Consider

$$u_t + f(u)_x = 0$$

with initial condition

$$t = 0 \qquad \begin{matrix} u_{-2} & & u_{-1} & & u_0 & & u_1 & \\ & x_{-1} & & x_0 & & x_1 & \end{matrix} \qquad x . \tag{4.6}$$

Consider the exact solution $u(x_0,t)$. Then

$$\frac{1}{T} \int_0^T f(u(x_0,t))\, dt = h^G(u_r(T), u_L(T))$$

with some u_r in the convex hull of $\{u_0, \ldots, u_{k-1}\}$

and some u_L in the convex hull of $\{u_{-k}, \ldots, u_{-1}\}$,

and where the domain of dependence of $u(x_0,t)$ for $0 \leq t \leq T$ is included in $[u_{-k}, u_{k-1}]$.

4.2. A formula for the solution of the Riemann problem

The following theorem is mentioned in [O], but our proof is different from that given there. We use the notion of convex hull of a function f between u_L and u_r as used in section 3, denoted by $CH_{u_L}^{u_r} f(u)$ and illustrated in Fig. 3.1.

Theorem 2 The solution u to the Riemann problem of $u_t + f(u)_x = 0$ with initial constant states $u_L < u_r$ is given by

$$u = u\left(\frac{x}{t}\right) = \Big\{ \tilde{u} \in \{[u_L, u_r] \cap \{u : f(u) = CH_{u_L}^{u_r} f(u)\}\} \text{ such that } \tag{4.7}$$
$$CH_{u_L}^{u_r} f(\tilde{u}) - \frac{x}{t}\tilde{u} \text{ is the minimum} \Big\} .$$

Proof: Minimizing $CH_{u_L}^{u_r} f(u) - \frac{x}{t} u$ means that the slope of $CHf(u) - \frac{x}{t} u$ is horizontal at the minimum value of $\tilde{u}$. Hence

$$\frac{d}{du}\left(CHf(u) - \frac{x}{t} u\right)\Big|_{u=\tilde{u}} = 0 ,$$

which implies $f'(\tilde{u}) = \frac{x}{t}$.

This is the definition of a characteristic, i.e. the value of u along a ray through the origin with slope $\frac{x}{t}$ is $\tilde{u}$. Notice that in case that there are two such minima in (4.7), then they are the left and right states bounding a contact discontinuity.

5. Some examples

Consider a special case $\mathcal{B}$ of flux functions as follows: let $f \in C^2$ be such that for all $u_L \in [-M,-1]$ and u_r $[1,M]$, $M > 1$, we have that $\min_{u \in u_L, u_r} \{CH_{u_L}^{u_r}(u) - su , -\varepsilon < s < \varepsilon\}$, remains unchanged. For an example of $f \in \mathcal{B}$ see Fig. 5.1.

As before consider

$$u_t + f(u)_x = 0, \quad \text{fix } f \in \mathcal{B} \tag{5.1}$$

with initial data piecewise constant with jumps as in (4.6), and

$$\begin{aligned} u_i &\in (-M,-1), & i &\in \{-k, ..., -1\} \\ u_i &\in (1,M), & i &\in \{0, ..., k\} . \end{aligned} \tag{5.2}$$

Then by theorem 1 one obtains

$$\frac{1}{T}\int_0^T f(u(x_0,t))\, dt = h^G(u_r, u_L)$$

for some $u_L \in [-M,-1]$
$u_r \in [1,M]$.

By theorem 2 we find that $h^G(u_r,u_L) = f(u_{min})$ where $u_{min} \in [-1,1]$ and u_{min} is the absolute minimum of f in $[-M,M]$. Note that by the choice of f, u_{min} is always the same, regardless of u_i in (5.2).

Now we make use of the assumption that not only CHf(u) always attains a minimum in (–1,1), but also CHf(u) – su, $-\varepsilon < s < \varepsilon$. By the change of variable $y = x - st$ equation (5.1) becomes

$$u_t + (f(u) - su)_y = 0\,. \tag{5.4}$$

Consider (5.4) with initial condition (5.2) as before. We may again conclude that

$$\frac{1}{T}\int_0^T \{f(u(st,t)) - su\}\,dt = f(u_{min}) - s\,u_{min}\,,$$

where u_{min} is the absolute minimum in [–M,M] of f-su.

Thus for (5.1) and initial data (5.2) we conclude that a fan wave dominates the solution for all time, see Fig. 5.1.

We may extend this result to an initial value problem to (5.1) with piecewise continuous initial data:

$$u(x,0) = \begin{cases} u_1(x) \text{ such that } -M < f_1 < 1, & x \in [x_{-k}, x_0] \\ u_2(x) \text{ such that } 1 < f_2 < M, & x \in [x_0, x_k] \\ f_1(x_{-k}) & x < x_{-k} \\ f_s(x_k) & x > x_k \end{cases} \tag{5.5}$$

with u_1 and u_2 continuous functions, for an example see Fig. 5.2.

Now approximate the initial data u_0 in (5.5) by piecewise constant data on intervals of length Δx. For this data the above conclusion holds. Since the flux function is in class $\mathcal{B}$, regardless of the size of Δx in the approximation of u_0 we obtain the same value of u along a fixed ray $y = x_0 - st$, $-\varepsilon < s < \varepsilon$. Thus when passing to the limit $\Delta x \to 0$, we find that the solution u to (5.1), (5.5) is also dominated by a fan wave for all time which depends only on the choice of f.

Finally we give a two dimensional example. Consider

$$u_t + f(u)_x + f(u)_y = 0 \qquad f \in \mathcal{B} \tag{5.6}$$

with initial condition

$$u(x,y,0) = \begin{cases} \varepsilon, & y > -x \\ -\varepsilon, & y < -x \end{cases} \tag{5.7}$$

This is a one dimensional Riemann problem with a fan as a solution. By the above examples we may now give a special perturbation of the constant states (5.7) s.th. the fan wave remains unchanged, but the rest of the solution changes. Let u(x,y,0) be constant on rays through the origin, i.e.

$$u(x,y,0) = u_0(\theta),\ \theta \in [0,2\pi]$$

$$u_0(\theta) = \begin{cases} f_1(\theta) \text{ such that } 1 < f(\theta) < M, & \theta \in \left[-\frac{\pi}{4}, \frac{3\pi}{4}\right] \\ f_2(\theta) \text{ such that } -M < f(\theta) < -1, & \theta \in \left[\frac{3\pi}{4}, \frac{7\pi}{4}\right] \end{cases}$$

with f_1 and f_2 continuous functions.

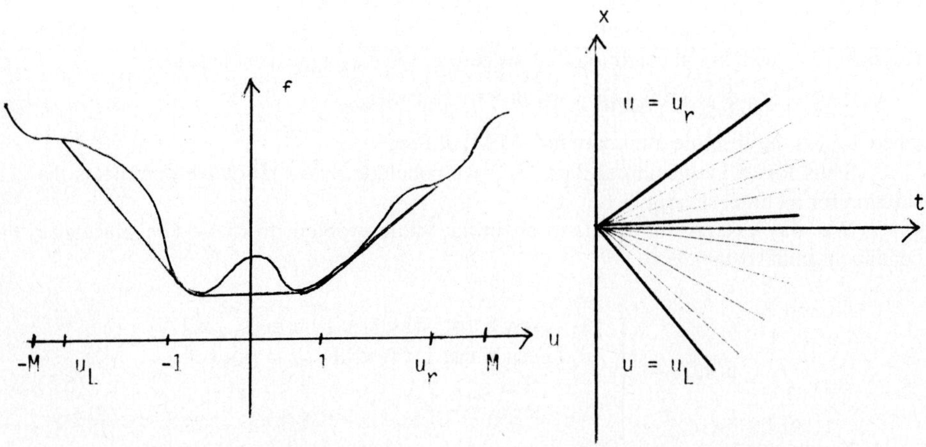

Fig. 5.1 An example of a flux function in class $\mathcal{B}$ together with a convex hull on the left. On the right the fan solution of the corresponding Riemann problem.

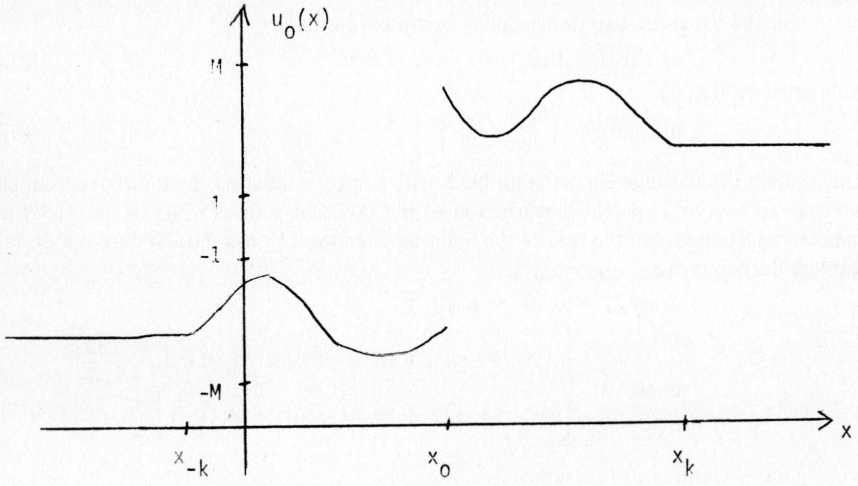

Fig. 5.2 Initial data for which the solution to

$$u_t + f(u)_x = 0\,, \qquad f \in \mathcal{B}$$

will contain the fan drawn in Fig. 5.1.

References

[CK] Chang, Klingenberg, "The Riemann Problem of a Scalar Conservation Law in Two Space Dimensions", preprint

[CS] Conway, Smoller, "Global Solutions to the Cauchy Problem for Quasilinear First Order Equations in Several Space Variables", Comm. Pure Appl. Math., 19 (1966)

[GK] Glimm, Klingenberg, McByran, Plohr, Sharp, Yaniv, "Front Tracking and Two Dimensional Riemann Problems", Adv. Appl. Math., 6 (1985)

[HK] Hsiao, Klingenberg, "The Construction of the Solution to a Non-convex Two Dimensional Riemann Problem", preprint

[G] Godunov, "A Finite Difference Method of the Numerical Computation of discontinuous Solutions of Equations of Fluid Mechanics", Mat. Sb., 47 (1959)

[K] Kruzkov, "First Order Quasilinear Equations in Several Independent Variables", Mat. Sb., 123 (1970)

[O] Osher, "Riemann Solvers, The Entropy Condition, and Difference Approximations", Siam J. Num. Anal., 21 (1984)

[W] Wagner, "The Riemann Problem in Two Space Dimensions for a Single Conservation Law", J. Math. Anal., 14 (1984)

Upwind Schemes for the Navier-Stokes Equations

Barry Koren
Centre for Mathematics and Computer Science
P.O. Box 4079, 1009 AB Amsterdam, The Netherlands

SUMMARY

A discretization method is presented for the full, steady, compressible Navier-Stokes equations. The method makes use of quadrilateral finite volumes and consists of an upwind discretization of the convective part and a central discretization of the diffusive part. In the present paper, the emphasis lies on the discretization of the convective part.

The applied solution method directly solves the steady equations by means of a Newton method, which requires the discretization to be continuously differentiable. For two upwind schemes which satisfy this requirement (Osher's and van Leer's scheme), results of a quantitative error analysis are presented. Osher's scheme appears to be more and more accurate than van Leer's scheme with increasing Reynolds number. A suitable higher-order accurate discretization of convection is chosen. Based on this higher-order scheme, a new limiter is constructed. Further, for van Leer's scheme, a solid wall - boundary condition treatment is proposed, which ensures a continuous transition from the Navier-Stokes flow regime to the Euler flow regime.

Numerical results are presented for a subsonic flat plate flow and a supersonic flat plate flow with oblique shock wave - boundary layer interaction. The results obtained agree with the predictions made.

Useful properties of the discretization method are that it allows an easy check of false diffusion and that it needs no tuning of parameters.

1980 Mathematics Subject Classification: 65N05, 65N30, 76N05, 76N10.
Key Words and Phrases: upwind schemes, Navier-Stokes equations
Note: This work was supported by the European Space Agency (ESA), via Avions Marcel Dassault - Bréguet Aviation (AMD-BA).

1. INTRODUCTION

1.1. Navier-Stokes equations

The equations considered are the full, steady, 2D, compressible Navier-Stokes equations

$$\frac{\partial f(q)}{\partial x}+\frac{\partial g(q)}{\partial y}-\frac{1}{Re}\{\frac{\partial r(q)}{\partial x}+\frac{\partial s(q)}{\partial y}\} = 0, \tag{1.1}$$

with $f(q)$ and $g(q)$ the convective flux vectors, Re the Reynolds number, and $r(q)$ and $s(q)$ the diffusive flux vectors. As state vector we consider the conservative vector $q=(\rho,\rho u,\rho v,\rho e)^T$, with for the total energy e the perfect gas relation $e=p/(\rho(\gamma-1))+\frac{1}{2}(u^2+v^2)$. The primitive flow quantities used are density ρ, pressure p, and the velocity components u and v. The ratio of specific heats γ is assumed to be constant. The convective flux vectors are defined by

$$f(q) = \begin{bmatrix}\rho u\\ \rho u^2+p\\ \rho uv\\ \rho u(e+p/\rho)\end{bmatrix},\quad g(q)=\begin{bmatrix}\rho v\\ \rho vu\\ \rho v^2+p\\ \rho v(e+p/\rho)\end{bmatrix}, \tag{1.2}$$

and the diffusive flux vectors by

$$r(q) = \begin{bmatrix}0\\ \tau_{xx}\\ \tau_{xy}\\ \tau_{xx}u+\tau_{xy}v+\frac{1}{\gamma-1}\frac{1}{Pr}\frac{\partial(c^2)}{\partial x}\end{bmatrix},\quad s(q) = \begin{bmatrix}0\\ \tau_{xy}\\ \tau_{yy}\\ \tau_{yy}v+\tau_{xy}u+\frac{1}{\gamma-1}\frac{1}{Pr}\frac{\partial(c^2)}{\partial y}\end{bmatrix}, \tag{1.3}$$

with Pr the Prandtl number, c the speed of sound (for a perfect gas: $c=\sqrt{\gamma p/\rho}$) and with τ_{xx}, τ_{xy} and τ_{yy} the viscous stresses. Assuming the diffusion coefficients to be constant and Stokes' hypothesis to hold, the viscous stresses are

$$\tau_{xx} = \frac{4}{3}\frac{\partial u}{\partial x} - \frac{2}{3}\frac{\partial v}{\partial y}, \tag{1.4a}$$

$$\tau_{xy} = \frac{\partial u}{\partial y} + \frac{\partial v}{\partial x}, \tag{1.4b}$$

$$\tau_{yy} = \frac{4}{3}\frac{\partial v}{\partial y} - \frac{2}{3}\frac{\partial u}{\partial x}. \tag{1.4c}$$

Here, we present a discretization method which allows an accurate (and efficient) computation of (steady) high-Reynolds number flows up to and including the Euler flow regime. The challenge in developing such a method is to find a discretization of the convective part which is accurate not only for typical Euler flows, but also for typical Navier-Stokes flows, like boundary layer flows. Finding a discretization for the diffusive part which satisfies the same requirements, is thought to be easy.

1.2. Discretization method

To still allow Euler flow solutions with discontinuities, the equations are discretized in integral form. A straightforward and simple discretization of the integral form is obtained by subdividing the integration region Ω into quadrilateral finite volumes $\Omega_{i,j}$ and by requiring that the conservation laws hold for each finite volume separately:

$$\oint_{\partial\Omega_{i,j}} (f(q)n_x + g(q)n_y)ds - \frac{1}{Re}\oint_{\partial\Omega_{i,j}} (r(q)n_x + s(q)n_y)ds = 0, \ \forall i,j \tag{1.5}$$

This discretization requires an evaluation of convective and diffusive fluxes at each volume wall.

1.2.1. Evaluation of convective fluxes. Based on experience with the Euler equations (see [5] for an overview), for the evaluation of the convective fluxes we prefer an upwind approach, following the Godunov principle [2]. So, along each finite volume wall, the convective flux is assumed to be constant and to be determined by a uniformly constant left and right state only. For the 1D Riemann problem thus obtained, an approximate Riemann solver is applied. The choice of the left and right state, to be used as input for the approximate Riemann solver, determines the accuracy of the convective discretization. First-order accuracy is simply obtained by taking the left and right state equal to that in the corresponding adjacent volume [6]. Higher-order accuracy is obtained by applying low-degree piecewise polynomial state interpolation (MUSCL-approach), using two or three adjacent volume states for the left and right state separately [4]. For this flux evaluation, we make use of the rotational invariance of the Navier-Stokes equations in order to reduce the number of these evaluations per finite volume wall from two to one. A more detailed discussion of the discretization of the convective part is given in section 2.

1.2.2. Evaluation of diffusive fluxes. For the evaluation of the diffusive fluxes, it is necessary to compute ∇u, ∇v and ∇c^2 at each volume wall. To compute for instance $(\nabla u)_{i+\frac{1}{2},j}$, where $i+\frac{1}{2}$ refers to the volume wall separating $\Omega_{i,j}$ and $\Omega_{i+1,j}$, we use Gauss' theorem

$$\nabla u_{i+\frac{1}{2},j} = \frac{1}{A_{i+\frac{1}{2},j}} \oint_{\partial\Omega_{i+\frac{1}{2},j}} u\mathbf{n}ds, \tag{1.6}$$

with $\partial\Omega_{i+\frac{1}{2},j}$ the boundary and $A_{i+\frac{1}{2},j}$ the area of a shifted quadrilateral finite volume $\Omega_{i+\frac{1}{2},j}$ which vertices $\mathbf{z}=(x,y)^T$ are defined by

$$\mathbf{z}_{i,j\pm\frac{1}{2}} = \frac{1}{2}(\mathbf{z}_{i-\frac{1}{2},j\pm\frac{1}{2}} + \mathbf{z}_{i+\frac{1}{2},j\pm\frac{1}{2}}), \tag{1.7}$$

and a similar expression for $\mathbf{z}_{i+1,j\pm\frac{1}{2}}$. The line integral in (1.6) is approximated by

$$\begin{aligned} \oint_{\partial\Omega_{i+\frac{1}{2},j}} u\mathbf{n}ds = \; & u_{i+1,j} && (\mathbf{z}_{i+1,j+\frac{1}{2}} - \mathbf{z}_{i+1,j-\frac{1}{2}}) + \\ & u_{i+\frac{1}{2},j+\frac{1}{2}} && (\mathbf{z}_{i,j+\frac{1}{2}} - \mathbf{z}_{i+1,j+\frac{1}{2}}) + \\ & u_{i,j} && (\mathbf{z}_{i,j-\frac{1}{2}} - \mathbf{z}_{i,j+\frac{1}{2}}) + \\ & u_{i+\frac{1}{2},j-\frac{1}{2}} && (\mathbf{z}_{i+1,j-\frac{1}{2}} - \mathbf{z}_{i,j-\frac{1}{2}}), \end{aligned} \tag{1.8}$$

with for $u_{i+\frac{1}{2},j\pm\frac{1}{2}}$ the central expression

$$u_{i+\frac{1}{2},j\pm\frac{1}{2}} = \frac{1}{4}(u_{i,j} + u_{i,j\pm1} + u_{i+1,j} + u_{i+1,j\pm1}). \tag{1.9}$$

Similar expressions are used for the other gradients and other walls. For sufficiently smooth grids this central diffusive flux computation is second-order accurate. Notice that by using central expressions, as (1.9), the directional dependence coming from the cross derivative terms is neglected. No significant gain in solution quality is expected from a biased approach as proposed in [1]. Given the fact that the present diffusive flux evaluation is rather cheap, here, use of rotational invariance is hardly advantageous.

1.3. Solution method

For a detailed description of the solution method, we refer to [7]. Here we give a brief summary.

For the nonlinear system of first-order discretized equations, symmetric point Gauss-Seidel relaxation is used. In here, one or more (exact) Newton steps are used for the collective updates of the four state vector components in each finite volume. Nonlinear multigrid is applied as an acceleration technique, the process is started by nested iteration.

For the higher-order accurate operator the same method leads to poor convergence or even divergence. As a remedy, we use iterative defect correction as an outer iteration for nonlinear multigrid applied, again, to the first-order discretized equations.

The application of the (exact) Newton method requires the convective and diffusive fluxes to be continuously differentiable (The diffusive fluxes as described in the previous section already fulfil this requirement.)

2. DISCRETIZATION OF CONVECTIVE PART

2.1. Approximate Riemann solver

As approximate Riemann solver for the Euler equations, we preferred Osher's scheme [12]. Reasons for this preference were: (i) its continuous differentiability, and (ii) its consistent treatment of boundary conditions. Here, the question arises whether it is still a good choice to use Osher's scheme when typical Navier-Stokes features such as shear, separation and heat conduction also have to be resolved. We should make a choice again.

Since continuous differentiability is an absolute requirement for the success of our solution method, and since we know no other approximate Riemann solvers with this property than Osher's and van Leer's [9], our choice is confined to these two only. So far, van Leer's scheme is more widespread in the field of Navier-Stokes than Osher's scheme [13, 14, 18]. Probably, the main reason for this is its greater conceptual and operational simplicity appealing from its first publications. However, recent publications on Osher's scheme, such as [6, 16], may help to reduce this difference.

With as next requirement the accurate modelling of physical diffusion, in fact, the definite choice can be made already. In [9], van Leer stated already that his flux vector splitter cannot preserve a steady contact discontinuity. Since a discrete shear layer may be interpreted as a layer of contact discontinuities, doubt rose already about the suitability of van Leer's scheme for Navier-Stokes codes. Recently, this doubt was confirmed in [11] where van Leer et al made a qualitative analysis (supplemented with numerical experiments) for various upwind schemes. There, Osher's scheme turned out to be better than van Leer's scheme, in particular for the resolution of boundary layer flows.

To shed some light on how large this difference in quality is, here results of a quantitative error analysis are presented for both Osher's and van Leer's scheme. The analysis is confined to the steady, 2D, isentropic Euler equations for a perfect gas with $\gamma=1$:

$$\frac{\partial f(q)}{\partial x} + \frac{\partial g(q)}{\partial y} = 0, \tag{2.1}$$

with

$$f(q) = \begin{bmatrix} \rho u \\ \rho(u^2+c^2) \\ \rho uv \end{bmatrix}, \quad g(q) = \begin{bmatrix} \rho v \\ \rho vu \\ \rho(v^2+c^2) \end{bmatrix}, \tag{2.2}$$

where c is constant. (The choice of 2D equations allows us to consider a boundary layer flow in the analysis.) In [8], for both upwind schemes, the system of modified equations is derived, considering (i) a first-order accurate, square finite volume discretization, and (ii) a subsonic flow with u and v positive, and $\rho \approx$ constant. Neglecting the density variation, the systems of modified equations are, for Osher:

$$\frac{\partial f(q)}{\partial x}+\frac{\partial g(q)}{\partial y}-h\frac{\rho}{2c}\left[\frac{\partial}{\partial x}\begin{bmatrix}\frac{1}{2}\frac{\partial(u^2)}{\partial x}\\ (u^2+c^2)\frac{\partial u}{\partial x}\\ uv\frac{\partial u}{\partial x}+uc\frac{\partial v}{\partial x}\end{bmatrix}+\frac{\partial}{\partial y}\begin{bmatrix}\frac{1}{2}\frac{\partial(v^2)}{\partial y}\\ vu\frac{\partial v}{\partial y}+vc\frac{\partial u}{\partial y}\\ (v^2+c^2)\frac{\partial v}{\partial y}\end{bmatrix}\right]=O(h^2), \tag{2.3}$$

and for van Leer:

$$\frac{\partial f(q)}{\partial x}+\frac{\partial g(q)}{\partial y}-h\frac{\rho}{2c}\left[\frac{\partial}{\partial x}\begin{bmatrix}\frac{1}{2}\frac{\partial(u^2)}{\partial x}\\ 2c^2\frac{\partial u}{\partial x}\\ \frac{1}{2}\frac{\partial((u^2+c^2)v)}{\partial x}\end{bmatrix}+\frac{\partial}{\partial y}\begin{bmatrix}\frac{1}{2}\frac{\partial(v^2)}{\partial y}\\ \frac{1}{2}\frac{\partial((v^2+c^2)u)}{\partial y}\\ 2c^2\frac{\partial v}{\partial y}\end{bmatrix}\right]=O(h^2). \tag{2.4}$$

In both first-order error terms, a typical Navier-Stokes flow solution is substituted, which clearly shows the differences between both schemes. As flow, we consider an incompressible semi-infinite flat plate flow. For simplicity, for this we do not use the exact Blasius solution, but Lamb's approximate solution which reads

$$\begin{bmatrix}\rho\\ u\\ v\end{bmatrix}=\begin{bmatrix}P\\ U\sin(\frac{\pi}{2}\frac{\sqrt{Re/x}}{5}y)\\ \frac{y}{2x}u(x,y)+\frac{5}{\pi}\frac{U}{\sqrt{Rex}}\{\cos(\frac{\pi}{2}\frac{\sqrt{Re/x}}{5}y)-1\}\end{bmatrix}, \tag{2.5}$$

with P and U constant. Substituting the solution vector (2.5) into the error vector of both (2.3) and (2.4), considering the boundary layer edge

$$y=\delta(x)\equiv\frac{5}{\sqrt{Re/x}} \tag{2.6}$$

at $x=1$, and taking the ratio of absolute values of both error vectors, using $Re\gg1$ (which is our interest), we find

$$\frac{\text{error Osher}}{\text{error van Leer}}=\begin{bmatrix}1\\ (1-\frac{2}{\pi})\frac{5}{\sqrt{Re}}\frac{U}{C}\\ 1/2\end{bmatrix}, \tag{2.7}$$

where we write $c=C$. From (2.7) it appears that van Leer's scheme deteriorates compared to Osher's scheme for increasing Re. Assuming the reliability of (2.7) for rather small Re, it appears that already for $Re>\{5(1-2/\pi)U/C\}^2$, where $U/C<1$, Osher's scheme definitely is to be preferred above van Leer's scheme.

To ensure a continuous transition along a solid wall boundary from the Navier-Stokes flow regime to the Euler flow regime, for van Leer's scheme it will be necessary to impose only the Euler boundary condition to the convective part. So, for a non-permeable solid wall this means that one should only impose a zero normal velocity component to the convective part (though all boundary conditions to the diffusive part, i.e. a zero normal and tangential velocity component, and some temperature condition). By not imposing the no-slip and temperature boundary condition to the convective part, we avoid that it 'feels' the severe contact discontinuity in the realistic case of a boundary layer flow on a very coarse grid and an outer flow with M not small. Such a contact discontinuity will be erroneously spread by van Leer's scheme, and cause that there is some finite, rather low value of Re above which the solution is insensitive to Re-variation.

Osher's scheme can preserve a steady contact discontinuity as long as it is aligned with the grid. Application of (commonly used) body-fitted grids guarantees this alignment along solid walls. Therefore, with a body-fitted grid, Osher's scheme does not need the careful solid wall - boundary condition treatment as proposed for van Leer's scheme.

2.2. Higher-order accuracy

As mentioned in section 1.2.1, higher-order accuracy is obtained by applying low-degree piecewise polynomial functions through two or three adjacent volume states. The polynomials are given by van Leer's κ-scheme [10]

$$q^l_{i+\frac{1}{2},j} = q_{i,j} + \frac{1+\kappa}{4}(q_{i+1,j}-q_{i,j}) + \frac{1-\kappa}{4}(q_{i,j}-q_{i-1,j}), \tag{2.8a}$$

$$q^r_{i+\frac{1}{2},j} = q_{i+1,j} + \frac{1+\kappa}{4}(q_{i,j}-q_{i+1,j}) + \frac{1-\kappa}{4}(q_{i+1,j}-q_{i+2,j}), \tag{2.8b}$$

with $\kappa \in [-1,1]$. For $\kappa = -1,0,1$, we have the fully one-sided upwind, the Fromm and the central scheme, respectively.

The aim now is to optimize κ. For this purpose, we consider the scalar model equation

$$\frac{\partial u}{\partial x} + \frac{\partial u}{\partial y} - \epsilon\left(\frac{\partial^2 u}{\partial x^2} + \frac{\partial^2 u}{\partial x \partial y} + \frac{\partial^2 u}{\partial y^2}\right) = 0. \tag{2.9}$$

On a square grid, a finite volume discretization which uses the κ-scheme for convection and the central scheme for diffusion, yields as modified equation

$$\begin{aligned}&\frac{\partial u}{\partial x} + \frac{\partial u}{\partial y} - \epsilon\left(\frac{\partial^2 u}{\partial x^2} + \frac{\partial^2 u}{\partial x \partial y} + \frac{\partial^2 u}{\partial y^2}\right) + \\ &+ h^2\left\{\frac{\kappa-1/3}{4}\left(\frac{\partial^3 u}{\partial x^3} + \frac{\partial^3 u}{\partial y^3}\right) - \frac{\epsilon}{12}\left(\frac{\partial^4 u}{\partial x^4} + 2\frac{\partial^4 u}{\partial x^3 \partial y} + 2\frac{\partial^4 u}{\partial x \partial y^3} + \frac{\partial^4 u}{\partial y^4}\right)\right\} = O(h^3).\end{aligned} \tag{2.10}$$

As optimal value for κ, we define: the value that gives the highest possible accuracy, i.e. third-order accuracy in this case. Assuming the reliability of the underlying Taylor series expansion, from (2.10), we find for this value

$$\kappa = \frac{1}{3}\left\{1+\epsilon\left(\frac{\partial^4 u}{\partial x^4} + 2\frac{\partial^4 u}{\partial x^3 \partial y} + 2\frac{\partial^4 u}{\partial x \partial y^3} + \frac{\partial^4 u}{\partial y^4}\right)\Big/\left(\frac{\partial^3 u}{\partial x^3} + \frac{\partial^3 u}{\partial y^3}\right)\right\}. \tag{2.11}$$

Since convection dominated problems, problems with $\epsilon \ll 1$, are our interest, we assume the above diffusion-dependence of κ to be negligible, which simply leads to $\kappa = 1/3$.

2.3. Monotonicity

To preserve monotonicity, we construct a limiter which is based on the $\kappa = 1/3$-scheme. For this, we use the monotonicity theory of Spekreijse [15], an extension of Sweby's theory [17], allowing more freedom in the limiter construction.

For the limited, higher-order, left and right state components, we write

$$q^{l(k)}_{i+\frac{1}{2},j} = q^{(k)}_{i,j} + \frac{1}{2}\phi(R^{(k)}_{i,j})(q^{(k)}_{i,j} - q^{(k)}_{i-1,j}), \tag{2.12a}$$

$$q^{r(k)}_{i+\frac{1}{2},j} = q^{(k)}_{i+1,j} + \frac{1}{2}\phi(1/R^{(k)}_{i+1,j})(q^{(k)}_{i+1,j} - q^{(k)}_{i+2,j}), \tag{2.12b}$$

with $k = 1,2,3,4$, $\phi(R)$ the limiter, and

$$R^{(k)}_{i,j} = \frac{q^{(k)}_{i+1,j} - q^{(k)}_{i,j}}{q^{(k)}_{i,j} - q^{(k)}_{i-1,j}}. \tag{2.13}$$

The limited $\kappa = 1/3$-scheme can be written in the one-sided form (2.12a-b) as

$$q^{l(k)}_{i+\frac{1}{2},j} = q^{(k)}_{i,j} + \frac{1}{2}\xi(R^{(k)}_{i,j})\left(\frac{1}{3} + \frac{2}{3}R^{(k)}_{i,j}\right)(q^{(k)}_{i,j} - q^{(k)}_{i-1,j}), \tag{2.14a}$$

$$q^{r(k)}_{i+\frac{1}{2},j} = q^{(k)}_{i+1,j} + \frac{1}{2}\xi(1/R^{(k)}_{i+1,j})\left(\frac{1}{3} + \frac{2}{3}/R^{(k)}_{i+1,j}\right)(q^{(k)}_{i+1,j} - q^{(k)}_{i+2,j}). \tag{2.14b}$$

Notice that for $\xi(R) = 1$ we have the (non-limited) $\kappa = 1/3$-scheme, and that $\xi(R)$ defines the limiter $\phi(R)$ by

$$\phi(R) = \xi(R)\left(\frac{1}{3} + \frac{2}{3}R\right). \tag{2.15}$$

General requirements to be fulfilled by $\xi(R)$ are: $\xi(1)=1$ to preserve higher-order accuracy, and: $\xi(0)=0$ and boundedness for large $|R|$ to preserve monotonicity. For the latter, we require that $\lim_{R\to\pm\infty} \xi(R)(\frac{1}{3}+\frac{2}{3}R)=1$. To make the limiter now a $\kappa=1/3$-limiter, we require that $\xi'(1)=0$. (This last requirement makes the limiter tangential to the $\kappa=1/3$-scheme in the monotonicity region [15].) Imposing these five requirements to the general form

$$\xi(R) = \frac{\alpha_1 R^2 + \alpha_2 R + \alpha_3}{\alpha_4 R^2 + \alpha_5 R + 1}, \tag{2.16}$$

we find with (2.15)

$$\phi(R) = \frac{2R^2+R}{2R^2-R+2}. \tag{2.17}$$

3. NUMERICAL RESULTS

3.1. Flow problems

To evaluate the discretization method, the following flow problems are considered: (i) a subsonic flat plate flow with $M=0.5$ and Re ranging from 10^2 up to 10^{100}, and (ii) a supersonic flat plate flow with oblique shock wave - boundary layer interaction at $M=2$, $Re=2.96\,10^5$. The latter problem stems from [3].

For the subsonic flow problem, the Blasius solution is used as a reference. The grids used for this flow problem are all composed of square finite volumes. As coarsest grid in all multigrid computations, we use a 4×2-grid.

For the supersonic flow problem, the experimental results from [3] are used as a reference. Here, in all multigrid computations a 5×2-grid is applied as coarsest grid. The grid was optimized for convection by introducing a stretching in flow direction, and in particular by aligning it with the impinging shock wave. A grid adaptation for diffusion was realized by introducing a stretching in crossflow direction.

For both flow problems, we use $\gamma=1.4$ and $Pr=0.71$. For further details about the implementation of both problems, we refer to [8].

3.2. Osher versus van Leer

To show at first the benefit of the solid wall - boundary condition treatment as proposed for van Leer's scheme in section 2.1, we consider the subsonic flat plate flow at $Re=10^{100}$. For both Osher's and van Leer's scheme we compute the flow on a 64×32-grid, using the first-order accurate discretization and imposing to the convective part, successively: (i) non-permeability, no-slip and no-heat-transfer, and - carefully - (ii) non-permeability only. The numerical results obtained are given in fig. 3.1. For the case with all Navier-Stokes boundary conditions imposed, it appears that van Leer's scheme severely thickens the thin layer, whereas Osher's scheme preserves it. With the careful approach, both schemes preserve the layer.

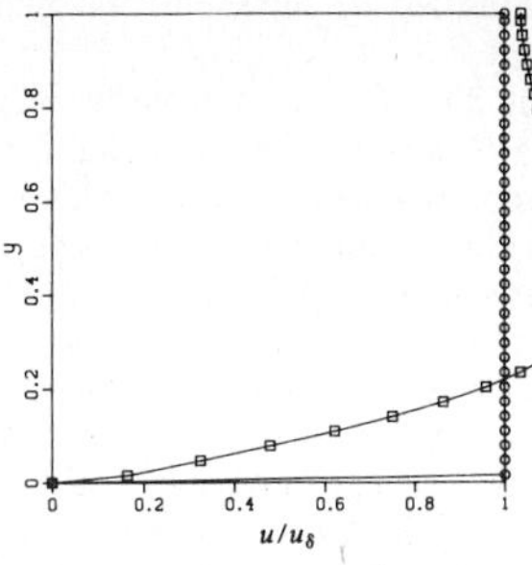

a. $u=v=0,\ \dfrac{\partial c}{\partial y}=0.$

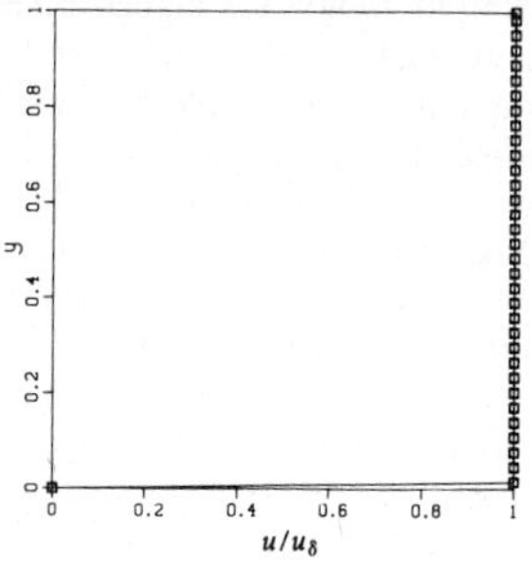

b. $v=0$ only.

Fig. 3.1. Velocity profiles at $x=0$ for the subsonic flat plate flow at $Re=10^{100}$ and $h=\frac{1}{32}$, for two solid wall - boundary condition treatments (○: Osher, □: van Leer).

Using the careful boundary condition treatment, for both schemes we perform an experiment with *h*- and *Re*-variation, using again the first-order accurate discretization. Numerical results obtained are given in fig. 3.2. The results show the superiority of Osher's scheme, in particular for high mesh Reynolds numbers. The deterioration of van Leer's scheme with respect to Osher's scheme which occurs in fig. 3.2b for increasing *Re*, is in agreement with the analytical results presented in section 2.1.

All numerical results presented hereafter were obtained with Osher's scheme only.

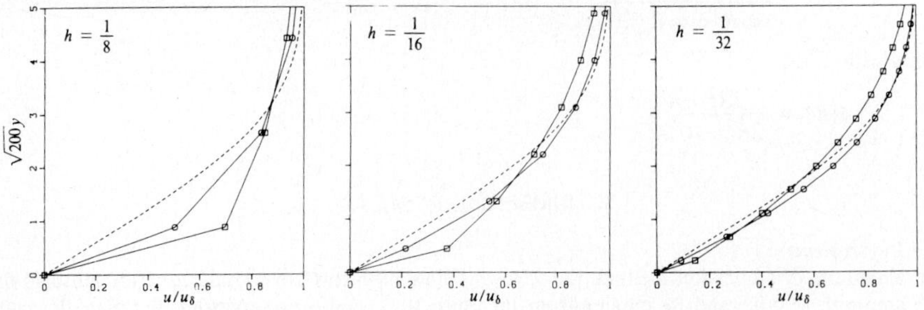

a. *h*-variation with $Re = 100$ (○ : Osher, □ : van Leer).

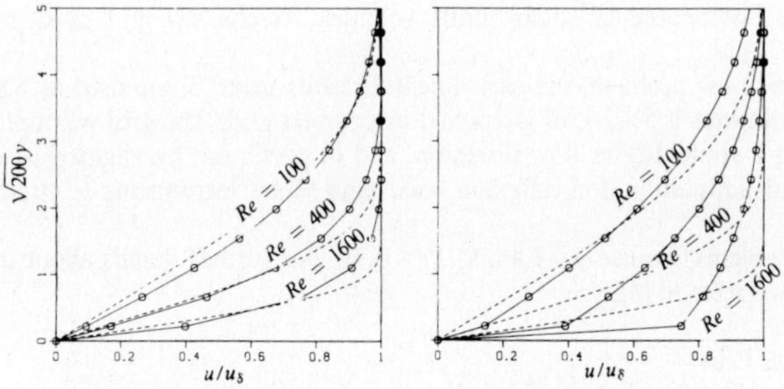

b. *Re*-variation with $h = 1/32$ (left: Osher, right: van Leer).

Fig. 3.2. Velocity profiles at $x = 0$ for the subsonic flat plate flow (-----: Blasius solution).

3.3. Monotone higher-order accuracy

To evaluate our monotone higher-order accurate scheme, we consider the supersonic flat plate flow. At first, we evaluate monotonicity, and next higher-order accuracy.

For monotonicity, we compute the Euler flow solution on the 80×32-grid given in fig. 3.3a, using the $\kappa = 1/3$-scheme with and without limiter. Numerical results obtained are given in fig. 3.3b. The results clearly show that the limiter does what it is supposed to do: making the solution monotone.

For higher-order accuracy, we compute on the same grid the Navier-Stokes solution, using now the limited $\kappa = 1/3$-scheme and the first-order scheme. A comparison is made with the experimental results from [3]. The results, given in fig. 3.3c, clearly show the need for higher-order accuracy. The first-order accurate surface pressure distribution lacks the plateau in the pressure distribution, which indicates that its solution has no separation bubble (i.e. no separation and no re-attachment). In agreement with the experimental results, the limited higher-order accurate surface pressure distribution does have a separation bubble. The quantitative differences still existing between the limited higher-order and measured surface pressure distribution must be due to uncertain influences in both the experiment and the computation. (As far as the experiment is concerned, this might be crossflow influences, non-observed though influential turbulence, some slight heat transfer through the wall, and so on. Concerning the computation, this might be for instance the neglect of temperature dependence in the diffusion coefficients.)

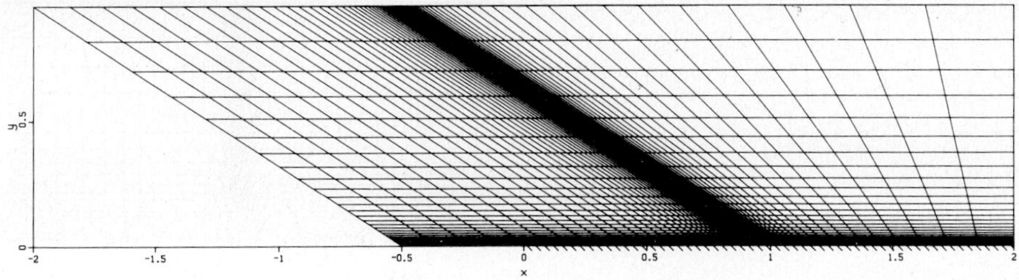

a. Finest grid applied (80 × 32).

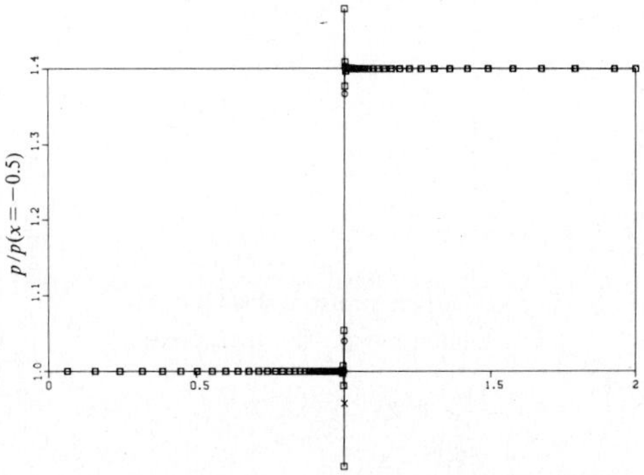

b. Inviscid surface pressure distributions
(○ : limited $\kappa=\frac{1}{3}$, □ : non-limited $\kappa=\frac{1}{3}$).

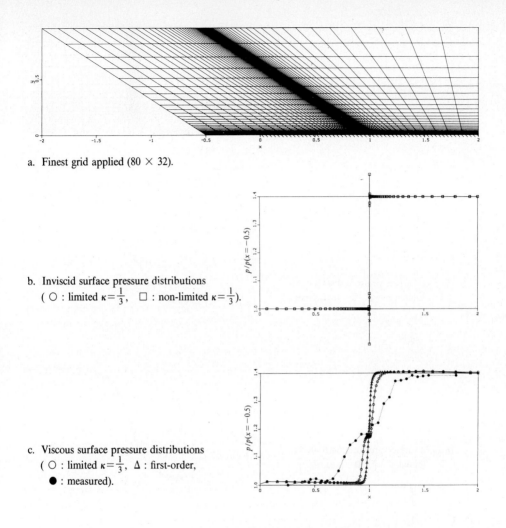

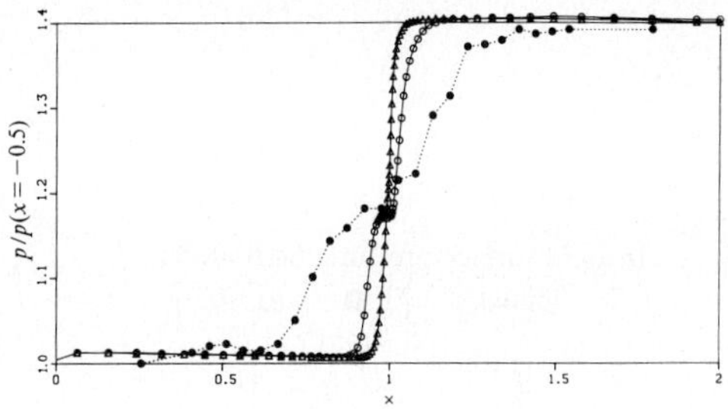

c. Viscous surface pressure distributions
(○ : limited $\kappa=\frac{1}{3}$, Δ : first-order,
● : measured).

Fig. 3.3. Results supersonic flat plate flow on oblique grid.

3.4. False diffusion

It should be noticed that by presenting for the supersonic flat plate flow, besides the viscous solution (obtained with the limited $\kappa=1/3$-scheme) also the corresponding inviscid solution, insight was given about the amount of false diffusion present in the viscous solution. The fact that the present method is hybrid in the sense that it can be used for both Navier-Stokes and Euler flows makes it easy to do this investigation. Omitting this investigation for the supersonic flat plate flow, when applying a commonly used rectangular grid, such as for instance the 80×32-grid shown in fig. 3.4a, leads to a viscous surface pressure distribution which seems to be very close to the experimental data (fig. 3.4b). However, the corresponding inviscid distribution indicates that this good resemblance is absolutely fake (fig. 3.4c).

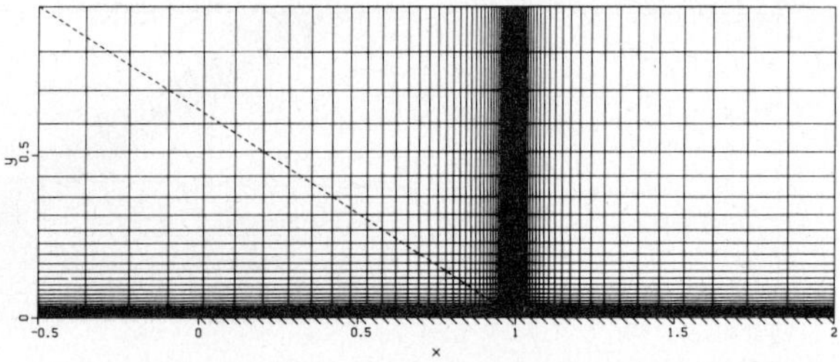

a. Finest grid applied (80 × 32).

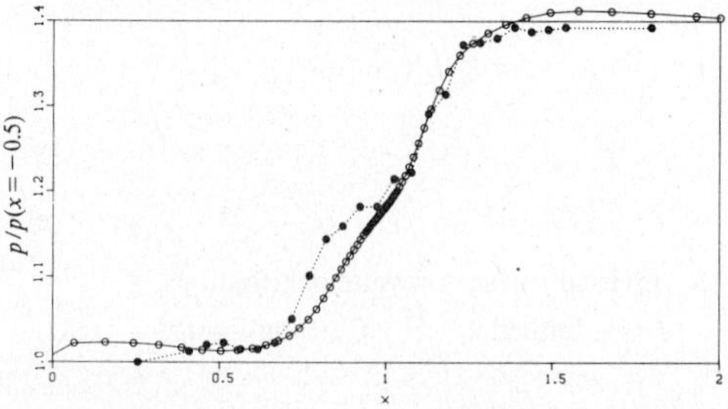

b. Viscous surface pressure distributions (O : limited $\kappa=\frac{1}{3}$, ● : measured).

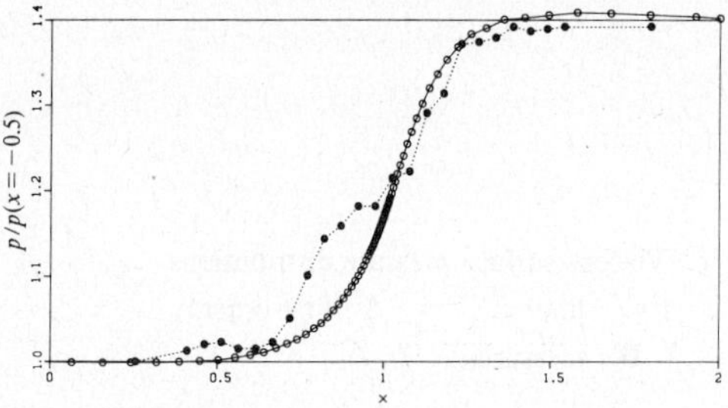

c. Inviscid surface pressure distributions (O : limited $\kappa=\frac{1}{3}$, ● : measured).

Fig. 3.4. Results supersonic flat plate flow on rectangular grid.

4. CONCLUSIONS

Theory and practice show that Osher's scheme leads to a more accurate resolution of boundary layers than van Leer's scheme. The difference in accuracy becomes larger with increasing Reynolds number. Already for rather low Reynolds numbers, the difference is such that, for engineering purposes, Osher's scheme is to be preferred above van Leer's scheme. An accidental circumstance is that Osher's scheme needs no special care in the application of solid wall - boundary conditions, whereas van Leer's scheme does. (To avoid, when still using van Leer's scheme, that there is some rather low Reynolds number above which the flow solutions are insensitive to Reynolds-variation, only the Euler boundary condition should be imposed to the convective part.)

It is important to investigate the reliability of any computed Navier-Stokes solution with respect to the numerical errors in the discretization of the convective part. The present code allows an easy check of false diffusion: the same code can be used for both viscous ($1/Re>0$) and inviscid ($1/Re=0$) flow computations.

The discretization method is parameter-free; it needs no tuning.

REFERENCES

1. S.R. CHAKRAVARTHY, K.Y. SZEMA, U.C. GOLDBERG, J.J. GORSKI and S. OSHER (1985). *Application of a New Class of High Accuracy TVD Schemes to the Navier-Stokes Equations.* AIAA-paper 85-0165.
2. S.K. GODUNOV (1959). *Finite Difference Method for Numerical Computation of Discontinuous Solutions of the Equations of Fluid Dynamics* (in Russian, also Cornell Aeronautical Lab. Transl.). Math. Sbornik 47, 272-306.
3. R.J. HAKKINEN, I. GREBER, L. TRILLING and S.S. ABARBANEL (1958). *The Interaction of an Oblique Shock Wave with a Laminar Boundary Layer.* NASA-memorandum 2-18-59 W.
4. P.W. HEMKER (1986). *Defect Correction and Higher Order Schemes for the Multi Grid Solution of the Steady Euler Equations.* Proceedings of the 2nd European Conference on Multigrid Methods (W. Hackbusch and U. Trottenberg, eds.), Cologne, 1985. Springer, Berlin.
5. P.W. HEMKER and B. KOREN (1988). *Defect Correction and Nonlinear Multigrid for the Steady Euler Equations.* Lecture Series on Computational Fluid Dynamics, von Karman Institute for Fluid Dynamics, Rhode-Saint-Genèse.
6. P.W. HEMKER and S.P. SPEKREIJSE (1986). *Multiple Grid and Osher's Scheme for the Efficient Solution of the Steady Euler Equations.* Appl. Num. Math. 2, 475-493.
7. B. KOREN (1988). *Multigrid and Defect Correction for the Steady Navier-Stokes Equations.* Proceedings of the 4th GAMM-Seminar Kiel on Robust Multi-Grid Methods, (W. Hackbusch, ed.), Kiel, 1988. Vieweg, Braunschweig (to appear).
8. B. KOREN (1988). *Upwind Discretization for the Steady Navier-Stokes Equations.* Report NM-R88xx, Centre for Mathematics and Computer Science, Amsterdam (to appear).
9. B. VAN LEER (1982). *Flux-Vector Splitting for the Euler Equations.* Proceedings of the 8th International Conference on Numerical Methods in Fluid Dynamics (E. Krause, ed.), Aachen, 1988. Springer, Berlin.
10. B. VAN LEER (1985). *Upwind-Difference Methods for Aerodynamic Problems governed by the Euler Equations.* Proceedings of the 15th AMS-SIAM Summer Seminar on Applied Mathematics (B.E. Engquist, S. Osher and R.C.J. Somerville, eds.), Scripps Institution of Oceanography, 1983. AMS, Rhode Island.
11. B. VAN LEER, J.L. THOMAS, P.L. ROE and R.W. NEWSOME (1987). *A Comparison of Numerical Flux Formulas for the Euler and Navier-Stokes Equations.* AIAA-paper 87-1104.
12. S. OSHER and F. SOLOMON (1982). *Upwind-Difference Schemes for Hyperbolic Systems of Conservation Laws.* Math. Comp. 38, 339-374.
13. W. SCHRÖDER and D. HÄNEL (1987). *An Unfactored Implicit Scheme with Multigrid Acceleration for the Solution of the Navier-Stokes Equations.* Computers and Fluids 15, 313-336.
14. G. SHAW and P. WESSELING (1986). *Multigrid Solution of the Compressible Navier-Stokes Equations on a Vector Computer.* Proceedings of the 10th International Conference on Numerical Methods in Fluid Dynamics (F.G. Zhuang and Y.L. Zhu, eds.), Beijing, 1986. Springer, Berlin.
15. S.P. SPEKREIJSE (1987). *Multigrid Solution of Monotone Second-Order Discretizations of Hyperbolic Conservation Laws.* Math. Comp. 49, 135-155.
16. S.P. SPEKREIJSE (1987). *Multigrid Solution of the Steady Euler Equations.* Ph.D.-thesis, Centre for Mathematics and Computer Science, Amsterdam.
17. P.K. SWEBY (1984). *High Resolution Schemes using Flux Limiters for Hyperbolic Conservation Laws.* SIAM J. Num. Anal. 21, 995-1011.
18. J.L. THOMAS and R.W. WALTERS (1985). *Upwind Relaxation Algorithms for the Navier-Stokes Equations.* AIAA-paper 86-1501.

NORMAL REFLECTION TRANSMISSION OF SHOCKS WAVES ON A PLANE INTERFACE BETWEEN TWO RUBBER-LIKE MEDIA

S. Kosiński

Technical University of Łódź, Institute of Construction Engineering I-32, Al. Politechniki 6, 93-590 ŁÓDŹ 40, POLAND

SUMMARY

Using a semi-inverse method proposed by Wright [1] a normal reflection and transmission of a finite elastic plane shock wave, at a plane interface of two rigidly coupled rubber-like elastic solids is examined. It is assumed that the material solids in front of the shock wave are unstrained and at rest. It is found that, depending on the material properties. the reflected wave is either a single simple wave or a shock wave; the transmitted wave is always a shock wave.

INTRODUCTION

The equations expressing balance of momentum for differentiable fields (stress and velocity) or in the case of the discontinuity surface Σ representing a shock wave are given by (1a,b), respectively.

$$T_{i\alpha,\alpha} = \rho_R \dot{u}_i, \qquad [\![T_{i\alpha}]\!] N_\alpha = -\rho_R V [\![u_i]\!], \tag{1a,b}$$

where $T_{i\alpha}$ - the first Piola-Kirchoff stress tensor, u_i - particle velocity, ρ_R - the density in the reference configuration B_R, V - shock wave speed along the normal, N_α - material unit normal to the discontinuity surface. The dot and comma notation signify partial differentiation with respect to the time and coordinates, respectively. The bold square brackets indicate the jump in the quantity enclosed across Σ. $\{x_\alpha\}$ and $\{x_i\}$ are the Cartesian coordinates of a particle in B_R and B, respectively. The jump conditions on Σ for deformation gradient and velocity are given by

$$[\![x_{i\alpha}]\!] = H_i N_\alpha, \qquad [\![u_i]\!] = -H_i V, \tag{2}$$

where H_i - amplitude vector of the jump.

The strength of the shock wave is defined by

$$m = \sqrt{H_i H_i}, \quad \text{and} \quad H_i = m d_i, \tag{3}$$

where d_i - vector of polarisation.

The equation of motion and the compatibility condition in the region of the simple wave are given by

$$(Q^*_{ij} - \rho_R U^2 \delta_{ij}) u'_j = 0, \quad U x'_{j\beta} + u'_j N_\beta = 0 \tag{4}$$

where $Q^*_{ij} = \tilde{Q}_{ij} - \tilde{Q}_{kj} n_k n_i$, $\tilde{Q}_{ij} = \sigma_{i\alpha j\beta} N_\alpha N_\beta$, $\sigma_{i\alpha j\beta} = \dfrac{\partial^2 \sigma}{\partial x_{i\alpha} \partial x_{j\beta}}$, σ - denotes the internal energy per unit mass in B_R, Q^*_{ij}, $\tilde{Q}_{ij}$ are the components of the acoustic tensor and the reduced acoustic tensor, respectively, the prime indicates differentiation with respect to λ - simple wave parameter, U denotes the normal speed of propagation. For expression $(4)_1$ to have a non zero solution in u'_j it is necessary that

$$\det (Q^*_{ij} - \rho_R U^2 \delta_{ij}) = 0, \qquad u'_j = k r_j, \tag{5}$$

where r_j is the right proper vector of the reduced acoustic tensor, k - - arbitrary parameter. Assuming $k = U$ in (5) we obtain

$$u'_j = U f(x_{i\alpha}) r_j, \qquad x'_j = -f(x_{i\alpha}) r_j N_\beta. \tag{6}$$

These are the ordinary differential equations for the deformation gradient and particle velocity in the region of the simple wave. They can be solved with the initial conditions taken from the region of constant state. The right proper vectors can be determined exact to an arbitrary scalar function of the deformation gradient $f(x_{i\alpha})$.

INCIDENT SHOCK WAVE

Consider an unbounded medium consisting of two elastic half-spaces of different material properties, joined along the plane $x_2 = 0$. Suppose that a plane shock wave of strength m_o, unit normal $\underset{\sim}{N}_o = (0,-1,0)$ and polarisation vector $\underset{\sim}{d}_o = (0,0,1)$ propagates in the half-space $x_2 > 0$ with speed V_o (Fig. 1). Such a wave has displacement component in the x_3 direction only. The material regions 0, $\hat{0}$ ahead of the incident shock wave are unstrained and at rest and are compatible with the interface conditions.

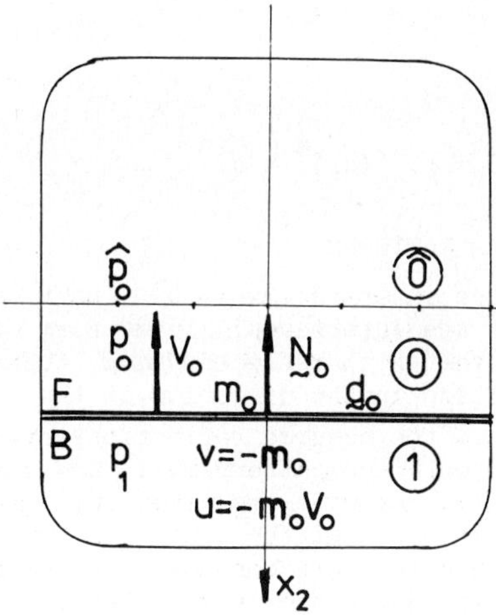

Fig.1. Incident shock wave

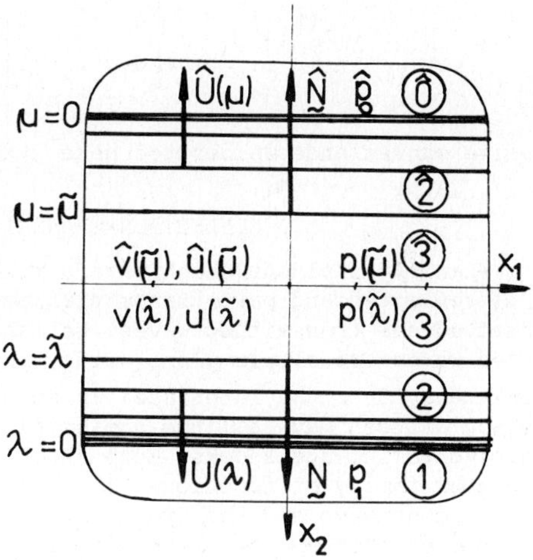

Fig.2. Assumed shock reflection-transmission pattern.

The jump conditions for deformation gradient and particle velocity are given by

$$[\![x_{32}]\!] = (x_{32})^B = -m_o, \qquad [\![u_3]\!] = (u_3)^B = -m_o V_o, \tag{7,8}$$

where $m_o = |\underset{\sim}{H}_o|$ - incident shock wave strength.

Both material solids are isotropic incompressible and are defined by the constitutive equations proposed by Zahorski [3].

$$W(I_1, I_2) = \rho_R \sigma(I_1, I_2) = C_1(I_1 - 3) + C_2(I_2 - 3) + C_3(I_1^2 - 9), \tag{9}$$

where I_1, I_2 - invariants of the left Cauchy-Green strain tensor $\underset{\sim}{B}$ and C_1, C_2, C_3 - elastic constants.

Using the momentum conservatioon law $(1)_1$ we obtain an equation for the shock wave speed

$$V^2 = \frac{c^2}{\rho_R}(1 + \eta m_o), \tag{10}$$

where $c^2 = 2(C_1 + C_2 + 6C_3) > 0; \quad \eta = 4C_3 c^{-2} > 0.$

For real elastic materials constants η and c are positive. The squared speed of the shock wave is a quadratic function of the shock wave strength. If the wave propagates in the medium which is unstrained and at rest, the results are independent of the direction of propagation and polarisation. The state behind the propagating shock wave (region 1) is now completely specified by the shock strength m_o. Eqs. (7,8) determine the deformation gradient and the particle velocity

$$(x_{i\alpha})^B = \begin{bmatrix} 1 & 0 & 0 \\ 0 & 1 & 0 \\ 0 & v & 1 \end{bmatrix}; \qquad \underset{\sim}{u} = (0,0,u).$$

For the convenience we denoted here $(x_{32})^B = v, \; (u_3)^B = u.$

REFLECTION-TRANSMISSION PATTERNS

When the incident shock wave strikes the interface $x_2 = 0$, part of it is reflected and part transmitted across the interfaces, in a form of reflected and transmitted waves. We assume that both reflected and transmitted waves are simple plane waves, travelling in the direction of the x_2- - axis away from the interface. In some cases the assumed reflection-transmission pattern may fail the admissibility test, the pattern must be then modified to include shocks. Both shock and simple waves are given by one parameter family of functions. It is also assumed that the state 1 behind the incident wave and the state 3 at the interface are connected by means of a sequence of one parameter families of reflected and transmitted waves, and constant state regions. The analogous connection exists between states $\hat{0}$, $\hat{3}$. The constraint of incompressibility restricts the propagating waves to transverse waves only. In general, two such simple waves families in every material are required. This means that there are four free parameters, with six interfacial continuity conditions for velocity and traction to be met. However, solution may exist in particular cases as presented here. We have chosen the coordinate system such that the direction of polarisation given by unit vector $\underset{\sim}{d}$ is parallel to x_3 - axis and the motion is restricted to x_3 direction. It appears that for this case the reflected and transmitted wave are single simple waves (regions 2, $\hat{2}$, Fig.2). The wavelets λ = const of the reflected wave are parallel planes with normals $N = (0,1,0)$; the wavelets u = const of the transmitted wave are planes with normals $\hat{N} = (0,-1,0)$. The reflected wave propagates into the just fixed state (region 1); the transmitted wave propagates into the "zero" state (region $\hat{0}$). All remain regions indicated in Fig.2 are regions of constant state. The reflection-transmission problem reduced then to an initial boundary value problem for differential equations governing the variation of the deformation gradient and velocity in the region of the simple wave. The problem now is to fit these waves so as to connect the states at the interface (region 3 and $\hat{3}$) that are compatible with the interfacial cocditions, with the states of region 1 and $\hat{0}$. The deformation gradient and velocity in the regions 2, $\hat{2}$ are of similar to (11) form. Since the components n_i (and $\hat{n}_i$) of the normals referred to the present configuration B remain the same: $n_i = N_i$ ($\hat{n}_i = \hat{N}_i$), the acoustic tensor Q^*_{ij} assumes a simpler form:

$$Q^*_{ij} = \rho_R \sigma_{i2j2} \quad \text{for} \quad i = 1,3, \qquad Q^*_{ij} = 0 \quad \text{for} \quad i = 2, \tag{12}$$

$$Q^*_{i3} = = 0 \quad \text{for} \quad i \neq 3 \quad \text{and} \quad Q^*_{33} = \rho_R \sigma_{3232}. \tag{13}$$

The simple form of the acoustic tensor, leads to the propagation condition $(4)_1$ for simple waves reduced to a single equation for reflected and transmitted wave

$$(\sigma_{3232} - U^2)u' = 0, \qquad (\hat{\sigma}_{3232} - \hat{U}^2)\hat{u}' = 0. \tag{14}$$

The characteristic root $U = \sqrt{\sigma_{3232}}$ is a real single valued function of v, and its represent the speed of the simple wave

$$U = c[\rho_R^{-1}(1 + 3\eta v^2)]^{\frac{1}{2}}; \tag{15}$$

the corresponding characteristic vector function $\tilde{u}'$ is given by

$$\tilde{u}' = (0,0,f(\lambda)) \tag{16}$$

where f is an arbitrary function of the wave parameter λ. Any particular choice of f affects only the parametrisation of the field quantities. Analogous for the transmitted wave.

The differential equations relating the particle velocity and the deformation gradient in the region of the simple wave is obtained form the compatibility condition $(4)_2$. We have in regions 2, $\hat{2}$, respectively

$$Uu' + v' = 0, \qquad \hat{U}\hat{u}' - \hat{v}' = 0 \tag{17}$$

where U is given by (15). It is convenient to assume $f = -U$ and $\hat{f} = \hat{U}$. From Eqs. (16), (17) it follows that

$$\text{in region 2: } u' = -U, \qquad v' = 1, \tag{18}$$

$$\text{in region 2: } \hat{u}' = \hat{U}, \qquad \hat{v}' = 1. \tag{19}$$

The deformation gradient and velocity are assumed to be continuous throughout regions 1, 2, 3 and throughout regions $\hat{0}$, $\hat{2}$, $\hat{3}$ (Fig.2). Thus the initial values for differential equations that describe region 2 are the constant values of region 1: $u(0) = -m_o V_o$, $v(0) = -m_o$, and the initial values for these equations in region $\hat{2}$ are the constant values of region $\hat{0}$: $\hat{u}(0) = \hat{v}(0) = 0$.

Integrating Eqs. (18), (19) we obtain

$$v(\lambda) = \lambda - m_o, \qquad u(\lambda) = -\int_o^{\lambda} U(\lambda)d\lambda - m_o V_o, \tag{20}$$

$$v(\mu) = \mu, \qquad u(\mu) = \int_o^{\mu} U(\mu)d\mu. \tag{21}$$

There are three conditions for stresses and one for velocity to consider at $x_2 = 0$:

$$T_{i2} = \hat{T}_{i2} \quad \text{for} \quad i = 1,2,3, \qquad u = \hat{u}. \tag{22}$$

On the part of the interface which is situated in front of the incident shock three sections are satisfied identically. The fourth one gives the relation between the static pressures of regions 0 and $\hat{0}$ across the interface

$$T_{22} = \hat{T}_{22} \Longrightarrow p_o - \hat{p}_o = 2 \cdot (\hat{C}_1 - C_1) + 2(\hat{C}_2 - C_2) + 6(\hat{C}_3 - C_3). \tag{23}$$

On the interface between regions 3 and $\hat{3}$ the condition $T_{12} = \hat{T}_{12}$ is satisfied identically, the condition $T_{22} = \hat{T}_{22}$ is equivalent to (23) and independent of the deformation gradient. Finally, substitution of $\sigma_{\underset{\sim}{i}\alpha\underset{\sim}{j}\beta}(x_{i\alpha})$ into Eqs. (22) leads to two nontrivial equations involving $\tilde{\lambda}, \tilde{\mu}$.

$$T_{32} = \hat{T}_{32} \Longrightarrow c^2(1 + \eta v^2(\tilde{\lambda}))v(\tilde{\lambda}) = \hat{c}^2(1 + \hat{\eta}\hat{v}(\tilde{\mu}))\hat{v}(\tilde{\mu}), \tag{24}$$

$$u(\tilde{\lambda}) = \hat{u}(\tilde{\mu}).$$

However, for $U(\lambda)$, $(\hat{U}(\mu))$, to represent a simple wave it is necessary that it is a monotonically decreasing function of the wave parameter λ, when λ changes from zero to its terminal value $\lambda = \tilde{\lambda}$. Whether the solution represent a simple wave depends on the sign of the derivative

$$\frac{dU}{dv} = \frac{3c^2\eta}{U\rho_R} v(\lambda), \qquad \frac{dU}{dv} < 0 \quad \text{if} \quad \lambda \in (0, m_o), \tag{25}$$

which is negative when λ belongs to the interval $(0, m_o)$. For this reason according to $(20)_1$ the final value $v(\tilde{\lambda})$ must be negative. The requirement that $v(\tilde{\lambda})$ is negative falls into two cases (26 a), (26 b).

$$-m_o < v(\lambda) < v(\tilde{\lambda}) < 0 \qquad \text{for} \qquad 0 < \lambda < \tilde{\lambda}, \tag{26 a}$$

$$v(\tilde{\lambda}) < v(\lambda) < -m_o \qquad \text{for} \qquad \tilde{\lambda} < \lambda < 0. \tag{26 b}$$

The counterpart of (26) for the transmitted wave are

$$\hat{v}(\tilde{\mu}) < \hat{v}(\mu) < 0 \qquad \text{for} \qquad \tilde{\mu} < \mu < 0. \tag{27}$$

We recall that each of the propagating wavelets is identified by a fixed value of the wave parameter λ changing from 0 to its final value $\tilde{\lambda}$. It follows that a wavelet λ precedes the wavelet $\lambda + d\lambda$. Consequently, if the wave speed U is a decreasing function of λ changing from 0 to $\tilde{\lambda}$, the wavelet $\lambda + d\lambda$ propagates at a lower speed than the wavelet λ and the reflected wave is a simple wave (26 a). If U is increasing with λ the wavelet $\lambda + d\lambda$ travels faster than the wavelet λ and in due course a shock is formed (26 b). It may happen that U is not a monotone function of λ. If such is the case, the reflected wave may be formed by a combination shock and simple wave.

We modify the solution pattern in case (26 b) assuming now, that the reflected wave is a shock propagating in direction $\underset{\sim}{N}$ (Fig.3)

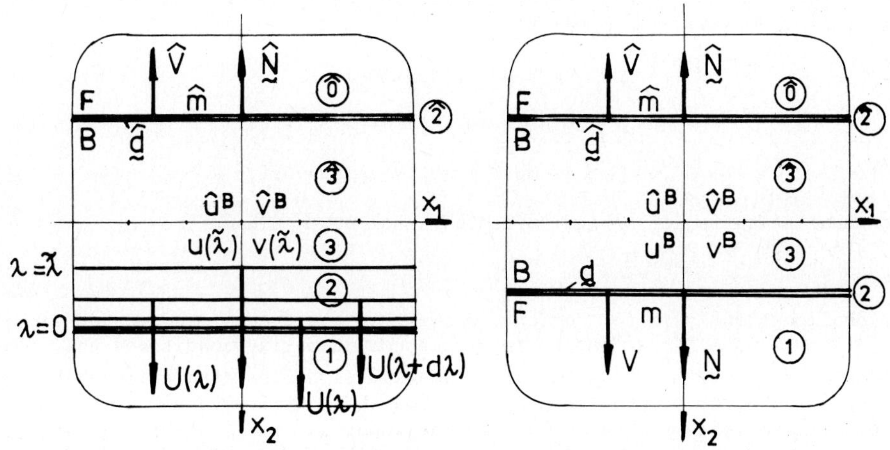

Fig.3. Shock reflection-transmission patterns

Equations of motion $(1)_1$ are now replaced by jump conditions (2) connecting the corresponding quantities in region 1 and 3 across the wave. The constant state ahead of the wave is given by (7), (8). We denote the constant values of the region behind the wave by u^B, v^B. Thus the jump of the deformation gradient and velocity across the wave are:

$$[\![v]\!] = v^B + m_o, \qquad [\![u]\!] = u^B + m_o V_o. \tag{28}$$

Eliminating the jump of velocity, from equation expressing balance of momentum on the discontinuity surface, we obtain the equation for the reflected shock wave speed:

$$V^2 = \frac{c^2}{\rho_R} (1 + \eta((v^B)^2 - v^B m_o + m_o)). \tag{29}$$

According to Lax [2], for (29) to represent an admissible shock it must also satisfy a stability criterion:

$$U^F \leqq V \leqq U^B \tag{30}$$

where U^B and U^F is the characteristic (acoustic) speed (15) in the material region behind and ahead of the shock, respectively. Thus

$$(U^B)^2 = \frac{c^2}{\rho_R}(1 + 3\eta(v^B)^2); \qquad (U^F)^2 = \frac{c^2}{\rho_R}(1 + 3\eta m_o^2). \tag{31}$$

From (30) it follows that

$$3(v^B)^2 > (v^B)^2 - v^B m_o + m_o^2 > 3m_o^2 \tag{32}$$

and the stability criterion (30) is satisfied for arbitrary positive value of the incident shock strength $m_o > 0$ if

$$v^B < -m_o \qquad \text{or} \qquad [\![v]\!] < 0. \tag{33}$$

Amplitude vector of the reflected shock is given by (34). The jump of the deformation gradient component is negative (35). For this reason d_3 must be equal -1. This means that the directions of polarisation for incident and reflected wave are opposite.

$$\underset{\sim}{H} = m\underset{\sim}{d} = (0,0,md_3), \tag{34}$$

$$[\![v]\!] = md_3N_2 = md_3 < 0 \implies d_3 = -1, \tag{35}$$

$$[\![u]\!] = -md_3V = mV \implies [\![u]\!] > 0. \tag{36}$$

Constant state region behind reflected shock wave is described by

$$v^B = -m_o - m, \qquad u^B = mV - m_oV_o. \tag{37}$$

An analogous analysis for region $\hat{2}$ shows that the transmitted wave cannot be a simple wave. According to interface compatibility conditions the components v^B and $\hat{v}^B$ have the same negative sign. The assumption that region $\hat{2}$ is a shock propagating in direction $\hat{\underset{\sim}{N}} = (0,-1,0)$ into region $\hat{0}$ of "zero" state leads to the following equations:

$$\hat{\underset{\sim}{H}} = \hat{m}\hat{\underset{\sim}{d}} = (0,0,\hat{m}\hat{d}_3), \tag{38}$$

$$\hat{v}^B = [\![v]\!] = \hat{m}\hat{d}_3\hat{N}_2 = -\hat{m} < 0 \quad \text{if} \quad d_3 = 1,$$

$$\hat{u}^B = [\![u]\!] = -\hat{m}\hat{d}_3\hat{V} = -\hat{m}\hat{V} < 0 \tag{39}$$

where $\hat{V}$ is the shock speed and $\hat{m}$ is the transmitted shock strength.

$$\hat{V}^2 = \frac{\hat{c}^2}{\hat{\rho}_R}(1 + \hat{\eta}(\hat{v}^B)^2). \tag{40}$$

Since the characteristic speed calculated in region 3 and $\hat{0}$ is

$$(\hat{U}^B)^2 = \frac{\hat{c}^2}{\hat{\rho}_R}(1 + 3\hat{\eta}(\hat{v}^B)^2), \qquad (\hat{U}^F)^2 = \frac{\hat{c}^2}{\hat{\rho}_R},$$

the stability criterion (30) is always satisfied. The results presented here were obtained under assumption that the system (24) has a solution. The question to be considered now is whether a combination of the parameters defining the two materials and the incident shock is possible for which such a solution exists.

REFLECTION-TRANSMISSION SOLUTION

Let us consider a case (26 a) when the reflected wave is a simple wave and transmitted wave is a shock. The terminal values of the simple wave are given by (41) and the constant values of region $\hat{3}$ behind the transmitted wave are given by (39)

$$v(\tilde{\lambda}) = \tilde{\lambda} - m_o, \qquad u(\tilde{\lambda}) = -\int_o^{\tilde{\lambda}} U(\lambda)d\lambda - m_o V_o. \tag{41}$$

Substituting Eqs. (39), (41) into Eqs. (24) we obtain two equations for two unknown values of parameters $\tilde{\lambda}, \hat{m}$.

$$\frac{c^2(1 + \eta(\tilde{\lambda} - m_o)^2)}{\hat{c}^2(1 + \hat{\eta}\hat{m}^2)} = \frac{-\hat{m}}{\tilde{\lambda} - m_o}$$
$$\hat{m}\hat{V} = \int_o^{\tilde{\lambda}} U(\lambda)d\lambda + m_o V_o. \tag{42}$$

Let us consider the case when both waves are shocks. For convenience in calculations, we have assumed that d has opposite direction as follows from (35). The parameter $\tilde{\lambda}$ must be replaced in $(42)_1$ by m - the reflected shock wave strength and $(m - m_o)$ replaces in this equation $(\tilde{\lambda} - m_o)$.

$$\hat{m}\hat{V} = mV + m_o V_o. \tag{43}$$

We can use (43) and $(42)_1$ with this modification to obtain unknown values of the reflected and transmitted shock wave strength $m, \hat{m}$.

PARTICULAR CASES

The shock is completely transmitted if $m = 0$ or $\tilde{\lambda} = 0$ and $\hat{m} \neq 0$. For equal mass densities in both materials we obtain

$$V = V_o, \qquad \hat{m} = m_o \tag{44}$$

provided the incident shock and the combined materials satisfy the condition:

$$m_o = \frac{\hat{c}^2 - c^2}{4(C_3 - \hat{C}_3)}. \tag{45}$$

Unfortunately, the possible value for m_o is outside of the admissibility interval. In cases considered here transmission is always associated with reflection and reflection without transmission is not possible.

NUMERICAL CALCULATIONS

A numerical analysis is conducted for six media composed of two different kinds of rubber. The sets of values for three kinds of rubber are given in [4], they are: I material $C_1 = 0.64$, $C_2 = 0.09$, $C_3 = 0.07$, II material $C_1 = 2.14$, $C_2 = 0.13$, $C_3 = 0.04$, III material $C_1 = 3.52$, $C_2 = 0.00$, $C_3 = 0.25$. The constants are expressed in KG/cm^2. For each of the following material combination A,B,C (Fig. 4a) and material combinations: D,E,F (Fig. 4b) exist a real positive solutions for $(\hat{m}, \tilde{\lambda})$ and

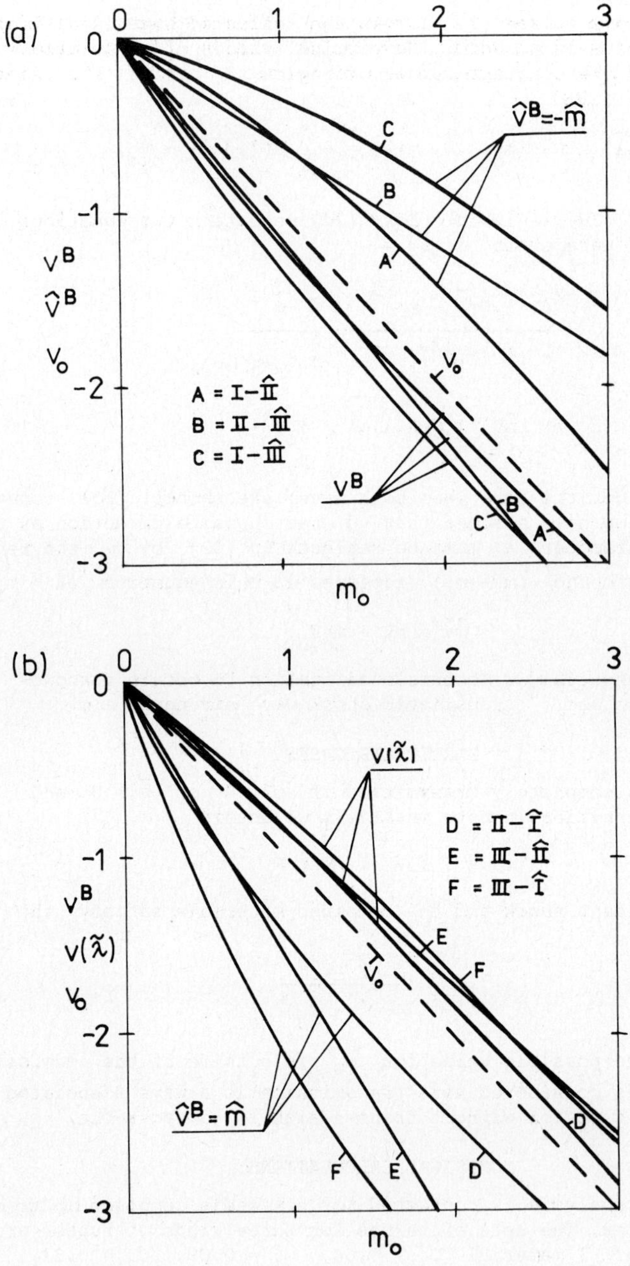

Fig.4. Components of the deformation gradient as function of m: (a) behind the reflected and transmitted shock, (b) behind the transmitted shock and reflected simple wave.

$(\hat{m},m)$, respectively, as a function of m_o. The results are displayed in Fig. 4. Components v^B and $\hat{v}^B$ of the deformation gradient in the region behind the transmitted and reflected shocks are plotted in Fig. 4a as functions on the incident shock parameter m_o for combinations A,B,C. components $\hat{v}^B$ and $v(\tilde{\lambda})$ of the deformation gradient in the regions behind the transmitted shock and reflected simple wave are plotted in Fig. 4b as functions of m_o for combinations D,E,F. The graphs on both Figures shows that the curves intersect for some value of m_o. Thus different material combinations are possible for which the transmitted (or reflected) waves are characterized by the same parameters while the corresponding reflected (or transmitted) waves have different parameters. The graphs in Fig.4 also show that $\hat{m} > m_o$ when the reflected wave is a shock and $\hat{m} < m_o$ when the reflected wave is a simple wave. The strength m of the reflected shock is comparatively small, and the range of variation of the reflected simple wave is also small. The type of reflected waves depends on the material properties of the composite medium, and through which of the two materials the incident shock propagates. In combinations A,B,C the reflected wave is a shock. In the reversed combinations D,E,F the reflected wave is a simple wave.

REFERENCES

[1] WRIGHT, T.W.: "Reflection of oblique shock waves in elastic solids", Internat. J. Solids and Structures, **7** (1971), pp. 161-181.

[2] LAX, P.: "Hyperbolic systems of conservation laws II", Comm. Pure Appl. Math. **10** (1957), pp. 537-566.

[3] ZAHORSKI, S.: "A form of the elastic potential for rubber-like materials", Arch. Mech. Stos. **5** (1959), pp. 613-617.

[4] ZAHORSKI, S.: "Experimental investigation of certain mechanical properties of rubber", Engineering Trans. **10** (1962), pp. 193-207, (in Polish).

[5] DUSZCZYK, B., KOSIŃSKI, S., WESOŁOWSKI, Z.: "Normal shock reflection in rubber like elastic material", Arch. Mech. Stos. **38** (1986), pp. 675-688.

ON THE CONCEPT OF WEAK SOLUTIONS IN THE BV-SPACE

Witold Kosiński

Polish Academy of Sciences
Institute of Fundamental Technological Research
ul. Świętokrzyska 21, PL-00-049 Warszawa, Poland

SUMMARY

A general system of quasi-linear first order equations written in the divergence form and constrained by a differential inequality (the second law of thermodynamics) is analysed. Compatibility conditions are presented. In the BV-space, which is a subspace of those regular distributions that are represented by functions of bounded variation, a weak solution to the system is defined. In the proposed definition, written as a kind of a variational inequality, the system of equations and the thermodynamic admissibility condition are accommodated. The application of this concept in the investigation of uniqueness of an admissible weak solution to a Cauchy problem in the BV-space is shortly presented.

INTRODUCTION

In the paper we shall be concerned with a general system of first order partial differential equations, written in the divergence form

$$\frac{\partial}{\partial t} f^{o}(U) + \mathrm{Div}\, f(U) = B(U, t, x). \tag{1}$$

Such systems, called systems of *balance laws*, are common features of individual theories of continuum physics. Particular *constitutive relations* characterizing the medium in question have supplied the system (1) with the explicite function relations between f^{o}, the flux f, the source term B and a vector field U. Here we assume that those relations are expressed by smooth functions. We should add at this stage that in a number of practical situations it is possible to choose the function f^{o} as the identity map ; then further calculations simplify. The variables (t, x) vary in some domain $\mathcal{P} \subseteq (0,T) \times \mathbb{R}^{n}$, and the vector field U is a map from $\mathcal{P}$ into $\mathcal{D} \subseteq \mathbb{R}^{m}$, in general $m \neq n > 1$.

The operator Div denotes the differantiation with respect to the *spatial* variable x. One should notice that in particular problems x may be the *material* variable, often denoted by X or y. Of importance is that t and x are regarded as independent varaiables in the description.

In order to make the notation more compact, we may introduce the (n+1) D divergence operator div acting on F: $\mathcal{D} \to \mathbb{R}^m \otimes \mathbb{R}^{n+1}$, defined by

$$F(U) := (f^0(U), f(U)) = (f^0(U), f^1(U), \ldots, f^n(U)),$$

as follows

$$\frac{\partial}{\partial t} f^0(U) + \mathrm{Div}\, f(U) = \frac{\partial}{\partial t} f^0 + \frac{\partial}{\partial x^1} f^1 + \cdots + \frac{\partial}{\partial x^n} f^n =: \mathrm{div}\, F(U).$$

Then the balance laws (1) can be written as follows

$$\mathrm{div}\, F(U) = B(U, \underline{x}), \text{ with } \underline{x} := (t.x) \ . \tag{2}$$

Physical observations show the irreversibility of processes in nature. Since solutions to the system of balance laws (1) have to describe real processes, in the sense they may take place in the nature, one needs a physical criterion which will rule out unrealistic processes. That criterion is the *second law of thermodynamics*, commonly written as the unilateral differential constraint

$$\frac{\partial}{\partial t} \eta(U) + \mathrm{Div}\, k(U) \geq r(U, t, x). \tag{3}$$

It states that within the medium and during any process, i.e. along any solution of the system (1), the production of the entropy must be non-negative. Here $\eta(U)$ represents the *entropy function* for the medium described by (1), k(U) is the *entropy flux function* (the entropy efflux, exactly), while r(U,t,x) represents the *entropy supply term* within the medium. Defining the vector function K: $\mathcal{D} \to \mathbb{R}^{n+1}$ by

$$K(U) := (\eta(U), k(U)) = (\eta(U), k^1(U), k^2(U), \ldots, k^n(U)),$$

we write the above constraint as

$$\mathrm{div}\, K(U) \geq r(U, \underline{x}). \tag{4}$$

Before consequences of (3) will be consider, we should point out that the system of balance laws (1) as well as the inequality (3) are the local counterparts of integral balance laws and an integral inequality, respectively, written for the whole material system considered. The local forms have been obtained by a localization procedure in which the essential assumptions are: the Lipschitz continuity of the field vector U, jointly in *t* and x, and the conjecture, that the integral laws have the same forms for the whole material system as well as for its every subsystem.

Hence in the above expressions the differentiation as well as solutions to the system are understood in the classical sense. In the case of non-smooth processes, when the Lipschitz continuity of the field U is lost, the local forms of the laws and the inequality are nonderivable from the integral forms. Consequently we are free in choosing their forms, in particular weak forms are acceptable.

The entropy production inequality is a constraint on particular constitutive relations appearing in (2) and (4) through the forms of the functions F and K. Hence to be compatible with the second law we ought to find conseqences of the constraint.

The following result, obtained with the help of I-Shih Liu's theorem [10] and Boillat's approach [1] applied to (2) and (4), gives the required consequences:

LEMMA 0. The system (1) is compatible with the inequality (3), i.e. every Lipschitz continuous solution U: $\mathcal{P} \to \mathcal{D}$ of the system (2) satisfies the constraint (4), iff

a) the constitutive functions η, k and f^{k}, $k = 0,1,..,n$, are such that, there exist a Lagrange multiplier-valued vector $\overline{U} \in \mathbb{R}^m$, as a function of U, i.e. $\overline{U} = W(U)$ and a function $\overline{K}: \mathbb{R}^m \to \mathbb{R}^{n+1}$; $\overline{K}(\overline{U}) = (\overline{k}^0(\overline{U}), \overline{k}^1(\overline{U}), .., \overline{k}^n(\overline{U}))$ such that

$$(\nabla_U f^{k})^T \overline{U} = \nabla_U k^{k}, \text{ and } f^{k} = -\nabla_{\overline{U}} \overline{k}^{k}, \; k = 0,1,..,n, \tag{5}$$

b) the source term B and the supply term r are such that

$$\overline{U} \cdot B(U,\cdot) - r(U,\cdot) \geq 0. \square \tag{6}$$

For the proof of that result consult the more general approach presented by Ruggeri and Strumia [12], when the case of a covariant vector field was considered (cf.also [2,11]). In their termonology $\overline{U}$ is called the main field.

The both differential relations (5) imply that the potentials $\overline{k}^{k}$ are related to k^{k} by a Legendre transformation

$$\overline{k}^{k} = k^{k} - \overline{U} \cdot f^{k}. \tag{7}$$

Moreover, the change of variables $\overline{U} = W(U)$ transforms (2), at least locally (where $\nabla_{\overline{U}} W$ is nonsingular), into a symmteric system of balance laws in $\overline{U}$

$$\operatorname{div} (\nabla_{\overline{U}} \overline{K}) - B = 0, \tag{8}$$

which forms a symmetric hyperbolic system (in the sense of Friedrichs), if the matrix $\overline{A}^0 := \nabla_{\overline{U}} \nabla_{\overline{U}} \overline{k}^0$ is positive definite.

We should notice that for a particular system (1), in which f^0 is the identity map, we have from $(5)_1$

$$\overline{U} = \nabla_U \eta(U), \tag{9}$$

and the *strict concavity* of the entropy function $\eta(U)$, or likewise - the strict convexity of the potential $\overline{k}^0$ as a function of $\overline{U}$, is sufficient for the positive definiteness of $\overline{A}^0$. We see that in this case thanks to Lemma 0, the compatibility of the system (1) with the thermodynamic inequality (3) is enough for (1) to be written as a symmetric hyperbolic system in the new variable $\overline{U}=\nabla_U\eta$, provided the entropy function is strictly concave. It means that in the class of first order systems of balance laws of the form (1), with $f^0(U) = U$ and with strictly concave entropy functions appearing in the unilateral constraint (3), the hyperbolicity and the symmetry are equivalent to the compatibility of (1) with the second law of thermodynamics.

COROLLARY. Any Lipschitz continuous solution of the system (2) satisfies automatically an additional balance law

$$\operatorname{div} K(U) = \overline{U}\cdot B(U,\underline{x}).\square \qquad (10)$$

The last result means that the system of balance laws (1) (or equivalently, (2)) possesses the suplementary balance law in the form of the balance law for the entropy (10). On the other hand, in the thermodynamics (6) is called the *internal dissipation inequality.* Hence it is natural to call any thermodynamic system non-dissipative, if the quantity p_η, defined by the left-hand side of (6), vanishes identically in U; for p_η the name an *entropy production density* is reserved [9]. Besides the system of equations of ideal (barotropic) compressible fluid the system of equations of finite thermoelasticity, both under adiabatic conditions, serve the examples of a non-dissipative system provided the absolute (thermodynamic) temperature is identified with the empirical one. For example, for the latter system we have

$$U = (\rho_0 v, \underline{F}, E)^T, \quad F = (U, -S, -v \otimes 1, -vS)^T, \quad K = (\eta, 0),$$

$$\overline{U} = (-v, -S, 1)\ \vartheta^{-1}, \quad \overline{K} = (\vartheta\eta + \rho_0 v^2 + S\cdot\underline{F} - E, -vS)\ \vartheta^{-1},$$

where $E = \rho_0(0.5v^2 + \varepsilon(\underline{F},\eta))$ and $S = \rho_0\nabla_{\underline{F}}\varepsilon$, $\vartheta = \rho_0\nabla_\eta\varepsilon$, with ε as the internal energy function. Here v is the particle velocity, $\underline{F}$ - the deformation gradient tensor, ρ_0 - the reference mass density; η is the specific entropy per unit mass and S - the 1st Piola-Kirchhoff stress.

We would like to point out that dissipative hyperbolic systems appear in thermodynamics with internal state variables, where dissipation results from a nondifferential type of a viscosity. It is manifested by a nonhomogeneity of the system (1), and may follow from non-elastic properties of the matter, for example (cf. [7,9]).

It is wortwhile to point out that in a dissipativeless continuum medium the entropy production may occure. However, it may only happen in non-smooth processes and on n-dimensional *hypersurfaces of discontinuity* of solutions. In such a case we face with a concentrated entropy production [1-4,9,11,12]. If a concentrated production of entropy appears in a non-smooth process, it cannot be negative, due to the thermodynamic principle. This observation prompts us the form of thermodynamic inequality for the non-smooth vector field U as follows

$$\operatorname{div} K(U) \geq \overline{U} \cdot B(U,\underline{x}). \tag{11}$$

This form plays the role of an *admissibility criterion* for non-smooth solutions, because the second law of thermodynamics has been already exploited and is not able to rule out all nonrealistic weak solutions of the system of balance laws. The inequality should be regarded in a generalized sense, e.g. in the sense of distributions or measures. To explain what we mean by this we have to characterize the space in which weak solutions to (2) will be defined.

WEAK SOLUTION IN THE BV-SPACE

Let $\mathcal{D}'(\mathcal{P})$ be the space of distributions, where $\mathcal{P}$ is open in $\mathbb{R}^{n+1}$. In $\mathcal{D}'(\mathcal{P})$ we are selecting $\mathcal{D}'(\mathcal{P})_{reg}$ the subspace of all regular distributions and $\mathcal{D}'(\mathcal{P})_{[0]}$-the subspace of all zero-order distributions. It is well-known that $L^1_{loc}(\mathcal{P}) \cong \mathcal{D}'(\mathcal{P})_{reg} \subset \mathcal{D}'(\mathcal{P})_{[0]}$, and $u \in \mathcal{D}'(\mathcal{P})_{[0]}$ if for each compact $\mathcal{A} \subset \mathcal{P}$ there exists $K_{\mathcal{A},u} > 0$, such that for $\forall\, \varphi \in C_0^\infty(\mathcal{P})$, $\operatorname{supp} \varphi \subset \mathcal{A}$

$$|\langle u,\varphi\rangle| \leq K_{\mathcal{A},u} \, ||\varphi||_\infty = K \sup_{\underline{x}\in\mathcal{P}} \{ |\varphi(\underline{x})| \}.$$

The characterization of $BV(\mathcal{P},\mathbb{R}^m)$ is as follows: $U:\mathcal{P} \to \mathbb{R}^m$ belongs to $BV(\mathcal{P},\mathbb{R}^m)$ if each component function U^α, $\alpha = 1,2,\ldots,m$, is an element of $BV(\mathcal{P})$, where

$$BV(\mathcal{P}) := \left\{ u \in \mathcal{D}'(\mathcal{P}) \; : \begin{array}{l} 1^0 \;\; u \in \mathcal{D}'(\mathcal{P})_{reg} \\ 2^0 \;\; D^1 u \in \mathcal{D}'(\mathcal{P})_{[0]} \end{array} \right\}, \tag{12}$$

where $D^1 u$ means a first order distributional derivatives of u. Hence the existence of a matrix-valued Borel regular measure $\mu_{\mathrm{grad}\, u}$ of a locally bounded variation (i.e. locally finite matrix-valued Radon measure) follows, such that for every $\varphi \in C_0^\infty(\mathcal{P})$

$$\int u \operatorname{grad} \varphi \, d\lambda^{n+1} = - \int \varphi \, d\mu_{\mathrm{grad}\, u} \,. \tag{13}$$

We shall write grad u for the measure $\mu_{grad u}$. If grad u is absolutely continuous with respect to the Lebesgue measure λ^{n+1}, then due to the Radon-Nikodym theorem we can write grad u = G λ^{n+1}, for some matrix-valued measurable function $G \in L^1_{loc}(\mathcal{P})$. Of importance is that for $u \in BV(\mathcal{P})$, in general, grad u is not absolutely continuous with respect to λ^{n+1}; it is often concentrated on an n D hypersurface, on which u is discontinuous. Hence the concept of the n D (surface) measure H^n well defined in $\mathbb{R}^{n+1}$ is essential. Due to the known properties [5-6,13-14] for any $U \in BV(\mathcal{P},\mathbb{R}^m)$ the splitting

$$\mathcal{P} = \mathbb{C}(U) \cup \Gamma(U) \cup \mathbb{A}(U) \text{ with } H^n(\mathbb{A}(U)) = 0 \tag{14}$$

is possible, where $\mathbb{C}(U) \cup \Gamma(U)$ forms the *set of regular points* of U, i.e. the union of two sets: $\mathbb{C}(U)$ - the set of all points of *approximate continuity* of U and $\Gamma(U)$ - the set of all points of *approximate jump discontinuity* of U, called the *jump set*, with $\lambda^{n+1}(\Gamma(U) = 0$.

Other properties of the class BV can be found for example in [6, Sec. 3.23] together with the evidence of the importance of *sets with finite perimeter*, which are characterized by the fact that their characteristic functions are elements of $BV(\mathbb{R}^{n+1})$. Since on those sets functions from BV possess their traces, formulating of a version of the Green-Gauss theorem is possible.

Now with the help of F and K, we define the main concept; it is a function of two vector variables given by the formula

$$S(U,V) := -(K(U) - K(V)) + \overline{V}(F(U) - F(V)), \tag{15}$$

where $\overline{V} = W(V)$ is associated with F and K by (5), i.e. $\overline{V}\,\nabla_{\overline{V}} F(V) = \nabla_{\overline{V}} K$. Hence we get immediately

$$S(U,V) = \overline{K}(\overline{V}) - \overline{K}(\overline{U}) - \nabla_{\overline{U}}\overline{K}(\overline{U})(\overline{V} - \overline{U}) =: \overline{S}(\overline{V},\overline{U}). \tag{16}$$

The properties of the function S are fundamental in the derivation of an evolutionary inequality for the stability analysis [2-4,7,8]. From (16) follows that each component of S(U,V) is of quadratic order in the difference U - V, Moreover the gradient of $\overline{V}$ $\nabla_V F(V)$ contracted with an arbitrary vector $d \in \mathbb{R}^{n+1}$ is a symmetric tensor, and consequently the tensor $\nabla_V W(V)\,\nabla_V F(V) d$ is symmetric. Now we introduce the main notion.

DEFINITION. A bounded (H^n-a.e.) function $U \in BV(\mathcal{P},\mathbb{R}^m)$ is said to be a *weak solution to the system* (2) if for any vector $C \in \mathbb{R}^m$ and a set $\mathcal{E}$ with finite perimeter, the following inequality holds:

$$\operatorname{div} S(U,C)(\mathcal{E}) \le -\int_{\mathcal{E}} (W(U) - W(C))\cdot B(U,\underline{x})\,d\lambda^{n+1}(\underline{x}). \tag{17}$$

We should point out that the left-hand side of (17) is a scalar Borel regular measure of the set $\mathcal{E}$, for the superposition $U \to S(U,C)$ is an element of $BV(\mathcal{P},\mathbb{R}^m)$. The following results support our definition.

LEMMA 1. If U is a weak solution and the function **W** is onto $\mathbb{R}^m$, then (2) and (11) hold in the sense of measures, i.e.

$$\operatorname{div} F(U)(\mathcal{E}) = \int_{\mathcal{E}} B(U,\underline{x})\,d\lambda^{n+1}(\underline{x}) , \tag{18}$$

$$\operatorname{div} K(U)(\mathcal{E}) \geq \int_{\mathcal{E}} \overline{U}\cdot B(U,\underline{x})\,d\lambda^{n+1}(\underline{x}) \tag{19}$$

for any set $\mathcal{E}$ with finite perimeter.

Proof. In view of (17), we have

$$0 \geq \operatorname{div} S(U,C)(\mathcal{E}) - \int_{\mathcal{E}} (W(C) - W(U))\cdot B(U,\underline{x})\,d\lambda^{n+1}(\underline{x}) =$$

$$= -\left(\operatorname{div} K(U)(\mathcal{E}) - \int_{\mathcal{E}} W(C)\cdot B(U,\underline{x})\,d\lambda^{n+1}(\underline{x}) \right) +$$

$$+ W(C)\cdot\left(\operatorname{div} F(U)(\mathcal{E}) - \int_{\mathcal{E}} B(U,\underline{x})\,d\lambda^{n+1}(\underline{x}) \right).$$

Now, since the map **W** is surjective, **W**(C) may take any value. Consequently (18) and (19) have to hold.□

COROLLARY. If $\Gamma(U)$ is the jump set of U, then for any $C \in \mathcal{D}$

$$S(U_n,C)(\underline{x})\cdot n \leq S(U_{-n},C)(\underline{x})\cdot n,\ H^n\text{- a.e. on } \Gamma(U) , \tag{20}$$

where n is the (Federer) normal vector to $\Gamma(U)$ at $\underline{x}$, while U_{-n} and U_n are distinct one-sided approximate limits of U, when the point $\underline{x}$ is approached from either halfspace determined by the hyperplane with its normal n. This inequality splits into the Rankine- Hugoniot relation

$$(F(U_n) - F(U_{-n}))(\underline{x})\cdot n = 0, \tag{21}$$

and the entropy increase inequality

$$(K(U_n) - K(U_{-n}))(y)\cdot n \geq 0. \tag{22}$$

Proof. From the known result (cf.[13, Th. 15.2]) it follows that there exists Borel set $\mathcal{C} \subset \mathbb{R}^{n+1}$ of class Γ such that $\Gamma(U) = \mathcal{C} \cup \mathcal{M}$, where $H^n(\mathcal{M}) = 0$. Since $\lambda^{n+1}(\mathcal{C} \cap \mathcal{E}) = 0$ for any Borel set $\mathcal{E}$, from (17) we get

$$\operatorname{div} S(U,C)(\mathcal{C}\cap\mathcal{E}) \leq 0.$$

Using the Green-Gauss theorem, we obtain

$$\operatorname{div} S(U,C)(\mathcal{C}\cap\mathcal{E}) = \int_{\mathcal{C}\cap\mathcal{E}} (S(U_n,C) - S(U_{-n},C)(\underline{x})\cdot n\,dH^n(\underline{x}) \leq 0.$$

The arbitrariness of $\mathcal{E}$ and the representation of $\Gamma(U)$ imply (21). To conclude the proof we substitute S from (15) into the last inequality, to get

$$(K(U_{-n}) - K(U_n))(\underline{x})\cdot n + W(C)(F(U_n) - F(U_{-n})(\underline{x})\cdot n \le 0$$

Hence (21) and (22) follow.□

If we normalize the normal vector to $\Gamma(U)$ at $\underline{x}$ in the space-time so that $n = (-s,N)$, with $N\cdot N = 1$ (with N as the space direction and s as the speed of propagation of the spatial surface of discontinuity of U, e.g. a shock wave), then the above inequality can be rewritten in the form

$$-s(f^0(U_{-n}) - f^0(U_n)) + (f(U_{-n}) - f(U_n))\, N = 0, \tag{23}$$

$$s(\eta(U_{-n}) - \eta(U_n)) - (k(U_{-n}) - k(U_n))\cdot N \ge 0 \tag{24}$$

The proposed definition is supported by the following result.

LEMMA 2. If U is a bounded element of BV, then conditions (18) and (19) imply (17), i.e. a necessary and sufficient condition for a H^n-bounded function U from BV to satisfy the system (2) in the sense of measures, and the condition (11), both in the region $\mathcal{P}$, is that (18) and (19) hold.□ (The proof is similar to that pesented in [8].)

To give the concept of a weak solution to the Cauchy problem we use the notion of symmetric mean value $\hat{U}$ of U from BV [13,14]. Note that at regular points $2\hat{U} = U_n + U_{-n}$. We say that U from $BV(\mathcal{P},\mathbb{R}^m)$ is a *solution of the Cauchy problem*

$$\operatorname{div} F(U) = B(U,t,x), \quad U(0,x) = g(x), \; x \in \mathbb{R}^n, \tag{25}$$

if U is a weak solution to $(25)_1$ and

$$\lim_{t\to 0,} \hat{U}(t,x) = g(x), \; \lambda^n\text{- a.e. on } \mathbb{R}^n, \; t > 0.$$

Note that the initial condition cannot be satisfied as an equality, for the function U is defined on the open set $\mathcal{P} \subseteq (0,T) \times \mathbb{R}^n$.

The concept of weak solution to (25) was used in [8], when a uniqueness result in BV was shown. It has been done by the method of a parabolic regularization of the initial system (1), i.e. instead of the function f in (1) into another one, namely $f'(U,\mathrm{Grad}U) = f(U) - \varepsilon 1\, \mathrm{Grad}U$, where ε is a positive parameter, was substituted. It was show that the unique weak solution has to be a limit, when ε tends to zero, of a sequence of regular (classical) solutions to the parabolized problem. The method used in [8] requires that the limit solution has a regularity of a function from $W^{1,\infty}_{loc}$. In the next paper the concept of a weak solution to the initial boundary-value problem will be discussed.

REFERENCES

[1] BOILLAT, G.: Sur une fonction croissante comme l'entropie et génératrice des chocs dans les systèmes hyperboliques. C. R. Acad. Sci. Paris, 283 A (1976), pp. 409-412.

[2] DAFERMOS, C.M.: Hyperbolic systems of conservation laws, in: Systems of Nonlinear Partial Differential Equations. J.M. Ball (ed.), D. Reidel Publ.Comp., Dordrecht, Holland, 1983, pp. 25-70.

[3] DAFERMOS, C.M. and KOSIŃSKI, W.: Continuous dependence results for quasi-linear hyperbolic systems. SFB 72-Preprint no. 666, Universität Bonn 1984.

[4] DIPERNA, R.J.: Uniqueness of solutions to hyperbolic conservation laws. Indiana U. Math. J. 28 (1979), pp. 137-188.

[5] FEDERER, H.: Geometric Measure Theory. Springer, Berlin, Heidelberg, New York 1969.

[6] HANYGA, A.: Mathematical Theory of Non-Linear Elasticity. Ellis Horwood Ltd., Chichester, Halsted Press:a division of John Wiley & Sons, and PWN-Polish Scientific Publishers, Warsaw 1985.

[7] KOSIŃSKI, W.: A note on stability of dissipative bodies. Arch. Mech., 34, 3 (1982), pp. 401-407.

[8] KOSIŃSKI, W.: Admissibility and uniqueness of weak solutions to hyperbolic systems of balance laws. Mathematical Methods in Applied Sciences, in print.

[9] KOSIŃSKI, W.: Field Singularities and Wave Analysis in Continuum Mechanics. Ellis Horwood Ltd., Chichester, Halsted Press: a division of John Wiley & Sons, and PWN-Polish Scientific Publishers, Warsaw 1986.

[10] LIU, I-Shih,: Method of Lagrange multipliers for exploitation of the entropy principle. Arch. Rat. Mech. Anal., 46 (1972), pp.131-148.

[11] RUGGERI, T.: Entropy principle, symmetric hyperbolic systems and shock waves. in: Wave Phenomena: Modern Theory and Applications. C. Rogers and T. Bryant Moodie (eds.), North-Holland, Amsterdam, New York, Oxford, 1984, pp.211-220.

[12] RUGGERI, T. and STRUMIA, A.: Main field and convex covariant density for quasi-linear hyperbolic systems. Relativistic fluid dynammics. Ann. Inst. H. Poincaré, 34, A (1981), pp. 65-84.

[13] VOL'PERT, A.I.: The space BV and quasilinear equations. Matem. Sborn. Akad. Nauk 73, 115 (1967), pp. 225-302 (in Russian). English translation: Mathem. USSR-Sbornik 2 (1967), pp. 225-267.

[14] VOL'PERT, A.I. and CHUDAYEV, S.I.: Analysis in a Class of Discontinuous Functions and Equations of Mathematical Physics (in Russian). Izdatelstwo "Nauka", Moskwa 1975.

Numerical Solution of the Euler Equations Used for Simulation of 2D and 3D Steady Transonic Flows

KAREL KOZEL, MIROSLAVA VAVŘINCOVÁ

Dept. of Comp. Techniques and Informatics, Faculty of Mechanical Engineering
TU Prague, Suchbátarova 4, 166 07 Prague 6
(ČSSR)

and NGUYEN VAN NHAC

Dept. of Mathematics , Faculty of Nuclear Engineering
TU Prague, Trojanova 13, 120 00 Prague 2
(ČSSR)

Summary

The work deals with the numerical solution of the system of Euler equations for the case of 2D steady transonic flows in a channel or through a cascade and 3D steady transonic flows in a channel.

The 2D weak solution of the problem is computed by the conservative finite volume formulation of the explicit MacCormack difference scheme with a nonlinear artificial dissipative term of second order and a linear dissipative term of fourth order. The steady solution is obtained by a time dependent method by integrating t to infinity and using appropriate steady boundary and periodical conditions.

The explicit MacCormack difference scheme in conservation form is used for computing the numerical solution of 3D transonic flows in a channel.

The presented 2D numerical results are compared with numerical results of Ron–Ho–Ni in the case of channel flows for $M_\infty < 1$ and with interferometric measurements of the Institute of Thermomechanics of Czechoslovak Academy of Sciences in the case of transonic flows through 8% DCA cascade for upstream Machnumbers $M_\infty < 1$ as well as $M_\infty > 1$.

The presented 3D numerical results of transonic flows in a channel are compared to numerical results computed by 1D theory and 2D theory.

1. 2D Steady Transonic Flows in a Channel and through a Cascade

Consider the 2D system of Euler equations in conservation form

$$W_t + F(W)_x + G(W)_y = 0 \tag{1.1}$$

where

$$\begin{aligned}
W &= \mathrm{col}\, \|\rho, \rho u, \rho v, e\|,\\
F &= \mathrm{col}\, \|\rho u, \rho u^2 + p, \rho uv, (e+p)u\|,\\
G &= \mathrm{col}\, \|\rho v, \rho uv, \rho v^2, (e+p)v\|,\\
p &= (\kappa - 1)\left[e - \frac{1}{2}\rho(u^2+v^2)\right].
\end{aligned}$$

We use dimensionless values of density $\rho = \widetilde{\rho}/\widetilde{\rho}_\infty$, velocity $(u,v) = (\widetilde{u}/\widetilde{a}_\infty, \widetilde{v}/\widetilde{a}_\infty)$, energy per unit volume $e = \widetilde{e}/(\widetilde{\rho}_\infty \widetilde{a}_\infty^2)$ and pressure $p = \widetilde{p}/(\widetilde{\rho}_\infty \widetilde{a}_\infty^2)$, $t = \widetilde{t}\,\widetilde{a}_\infty/c$, where $\widetilde{\rho}$, $(\widetilde{u},\widetilde{v})$, $\widetilde{e}$, $\widetilde{p}$, $\widetilde{t}$, $\widetilde{a}$ are density, velocity vector, ernergy per unit volume, pressure, time, sonic velocity; $\widetilde{\rho}_\infty$ is the upstream density etc. and c is the length of the chord of the given profile. Also $x = \widetilde{x}/c$, $y = \widetilde{y}/c$, $z = \widetilde{z}/c$ where $\widetilde{x}$, $\widetilde{y}$, $\widetilde{z}$ are space coordinates in the physical plane.

A piecewise smooth function $W(x,y,t)$ is called a weak solution of (1.1) if it satisfies

$$\iint_D W|_{t_1}^{t_2}\, dxdy = -\int_{t_1}^{t_2} \left\{ \oint_{\partial D} F dy - G dx \right\} dt \tag{1.2}$$

for every suitable Jordan's curve $\partial D \subset \Omega$, $D = \mathrm{Int}\,\partial D$, $\forall t_1, t_2 > 0$. Let the domain D be sufficiently small, then we can rewrite (1.2) using the mean value theorem in D on the left hand side of (1.2) and in interval $]t_1, t_2[$ on the right hand side of (1.2):

$$[W(\overline{x},\overline{y},t_2) - W(\overline{x},\overline{y},t_1)]\,\mu = -\left[\oint_{\partial D} F(x,y,\overline{t})\,dy - G(x,y,\overline{t})\,dx\right]\cdot(t_2 - t_1), \tag{1.3}$$

$(\overline{x},\overline{y}) \in D$, $\overline{t} \in]t_1, t_2[$, $\mu = \iint_D dxdy$. We consider the finite volume formulation of the numerical scheme. The difference scheme satisfies relation (1.3) for each computational cell (see Fig. 1.1)

$$D_{ij} \equiv D_m^{(1)} D_m^{(2)} D_m^{(3)} D_m^{(4)}, \quad D_m^{(k)} = (x_k^m, y_k^m), \quad \mu_{ij} = \iint_{D_{ij}} dxdy.$$

$$W_m^{n+1} - W_m^n = -\frac{\Delta t}{\mu_m} \oint_{\partial D_m} (\mathcal{F}\,dy - \mathcal{G}\,dx), \tag{1.4}$$

where W_m is the mean value of W in the computationial cell D_m; $\mathcal{F}$, $\mathcal{G}$ are approximations of $F(W)$, $G(W)$ along ∂D_m, $\Delta t = t_2 - t_1 = t^{n+1} - t^n$, $\mu_m =$

$\int\int_{D_m} dx dy$. The steady state solution of our problem has to satisfy the integral relation (1.4) for $W_m^{n+1} - W_m^n = 0$ for all $D_m \subset \Omega$.

To obtain the numerical solution we use the time dependent method. Steady state is reached by letting t tend to infinity and using steady boundary conditions. MacCormack's explicit difference scheme [7] in finite volume formulation is used.

The **predictor** step has the form

$$\begin{aligned} W_{ij}^{n+\frac{1}{2}} &= W_{ij}^n - \frac{\Delta t}{\mu_{ij}} \oint_{\partial D_{ij}} (F^n dy - G^n dx) \\ &= W_{ij}^n - \frac{\Delta t}{\mu_{ij}} \left[F_{i+1,j}^n (x_2 - x_1) - G_{i+1,j}^n (y_2 - y_1) + \right. \\ &\qquad F_{i,j+1}^n (x_3 - x_2) - G_{i,j+1}^n (y_3 - y_2) + \\ &\qquad F_{ij}^n (x_4 - x_3) - G_{ij}^n (y_4 - y_3) + \\ &\qquad \left. F_{ij}^n (x_1 - x_4) - G_{ij}^n (y_1 - y_4)\right] \\ &= W_{ij}^n - \frac{\Delta t}{\mu_{ij}} \operatorname{Res} W_{ij}^n, \end{aligned} \tag{1.5a}$$

where $F_1^n = F(W_{i+1,j}^n)$, $F_2^n = F(W_{i,j+1}^n)$, $F_3^n = F_4^n = F(W_{i,j}^n)$, analogous for G_k^n; $\Delta x_k = x_{k+1}^m - x_k^m$, $\Delta y_k = y_{k+1}^m - y_k^m$ and $x_5^m = x_1^m$, $y_5^m = y_1^m$. A similar form is used for the **corrector** step

$$\begin{aligned} \overline{W}_{ij}^{n+1} &= \frac{1}{2}\left(W_{ij}^n + W_{ij}^{n+\frac{1}{2}}\right) - \frac{1}{2}\frac{\Delta t}{\mu_{ij}} \oint_{\partial D_{ij}} \left(F^{n+\frac{1}{2}} dy - G^{n+\frac{1}{2}} dx\right) \\ &= \frac{1}{2}\left(W_{ij}^n + W_{ij}^{n+\frac{1}{2}}\right) - \frac{1}{2}\frac{\Delta t}{\mu_{ij}} \left[F_{ij}^{n+\frac{1}{2}} (x_2 - x_1) - G_{ij}^{n+\frac{1}{2}} (y_2 - y_1) + \right. \\ &\qquad F_{ij}^{n+\frac{1}{2}} (x_3 - x_2) - G_{ij}^{n+\frac{1}{2}} (y_3 - y_2) + \\ &\qquad F_{i-1,j}^{n+\frac{1}{2}} (x_4 - x_3) - G_{i-1,j}^{n+\frac{1}{2}} (y_4 - y_3) + \\ &\qquad \left. F_{i,j-1}^{n+\frac{1}{2}} (x_1 - x_4) - G_{i,j-1}^{n+\frac{1}{2}} (y_1 - y_4)\right] \\ &= \frac{1}{2}\left(W_{ij}^n + W_{ij}^{n+\frac{1}{2}}\right) - \frac{1}{2}\frac{\Delta t}{\mu_{ij}} \operatorname{Res} W_{ij}^{n+\frac{1}{2}}, \end{aligned} \tag{1.5b}$$

where $F_ij^n = F(W_{ij}^n)$. The final value W_{ij}^{n+1} is corrected by an artificial dissipative term.

$$W_{ij}^{n+1} = \overline{W}_{ij}^{n+1} + DW_{ij}^n, \tag{1.5c}$$

where

$$DW_{ij}^n = D_x W_{ij}^n + D_y W_{ij}^n \tag{1.6}$$

and

$$D_x W_{ij}^n \sim \varepsilon_x^{(1)} \frac{\Delta t}{\Delta x} \frac{\partial}{\partial x}(|W_x| W_x) + \varepsilon_x^{(2)} \frac{\Delta t}{\Delta x} W_{xxxx},$$

$$\varepsilon_x^{(1)} \sim \varepsilon_x^{(2)} = O(\Delta t \Delta x^2). \tag{1.7}$$

A similar expression is used in conservative finite volume formulation for $D_y W_{ij}^n$.

Boundary conditions along a profile surface (wall) are treated as follows. We consider a fictitious computational cell e.g. D_{ij-1} (inside a profile) with fictitious values $F_{i,j-1}$, $G_{i,j-1}$. Let

$$F_{i,j-\frac{1}{2}} = -(F_{ij} + F_{i,j-1}), \quad G_{i,j-\frac{1}{2}} = -(G_{ij} + G_{i,j-1}).$$

Then we can express $F_{i,j-1}$, $G_{i,j-1}$ and use this in (1.5b) together with ($v/u = f'(x)$, $f(x)$ describes the profile surface) the following relation which is valid along the boundary

$$\int\limits_{D_{ij}^{(4)} D_{ij}^{(1)}} F dy - G dx = F_4 \Delta y_4 - G_4 \Delta x_4 = \mathrm{col}\, \|0, \Delta y_4, -\Delta x_4, 0\| \cdot p_4 \tag{1.8}$$

if $D_{ij}^{(4)} D_{ij}^{(1)}$ is a line. If $D_{ij}^{(4)} D_{ij}^{(1)}$ is not a line one has to replace (1.8) by the expression

$$\int\limits_{D_{ij}^{(4)} D_{ij}^{(1)}} F dy - G dx = \int\limits_{D_{ij}^{(4)} D_{ij}^{(1)}} p_4 \cdot \mathrm{col}\, \|0, dy, -dx, 0\|. \tag{1.9}$$

Pressure p is extrapolated by double quadratic extrapolation or computed using the relation

$$\frac{\partial p}{\partial n} = \rho q^2 / R \tag{1.10}$$

in difference form, $q^2 = u^2 + v^2$, R is the radius of curvature of the boundary streamline, n is the outer normal.

Periodical conditions are considered in the usual way. For steady state computation of 2D channel flows we use 1D theory to fulfill upstream and downstream boundary conditions. For 2D cascade flows all components of W are considered along the upstream boundary; along the downstream boundary the first three components of W are extrapolated using $W_{ss} = 0$ (meaning of s–directionis given below) and e is computed using the given downstream pressure p_2 and extrapolated values ρ_2, $(\rho u)_2$, $(\rho v)_2$. The grid directions in the case of channel flows are aligned with the y–direction and the s–direction (approximated streamline direction), in the case of cascade flows with the s–direction and the pitch direction.

In the next part we present several numerical results. A good agreement of our results and Ron–Ho–Ni's numerical results for 2D channel flows with $M_\infty = 0.675$ is presented in [3]. The first part of our numerical results dealing with cascade flows is devoted to a comparison of our results and interferometric measurements published in [2]. We can compare our numerical results using lines $M =$ const. not only qualitatively (shape of lines $M =$ const. and shape of black and white stripes in the interferogram) but also quantitatively because the sonic line is denoted in the interferogram by a dotted (broken) line and difference 2.5 strips in the interferogram corresponds to $\Delta M = 0.05$ in our numerical results. Fig. 1.2a–f show a comparison of interferometric measurements and computed results for increasing M_∞. The downstream pressure for numerical results is observed to be approximately the same as for experimental results. The same is true for the angle of attack α. Constant difference ΔM_∞ in experimental (e) and computational (c) results is considered $\Delta M_\infty = M_\infty^c - M_\infty^e = 0.06$; $M_\infty^e \in \langle 0.91; 1.13 \rangle$.

We can observe the behaviour of the first sonic line (near leading edge) for increasing upstream Mach numbers in the experimental and numerical results as well as the appearance of a bowed detached shock wave. The reflection of this shock wave and the behaviour of the strong shock wave near the trailing edge of the upper profile is also similar for experimental and numerical results.

The next results (Fig. 1.3a) show the back pressure (downstream pressure) effect in transonic cascade flows for ČKD1 compressor cascade [3], $M_\infty = 0.87$, $\alpha = 22.82°$. The back pressure is considered in relation to $p_\infty = p_1$ by constant k: $p_2 = k \cdot p_1$ ($k > 1$ for compressor cascade). We can observe, that for increasing k the shock wave is moving in the direction to the leading edge, the jump in the shock wave is decreasing and the maximal Mach number on the upper profile surface is also decreasing. Fig. 1.3b shows a comparison of numerical results of transonic flows through ČKD1 compressor cascade for $M_\infty = 0.87$, $\alpha = 22.82°$, $p_2 = 1.285 p_1$, using a Mach number distribution along the upper and lower profile surface computed by the numerical solution of the full potential equation [4] and by the finite volume solution of the Euler equations. All numerical results were achieved by an ICL–4–72 computer using 3000 iterations for channel flows and 1200 – 2000 iterations for cascade flows.

2. Numerical Solution of 3D Transonic Flows in a Channel

Consider the 3D system of Euler equations in conservation form

$$W_t + F(W)_x + G(W)_y + H(W)_z = 0 \tag{2.1}$$

where

$$W = \text{col}\,\|\rho, \rho u, \rho v, \rho w, e\|,$$

$$F = \text{col}\,\|\rho u, \rho u^2 + p, \rho uv, \rho uw, (e+p)u\|,$$

$$G = \text{col}\,\|\rho v, \rho uv, \rho v^2 + p, \rho vw, (e+p)v\|,$$

$$H = \text{col}\,\|\rho w, \rho wu, \rho wv, \rho w^2 + p, (e+p)w\|,$$

$$p = (\kappa - 1)\left[e - \frac{1}{2}\rho(u^2 + v^2 + w^2)\right].$$

(u, v, w) is the velocity vector

In this case a finite method is use for the numerical solution. The cross–section of the considered 3D channel is oblong (see Fig. 2.1). The governing curve of the channel is given by $x = f_1(y)$, $z = f_2(y)$ and the cross–section is considered to be oblong with the sides $a(y)$ and $b(y)$, where $f_1(y), f_2(y), a(y), b(y) \in C^1(I)$, $I \equiv \langle 0, Y_0 \rangle$.

The transformation $(x, y, z) \longrightarrow (x, s, z)$ is used. s is the approximated streamline direction. Then, system (2.1) is transformed to the following form

$$W_t + \widetilde{F}(W)_x + \widetilde{G}(W)_s + \widetilde{H}(W)_z = 0 \tag{2.2}$$

which is used for our numerical computation. We do not consider grid points on the walls. Fig. 2.2a–b show a grid used in the cross–section $y = y_j = j \cdot \Delta y$, $z = z_k = (k + \frac{1}{2})\Delta z + f_2(y_j)$. A similar grid is used in the cross–section $x = x_i = (i + \frac{1}{2})\Delta x + f_1(y_j)$. Similar to the 2D case we use MacCormack's explicit difference scheme in the following form:

Predictor step

$$\begin{aligned} W_{ijk}^{n+\frac{1}{2}} = W_{ijk}^{n} &- \frac{\Delta t}{\Delta x}\left(\widetilde{F}_{i+1,jk}^{n} - \widetilde{F}_{ijk}^{n}\right) \\ &- \frac{\Delta t}{\Delta s}\left(\widetilde{G}_{i,j+1,k}^{n} - \widetilde{G}_{ijk}^{n}\right) \\ &- \frac{\Delta t}{\Delta z}\left(\widetilde{H}_{ij,k+1}^{n} - \widetilde{H}_{ijk}^{n}\right) \end{aligned} \tag{2.3a}$$

Corrector step

$$\begin{aligned} \overline{W}_{ijk}^{n+\frac{1}{2}} = \frac{1}{2}\left(W_{ijk}^{n} + W_{ijk}^{n+\frac{1}{2}}\right) - \frac{1}{2}\Bigg[&\frac{\Delta t}{\Delta x}\left(\widetilde{F}_{ijk}^{n+\frac{1}{2}} - \widetilde{F}_{i-1,jk}^{n+\frac{1}{2}}\right) \\ &+ \frac{\Delta t}{\Delta s}\left(\widetilde{G}_{ijk}^{n+\frac{1}{2}} - \widetilde{G}_{i,j-1,k}^{n+\frac{1}{2}}\right) \\ &- \frac{\Delta t}{\Delta z}\left(\widetilde{H}_{ijk}^{n+\frac{1}{2}} - \widetilde{H}_{ij,k-1}^{n+\frac{1}{2}}\right)\Bigg]. \end{aligned} \tag{2.3b}$$

W_{ijk}^{n+1} is corrected by an artificial dissipative term similar to the 2D case

$$W_{ijk}^{n+1} = \overline{W}_{ijk}^{n+1} + DW_{ijk}^{n} \,. \tag{2.3c}$$

Boundary conditions along the walls are satisfied as in the 2D case of the finite volume form. Extraplolated values of the pressure p along the walls are computed by double quadratic extrapolation. The upstream and downstream boundary conditions are similar to the ones in the 2D calculation.

The presented numerical results are computed for a channel given by

$$\begin{aligned} x &= f_1(y) = \frac{1}{2}\left[1 - b(y)\right], \\ z &= f_2(y) = \frac{1}{2}\left[1 - a(y)\right], \\ a(y) = b(y) &= \begin{cases} \left[\frac{1+1.5\cdot\left(1-(y/5)\right)^2}{2.5}\right]^{\frac{1}{2}} & , y \in \langle 0; 5) \\ \left[\frac{1+0.5\cdot\left(1-(y/5)\right)^2}{2.5}\right]^{\frac{1}{2}} & , y \in \langle 5; 10\rangle . \end{cases} \end{aligned} \tag{2.4}$$

Because we have no other suitable 3D numerical or experimental results, we compare our 3D results with our 1D and 2D numerical results. Fig. 2.3 shows a comparison of our numerical results using Mach number distribution in the midpoints of the channel and in the points near the wall with numerical results computed by 1D theory for $p_2 = 0.45 \cdot p_1$. We can only compare our 1D and 2D numerical results qualitatively with other 3D and 1D results [1]. Fig. 2.4 shows a comparison of 3D and 2D results. The 2D ones were computed using cross–section $A(x) = a(x)^2$, not $A(x) = a(x)$ as it is used in many cases. Fig 2.5 shows 3D and 2D results mapped by lines $M = const.$, the 3D ones for cross–section $z = z_3$ (mid–cross–section). We use an ICL–4–72 computer (with double precission) and 3000 iterations.

References

[1] Daguji, H., Motohashi, Y., Yamamoto, S.: An Implicit Time–Marching Method for Solving the 3D Compressible Euler Equations, Proceedings of 10. ICNMFD, Lecture Notes in Physics 264, Springer–Verlag 1987.

[2] Dvořák, R.: On the Development and Structure of Transonic Flow in Cascades, Proceedings of Symposium Transsonicum II, Göttingen, 1975.

[3] Huněk, M., Kozel, K., Vavřincová, M.: Numerical Solution of 2D Transonic Flow Problem in Compressor Cascades Using Full Potential Equation and Multiple–Grid Techniques. Numerical Solution of Boundary Layer and

Transonic Euler Equations, Technical Report o. p. ČKD1 Prague Compressor, No KKS–Tk 2.7–285, November 1987.

[4] Huněk, M., Kozel, K., Vavřincová, M.: Numerical Solution of Transonic Potential Flow in 2D Compressor Using Multiple–Grid Techniques, Proceedings of IV. GAAM Seminar: Robust Multiple–Grid Methods (Kiel, January 1988).

[5] Kozel, K., Vavřincová, M.: Finite Volume Solution of the Euler Equations, Proceedings of 2nd ISNA Conference, Prague, August 1987.

[6] Lerat, A.: Implicit Method of Second–Order Accuracy for the Euler Equations, AIAA Journal, Vol. 23, No. 1, 1985.

[7] MacCormack, R. W.: The Effect of Viscosity in Hypervelocity Impact Cratering, AIAA Paper 69–354, 1969.

[8] Ron–Ho–Ni: A Multiple–Grid Scheme for Solving the Euler Equations, AIAA Journal, Vol. 20, No. 11, 1982.

[9] Thomkins, W. T. Jr.: Analysis of Pseudo–Time–Marching Schemes for Application to Turbomachinery Cascade Calculations, in Advances in Computational Transonics, Vol. 4, Pineridge Press, 1985, ed. W. G. Habashi.

Figures

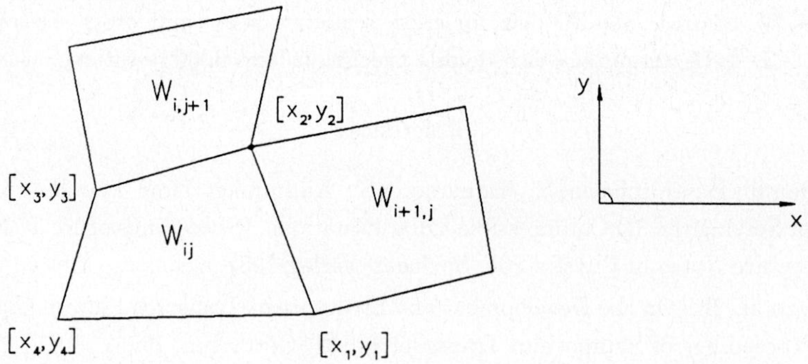

Fig. 1.1 Computational cell used for equation (1.4)

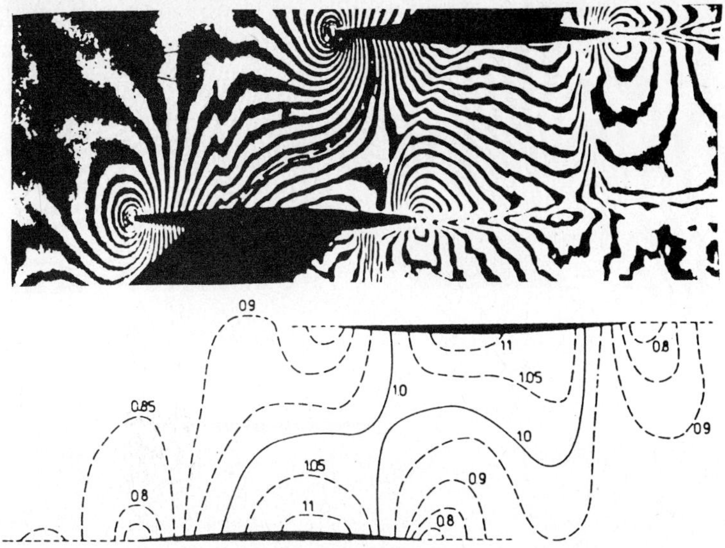

Fig. 1.2a $M_\infty^c = 0.91,\ \alpha = 0.2°,\ p_2 = 0.98 p_1$

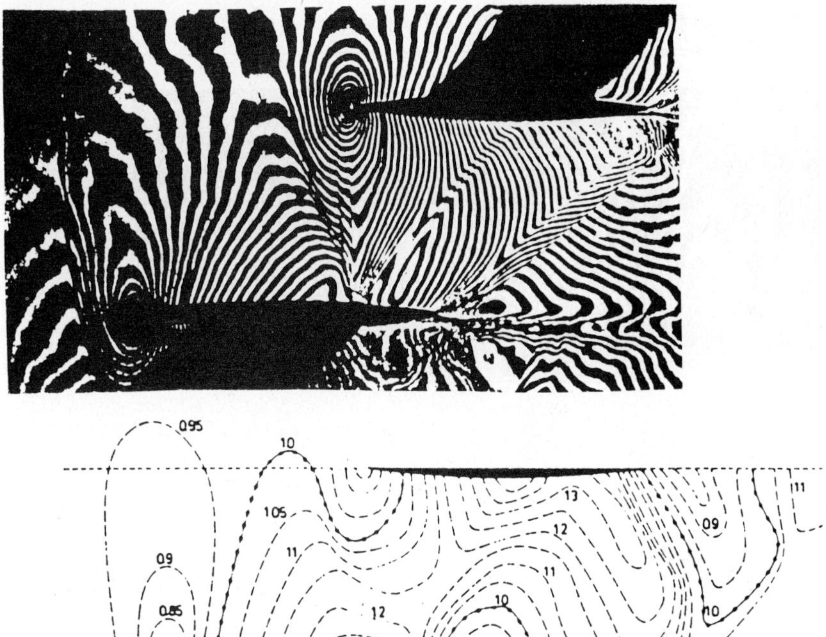

Fig. 1.2b $M_\infty^c = 1.013,\ \alpha = 0.2°,\ p_2 = 0.95 p_1$

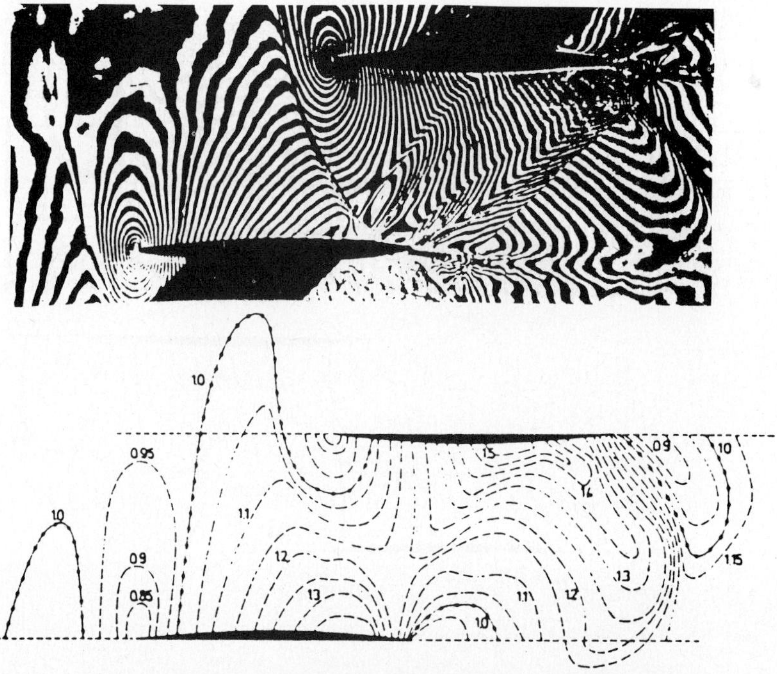

Fig. 1.2c $M^c_\infty = 1.05,\ \alpha = 0.2°,\ p_2 = 0.95 p_1$

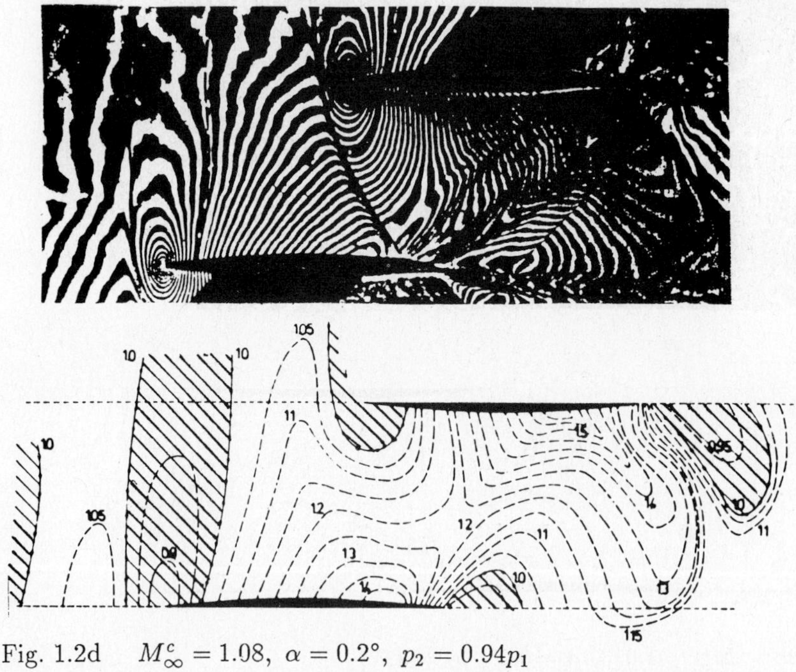

Fig. 1.2d $M^c_\infty = 1.08,\ \alpha = 0.2°,\ p_2 = 0.94 p_1$

Fig. 1.2e $M_\infty^c = 1.1,\ \alpha = 0.2°,\ p_2 = 0.95p_1$

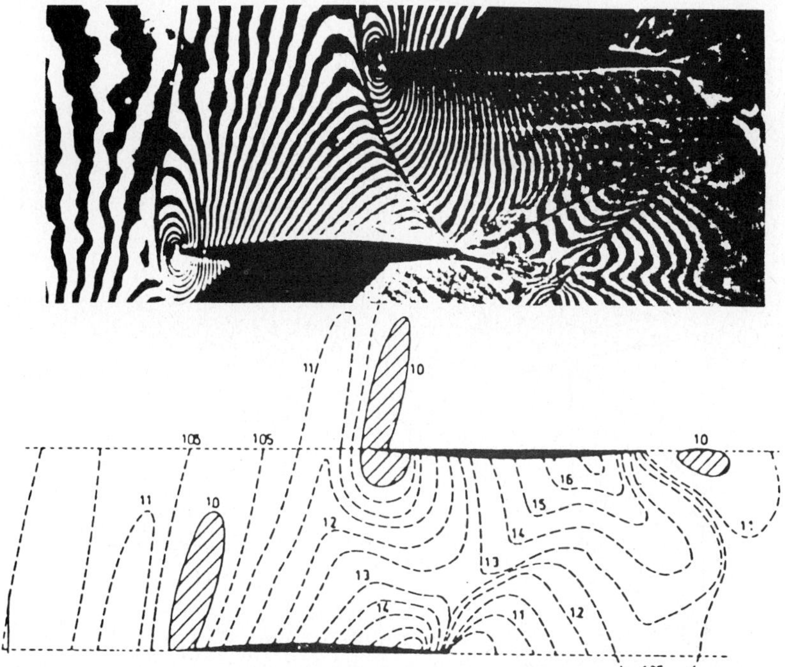

Fig. 1.2f $M_\infty^c = 1.13,\ \alpha = 0.3°,\ p_2 = 0.9p_1$

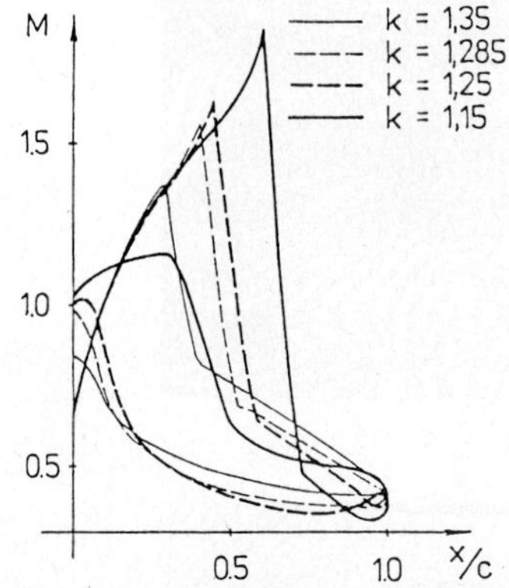

Fig. 1.3a Back pressure effect; ČKD1 compressor cascade
$M_\infty = 0.87,\ \alpha = 22.82°$

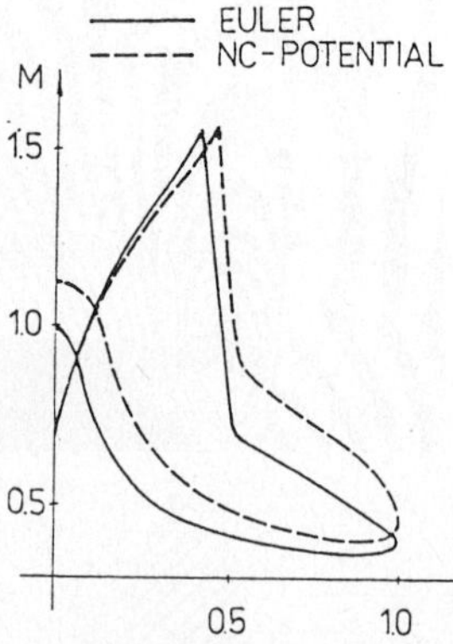

Fig. 1.3b Comparison of numerical solution using full potential eq. and Euler eq.; ČKD1 compressor cascade

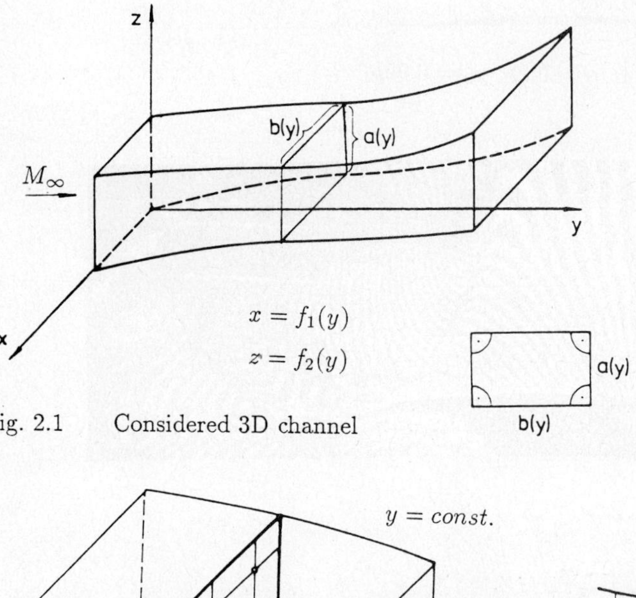

Fig. 2.1 Considered 3D channel

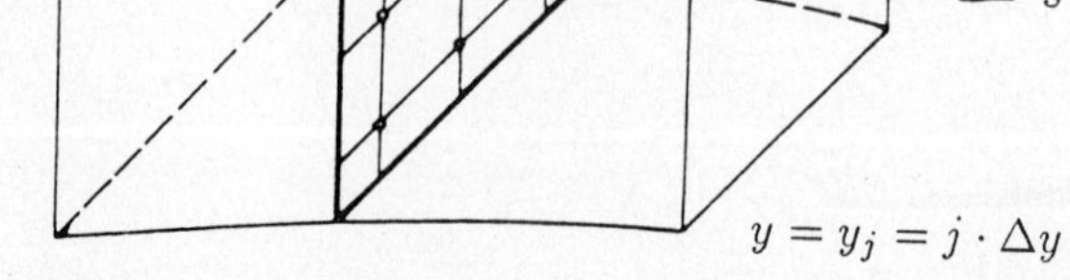

Fig. 2.2a Grid considered in $y = const.$

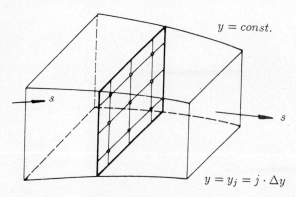

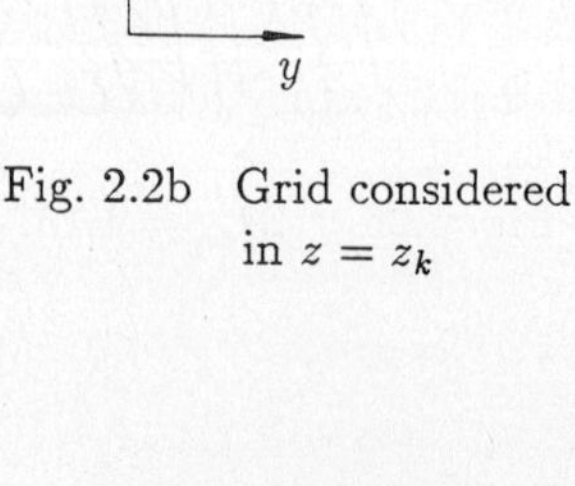

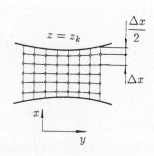

Fig. 2.2b Grid considered in $z = z_k$

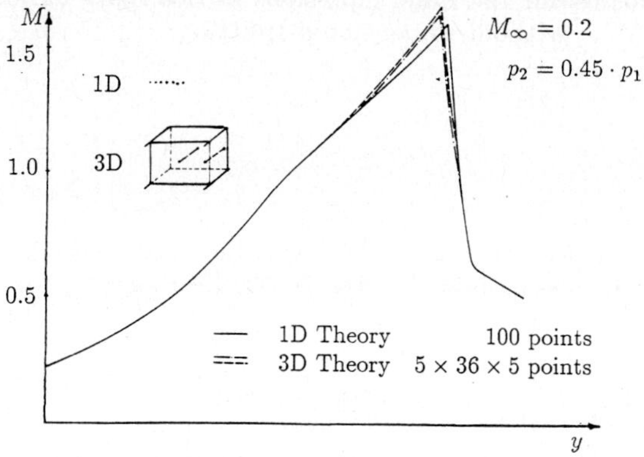

Fig. 2.3 Comparison of 1D and 3D results

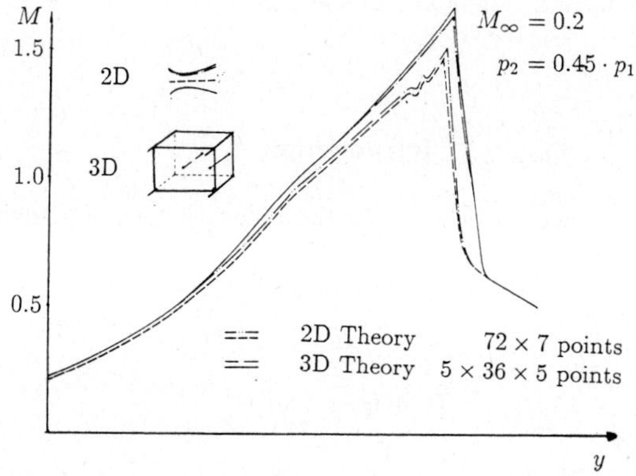

Fig. 2.4 Comparison of 2D and 3D results using Mach number distribution

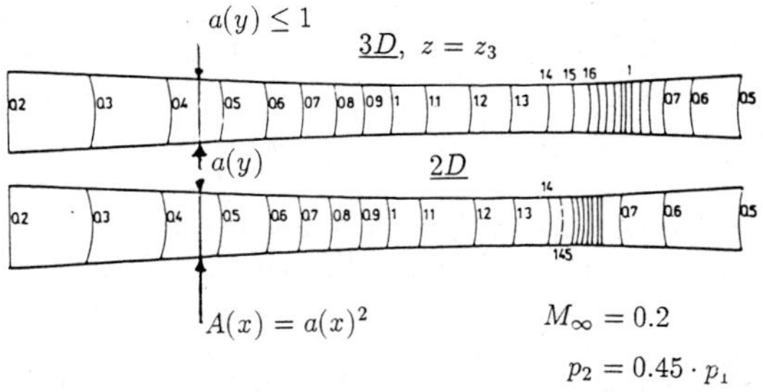

Fig. 2.5 Comparison of 2D and 3D results using lines $M = const.$

Numerical schemes for the Euler equations in two space dimensions without dimensional splitting.

Dietmar Kröner
Universität Heidelberg, Institut für Angewandte Mathematik
Im Neuenheimer Feld 294, D-6900 Heidelberg

Abstract

We consider a numerical scheme for the nonlinear Euler equations of gasdynamics in 2-D. The algorithm doesn't use any dimensional splitting. It is a generalization of a scheme, which was developped by Roe for the linear Euler equations. In 1-D perturbations can propagate only in two directions but in 2-D there are infinite many directions of propagation. Therefore the algorithm should be able to select the most important directions and to ensure that the scheme takes this fact into account. In this paper we shall describe some details of this algorithm and we shall present some numerical results in 1-D and 2-D.

Introduction

In this paper we consider numerical schemes for the Euler equation of gasdynamics

$$\partial_t U + \partial_x F(U) + \partial_y G(U) = 0, \tag{1}$$

where

$$U := (p, u, v, \rho)^t,$$

$$F(U) := (u\partial_x p + \rho a^2 \partial_x u, \frac{1}{\rho}\partial_x p + u\partial_x u, u\partial_x v, \rho\partial_x u + u\partial_x \rho),$$

$$G(U) := (v\partial_y p + \rho a^2 \partial_y v, v\partial_y u, \frac{1}{\rho}\partial_y p + v\partial_y v, \rho\partial_y v + v\partial_y \rho).$$

Here ρ denotes the density, u and v the components of the velocity with respect to x and y, e the energy density and p the pressure. For an ideal gas we have the following equation of state

$$p = (\gamma - 1)(e - 0.5\rho(u^2 + v^2)). \tag{2}$$

Up to now there is no general existence result in particular no convergence result for a numerical scheme for the Euler equation in two space dimension . On the contrary to one-dimensinal problems in two dimensions , perturbations can spread into infinite many directions. The schemes using dimensional splitting don't take care of this fact and there are examples for which they do not work (see Roe [9]). Recently Roe [6]and [7]has published some basic ideas for schemes which use the direction of propagation and which work without dimensional splitting. For other schemes which do not use dimensional splitting in two dimensions we refer to Hirsch et al. [3],

Colella [1], Davis [2], LeVeque [4], [5]. Now let me describe Roe's scheme for the linear Euler equations.

Roe's scheme for the linear Euler equation:

In the papers [6]and [7]Roe has developped a numerical scheme for the linearized Euler equation in 2-D .The algorithm is able to select the most significant directions and to ensure that the information propagates numerically in the same directions as it would propagate physically. The basic idea of Roe consists in using the gradients of the last time step to approximate locally the unknown solution by a modelflow of travelling waves of the form

$$v(x cos(\Theta) + y sin(\Theta) - St)$$

where v is some smooth function and Θ is related to the direction of propagation. Then Roe constructs a monotone scheme of first order for scalar equations in two space dimensions and generalizes it to linear systems. In particular the Roe scheme works as follows for the linear Euler equation

$$\partial_t U + P\partial_x U + Q\partial_y U = 0, \tag{3}$$

where U is as above,

$$P = \begin{pmatrix} u & \rho a^2 & 0 & 0 \\ \frac{1}{\rho} & u & 0 & 0 \\ 0 & 0 & u & 0 \\ 0 & \rho & 0 & u \end{pmatrix} \tag{4}$$

$$Q = \begin{pmatrix} v & 0 & \rho a^2 & 0 \\ 0 & v & 0 & 0 \\ \frac{1}{\rho} & 0 & v & 0 \\ 0 & 0 & \rho & v \end{pmatrix} \tag{5}$$

and $a = \sqrt{\gamma p/\rho}$ is the sound velocity. Since for solutions of the form

$$v(x cos(\Theta) + y sin(\Theta) - S(\Theta)t)$$

it turns out that v' is an eigenvector of $P cos(\Theta) + Q sin(\Theta)$, let us start to compute the eigenvectors of $cos(\Theta)P + sin(\Theta)Q$. They are

$$R_1(\Theta) = \begin{pmatrix} \rho a^2 \\ a cos\Theta \\ a sin\Theta \\ \rho \end{pmatrix}, R_2(\Theta) = \begin{pmatrix} \rho a^2 \\ -a cos\Theta \\ -a sin\Theta \\ \rho \end{pmatrix}, R_3(\Theta) = \begin{pmatrix} 0 \\ 0 \\ 0 \\ \rho \end{pmatrix}, R_4(\Theta) = \begin{pmatrix} 0 \\ -a sin\Theta \\ a cos\Theta \\ 0 \end{pmatrix} \tag{6}$$

and the corresponding eigenvalues

$$\lambda_1 = u cos\Theta + v sin\Theta + a,$$

$$\lambda_2 = -u cos\Theta - v sin\Theta + a, \tag{7}$$

$$\lambda_3 = ucos\Theta + vsin\Theta,$$

$$\lambda_4 = ucos\Theta + vsin\Theta.$$

The first and the second eigenvector are related to acoustic waves, the third one to an entropy and the fourth one to a shear wave. Roe chooses the following six eigenvectors

$$R_1(\Theta), R_1(\Theta + \frac{\pi}{2}), R_2(\Theta + \pi), R_1(\Theta - \frac{\pi}{2}), R_3(\phi), R_4(0), \tag{8}$$

where ϕ is an additional unknown parameter. Now let us assume that we have already an approximation w_A^n of U at time $t_n := n\Delta t$ at the point A of a given triangulation (see Fig.1).

Then we have to describe how to compute w_A^{n+1}. Because of the travelling-wave-Ansatz, for each neighbour point B of A we have to compute $\beta_k^B, k = 1, ..., 6, \Theta$ and ϕ, such that

$$\sum_{k=1}^{6} \beta_k^B \begin{pmatrix} c_k \\ s_k \end{pmatrix} r_k(\Theta_k) = \nabla w_A^n, \tag{9}$$

where

$$\Theta_1 = \Theta, \Theta_2 = \Theta + \pi/2, \Theta_3 = \Theta - \pi/2, \Theta_4 = \Theta + \pi, \Theta_5 = \phi, \Theta_6 = 0,$$

$c_k = cos(\Theta_k), s_k = sin(\Theta_k)$ and r_k are the eigenvectors (8). This is a system of eight equations for the unknowns $\beta_k^B, k = 1, ..., 6, \Theta$ and ϕ. Let us assume that it is solvable. Actually by this special choice of eigenvectors there is an explicit formular to solve it. Then we set

$$w_{AB} := 1/2(w_A^n + w_B^n) - 1/2 \sum_{k=1}^{6} sign(n_B \begin{pmatrix} c_k \\ s_k \end{pmatrix} \lambda_k) \beta_k^B \begin{pmatrix} c_k \\ s_k \end{pmatrix} (A - B) r_k, \tag{10}$$

where n_B is the inner normalvector to the dual mesh (see Fig.1, cell with dotted line) with respect to B. Then for w_A^{n+1} we define

$$w_A^{n+1} := w_A^n - \Delta t/(\Delta x^2) \sum_B n_B \begin{pmatrix} P w_{AB} \\ Q w_{AB} \end{pmatrix}. \tag{11}$$

(10) and (11) define a two-dimensional upwind scheme of first order, which is monotone and in conservation form for scalar linear equations (see Roe [6]and [7].)
As far as we know there are no numerical experience for this algorithm even in the linear case. Therefore it remains to test this scheme, to generalize it to nonlinear systems and to compare it with other existing schemes. Therefore we started to do this. The most important question is how to generalize it to nonlinear systems. We have tried two different ideas which we are now going to describe.

Generalization to nonlinear systems,first version:

Consider a local part of the grid (see Fig.1). In order to update the solution in A for the timestep t^{n+1} we have to take into account the points $B_i, i = 1, ..., 6$. Let me describe the procedure for B_1 which we have to repeat successively for B_i, $i = 2, ..., 6$. In order to get an approximation for ∇w_A^n in formular (9) we use the values of w^n in the neighbour points B_2, A, B_6 of B_1. For defining the eigenvectors $r_k(\Theta), k = 1, ...6$ we need u, v,ρ and p, evaluated at a suitable point or some meanvalues of $w^n_{B_1}, w^n_{B_2}, w^n_A, w^n_{B_6}$. In our case we choose the Roe-mean-value (see [8]) of $w^n_{B_1}, w^n_A$ which is defined for instance for u as

$$u_{mean} = \frac{\sqrt{(\rho_{B_1})}u_{B_1} + \sqrt{(\rho_A)}u_A}{\sqrt{(\rho_{B_1})} + \sqrt{(\rho_A)}} \quad (12)$$

We solve (9) in order to get $\beta^B_k, k = 1, ...6, \Theta$ and ϕ using an explicit formular. Then we compute w_{AB} as defined in (10). In the same way we treat the other points $B_i, i = 2, ..., 6$. The new value w^{n+1}_A is then defined as

$$w^{n+1}_A := w^n_A - \Delta t/(\Delta x^2) \sum_B n_B \begin{pmatrix} F(w_{AB}) \\ P(w_{AB}) \end{pmatrix}. \quad (13)$$

Since the eigenvectors change with each B_i, this scheme applied to a scalar equation is in general not monotone.

Testproblem 1

Shock-tube problem: As a first testproblem for this version of the 'Roe-scheme' we consider the flow through a two dimensional tube with one-dimensional Riemann data. The basic domain Ω for our calculations is $\Omega = (0, 0.3)x(0, 0.1)$. In particular we choose the following data:

$$grid : 20x60, \Delta x = 0.005, \Delta y = 0.005, \Delta t = 0.0025.$$

Initial conditions:

$$u_0 = 0, v_0 = 0 in \Omega$$

and

$$\rho_0(x, y) = \begin{cases} 1.0, & \text{if } y \in (0, 0.1), x \in (0, 0.015); \\ 0.125, & \text{if } y \in (0, 0.1), x \in (0.015, 0.03); \end{cases}$$

$$e_0(x, y) = \begin{cases} 2.5, & \text{if } y \in (0, 0.1), x \in (0, 0.015); \\ 0.25, & \text{if } y \in (0, 0.1), x \in (0.015, 0.03). \end{cases}$$

These initial conditions are known as 'Sod's testproblem' (see Sod[11]). On the boundary we use mixed Dirichlet and Neumann boundary conditions.

These data are of 1-D structure but we use it as a testproblem for the 2-D scheme of Roe. For the result we expect the functions shown in Fig.2 (see Sod[11]). The density is plotted against the length x of the tube. The dotted lines refer to the numerical solution computed with the Godunov scheme and the solid line to the exact solution. The solution is a shockwave going to the right-hand side, followed by a contactdiscontinuity and a rarefaction wave going to the left-hand side.

The results of this scheme are shown in Fig.3, where we have plotted the density ρ as a function of x for T=0.15 There arise some problems concerning the stability. After some time iterations we obtain oscillations near the contactdiscontinuity which increases in time. We believe that this is due to the fact that the scheme is not monotone (for the scalar equation). Therefore let us study a second version of this scheme, which avoids this oscillations.

Generalization to nonlinear systems,second version:

Now let us consider the Euler equations in conservative variables.

$$\partial_t U + \partial_x F(U) + \partial_y G(U) = 0, \tag{14}$$

where
(15)
$$U := (\rho, \rho u, \rho v, e)^t, F(U) := (\rho u, \rho u^2 + p, \rho uv, u(e+p))^t, G(U) := (\rho v, \rho uv, \rho v^2 + p, v(e+p))^t. \tag{15}$$

Again we write the equations in the form

$$\partial_t U + P\partial_x U + Q\partial_y U = 0, \tag{16}$$

with suitable matrices P and Q.

We consider again a local part of the grid as in Fig.1 . Now in the first step we compute the local direction of propagation by approximating the discontinuity by travelling waves. Assume that for instance

$$\rho(t,x,y) = \rho_0(x\cos(\Theta) + y\sin(\Theta) - \lambda t)$$

for some unknown function $\rho_0 \in C^1(R,R)$ and $\Theta, \lambda \in R$. Then

$$\partial_x \rho = \rho_0' \cos\Theta, \partial_y \rho = \rho_0' \sin\Theta$$

and

$$\tan\Theta = \frac{\partial_y \rho}{\partial_x \rho}.$$

This gives a condition for Θ if $\partial_x \rho \neq 0$. Otherwise we choose $\Theta = \frac{\pi}{2}$ or the solution is smooth and the approximating solution should not depend on Θ.

Now let me explain the local linearization using geometrical arguments. We draw a line g orthogonal to the direction given by Θ through the point A (see Fig.1). Then on each side of this line we compute the meanvalues M_l and M_r of the state vectors in all neighbour points and afterwards we take the Roe-mean-value R of M_l and M_r (see (12)). Now for evaluating P and Q in (16) we use the values of R. Before going to more details of the numerical scheme let us derive the local equation which we shall approximate later on by the numerical scheme. Define

$$V(t,\xi,\eta) := U(t,x,y),$$

where

$$\xi = \xi(\Theta) = x\cos\Theta + y\sin\Theta, \eta = \eta(\Theta) = -x\sin\Theta + y\cos\Theta.$$

For V we obtain

$$\partial_t V + D(\Theta)\partial_\xi V + D(\Theta + \frac{\pi}{2})\partial_\eta V = 0, \tag{17}$$

where $D(\Theta) = P\cos\Theta + Q\sin\Theta$. While Roe chooses six eigenvectors (see (8)) we take only four linear independent eigenvectors of $D(\Theta)$:

$$R_1(\Theta) := \begin{pmatrix} 1 \\ u + a\cos\Theta \\ v + a\sin\Theta \\ H + a(u\cos\Theta + v\sin\Theta) \end{pmatrix}, R_2(\Theta) := \begin{pmatrix} 1 \\ u - a\cos\Theta \\ v - a\sin\Theta \\ H - a(u\cos\Theta + v\sin\Theta) \end{pmatrix},$$

$$R_3(\Theta) := \begin{pmatrix} 1 \\ u \\ v \\ \frac{1}{2}(u^2+v^2) \end{pmatrix}, R_4(\Theta) := \begin{pmatrix} 0 \\ -asin\Theta \\ acos\Theta \\ a(vcos\Theta - usin\Theta) \end{pmatrix} \tag{18}$$

where $H = (e+p)/\rho$ is the enthalpy. The corresponding eigenvalues are the same as in (7). Then there exists $\alpha_j(t,\xi,\eta), j = 1,...4$ such that

$$V(t,\xi,\eta) = \sum_{j=1}^{4} \alpha_j(t,\xi,\eta) R_j. \tag{19}$$

Because of (17) we obtain for α_j:

$$\sum_j (\partial_t \alpha_j + D(\Theta)\partial_\xi \alpha_j + D(\Theta + \frac{\pi}{2})\partial_\eta \alpha_j) R_j = 0\,. \tag{20}$$

Now we assume that the derivative of V tangential to the discontinuity vanishes: $\partial_\eta V = 0$. This is satisfied for instance, if the states in front and behind of the discontinuity are constant or if the problem is rotationally symmetric. Therefore we obtain for $\alpha_1, ..., \alpha_4$ the following equations:

$$\partial_t \alpha_j + \lambda_j \partial_\xi \alpha_j = 0 \tag{21}$$

or in the original coordinate-system

$$\partial_t \alpha_j + \lambda_j \cos(\Theta_j) \partial_x \alpha_j + \lambda_j \sin(\Theta_j) \partial_y \alpha_j = 0. \tag{22}$$

Assume we have already computed V^n (this means an approximation of V at time $n\Delta t$). Then define α_j^n by solving a system of linear equations. For any j we have

$$\sum_{j=1}^{4} \alpha_{j,A}^n r_j = V_A^n \tag{23}$$

for any point A of the grid. Then for any j we compute an approximation $\alpha_{j,A}^{n+1}$ of (22). with respect to the initial conditions

$$\alpha_j^{n+1}(0,.) = \alpha_j^n. \tag{24}$$

(22) is a scalar equation and we solve it using Roe's scheme. If one applies it to (22),(24) we get:

$$\alpha_{j,AB} := \frac{1}{2}(\alpha_{j,A}^n + \alpha_{j,B}^n) - \frac{1}{2} sign(n_B \begin{pmatrix} c_j \\ s_j \end{pmatrix} \lambda_j)(\alpha_{j,A}^n - \alpha_{j,B}^n)\,. \tag{25}$$

Depending on the flow direction, $\alpha_{j,AB}$ is either equal to $\alpha_{j,A}^n$ or equal to $\alpha_{j,B}^n$. For $\alpha_{j,A}^{n+1}$ we obtain

$$\alpha_{j,A}^{n+1} := \alpha_{j,A}^n - \frac{\Delta t}{(\Delta x^2)} \sum_B n_B \begin{pmatrix} c_j \\ s_j \end{pmatrix} \lambda_j \alpha_{j,AB}^n \tag{26}$$

and therefore

$$V_A^{n+1} := V_A^n - \Delta t/(\Delta x^2) \sum_j \sum_B n_B \begin{pmatrix} c_j \\ s_j \end{pmatrix} \lambda_j \alpha_{j,AB}^n R_j. \tag{27}$$

This is again a first order upwind scheme for (14). It can be seen very easily that the scheme (25), (26) is monotone.

Testproblem 2

Shock-tube problem: Again we have tested this scheme with Sod's data as in Testproblem 1 and obtained the results shown in Fig.4. Now the oscillation obtained by the first version disappeared and the qualitative behaviour agree with the results computed with the Godunov scheme(see Fig.2).

Testproblem 3

Comparison with an exact solution: In order to get some informations about the correctness of the solution we have solved the Euler equation (14) with respect to initial conditions, satisfying the Rankine-Hugoniot conditions. For the values of the left side we choose $\rho_l = 1$, $p_l = 1$, $u_l = 0$, $v_l = 0$, and for the pressure on the right side we choose $p_r = 5$. Then we compute ρ_r, u_r, v_r and the shock velocity, such that the Rankine-Hugoniot conditions are satisfied (see Smoller [10]). It turn out that the exact shock velocity is $\sigma = -2.4899....$ The numerical experiments for $\Delta t = 0.003$, $\Delta x = 0.01$, and $T = 0.15$ show (see Fig.5,6), that $\sigma \approx -2.4$ and that the error in the descret L^1-Norm is equal to 0.05975. For $\Delta t = 0.006$, $\Delta x = 0.02$, and $T = 0.15$ we obtain for this error 0.16160.

Testproblem 4

Interacting blast waves; strong shocks: We have applied our scheme to the problem of two interacting blast wave in a tube of finite length with strong shocks (see Woodward, Colella [12]). As the input data we have used the same one as in Woodward, Colella [12], i.e. reflecting walls and for the initial conditions: $p = 1000$ in the leftmost tenth of the tube, $p = 100$ in the rightmost tenth and in between $p = 0.01$. The density ρ is equal to 1 everywhere, and the velocities are $u = 0$, $v = 0$. The numerical results for $\Delta t = 0.00002$, $\Delta x = 0.001$, and $T = 0.016$ and $T = 0.026$ are shown in Fig.7,8. If one compares the result with those of Woodward, Colella [12]it turns out that the shock velocity of the left shock is a little bit too small.

Testproblem 5

Converging cylindrical shockwaves in 2-D: As initial conditions we choose radial symmetric values:Within an interior circel we prescribe the density and the energy density equal to 1 and in the exterior domain equal to 4. Initially the velocities are equal to zero everywhere. On the boundaries of the basic domain we use reflecting boundary conditions. Then we should expect a radialsymmetric solution. In our experiments we have used a grid of $100x100$ points and we have chosen $\Delta t = 0.05$, $\Delta x = 0.1$. The results for the time $T = 1.25$ are shown in Fig. 9 for a 3-D-view, in Fig.10 for a cross section through the center, and in Fig.11 for level-lines. We have plotted the density ρ as a function of x and y. We should mention that it would be more convenient to solve this problem in polarcoordinates because of its radial symmetry. But on the other hand it seems to be a good testproblem for our scheme. The obtained results are of the same structure as the exact solution. At the moment the algorithm does not compute the angle Θ automatically. Up to know we compute it explicitly using the radial symmetry of the problem.

All the computations have been done on an IBM 3090. The CPU-time can be used only for internal comparisons since the program was not optimized with respect to the CPU-time.

Acknowledgements

The author wishes to thank Professor Jäger and Dr.Klingenberg for their support and valuable discussions. Furthermore the author would like to acknowledge the financial support of the Deutsche Forschungsgemeinschaft.

References

[1]P.Colella, Multidimensional upwind methods for hyperbolic conservation laws. Report Lawrence Berkeley Laboratory 17023,1984.

[2]S.F.Davis, A rotationally biased upwind difference scheme for the Euler equations.J.of Comp. Physics 56(1984), 65-92.

[3]Ch.Hirsch, C.Lacor, H.Deconinck, Convection algorithm on a diagonal procedure for the multidimensional Euler equation. AIAA, Proceedings of the 8-th Computational Fluid Dynamics Conference 1987.

[4]R.J.LeVeque, High resolution finite volume methods on grids via wave propagation. ICASE REPORT NO.87-68, 1987

[5]R.J.LeVeque, Cartesian grid methods for flow in irregular regions. To appear in the Proceedings of the Oxford Conference on Numerical Methods in Fluid Dynamics, 1988.

[6]P.Roe, A basis for upwind differentiating of the two dimensional unsteady Euler equations. Numerical methodes for fluid dynamics II, Eds.: Morton, Baines, Oxford Univ. Press 1986.

[7]P.Roe, Discrete models for the numerical analysis of time dependent multidimensional gas dynamics. Icase report 85-18,1985.

[8]P.Roe, Approximate Riemann solvers, parameter vectors difference schemes. J.of Comp. Physics 43 (1981), 357-372 .

[9]P.Roe, Discontinuous solutions to hyperbolic systems under operator splitting. Manuscript.

[10]J.Smoller, Shockwaves and reaction-diffusion equations. New York Heidelberg Berlin.

[11]G.A.Sod, A survey of several finite difference methods for systems of nonlinear hyperbolic conservation laws. J.of Comp. Physics 27, (1978), 1-31 .

[12]P. Woodward, P. Colella, The numerical simulation of two-dimensional fluid flow with strong shocks. J.Comp.Phys.54(1984), 115-173.

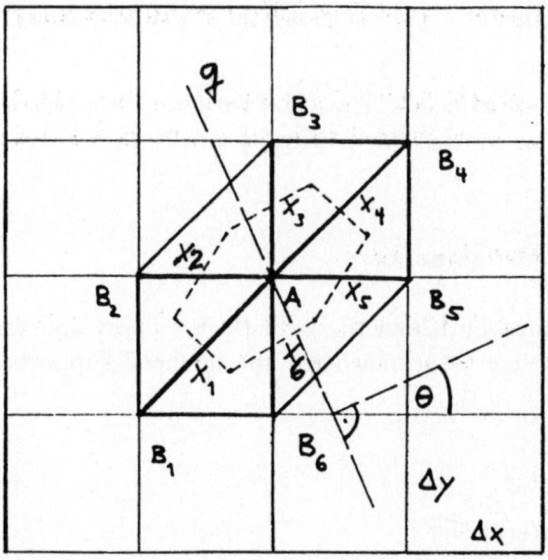

Figure 1

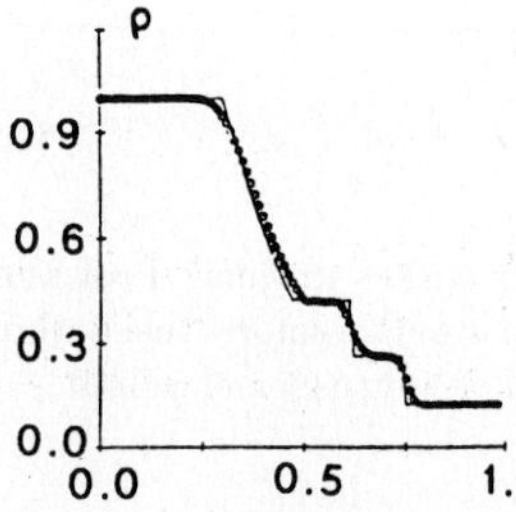

Figure 2

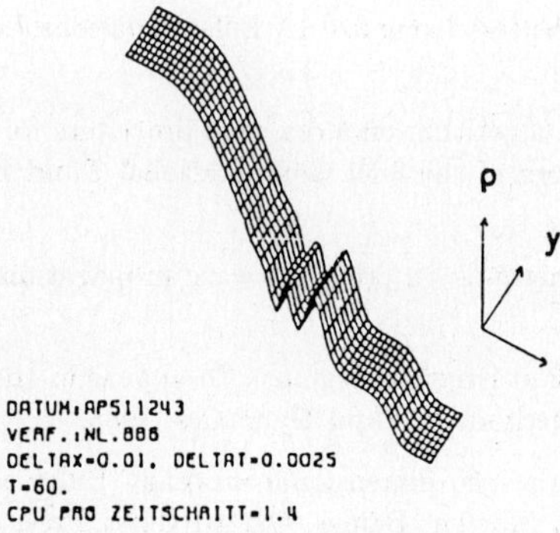

Figure 3

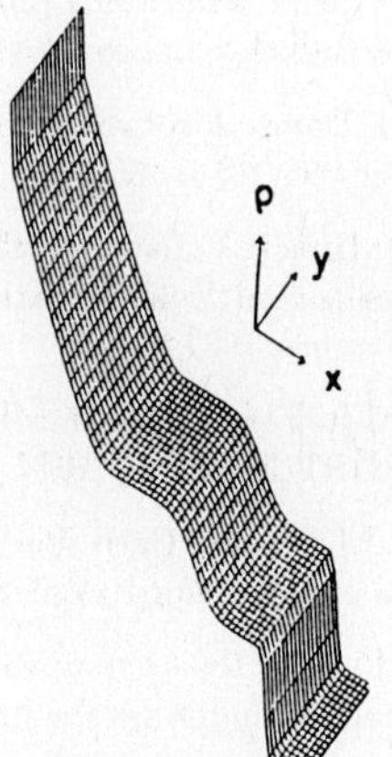

Figure 4

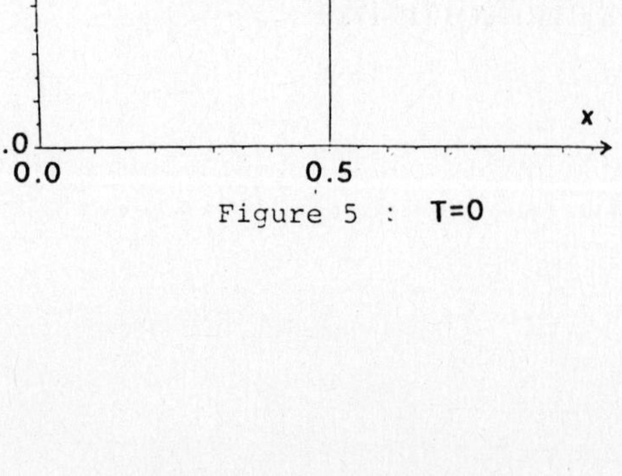

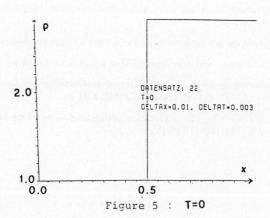

Figure 5 : T=0

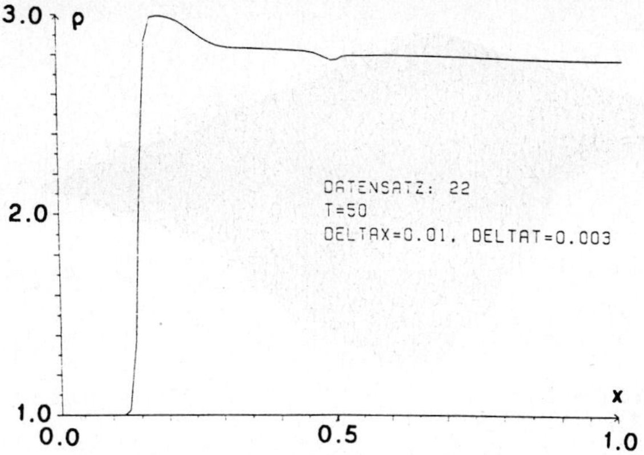

Figure 6 : **T=0.15**

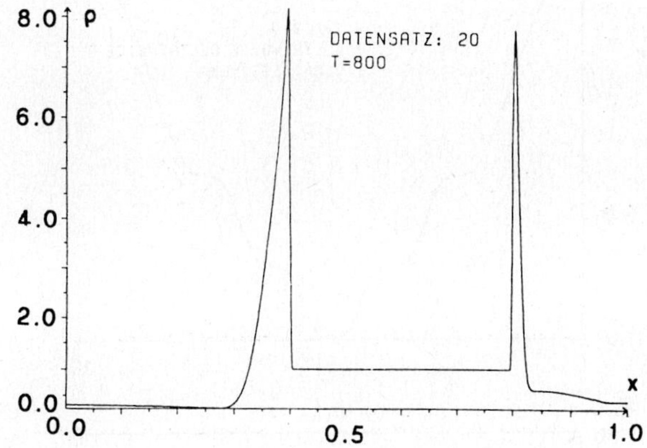

Figure 7

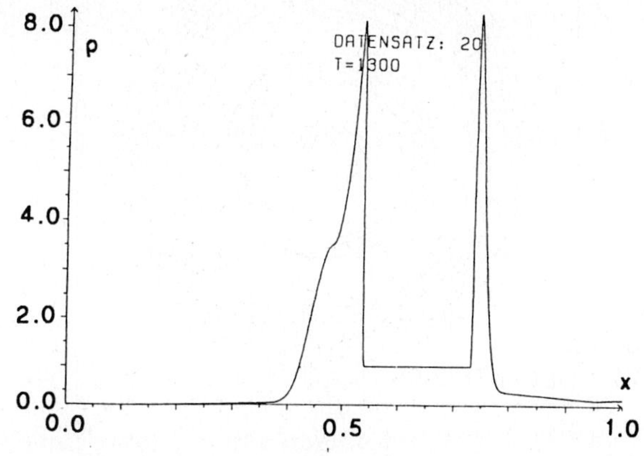

Figure 8

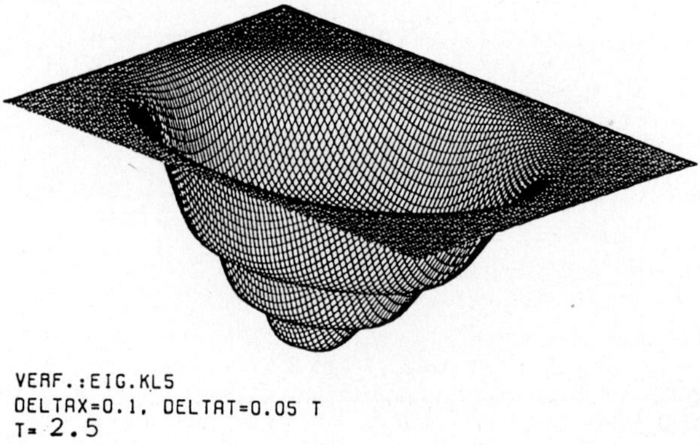

Figure 9

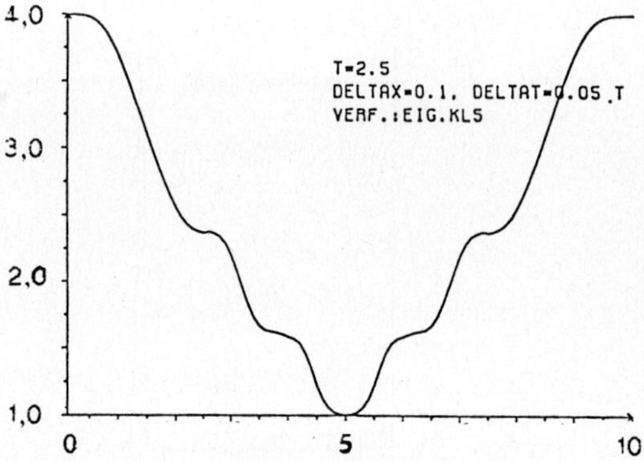

Figure 10

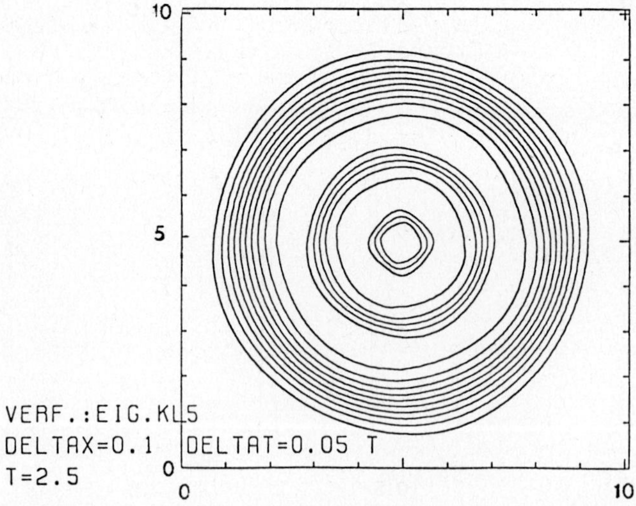

Figure 11

INITIAL-BOUNDARY VALUE PROBLEMS FOR TRANSONIC EQUATIONS IN THE UNBOUNDED DOMAIN

N.A.Lar'kin
The Institute of Theoretical and Applied Mechanics. Novosibirsk-9o, 63oo9o, USSR.
(This work was accomplished during author's working in the Mathematical Institute 1 of Karlsruhe as Alexander von Humboldt Fellow)

Summary

Initial-boundary value problems for nonlinear equations, modelling nonsteady 3-D transonic gas flows near a body, which differs only sligtly from a slender cylinder, are considered. Local in time existence and uniqueness of classical solutions for viscous and inviscous flows are proved.

The Hyperbolic Problem

When studying a transonic flow, model equations governing the development of perturbations near a known solution are widely used. These equations are derived under various assumptions from Navier-Stokes equations for a compressible heatconducting gas. The Lin-Reissner-Tsieghn equation

$$u_{xt} + u_x u_{xx} - \Delta_y u = 0 , \qquad (1)$$

where $\Delta_y = \partial^2/\partial y_1^2 + \partial^2/\partial y_2^2$, u is a potential of disturbances, simulates the development of perturbations in nonsteady non-viscous transonic flow near a body, which differs only slightly from a slender cylinder [1]. It is easy to verify, that (1) is hyperbolic for all finite values of u_x. (1) is considered in the domain $G = D \times (0,T)$, where $D = R^2 \setminus \Omega$. Here $y \in \Omega \subset R^2$, $x \in R^1 = R$. Ω is a domain with a boundary $\partial\Omega$, which is sufficiently smooth. Denote $S = \partial\Omega \times R$, $S_T = S \times (0,T)$; n is an outward normal vector on $\partial\Omega$, then on S_T Neuman's condition is given

$$\partial u/\partial n|_{S_T} = \partial\Phi/\partial n|_{S_T} , \qquad (2)$$

where $\Phi(x,y,t)$ is a known function. We recall, that (2) is a linear version of the impermeability condition. At $t = 0$ the initial data are given

$$u(x,y,0) = u_o(x,y) . \qquad (3)$$

At the infinity the decay of perturbations is given

$$\lim_{|x|+|y|\to\infty} u_x = 0, \qquad \lim_{x\to-\infty} u_y = 0, \qquad \lim_{|y|\to\infty} u_y = 0. \tag{4}$$

Theorem 1. Let be $\Phi \in W_2^6(G)$, $u_o \in W_2^5(D)$, $\int_{-\infty}^{x} u_o(t,y)dt \in W_2^5(D)$. $\Phi = 0$, if $|x| + |y| \geq r$; $\partial u_o/\partial n|_s = \partial\Phi/\partial n|_s(x,y,0)$. Then one can find such a $T_o \in (0,T)$, that in $G_o = D \times (0,T_o)$ there exists a unique solution of (1) -(4): $\partial_t^i u_x \in L_\infty(0,T_o; W_2^{3-i}(D))$, $\partial_t^j \Delta_y u \in L_\infty(0,T_o;W_2^{2-j}(D))$; $(i = 0,1,2;\ j = 0,1)$.

Here and further on $W_p^l(D)$, $L_q(0,T;W_p^l(D))$ denote respectively isotropic and anisotropic spaces of S.L.Sobolev [3] .

Remark. We mean it in all our assertions on uniqueness, that $u(x,y,t)$ is defined except up to a constant.

In order to prove Theorem 1 we consider the following auxiliary problem.

The Viscous Problem

Consider in G the equation

$$u_{\mu xt} - \mu\partial_x^3 u_\mu + u_{\mu x}u_{\mu xx} - \Delta_y u_\mu = 0 , \tag{5}$$

where μ is a positive small number. (5) is also a physical equation. It describes the development of nonstationary perturbations in a viscous heatconducting transonic flow [2] . Unlike (1), it is nonclassical one. We study for (5) the following initial-boundary value problem

$$\partial u_\mu/\partial n|_{S_T} = \partial\Phi_\mu/\partial n|_{S_T}, \quad u_\mu(x,y,0) = u_{\mu o}(x,y),$$

$$\partial u_{\mu o}/\partial n|_S = \partial\Phi_\mu(x,y,0)/\partial n|_S, \quad \lim_{|x|+|y|\to\infty} u_{\mu x} = 0, \quad \lim_{x\to-\infty} u_{\mu y} = 0,$$

$$\lim_{|y|\to\infty} u_{\mu y} = 0 . \tag{6}$$

Here Φ_μ, $u_{\mu o}$ are smooth approximations of Φ, u_o. The following assertion holds.

Theorem 2. Let be $\Phi_\mu \in W_2^7(G)$, $u_{\mu o} \in W_2^6(D)$, $\int_{-\infty}^{x} u_{\mu o}(t,y)dt \in W_2^6(D)$; $\lim_{\mu\to o} ||\Phi-\Phi_\mu||_{W_2^6(G)} = 0$, $\lim_{\mu\to o} (||u_0 - u_{\mu 0}||_{W_2^5} + ||\int_{-\infty}^{x}(u_{\mu o}(t,y)-$

$u_o(t,y))dt||_{W_2^5(D)}) = 0$. Then for any $\mu > 0$ such $T_1 \in (0,T)$, $T_1 \geq T_o$, can be found, that there exists in $G_1 = D \times (0,T_1)$ the unique solution of (5),(6): $\partial_t^i u_{\mu x} \in L_\infty(0,T_1;W_2^{3-i}(D))$ $(i = 0,1,2)$; $\partial_{t,x}^{j+3} u_\mu \in L_\infty(0,T_1;W_2^{2-j}(D))$, $\partial_t^j \Delta_y u_\mu \in L_\infty(0,T_1;W_2^{2-j}(D))$ $(j = 0,1)$, and the following inequality holds

$$\sum_{i=o}^{2} ||\partial_t^i u_{\mu x}||_{L_\infty(0,T_1;W_2^{3-i}(D))} + \sum_{j=o}^{1} \mu^{1/2} |\partial_{t,x}^{j+3} u_\mu||_{L\infty(0,T_1;W_2^{2-j}(D))}$$

$$+ \quad ||\partial_t^j \Delta_y u_\mu||_{L\infty(0,T_1;\, W_2^{2-j}(D))} \leq C,$$

where a constant C does not depend on μ; $\partial_{t,x}^{i+j} = \partial^{i+j}/\partial t^i \partial x^j$.

It is clear, that existence assertion in Theorem 1 follows from Theorem 2 and from a *-weak convergence of sequences $\{u_{\mu x}\}$, $\{\Delta_y u_\mu\}$ if one pass to the limit in (5),(6) as μ tends to zero. A crucial point therefore is to prove Theorem 2. To do it we investigate at first a linear version of (5),(6) in a sequence of domains G_m, bounded in y-variables. We will construct solutions of these problems by Galerkin's method and prove à priori estimates, which make it possible: 1) to prove convergence of Galerkin's approximations, 2) to prove solvability of a linear problem in G_T, 3) to prove Theorem 2 with the help of the Contracted mapping theorem and hence Theorem 1.

<u>The linear problem.</u> 1) Define G_m as follows: $G_m = D_m \times (0,T)$, $D_m = \Omega_m \times R$, $\Omega_m = K_m \setminus \Omega$, $K_m := \{|y| < m\}$, $m > r$. $S_m = \partial\Omega_m \times R \times (0,T)$. We construct in D a sequence of functions $v_{mo}(x,y)$: $v_{mo} \in W_2^6(D)$, $\int_{-\infty}^{x} v_{mo}(t,y)dt \in W_2^6(D)$, $v_{mo} = 0$, if $|y| > m$, $\partial v_{mo}/\partial n|_S = 0$; $\lim_{m\to\infty}(||v_{mo} - v_o||_{W_2^6(D)} + ||\int_{-\infty}^{x}[v_{mo}(t,y) - v_o(t,y)] \times dt||_{W_2^6(D)}) = 0$. Consider in G_m the linear problem

$$L_\mu v = v_{xt} - \mu v_{xxx} + K(x,y,t)v_{xx} + \alpha(x,y,t)v_x -$$

$$\Delta_y v = f(x,y,t), \tag{7}$$

$$\partial v/\partial n|_{S_m} = 0, \quad v(x,y,0) = v_{mo}(x,y), \quad \lim_{|x|\to\infty} v_x = 0,$$

$$\lim_{x\to -\infty} v_y = 0. \tag{8}$$

Here $K(x,y,t)$, $\alpha(x,y,t)$, $f(x,y,t)$ are known functions.
2) We fix $m < \infty$ and construct approximate solutions of (7), (8) by Galerkin's method

$$V^N(x,y,t) = \sum_{j=1}^{N} g_j(x,t)w_j(y), \quad \Delta_y w_j + \lambda_j w_j = 0 \quad \text{in } \Omega_m;$$

$$\partial w_j/\partial n\big|_{\partial\Omega m} = 0; \quad (w_i,w_j) = \delta_{ij}, \quad (u,v) = \int_{\Omega_m} u(x,y,t)v(x,y,t)dy.$$

We find unknown functions $g_j(x,t)$ as solutions of the following evolution problem

$$(L_\mu v^N, w_j) = (f,w_j) \equiv f_j \quad (j = 1,\dots,N). \tag{9}$$

$$g_j(x,0) = g_{jo}(x) = (v_{mo},w_j), \quad \lim_{|x|\to\infty} g_{jx} = 0,$$

$$\lim_{x\to -\infty} \lambda_j g_j = 0. \tag{10}$$

<u>Lemma 1</u>. Let be $\partial_t^i K$, $\partial_t^i \alpha \in L_\infty(0,T;W_2^{3-i}(D))$, $\partial_t^i f \in L_2(0,T;W_2^{3-i}(D))$, $i = 0,1,2$. Then for any fixed $\mu > 0$, $m<\infty$, $N<\infty$ there exists the unique solution of (9),(10): $\partial_t^i g_{jx} \in L_\infty(0,T;W_2^{4-i}(R)) \cap L_2(0,T;W_2^{5-i}(R))$, $\partial_t^3 g_{jxx} \in L_2(0,T;L_2(R))$, $g_j \in W_\infty^2(0,T;L_2(R))$. $i = 0,1,2$; $j = 1,\dots,N$.

<u>Proof of Lemma 1</u>. Assume, without a loss of generality, that in G_m the following inequality is fulfilled

$$2\alpha - 5|K_x| > \delta > 0 .$$

Indeed, we have after changing in (7) the unknown function as $v = e^{\lambda t}u$:

$$L_\mu u = u_{xt} - \mu\partial_x^3 u + K(x,y,t)u_{xx} + (\alpha(x,y,t) + \lambda)u_x -$$

$$\Delta_y u = e^{-\lambda t} f(x,y,t).$$

We obtain the desired inequality choosing $\lambda > 0$ big enough. Divide segment $[0,T]$ in L equal parts: $h = [0,T]/L$. Denote $g^l(x) = g(x,lh)$, $l = 0,\dots,L$, and consider the discretization of (9) in t:

$$L_h g_j^l = (g_{jx}^l - g_{jx}^{l-1})/h - \mu\partial_x^3 g_j^l + (K^l u_{xx}^{Nl}, w_j) + (\alpha^l u_x^{Nl}, w_j) +$$

$$\lambda_j g_j^l = f_j^l , \quad (j = 1,\dots,N; \quad l = 1,\dots,L). \tag{11}$$

$$g_j^o(x) = g_{jo}(x), \quad \lim_{|x|\to\infty} g_{jx}^l = 0, \quad \lim_{x\to-\infty} \lambda_j g_j^l = 0. \tag{12}$$

Lemma 2. Let conditions of Lemma 1 be fulfilled. Then for any fixed $\mu > 0$, $N < \infty$ and for any $1 \le l \le L$ there exists in R the unique solution of (11),(12):

$$g_{jx}^l \in W_2^5(R), \quad \lambda_j g_j^l \in L_2(R) \quad (j = 1,\dots,N).$$

Proof of Lemma 2. Rewrite (11) in the form

$$-\mu\partial_x^3 g_j^l + (K^l u_{xx}^{Nl}, w_j) + ((\alpha^l + 1/h)u_x^{Nl}, w_j) + \lambda_j g_j^l =$$

$$f_j^l + g_{jx}^{l-1}/h = f_{j1}^l. \tag{13}$$

It is easy to verify, that for any smooth solution of (13), (12) the following estimate holds

$$\sum_{j=1}^N ||g_{jx}^l||_{W_2^2(R)} + ||\lambda_j g_j^l||_{L_2(R)} \le C\sum_{j=1}^N ||f_{j1}^l||, \quad 1\le l \le L,$$

which imply a weak solvability of (13),(12). One can show the regularity of solutions by standard methods. Lemma 2 is proved. Now we can continue to prove Lemma 1. Let us define:

$v_{xh}^l = (v_x^l - v_x^{l-1})/h$, $v_{xhh}^l = (v_{xh}^l - v_{xh}^{l-1})/h$, $L_{2h}g^l = (L_h g^l - L_h g^{l-1})/h$, $L_{3h}g^l = (L_{2h}g^l - L_{2h}g^{l-1})/h$ and consider integrals: $\int_R (L_h g_j^l - f_j^l) g_{jx}^l dx = 0$, $\int_R \partial_x^i (L_h g_j^l - f_j^l)\partial_x^{i+3} g_j^l dx = 0$,

$i = 1,2,3$; $j = 1,\dots,N$. After some calculations one can get the estimate

$$\sum_{s=1}^5 ||\partial_x^s v^l||^2 + \sum_{n=1}^l ||\partial_x^6 v^n||^2 h \le C(\sum_{s=1}^5 ||\partial_x^s v^o||^2 + \sum_{n=1}^{lN}\sum_{j=1} ||f_j^n||^2_{W_2^3(R)} h). \tag{14}$$

Here $||u||^2 = \langle u,u\rangle$, $\langle u,v\rangle = \int_{D_m} u(x,y,t)v(x,y,t)dxdy$; a constant C in (14) does not depend on h. To be sure, that g_j^l are regular functions of h, consider integrals:

$\int_R g_{jhx}^l L_{2h} g_j^l dx$, $\int_R \partial_x^{i+3} g_{jh}^l \partial_x^i L_{2h} g_j^l dx$, $i=0,1,2$; $\int_R g_{jhhx}^l L_{3h} g_j^l dx$,

$\int_R \partial_x^{s+3} g_{jhh}^l \partial_x^s L_{3h} g_j^l dx$, $(s = 0,1)$, whence one come to inequalities

$$\sum_{i=1}^{4}||\partial_x^i v_h^1||^2+\sum_{n=1}^{1}||\partial_x^5 v_h^n||^2 h \le C(\sum_{i=1}^{4}||\partial_x^i v_h^o||^2+\sum_{n=1}^{1}\sum_{j=1}^{N}||f_{jh}^n||^2_{W_2^2(R)} h),$$

$$\sum_{s=1}^{3}||\partial_x^s v_{hh}^1||^2+\sum_{n=1}^{1}||\partial_x^4 v_{hh}^n||^2 h \le C(\sum_{s=1}^{3}||\partial_x^s v_{hh}^o||^2 +$$

$$\sum_{n=1}^{1}\sum_{j=1}^{N}||f_{jhh}^n||^2_{W_2^1(R)} h). \qquad (15)$$

We define values for v_h^o, v_{hh}^o as follows: $v_{hx}^o = v_{xt}(x,y,0)$, $v_{hhx}^o = v_{xtt}(x,y,0)$, and calculate values of $v_{xt}(x,y,0)$, $v_{xtt}(x,y,0)$ with the help of $v_o(x,y)$ and (7). Constants C in (14), (15) do not depend on $h > 0$, that makes it possible to pass to the limit in (11) as h tends to 0, that means, to prove Lemma 1. The next step is most complicated in technical sense. It consists of finding à priori estimates for solutions of (9),(107 in suitable Sobolev spaces.

<u>À priori estimates</u>. The structure of domains G_m let it possible to differentiate (9) in t and x. The differentiation in y is created with the help of the equality $\lambda_j w_j = -\Delta_y w_j$. In order to prove existence of a solution for (9),(10) in a classical sense, it is sufficient to show,that $u_{xt}, u_{xx}, \Delta_y u \in L_\infty(0,T;W_2^2(D_m))$. Then smoothness of a solution will follow from the embedding theorems [3] . We consider with this purpose scalar products of the form

$$<\partial_{x,t}^{i+j}(L u - f),\partial_{x,t}^{i+j}u_x>= 0 \quad (i + j = 0,1,2,3; \ j = 0,1,2),$$

$$<\partial_{x,t}^{k+l}(L u - f),\Delta_y\partial_{x,t}^{k+l}u_x> = 0, \quad k + l = 0,1,2; \ l = 0,1.$$

Now we will show, for example, how posessing the estimate

$$||u_x||(t)_{W_2^1(D_m)}+ \mu||\partial_x^3 u||(t) +||u_{xt}||(t) +$$

$$||\Delta_y u||(t) \le C , \qquad \forall t \in (0,T) \qquad (16)$$

to obtain the inequality

$$||u_x||(t)_{W_2^2(D_m)} +\mu||\partial_x^3 u||(t)_{W_2^1(D_m)} +||\Delta_y u||(t)_{W_2^1(D_m)} +$$

$$||u_{xt}||(t)_{W_2^1(D_m)} \le C, \qquad (17)$$

where constants C do not depend on μ , N,m; $||u||^2(t) = \int_{D_m} u^2(x,y,t)dxdy$. Indices μ ,N,m will be omitted. Consider the equality $\int_o^t\{<\partial_x^2(L_\mu u - f),\partial_x^3 u>- <(L_\mu u - f)_x,\Delta_y\partial_x^2 u>\} d\tau= 0.$

After integrating by parts, using (16) and the multiplicative inequality $||g||_{L_4(D)}(t) \leq C||g||^{1/4}(t)||\nabla g||^{3/4}(t)$, [3] , we come to the expression

$$||\partial_x^2 u||^2_{W_2^1(D_m)}(t) + \mu\int_0^t||\partial_x^3 u||^2_{W_2^1(D_m)}(\tau)d\tau \leq C(1+ \int_0^t||\partial_x^2 u||^2_{W_2^1(D_m)}(\tau)d\tau).$$

The constant C does not depend on μ, N, m. From here, according to Gronwall's Lemma

$$||\partial_x^2 u||_{W_2^1(D_m)}(t) \leq C \qquad \forall\, t \in (0,T).$$

Consider further the equality

$$<(L_\mu u - f)_{xt}, u_{xxt} - (L_\mu u - f)_t, \Delta_y u_{xt}>(t) = 0.$$

After some transformations with the help of yearlier obtained estimates for $\partial_x^2 u$, we come again to the inequality

$$||u_{xt}||^2_{W_2^1(D_m)}(t) \leq C(1 + \int_0^t||u_{x\tau}||^2_{W_2^1(D_m)}(\tau)d\tau), \forall t \in(0,T).$$

From here $||u_{xt}||_{W_2^1(D_m)}(t) \leq C$. Now we may differentiate (9) in x and, using the fact, that $<\Delta_y u_x, {}_x^4 u> = 0$, to find:

$$||\Delta_y u_x||(t) \leq ||u_{xxt} + (Ku_{xx})_x + (\alpha u_x)_x - f_x||(t) \leq C,$$

$$\mu||\partial_x^3 u||_{W_2^1(D_m)}(t) \leq ||u_{xt} + Ku_{xx} + \alpha u_x - f||_{W_2^1(D_m)}(t) \leq C.$$

At last, one can see, rewriting (9) in the form

$$\Delta_y u = u_{xt} + Ku_{xx} + \alpha u_x - f - \mu u_{xxx},$$

that $\Delta_y u \in L_\infty(0,T;W_2^1(D_m))$. Proof of (17) is accomplished. Proceeding successively in the same manner, one may be sure, that the following assertion holds.

<u>Lemma 3.</u> A solution of (9),(10) satisfies the inequality

$$||u_x||_{W_2^3(D_m)}(t) + ||u_{xt}||_{W_2^2(D_m)}(t) + ||\Delta_y u||_{W_2^2(D_m)}(t) + \mu||\partial_x^3 u||_{W_2^2(D_m)}(t) \leq C, \qquad \forall\, t \in (0,T)$$

and a constant C does not depend on μ, N, m. Now, having sufficient estimates, we can prove Theorem 2. The sketch of the proof consists of Lemmas 1,3 and making use of the Contracted mapping Theorem.

Corrolary 1. Let Lemmas 1-3 hold. Then for any fixed $m < \infty$, $\mu > 0$ there exists in G_m the unique solution of (7),(8):

$$\partial_t^i u_x \in L_\infty(0,T;W_2^{3-i}(D_m)) \quad (i = 0,1,2); \quad \Delta_y \partial_t^j u \in L_\infty(0,T;$$

$$W_2^{2-j}(D_m)) \quad (j = 0,1); \quad \mu\partial_x^3 u \in L_\infty(0,T;W_2^2(D_m)), \quad \mu\partial_x^6 u \in L_\infty(0,T;$$

$L_2(D_m))$, and the inclusions do not depend on m, μ.

To prove Corrolary 1 it is sufficient to pass to the limit in (9) as N tends to ∞.

Corrolary 2. Let Lemmas 1-3 hold. Then for any fixed $\mu > 0$ there exists in G the unique solution of (7), satisfying the following initial and boundary conditions

$$v_\mu(x,y,0) = v_{\mu o}(x,y), \quad \lim_{|x|+|y|\to\infty} v_{\mu x} = 0, \quad \lim_{x\to-\infty} v_{\mu y} = 0, \quad \lim_{|y|\to\infty} v_{\mu y} = 0.$$

Proof. We continue solutions $v_{\mu m}(x,y,t)$, obtained in Corrolalary 1, from G_m in G, conserving smoothness properties, and pass to the limit in sequence $\{v_{\mu m}\}$ as $m \to \infty$. It is possible due to Lemma 3.

Proof of Theorem 2. Introduce a new unknown function $z=u-\Phi$ in (5),(6) and define in G compact set $S_M := \{v_x(x,y,t):$

$$\partial_t^i v_x \in L_\infty(0,T;W_2^{3-i}(D)) \quad (i=0,1,2), \quad \partial v_x/\partial n|_{S_T} = 0;$$

$$\sum_{i=o}^{2} ||\partial_t^i v_x||L_\infty(0,T;W_2^{3-i}(D)) \leq M, \quad v_x(x,y,0) = v_{ox}(x,y) \}.$$

We substitute an arbitrary function from S_M into the nonlinear term in (5), rewritten in terms of z. Thus we come to linear problem (7),(8). For any fixed $M < \infty$ all conditions of Lemma 1 will be fulfilled, therefore one can define an operator $P: z_x = Pv_x$. It is easy to verify by choosing $T_1 \in (0,T)$ sufficiently small, that Contracted mapping Theorem takes a place. It permits us to prove Theorem 2. Uniqueness of solutions of problems (1),(2); (5),(6) etc is proved in a standard way. Let u_1, u_2 be two solitions, then $z = u_1 - u_2$ satisfies the following homogeneous linear problem

$$Lz = z_{xt} - \mu z_{xxx} + u_{1x} z_{xx} + u_{2xx} z_x - \Delta_y z = 0.$$

$$\partial z/\partial n|_{S_T} = 0, \quad z(x,y,0) = 0.$$

Considering the scalar product $\langle Lz, z_x \rangle = 0$ and integrating it in τ from 0 till t, come to the inequality

$$||z_x||(t) \le C\int_0^t ||z_x||(\tau)d\tau , \qquad \forall\, t \in (0,T_0),$$

where C does not depend on $\mu > 0$. Hence, $z_x \equiv 0$ in G_0, and $z(x,y,t)$ is defined up to a constant, that is typical for gasdynamics.

Literature

[1] Lin, C., Reissner, E., Tsieghn.: "On two-dimensional non-steady motion of a slender body in a compressible fluid", J. Mathematics and Physics, 27, n3 (1948).

[2] Ryzhov, O.S., Shefter, G.: " O vliyanii vyazkosti i teploprovodnosti na structuru szhimaemych techenij", Prikladnaya Matematika i Mechanika, (Russian), 28, n6 (1964).

[3] Besov, O. V., Il'in, V. P., Nikol'skij, S.M.: "Integralnye predstavleniya funktsij i teoremy vlozheniya", Moscow, Nauka (1975).

ENTROPY WEAK SOLUTIONS TO NONLINEAR HYPERBOLIC SYSTEMS IN NONCONSERVATION FORM

Philippe Le Floch
Centre de Mathématiques Appliquées, Ecole Polytechnique,
91128 Palaiseau Cédex, France.

ABSTRACT

For nonlinear hyperbolic systems in nonconservation form, we consider weak solutions in the class of bounded functions of bounded variation . A generalized global entropy inequality is proposed and studied. In this mathematical framework, we solve the Riemann problem and prove, for the Cauchy problem, the consistancy of the random choice method for systems in nonconservation form. Our theory of entropy weak solutions is applied to nonconservative systems of elastodynamics and gasdynamics. In particular, we give here a nonconservation form of the system of conservation laws of gasdynamics, which is equivalent for weak solutions in BV.

1. INTRODUCTION

We are interested in nonlinear hyperbolic systems in nonconservation form :

$$A_0(u)\,\partial_t u + A(u)\,\partial_x u = 0, \quad u(x,t) \in \mathbf{U}, \quad x \in \mathbb{R}, \quad t > 0. \qquad (1.1)$$

Here, $\mathbf{U}$ is an open subset of $\mathbb{R}^p$; A_0 and A are continuously differentiable functions defined from $\mathbf{U}$ into the space of $p\times p$ matrix. For each u in $\mathbf{U}$, we assume that $A_0(u)$ is invertible, and for the sake of simplicity the matrix $A_0(u)^{-1}.A(u)$ has p distinct eigenvalues :

$$\lambda_1(u) < \lambda_2(u) < \dots < \lambda_p(u),$$

with a corresponding basis of right eigenvectors $r_1(u), r_2(u), \dots, r_p(u)$. Each i-characteristic field is supposed to be globally either genuinely nonlinear or linearly degenerate ([12]).

Generally, the nonlinear hyperbolic system (1.1) is not a system of conservation laws, i.e. of the form

$$\partial_t f_0(u) + \partial_x f(u) = 0, \quad u(x,t) \in \mathbf{U}, \quad x \in \mathbb{R}, \quad t > 0, \qquad (1.2)$$

with C^2-functions f_0 and $f : \mathbf{U} \subset \mathbb{R}^p \to \mathbb{R}^p$, so that the theory of conservation laws (Lax [5], Glimm [4]) does not apply : the notions of weak solutions and entropy conditions have no sense for (1.1). But such nonconservative systems appear in some applications in elastodynamics (where the evolution of the stress into an elastic medium is given by physicists by a nonconservative equation) or in gasdynamics (where it may be very useful to work with nonconservation forms of the well known conservation laws for the construction of new efficient numerical schemes); see [6-10] and Section 4. below. Thus a theory of entropy weak solutions for systems in nonconservation form is needed.

Here we define a notion of entropy weak solution for (1.1) in the space of bounded functions of (locally) bounded variation (BV). Let us recall that the relevance of this space BV for studying systems of conservation laws is recognized by many authors as Glimm [4], DiPerna [2], DiPerna-Majda [3],...In this paper, we extend the usual definition of entropy weak solutions for (1.2) to the systems (1.1).

First, we have to define in which weak sense the equations must be understood. To seek weak solutions to (1.1) in the space $L^\infty \cap BV(\mathbb{R}\times\mathbb{R}^+)$ of bounded functions u of (locally) bounded variation, the main tool here is the notion of "functional superposition" introduced by Volpert in [13] : roughly speaking, to make sense to products as those appearing in (1.1) for discontinuous functions, the idea is to complete a discontinuous function by specifying "its" value at points of discontinuity. In fact, the way to complete is not trivial at all and is contained into the definition of functional superposition of Volpert (Section 2., below).

Second, for the sake of uniqueness of weak solutions, it is necessary as in the context of systems of conservation laws ([5],[12]...) to add a so called entropy condition. Here, we propose such a condition for systems in nonconservation form. It takes the form of a global entropy inequality, in general in nonconservation form,

$$(\phi^T A_0)(u)\ \partial_t u + (\phi^T A)(u)\ \partial_x u \leq 0, \tag{1.3}$$

where the function $\phi : \mathbb{R}^p \to \mathbb{R}^p$ is assumed to satisfy some positivitness and compatibility (with respect to A_0 and A) properties. We emphasize that a lot of properties, well known for conservation laws, may be generalized to (1.1) thanks to this new notion of entropy condition (Le Floch [6-9]).

The coherence of our definitions is shown by the following result : if the matrix A_0 and A in (1.1) are Jacobian matrix of some fluxes f_0 and f, so that the system in nonconservation form (1.1) may be written in the conservation form (1.2), then our new notion of entropy weak solution for (1.1) and the usual notion (in the sense of distributions, see for instance Glimm) are equivalent. At the contrary, let us recall that a completely mathematical framework is proposed by Leroux-Colombeau in [10] to define products of distributions as those appearing in (1.1) : their solutions are more general than distributions and thus are not at all "classical" solutions.

In the proposed mathematical framework, we find the entropy weak solution of the Riemann problem associated with (1.1) for small initial data. Our result is a generalization of the Lax 's theorem for systems of conservation laws [5]. We also establish the equivalence between our nonconservative entropy condition (1.3) and the usual Lax entropy criterium for speeds of shocks (Lax [5]).

Section 3. is devoted to the Cauchy problem for (1.1). Because we are now able to solve Riemann problems for a nonconservative system, it is a simple matter to follow the construction of the random choice method of Glimm [4] and hence to define a sequence u^h of approximate solutions of a Cauchy problem with small data in uniform and BV norms. We are concerned with the convergence of this sequence to an entropy weak solution of (1.1),(1.3). First, we prove that the sequence u remains uniformly bounded in sup and BV norms and uniformly Lipschitz continuous in time. By a standart compactness argument, it results that a subsequence tends to a limit function u almost everywhere in the sense of the Lebesgue measure. Using the previous estimations we give a proof of the consistancy of the sequence u^h with respect to both the system (1.1) and the entropy inequality (1.3). Finally the last point is to pass to the limit into the product $A_0(u^h)\,\partial_t u^h$ and $A(u^h)\,\partial_x u^h$. To this purpose, we note that the convergence almost everywhere is not sufficient in general ! However, being the results of sharp convergence of the random choice method proved theoretically for instance by Liu [11], we may conjecture that

$$A_0(u^h)\,\partial_t u^h \rightarrow A_0(u)\,\partial_t u \qquad \text{and} \qquad A(u^h)\,\partial_x u^h \rightarrow A(u)\,\partial_x u \tag{1.4}$$

weakly in the sense of measures. If (1.4) holds, then the limit-function u is an entropy weak solution of (1.1)(1.3).

In Section 4., our theory is applied to nonconservative systems issued of elastodynamics and gasdynamics. The modelisation of an elastic medium (tridimensional but with propagation in only one direction) yields a system of four equations: the usual three conservation laws of mass, momentum and total energy, plus an equation for the evolution of the stress deviator of the material. This latter is given by physicists in nonconservation form [10], as a consequence of the Hooke's law. It seems that it is not possible to express it in conservation form, so that a direct "nonconservative" study is needed. Here, we use our theory of Section 1. to analyse a simplified version of this system assuming the pressure is constant. The study of the complete system will be published later in [8]. For this system of 3 equations (conservation of mass and momentum and the equation for the deviator of stress), we solve the Riemann problem -for non necessarily small initial data- and we find the entropy inequalities (1.3) for this system.

Moreover, we are also interested in the conservation laws of gas dynamics. For this system we get a nonconservative form which is equivalent for weak solutions in the sense of Section 1. As noted in a different mathematical context by [10], such nonconservation forms of systems of conservation laws may be very useful to construct new numerical finite difference schemes.

2. AN ORIGINAL DEFINITION OF ENTROPY WEAK SOLUTIONS

Let us briefly recall a regularity property of BV functions, i.e. functions whose partial derivatives are locally bounded Borel measures (Volpert [13]). For an element u in $L^\infty \cap BV(\mathbb{R}\times\mathbb{R}^+;\mathbb{R}^p)$, it turns out that -with the possible exception of a set with zero 1-dimensional Hausdorff measure -each point (x,t) of $\mathbb{R}\times\mathbb{R}^+$ is regular, that is : either a point of approximate continuity $(u(x,t) = u_-(x,t) = u_+(x,t))$ or a point of approximate jump where one may define two distinct values $u_-(x,t)$ and $u_+(x,t)$. So we may consider representants of BV-functions modulo 1-dimensional Hausdorff measure.

Following Volpert [13], we define the "functional superposition" of a BV function. Consider a continuous function f in $C^0(\mathbb{R}^p;\mathbb{R})$ and an element u of $L^\infty\cap BV(\mathbb{R}\times\mathbb{R}^+;R^p)$. <u>The functional superposition</u> of u by f, denoted by $\hat{f}(u)$, is the function of $L^\infty\cap BV(\mathbb{R}\times\mathbb{R}^+;\mathbb{R}^p)$ given by the formula

$$\hat{f}(u)(x,t) = \int_0^1 f(\ u_-(x,t) + \alpha\ (u_+(x,t) - u_-(x,t)))\ d\alpha, \tag{2.1}$$

valid for each (x,t) in $\mathbb{R}\times\mathbb{R}^+$ without a set of zero 1-dimensional Hausdorff measure. If A is a matrix valued function, A(u) is defined similarly. The main result of Volpert we need here is : for each arbitrary bounded BV function v, the function $\hat{f}(u)$ given by (2.1) is measurable and locally integrable with respect to the Borel measure defined by a partial derivative $\partial v/\partial t$ or $\partial v/\partial x$. Thus a product $\hat{f}(u).\partial v/\partial t$ or $f(u).\partial v/\partial x$ makes sense as a locally finite Borel measure. Indeed, for systems of conservation laws, this concept of superposition is known to be very useful by many authors (DiPerna [2], DiPerna-Majda [3]). Using the notion of functional superposition, we propose:

<u>Definition 2.1</u> A function u in $L^\infty\cap BV(\mathbb{R}\times\mathbb{R}_+;U)$ is a <u>weak solution</u> to the nonlinear hyperbolic system (1.1) if the equality

$$\hat{A}_0(u)\ \partial_t u + \hat{A}(u)\ \partial_x u = 0 \tag{2.2}$$

holds in the sense of Borel measures.

Let us apply this definition of weak solutions to discontinuous functions consisting of two constant states and get a practical formula for computing jump relations for systems in nonconservation form.

Theorem 2.1 **The discontinuous function u given by:**

$$u(x,t) = u_L \quad \text{for } x-\sigma t<0, \quad u_R \quad \text{for } x-\sigma t>0, \tag{2.3}$$

with u_L and u_R in U, σ in $\mathbb{R}$, is a weak solution to the system (1.1) if and only if the following generalized Rankine-Hugoniot jump relation holds

$$\int_0^1 \{ -\sigma A_0(u_L+\alpha(u_R-u_L)) + A(u_L+\alpha(u_R-u_L)) \} \, d\alpha \, (u_R-u_L) = 0. \tag{2.4}$$

Then, for the sake of uniqueness, we need a so called entropy condition which applies to systems in nonconservation form. Here let us define for the system (1.1) a notion of (global) entropy inequality which generalizes the well known Lax entropy inequalities [5] for conservation laws. We set

Definition 2.2 **A p-vector valued function $\phi : U \to \mathbb{R}^p$ of C^1-class is an admissible function for the system (1.1) if it is increasing and satisfies the compatibility property**

$$D\phi^T A_0 = A_0^{\,T} D\phi, \qquad D\phi^T A = A^T D\phi. \tag{2.5}$$

In general, there do not exist admissible functions for an arbitrary nonlinear hyperbolic system. However, as for entropy -entropy flux of conservations laws, we hope that physically meaningfull systems admit admissible functions. Namely, examples of admissible functions for nonconservative systems are presented in [7,8] ; see also Section 4. Finally, our definition of entropy weak solution to (1.1) is :

Definition 2.3 **Suppose there exists an admissible function ϕ for (1.1). A function u in $L^\infty \cap BV(\mathbb{R} \times \mathbb{R}_+)$ which is a weak solution to the system in nonconservation form (1.1) (in the sense of Definition 2.1) is an entropy weak solution to (1.1) (with respect to the admissible function ϕ) if the generalized entropy inequality**

$$(\phi^T A_0)\hat{}(u)\ \partial_t u + (\phi^T A)\hat{}(u)\ \partial_x u \ \leq 0 \tag{2.6}$$

holds in the space of measures.

Inequalities (2.6) are really a generalization of the usual entropy inequalities derived for conservation laws by the viscosity method. An important fact is that the usual notion of entropy weak solution to conservation laws is contains into our new definition: suppose A_0 and A are Jacobian matrix and thus (1.1) is equivalent to (1.2) for smooth solutions; then we prove in [7] that a BV function is a solution of the system in nonconservation form (1.1) in the sense of definitions 2.1-2.3 if and only if it is a solution of the conservation laws (1.2) in the sense of distributions. Henceforth, in that case, our notion of entropy weak solution reduces exactly to the usual notion of Glimm [4] and Volpert [13].

3. RIEMANN PROBLEM AND RANDOM CHOICE METHOD

To get existence of entropy weak solutions of the Cauchy problem for systems in nonconservation form, we use the random-choice method introduced by Glimm [4] for systems of conservation laws.

First, concerning the Riemann problem for (1.1) (which is the basis of the random choice method), we give briefly our main results. It is a Cauchy problem with a piecewise constant initial data u_0 of the form:

$$u_0(x) = u_L \quad \text{if } x<0, \quad u_R \quad \text{if } x>0, \tag{3.1}$$

with u_L and u_R in **U**. On one hand, the usual notion of rarefaction waves [5] is clearly still valid for (1.1). On the other hand, we remark that the Lax admissibility criterion on the speed of a discontinuity [5] makes also sense for (1.1); thus, using the definition 2.1 of weak solution to systems in nonconservation form, we have defined in [7] the notion of shock waves and contact discontinuities for (1.1). Then, as in Lax [5], it is a simple matter to get:

Theorem 3.1 **For an initial jump $|u_R - u_L|$ small enough, there exists a unique weak solution (Definition 2.1) of the Riemann problem for the nonconservative system (1.1) in the class of self-similar functions composed with at most p rarefaction waves, shock waves or contact discontinuities.**

Then, we prove that this solution which an entropy solution in the sense of the Lax criterium is also an entropy solution in the sense of our definitions 2.2-2.3.

Theorem 3.2 **Suppose there exists an admissible function ϕ for (1.1) (Definition 2.2). For weak shock waves (associated with genuinely characteristic fields) of the system in nonconservation form, the entropy inequality (2.6) is equivalent to the Lax admissibility criterion. Thus, when all the characteristic fields of (1.1) are genuinely nonlinear, the weak solution of the Riemann problem given by Theorem 3.1 is an entropy solution in the sense of Definition 2.3.**

In fact, in the case of linearly degenerate fields, the signification of (2.6) is not clear in general. Nevertheless, refer to Section 4. where we analyse a physically meaningfull system of 3 equations with a linearly degenerate field.

Then, let us pass to the Cauchy problem for (1.1). Glueing together solutions of different Riemann problems as in the random-choice method of Glimm [4], we easily contruct approximate solutions $\{ u^h \}$ of the problem (1.1) and

$$u(x,0) = u_0(x). \tag{3.2}$$

The initial data u_0 is assumed to sufficiently small in sup norm and BV norm. We refer to [] for the precise definition of u^h: recall only that this construction needs an (equidistributed) sequence $a=(a_n)$. As usual in the frame of systems of conservation laws, this family of approximate solutions is uniformly bounded in norms L^∞ and BV and is uniformly Lipschitz continuous in time. Hence, by the Helly compactness theorem, it (or a subsequence) converges almost everywhere with respect to the Lebesgue measure to a boundedand Lipschitz continuous BV-function $u = u(x,t)$.

We are able to prove the consistancy of the family of approximate solutions $\{ u^h \}$ with both the system in nonconservation form (1.1) and our nonconservative entropy condition (2.6), in the following sense :

Theorem 3.3 **Consider $E=[-1,+1]^{\mathbb{N}}$ with the equidistributed measure. There exists a subset E_0 of E with zero measure and a subsequence $(h_n)_{n \in \mathbb{N}}$ tending to zero, such that**

$$\int_{\mathbb{R}} \int_{\mathbb{R}_+} \theta(x,t) \{ A_0(u^h)\, \partial_t u^h + A(u^h)\, \partial_x u^h \}\, dx\, dt \to 0, \quad h = h_n \to 0, \tag{3.3}$$

for every function θ in $C^0(\mathbb{R} \times \mathbb{R}_+, \mathbb{R})$ with compact support and each equidistributed sequence a of $E \backslash E_0$. Moreover, assume that the system (1.1) admits an admissible function ϕ (Definition 2.2), and that all the characteristic fields of (1.1) are genuinely nonlinear. Then, the family of approximate solutions $\{u^h\}$

is also consistant with the entropy inequality (2.6) in the sense:

$$\lim \int_{\mathbb{R}} \int_{\mathbb{R}_+} \theta(x,t) \{ (\phi^T A_0)\hat{~}(u^h)\, \partial_t u^h + (\phi^T A)\hat{~}(u^h)\, \partial_x u^h \}\, dx\, dt \leq 0, \qquad (h=h_n \to 0). \tag{3.4}$$

4. NONCONSERVATION FORMS OF SYSTEMS OF GASDYNAMICS AND ELASTODYNAMICS

In this section, we show how our general theory applies to both gasdynamics and elastodynamics equations.

4.1. The modelisation of an elastic medium provides a system of nonlinear hyperbolic equations which is given in nonconservation form. In Euler coordinates, the density ρ, the velocity u, the energy E, the internal energy e, the pressure p and the stress deviator σ of an one dimensional homogenous elastic medium satisfy [10] :

$$\partial_t \rho + \partial_x(\rho u) = 0, \quad \partial_t(\rho u) + \partial_x(\rho u^2 + p - \sigma) = 0, \quad \partial_t(\rho E) + \partial_x(\rho u E + (p-\sigma)u) = 0,$$
$$\partial_t \sigma + u\partial_x \sigma - k^2 \partial_x u = 0, \qquad \text{where } E = e + u^2/2, \quad p=p(\rho,e) \text{ and } k>0. \tag{4.1}$$

Here the medium is supposed to be tridimensional but the propagation is only in one direction. This system is composed of the three conservation laws of mass, momentum and total energy, plus an equation for the evolution of the stress. The fourth equation in (4.1) is in nonconservation form because of the advection term $u\partial_x\sigma$, and is a consequence of the Hooke's law expressed in Lagrangian coordinates.

For the sake of simplicity, we study a simplified version of (4.1), assuming the pressure p is constant. For this model of 3 equations

$$\partial_t \rho + \partial_x(\rho u) = 0, \quad \partial_t(\rho u) + \partial_x(\rho u^2 - \sigma) = 0, \quad \partial_t \sigma + u\partial_x \sigma - k^2 \partial_x u = 0, \tag{4.2}$$

we solved in [7] the Riemann problem <u>without restriction</u> on the size of the initial data. Let us transform (4.2) by using the mass Lagrangian coordinates (y(x,t),t) defined for smooth solutions by : $\partial_t y(x,t) = (\rho\ u)(x,t), \quad y(x,0) = x.$

We get a system equivalent to (4.2) for smooth solutions :

$$\partial_t v - \partial_x u = 0, \quad \partial_t u - \partial_y \sigma = 0, \quad v\partial_t \sigma - k^2 \partial_y u = 0, \tag{4.3}$$

where here v denotes the specific volume : $v= 1/\rho$. Moreover, we may verify that (4.2) and (4.3) are even equivalent for weak solutions.

Theorem 4.1 The shock curves of the two systems (4.2) and (4.3) are equivalent.

We begin with the entropy inequalities (2.6) for the system (4.3). Writting the positivity and compatibility properties (2.5).

Theorem 4.2 Define the function $S(v,\sigma)$ **by :** $S(v,\sigma) = v^{-k^2} e^{\sigma}$. **Then, the entropy inequalities (2.6) for the nonconservative system (4.3) are :**

$$v\partial_t g(S(v,\sigma)) + K\{k^2 u\partial_t u + v\partial_y\sigma - k^2\partial_y(\sigma u)\} \leq 0, \quad (4.4)$$

for each convex function $g : \mathbb{R} \to \mathbb{R}$ **and each positive constant** K.

We consider now the system in Eulerian coordinates. A surprising fact is that the system (4.2) does not admits any admissible function : the relations (2.5) yield linear partial differential equations which are incompatible ! However, we know that our notion of both weak and entropy solution is not stable by a change of variable. So that, we may hope that choosing different unknown will correspond to a "better" system. Namely, we prove

Theorem 4.3 The nonconservative system (4.2) with the unknown (ρ,u,σ) **is equivalent for weak solutions in BV (Definition 2.1) to the following system with the unknown** (v,u,σ) :

$$\begin{aligned} &v^{-1}\partial_t v + v^{-1} u\partial_x v - \partial_x u = 0,\\ &v^{-1}\partial_t u + u v^{-1}\partial_x u - \partial_x\sigma = 0,\\ &\partial_t\sigma + u\partial_x\sigma - k^2\partial_x u = 0. \end{aligned} \quad (4.5)$$

And finally, we find the entropies for the system of elastodynamics in Eulerian coordinates :

Theorem 4.4 The entropy inequalities for the system (4.5) are

$$\partial_t g(S(v,\sigma)) + u\partial_x g(S(v,\sigma)) + K\{k^2 v^{-1} u(\partial_t u + u\partial_x u) + v\partial_x\sigma - k^2\partial_x(\sigma u)\} \leq 0,$$

where g is a convex function and K a positive constant. (4.6)

Broadly speaking, we remark that there exists in general a lot of different nonconservation forms of a given nonconservative system, which are equivalent for weak solutions. However, we think that some of them have better stability properties (for instance of the point of view of the construction of numerical schemes). Hence, we hope that our notion of entropy inequality -which relies on symetry properties of the system- provides a criterium to select the "good" nonconservation forms of a nonlinear hyperbolic system.

4.2. We now consider the system of conservation laws of gasdynamics. Following some previous ideas of [10] (in a completely different mathematical framework), we derive a nonconservation form of this system equivalent for weak solutions. Let us begin by the system in mass Lagrangian coordinates [12] :

$$\partial_t v - \partial_x u = 0, \quad \partial_t u + \partial_x p = 0, \quad \partial_t E + \partial_x (pu) = 0, \tag{4.7}$$

where v, u, p, E and $e=E-u^2/2$ are the specific volume, the velocity, the pressure, the total energy and the internal energy of the gas respectively. We assume that the equation of state for the pressure

$$p = \mathbf{P}(e,v)$$

may be equivalently written as a function for the internal energy

$$e = \mathcal{E}(v,p).$$

We know [13] that (v,u,p) are in some sense "natural variables" for the resolution of the Riemann problem for (4.7). And that gives the idea to look for a system for (v,u,p).

<u>Theorem 4.5</u> The system (4.7) is equivalent to the following system in nonconservation form with the unknown (v,u,p) :

$$\partial_t v - \partial_x u = 0, \tag{4.8a}$$

$$\partial_t u + \partial_x p = 0, \tag{4.8b}$$

$$\partial_p \mathcal{E}(v,p)\, \partial_t p + (p + \partial_v \mathcal{E}(v,p))\, \partial_x u = 0, \tag{4.8c}$$

for weak solutions in $L^\infty \cap BV$ (Definition 2.1).

For the proof, we need a lemma which displays the importance of a property of <u>linearity</u> of an arbitrary system (1.1) :

<u>Lemma 4.6</u> Consider a nonlinear hyperbolic system : $A_0(U)\partial_t U + A(U)\partial_x U = 0$, and set $A_0(U)=(a_{0,ij}(U))$ and $A(U)=(a_{ij}(U))$. Let us assume that there exists an integer $q \leq p$ such that the functions $a_{0,ij}(U)$ and $a_{ij}(U)$ are constant for $i \leq q$ and $j \leq p$. Then, let $C(U)=(c_{ij}(U))$ be a matrix satisfying the properties

$$\int_0^1 C(U_L+\alpha(U_R-U_L))\, d\alpha \text{ is invertible for each } U_L,\, U_R \text{ in } \mathbf{U}, \tag{4.9}$$

and

$c_{ij}(U)$ = Cst., **for each $i \le p$ and $j \ge q$.** (4.10)

Then, the two nonconservative systems :

$$A_0(U)\, \partial_t U + A(U)\, \partial_x U = 0, \qquad \textbf{and} \qquad C(U)\, A_0(U)\, \partial_t U + C(U)\, A(U)\, \partial_x U = 0,$$

are equivalent for weak solutions in $L^\infty \cap BV$.

For the proof of Theorem 4.5, it then suffices to use the linearity of the two first composants of the flux (-u,p,p.u) with respect to the variables (v,u,p). In the case of a polytropic perfect gas, the functions **P** and $\mathcal{E}$ are given by

$$\mathbf{P}(v,e) = (\gamma-1)\, e/v, \qquad \mathcal{E}(v,p) = p\, v\, /\, (\gamma-1), \qquad \text{with } \gamma > 1,$$

the equation (4.8c) becomes

$$v\, \partial_t p + \gamma\, p\, \partial_x u = 0. \tag{4.11}$$

We now turn to the system of gas dynamics in Eulerian coodinates:

$$\partial_t \rho + \partial_x(\rho u) = 0, \quad \partial_t(\rho u) + \partial_x(\rho u^2 + p) = 0, \quad \partial_t(\rho e) + \partial_x(\, (\rho e + p) u\,) = 0. \tag{4.12}$$

Here $\rho = 1/v$ is the density and the other variables have the same signification as in (4.12). Using again the variables (v, u, p), we get an equivalent nonconservation form of (4.12) :

<u>Theorem 4.7</u> **The system of conservation laws (4.12) and the following system in nonconservation form with the unknown** (v,u,p,):

$$v^{-1}\, \partial_t v + v^{-1}\, u\, \partial_x v - \partial_x u = 0, \tag{4.13a}$$

$$v^{-1} \partial_t u + u\, v^{-1} \partial_x u + \partial_x p = 0, \tag{4.13b}$$

$$\partial_t p + u \partial_x p + \gamma\, p\, \partial_x u = 0, \tag{4.13c}$$

are equivalent for weak solutions in $L^\infty \cap BV$ in the sense of Definition 2.1.

As in paragraph 4.1, we may prove that the systems (4.8) or (4.13) with the unknown (v,u,p) do not admit any entropy inequality in the sense of definition 2.2-2.3. Again our interpretation is that the systems (4.7) and (4.12) have probably better properties of (numerical) stability than the new systems (4.8) and (4.13).

REFERENCES

[1] Dafermos, C.M., Hyperbolic systems of conservation laws, J.M. Ball ed., Systems of nonlinear P.D.E. (1983), 25-70.

[2] DiPerna, R.J., Uniqueness of solutions of hyperbolic conservation laws, Ind. Univ. Math. J., 28 (1979), 137-188 .

[3] DiPerna, R.J., Majda, A., The validity of nonlinear geometric optics for weak solutions of conservation laws, Comm. Math. Phys. 98 (1985), 313-347.

[4] Glimm, J., Solutions in the large for nonlinear hyperbolic systems of equations, Comm.Pure Appl. Math. 18 (1965),697-715.

[5] Lax, P.D., Hyperbolic systems of conservation laws and the mathematical theory of shocks waves, CBMS Monograph N°11, SIAM (1973).

[6] Le Floch, Ph., Solutions faibles entropiques des systèmes hyperboliques nonlinéaires sous forme non conservative, Note aux C.R. Acad. Sc. Paris, t.306, Série1, p. 181-186,1988.

[7] Le Floch, Ph., Nonlinear hyperbolic systems in nonconservation form, to appear in Comm. in Part. Diff. Equa.(1988); and Thesis of the Ecole Polytechnique Palaiseau, France.

[8] Le Floch, Ph., Nonconservation forms of systems of gasdynamics and elastodynamics, to appear.

[9] Le Floch, Ph., Raviart,P.A., An asymptotic expansion for the solution of the generalized Riemann problem, Ann. Inst.H. Poincaré, Non Linear Analysis, in press.

[10] LeRoux,A.Y., Colombeau,J.F., Techniques numériques en élasticité dynamique, Communication to the First International Conference on Hyperbolic Problems, Saint-Etienne (France), january 1986.

[11] Liu, Tai Ping, Admissible solutions to systems of conservation laws, A.M.S. Memoirs (1982).

12[] Smoller, J.S., Shock waves and reaction diffusion Equations, Springer Verlag, New York (1983).

[13] Volpert, A.I., The space BV and quasilinear equations, Math. USSR Sb.2, 257-267 (1967).

[14] Wagner, D.H., Equivalence of the Euler and Lagrangian Equations of gas dynamics for weak solutions, J. Diff. Equa., 68 (1987), 118-136.

A VELOCITY-PRESSURE MODEL FOR ELASTODYNAMICS

A.Y. LE ROUX
UER Mathématiques
Université de Bordeaux 1
F- 33405 TALENCE

P. DE LUCA
Centre d'études de Gramat
F- 46500 GRAMAT

SUMMARY

The experimental construction of the shock polar of a material leads to a curve $u = g(p)$ in a velocity-pressure diagram. This corresponds to the rankine Hugoniot condition for any shock linked to the state (0,0) . Since the Volpert rule for the multiplication of distributions works for a non conservative model involving the velocity and the pressure (this means that usual shock waves are to be found with this rule), we build such a model, the Rankine Hugoniot curves of which correspond to the experimental shock polar. Then a transport projection method, as for the Godunov scheme, is described. The transport step uses a Riemann solver and the projections are performed in such a way to conserve the same quantities as for a conservative model. Here the variations of the density are neglected for this model is to be used mainly for solids. Some remarks are added for the case of a density variation taken in account, as for a several material case.

THE EXPERIMENTAL CONSTRUCTION OF THE SHOCK POLAR

We denote by A the material for which the shock polar is to be constructed and by B an other material the shock polar of which is known. A cylindrical block of B is thrown against a target which is a cylindrical block of A. The bases of both cylinders are parallel and large enough to enable the use of a one dimension model, along the axe of symmetry.

Immediately after the impact a shock wave propagates across A and reaches the opposite face. Then a rarefaction wave comes back into A and makes the target to start with some velocity. In the same way another wave runs across B. The initial velocity of B is known before the impact and is denoted by v. The initial velocity of A is zero. The initial pressures in A and in B are both supposed to be zero.

Then the state of A behind the shock wave is some velocity u and some pressure p . This state (u,p) is a point of the shock polar of A, which is the shock curve passing through the state (0,0). This is usually derived from the Rankine Hugoniot conditions when these conditions are known.

Since the two shocks after the impact are separated by a contact discontinuity, the state of B is also (u,p) . This state can be linked to the state (v,0) by a shock curve of B . This gives a first condition to determine the two values u and p . A second condition is obtained by considering the instant when the shock wave reaches the opposite face of A . The rarefaction wave which appears at that time links the state (u,p) to the state (w,0) . Here w is the velocity of the target after the rarefaction wave has crossed it. This means that we have assimilated the ambient medium to the vaccuum. The value of w is measured during the experiment.

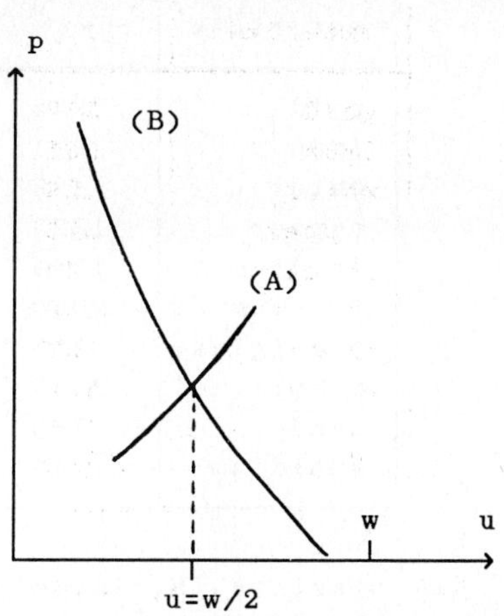

Now we claim that the curve linking the value (u,p) to (w,0) is closed to the symmetric of the shock polar of A . This assumption corresponds to identify the shock curve to the Riemann invariant and leads to set w = 2 u .

Next the value (u,p) is obviously found at the intersection of the line u = w / 2 and the shock curve of B going down through (v,0). We get this way one point of the shock polar of A. By doing the same experiment with other values of the initial velocity v of B we get other points of this shock polar, and then build the curve.

We have found this way a curve u = g(p) with g increasing and satisfying g(0) = 0 . In practice g is a concave function

In practice such curves are to be found in this form,

$$p = \alpha u + \beta u^2$$

with given positive real parameters α and β (see [4,5]). Then for the velocity we get the form

$$u = \alpha_0 \left(\sqrt{1 + \beta_0 p} - 1 \right) \quad . \qquad (1)$$

We give here a few examples of the corresponding values α_0 and β_0 for several materials in m-k-s- units and with p given in pascals. These values are obtained from [5]. We give also the values of the density ρ_0 and of the soundspeed c_0 for the same materials (at rest and for a given temperature).

materials	α_0	β_0	ρ_0	c_0
gold	972	$3.5\ 10^{-11}$	19240	3056
iron	931	$7.66\ 10^{-11}$	7850	3574
water	429	$2.84\ 10^{-9}$	998	1647
copper	1323	$4.3\ 10^{-11}$	8930	3940
iridium	1344	$1.7\ 10^{-11}$	22484	3916
lithium	2050	$3.96\ 10^{-10}$	530	4645
plexiglas	857	$7.57\ 10^{-10}$	1186	2598
beryllium	3558	$3.8\ 10^{-10}$	1851	7998
lead	702	$1.22\ 10^{-10}$	11350	2051
rubidium	445	$2.59\ 10^{-9}$	1530	1134

In practice this model is valid for a pressure of the order of 10 kbars which leads to a value of $\beta_0\, p$ of the order of a few units.

We may remark that g(p) is a concave inceasing function.

THE VELOCITY PRESSURE MODEL

This experimental method does not give directly a state law of the form

$$p = F(\rho, I)$$

that is the pressure p as a function of the density ρ and the internal energy I. We use to find such a state law in the conservative equations of mass, momentum and total energy. This is the case for example in hydrodynamics.

The system we shall derive from this shock polar will not have a conservation form, but this does not mean that mass, momentum and total energy will not be conserved.

We choice to work with the two parameters involved in the experiment: the velocity and the pressure. The density can be also introduced; this is discussed later. We aim to get the same shock curves than the one described by the shock polar we have built and which can be seen as a curve

$$u = g(p) \ . \tag{2}$$

We give our model the following form, where v_0 is the specific volume

$$u_t + u u_x + v_0\, p_x = 0 \qquad , \tag{3}$$

$$p_t + u\, p_x + \frac{\varphi(p)}{v_0}\, u_x = 0 \ . \tag{4}$$

This comes from the following arguments. Equation (3) corresponds to the conservation of the momentum. The effects due to the variation of the density have been neglected. The velocity of the material is taken in account in the inerty terms $u\, u_x$ and $u\, p_x$, what we call the convection terms. The function $\varphi(p)$ corresponds to the square of the soundspeed which is only depending on the pressure and not on the velocity. This function $\varphi(p)$ will be derived from (1) and by using some arguments allowing to compute the two products of distributions $u\, p_x$ and $\varphi(p)\, u_x$. From [3] we know that the Volpert rule (see [6]) for the multiplication of distributions can be applied to the couple of parameters u and p .

Let us consider now a shock wave which propagates along a line $x = c\, t$. We denote respectively by (u_1, p_1) and (u_2, p_2) the states of the material on the left and on the righ hand side of the shock. Then we set

$$\Delta u = u_2 - u_1 \quad , \quad \overline{u} = \frac{u_1 + u_2}{2} \quad ,$$

and define Δp, $\overline{p}$,.. in a same way.

By denoting Y the Heaviside function we have

$$u = u_1 + \Delta u\ Y(x-ct) \quad , \quad p = p_1 + \Delta p\ Y(x-ct) \quad ,$$

and denoting $\xi = x-ct$, we get

$$-c\ \Delta u\ Y'(\xi) + (u_1 + \Delta u Y(\xi))\ Y'(\xi) + v_0\ \Delta p\ Y'(\xi) = 0\ .$$

Since from the Volpert rule we have $2\, Y\, Y' = Y'$, we get the condition

$$(\overline{u} - c)\ \Delta u + v_0\ \Delta p = 0 \quad . \qquad (5)$$

In a same way, we have

$$-c\ \Delta p\ Y'(\xi) + (u_1 + \Delta u Y(\xi))\ Y'(\xi) + \frac{1}{v_0}\ \varphi(p_1 + \Delta p Y(\xi))\ \Delta u\ Y'(\varphi) = 0$$

and since from the Volpert rule,

$$\varphi(p_1 + \Delta p Y(\xi))\ Y'(\xi) = \frac{\Delta \Phi(p)}{\Delta p}\ Y'(\xi)$$

with
$$\Phi(p) = \int_0^p \varphi(\eta)\ d\eta \quad .$$

We get the condition

$$(\bar{u}-c)\,\Delta p + \frac{\Delta\Phi(p)}{v_0\,\Delta p}\,\Delta u = 0\,. \qquad (6)$$

From (5) and (6) we obtain

$$\frac{\Delta\Phi(p)}{\Delta p} = \left[v_0 \frac{\Delta p}{\Delta u} \right]^2 \qquad (7)$$

Now we consider the particular case of the experiment where $u_2=0$, $p_2=0$, $u_1=u=g(p)$ and $p_1=p$. This gives

$$\Phi(p) = \frac{v_0^{\,2}\, p^3}{g(p)^2}\,.$$

Thus we get $\varphi(p)$ from $g(p)$ by

$$\varphi(p) = v_0^{\,2} \frac{d}{dp} \left\{ \frac{p^3}{g(p)^2} \right\}. \qquad (8)$$

However we need the system to be hyperbolic, that is $\varphi(p) \geqslant 0$, which is here the same as

$$3\,g(p) - 2\,p\,g'(p) \geqslant 0\,. \qquad (9)$$

This is true for any increasing concave function $g(p)$. Moreover this system is genuinely nonlinear.

In conclusion, by taking $\varphi(p)$ as defined in (8) we get a hyperbolic system whose shocks are ruled by the Rankine Hugoniot condition (7), which corresponds exactly to the experimental shock polar.

THE RIEMANN SOLVER

The numerical scheme will use a splitting of the system (3) (4) into two parts: a convection part corresponding to solve the equations

$$u_t + u\,u_x = 0 \qquad (10)$$

$$p_t + u\,p_x = 0 \qquad (11)$$

for a time increment,

and a propagation part corresponding to

$$u_t + v_0 \, p_x = 0 \tag{12}$$

$$v_0 \, p_t + \varphi(p) \, u_x = 0 \tag{13}$$

also for a time increment.

For each part, the scheme corresponds to a transport projection method. The transport step uses a Riemann solver in each case.

For the convection part (10),(11), and starting with the initial value

$$(u(x,0),p(x,0)) = \begin{cases} (u_l , p_l) & \text{for } x < 0 \ , \\ (u_r , p_r) & \text{for } x > 0 \ , \end{cases}$$

the Riemann solver gives the value (u,p) of the solution for $x = 0$ and $t > 0$. We get

$$(u,p) = \begin{cases} (u_l , p_l) & \text{for } u_l > 0, \ u_r + u_l > 0, \\ \left(0, \dfrac{u_r p_l - u_l p_r}{u_r - u_l}\right) & \text{for } u_r > 0, \ u_l < 0, \\ (u_r , p_r) & \text{for } u_r < 0, \ u_r + u_l < 0 \ . \end{cases} \tag{14}$$

Now for the propagation part we shall approximate any wave by a shock wave. This is argued by the fact that for a rarefaction wave, (u_l , p_l) and (u_r , p_r) are close to one another in practice and then to take the shock wave instead of the Riemann invariant is a good approximation. Then we have to approximate the value at $x=0$ for $t>0$ of the solution of (12), (13) by (u,p) solution of the nonlinear system

$$\begin{aligned} p - p_l &= - Z(p,p_l) \ (u - u_l) \\ p - p_r &= Z(p,p_r) \ (u - u_r) \end{aligned} \tag{15}$$

with

$$Z(p,q) = \frac{1}{v_0} \sqrt{\frac{\Phi(p)-\Phi(q)}{p-q}}$$

which is known as the dynamical impedance, equal to $\rho \, c$, the product of the density by the soundspeed c .

This is solved very rapidly by starting with $p^0=\overline{p}$ and iterating

$$Z_r = Z(p^\nu, p_r) \quad , \quad Z_l = Z(p^\nu, p_l)$$

$$p^{\nu+1} = \frac{Z_r p_l + Z_l p_r - Z_l Z_r (u_r - u_l)}{Z_l + Z_r} \tag{16}$$

which converges in a few iterations, and we set

$$u = \frac{Z_l u_l + Z_r u_r - (p_r - p_l)}{Z_l + Z_r} . \tag{17}$$

THE NUMERICAL SCHEME

We describe the numerical scheme for the two parts, the construction of which is the same as for the Godunov scheme in each case. We denote by h the meshsize and by Δt the time increment, and we set $r = \Delta t/h$. On any cell

$$M_i = \left]\,(i-1/2)h, (i+1/2)h\,\right[\qquad \text{with } i \in \mathbb{Z}$$

the approximate values of the velocity and of the pressure are constant and respectively denoted by u_i^n and p_i^n at $t = n\,\Delta t$.

The convection part is performed as follows. We compute $(u^n_{i+1/2}, p^n_{i+1/2})$ as in (14) with $u_l=u_i^n$, $p_l=p_i^n$, $u_r=u^n_{i+1}$, $p_r=p^n_{i+1}$. Then for m=n+1/2, we compute

$$u_i^m = u_i^n - \frac{r}{2}\left((u^n_{i+1/2})^2 - (u^n_{i-1/2})^2 \right) \tag{18}$$

which corresponds to the well known Godunov scheme for the Burgers equation. Then we compute the pressue by

$$p_i^m = \frac{p_i^n - r\,(p^n_{i+1/2}u^n_{i+1/2} - p^n_{i-1/2}u^n_{i-1/2})}{1 - r\,(u^n_{i+1/2} - u^n_{i-1/2})} . \tag{19}$$

This last scheme is built as follows. Since (11) has not a conservative form and since the velocity field is known by (1!), we introduce a function α which is the solution of $\alpha_t + (\alpha u)_x = 0$ with the initial data $\alpha(x,0)=1$. Then αp is also solution of $(\alpha p)_t + (\alpha p u)_x = 0$. Here (19) corresponds to the computation of the rate of the projections of α and (αp) on M_i after a time step Δt.

The schemes (18) and (19) are stable and preserve the variation of u and p if r is such that the stability condition

$$r \max_{i,n}(|u_i^n|) \leqslant 1 . \qquad (20)$$

For the propagation part we start with the values u_i^m and p_i^m on each cell and compute $(u_{i+1/2}^m, p_{i+1/2}^m)$ as the solution of (15) with $(u_l, p_l) = (u_i^m, p_i^m)$ and $(u_r, p_r) = (u_{i+1}^m, p_{i+1}^m)$. Then we project the solution on the cells. this can be done by computing successively

$$\begin{aligned} \beta_i^m &= r\ v_0\ Z(p_{i-1/2}^m, p_i^m) \\ \gamma_i^m &= r\ v_0\ Z(p_{i+1/2}^m, p_i^m) \end{aligned} \qquad (21)$$

$$u_i^{n+1} = \beta_i^m\ u_{i-1/2}^m + \gamma_i^m\ u_{i+1/2}^m + (1-\beta_i^m-\gamma_i^m)\ u_i^m \qquad (22)$$

$$p_i^{n+1} = \beta_i^m\ p_{i-1/2}^m + \gamma_i^m\ p_{i+1/2}^m + (1-\beta_i^m-\gamma_i^m)\ p_i^m . \qquad (23)$$

This method is stable for

$$\beta_i^m + \gamma_i^m \leqslant 1 , \qquad (24)$$

This condition can be improved (and replaced by $\beta_i^m \leqslant 1$, $\gamma_i^m \leqslant 1$) by solving an additional Riemann problem on each cell where (24) is not satisfied.

Note that the computations already done by iterating (16) can be used to obtain β_i^m and γ_i^m in (21).

SOME REMARKS AND CONCLUSION

The formula of projection used in (22) corresponds to the conservation of the momentum since the variation of the density has neen neglected. The projection of the pressure by (23) uses the same formulabut this does not correspond to a conservation law. However the value of the pressure is large in practice (more than one kilobar, when $u \neq 0$)and we find here a good behaviour of the scheme and a good position of the shock waves. This has been tested in a simulation of a shock "iron against iron" and compared with experimental results. Other numerical simulations have been performed . They give results of the same quality as the one expected for the Godunov scheme. An antidiffusion technique has been also tested which improved efficiently the numerical results. Such a method was described in [2].

The quality of this scheme (with the projection of the pressure as in (23)) is not so good for lower values of the pressure. For example in hydrodynamics, the value p_i^{n+1} needs to be corrected by adding a non negative term of the order of $(\Delta u)^2$.

The introduction of the density can be performed as follows. the system becomes

$$v_t + u\, v_x - v\, u_x = 0$$

$$u_t + u\, u_x + v\, p_x = 0$$

$$p_t + u\, p_x + \frac{\varphi(p)}{v}\, u_x = 0$$

and the Riemann solver for the convection part is the same as above sinc v is also solution of (11). Then we compute the propagation part. Here the Volpert rule cannot work, for another rankine Hugoniot than the expected one will be derived. So we give the following sense to the product

$$\frac{\varphi(p_1 + \Delta p\, H(\xi))}{v_1 + \Delta v\, H(\xi)}\, \Delta u\, H'(\xi) = \frac{\Delta\, \Phi(p)}{\overline{v}\, \Delta p}\, \Delta u\, H'(\xi))$$

which leads to the same rankine Hugoniot condition as the shock polar. This product is meaning full in the algebra introduced in [1]. Now a Riemann solver can be constructed which generalized the one above. The projection must be performed as follows

$$\frac{1}{v_i^{n+1}} = \frac{\beta_i^m}{v_{i-1/2}^m} + \frac{\gamma_i^m}{v_{i+1/2}^m} + \frac{1-\beta_i^m-\gamma_i^m}{v_i^m}$$

to ensure mass conservation. To preserve the momentum we set

$$u_i^{n+1} = v_i^{n+1}\left(\beta_i^m \frac{u_{i-1/2}^m}{v_{i-1/2}^m} + \gamma_i^m \frac{u_{i+1/2}^m}{v_{i+1/2}^m} + (1-\beta_i^m-\gamma_i^m)\, \frac{u_i^m}{v_i^m}\right) .$$

Note that (22) still works for a small variation of v which is often the case in practice.

The same scheme as in (23) can be applied for the pressure, and works. This method allows to get a good behaviour of the; velocity and the pressure near a contact discontinuity. As a matter of fact these parameters are then computed without oscillations. This was not the case when the projection uses the total energy, by mean of the specific volume, to compute the pressure. This can be explained by the fact that a very small variation of the specific volume leads to a large variation of the pressure in elastodynamics. Then any perturbation of the specific volume, which appears necessary near a contact discontinuity, produces oscillations in the pressure profile.

REFERENCES

[1] J.F. Colombeau - *Elementary introduction to new generalized functions*, North Holland - 1978.

[2] J.F.Colombeau, A.Y.Le Roux - Numerical method for hyperbolic systems in non conservative form using product of distributions, *Advances in computer methods for partial differential equations VI*- R.Vichnevetsky and R.S. Stepleman ed, Publi. IMACS, 1987.

[3] J.F.Colombeau, A.Y.Le Roux, B.Perrot - Multiplication de distribution et ondes de choc élastiques ou hydrodynamiques en dimension un. Comptes Rendus Acad. Sciences, 305,pp453-456(1987).

[4] M.Desfourneaux - Théorie et mesure des ondes de choc dans des solides, Cours CP32, ENSTA Paris-1975.

[5] Selected Hugoniot- Group GMX-6, Los Alamos Scientific Laboratory, New Mexico, 1969.

[6] A.I.Volpert - The space BV and quasi linear equations. Math Sb. 73(115), pp 255-302,1967.

HIGHER ORDER ACCURATE KINETIC FLUX VECTOR SPLITTING METHOD FOR EULER EQUATIONS

J. C. Mandal and S. M. Deshpande

Department of Aerospace Engineering,
Indian Institute of Science, Bangalore 560012,
India

SUMMARY

A new upwind method called Kinetic Flux Vector Splitting (KFVS) has been developed for the solution of the Euler equations of gas dynamics. This method is based on the fact that the Euler equations are the moments of the Boltzmann equation when the velocity distribution is a Maxwellian. It is shown that the KFVS is a suitable moment of the Courant-Isaacson-Rees (CIR) scheme applied to the Boltzmann equation and further that it is equivalent to the flux-difference splitting approach. It can also be regarded as a Kinetic Theory based Riemann solver. The KFVS has been combined with the TVD and UNO formalisms and its application to the test case of one-dimensional shock propagation has been shown to yield accurate wiggle-free solution with high resolution.

INTRODUCTION

The one-dimensional unsteady Euler equations are

$$\partial U/\partial t + \partial G/\partial x = 0 , \tag{1}$$

where $U = [\rho, \rho u, \rho e]^T$, $G = [\rho u, p+\rho u^2, u(\rho e+p)]^T$ (2)
ρ = mass density, u = fluid velocity, p = pressure and e is the total energy per unit mass. These equations can be obtained as the moments of the Boltzmann equation, that is,

$$\langle\psi, \partial F/\partial t + v\, \partial F/\partial x\rangle = 0 , \tag{3}$$

where ψ = moment function vector = $[1, v, I+v^2/2]^T$,
F = Maxwellian velocity distribution
$= (\rho/I_0\sqrt{2\pi RT}) \exp[(-(v-u)^2/2RT)-I/I_0]$, (4)
v = molecular velocity, I = internal energy varible corresponding to the nontranslational degrees of freedom, T = temperature, R = Gas constant per unit mass, and $I_0 = [(3-\gamma)/((\gamma-1)2)]RT$, γ = ratio of specific heats, and the inner product $\langle\psi,F\rangle$ is defined by

$$\langle\psi,F\rangle = \int_{-\infty}^{+\infty}dv \int_0^{\infty}dI\, \psi F . \tag{5}$$

The vectors U and G in terms of the inner product are

$$U = \langle \psi, F \rangle, \qquad G = \langle \psi, vF \rangle = \langle v\psi, F \rangle . \tag{6}$$

The equation (3) is the basis of the KFVS method.

ANALYSIS

In case of one-dimensional unsteady flows, the Maxwellian can be split into two parts corresponding to $v>0$ and $v<0$. The flux vector G therefore splits as

$$G^{+} = \langle \psi, ((v+|v|)/2)F \rangle, \qquad G^{-} = \langle \psi, ((v-|v|)/2)F \rangle . \tag{7}$$

The split flux vectors G^+ and G^- are integrals of $vF\psi$ over positive and negative half spaces in velocity varible respectively. They can be evaluated in closed form in terms of error functions as

$$\begin{aligned} G_1^{\pm} &= \rho u(1 \pm \mathrm{erf} S)/2 \pm \rho e^{-S^2}/(2\sqrt{\pi\beta}) , \\ G_2^{\pm} &= (p+\rho u^2)(1 \pm \mathrm{erf} S)/2 \pm \rho u\, e^{-S^2}/(2\sqrt{\pi\beta}) , \\ G_3^{\pm} &= (\gamma p u/(\gamma-1) + \rho u^3/2)(1 \pm \mathrm{erf} S)/2 \\ &\quad \pm((\gamma+1)p/(2(\gamma-1)) + \rho u^2/2)\, e^{-S^2}/(2\sqrt{\pi\beta}) , \end{aligned} \tag{8}$$

$$S = u\sqrt{\beta} , \qquad \beta = 1/(2RT) . \tag{9}$$

In terms of split flux vector (1) can be written in the form

$$\partial U/\partial t + \partial G^{+}/\partial x + \partial G^{-}/\partial x = 0 . \tag{10}$$

Upwind differencing the split-flux terms in (10) we obtain the first-order accurate KFVS

$$(\partial U/\partial t)_j^n + (G_j^{+n} - G_{j-1}^{+n})/\Delta x + (G_{j+1}^{-n} - G_j^{-n})/\Delta x , \tag{11}$$

where the superscript n corresponds to time level t_n and j is any mesh point along x-axis. Substituting for U, G^+ and G^- from (6) and (7) we obtain

$$\begin{aligned} \partial\langle \psi, F_j^n \rangle/\partial t &+ \langle \psi, ((v+|v|)/2)(F_j^n - F_{j-1}^n) \rangle/\Delta x \\ &+ \langle \psi, ((v-|v|)/2)(F_{j+1}^n - F_j^n) \rangle/\Delta x = 0 , \end{aligned}$$

from which it is evident that (11) is a ψ-moment of the CIR differenced Boltzmann equation

$$\begin{aligned} (\partial F/\partial t)_j^n &+ ((v+|v|)/2)(F_j^n - F_{j-1}^n)/\Delta x \\ &+ ((v-|v|)/2)(F_{j+1}^n - F_j^n)/\Delta x = 0 . \end{aligned} \tag{12}$$

Now an interesting question arises whether the KFVS scheme

(11) which is obtained from (12) remains an upwind scheme after the moments are taken. It can be very easily verified that the Jacobians $\partial G^+/\partial U$, $\partial G^-/\partial U$ have complex eigenvalues having real positive and real negative parts respectively. However, using the theory of [1] and [2] it can be shown that the split-flux Euler equations (10) can be transformed to the symmetric hyperbolic form

$$P\ \partial q/\partial t + B^+\ \partial q/\partial x + B^-\ \partial q/\partial x = 0\ , \tag{13}$$

where q is a transformed vector, P is a positive symmetric matrix, B^+ and B^- are positive and negative symmetric matrices respectively. In [2] $P^{-1}B^+$ and $P^{-1}B^-$ have been shown to have real positive and real negative eigenvalues thus confirming the upwinding property of the scheme (11). It is also observed that the eigenvalues are smooth functions of the Mach number M and have no sonic glitches which are present in Steger and Warming flux splitting [3]. The eigenvalues of $P^{-1}B^-$ (see Fig.1) decrease to very small values as M becomes increasingly supersonic, while the eigenvalues of $P^{-1}B^+$ (see Fig.2) tend to those of $P^{-1}B$ (see Fig.3) as $M \to 1$. It is interesting to note that though this splitting also leads to split fluxes whose Jacobians (in symmetric hyperbolic form) have positive and negative eigenvalues, it has been performed in a completely different manner compared to that of Steger and Warming [4].

It is interesting to observe that (12) can be written as

$$(\partial F/\partial t)^n_j + v(F^n_{j+1/2} - F^n_{j-1/2})/\Delta x = 0\ , \tag{14}$$

where $F^n_{j+1/2} = F^L_{j+1/2}$ for $v>0$, and $F^n_{j+1/2} = F^R_{j+1/2}$ for $v<0$, and $F^L_{j+1/2}$, $F^R_{j+1/2}$ are the velocity distribution functions immediately to the left and the right of the interface $j+1/2$ respectively. The equation (14) reduces to (12) by taking

$$F^L_{j+1/2} = F_j\ , \quad F^R_{j+1/2} = F_{j+1}\ . \tag{15}$$

The ψ-moment of (14) gives the corresponding differenced Euler equations as

$$(\partial U/\partial t)^n_j + (G^n_{j+1/2} - G^n_{j-1/2})/\Delta x = 0\ . \tag{16}$$

Obviously, $G^n_{j+1/2} = G(U^n_j, U^n_{j+1}) = G^+(U^n_j) + G^-(U^n_{j+1})$.

The KFVS scheme is therefore a ψ-moment of the Riemann solver for the 1-D wave equation

$$\partial F/\partial t + v\ \partial F/\partial x = 0\ .$$

Another interesting property of the KFVS is that it is equivalent to the flux-difference splitting approach. For demonstrating this equivalence we observe that

$$\begin{aligned} vF_{j+1/2} &= ((v+|v|)/2)F_{j+1/2} + ((v-|v|)/2)F_{j+1/2} \\ &= ((v+|v|)/2)F_j + ((v-|v|)/2)F_{j+1}\ , \text{ using (15)} \end{aligned}$$

Therefore

$$vF_{j+1/2} = v\,(F_j+F_{j+1})/2 - |v|(F_{j+1}-F_j)/2$$
$$= v(F_j+F_{j+1})/2 + ((v-|v|)/2)(F_{j+1}-F_j)/2$$
$$- ((v+|v|)/2)(F_{j+1}-F_j)/2 \,. \quad (17)$$

Taking ψ-moment of (17) we obtain

$$G_{j+1/2} = (G_j+G_{j+1})/2 + (DG^-_{j+1/2}-DG^+_{j+1/2})/2 \,, \quad (18)$$

where the flux-differences are defined by

$$DG^{\pm}_{j+1/2} = \langle\psi,((v\pm|v|)/2)(F_{j+1}-F_j)\rangle. \quad (19)$$

The extension of the first-order KFVS to higher order schemes can be done in many ways. A method adopted here is based on the analysis of Chakravarthy and Osher [5], that is, define the flux vector $G_{j+1/2}$ by

$$G_{j+1/2} = \text{EFS} + [(1+\phi)(DG^+_{j+1/2}-DG^-_{j+1/2})$$
$$+ (1-\phi)(DG^+_{j-1/2}-DG^-_{j+3/2})]/4 \,. \quad (20)$$

where EFS stands for <u>E</u>xpression for <u>F</u>irst-order <u>S</u>cheme and is equal to the right hand side of (18). The parameter ϕ takes on respectively the values -1 and 1/3 for second- and third-order accurate upwind schemes. As higher order schemes are known to have spurious wiggles in the solutions, modified differences [5] , [6] can be introduced to suppress the wiggles. The modified differences are given by

$$\tilde{D}G^{\pm}_{j+1/2} = \text{minmod}[DG^{\pm}_{j+1/2}, R.DG^{\pm}_{j-1/2}] \,, \quad (21)$$
$$\tilde{\tilde{D}}G^{\pm}_{j+1/2} = \text{minmod}[DG^{\pm}_{j+1/2}, R.DG^{\pm}_{j+3/2}] \,, \quad (22)$$

where $\text{minmod}[a,b] = 0.5[\text{sign}(a)+\text{sign}(b)]\min[|a|,|b|]$, (23)

$$\text{sign}(a) = \begin{cases} +1 & \text{for } a>0 \\ -1 & \text{for } a<0 \\ 0 & \text{for } a=0 \end{cases} ,$$
$$0 \leqq R \leqq (3-\phi)/(1-\phi) \,.$$

The modified expression for $G_{j+1/2}$ to be used instead of (20) is therefore given by

$$G_{j+1/2} = \text{EFS} + [(1+\phi)(\tilde{D}G^+_{j+1/2}-\tilde{\tilde{D}}G^-_{j+1/2})$$
$$+(1-\phi)(\tilde{\tilde{D}}G^+_{j-1/2}-\tilde{D}G^-_{j+3/2})]/4 \,. \quad (24)$$

In terms of $F^L_{j+1/2}$, $F^R_{j+1/2}$ the various order accurate KFVS methods correspond to the choices :

(a) <u>First-order</u> <u>scheme</u>

$$F^L_{j+1/2} = F_j, \quad F^R_{j+1/2} = F_{j+1} \,.$$

(b) Second-order(ϕ=-1) and third-order(ϕ=1/3) schemes

$$\left.\begin{aligned} F^L_{j+1/2} &= F_j + [(1+\phi)DF_{j+1/2} + (1-\phi)DF_{j-1/2}]/4 , \\ F^R_{j+1/2} &= F_{j+1} - [(1+\phi)DF_{j+1/2} + (1-\phi)DF_{j+3/2}]/4 , \end{aligned}\right\} \quad (25)$$

where $DF_{j+1/2} = F_{j+1} - F_j$.

(c) Schemes with modified differences

$$\left.\begin{aligned} F^L_{j+1/2} &= F_j + [(1+\phi)\tilde{D}F_{j+1/2} + (1-\phi)\tilde{\tilde{D}}F_{j-1/2}]/4 , \\ F^R_{j+1/2} &= F_{j+1} - [(1+\phi)\tilde{\tilde{D}}F_{j+1/2} + (1-\phi)\tilde{D}F_{j+3/2}]/4 , \end{aligned}\right\} \quad (26)$$

where

$$\left.\begin{aligned} \tilde{D}F_{j+1/2} &= \text{minmod}[DF_{j+1/2}, R.DF_{j-1/2}] , \\ \tilde{\tilde{D}}F_{j+1/2} &= \text{minmod}[DF_{j+1/2}, R.DF_{j+3/2}] . \end{aligned}\right\} \quad (27)$$

TVD and UNO interpolation

At this stage it is important to make a few remarks about the TVD property of the above schemes. Defining the TV norm of F by

$$TV(F) = \int_{-\infty}^{+\infty} |\partial F/\partial x| \; dx .$$

We can easily prove that the first-order scheme described by (14) and (15) is TVD if the Courant number $|v \Delta t/\Delta x|$ is less than unity . This condition introduces a cut-off $-\Delta x/\Delta t \leq v \leq \Delta x/\Delta t$ in the velocity space. For taking moments of F (and this is always done in the present formulation) the velocity has to vary from $-\infty$ to $+\infty$, and hence the TVD condition is violated for velocities beyond the cut-off bounds. However, the contributions from these velocities to the ψ-moments become negligibly small as $\Delta t \to 0$ because of the exponential decay of F. It is therefore reasonable to expect that the moments will still be wiggle-free if Δt is kept sufficiently small. Similar remarks will apply to the higher order accurate schemes (25) and (26).

Since TVD formalism has inherent mechanism to degenerate to first-order accuracy at extrema, another technique called UNO (Uniformly accurate essentially Non-Oscillatory) interpolation [7] has also been adopted together with the KFVS method. UNO interpolation is based on the Newton polynomial. Here instead of using fixed stencils as in TVD, adaptive stencils are used and this way higher order accuracy right upto the discontinuity is possible [7]. Suppose n-th order interpolation is used. Then there are n stencils possible for the n-th order polynomial to be fitted to data including the j and j+1 mesh points depending on the upwinding requirement. Choices are made among these stencils that give the smoothest polynomial using the criterion given in [7]. Advantages of the use of this sort of interpolation is that divided differences can be constructed recursively which will be consistent with the above criterion and the numerical scheme can be extended to any order of accuracy. The UNO interpolation has been used along with the KFVS method at the Euler level.

RESULTS AND DISCUSSION

Fig.4 shows the results of the shock-tube problem solved by using various order KFVS schemes. The higher order accurate KFVS schemes with modified differences show progressive improvement over the results of the lower order schemes and have very good agreement with the exact results. The near absence of wiggles do indeed confirm the validity of the arguments given before. Fig.5 shows the results of the third-order KFVS with UNO interpolation. Also on the same figure the results of the third-order TVD scheme have been plotted for the purpose of comparison. One can notice that the UNO method gives comparatively higher resolution as expected.

REFERENCES

[1] DESHPANDE, S. M. : "On the Maxwellian distribution, symmetric form, and entropy conservation for the Euler equations", NASA-TP-2583, 1986.

[2] DESHPANDE, S. M. and MANDAL, J. C. : "Kinetic Flux Vector Splitting (KFVS) for the Euler Equation", Fuid Mech. Rep. 87 FM 2, Dept. of Aerospace Engg., Indian Institute of Science, Bangalore, India.

[3] VAN LEER, B. : "Flux Vector Splitting for the Euler equations", ICASE Report No. 82-30, Sept. 1982.

[4] STEGER, J. L. and WARMING, R. F. : "Flux Vector Splitting of the inviscid Gasdynamic equations with applications to Finite-difference methods", J. Computational Phys., 40 (1981), pp.263.

[5] CHAKRAVARTHY, S. R. and OSHER, S. : "High resolution applications of the Osher upwind scheme for the Euler equations", AIAA Paper 83-1943 (1983).

[6] DESHPANDE, S. M. and MANDAL, J. C. : "Kinetic Theory based new upwind methods for inviscid compressible flows", Paper presented at the Euromech Colloquim 224 on Kinetic Theory aspects of evaporation-condensation phenomena held at Kardijali, Bulgaria during July 6-10, 1987.

[7] CHAKRAVARTHY, S.R., HARTEN, A. and OSHER, S. : "Essentially non-oscillatory shock-capturing schemes of arbitrarily-high accuracy", AIAA Paper 86-0339 (1986).

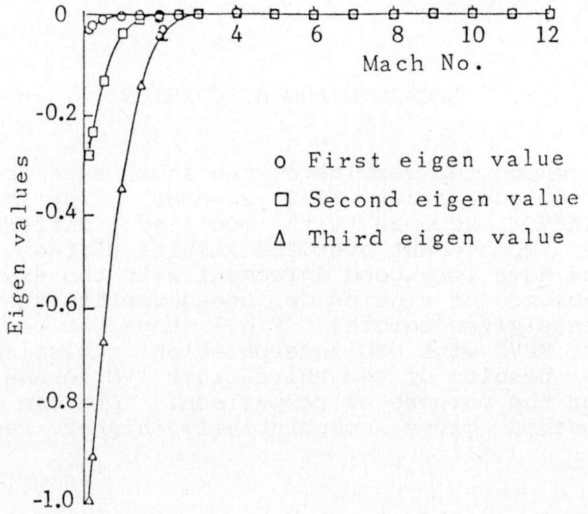

Fig. 1 Normalized (with respect to wave speed) Eigen values of $P^{-1}B^{-}$ matrix.

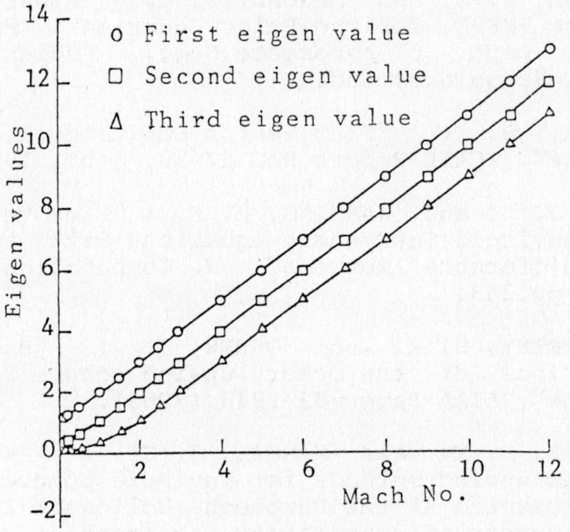

Fig.2 Normalized (with respect to wave speed) Eigen values of $P^{-1}B^{+}$ matrix.

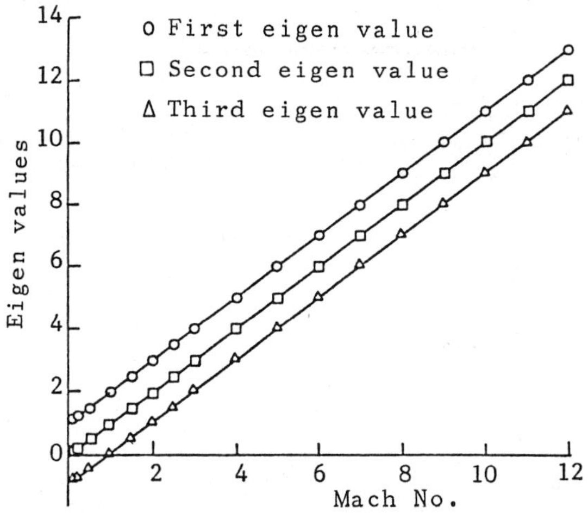

Fig.3 Normalized (with respect to wave speed) Eigenvalues of $P^{-1}B$ matrix.

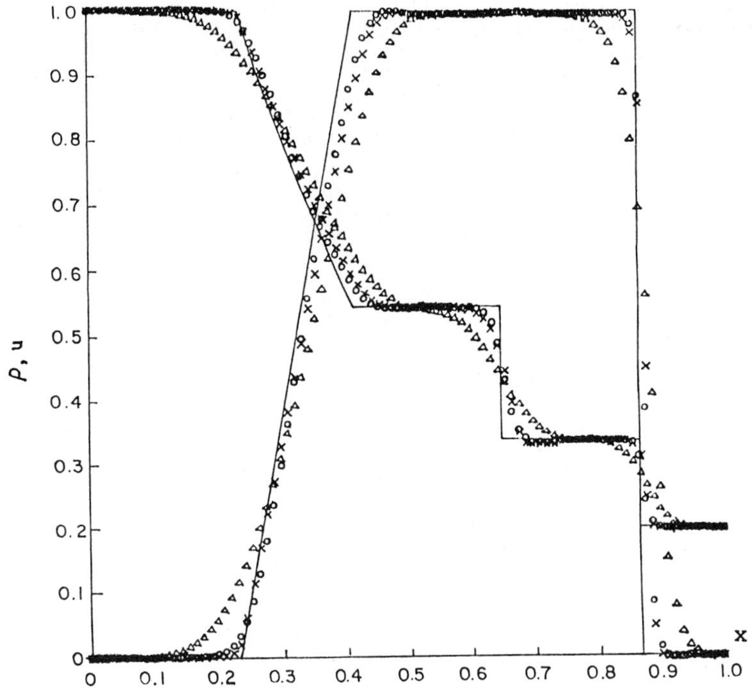

Fig.4 Computed ρ,u and exact(-) profiles for shock tube problems for KFVS. First-order (Δ), second-order (×) and third-order (o).

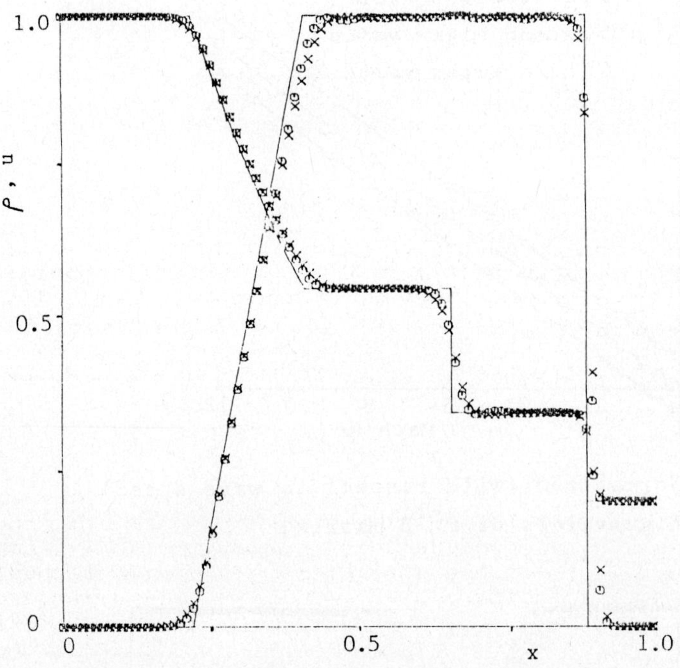

Fig.5 Computed ρ ,u and exact (-) profiles for shock tube problems for third-order KFVS using UNO (o) and TVD (×).

MONTE CARLO FINITE DIFFERENCE METHODS FOR THE SOLUTION OF HYPERBOLIC EQUATIONS

Guillermo Marshall
EPFL, GASOV Group
1015 Lausanne, Switzerland.

ABSTRACT

A new class of stochastic finite difference methods for the solution of hyperbolic partial differential equations is introduced. They are monotonicity preserving, unconditionally stable and grid free. The numerical results presented show the convergence of these methods. They also evidence the simplicity, robustness and universality of the Monte Carlo approach.

INTRODUCTION

The aim of this work is to extend the use of probabilistic models to the numerical solution of nonlinear hyperbolic partial differential equations (for nonlinear parabolic and elliptic equations see [5]). A probabilistic model related to a differential equation is any procedure which involves the use of sampling devices based on probabilities to approximate its solution. A probabilistic model uses stochastic processes, that is, a sequence of states whose transition is governed by random events. The type of stochastic process under consideration is commonly called a random walk.

The conexion between probabilistic models and certain linear differential equations of mathematical physics dates back to the works of Lord Raleigh, Einstein, Langevin and many others, it originated the models at present known as random walk or pedestrian models . In particular, the relation of random walks to linear hyperbolic equations was established in 1938 by Polya who introduced a random walk method for an hyperbolic equation studied by Albert Einstein Junior. Mathematical interest was further stimulated by the work of Kolmogorov who found the relation between stochastic processes of the Markov type and certain integro differential equations. The name Monte Carlo was given to these probabilistic models by von Neumann and coworkers at Los Alamos Scientific Laboratory in the mid forties, but became widely known only after the article of Metropolis and Ulam in 1949 (see references in [5]).

The main advantages of linear random walk models are that the solution at a point can be estimated independently of the solution at other points, and that the dependency of computational time on dimensionality is weak. The main disadvantage is that the convergence is proportional only to 1/sqrt(N), where N is the number of random walks.

The conexion between probabilistic models and nonlinear

hyperbolic systems was established in [3]. A new class of stochastic methods for the solution of nonlinear hyperbolic equations based on [3], the Random Choice Method (RCM), was introduced in [1]. The RCM is a numerical technique consisting in sampling local exact solutions of Riemann problems. To avoid the inherent difficulty in finding local exact solutions, an approximate Riemann solver was presented in [4] and [2], these methods however, are only valid for scalar conservation laws. The main advantages of these type of stochastic methods are that they are grid free, unconditionally stable and give high resolution near sharp fronts at a modest price.

The stochastic methods for nonlinear hyperbolic equations introduced here are a natural extension of Monte Carlo linear random walk models. They consist in sampling local exact solutions of finite difference schemes; in this context they are located between the RCM and plain deterministic finite difference schemes. Following this reasoning we have generically called them Monte Carlo Finite Difference Methods (MCFDM in short) rather than random walk models.

One of the main advantages of the MCFDM for linear hyperbolic equations, namely, that the solution at a point can be estimated independently of the solution at other points, is lost for nonlinear problems due to the inherent global nature of the problem. However, other advantages become extremely important: grid independency, unconditional stability and oscillation free. In brief, MCFDM share many of the good properties of RCM. The rate of convergence of MCFDM can be greatly improved by appropriate variance reduction techniques. Moreover, the advent of massively parallel processors stimulated the application of these methods because the parallel environment architecture is ideally suited for the Monte Carlo approach.

MONTE CARLO FINITE DIFFRENCE METHODS FOR NONLINEAR SCALAR HYPERBOLIC EQUATIONS

In this section we introduce the MCFDM for the solution of nonlinear scalar hyperbolic equations using as a model problem the Burgers inviscid equation. Consider the hyperbolic equation

$$u_t + F(u)_x = 0, \; 0 \le x \le 1, \; t > 0 \tag{1}$$

with $F(u) = u^2/2$, $u > 0$, and with initial conditions

$$u(x_i,0)=g(Q), \quad 0\le x\le 1, \; t=0, \tag{2}$$

and boundary conditions

$$u(0,t)=f(Q), \; u(1,t)=f(Q), \; t>0. \tag{3}$$

For solving this problem with a deterministic model, for instance a finite difference method, we discretize the domain and the boundaries with a rectangular grid of width h and height k, h and k denoting the space and time steps, respectively. We denote by P_i internal nodes and by Q_j boundary nodes (by boundary nodes we mean initial or boundary nodes). The approximate solution at node $P_i=P(x_i)$ and time $t_n=nk$ is denoted $u^n(P_i)$. An explicit finite difference analogue of equation (1) at an internal node P_0 (with P_1 and P_2 its left and right neighbours, respectively) of the domain, reads

$$u^{n+1}(P_0) = \sum_{i=0}^{i=2} a_i(P_0)\, u^n(P_i) \tag{4}$$

where the coefficients $a_i(P_0)$ depend on the particular scheme being chosen. For the Lax scheme, for instance, $a_0(P_0)=0$, $a_1(P_0)=(1+q_0/2)$, $a_2(P_0)=(1-q_0/2)$, where $q_0=u_0k/h$, and for the Godunov scheme, $a_0(P_0)=1-q_0$, $a_1(P_0)=q_0$, $a_2(P_0)=0$, where $q_0=0.5(u_0^n+u_1^n)k/h$. Similar expressions can be obtained for the Wendroff implicit scheme. Equation (4) can be written in matrix form as $U^{n+1}=P\,U^n+D$, where P is a tridiagonal matrix readily derived from (4). Stability in the L_2 norm is satisfied if the spectral radius $\rho(P)$ of the matrix P lies inside the unit disk; this is ensured if the CFL condition is fulfilled. Convergence of the method follows.

We now consider the stochastic model for (1) in the realm of finite Markov processes. For this we impose the following restrictions

$$\sum_{i=0}^{i=2} a_i(P_0) = 1 \quad \text{and} \quad a_i(P_0) \geq 0 \tag{5}$$

which are necessary for probability assignment. It is worth observing that in principle it is not possible to associate a Markov process to an hyperbolic equation since its difference approximation yields negative coefficients not allowing probability assignement; however, since every monotone finite difference approximation to a hyperbolic equation is first order accurate and hence, is a second order approximation to a parabolic equation, the MCFDM is then justified. Moreover, in nonlinear problems the transition probabilities are unknown apriori since they are function of the unknown solution, thus we are forced to estimate the latter to calculate the former. A discrete Markov process or Markov chain is defined to be a system S consisting of a finite set of states S_i , at each of a sequence of discrete times t=0, 1, 2,.., n the system S is in one of the states S_i. The state S_i determines a set of conditional probabilities p_{ij}, the quantity p_{ij} is the probability that the system which, at the time n is in the state S_i, will be in the state S_j at the n+1 time, clearly, p_{ij}

is the probability of the transition S_i to S_j. The characteristic property of a Markovian process is that p_{ij} only depends on the current state S_i and is independent of the previous states of the system. The set of all conditional probabilities p_{ij} forms a stochastic matrix P which completely determines the properties of the given chain. The state S_i is said to be an absorbing state if the system S remains in this state with probability one, the states S_i and S_j are said to be linked if there is a non-zero probability that S_i may attain S_j in a finite number of time steps and finally, a Markov chain is said to be terminating if each of its states is linked to an absorbing state. It follows that if a finite Markov process is terminating it attains an absorbing state with probability one, in a finite number of states. We can now associate the set of states S_i of a finite Markov chain with the approximate solution at the nodes of the domain on which the finite difference is defined, the internal nodes corresponding to transient states and boundary nodes to absorbing states. The coefficients $a_i(P_0)$ of the difference equation (4) can be associated with transition probabilities p_{ij} in the Markov chain. Consider now the following random walk procedure. Let P_0 at time t_{n+1} be the current state of an hypothetical particle and P_i, i=1,2, the next possible states at t_n, reached in the unit time. The transition from P_0 at t_{n+1} to P_i at t_n occurs with probability $a_i(P_0)$ according to the following formula

$$u^{n+1}(P_0) = \{ \begin{array}{lll} u^n(P_1) & \text{if} & 0 \le \xi_i \le a_1(P_0) \\ u^n(P_0) & \text{if} & a_1(P_0) \le \xi_i \le a_1(P_0)+a_2(P_0) \\ u^n(P_2) & \text{if} & a_1(P_0)+a_2(P_0) \le \xi_i \le 1 \end{array} \qquad (6)$$

where a point falls at random in the interval [0,1] with coordinate ξ_i picked from a uniform distribution in the range [0,1]. We call expression (6) a Monte Carlo Finite Difference Scheme, it satisfies the finite difference equation (4) but not its boundary conditions. For this we need the following considerations. If $u(P_0,Q_j)$ denotes the probability of ending a random walk at a boundary Q_j having started at P_0, the expectation of the boundary values reached is given by

$$V(P_0) = \sum_{j=1}^{j=s} u(P_0,Q_j)\ f(Q_j) \qquad (7)$$

where s is the total number of boundary nodes. It can be shown that $V(P_0)$ satisfies the finite difference equation (4) and its boundary conditions (see details in [5]). For estimating

$u(P_0,Q_j)$ we simulate N times the random walk starting at P_0 and counting the number of times n_j in which a boundary node Q_j is reached. An approximation of (7) is given by

$$V^{n+1}(P_0) \cong \sum_{j=1}^{j=s} n_j/N\ f(Q_j) = 1/N \sum_{j=1}^{j=s} n_j\ f(Q_j). \qquad (8)$$

The last summation is the average of all the boundaries reached after N random walks. It can be shown that the expectation of this average is $u(P_0)$ and by the law of large numbers this average converges to the exact solution of (4) for increasing values of N. For the construction of the stochastic model the random walk begins at a node P_0 of a time level t_{n+1} for which the solution, at the previous time level and for all grid nodes, is known. Here a random walk consists in one random step since after it a boundary node or absorbing state is inevitably encountered. Implicit in this procedure is the fact that the unknown transition probabilities have been estimated using the solution at the previous time level; obviously the solution is calculated for all nodes at each time level.

Formula (6) can be written in matrix notation as $U^{n+1}=PU^n+D$ where P is a tridiagonal matrix whose elements are the set of all probability transitions $a_i(P_i)$. Stability of the MCFDS can be established in the L_2 norm if the spectral radius $\rho(P)$ of the matrix P is inside the unit disk. Since the matrix P is a stochastic matrix and by definition its spectral radius is equal to one, it follows that the MCFDS is unconditionally stable.

To reduce the variance we have used the following strategy introduced in [1]. The interval [0,1] is subdivided into m_2 subintervals, ξ_1 is picked in the first subinterval, ξ_2 in the second, etc., ξ_{m_2+1} in the first subinterval, i.e., $\xi_i^*=(\xi_i+\eta_{i+1})/m_2$. The subinterval ordering is obtained with $\eta_{i+1}=(\eta_i+m_1) \bmod m_2$, where m_1, m_2, $m_1<m_2$ are prime integers, $\eta_0<m_2$ and η_0 given. It is clear that since only one ξ_i^* is picked per random walk, after m_2 random walks, m_2 random coordinates ξ_i^* have been picked and each one on a different subinterval. With this procedure the sequence of samples ξ_i^* reach approximate equidistribution over [0,1] at a faster rate. Numerical experiments presented below show a significant improvent using this technique as compared with simple sampling.

The MCFDM can be easily extended to hyperbolic partial differential equations in any number of dimensions, the only constraint being the satisfaction of the stochasticity conditions given by expression (5).

MONTE CARLO FINITE DIFFERENCE METHODS FOR HYPERBOLIC SYSTEMS OF CONSERVATION LAWS

In this section we introduce the MCFDM for the solution of one dimensional homogeneous hyperbolic system of conservation laws. This system can be written as

$$U_t + F(U)_x = 0, \tag{9}$$

where F(U) is a flux density, U is a vector valued funcion and x and t are the space and time coordinates. A simple example is provided by the one dimensional shallow water system (or isentropic gas flow) which is given by (9), where U = { uh,h }, $F(U)=\{u^2+gh^2/2\}$, here u is the water velocity, h is the depth and g is the gravity. System (9) can be written in characteristic form as

$$W_t + A\, W_x = 0, \tag{10}$$

where now W = { r, s } and A = diag { (3r+s)/4, (r+3s)/4 }, r and s are the Riemann invariants: r= u+2C, s=u-2C and C=sqrt(gh) (see details in [6]). An explicit finite difference scheme is given by

$$W^{n+1}(P_0) = \sum_{i=0}^{i=2} a_i(P_0)\, W^n(P_i)\,. \tag{11}$$

Using a Godunov type scheme and assuming supercritical flow (for instance), $a_0(P_0)=I-k/2\ A(P_0)$, $a_1(P_0)=k/2\ A(P_0)$ and $a_2(P_0)=0$. The numerical method consists in solving the shallow water system for the Riemann invariants, recovering then the primitive variables with the formulae: $h=(r-s)^2/(16g)$ and $u=(r+s)/2$. Similar arguments as those used in the scalar case show that formula (11) is stable if the CFL condition is satisfied.

For the construction of the stochastic analogy we need to satisfy the stochasticity conditions given by expression (5) but now for vector valued functions. Thus a Monte Carlo Godunov Scheme for system (10) can be written as

$$W^{n+1}(P_0) = \begin{cases} W^n(P_1) & \text{if} \quad 0 \le \eta_i \le a_1(P_0) \\ W^n(P_0) & \text{if} \quad a_1(P_0) \le \eta_i \le a_1(P_0)+a_2(P_0) \\ W^n(P_2) & \text{if} \quad a_1(P_0)+a_2(P_0) \le \eta_i \le I \end{cases} \tag{12}$$

where now $\eta_i = \{\xi_1, \xi_2\}_i$ and ξ_1 and ξ_2 are picked from a uniform distribution in the range [0,1]. With similar arguments as those given for the scalar case it can be shown that the MCFDM just described is unconditionally stable.

NUMERICAL RESULTS

In this section we present some results obtained in the application of the MCFDM to scalar and hyperbolic system of conservation laws. In figures 1 and 2 we present the results obtained with the Monte Carlo Godunov method in the solution of the Riemann problem constituted by the Burgers inviscid equation (1) with the following initial data: $u(x,0)=u_l=2$ for $x<0$ and $u(x,0)=u_r=1$ for $x>0$. The exact solution is given by

$$u(x,t)=\begin{cases} u_l & \text{for } x < s \\ u_r & \text{for } x > s \end{cases}$$

where $s=1/2(u_l+u_r)=3/2$ is the shock propagation speed. The local CFL conditions at the left and right of the shock are $CFL=k/h<1/2$ and $CFL=k/h<1$, respectively, while at the shock is $CFL=k/h<3/2$; the critical CFL is then $CFL_{cr}=1/2$. Figure 1 shows the exact and Monte Carlo Godunov solutions for $t=0$ and $t=0.1172$ using the CFL_{cr}. Characteristics at the left of the shock travel at the correct speed but the shock is smeared. The results of the Monte Carlo method coincide with those obtained with the Godunov deterministic scheme. In figure 2 we present the results of the Monte Carlo Godunov method for the same problem but now using a $CFL=3/2=1.333\ CFL_{cr}$. For this value the stochastic solution coincides with the exact solution while the deterministic scheme is unstable. This remarkable result has been obtained violating the stochasticity condition and thus the consistency of the Markov chain (except at the discontinuity) and is due to the fact that the Monte Carlo method is unconditionally stable. More specifically: the Monte Carlo Godunov scheme is given by

$$u^{n+1}(P_0)=\begin{cases} u^n(P_1)\ , & 0 < \xi_i < a_1(P_0) & (13.a) \\ u^n(P_0)\ , & a_1(P_0) < \xi_i < a_1(P_0)+a_2(P_0) & (13.b) \end{cases}$$

$a_1(P_0)=1$, $a_2(P_0)=0$ and $u^{n+1}(P_0)=u^n(P_1)$ with probability one. At the left of the shock $q_0=4/3$, $a_1(P_0)=4/3 > 1$ and $a_2(P_0)=-1/3 < 1$, which is not consistent. However, since the condition (13.a) is checked first and because $a_1(P_0) > 1$, $u^{n+1}(P_0)=u^n(P_1)$ with probability one. This has no consequences here because of the initial data involved; had we been dealing with initial data constituted by a rarefaction wave, the violation of the CFL condition would have produced a spurious constant state. For the particular problem discussed here the Monte Carlo Godunov method has infinite resolution.

Figures 3 and 4 show the results obtained with the Monte Carlo Godunov method (12) in the solution of the Riemann problem constituted by the shallow water system (10) with the following initial data: $r(x,0)=r_l=20$, $s(x,0)=s_l=-20$, for $x<0$ and $r(x,0)=r_r=8$, $s(x,0)=s_r=-8$, for $x>0$; these data corresponds to the dam breaking problem with initial conditions $h_l=10.2$, $h_r=1.63$ and $u_l=u_r=0$. The exact solution in the primitive variables is given by a constant state separated by a right shock and a left rarefaction (see details in [6]). Figure 3 shows the Monte Carlo Godunov solution for the Riemann invariants r and s, initially and 40 time steps later. Figure 4 shows the corresponding results for the depth and velocity. These run were made with a time step corresponding to the critical CFL condition, a grid of $h=1/256$, 110 random walks per node and stratified sampling. The Monte Carlo results coincide with those obtained with the exact solution (not shown here). The sharpness and non oscillatory character of the solution are noticeable.

CONCLUSIONS

A new class of stochastic finite difference schemes for the solution of nonlinear hyperbolic systems of conservation laws has been introduced. It is based on exact solutions of finite difference schemes and sampling techniques . Its main advantages are that it is unconditionally stable and grid and oscillation free. Numerical results show the convergence of these methods. For certain data, the MCFDM possess infinite resolution.

REFERENCES

[1] A. Chorin, "Random Choice Solutions of Hyperbolic Systems", J. Comp. Phys.V 22, p. 517-533, 1976.

[2] J. S. Cohen and J. A. Lavita,"A Simple Approximate Random Choice Method for Scalar Conservation Laws", SIAM J. Sci. Stat. Comput., V. 7, p. 1350-1359, 1986.

[3] J. Glimm,"Solutions in the Large for Nonlinear Hyperbolic Systems of Equations", Comm. Pure and Appl. Math., V 18, p. 695-715, 1965.

[4] A. Harten and P. D. Lax, "A Random Choice Finite Difference Scheme for Hyperbolic Conservation Laws", SIAM J. Numer. Anal. 18, p. 289-315, 1981.

[5] G. Marshall, "Monte Carlo Finite Difference Methods for the Solution of Differential Equations", Report UGVA-DPT 1988/3-567, Department of Theoretical Physics, University of Geneve.

[6] G.Marshall and A. Menendez, "Numerical Treatment of Nonconservations Forms of the Equations of Shallow Water Theory", J. Comput. Phys. 44, 167-188, 1981.

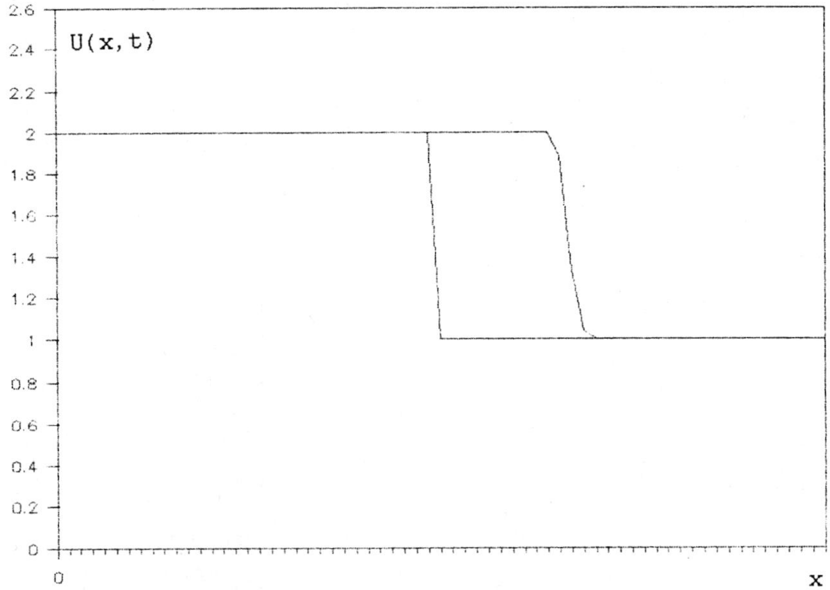

Fig. 1. Monte Carlo Godunov solution for Burgers inviscid at t=0 and t=15k ($CFL=CFL_{cr}$).

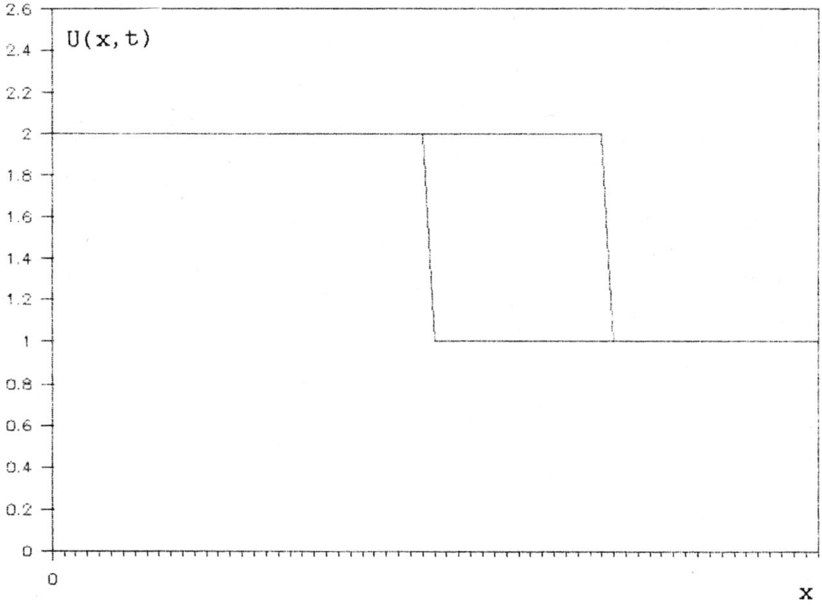

Fig. 2. Monte Carlo Godunov solution for Burgers inviscid at t=0 and t=15k ($CFL=1.33CFL_{cr}$).

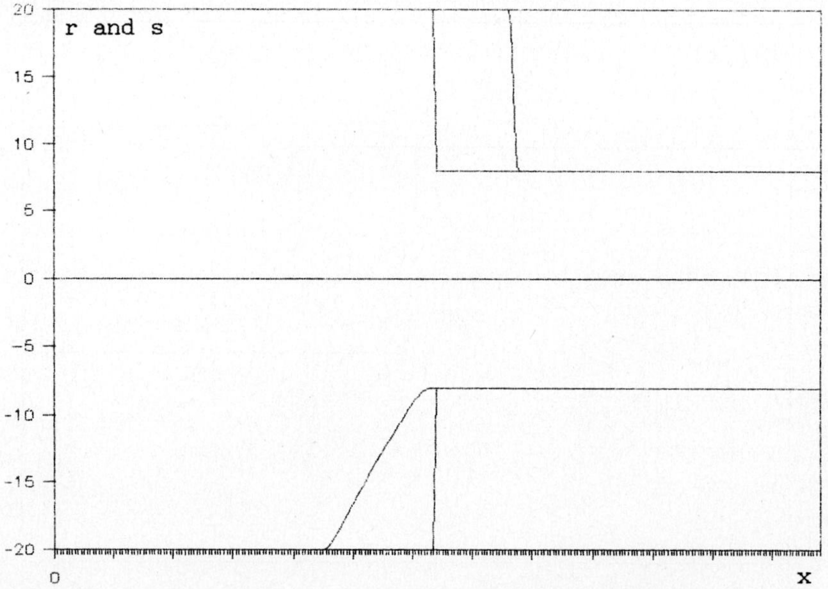

Fig. 3. Monte Carlo Godunov solution for the shallow water system at t=0 and t=40k, Riemann invariants: r top, s bottom.

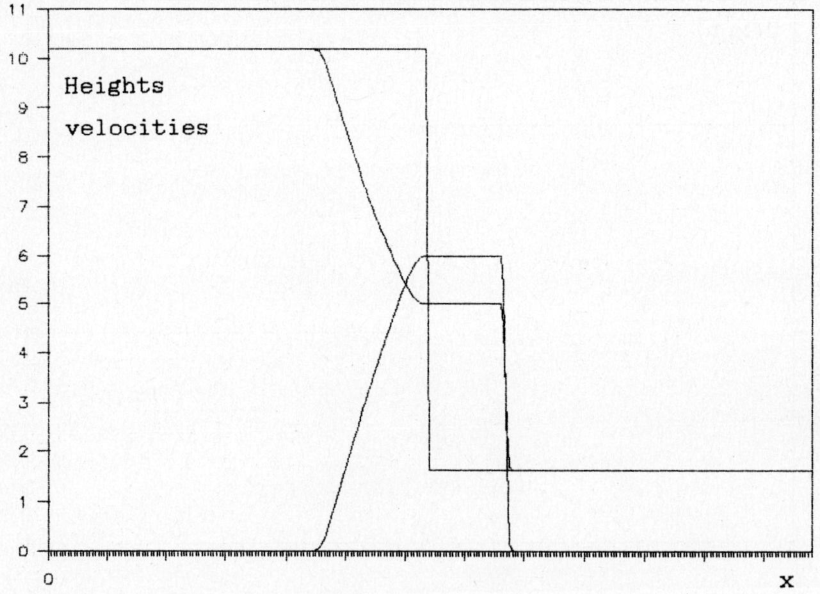

Fig. 4. Monte Carlo Godunov solution for the shallow water system at t=0 and t=40k, heights and velocities superimposed.

NUMERICAL SOLUTION OF FLOW EQUATIONS AN AIRCRAFT DESIGNER'S VIEW

Josef Mertens, Klaus Becker
Department of Theoretical Aerodynamics
MBB-UT, TE 212, Hünefeldstr. 1-5, D-2800 Bremen 1

SUMMARY

Today the most accurate and cost effective industrial codes used in aircraft design are based on the full potential equation coupled with boundary layer equations. However, these are not capable to solve complicated three-dimensional problems of vortical flows and shocks. On the other hand Euler and Navier-Stokes codes are too expensive and not accurate enough for design purposes, especially in regard of drag and interference prediction. The reasons for these deficiencies are investigated and a way to overcome them by future developments is demonstrated.

NOMENCLATURE

a	speed of sound	p	pressure
a_o	stagnation speed of sound	R	special gas constant
dA	element of surface ∂V	s	specific entropy
$div_n \cdot$	div. in plane normal to $\underset{\sim}{n}$	t	time
$D_M \cdot$	wave operator at Mach cone	T	temperature
$D_P \cdot$	wave operator at path line	$\underset{\sim}{v}^*$	space-time velocity, stationary: $\underset{\sim}{v}^* = \underset{\sim}{v}$
$\frac{D\cdot}{Dt}$	substantial derivative	$\underset{\sim}{v}$	space velocity
		v_n	normal component of velocity
∂V	surface of V	$v_{n_{rel}}$	$:= (\underset{\sim}{v} - \underset{\sim}{v}_{shock}) \cdot \underset{\sim}{n}_{shock}$
e	specific inner energy	$\underset{\sim}{v}_t$	tangential part of velocity
$\underset{\sim}{F}$	flux	V	control volume
h_o	specific stagnation enthalpy	γ	ratio of specific heats
$\underset{\sim}{n}^*$	space-time like normal vector	λ	Riemann invariant
$\underset{\sim}{n}$	space like part of $\underset{\sim}{n}^*$	ρ	density

DEFICIENCIES OF MODERN NUMERICAL METHODS

In the aerodynamic design of modern aircraft, especially transonic transport aircraft, numerical methods became one of the most important design tools. The mayority of the codes used nowadays relies heavily on the experience gained with the elliptic subsonic potential equation. To make possible a solution of transonic flows, conditions were introduced to enforce numerical stability. Even today, the most accurate codes for drag prediction are full potential codes coupled with a boundary layer method. Often viscous effects strongly influence the solution (Fig. 1: shock/ boundary layer interaction, rear loaded profiles, transonic wakes). The H-type grid enables an accurate coupling of the inviscid and viscous solution including the wake, and an easy capture of normal shocks [4].

These codes are restricted to two-dimensional or nearly two-dimensional flow problems because they cannot capture the typical three-dimensional effects (Fig. 2: unknown three-dimensional shocks, free vortices, wake interferencies, nacelle and jet interferencies, rotational flow fields).

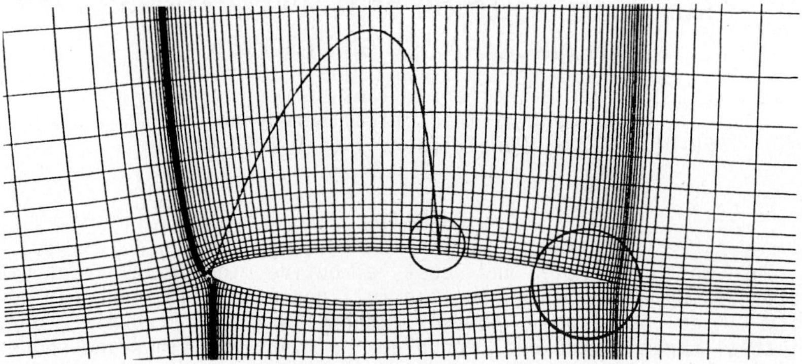

Fig. 1: Transonic airfoil calculation

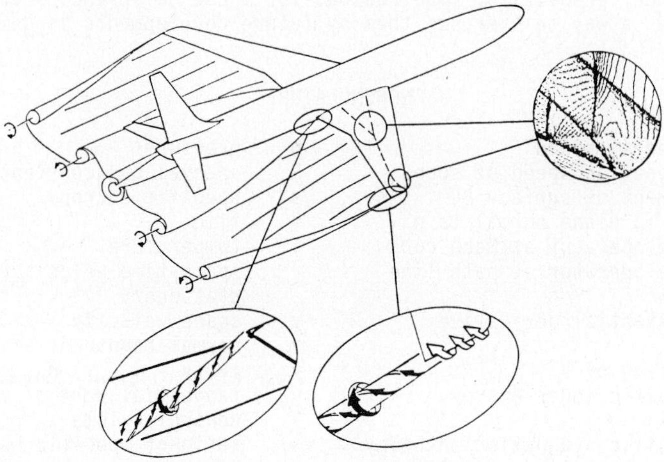

Fig. 2: 3D-flow problems around transport aircraft

For aircraft having low aspect ratio wings the older methods are completely insufficient: the flow field is dominated by vortex systems; at higher Mach numbers the steep entropy gradients and real gas effects do not allow a potential approximation.

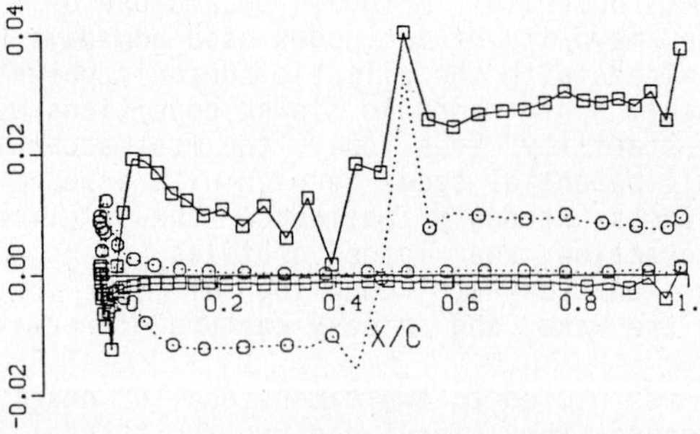

Fig. 3: Relative loss of total pressure on a midwing airfoil (Solutions of two different 3D Euler codes)

On the other hand Euler and Navier-Stokes (NS) codes are available, with it should possible to solve these problems [8]. Figs. 3-8 show some typical 3D-results of two modern Euler codes representing the state of the art. Due to smearing and wiggles shock location and strength cannot be determined accurately (Fig. 3). At the leading edge spurious pressure raises resp. entropy losses occur which partly vanish downstream; the trailing edge solution produces similar errors.

Inviscid wave and induced drag can be determined by pressure integration in the direction perpendicular to the free stream. It is given as the small difference of the large areas enclosed by the pressure curves (Fig. 4); errors result mainly from incorrect pressure computation at leading and trailing edges.

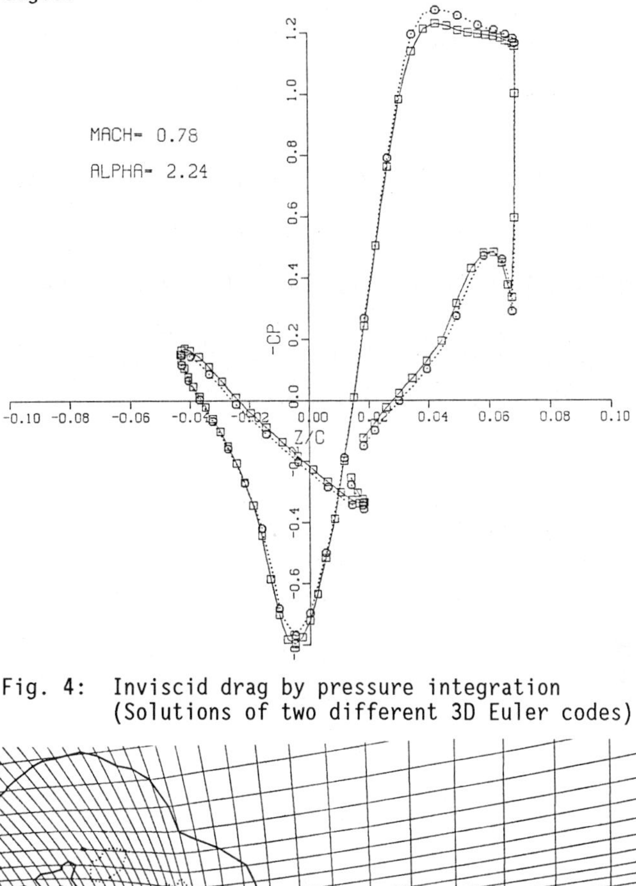

Fig. 4: Inviscid drag by pressure integration (Solutions of two different 3D Euler codes)

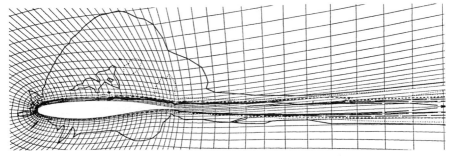

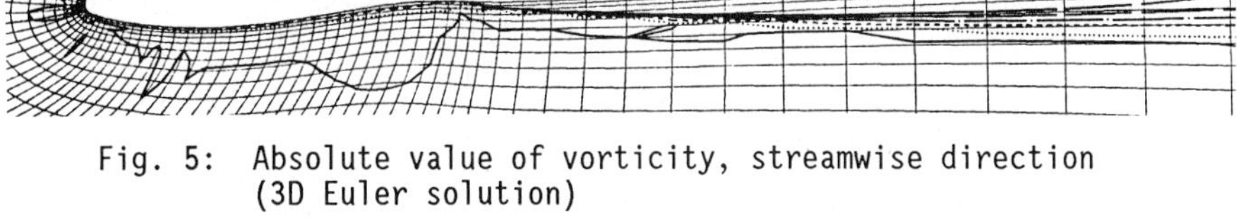

Fig. 5: Absolute value of vorticity, streamwise direction (3D Euler solution)

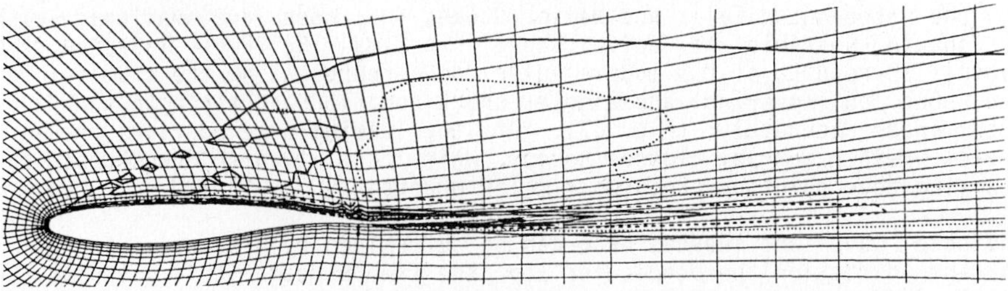

Fig. 6: Relative loss of total pressure, streamwise direction (3D Euler solution, values ≥ 0.005)

Another possibility is to calculate wave drag from the entropy rise at the shock and induced drag in the Trefftz plane. This requires accurate shock determination and vorticity transport; both are not sufficiently resolved by current codes (Figs. 5/6).

Another important problem is interference caused by vortex systems, the computation of which should also be possible by Euler and NS solvers.

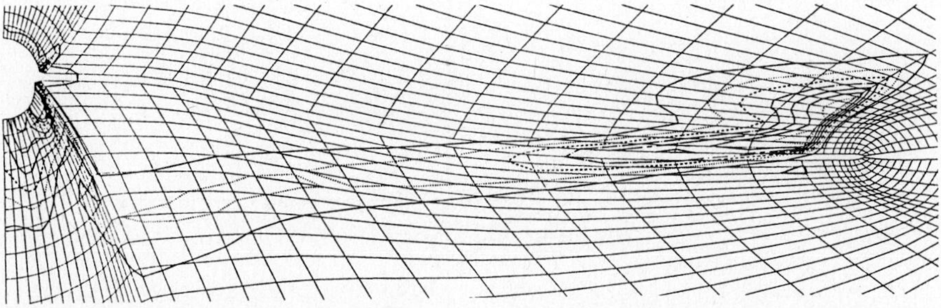

Fig. 7: Absolute value of vorticity, wake behind wing (3D Euler solution)

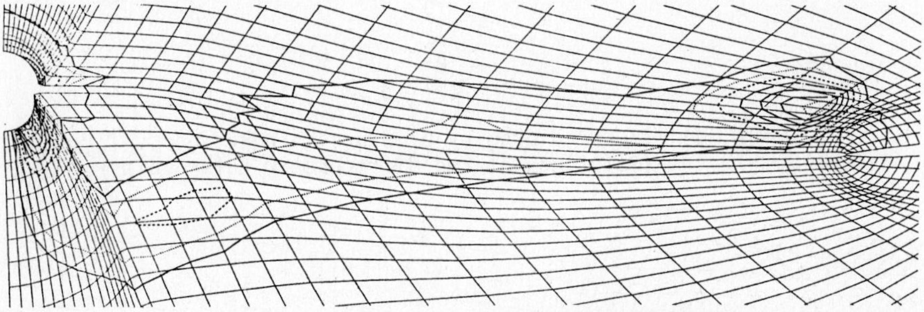

Fig. 8: Total pressure loss, wake behind wing (3D Euler solution)

Figs. 7,8 were to show only the inviscid wake and shock vorticity of an Euler solution. Partly vorticity occurs at the physically known places. But obviously, additional vorticity is generated by grid properties, is smeared out, and in the downstream direction the vorticity content [3] diminishes rapidly. So these codes are not yet able to solve this problem with the accuracy needed for effective aircraft design.

Looking at computing costs the Euler (and NS) codes are surprisingly expensive: Today's Euler codes still need grid sizes comparable to those of full potential codes. But the required pressure or velocity values are directly obtained as solutions of the Euler equations. Therefore a much coarser mesh should yield the same accuracy as for potential solutions, which produce an order one loss of accuracy by numerical differentiation of the potential function. Moreover - because of the required degree of continuity - additional difficulties occur for potential solvers in regions of strongly varying solutions. Therefore important accuracy and cost improvements of future Euler (and NS) codes can be expected.

The viscous nature of real flow may lead to assume that difficulties could be overcome by using the NS equations. However, since the errors mentioned are generated by the numerical method and not by the equations, an improvement appears possible only by proper formulation of the algorithms. Especially for high Reynolds number flows the Euler terms remain essential in the NS equations; they completely describe the flow field away from body surfaces or wakes. Therefore solving the NS equations will only be possible by accurately solving the intrinsic hyperbolic Euler equations; their numerical treatment by field methods will be discussed here.

It is well known, that the errors in numerical solutions are generated by numerical smearing, amplification and artificial damping, but it is difficult to localize the hidden sources of these effects. We will try to identify some of them and to show possibilities to overcome them. The facts presented are well known, but not fully respected in numerical methods.

The development of numerical field methods is made in five steps:
- Selection and analysis of the governing equations,
- Selection of a point distribution or grid to represent the flow field,
- Approximation of solution values between grid points,
- Formulation of the boundary conditions defining the special problem,
- Mathematical solution algorithm.

In the next sections only the first three points are discussed for the Euler equations as an example of systems of nonlinear hyperbolic equations.

GOVERNING EQUATIONS

The well known Euler equations are

$$\text{C} \qquad \frac{D\rho}{Dt} + \rho \operatorname{div} \underset{\sim}{v} = \frac{\partial \rho}{\partial t} + \operatorname{div}(\rho \underset{\sim}{v}) = 0 \,,$$

$$\text{M} \qquad \rho \frac{D\underset{\sim}{v}}{Dt} + \operatorname{grad} p = \underset{\sim}{0} \,, \tag{1}$$

$$\text{E} \qquad \rho \frac{De}{Dt} + p \operatorname{div} \underset{\sim}{v} = \rho \frac{De}{Dt} - \frac{p}{\rho} \frac{D\rho}{Dt} = \rho T \frac{Ds}{Dt} = 0 \,.$$

Hyperbolic differential equations have real directions with undefined derivatives. For the Euler equations (1) these directions are the directions $\underset{\sim}{n}^*$ normal to the path lines and normal to the Mach cone:

$$\begin{aligned} &\text{path line:} && \underset{\sim}{v}^* \cdot \underset{\sim}{n}^* = 0 \,, \\ &\text{Mach cone:} && \underset{\sim}{v}^* \cdot \underset{\sim}{n}^* = -a \,. \end{aligned} \tag{2}$$

All solutions of hyperbolic systems, except the trivial ones, are defined by jumps of (sometimes higher order) derivatives. The possible discontinuities are (depending on the selected set of dependant variables) e.g.

$$\begin{aligned}
&\text{across the path line: all variables except} && p,\ v_n\ , \\
&\text{across the Mach cone: 1. derivatives of} && p,\ (\underset{\sim}{v}\cdot\underset{\sim}{n})\ , \\
&\text{across shocks: all variables except} && \underset{\sim}{v}_t\,,\ \rho\, v_{n_{rel}}\ ,\ p + \rho\, v_{n_{rel}} \cdot v_n\ , \\
& && \rho\,(e + \tfrac{v^2}{2})\ v_{n_{rel}} + p\ v_n\ .
\end{aligned} \tag{3}$$

The path line and wake discontinuities are connected with vortices and occur even in steady subsonic flow. The Euler equations can generate wake discontinuities only at shocks or boundaries, but they completely describe their transportation along path lines.

In elliptic problems the polynomial order of the Taylor approximation is a quality measure for the discretization. For hyperbolic problems this is only true for regions with very smooth solutions. In the physically more interesting zones different kinds of discontinuities are significant; especially here the solution cannot be expanded into Taylor series.

Normal to the directions $\underset{\sim}{n}^*$ of possible jumps there exist corresponding directions of wave propagation with the associated wave operators:

$$\begin{aligned}
&\text{path line}\ \ \underset{\sim}{v}^* : && D_P\,.\ := \frac{D.}{Dt} := \frac{\partial .}{\partial t} + (\underset{\sim}{v}\cdot\text{grad})\,.\ , \\
&\text{Mach cone}\ \ \underset{\sim}{v}^* + a\,\underset{\sim}{n} : && D_M.\ := \frac{D.}{Dt} + a\ (\underset{\sim}{n}\cdot\text{grad})\,.\ .
\end{aligned} \tag{4}$$

The continuous part of the solution is defined by the set of compatibility conditions of characteristics theory, e.g. (depending on variables)

$$\begin{aligned}
&\text{along path line:}\ \ E && D_P\ s = 0\ , \\
&\qquad\qquad\qquad \underset{\sim}{v}\cdot M && D_P\ (\tfrac{v^2}{2}) = -\ \tfrac{1}{\rho}\ \underset{\sim}{v}\cdot \text{grad}\ p\ , \\
&\text{along Mach cone:} && D_M\ (\underset{\sim}{v}\cdot\underset{\sim}{n}) + \frac{1}{\rho a}\ D_M\, p + a\ \text{div}_n\ \underset{\sim}{v} = 0
\end{aligned} \tag{5}$$

($\text{div}_n\ \underset{\sim}{v}$: div $\underset{\sim}{v}$ taken in the plane normal to $\underset{\sim}{n}$).

These conditions are special combinations of the governing equations (1) which are valid everywhere except across shocks. But only in the directions of the corresponding characteristics they describe continuous wave propagation although continuity is not required for each single term.

Along the path line entropy (and in steady flow stagnation enthalpy) is convected, without any continuity required in the transverse direction. For vortical flows the correct calculation of vorticity transport is substantial; but this is described only indirectly by convection.

Important for numerical treatment is, that the hyperbolic solution is exclusively defined by jumps of derivativs across the characteristics and in certain cases as jumps of the solutions themselves.

Referring to integral formulations instead of differential equations, it is possible to capture all discontinuities within one cell for one-dimensional problems, because fluxes may be treated as unknowns.

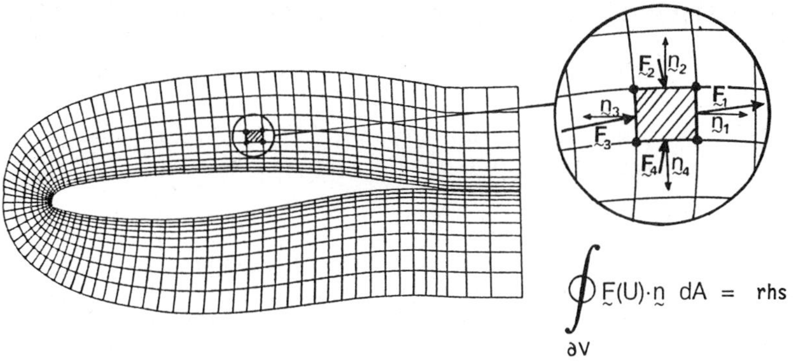

Fig. 9: Finite volume

$$\text{C} \qquad \int_V \frac{\partial \rho}{\partial t}\, dV = -\oint_{\partial V} (\rho\, \underset{\sim}{v}\cdot \underset{\sim}{n})\, dA\ ,$$

$$\text{M} \qquad \int_V \frac{\partial}{\partial t}(\rho \underset{\sim}{v})\, dV = -\oint_{\partial V} [\rho \underset{\sim}{v}\ (\underset{\sim}{v}\cdot \underset{\sim}{n}) + p\ \underset{\sim}{n}]\ dA\ , \qquad (6)$$

$$\text{E} \qquad \int_V \frac{\partial}{\partial t}[\rho(e+\frac{v^2}{2})]dV = -\oint_{\partial V} [\rho(e+\frac{v^2}{2} + \frac{p}{\rho})(\underset{\sim}{v}\cdot \underset{\sim}{n})]\ dA\ ,$$

This normally fails in multidimensional cases since the fluxes are tensors one degree higher than the flow variables. Therefore fluxes are computed by surface integration and their values have to be calculated from a usual set of variables. However, the values of these variables are only known at distinct points and not at the whole boundary. This must be overcome by interpolation assumptions which often are inconsistent with the discontinuities. Therefore a combination with other techniques is recommended which will be described later.

The selection of unknowns has a strong influence on numerical properties. For the Euler equations the so-called conservative variables yield shock capturing capability to finite difference schemes, but are normally working well only for one-dimensional cases. For finite volume schemes based on the surface integral equations (6) instead of the differential equations, it is not necessary to use the conservative variables for shock capturing. The only important condition is, that the equations do not contain production terms.

Using conservative or primitive variables, the set of differential equations is strongly coupled. For numerical reasons and for consistency it is desirable to decouple the system of equations. This is at least partly possible by using a different set of variables. A complete decoupling is provided by Riemann invariants if they exist; unfortunately they usually do not exist. But often it is possible to construct a system of equations with weaker coupling and weaker nonlinearities, using the knowledge of Riemann properties of simpler cases. Choosing velocity, speed of sound and entropy as variables leads to a set of only mildly coupled and nearly linear compatibility conditions. Wave transportation is described by the wave operators **D** (4). The nonlinearity is restricted to the determination of the wave operator's characteristic direction and the right hand terms:

along path line: E $\quad D_P\, s = 0\,,$

$\underset{\sim}{v}\cdot M$ (transient) $\quad D_P\left(\frac{v^2}{2}\right) = \underset{\sim}{v}\left[T \operatorname{grad} s - \frac{1}{\gamma-1}\operatorname{grad}(a^2)\right],$

$\underset{\sim}{v}\cdot M$ (quasi stat.) $\quad D_P\,(a_o^2) = \frac{\partial}{\partial t}(a^2) - (\gamma-1)T\,\frac{\partial s}{\partial t}\,,$ (7)

along Mach cone: $\quad D_M\left(\frac{2}{\gamma-1}\,a + \underset{\sim}{v}\cdot\underset{\sim}{n}\right) = -\frac{a}{\gamma R}\,D_M\, s - a\,\mathrm{div}_n\,\underset{\sim}{v}\,.$

If we locally construct for each grid plane Riemann invariants by combining velocity and speed of sound, the weakest coupling of compatibility equations is obtained [1, 5]. This is implemented by using two differing normal vectors [7].

$$\lambda(\underset{\sim}{n}) := \frac{2}{\gamma-1}\,a + (\underset{\sim}{v}\cdot\underset{\sim}{n})\,, \quad \underset{\sim}{n}_1 = -\underset{\sim}{n}_2\,. \tag{8}$$

It is well known that for isentropic plane waves the equations are completely decoupled. However, this set has no shock capturing capability.

SELECTION OF REPRESENTATIVE POINTS

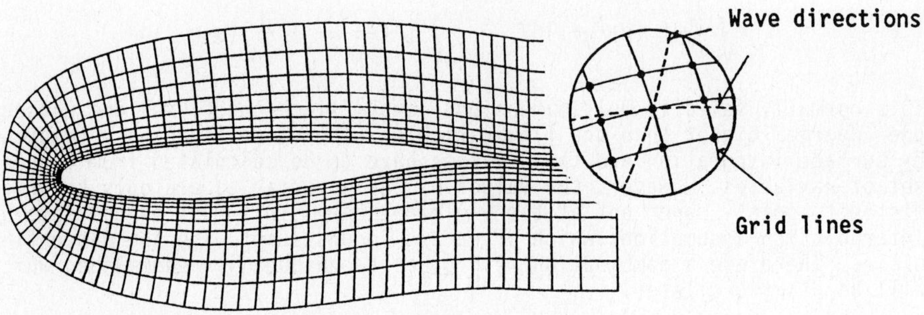

Fig. 10: Numerical grid

In most numerical field methods the solution is represented by a distinct number of grid points. Between the grid points the solution values are distributed by some kind of interpolation. These interpolation functions are defined locally, their definition changes at grid lines. Therefore grid lines introduce numerical discontinuities. The best results are obtained, if these numerical discontinuities coincide at least with the most important physical discontinuities.

For the Euler equations the most important discontinuities are:
- *At shocks:* shocks and path lines,
- *In nonisentropic, shock-free regions:* path lines and characteristics, especially "main characteristics" (The "main characteristic" is the downstream characteristic in the plane spanned by the boundary normal vector and the velocity vector.),
- *In the isentropic region:* characteristics, most important the "main characteristics",
- *Steady subsonic vortex flow:* path lines which here are stream lines.

This constitutes a great challenge on grid construction, but it is a way to get sufficiently accurate solutions with a restricted number of grid points (e.g. Fig. 11).

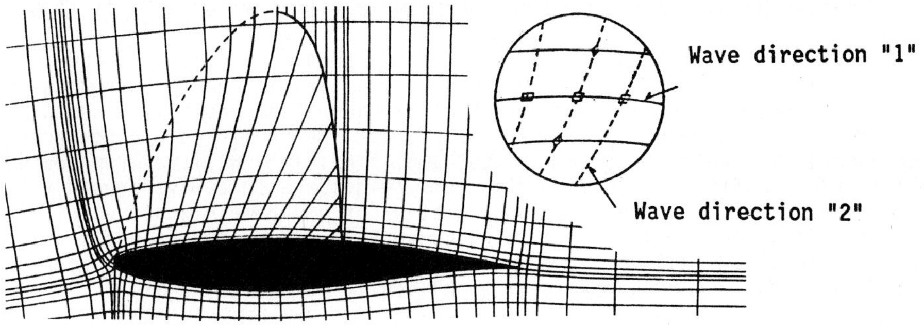

Fig. 11: Example of a physically motivated numerical grid

APPROXIMATION OF SOLUTION VALUES BETWEEN GRID POINTS

Most numerical field methods need some kind of solution distribution between grid points which generate numerical discontinuities along grid lines. If the grid lines do not coincide with the characteristics, part of the information transport changes direction from that of the characteristics to that of the grid lines due to the redistribution of discontinuities. This produces numerical dispersion.

On the other hand, the interpolation functions often must be continuous across grid lines. Then continuity is introduced numerically whereas physical nature may be discontinuous. This smears out solutions, amplifies disturbances and produces the well known wiggles.

Normally interpolation of different variables between the grid points is treated independently from each other, e.g. linearly for density, momentum and energy. But the equations as well as the physical distributions of values are strongly coupled. In the zones of steep gradients or even strongly varying gradients, the solution is affected by inconsistent interpolation. A subsequent computation, especially of sensitive functions like pressure or entropy, amplifies these errors due to the nonlinear combination of inconsistent values; spurious entropy often disappearing further downstream is generated. Moreover, truncation errors increase with nonlinearity and stronger coupling of equations. So it becomes impossible to accurately calculate wave or induced drag by pressure integration, because the most important regions have strongly varying gradients.

REQUIREMENTS FOR NUMERICAL FIELD METHODS

In contrast to most numerical schemes which stay in the tradition of elliptic solvers or one-dimensional approximations, an accurate numerical scheme requires proper modeling of the hyperbolic character. Therefore physical and numerical discontinuities should coincide as much as possible. For more than two variables it is necessary to select the most important directions: wave and discontinuity lines which transport the main information and the strongest discontinuities.

From the aircraft designer's view, a combination with a viscous solution is absolutely necessary. It is facilitated by the selection of grids well adapted to stream lines [2]. To facilitate code construction, stream line adaption can be used as a construction principle of the code itself [1,6].

To achieve accurate solutions the scheme should be of second order in smooth regions, but at physical discontinuities it should introduce as little numerical smoothness as possible. Therefore interpolation should be restricted to one mesh or cell surface. Then numerical inaccuracies introduced by approximation are not amplified; they will be transported along the grid lines. If the grid lines coincide with a wave propagation direction, all corresponding approximation errors remain confined within the neighbouring grid lines, with no further dispersion. A realization seems possible, e.g. with characteristics oriented schemes [1] or node oriented finite volume schemes using physically motivated grids.

To overcome the difficulties with inconsistent interpolation and truncation errors, it is possible to use sets of variables which better decouple the equations. These sets normally have no shock capturing capability. So they can only be used in a nearly converged state to improve accuracy; shock fitting must be performed. Another possibility is to use better interpolation functions, based on local approximations of the flow field, e.g. locally linearized potential solutions.

The methods mentioned above will give the possibility of cost effective and accurate solutions. But programming work will be more arduous, especially when versatility is to be maintained. At comparable accuracy for Euler codes the goal is to achieve

- *computing time in the order of that of full potential codes,*
- *with coarse grids as known from the method of characteristics.*

REFERENCES

[1] AGRAWAL, S., VERMELAND, R. E., VERHOFF, A., LOWRIE, R. B.: "Euler Transonic Solutions over Finite Wings", AIAA Paper 88-0009.

[2] HAASE, W.: "Influence of Trailing-Edge Meshes on Skin Friction in Navier-Stokes Calculations", AIAA J., 24, 9 (1986), pp. 1557-1559.

[3] HIRSCHEL, E. H., RIZZI, A.: "The Mechanism of Vorticity Creation in Euler Solutions for Lifting Wings", MBB-LKE122-AERO-MT-753 (1987), Presented at the Int. Vortex-Flow Symp., Stockholm, Oct. 1-3, 1987.

[4] MERTENS, J., KLEVENHUSEN, K.-D., JAKOB, H.: "Accurate Transonic Wave Drag Prediction Using Simple Physical Models", AIAA J., 25, 6 (1987), pp. 799-805.

[5] MORETTI, G., ZANETTI, L.: "A New and Improved Computational Technique for Two-Dimensional, Unsteady, Compressible Flows", AIAA J., 22, 6 (1984), pp. 758-765.

[6] OWEN, D. R., PEARSON, C. E.: "Numerical Solution of a Class of Steady-State Euler Equations by a Modified Streamline Method", AIAA paper 88-0625 (1988).

[7] SAUER, R.: "Nichtstationäre Probleme der Gasdynamik", Springer, 1966.

[8] "Accuracy Study of Transonic Flow Computations for Three Dimensional Wings", Study by GARTEUR AD(AG-05), August 1987.

HYPERSONIC BLUNT BODY COMPUTATIONS INCLUDING REAL GAS EFFECTS

J.-L. Montagné†
ONERA, B.P. 72, 92322 Chatillon Cedex, France
H.C. Yee‡
NASA Ames Research Center, Moffett Field, CA 94035 USA
G.H. Klopfer*
NEAR Inc., Mountain View, CA 94043 USA
and
M. Vinokur**
Sterling Software, Palo Alto, CA 94030 USA

I. Motivation and Objective

The recently developed second-order explicit and implicit total variation diminishing (TVD) shock-capturing methods of the Harten and Yee [1,2], Yee [3,4], and van Leer [5,6] types in conjunction with a generalized Roe's approximate Riemann solver of Vinokur [7] and the generalized flux-vector splittings of Vinokur and Montagné [8] for two-dimensional hypersonic real gas flows are studied. A previous study [9] on one-dimensional unsteady problems indicated that these schemes produce good shock-capturing capability and that the state equation does not have a large effect on the general behavior of these methods for a wide range of flow conditions for equilibrium air. The objective of this paper is to investigate the applicability and shock resolution of these schemes for two-dimensional steady-state hypersonic blunt body flows.

The main contribution of this paper is to identify some of the elements and parameters which can affect the convergence rate for high Mach numbers or real gases but have negligible effect for low Mach number cases for steady-state inviscid blunt body flows. In order to investigate these different points, two kinds of flows are considered. The blunt body calculations at Mach numbers higher than 15 allow significant real gas effects to occur, while the case of an impinging shock provides a test on the treatment of slip surfaces and complex shock structures. In separate papers, a temporally second-order, implicit, time-accurate TVD-type algorithm for viscous steady and unsteady flows is studied. Studies show that the behavior of the schemes with various temporal differencing but similar spatial discretization for inviscid and viscous flows are very different — in terms of stability and convergence rate. This point will be addressed in reference [10]. However, this paper only concerns itself with steady-state inviscid computations.

In the following section, the generalized Roe's approximate Riemann solver and flux-vector splittings for real gases are reviewed. Due to space limitation, only the ADI linearized conservative implicit version of the Harten and Yee schemes [11] and Yee [3] is reviewed here since most of the illustrations are computed with this particular algorithm. The findings concerning the various aspects in improving the convergence rate and numerical examples are discussed in the subsequent sections.

†Research Scientist, Theoretical Aerodynamics Division, this work was performed while on leave as an Ames Associate at NASA Ames Research Center, Moffett Field, CA 94035 USA.

‡Research Scientist, Computational Fluid Dynamics Branch

*Research Scientist

**Principal Analyst

II. Description of the Numerical Algorithm

The conservation laws for the two-dimensional Euler equations can be written in the form

$$\frac{\partial U}{\partial t} + \frac{\partial F(U)}{\partial x} + \frac{\partial G(U)}{\partial y} = 0. \tag{1}$$

where $U = [\,\rho,\ m,\ n,\ e\,]^T$, $F = [\,\rho u,\ mu{+}p,\ nu,\ eu{+}pu\,]^T$, and $G = [\,\rho v,\ mv,\ nv{+}p,\ ev{+}pv\,]^T$. Here ρ is the density, $m = \rho u$ is the x-component of the momentum per unit volume, $n = \rho v$ is the y-component of the momentum per unit volume, p is the pressure, $e = \rho[\epsilon + \frac{1}{2}(u^2 + v^2)]$ is the total internal energy per unit volume, and ϵ is the specific internal energy.

A generalized coordinate transformation of the form $\xi = \xi(x,y)$ and $\eta = \eta(x,y)$ which maintains the strong conservation-law form of equation (1) is given by

$$\frac{\partial \widehat{U}}{\partial t} + \frac{\partial \widehat{F}(\widehat{U})}{\partial \xi} + \frac{\partial \widehat{G}(\widehat{U})}{\partial \eta} = 0, \tag{2}$$

where $\widehat{U} = U/J$, $\widehat{F} = (\xi_x F + \xi_y G)/J$, $\widehat{G} = (\eta_x F + \eta_y G)/J$, and $J = \xi_x\eta_y - \xi_y\eta_x$, the Jacobian transformation. Let $A = \partial F/\partial U$ and $B = \partial G/\partial U$. Then the Jacobians $\widehat{A} = \partial\widehat{F}/\partial\widehat{U}$ and $\widehat{B} = \partial\widehat{G}/\partial\widehat{U}$ can be written as

$$\widehat{A} = (\xi_x A + \xi_y B) \tag{3a}$$

$$\widehat{B} = (\eta_x A + \eta_y B). \tag{3b}$$

2.1. Riemann Solvers

Here the usual approach of applying the one-dimensional scalar TVD schemes via the so called Riemann solvers for each direction in multidimensional nonlinear systems of hyperbolic conservation laws (see for example reference [2]) is used. The eigenvalues and eigenvectors of the Jacobian matrices $\widehat{A}$ and $\widehat{B}$ are used in approximate Riemann solvers. Given two states whose difference is ΔU, Roe [12] obtained an average $\overline{A}$ in the ξ-direction, for example, satisfying $\Delta\widehat{F} = \overline{A}\Delta U$ for a perfect gas. The generalization by Vinokur [7] for an arbitrary gas involves the pressure derivatives $\chi = (\partial p/\partial\rho)_{\tilde{\epsilon}}$ and $\kappa = (\partial p/\partial\tilde{\epsilon})_{\rho}$, where $\tilde{\epsilon} = \rho\epsilon$. The relation $c^2 = \chi + \kappa h$ then gives the speed of sound, where $h = \epsilon + p/\rho$. Introducing $H = h + (u^2 + v^2)/2$, Vinokur found the same expressions for $\overline{u}, \overline{v}$ and $\overline{H}$ as for the perfect gas, and that $\overline{\chi}$ and $\overline{\kappa}$ must satisfy

$$\overline{\chi}\Delta\rho + \overline{\kappa}\Delta\tilde{\epsilon} = \Delta p. \tag{4}$$

Unique values of $\overline{\chi}$ and $\overline{\kappa}$ are obtained by projecting the arithmetic averages of the values for the two states into this relation (see references [7] and [2] for the exact formulas).

Flux-vector splitting methods divide the flux $\widehat{F}$ into several parts, each of which has a Jacobian matrix whose eigenvalues are all of one sign. The approach by Steger and Warming [13] made use of the relation $F = AU$, valid for a perfect gas. Van Leer [6] constructed a different splitting in which the eigenvalues of the split-flux Jacobians are continuous and one of them vanishes leading to sharper capture of transonic shocks. Vinokur and Montagné [8] showed that the expressions for both these splittings can be generalized to an arbitrary gas by using the variable $\gamma = \rho c^2/p$, and adding to the split energy flux a term equal to the product of the split mass flux and the quantity $\epsilon - c^2/[\gamma(\gamma-1)]$ (see references [8] and [2] for the exact formulas).

2.2 Description of the Implicit TVD schemes

Let Δt be the time step and let the grid spacing be denoted by $\Delta\xi$ and $\Delta\eta$ such that $\xi = j\Delta\xi$ and $\eta = k\Delta\eta$. An implicit second-order in space, first-order in time TVD algorithm in generalized coordinates of Yee and Harten for two-dimensional systems (1) [2-4] can be written as

$$\widehat{U}^{n+1}_{j,k} + \frac{\Delta t}{\Delta \xi}\left[\widetilde{F}^{n+1}_{j+\frac{1}{2},k} - \widetilde{F}^{n+1}_{j-\frac{1}{2},k}\right] + \frac{\Delta t}{\Delta \eta}\left[\widetilde{G}^{n+1}_{j,k+\frac{1}{2}} - \widetilde{G}^{n+1}_{j,k-\frac{1}{2}}\right] = \widehat{U}^{n}_{j,k}. \tag{5}$$

The functions $\widetilde{F}_{j+\frac{1}{2},k}$ and $\widetilde{G}_{j,k+\frac{1}{2}}$ are the numerical fluxes in the ξ- and η-directions evaluated at $(j+\frac{1}{2},k)$ and $(j,k+\frac{1}{2})$, respectively. Typically, $\widetilde{F}_{j+\frac{1}{2},k}$ can be expressed as

$$\widetilde{F}_{j+\frac{1}{2},k} = \frac{1}{2}\left(\widehat{F}_{j,k} + \widehat{F}_{j+1,k} + R_{j+\frac{1}{2}}\Phi_{j+\frac{1}{2}}\right). \tag{6}$$

Here $R_{j+\frac{1}{2}}$ is the eigenvector matrix for $\partial\widehat{F}/\partial U$ evaluated at some symmetric average of $U_{j,k}$ and $U_{j+1,k}$ (for example, Roe average [12] for a perfect gas and generalized Roe average of Vinokur [7] for real gases). Similarly, one can define the numerical flux $\widetilde{G}_{k+\frac{1}{2}}$ in this manner. For viscous steady and unsteady flows, a fully implicit second-order in time and space algorithm (with the same spatial differencing for the convection terms) appears to be more stable and efficient (in terms of convergence rate) than (5). See references [10,14,15] for details.

Second-order Symmetric TVD Scheme: The elements of the $\Phi_{j+\frac{1}{2}}$ in the ξ-direction denoted by $(\phi^l_{j+\frac{1}{2}})^S$ for a spatially second-order symmetric TVD scheme [3,4] are

$$(\phi^l_{j+\frac{1}{2}})^S = -\psi(a^l_{j+\frac{1}{2}})\left[\alpha^l_{j+\frac{1}{2}} - \widehat{Q}^l_{j+\frac{1}{2}}\right]. \tag{7a}$$

The value $a^l_{j+\frac{1}{2}}$ is the characteristic speed a^l for $\partial\widehat{F}/\partial U$ evaluated at some average between $U_{j,k}$ and $U_{j+1,k}$. The function ψ is

$$\psi(z) = \begin{cases} |z| & |z| \geq \delta_1 \\ (z^2+\delta_1^2)/2\delta_1 & |z| < \delta_1 \end{cases}. \tag{7b}$$

Here $\psi(z)$ in equation (7b) is an entropy correction to $|z|$ where δ_1 is a small positive parameter. For steady-state problems containing strong shock waves, a proper control of the size of δ_1 is very important, especially for hypersonic blunt-body flows. See reference [2] or section III for a discussion. An example of limiter function $\widehat{Q}^l_{j+\frac{1}{2}}$ used in calculations is:

$$\widehat{Q}^l_{j+\frac{1}{2}} = \text{minmod}\left[\alpha^l_{j-\frac{1}{2}}, \alpha^l_{j+\frac{1}{2}}, \alpha^l_{j+\frac{3}{2}}\right]. \tag{7.c}$$

The minmod function of a list of arguments is equal to the smallest number in absolute value if the list of arguments is of the same sign, or is equal to zero if any arguments are of opposite sign. Here $\alpha^l_{j+\frac{1}{2}}$ are elements of

$$\alpha_{j+\frac{1}{2}} = R^{-1}_{j+\frac{1}{2}}(U_{j+1,k} - U_{j,k}). \tag{8}$$

Second-Order Upwind TVD Scheme: The elements of the $\Phi_{j+\frac{1}{2}}$ in the ξ-direction denoted by $(\phi^l_{j+\frac{1}{2}})^U$ for a spatially second-order upwind TVD scheme [11,2] are

$$(\phi^l_{j+\frac{1}{2}})^U = \frac{1}{2}\psi(a^l_{j+\frac{1}{2}})(g^l_{j+1} + g^l_j) - \psi(a^l_{j+\frac{1}{2}} + \gamma^l_{j+\frac{1}{2}})\alpha^l_{j+\frac{1}{2}}. \tag{9a}$$

where

$$\gamma^l_{j+\frac{1}{2}} = \frac{1}{2}\psi(a^l_{j+\frac{1}{2}})\begin{cases} (g^l_{j+1} - g^l_j)/\alpha^l_{j+\frac{1}{2}} & \alpha^l_{j+\frac{1}{2}} \neq 0 \\ 0 & \alpha^l_{j+\frac{1}{2}} = 0 \end{cases}. \tag{9b}$$

An example of limiter function g^l_j used in calculations is

$$g^l_j = \text{minmod}\left[\alpha^l_{j-\frac{1}{2}}, \alpha^l_{j+\frac{1}{2}}\right]. \tag{9c}$$

A Conservative Linearized ADI Form for Steady-State Applications: A conservative linearized ADI form of equation (5) used mainly for steady-state applications as described in detail in references [3,11], can be written as

$$\left[I + \frac{\Delta t}{\Delta\xi} H^{\xi}_{j+\frac{1}{2},k} - \frac{\Delta t}{\Delta\xi} H^{\xi}_{j-\frac{1}{2},k}\right] E^{*} = -\frac{\Delta t}{\Delta\xi}\left[\widetilde{F}^{n}_{j+\frac{1}{2},k} - \widetilde{F}^{n}_{j-\frac{1}{2},k}\right] - \frac{\Delta t}{\Delta\eta}\left[\widetilde{G}^{n}_{j,k+\frac{1}{2}} - \widetilde{G}^{n}_{j,k-\frac{1}{2}}\right], \tag{10a}$$

$$\left[I + \frac{\Delta t}{\Delta\eta} H^{\eta}_{j,k+\frac{1}{2}} - \frac{\Delta t}{\Delta\eta} H^{\eta}_{j,k-\frac{1}{2}}\right] E = E^{*}, \tag{10b}$$

$$\widehat{U}^{n+1} = \widehat{U}^{n} + E, \tag{10c}$$

where

$$H^{\xi}_{j+\frac{1}{2},k} = \frac{1}{2}\left[\widehat{A}_{j+1,k} - \Omega^{\xi}_{j+\frac{1}{2},k}\right]^{n}, \tag{10d}$$

$$H^{\eta}_{j,k+\frac{1}{2}} = \frac{1}{2}\left[\widehat{B}_{j,k+1} - \Omega^{\eta}_{j,k+\frac{1}{2}}\right]^{n}. \tag{10e}$$

The nonstandard notation

$$H^{\xi}_{j+\frac{1}{2},k} E^{*} = \frac{1}{2}\left[\widehat{A}^{n}_{j+1,k} E^{*}_{j+1,k} - \Omega^{\xi}_{j+\frac{1}{2},k} E^{*}\right]^{n} \tag{10f}$$

is used, and $\Omega^{\xi}_{j+\frac{1}{2},k}$, $\Omega^{\eta}_{j,k+\frac{1}{2}}$ can be taken as

$$\Omega^{\xi}_{j+\frac{1}{2},k} E^{*} = R_{j+\frac{1}{2},k} \mathrm{diag}\left[\psi(a^{l}_{j+\frac{1}{2}})\right] R^{-1}_{j+\frac{1}{2},k} (E^{*}_{j+1,k} - E^{*}_{j,k}) \tag{10g}$$

$$\Omega^{\eta}_{j,k+\frac{1}{2}} E = R_{j,k+\frac{1}{2}} \mathrm{diag}\left[\psi(a^{l}_{k+\frac{1}{2}})\right] R^{-1}_{j,k+\frac{1}{2}} (E_{j,k+1} - E_{j,k}). \tag{10h}$$

Here $\widehat{A}_{j+1,k}$ and $\widehat{B}_{j,k+1}$ are (3) evaluated at $(j+1,k)$ and $(j,k+1)$, respectively. The nonconserverative linearized implicit form suitable for steady-state calculations [2] is also considered. Numerical study indicated that the latter form appears to be slightly less efficient in terms of convergence rate than the linearized conservative form.

III. Enhancement of Convergence Rate for Hypersonic Flows

The current study indicated that the following three elements can affect the convergence rate at hypersonic speeds: (a) the choice of the entropy correction parameter δ_1, (b) the choice of the dependent variables on which the limiters are applied, and (c) the prevention of unphysical solutions during the initial transient stage.

(a). For blunt-body steady-state flows with $M > 4$, the initial flow conditions at the wall are obtained using the known wall temperature in conjunction with pressures computed from a modified Newtonian expression. Also, for implicit methods a slow startup procedure from freestream boundary conditions is necessary. Most importantly, it is advisable to use δ_1 in equation (7b) as a function of the velocity and sound speed. In particular

$$(\delta_1)_{j+\frac{1}{2}} = \widetilde{\delta}(|u_{j+\frac{1}{2}}| + |v_{j+\frac{1}{2}}| + c_{j+\frac{1}{2}}) \tag{11a}$$

$$(\delta_1)_{k+\frac{1}{2}} = \widetilde{\delta}(|u_{k+\frac{1}{2}}| + |v_{k+\frac{1}{2}}| + c_{k+\frac{1}{2}}) \tag{11b}$$

with $0.05 \le \widetilde{\delta} \le 0.25$ appear to be sufficient for the blunt-body flows for $4 \le M \le 25$. Equation (11) is written in Cartesian coordinates. In the case of generalized coordinates, the u and v should be replaced by the contravariant velocity components, and one half of the sound speed would be from the ξ-direction and the other half would be from the η-direction. For implicit methods, it is very important to use (11) in $\psi(z)$ on both the implicit and explicit operators. For the implicit operator, numerical experiments showed that the linearized conservative form (10) converges slightly faster than the linearized nonconservative form [11]. It seems also that when the freestream Mach number increases, the convergence rate of the linerarized conservative form (10) is slightly better than a simplified version which replaces $\Omega^{\xi}_{j+\frac{1}{2},k}$ and $\Omega^{\eta}_{j,k+\frac{1}{2}}$ of (10g,h) by $\max_l\ \psi(a^l_{j+\frac{1}{2}})$ and $\max_l\ \psi(a^l_{k+\frac{1}{2}})$ times the identity matrix.

(b). Higher-order TVD schemes in general involve limiter functions. However, there are options in choosing the types of dependent variables in applying limiters for system cases, in particular for systems in generalized coordinates. The choice of the dependent variables on which limiters are applied can affect the convergence process. In particular, due to the nonuniqueness of the eigenvectors $R_{j+\frac{1}{2}}$, the choice of the characteristic variables on which the limiters are applied play an important role in the convergence rate as the Mach number increases. For moderate Mach numbers, the different choice of the eigenvectors have negligible affect on the convergence rate. However, for large Mach number cases, the magnitudes of all the variables at the jump of the bow shock are not the same. In general, the jumps are much larger for the pressures than for the densities or total energy. Studies indicated that employing the form $R_{j+\frac{1}{2}}$ such that the variation of the α are of the same order of magnitude as for the pressure would be a good choice for hypersonic flows. The form similar to the one used by Gnoffo [16] or Roe and Pike [17] can improve the convergence rate over the ones used in references [4,18].

(c). Due to the large gradients and to the fact that the initial conditions are far from the steady-state physical solution, the path used by the implicit method can go through states with negative pressures if a large time step is employed. A convenient way to overcome the difficulties is to fix a minimum allowed value for the density and the pressure. With this safety check, the scheme allows a much larger time step and converges several times faster. In addition, since the Roe's average state allows the square of the average sound speed $c^2_{j+\frac{1}{2}}$ to lie outside the interval between c^2_j and c^2_{j+1}, $c^2_{j+\frac{1}{2}}$ might be negative even though c^2_j and c^2_{j+1} are positive during the transient stage when the initial conditions are far from the steady-state physical solution. In this case, we replace $c^2_{j+\frac{1}{2}}$ by $\max(c^2_{j+\frac{1}{2}}, \min(c^2_j, c^2_{j+1}))$. This later safety check is in particular helpful for the symmetric TVD algorithm (7).

IV. Numerical Results

The current study on the shock resolution of the various schemes for two-dimensional steady-state blunt-body computations indicates similar trends as the one dimensional study [9]. The main issue appears to be their relative efficiency. Due to extra evaluations per dimension in the curve fitting between the left and right states in a real gas for the van Leer formulation, additional computation is required for the van Leer type schemes than the Harten and Yee [1,2], and Yee [3,4] types of TVD schemes. Here van Leer type schemes refer to the use of the MUSCL approach in conjunction with Roe type approximate Riemann solver [12] or flux-vector splittings [6,13]. Moreover, for steady-state applications, implicit methods are preferred over explicit methods because of the faster convergence rate. In addition, it is easier to obtain a noniterative linearized implicit operator for the Harten and Yee, and Yee type schemes than for the van Leer type schemes. For these reasons, the linearized implicit versions of Harten and Yee [11] and Yee [3] are preferred over the van Leer type schemes.

Resolution of First- and Second-Order schemes: For problems containing complex shock structures, first-order upwind TVD schemes are too diffusive unless extremely fine grids are used. For a blunt-body flow containing a single steady bow shock only, the shock-capturing capability of a first-order upwind TVD scheme seems to be quite adequate if one is interested in the shock resolution only. However, a careful examination of the overall flow field of the density and Mach number contours of the first- and second-order TVD schemes compared with the exact solution (shock-fitting solution) reveals the inaccuracy of the first-order scheme. Figure 1 compares the resolution of the first-order (setting $g_j^l = 0$) and second-order upwind TVD schemes (10) using the Roe approximate Riemann solver [12] with the "exact solution" for a perfect gas ($\gamma = 1.4$) at a freestream Mach number of 10. The computations are performed on a 61×33 adapted grid for the full (half) cylinder, which yields a fairly good bow shock resolution by both schemes. However, the contour levels near the body are significantly shifted with the first-order scheme, while the second-order scheme reproduces almost identical results as the exact solution.

Convergence Rate of Explicit and Implicit TVD Schemes at Hypersonic Speed: The five different second-order TVD methods previously studied [9] in one dimension yield very similar shock-resolution for the blunt-body problem. In particular, for an inviscid blunt-body flow in the hypersonic equilibrium real gas range, the explicit second-order Harten and Yee, and Yee-Roe-Davis type TVD schemes [2-4] using the generalized approximate Riemann solver [7] produce similar shock-resolution but converge slightly faster than an explicit second-order van Leer type scheme using the generalized van Leer flux-vector splitting [8].

The freestream conditions for the current study are $M_\infty = 15$ and 25, $p_\infty = 1.22 \times 10^3\ N/m^2$, $\rho_\infty = 1.88^{-2}\ kg/m^3$, and $T_\infty = 226°$K. The grid size is 61×33 for the full (half) cylinder (figure 2). For the $M_\infty = 25$ case, the shock stand off distance is at approximately fourteen points from the wall on the symmetry axis. The relaxation procedure for the explicit methods employs a second-order Runge-Kutta time-discretization with a CFL of 0.5 (solution not shown). The parameter $\widetilde{\delta}$ is set to a constant value of 0.15. Pressure and Mach number contours converge and stabilize after 3000-4000 steps but the convergence rate is much slower for the density (with a 2-3 order of magnitude drop in L_2-norm residual). The bow shock is captured in two to three grid points. The curve fits of Srinivasan et al. [19] are used to generate the thermodynamic properties of the gas.

The same flow condition was tested on the implicit scheme (10). The convergence rate is many times faster. Figures (3) and (4) show the Mach number, density, pressure and κ contours computed by the linearized conservative ADI form of the upwind scheme (10) for Mach numbers 15 and 25. Figure 5 shows the slight advantage of the convergence rate of the linearized conservative implicit TVD scheme (10) over the linearized nonconservative implicit TVD scheme suggested in reference [11]. The convergence rate and shock resolution for the symmetric TVD scheme (10,7) behave similarly. For $M_\infty = 15$ case, the L_2-norm residual stagnated after a drop of four orders of magnitude. In general, for a perfect gas with $10 \leq M_\infty \leq 25$ and a not highly clustered grid, steady-state solutions can be reached in 800 steps with 12 orders of magnitude drop in the L_2-norm residual. However, the convergence rate is at least twice as slow for the real gas counter-part. An important observation for the behavior of the convergence rate for the Mach 15 real gas case is that the discontinuities of the thermodynamic derivatives which exist in the curve fits of Srinivasan et al. [19] might be the major contributing factor. This is evident from figures (3d) and (4d) and from comparing with the convergence rate for the perfect gas result.

Computations of impinging shocks: Figure (6) shows the Mach contours computed by the implicit upwind TVD scheme (10) of an inviscid shock-on-shock interaction on a blunt body in the low hypersonic range. Extensive study on flow fields of this type were reported in references [20-22] for the viscous case. This flow field is typical of what will be experienced by the inlet cowl of the National Aerospace Plane (NASP). The freestream conditions for this flow field are $M_\infty = 4.6$, $p_\infty = 14.93$ N/m^2, $T_\infty = 167°$K, $T_w = 556°$K, and $\gamma = 1.4$ for a perfect gas. An oblique shock with an angle of 20.9° relative to the free stream impinges on the bow shock. Various types of interactions occur depending on where the impingement point is located on the bow shock. As shown by the Mach contours, the impinging shock has caused the stagnation point to be moved away from its undisturbed location at the symmetry line. The surface pressures at the new stagnation point can be several times larger than those at the undisturbed location of the stagnation point. In addition, a slip surface emanates from the bow shock and impinging shock intersection point and is intercepted by a shock wave which starts at the upper kink of the bow shock. The interacting shock waves and slip surfaces are confined to a very small region and must be captured accurately by the numerical scheme if the proper surface pressures and heat transfer rates are to be predicted correctly. The 77×77 grid used and the convergence rate computed by the implicit scheme (10) are shown in figure

(6). Though the pattern of the flow is significantly more complicated than for the previous cases, the convergence rate remains quite satisfactory. Detailed study of viscous steady and unsteady flow fields of this type using a fully implicit second-order time-accurate scheme [10] of the same numerical flux (6-9) for the convection terms are reported in [10,14,15]. It was found that for viscous computations, the scheme suggested by Yee et al. [10] is more robust than equation (10) which is best suited for steady-state inviscid flows.

IV. Concluding Remarks

Some numerical aspects of the TVD schemes that can affect the convergence rate for hypersonic Mach numbers or real gas flows but have negligible effect on low Mach number or perfect gas flows are identified. Improvements have been made to the various TVD algorithms to speed up the convergence rate in the hypersonic flow regime. Even with the improvement though, the convergence is in general slightly slower for a real gas than for a perfect gas. The nonsmoothness in the curve fits of Srinivasan et al. may be a major contributing factor in slowing down the convergence rate. Due to extra evaluations per dimension in the curve fitting between the left and right states in a real gas for the van Leer formulation, more computation is required for the van Leer type schemes than for the Harten and Yee, and Yee types of TVD schemes.

Aside from the difference in convergence rate, the numerical results confirm the findings of the one dimensional study. The different methods yield very similar shock-resolution on the blunt body problem with freestream Mach numbers up to 25, and the state equation does not have a large effect on the general behavior of these methods. Further improvements on the ADI relaxation algorithm could speed up the convergence rate even more.

References

[1] A. Harten, On a Class of High Resolution Total-Variation-Stable Finite-Difference Schemes, SIAM J. Num. Anal, **21**, 1-23 (1984).

[2] H.C. Yee, Upwind and Symmetric Shock-Capturing Schemes, NASA TM-89464, May 1987; also to appear, proceedings of the "Seminar on Computational Aerodynamics," Dept. Mech. Engin., University of Calif., Davis, Spring Quarter, 1986.

[3] H.C. Yee, Construction of Explicit and Implicit Symmetric TVD Schemes and Their Applications, J. Comput. Phys., **68**, 151-179 (1987); also NASA TM-86775, July 1985.

[4] H.C. Yee, Numerical Experiments with a Symmetric High-Resolution Shock-Capturing Scheme, Proc. 10th Int. Conf. on Numerical Methods in Fluid Dynamics, June 1986, Beijing, China; also NASA TM-88325, June 1986.

[5] B. van Leer, Towards the Ultimate Conservation Difference Scheme V, A Second-Order Sequel to Godunov's Method, J. Comp. Phys., **32**, 101-136 (1979).

[6] B. van Leer, Flux-Vector Splitting for the Euler Equations, ICASE Report 82-30; Sept., 1982.

[7] M. Vinokur, Generalized Roe Averaging for Real Gas, NASA Contractor Report, in preparation.

[8] M. Vinokur and J.-L. Montagné, Generalized Flux-Vector Splitting for an Equilibrium Gas, NASA contractor report, in preparation.

[9] J.-L. Montagné, H.C. Yee and M. Vinokur, Comparative Study of High-Resolution Shock-Capturing Schemes for Real Gas, NASA TM-86839, July 1987.

[10] H.C. Yee, G.H. Klopfer and J.-L. Montagné, High-Resolution Shock-Capturing Schemes for Inviscid and Viscous Hypersonic Flows, Proceedings of the BAIL V conference, June 20-24, 1988, Shanghai, China, also NASA-TM, April 1988.

[11] H.C. Yee, Linearized Form of Implicit TVD Schemes for Multidimensional Euler and Navier-Stokes Equations, Computers and Mathematics with Applications, **12A**, 413-432 (1986).

[12] P.L. Roe, Approximate Riemann Solvers, Parameter Vectors, and Difference Schemes, J. Comp. Phys., **43**, 357-372 (1981).

[13] J.L. Steger and R.F. Warming, Flux-Vector Splitting of the Inviscid Gasdynamic Equations with Application to Finite Difference Methods, J. Comput. Phys., **40**, 263-293 (1981).

[14] G. Klopfer and Yee, H.C., Viscous Hypersonic Shock on Shock Interaction on Blunt Cowl Lips, AIAA-88-0233, AIAA 26th Aerospace Sciences Meeting, Jan. 11-14, 1988, Reno, Nevada.

[15] G. Klopfer, Yee, H.C. and P. Kutler, Numerical Study of Unsteady Viscous Hypersonic Blunt Body Flows With An Impinging Shock, Proceeding of the 11th International Conference on Numerical Methods in Fluid Dynamics, June 27 - July 1, 1988, Williamsburg, Virginia.

[16] P.A. Gnoffo, R.S. McCandless and H.C. Yee, Enhancements to Program LAURA for Efficient Computation of Three-Dimensional Hypersonic Flow, AIAA Paper 87-0280, Jan. 1987.

[17] P.L. Roe and J. Pike, Efficient Construction and Utilisation of Approximate Riemann Solutions, Computing Methods in Applied Sciences and Engineering, ed. R. Glowinski, North-Holland, Amsterdam, J.-L. Lions, 499-518 (1984).

[18] S. Chakravathy and K.Y. Szema, An Euler Solver for Three-Dimensional Supersonic Flows with Subsonic Pockets, AIAA Paper 85-1703, June 1985.

[19] S. Srinivasan, J.C. Tannehill, K.J. Weilmunster, Simplified Curve Fit for the Thermodynamic Properties of Equilibrium Air, ISU-ERI-Ames 86401; ERI Project 1626; CFD15.

[20] T.L. Holst and J.C. Tannehill, Numerical Computation of Three-Dimensional Viscous Blunt Body Flow Fields with an Impinging Shock, ERI Rept. 75169, 1975, Iowa State University, Ames, Iowa.

[21] J.C. Tannehill, T.L. Holst and J.V. Rakich, Numerical Computation of Two-Dimensional Viscous Blunt Body Flows with an Impinging Shock, AIAA J., **14**(2), 204-211 (1976).

[22] J.C. Tannehill, T.L. Holst and J.V. Rakich, Comparison of a Two-Dimensional Shock Impingement Computation with Experiment, AIAA J., **14**(4), 539-541 (1976).

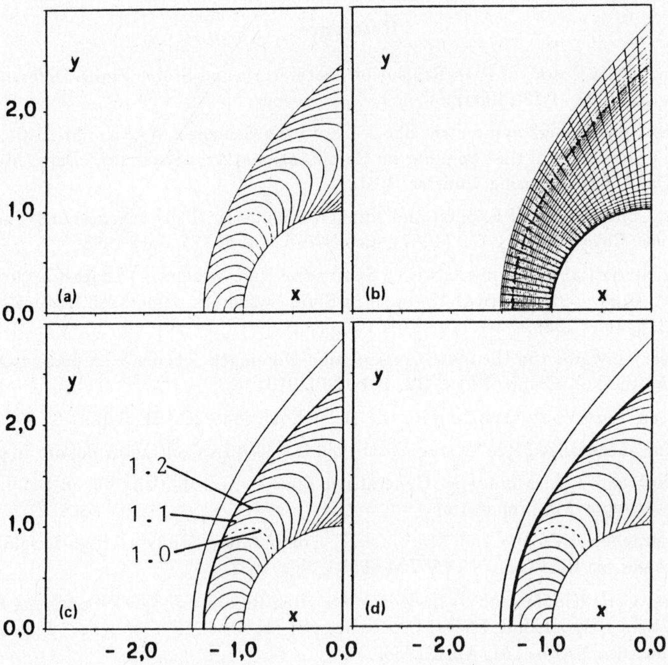

Fig. 1 Comparison among the Mach contours of a second-order implicit upwind TVD scheme (c), a first-order TVD scheme (d) and the shock fitting "exact" solution (a) using the adapted grid (b) for a perfect gas at $M_\infty = 10$.

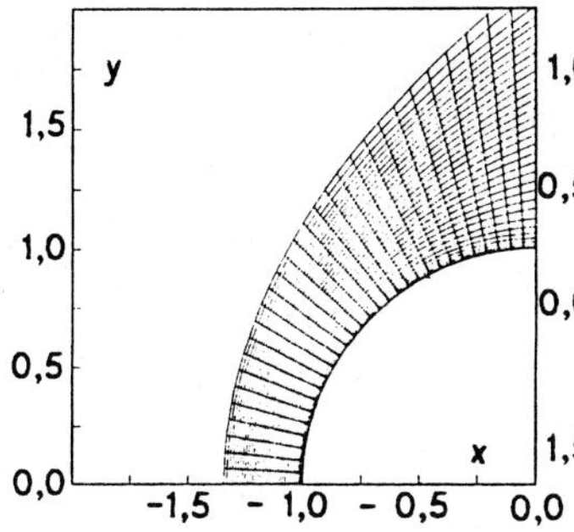

Fig. 2 The 31 × 33 grid.

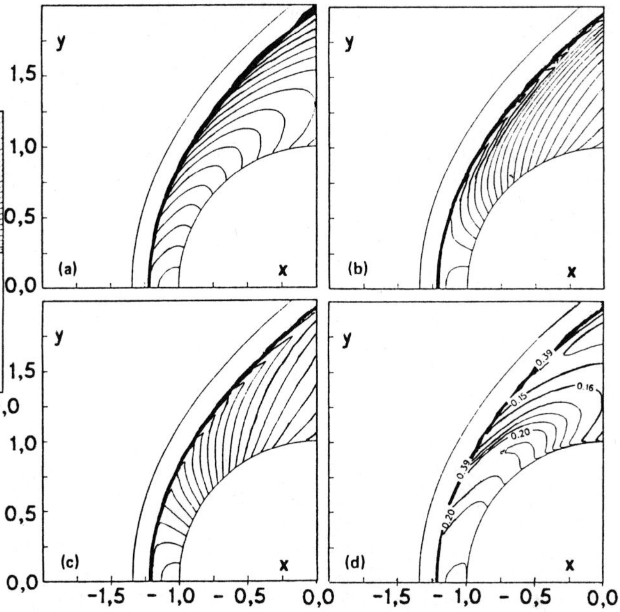

Fig. 3 The Mach contours (a), density contours (b), pressure contours (c) and κ (d) computed by a second-order implicit TVD scheme for an equilibrium real gas at $M_\infty = 15$.

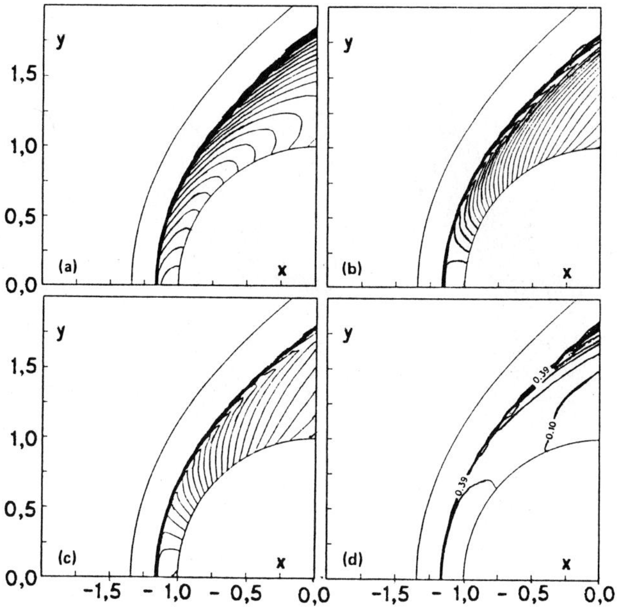

Fig. 4 The Mach contours (a), density contours (b), pressure contours (c) and κ (d) computed by a second-order implicit TVD scheme for an equilibrium real gas at $M_\infty = 25$.

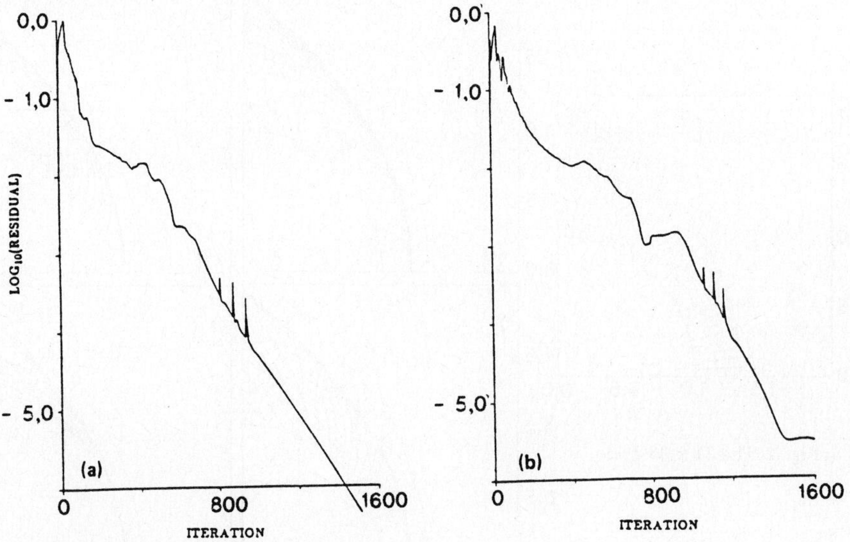

Fig. 5 Comparison of the L_2-norm residual of a linearized conservative implicit operator (a) and a linearized nonconservative implicit operator (b) for an equilibrium real gas at $M_\infty = 25$.

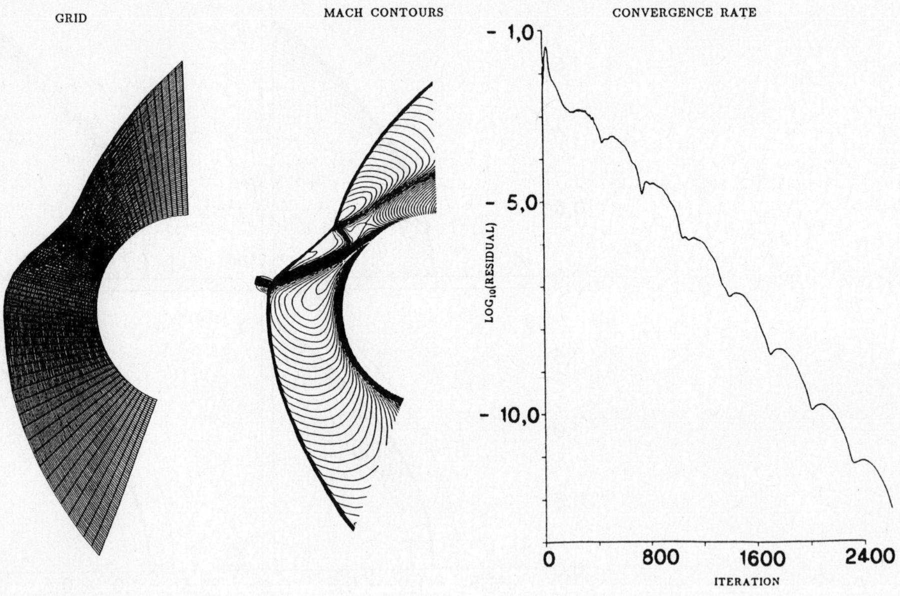

Fig. 6 Two-dimensional inviscid blunt-body flow computed by a second-order implicit scheme for a perfect gas at $M_\infty = 4.6$.

AIRFOIL CALCULATIONS IN CARTESIAN GRIDS

Gino Moretti
G.M.A.F., Inc., Freeport, NY, USA
and
Andrea Dadone
University of Bari, Italy

ABSTRACT

A technique to compute two-dimensional flows on Cartesian grids in the presence of bodies of arbitrary shape is presented. The basic technique is patterned on the λ-scheme and shock-fitting procedures. The special handling of boundary conditions on rigid bodies is described in detail. A description of the application of the technique to the calculation of transonic flows past a NACA 0012 airfoil at no incidence is made.

INTRODUCTION

Numerical analysis of Euler equations is generally performed using computational grids which are chosen to be as close to orthogonal as possible, and to have all rigid boundaries on grid lines. Orthogonality of the grid enhances accuracy. The handling of boundary conditions is simplified if boundaries coincide with grid lines.

Generation of suitable grids, however, becomes a major problem when the geometry of the bodies is complicated, when there are many bodies in the field, and in three-dimensional problems.

In the light of recent improvements, both in computing machines and in numerical techniques, the importance of a choice of a grid can be challenged. The possibility of using a Cartesian grid all throughout has already been explored. Successful attempts have been made using a finite volume method as the basic integration technique [1]. Here we present and discuss results obtained using the λ-scheme and shock-fitting. The reason for our attempt is that the latter technique has been proved to be accurate and efficient in all cases analyzed so far and we want to see whether its good qualities can be retained when a Cartesian grid is used to compute the flow about an arbitrary body.

GENERAL OUTLINE OF THE TECHNIQUE

The λ-scheme has been described in detail in [2], for general grids (orthogonal and not orthogonal). When a Cartesian grid is used, the scheme takes on the simplest possible form. For clarity's sake, we repeat here the equations valid for a Cartesian grid.

Let a, S, and **q** be speed of sound, entropy and velocity, respectively, γ the ratio of specific heats, and $\delta=(\gamma-1)/2$. Let x, y and t be the Cartesian coordinates, and time. All quantities are non-dimensionalized. The unit of speed is the speed of sound at infinity, divided by $\sqrt{\gamma}$, the unit of length is chosen according to the problem, and the unit of time is the ratio of the units of length and speed. In addition, the entropy is divided by γR, where R is the gas constant.

The general equations of motion (Euler equations):

$$
\begin{aligned}
&(a_t+ua_x+va_y)/\delta+a(u_x+v_y)-aS_t-a(uS_x+vS_y)=0\\
&u_t+uu_x+vu_y+aa_x/\delta-a^2S_x=0\\
&v_t+uv_x+vv_y+aa_y/\delta-a^2S_y=0\\
&S_t+uS_x+vS_y=0
\end{aligned}
\tag{1}
$$

are recast into the form:

$$
\begin{aligned}
&a_t=\delta(f_1{}^x+f_2{}^x+f_1{}^y+f_2{}^y)+\delta aS_t\\
&u_t=f_1{}^x-f_2{}^x+f_3{}^y\\
&v_t=f_1{}^y-f_2{}^y+f_3{}^x\\
&S_t=f_4{}^x+f_4{}^y
\end{aligned}
\tag{2}
$$

where

$$
f_i^x=-\lambda_i^x[(R_i^x)_x-aS_x]/2, \quad f_i^y=-\lambda_i^y[(R_i^y)_y-aS_y]/2 \quad (i=1,2)
$$

$$
\begin{aligned}
&f_3{}^x=-uv_x, \qquad f_3{}^y=-vu_y\\
&f_4{}^x=-uS_x, \qquad f_4{}^y=-vS_y
\end{aligned}
\tag{3}
$$

and

$$
\lambda_i^x=u\pm a, \quad \lambda_i^y=v\pm a, \quad R_i^x=a/\delta\pm u, \quad R_i^y=a/\delta\pm v \; . \tag{4}
$$

The derivatives are approximated by forward or backward differences according to the coefficients (λ, u, or v) being negative or positive.

The code may be formally first-order or second-order accurate. The latter consists of the former repeated twice; at the first level, the quantities defined by (3) are divided by 2; at the second level, each quantity is used as defined by (3), but the value obtained at the first level in a neighboring point is subtracted. The "neighboring" point is the node next to the point in question, in the forward or backward direction according to λ, u or v being negative or positive.

THE COIN TECHNIQUE

To minimize errors in the leading edge region and a consequent decay of total temperature on the surface of the body, we reformulate the equations of motion in the spirit of [3]. Let us split **q** and a into sums of two terms, of which the ones denoted by ^{0} are computed at the start of the calculation and never changed again, and the ones denoted by a prime are the unknowns to be computed: The velocity $\mathbf{q}^0$ is the velocity of an incompressible, irrotational flow about the profile; therefore, it satisfies the conditions:

$$
\nabla\cdot\mathbf{q}^0 = 0 \; , \qquad \nabla\times\mathbf{q}^0 = 0 \; . \tag{5}
$$

The term a^0 is related to $\mathbf{q}^0$ by the condition of conservation of total temperature:

$$\delta(q^0)^2 + (a^0)^2 = a_0^2 = a_\infty^2(1+\delta M_\infty^2) \quad (6)$$

where a_0 is the stagnation speed of sound, and a_∞, M_∞ are the values of a and M at infinity. As said above, neither $\mathbf{q}^0$ nor a^0 depend on time:

$$\mathbf{q}^0_t = 0 , \qquad a^0_t = 0 . \quad (7)$$

For brevity, we denote the following technique by the acronym COIN (Compressible Over INcompressible). In the vicinity of the leading edge, where the flow stagnates, a compressible flow behaves as incompressible; therefore, better accuracy will be obtained in discretizing the primed terms since their gradients are small [3].

All "incompressible" velocities are obtained by solving (5) and prescribing a velocity at infinity, $V_\infty{}^0$. The actual velocity at infinity is $V_\infty = a_\infty M_\infty$ but, instead of letting $V_\infty{}^0 = V_\infty$ (which, with the current non-dimensionalizing parameters, would make $V_\infty{}^0 = \sqrt{\gamma}M_\infty$), we first compute the ratio, density/stagnation density at infinity [which, in this context, is $1/(1+\delta M_\infty^2)^{1/(\gamma-1)}$], and then define

$$V_\infty{}^0 = \sqrt{\gamma}M_\infty/(1+\delta M_\infty^2)^{1/(\gamma-1)} . \quad (8)$$

By so doing, mass-flows in the stagnation region are well represented by the "incompressible" solution.

Taking (5), (6), and (7) into account, Eqs. (2) can be replaced by:

$$\begin{aligned} a'_t &= \delta(f_1{}^x + f_2{}^x + f_1{}^y + f_2{}^y) + \delta a S_t + f_1{}^P \\ u'_t &= f_1{}^x - f_2{}^x + f_3{}^y + f_2{}^P \\ v'_t &= f_1{}^y - f_2{}^y + f_3{}^x + f_3{}^P \\ S_t &= f_4{}^x + f_4{}^y \end{aligned} \quad (9)$$

where the coefficients of (3) remain unaltered, but the terms being differentiated are primed (with $S'=S$), and

$$\begin{aligned} f_1{}^P &= -(u a_x{}^0 + v a_y{}^0) \\ f_2{}^P &= -(u' u_x{}^0 + v' u_y{}^0 + a' a_x{}^0/\delta) \\ f_3{}^P &= -(u' v_x{}^0 + v' v_y{}^0 + a' a_y{}^0/\delta) . \end{aligned} \quad (10)$$

The integration technique, thus, is the same described above, with the addition of terms, locally defined by (10).

BOUNDARY CONDITIONS

Boundary conditions at outside boundaries are easily enforced. On an inlet boundary (x=constant), the entropy and the stagnation speed of sound are assumed to be constant, and the ratio, $\sigma = v/u$ to be known. It follows that

$$a a_t + \delta u u_t (1+\sigma^2) = 0 . \quad (11)$$

Therefore, using the first two Eqs. (2),

$$a(f_1^x+f_2^x+f_1^y+f_2^y)+(1+\sigma^2)u(f_1^x-f_2^x+f_3^y) = 0 \qquad (12)$$

from which f_1^x may be determined, since all the other quantities are known from inside or along the boundary. On an exit boundary (x=constant again), the pressure is assumed to be constant in time. This yields the condition:

$$f_1^x+f_2^x+f_1^y+f_2^y = 0 \qquad (13)$$

from which f_2^x may be determined.

At the upper and lower boundary, far from the body, the incompressible solution is accepted and no compressible perturbation has to be computed.

TREATMENT OF THE BODY

So far, the calculation is simple, straightforward and fast. The code is vectorizable without difficulty. The results are more accurate than using any

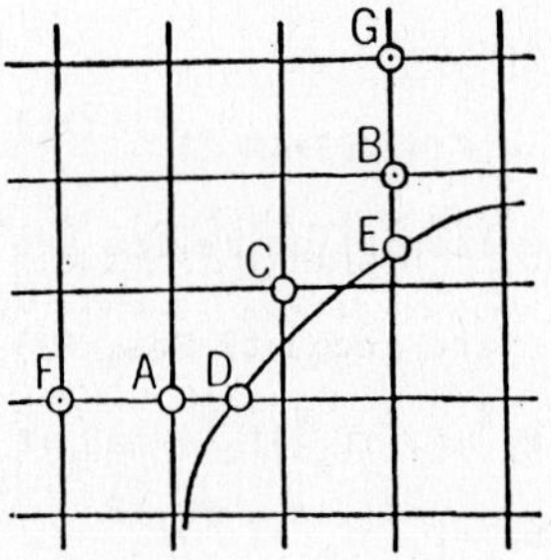

Fig. 1

other grid.

The novel feature is the enforcement of boundary conditions on a rigid body, the contour of which is not a coordinate line. Accuracy in enforcing such a condition is crucial. In using upwind schemes, particularly the λ-scheme, on an orthogonal grid wrapped around a body, the procedure does not introduce arbitrary elements; accuracy is not impaired. In our new attempt we must maintain spirit and accuracy of the above approach. To this effect, we first focus our attention on all grid points in the immediate vicinity of the body (such as points A, B, and C in Fig. 1). The boundary condition is easily enforced at A and B. At point A, according to the rules of the λ-scheme, there is only one quantity which cannot be evaluated from grid values, i.e. f_2^x, which contains the forward difference approximating one of the x-derivatives. Similarly, at point B, only f_1^y, which contains the backward difference approximating one of the y-derivatives cannot be evaluated. In either case, however, one boundary condition is available, i.e. the direction of the velocity vector at point D or point E. To use such a condition at A or B, the direction of the velocity vector at A or B is interpolated from F or G (where it has been computed) and D or E. At C, where none of the two above differences is computable, arbitrariness seems to be unavoidable, but there are ways to circumvent the difficulty.

In brief, one has to express derivatives in the directions normal and tangential to the body as functions of derivatives along Cartesian lines, without violating the domains of dependence. This can be accomplished in different ways, two of which have been explored in the present work.

In the first, all points at intersections of the body with a grid line are marked, the values of u, v, a, and S at such points are stored and the normal and tangential components of the velocity can be evaluated. Using them and some value interpolated from the Cartesian grid, Eqs. (1) can be reformulated in a local frame, normal and tangent to the body, in which only one f is unknown, with no discrimination between points A, B, or C of Fig. 1. By so doing, all points on the body are actually computed, and the neighboring grid points are interpolated.

In the second, points A and B are treated as explained above. In the vicinity of points such as C, the angle, α, between one grid line and the body is smaller than the similar angle for the other grid line (Fig. 2). The unknown f-term relative to the former line is extrapolated to C from M and N, and the other unknown is then obtained by imposing the boundary conditions. When all grid points are computed, the values at the body are extrapolated from the neighboring grid values.

Each technique has its advantages and disadvantages. The former is more appealing from a theoretical viewpoint, essentially for two reasons: a) it computes points on the body, without having to resort to extrapolations, and b) it has only one unknown to determine at each point, consistently with the availability of one boundary condition only. The latter is simpler to implement, but it uses more extrapolations.

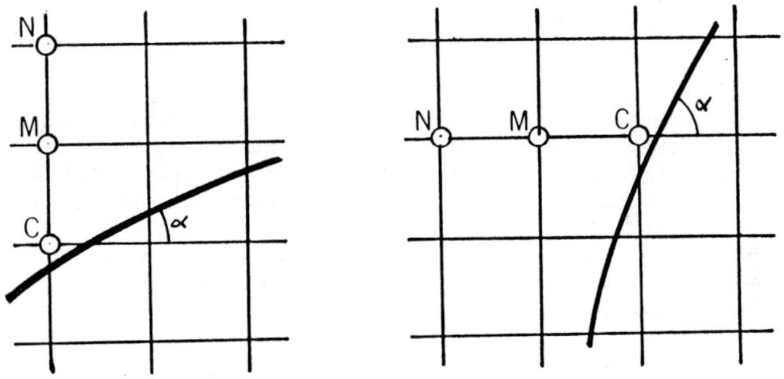

Fig. 2

The results presented here are obtained using the second option.

MULTIPLE GRIDS

High resolution is needed in the vicinity of certain portions of the rigid body. The mesh, instead, can be rather coarse at some distance from the body. Refinement of the mesh can be achieved by stretching the coordinates. It is well-known, however, that separate stretchings of the two coordinates create cells of high aspect ratios in certain regions. This results in a waste of computational work and a loss of accuracy. Better stretchings bring back either non-orthogonal coordinates or curvilinear grids, or both. In order to maintain simplicity and accuracy, and in view of extensions to complicated geometries and three-dimensional flows, we opted for using a number of rectangular regions, of increasing sizes, contained inside one another. One of the regions contains the rigid body. For example, to compute the flow past an airfoil, we use a maximum of four regions, as shown in Fig. 3. Each region is covered by a Cartesian grid. The fineness of the mesh varies from region to region, increasing toward the body. In going from one region to the surrounding one, the mesh intervals can be doubled, tripled or quadrupled.

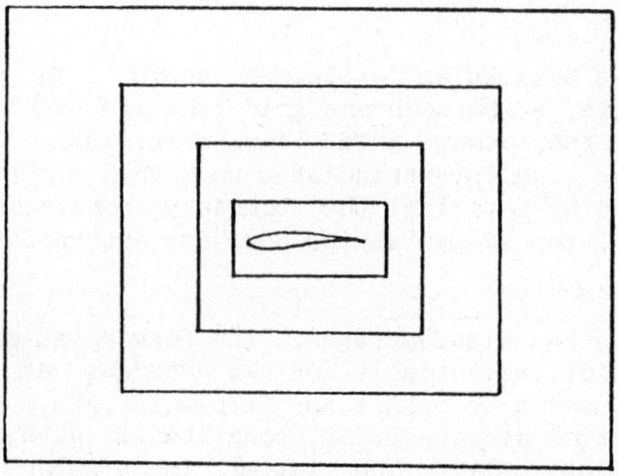

Fig. 3

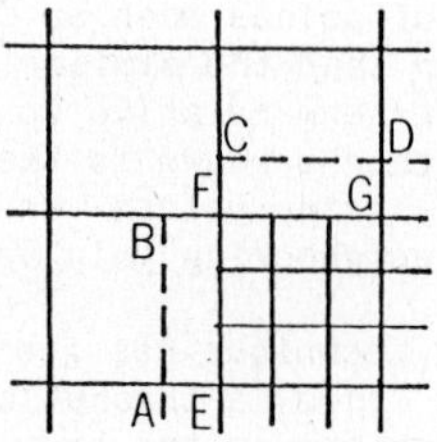

Fig. 4

Matching of regions is performed as follows (Fig. 4). Rows of values used in the λ-scheme (Riemann variables) are linearly interpolated along AB and CD from the outer (coarse) mesh, and used to generate certain normal derivatives along EF and FG, when needed. The integration of the inner region is performed including the lines EF and FG, but not along AB, CD. The values in the outer region along EF and FG are transferred from the inner region. This procedure is correct for a first-order accurate calculation. If a two-level scheme is used (to achieve second-order accuracy), some additional manipulation is needed, which we will not outline here.

We ran preliminary tests for a circle. The circle is centered in a square, the side of which is twice the diameter of the circle. This square, in turn, is contained within another square, the side of which is 12 times the diameter. The inner square is covered by a 30x30 mesh, the outer square by a 60x60 mesh. Both meshes are Cartesian, with equal spacing in x and y. The spacing in the outer mesh is three times the spacing in the inner mesh. The Mach number at infinity is 0.4. At convergence, the maximum Mach number is 0.989. This result compares well with results obtained by other Authors [4]. It is important to note that the total number of points on the circle is only 46.

The accuracy in the frontal section of the circle is very good. This means that the technique can be very accurate and no restrictions should be

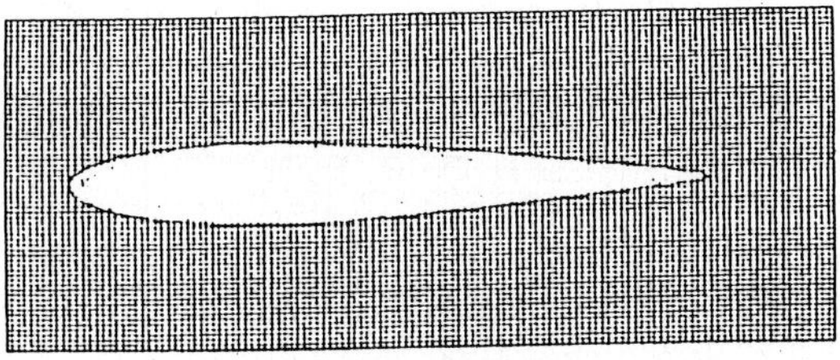

Fig. 5

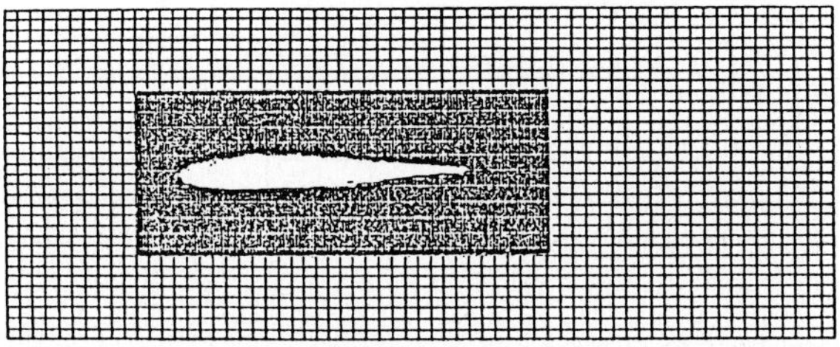

Fig. 6

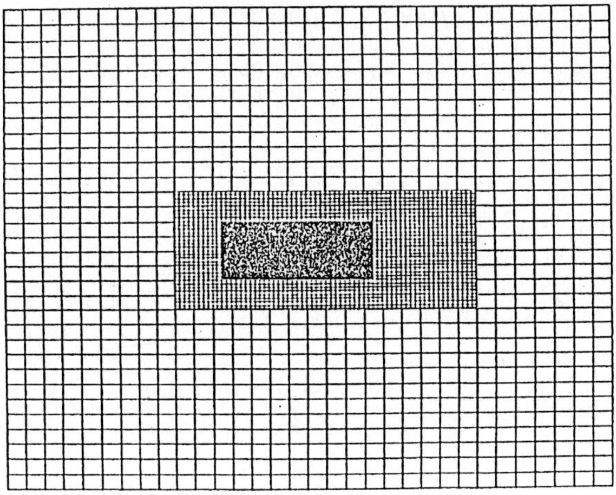

Fig. 7

anticipated on the use of Cartesian grids for bodies of any shape. The calculation is very fast. Using the two grids mentioned above, it takes about 30 seconds to advance 1000 steps on a CRAY X-MP computer; most of the time, however, is spent on the coarse grid and it could be reduced by making the grid even coarser. The convergence is good. The residual, defined as the mean square value of the difference between the moduli of the velocities at two successive steps, reaches machine zero at step 1000.

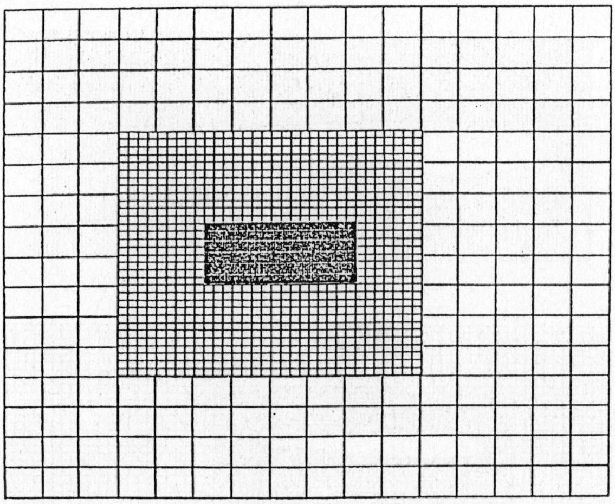

Fig. 8

CALCULATION OF A NACA 0012 AIRFOIL

Calculations have been made for a NACA 0012 airfoil, at no incidence. Four grids have been used, testing effects of different mesh sizes and different overall sizes. The regions sketched in Fig. 3 are shown in better detail in Figs. 5 through 8. Fig. 5 presents the inner grid, containing the airfoil. In Fig. 6 we see the two innermost grids; in Fig. 7 three grids are shown and finally the three outermost grids appear in Fig. 8.

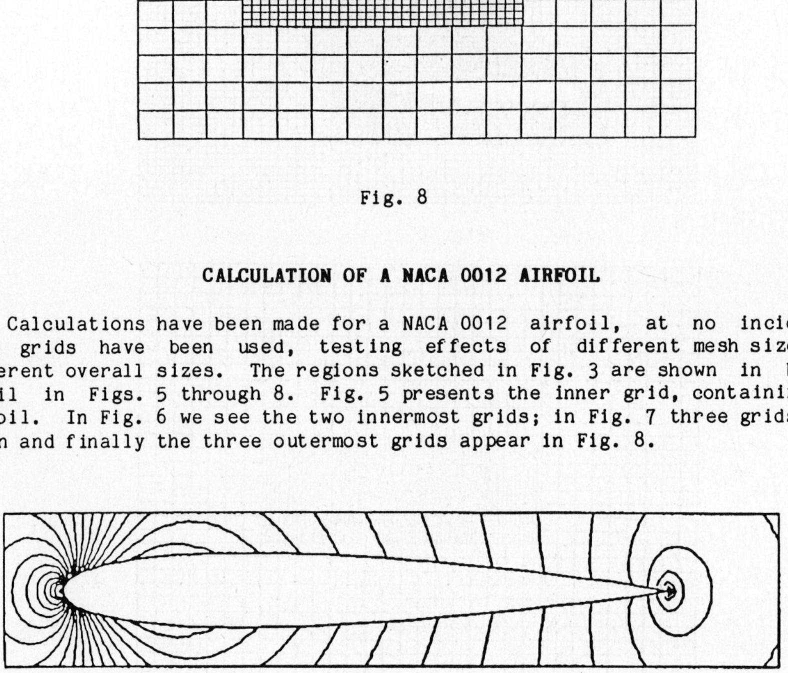

Fig. 9

SUBSONIC CASE

The first calculation presented here has been made for a Mach number at infinity equal to 0.72. The results shown in Figs. 9 (isobar plot in the innermost region), 10 (Mach number distribution on the profile), and 11 (C_p distribution on the profile) prove that an accuracy comparable with that of the most reliable codes can be reached. Indeed, the mesh used around the body is still coarse in the leading edge region, when compared with current C-grids or O-grids. A fair comparison can be made between the present results and results obtained using our fast solver code [5] with a 64x16 mesh; they are identical.

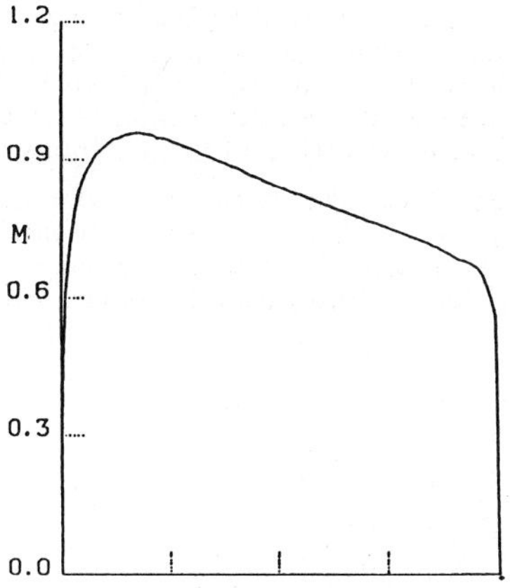

Fig. 10

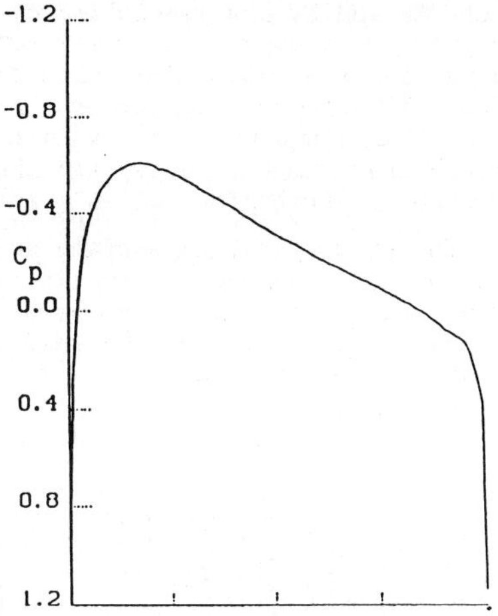

Fig. 11

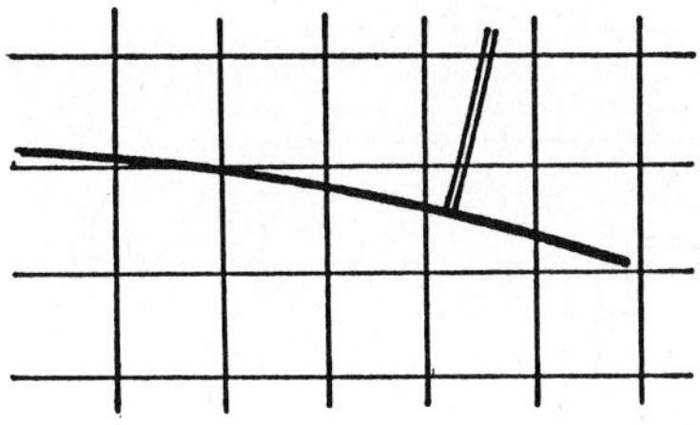

Fig. 12

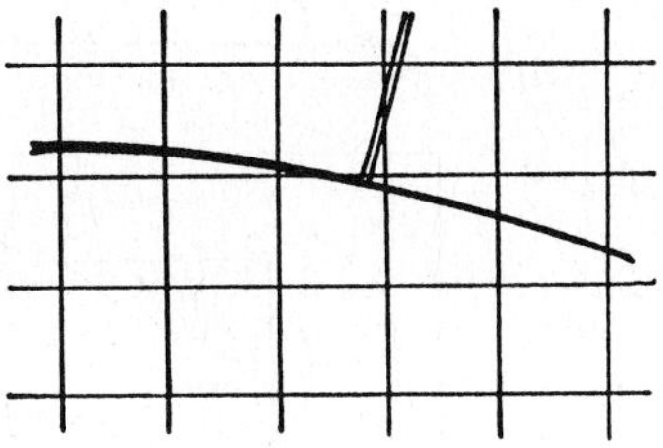

Fig. 13

TRANSONIC CASE

For the transonic case, provision has to be made for the appearance of shocks on either side of the airfoil. The analysis of the shocks has been outlined in [2] and explained in detail in [5] for a C-grid. Here, the same principles are used, with some simplification due to the simplicity of the grid, and some additional manipulation, made necessary by the obliquity of the airfoil surface over the Cartesian grid.

If the shock is located as shown in Fig. 12, the calculation can proceed as explained in [5]. Minor variations are needed if the shock is located as in Fig. 13. Substantially, certain extrapolations from upstream cannot be carried on and a simpler definition of values in front of the shock must be used.

Typical results are shown in Figs. 14 through 19. The free stream Mach number is .805; the inner region has 128x64 intervals; the other regions are, in order, 64x32, 32x32 and 16x16. Moving from one grid to the next, the values of Δx and Δy are multiplied by 4. In Fig. 14, isobars are shown for the inner region (above) and again for the inner region and the region surrounding it (below). The matching of isobars and their slopes is very good. The fitted shock is shown by a row of x's. Note that the shock is fitted in the inner region but not in the next one. In Fig. 15, lines of constant entropy, and the shock, are shown in the inner region. Figs. 16 and 17 show the distribution of C_p and M along the airfoil. Fig. 18 shows a detail of the grid, the body surface and the shock, to demonstrate that the fitted shock is actually oblique with respect to the grid, and that it crosses grid lines. Fig. 19 presents the location of the shock root as a function of computational steps. It is clear that the location of the shock is stabilized before 3000 steps.

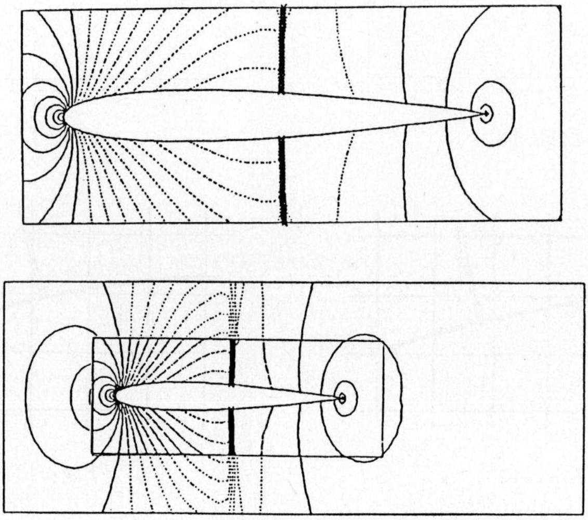

Fig. 14

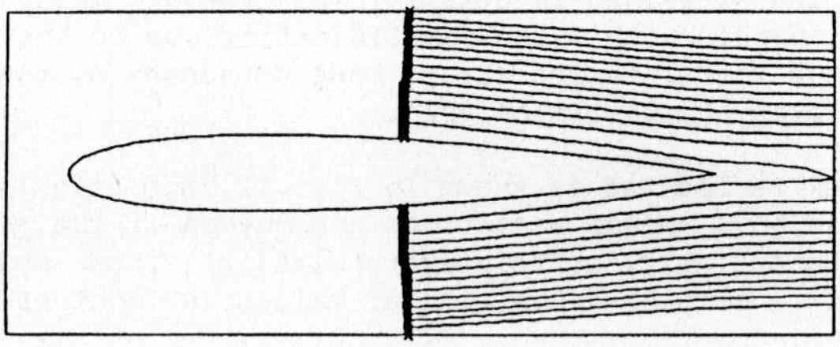

Fig. 15

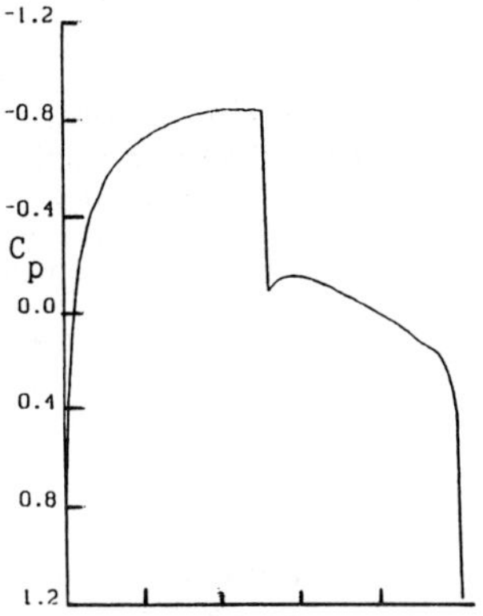

Fig. 16

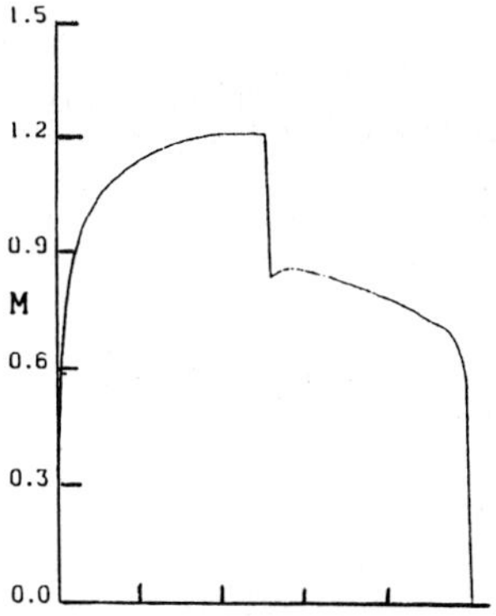

Fig. 17

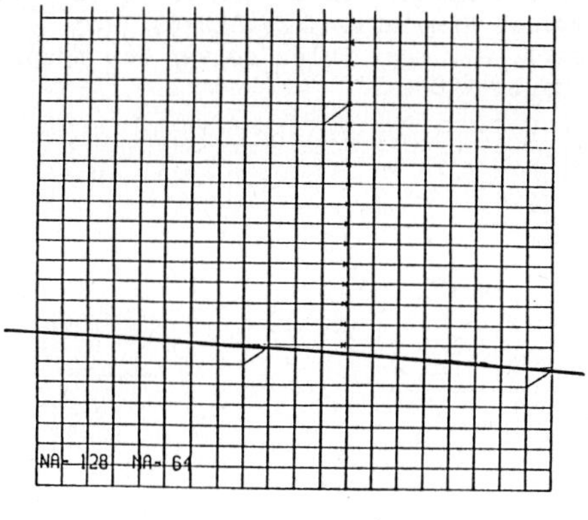

Fig. 18

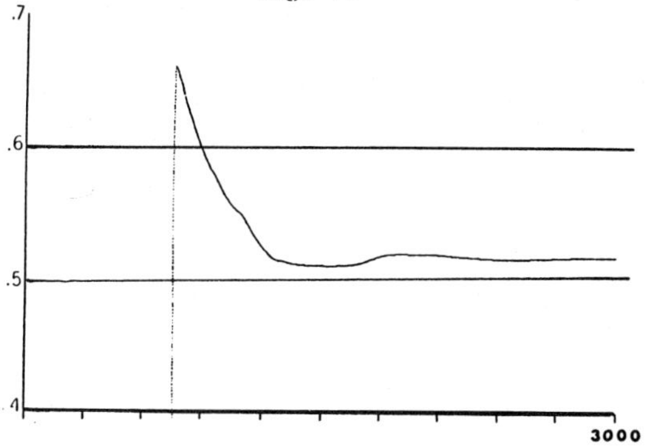

Fig. 19

CONCLUSIONS

Calculations made on circles and airfoils (in the latter case, for subsonic as well as supersonic flows) show that, with minor modifications, a technique based on the λ-scheme and shock-fitting over Cartesian grids can be used to evaluate flow fields about bodies of arbitrary shapes, without impairing accuracy. The present work should be considered as a first step towards generalization to multiple bodies and three-dimensional flows.

This work has been sponsored by the Science and Technology Foundation of New York State, under contract SBIR (87)-91.

REFERENCES

1. Clarke, D.K. et al., Euler calculations for multielement airfoils using Cartesian grids, AIAA J. 24:353-58, 1986
2. Moretti, G., A technique for integrating two-dimensional Euler equations, Comp.Fl., 15:59-75, 1987
3. Dadone, A., A quasi-conservative COIN lambda formulation, Lect. Notes in Phys. 264:200-4, 1986
4. Salas, M.D., Recent developments in transonic Euler flow over a circular cylinder, Math. and Comp. in Simulation XXV:232-6, 1983.
5. Dadone, A. and Moretti, G., Fast Euler solver for transonic airfoils, to appear in the AIAA J., March 1988.

CHARACTERISTIC GALERKIN METHODS FOR HYPERBOLIC SYSTEMS

K.W. Morton & P.N. Childs
ICFD, Oxford University Computing Laboratory
8-11 Keble Road, Oxford OX1 3QD

SUMMARY

Explicit use of characteristics together with the Galerkin projection proves to be a powerful combination in the approximation of hyperbolic equations. A basic scheme uses piecewise constant functions to produce first order approximations which are unconditionally stable, conservative, entropy-satisfying and monotonicity-preserving. An adaptive recovery technique at each time step then gives higher accuracy where warranted by the smoothness without affecting these properties.

1. INTRODUCTION

The finite element method needs some sort of global principle to be truly effective: for elliptic problems this is provided by variational principles; but for hyperbolic problems it is natural to look to the existence of Riemann invariants to play this rôle. Thus it is no surprise that as the scope of finite elements developed, the idea of combining the use of characteristics with the Galerkin projection occurred to many people at a similar time in the early 1980's, including Benqué and Ronat [1981], Bercovier et al [1982], Douglas and Russell [1982], Morton and Stokes [1982] and Pironneau [1982].

Suppose an evolutionary problem for $u(t)$ is characterised by the evolution operator $E(\cdot)$:

$$E(\Delta t): u(t) \mapsto u(t+\Delta t), \tag{1.1}$$

and that this is approximated by E_Δ. Suppose further that at the discrete times $\{t_n\}$, the spatial variation of $u(t_n) = u(x,t_n)$ is approximated by the finite element expansion

$$U^n(x) := \sum_{(j)} U_j^n \, \phi_j(x) \tag{1.2}$$

in basis functions $\{\phi_j\}$. Then the Galerkin projection leads to the time-stepping algorithm, for approximating the evolution of the approximation from t_n to $t_{n+1} = t_n + \Delta t$,

$$\left\langle U^{n+1},\phi_i\right\rangle = \left\langle E_\Delta U^n,\phi_i\right\rangle \qquad (1.3)$$

where $<\cdot,\cdot>$ denotes the L^2 inner product over the spatial variable(s). For example, use of a Taylor expansion to define E_Δ leads directly to the Taylor-Galerkin algorithms introduced by Donea [1984]. On the other hand, use of characteristic paths to define the mapping E_Δ leads to characteristic Galerkin methods, with Lagrange-Galerkin methods being special cases in which only particle paths are used.

Consider the scalar problem in two dimensions

$$u_t + \operatorname{div}(f,g) = 0, \qquad (1.4a)$$

written in the form

$$u_t + \underline{a}\cdot\underline{\nabla}u = 0 \qquad (1.4b)$$

where $\underline{a} = (\partial f/\partial u, \partial g/\partial u)$. If the characteristic paths $\underline{X}(\underline{x},t;t')$, given by

$$\frac{d\underline{X}}{dt'} = \underline{a}, \quad \underline{X}(\underline{x},t;t) = \underline{x}, \qquad (1.5)$$

are approximated by straight-line sections, we are led to the evolution operator

$$(\hat{E}u)(\underline{y}) := u(\underline{x}), \qquad (1.6a)$$

where

$$\underline{y} = \underline{x} + \underline{a}(u(\underline{x}))\Delta t \approx \underline{X}(\underline{x},t;t+\Delta t), \qquad (1.6b)$$

and thence to the Euler characteristic Galerkin (ECG) method

$$\left\langle U^{n+1},\phi_i\right\rangle = \left\langle \hat{E}U^n,\phi_i\right\rangle = \int U^n(\underline{x})\phi_i(\underline{y})d\underline{y}. \qquad (1.7)$$

Note that there are two approximations involved in this Euler time-stepping scheme: firstly the characteristics are approximated by (1.6b); and secondly the evolution of (1.4a) is approximated by constancy of the solution along the characteristics, (1.6a). Both of these may be improved upon in more sophisticated schemes.

Clearly, even in one dimension, a single-valued solution curve $u(x,t)$ at time t can lead to a multi-valued $u(y,t+\Delta t)$ through (1.6) being applied where the characteristics envelope.

In that case the integral (1.7) is interpreted as being along the graph $[y,\hat{E}U^n]$ as in the definition of the transport-collapse operator by Brenier [1984].

An alternative formulation of (1.7), introduced in Morton and Stokes [1982] and used extensively thereafter, makes more explicit the rôle of the flux functions in (1.4a) and in one dimension has the form

$$\left\langle U^{n+1} - U^n, \phi_i \right\rangle + \Delta t \left\langle \partial_x f(U^n),\ \Phi_i^n \right\rangle = 0, \qquad (1.8a)$$

where

$$\Phi_i^n(x) = \frac{1}{a(U^n(x))\Delta t} \int_x^{x+a\Delta t} \phi_i(z)dz. \qquad (1.8b)$$

Note that the special test function Φ_i^n is an average of the basis function ϕ_i over the interval covered by a characteristic in the time step, and is just applied to the spatial differential operator in (1.4a). This form implicitly assumes that a fixed approximation space is used at all levels and is easily derived from (1.7) by integrating by parts along the graphs $G := [x,U^n]$, $\hat{G} := [y,\hat{E}U^n]$, where $y = x + a\Delta t$:

$$\begin{aligned}\left\langle U^{n+1},\phi_i \right\rangle &= \int U^n(x)\phi_i(y)dy \\ &= \int U^n(x)d\int^y \phi_i(z)dz \\ &= -\int_{\hat{G}} \int^y \phi_i(z)dzdu; \end{aligned} \qquad (1.9a)$$

while we can write

$$\begin{aligned}\left\langle U^n,\phi_i \right\rangle &= \int U^n(x)\phi_i(x)dx \\ &= \int U^n(x)d\int^x \phi_i(z)dz \\ &= -\int_G \int^x \phi_i(z)dzdu; \end{aligned} \qquad (1.9b)$$

and adu can be replaced by df.

When an adaptive mesh is used that changes from time step to time step, the double integral form in (1.9) has to be used so that we write

$$\left\langle U^{n+1},\phi_i^{n+1} \right\rangle = -\int_{\hat{G}} \int_{-\infty}^y \phi_i^{n+1}(z)dzdu. \qquad (1.10)$$

We shall give results below that show the effectiveness of an adaptive mesh refinement strategy based on this form.

For the linear, constant coefficient, advection problem $u_t+au_x = 0$, which is the universal first model problem, the choice of piecewise constant basis functions on a uniform mesh leads to the first order upwind difference scheme

$$U_i^{n+1} = (1-\hat{\nu}\Delta_-)U_{i-p}^n, \qquad (1.11)$$

where it is assumed $a>0$, the CFL number $a\Delta t/\Delta x$ has integral part p and fractional part $\hat{\nu}$, and Δ_- is the backward difference operator defined by $\Delta_- U_i := U_i-U_{i-1}$. If piecewise linear basis functions are used, the well-known third-order accurate scheme

$$(1+\tfrac{1}{6}\delta^2)U_i^{n+1} = \left[(1+\tfrac{1}{6}\delta^2) - \hat{\nu}\Delta_0 + \tfrac{1}{2}\hat{\nu}^2\delta^2 - \tfrac{1}{6}\hat{\nu}^3\delta^2\Delta_-\right]U_{i-p}^n \qquad (1.12)$$

is obtained. The sequence can be continued indefinitely with splines of order s giving accuracy of order $2s-1$: and replacement of the Euler time-stepping by central differencing raises this order of accuracy to $2s$.

All these schemes are of course conservative and unconditionally stable: and (1.12) is highly accurate for smooth problems, even on nonuniform meshes and when extended into two dimensions - so long as the inner products are adequately approximated (see Morton, Priestley and Süli [1988]). However, only (1.11) is monotonicity preserving, and with the higher order schemes non-physical oscillations are introduced when nonlinear problems, typified by compressible gas dynamics, are approximated. Thus it is advantageous to regard the piecewise constant approximation as basic and to introduce an intermediate recovery stage at each time step to increase the accuracy to the level warranted by the smoothness of the solution. That is, the recovery stage is adaptive and may include explicit representation of shocks and other discontinuities. This recovered approximation is then substituted into the evolution equation (1.7) or (1.8), instead of the piecewise constant U^n, in order to produce a more accurate piecewise constant U^{n+1} at the next time step.

Algorithms developed along these lines by Morton and Sweby [1985, 1987] and by Childs and Morton [1986] are described in the next section, and convergence results for the scalar case given in section 3. New results are given in section 4 for the development and application of these methods for systems of equations in one dimension, using flux vector splitting and flux difference splitting techniques developed for finite difference methods. Finally, in section 5, we consider some of the special features which arise when these ideas are extended to hyperbolic systems in two space dimensions.

2. ECG ALGORITHMS BASED ON RECOVERY TECHNIQUES

Suppose $U^n(x)$ is piecewise constant on a nonuniform mesh, equalling U_i^h on the interval $(x_{i-½}, x_{i+½})$ of length Δx_i and mid-point x_i. As is implied by the Galerkin procedure, it is supposed to be a good approximation to the L^2 best fit of the true solution u^n by piecewise constants: that is, each U_i^n is interpreted as an average over the ith interval. The recovery stage seeks to construct an improved approximation to u^n from these averages. It is to a large degree quite separate from the evolution stage and is essentially an exercise in approximation theory.

We denote the recovered approximation by $\tilde{u}^n$ and construct it through the use of three types of information by means of

(i) combining several neighbouring values $\{U_i^n\}$;

(ii) exploiting a priori information regarding the underlying solution, e.g. positivity, monotonicity, smoothness; and

(iii) insisting on the projection property

$$\left\langle \tilde{u}^n - U^n, \phi_i \right\rangle = 0 \qquad \forall\ i, \tag{2.1}$$

where here the basis function ϕ_i is the characteristic function of the ith interval. Then the resulting ECG algorithm for the time step $t_n \to t_{n+1} = t_n + \Delta t$ replaces (1.7) by

$$\left\langle U^{n+1}, \phi_i \right\rangle = \int \tilde{u}^n(x)\phi_i(y)dy = \int_{x_{i-½}}^{x_{i+½}} \tilde{u}^n(x)dy \qquad \forall\ i, \tag{2.2}$$

with $y = x + a(\tilde{u}^n)\Delta t$. Similarly, because of (2.1), (1.8) can be replaced by merely substituting $\tilde{u}^n$ for U^n everywhere.

The choice of form for $\tilde{u}^n$ can be made in many different ways. For the linear advection equation, it was shown by Morton [1983] and Childs and Morton [1986] that, if U^n is a spline of order s on a uniform mesh and recovery is by a similar spline of order $s + p$, then the resulting order of accuracy for (2.2) is $2s + p - 1$: indeed, if p is even the same scheme is obtained as if a spline of order $s + \frac{1}{2}p$ were used without recovery. For example, recovery with quadratic splines from piecewise constants reproduces the third order scheme (1.12); while recovery with piecewise linears yields a new second order accurate scheme.

This is one attraction of basing the recovery procedure on splines. Another is that they have the same number of free parameters, the higher order polynomial forms being used to give greater smoothness. A disadvantage is that the recovery process is a global process: thus while (1.11) is explicit, reflecting the diagonal mass matrix obtained with piecewise constants, recovery introduces a banded mass matrix corresponding to the implicitness exhibited in (1.12). In the algorithms given here we shall limit the recovery to piecewise linears, which have their nodal values at the points $\{x_i\}$, the centre-points of the intervals introduced above: that is, as with the spline family, the corresponding finite elements for the piecewise linear representation are staggered relative to those for the piecewise constant representation.

Consider then the scalar conservation law in one dimension

$$u_t + f_x = 0, \qquad (2.3)$$

where the flux function $f(u)$ we suppose for simplicity is convex, so that its derivative $a(u)$ has only the one sonic point $\bar{u}$ at which $a(\bar{u}) = 0$. Then (1.8) for piecewise constants and without recovery is easily seen to yield a generalisation of the well-known algorithm of Engquist and Osher [1981]. We clearly have the relations

$$\sum_{(i)} \Phi_i^n(x) \equiv 1 \qquad (2.4a)$$

and

$$\left\langle \partial_x f(U^n), 1 \right\rangle = \sum_{(k)} \left\{ \left[f(U_k^n) - f(\bar{u}) \right] + \left[f(\bar{u}) - f(U_{k-1}^n) \right] \right\}. \qquad (2.4b)$$

Thus the update process can be carried out by dealing with each discontinuity in $f(U^n)$ in turn, cycling over k, and allocating some proportion of it to $\left\langle U^{n+1}-U^n,\phi_i\right\rangle$, cycling over i. Moreover, because of the assumed convexity of $f(u)$, the characteristic speed between U_k^n and $\bar{u}$ has the same sign: thus when the CFL number is less than unity all of the contribution from $[f(U_k^n)-f(\bar{u})]$ goes to the kth interval, to the right of the discontinuity at $x_{k-\frac{1}{2}}$, or to the $(k-1)^{th}$ interval to the left, according to whether $a(U_k^n)$ is positive or negative. Hence in this case we have the following simple result.

Algorithm: *set* $S_i := U_i^n \Delta x_i$; *for each* k

$$\textit{transfer} \quad -\Delta t[f(U_k^n)-f(\bar{u})]$$

$$\textit{from} \begin{cases} S_{k+1} \\ S_k \end{cases} \textit{to} \begin{cases} S_k \\ S_{k-1} \end{cases} \textit{if} \quad a(U_k^n) \begin{cases} >0 \\ <0 \end{cases}; \tag{2.5}$$

$$\textit{then} \quad U_i^{n+1}\Delta x_i := S_i .$$

Conservation is obvious from this form. Also the generalisation to larger time steps, in which a characteristic starting from the discontinuity at $x_{k-\frac{1}{2}}$ can reach beyond the interval on either side, is also clear. Details are given in Childs and Morton [1986].

This has been given in detail because after recovery essentially the same algorithm may be used. In order to achieve adaptability, each discontinuity is resolved by a linear section whose length is controlled by a parameter θ: for $\theta_{k-\frac{1}{2}} = 0$, there is no resolution of the discontinuity at $x_{k-\frac{1}{2}}$; and for $\theta_{k-\frac{1}{2}} = 1$, it is spread to the centre of the interval on either side. Then in order to implement (2.2) we need only to modify the flux function locally before carrying out (2.5): specifically, we introduce $A_{k-\frac{1}{2}}(u)$ to satisfy

$$A_{k-\frac{1}{2}}(u)\Delta t = a(u)\Delta t + (x-x_{k-\frac{1}{2}}) \tag{2.6a}$$

for u between $\tilde{u}_{k-1}^n$ and $\tilde{u}_k^n$, and x between $x_{k-\frac{1}{2}} - \frac{1}{2}\theta_{k-\frac{1}{2}}\Delta x_{k-1}$ and $x_{k-\frac{1}{2}} + \frac{1}{2}\theta_{k-\frac{1}{2}}\Delta x_k$; then we set

$$F_{k-\frac{1}{2}}(u) = \int A_{k-\frac{1}{2}}(u)du. \tag{2.6b}$$

In effect, for example, the characteristic starting at $x_{k-½} + \frac{1}{2}\theta_{k-½}\Delta x_k$ with speed $a(\tilde{u}_k^n)$ is started from $x_{k-½}$ with an appropriately increased speed. The limitation on Δt for the simple form (2.5) to hold is of course more severe, but this is of little consequence.

Once the parameters $\{\theta_{i-½}\}$ have been chosen, the $\{\tilde{u}_i^n\}$ are given by solving the projection equation (2.1) which yields the tridiagonal system

$$\frac{1}{4}\left[\theta_{i-½}\,\frac{\Delta x_i}{\Delta x_{i-1}+\Delta x_i}\,\Delta_-\tilde{u}_i^n + \theta_{i+½}\,\frac{\Delta x_i}{\Delta x_i+\Delta x_{i+1}}\,\Delta_+\tilde{u}_i^n\right] + \tilde{u}_i^n = U_i^n. \tag{2.7}$$

This is relatively straightforward: what is less so, is the choice of the $\{\theta_{i-½}\}$. It is convenient at this stage to introduce the possibility of explicitly fitting shocks or other discontinuities, because this breaks up the sequence of $\{\theta_{i-½}\}$ to be chosen to a set of subsequences.

Because of the nature of the information being used, the recovery procedure is inevitably subjective and problem-dependent. When it is known that a true solution $u^n(x)$ may have a discontinuity, its presence is recognised by two sequences of smoothly varying averages $\{U_i^n\}$, separated by one intermediate value, representing the average across the jump. For this purpose it is useful to introduce the ratios

$$r_i := \frac{\Delta_- U_i^n}{\Delta_+ U_i^n}\,\frac{\Delta_+ x_i}{\Delta_- x_i}. \tag{2.8}$$

Then, with a suitable choice of program parameters, a jump is deemed to lie in the jth interval if:

$$\text{(i) } r_j > 0 \text{ with } |r_{j-1}| \ll 1,\ |r_{j+1}| \gg 1; \tag{2.9a}$$

and $$\text{(ii) } a(U_{j-1}^n) \geq a(U_{j+1}^n). \tag{2.9b}$$

The second condition is imposed to prevent the development of an expansion wave being inhibited.

Suppose a jump has been recognised in interval j (and not in those on either side): then we set $\theta_{j-½} = \theta_{j+½} = 0$ before solving (2.7). All other parameters $\{\theta_{i-\frac{1}{2}}\}$ are chosen "as large as possible" in the interval [0,1] consistent with the monotonicity-preservation condition

$$\text{signum}\ \{\Delta_- \tilde{u}_i^n\} = \text{signum}\ \{\Delta_- U_i^n\}. \tag{2.10}$$

This choice and the solution of (2.7) has to be carried out iteratively: suitable algorithms are given by Morton and Sweby [1987] and Childs and Morton [1986]. Generally it will be necessary to have $\theta<1$ only at severe changes of gradient: see Fig.1 for a typical example, where initial data has been first projected onto piecewise constants and then recovered by piecewise linears after detection of the jump on the right.

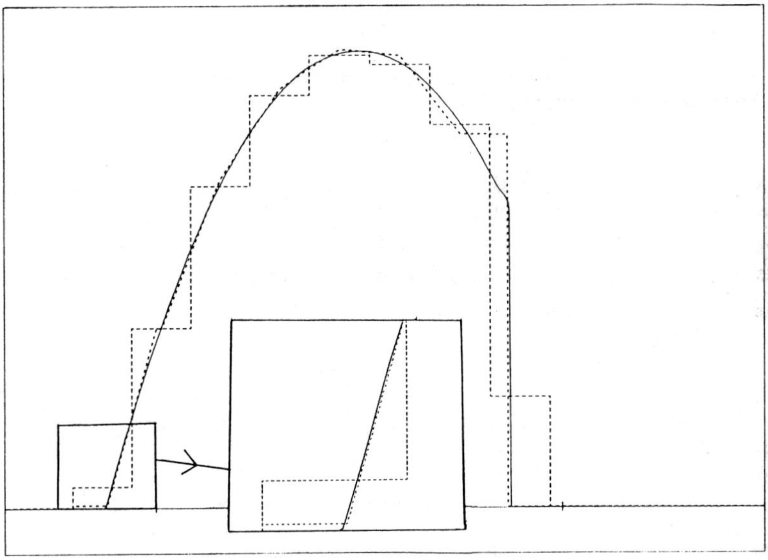

Fig.1 Recovery from piecewise constant projection.

Finally, in the subsequent update process, the shock in the jth interval has a starting position given by the projection equation $\left\langle \tilde{u}^n - U^n, \phi_j \right\rangle = 0$, namely

$$x_s := (1-\eta)x_{j-\frac{1}{2}} + \eta x_{j+\frac{1}{2}} \tag{2.11a}$$

where

$$\eta = (\Delta_+ \tilde{u}_j^n)/(\Delta_+ \tilde{u}_j^n + \Delta_- \tilde{u}_j^n); \tag{2.11b}$$

and it moves with the "shock speed"

$$a_s := \frac{f(\tilde{u}^n_{j+1})-f(\tilde{u}^n_{j-1})}{\tilde{u}^n_{j+1}-\tilde{u}^n_{j-1}}. \qquad (2.11c)$$

Otherwise the update is as before.

3. CONVERGENCE RESULTS IN THE SCALAR CASE

We denote by G^* the graph $[x,\tilde{u}^n]$ of the recovered solution, including jump recovery, and by $a^*(\tilde{u}^n)$ the characteristic speed modified to be the shock speed as given by (2.11c) over that part of the graph corresponding to a shock: E^* will be used to denote the evolution operator in this case. Then, as in (1.9) and (1.8), one can still write the update process in the compact forms

$$\left\langle U^{n+1},\phi_i\right\rangle = -\int_{G^*}\int_{-\infty}^{x+a^*\Delta t} \phi_i(z)dzdu \qquad (3.1a)$$

$$= \left\langle U^n,\phi_i\right\rangle - \Delta t\int_{G^*}\Phi_i^* df, \qquad (3.1b)$$

where Φ_i^* is defined as in (1.8b) with $a^*(\tilde{u}^n)$ replacing $a(U^n)$. It is these forms that are most useful for the analysis leading to the theorems given below. Proofs of all the results can be found in Childs and Morton [1986].

We begin with a result giving a key entropy inequality.

<u>Theorem 1</u> On a quasi-uniform mesh and for a uniformly bounded $a^*(\cdot)$, we have

(i) $\mathrm{Var}(U^{n+1}) \leq \mathrm{Var}(\tilde{u}^n)$;

and, for any convex entropy function $V(\cdot) \in W^{1,\infty}(\mathbb{R})$,

$$\text{(ii)}\quad V(BE^*\tilde{u}^n(z)) \leq -\int_{G^*} V'(u)H(x+a^*\Delta t-z)du, \qquad (3.2)$$

where E^* is the evolution operator defined as in (1.6) but using a^*, B is Brenier's collapse operator and $H(\cdot)$ is the Heaviside function.

With no shock recovery, sharper results are possible. Thus one can show that the operator $\hat{BE}$ produces no new extrema for any continuous single-valued graph. One can also in this case obtain an approximation result which indicates some of the constraints that need to be placed on the choice of the Θ-parameters.

<u>Lemma 1</u> Suppose that in the linear recovery from U to $\tilde{u}$, with no jump recovery, there is a constant C such that, for $h = \max(\Delta x_i)$,

$$\Theta_{i-\frac{1}{2}}|\Delta_- U_i| \leq Ch \qquad \forall\ i. \tag{3.3}$$

Then

$$\|\tilde{u}-U\|^2_{L^2} \leq \frac{2}{3}Ch^2 \mathrm{Var}(\tilde{u}). \tag{3.4}$$

[Note that this would be a straightforward result if we assumed $\tilde{u} \in H^1(\mathbb{R})$.] Using this, we have the following result.

<u>Theorem 2</u> On a quasi-uniform mesh (for both $\{x_i\}$ and $\{t_n\}$), with linear recovery satisfying (3.3) and the monotonicity criterion (2.10), but without jump recovery, suppose that the initial data u^0 is of bounded variation, with a finite number of extrema, and of compact support. Then as $h \to 0$, the ECG approximation $\{U^n\}$ converges in $L^\infty(L^1(\mathbb{R});[0,T])$ to the unique entropy-satisfying solution of the equation (2.3).

A similar theorem holds when jump recovery is included, but the hypotheses need to be tightened in several respects. In particular, undershoots may develop when too large a time step is used in conjunction with a recovered shock. The situation is well illustrated by the results shown in Fig.2 for the inviscid Burgers' equation, that is with $f(u) \equiv \frac{1}{2}u^2$: these also show well the effectiveness of the ECG scheme, particularly the value of being able to use a large time step. The initial data shown in Fig.2a provides a severe test on a uniform mesh of size $h = 0.02$ or 0.01: results for mesh ratios $\Delta t/h$ equal to 0.3125, 1.25, 2.5 and 7.5 are shown in Figs.2b,c,d. In Fig.2b the two smaller ratios are compared at time $t = 0.3$, $h = 0.02$, which show accuracy can be lost if too many small time steps are used, because of the excessive number of projections. In Fig.2c,d the two larger ratios are compared at times $t = 0.15$ and $t = 0.3$ respectively with $h = 0.01$. The results for mesh ratio 2.5 (effectively the maximum CFL number) are very good indeed but the mesh ratio of 7.5 is too large, causing an undershoot to occur to the right of the second shock at the earlier time, even though this has been eliminated by the later time.

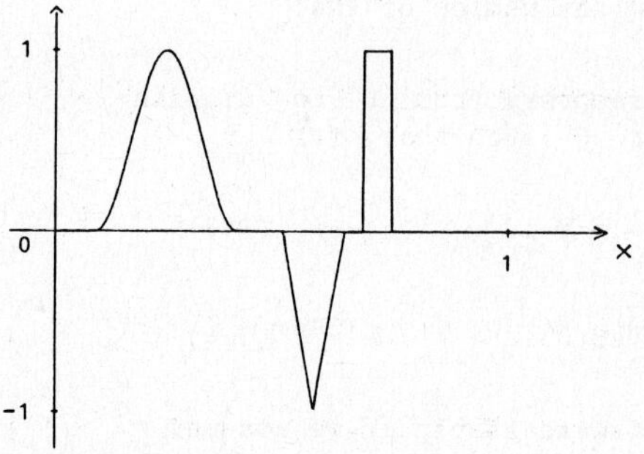

(a) Initial data

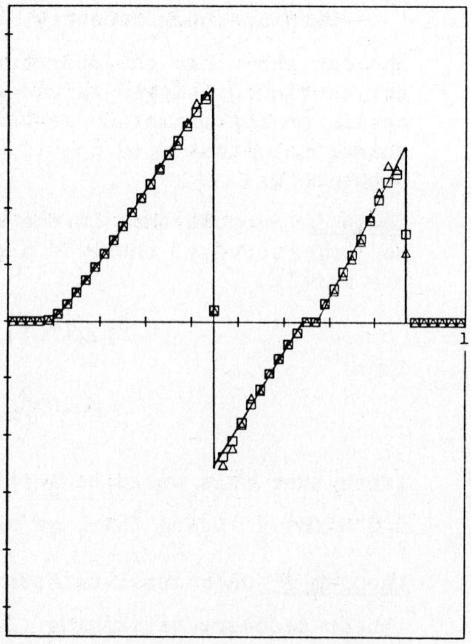

(b) t = 0.3, Δx = 1/50
Δ λ = 0.3125
□ λ = 1.25

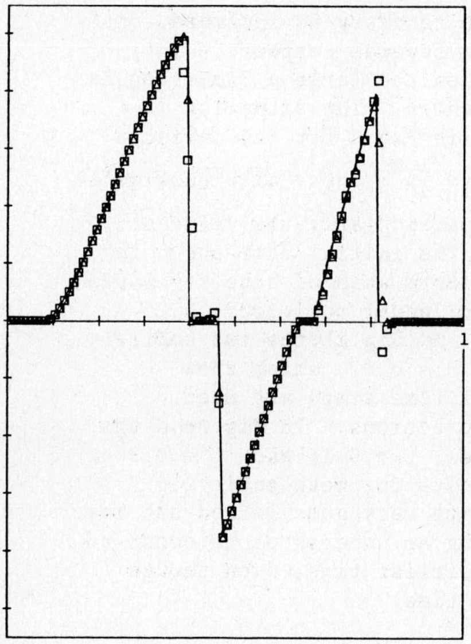

(c) t = 0.15, Δx = 1/100
Δ λ = 2.5
□ λ = 7.5

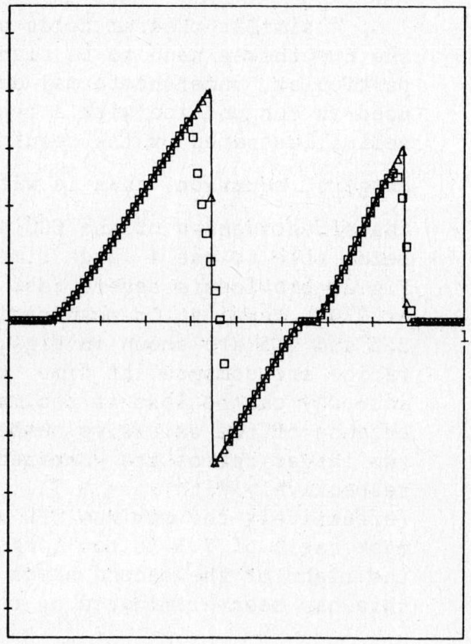

(d) t = 0.3, Δx = 1/100
Δ λ = 2.5
□ λ = 7.5

Fig. 2 Initial data and solution for Burgers' equation.

The undershoots occur because a^* in (3.1a) is discontinuous and the graph E^*G^* contains a closed loop. The possibility of new extrema arising from this cause requires more stringent hypotheses in the convergence theorem. The increase of entropy that results when a shock is recovered also requires modification to the hypotheses: details can be found in Childs and Morton [1986].

4. HYPERBOLIC SYSTEMS IN ONE DIMENSION

A characteristic based method for a system of equations must make some use of the Jacobian matrix $A := \partial \underline{f}/\partial \underline{u}$. Perhaps the most direct is to use its eigenvectors to convert the system to characteristic normal form: then an algorithm might consist of changing to characteristic variables at each time step so that the scalar algorithm might be applied to each. The flux vector splitting techniques of Steger and Warming [1981] and van Leer [1982] are in effect based on this approach; and, for the Euler equations of gas dynamics, exploit for this purpose the fact that the flux functions are homogeneous of degree one in the conserved variables. Some examples of applying the former technique to the ECG schemes given above have been presented in Morton and Sweby [1987].

However, the algorithm given in (2.5) lends itself more naturally to the use of techniques based on splitting the flux differences into components corresponding to the various waves supported by the system. That is, they approximately solve the Riemann problem for the jumps in $\underline{u}$ and $\underline{f}$ at each interval boundary $x_{i-\frac{1}{2}}$. The two most widely used are due to Osher and Solomon [1982] and to Roe [1981]: the former is based on the use of simple waves; the latter introduces a linearised problem for which the Jacobian satisfies $\hat{A}\,\Delta\underline{u} = \Delta\underline{f}$ and gives the correct shock speeds in the case of pure shocks. We have found this last technique both the simplest to apply and the most effective in the problems that we have studied, although it needs an "entropy fix" as in Harten and Hyman [1983] to preclude entropy violating solutions. Thus it is the only one we shall describe here. Note that we are generally limited to second order accuracy with ECG schemes for systems, as no account is taken of the curving of the characteristics: thus limiting the recovery to piecewise linears is entirely justified in this case.

If the vectors $\underline{u}$, $\underline{f}$ in the hyperbolic system $\underline{u}_t + \underline{f}_x = \underline{0}$ have dimension p, the graph $G \equiv [x, \underline{U}^n]$, or $G^* \equiv [x, \underline{\tilde{u}}^n]$ after recovery, lie in the space $\mathbb{R}^{p+1}$ and the Roe decomposition breaks them into linear sections, each parallel to a right eigenvector of the mean Jacobian $\hat{A}$. For the linear recovery we have used a single set of Θ-parameters so that the recovery equation (2.7) is applicable, with $\tilde{u}_i^n$ replaced by $\tilde{\underline{u}}_i^n$ and U_i^n by $\underline{U}_i^n$: the Θ's are chosen so that the condition (2.10) applies to all components, although of course for the differential system we no longer have the monotonicity-preservation properties which motivated this criterion. Each of the sub-graphs, associated with the mth characteristic field and the interval boundary $x_{k-\frac{1}{2}}$, is taken to run from $x_{k-\frac{1}{2}} - \frac{1}{2}\Theta_{k-\frac{1}{2}}\Delta x_{k-1}$ to $x_{k-\frac{1}{2}} + \frac{1}{2}\Theta_{k-\frac{1}{2}}\Delta x_k$ and has a length $\hat{\alpha}_{k-\frac{1}{2}}^{(m)} \hat{\underline{r}}_{k-\frac{1}{2}}^{(m)}$ in the p dimensions of $\underline{u}$-space, $m = 1,2,\ldots,p$. That is,

$$\Delta_- \tilde{\underline{u}}_k = \sum_{m=1}^{p} \hat{\alpha}_{k-\frac{1}{2}}^{(m)} \hat{\underline{r}}_{k-\frac{1}{2}}^{(m)} \tag{4.1}$$

where $\hat{\underline{r}}_{k-\frac{1}{2}}^{(m)}$ is the right eigenvector (corresponding to eigenvalue $\hat{\lambda}_{k-\frac{1}{2}}^{(m)}$) of the Roe decomposition matrix $\hat{A}_{k-\frac{1}{2}}$, for which

$$\Delta_- \underline{f}(\tilde{u}_k^n) = \hat{A}_{k-\frac{1}{2}} \Delta_- \tilde{\underline{u}}_k^n. \tag{4.2}$$

Then the characteristic speeds $\hat{\lambda}_{k-\frac{1}{2}}^{(m)}$ are modified as in (2.6a) so that the updates can be computed from the recovered approximation $\tilde{u}$ through an algorithm based on (3.1) which allows arbitrary time steps.

Boundary value problems are transformed to pure initial value problems to which (3.1) can be applied directly. For example, at a solid wall, boundary conditions for the Euler equations are implemented by reflection, forcing the recovered pressure and density to be symmetric and the velocity antisymmetric. This can be shown to preserve key properties of the scheme. As an illustration, results are given in Fig.3 for the problem of Woodward and Collela [1984] in which two strong blast waves collide in a shock tube closed at both ends. The mesh is uniform with 400 points, the mesh ratio is 0.1 giving a maximum CFL number of 5.4 and a very fine mesh result is given as a full line for comparison.

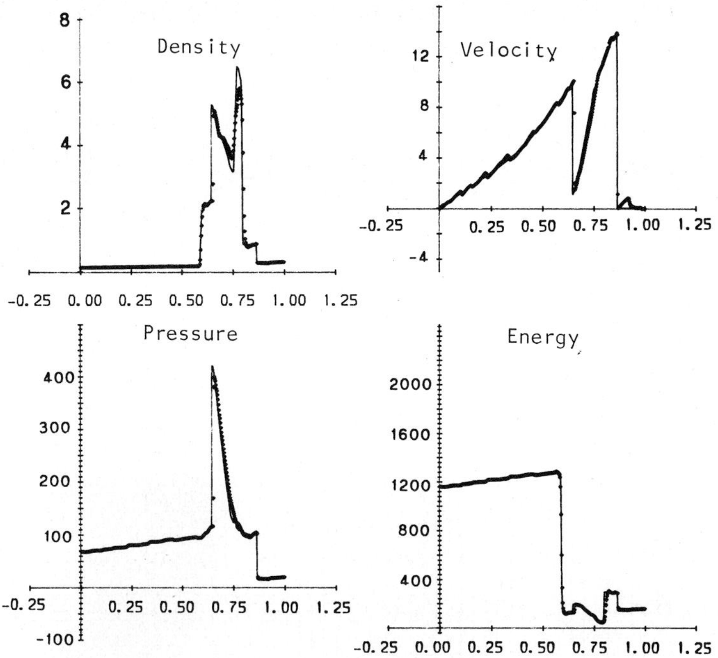

Fig.3 Woodward and Colella's blast wave problem at time t = 0.038

Moving and adaptive grids are readily incorporated in the ECG schemes, and with the lack of a time step restriction, local mesh refinement is an attractive option. We use an error monitor which takes account of the decomposed waves and typical results are shown in Figs.4 and 5. This is for the familiar shock tube problem of Sod [1978] but with closed ends: results are shown in Fig.4 and the corresponding mesh with its adaptive refinement (in x only) in Fig.5.

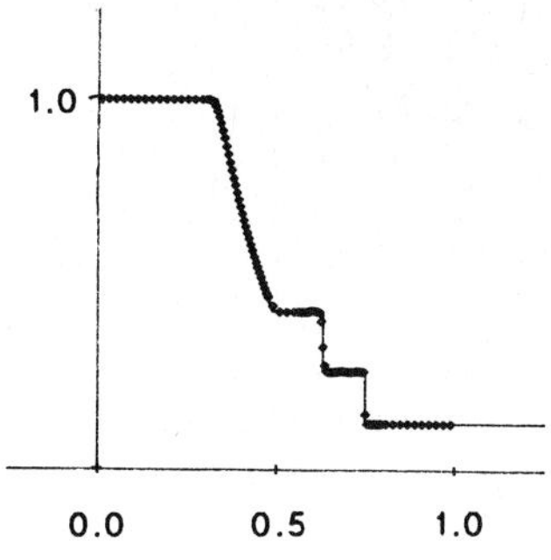

Fig.4a Density plot for shocktube problem at time 0.144

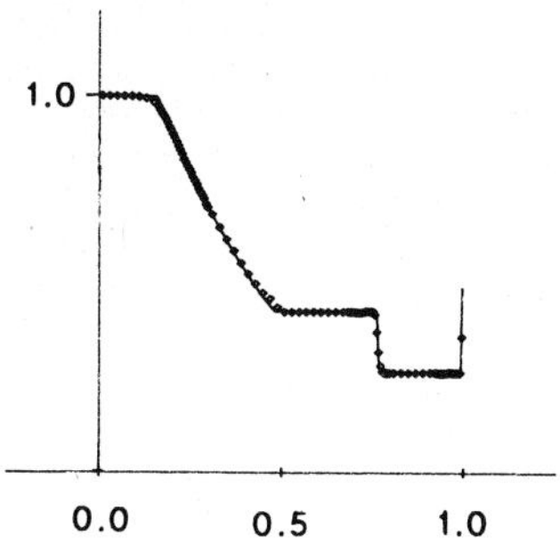

Fig.4b As Fig.41, but after time 0.288

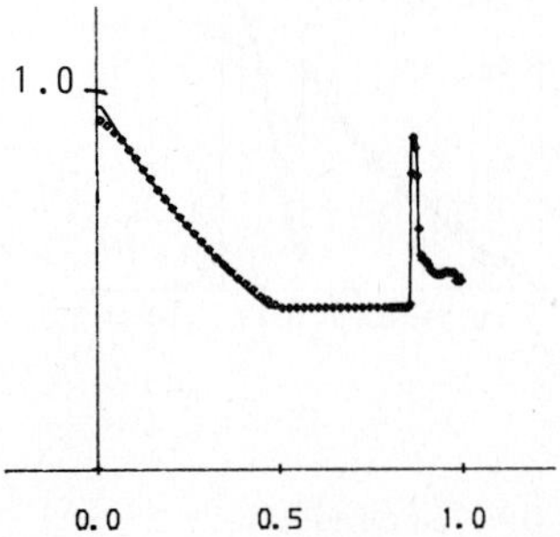

Fig.4c As Fig.4a, but after time 0.432

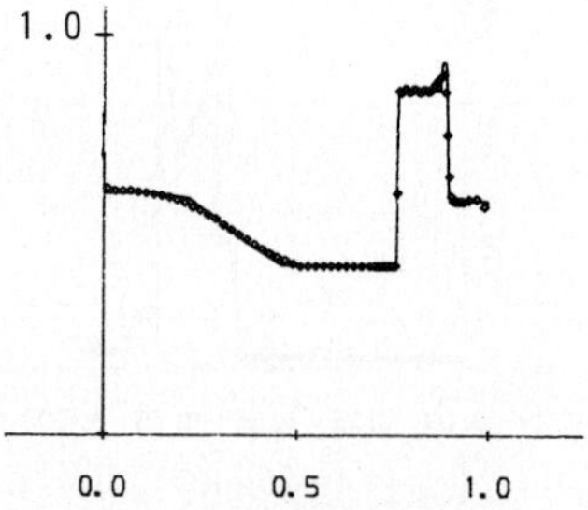

Fig.4d As Fig.4a, but after time 0.576

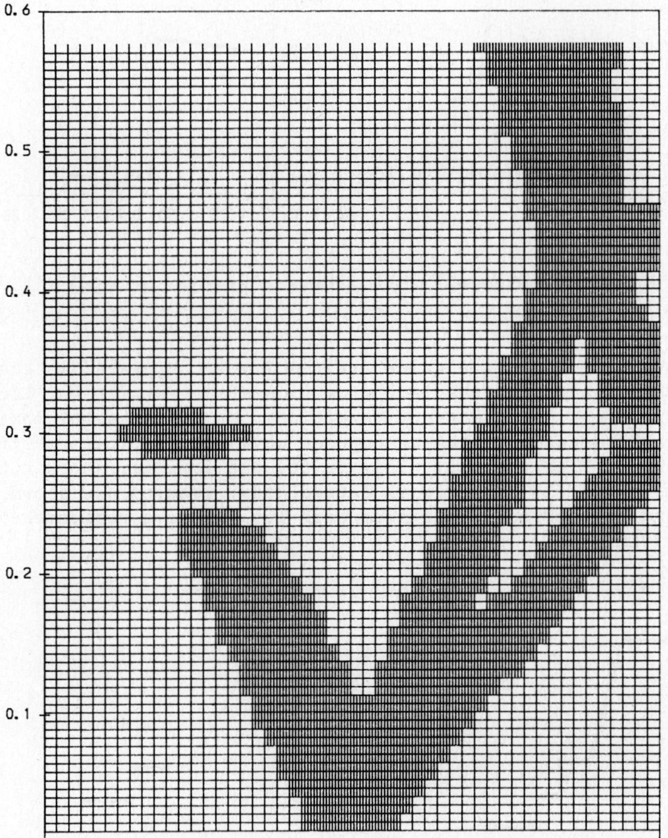

Fig.5 Adaptive mesh used for shocktube problem.

5. EXTENSIONS TO TWO DIMENSIONS

Much still remains to be done here to properly exploit both the flexibility of the finite element method in multi-dimensions and the key properties of the characteristic Galerkin techniques. Some preliminary results from a fairly straightforward extension of the one-dimensional algorithms to two dimensions were given in Fletcher and Morton [1986]. Here we confine our observations to just three points.

The first is that the inner products such as in (1.7) are very much more difficult to evaluate in multi-dimensions than in one dimension. However, there is a considerable literature on Lagrange Galerkin methods used for approximating the Navier Stokes equations and problems of flow in porous media. These usually use piecewise linear or quadratic elements and much is now known regarding how the corresponding inner products should be approximated. As pointed out in Morton, Priestley and Suli [1988] a straightforward use of standard quadrature formulae will often introduce instabilities which can be difficult to control through the choice of time step. On the other hand, if the tracing of the characteristic paths is approximated in such a way that the resulting inner products can be evaluated exactly, then the resulting scheme can be both accurate and stable. For example, with bilinear elements on a rectangular mesh one can move the centroid of each element according to (1.6b), but then consider the whole element to make this translation undistorted and unrotated: the resulting integrals are then quite straightforward. The stability of this scheme for linear advection is established in the cited paper where its accuracy on the commonly used "rotated cone" problem is also demonstrated. These results seem highly relevant to the task of extending our present algorithms into two dimensions.

Secondly, there is a question regarding the actual characteristic velocities that should be used at a discontinuity. Suppose a basic piecewise constant approximation over some mesh of elements is used. It is argued in Osher [1980] that the natural generalisation of the algorithm (2.5) is to use, at each element edge, the normal component of the jump in flux to effect the transfers. Thus contributions are made to the updates for the elements on either side of the edge but to no others. That is, corner effects are not taken into account or, equivalently, characteristic velocities in a tangential direction. As pointed out by LeVeque [1987] this is a serious inaccuracy when reasonably large time steps are used.

It is therefore useful to record the precise form that the update procedure for piecewise constants should take in a simple case. Consider the scalar conservation law

$$u_t + f_x + g_y = 0 \qquad (5.1a)$$

i.e.
$$u_t + au_x + bu_y = 0, \qquad (5.1b)$$

where $a = \partial f/\partial u$ and $b = \partial g/\partial u$: and suppose we have a uniform rectangular mesh of spacing $\Delta x, \Delta y$. Then if $a,b \geq 0$ and $a\Delta t \leq \Delta x$, $b\Delta t \leq \Delta y$, one can show that the correct generalisation of the first order upwind difference scheme in one dimension which is defined by (1.6), (1.7) is given by

$$\Delta x\Delta y(U_{ij}^{n+1} - U_{ij}^{n}) + \Delta t\Delta y\Delta_{-x} f(U_{ij}^{n}) + \Delta t\Delta x\Delta_{-y} g(U_{ij}^{n}) + (\Delta t)^2 \Delta_{-x}\Delta_{-y} h(U_{ij}^{n}) \qquad (5.2)$$

where $h(u) := \int ab\,du$. The last term in (5.2) gives the corner effects arising from tangential velocities at each element edge. This form is readily generalised to the case where both f and g possess sonic points.

Finally, we wish to point out the close link that exists between the methods described here for unsteady problems and the widely used cell vertex methods for steady gas dynamic flows. Suppose we have a structured mesh of quadrilateral elements on which we would have a piecewise constant approximation parameterised by $\{\underline{U}_{ij}\}$. After recovery by piecewise bilinears (generalising the Θ-recovery of (2.7)), attention is focussed on the variation between the values $\{\underline{\tilde{u}}_{ij}\}$ at the centres of the elements: and joining up neighbouring centres creates a staggered quadrilateral mesh of what we shall call cells, see Fig.6. An update of the form (1.8) will be constructed from inner products

$$\left\langle \partial_x \underline{f}(\underline{\tilde{u}}) + \partial_y \underline{g}(\underline{\tilde{u}}), \underline{\tilde{\Phi}}_{ij} \right\rangle, \qquad (5.3)$$

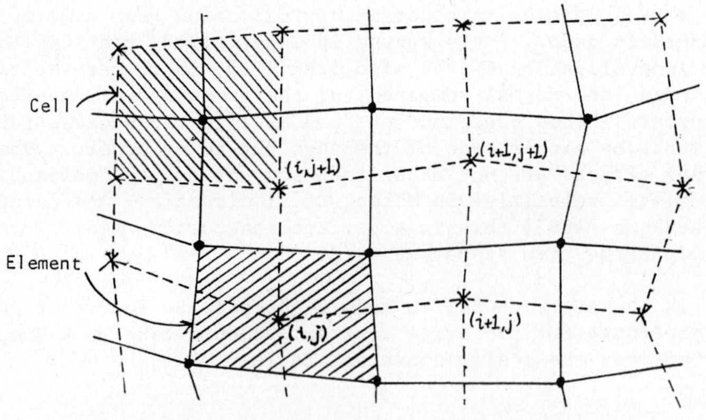

Fig.6 Staggered meshes of "cells" and "elements".

where each component of the test function vector $\tilde{\underline{\Phi}}_{ij}$, is an average of the characteristic function for the (i,j) element over an appropriate characteristic path. The steady state solution is obtained by setting to zero all these updates, for any choice of Δt, which means that it is not necessary to include the last term in (5.2). There is then clearly a good deal of flexibility available in how this might be done. But in algorithm (2.5), and its generalisation after recovery, the emphasis was on decomposing contributions into those arising from integrations between cell centres. In two dimensions this generalises into exploiting the fact that

$$\sum_{(i,j)}\tilde{\underline{\Phi}}_{ij}(x,y) \equiv \underline{1}, \qquad (5,4)$$

and using a partition of unity based on the characteristic functions of the cells whose vertices are the element centres (i,j). Denoting by Ω_{kl} such a cell, we hence obtain by means of Gauss' theorem and integration by parts around the cell perimeter $\partial\Omega_{kl}$

$$0 = \int_{\Omega_{kl}} \mathrm{div}(\underline{f},\underline{g})d\Omega = \int_{\partial\Omega_{kl}} \underline{f}dy-\underline{g}dx = \int_{\partial\Omega_{kl}} xd\underline{g}-yd\underline{f}, \qquad \forall\ k,l. \qquad (5.5)$$

The central form here is that which is used to implement the cell vertex method, the integrals between vertices being approximated by the trapezoidal rule. The last form is an appropriate one for generalising the algorithms of section 2 to two dimensions.

6. REFERENCES

1. J.P. Benqué & J. Ronat (1981), Quelques difficultés des modéles numérique en hydraulique, 5th Int. Symp. on Comp. Meths. in Appl. Sciences and Engng., INRIA, Versailles, France.

2. M. Bercovier, O. Pironneau, Y. Hasbani & E. Livne (1982), Characteristic and finite element methods applied to the equations of fluids, Proc. MAFELAP 1981, J.R. Whitemen, ed., Academic Press, London, pp. 471-478.

3. Y. Brenier (1984), Average multivalued solutions for scalar conservation laws, SIAM J. Numer. Anal., 21, pp. 1013-1037.

4. P.N. Childs & K.W. Morton (1986), Characteristic Galerkin methods for scalar conservation laws in one dimension, Oxford University Computing Laboratory, Report 86/5.

5. J. Donea (1984), A Taylor-Galerkin method for convective transport problems, Int. J. Numer. Meths. Engng., 20, pp. 101-119.

6. J. Douglas Jnr. & T. Russell (1982), Numerical methods for convection dominated diffusion problems based on combining the method of characteristics with finite element or finite difference procedures, SIAM J. Numer. Anal., 19, pp. 871-885.

7. B. Enquist & S. Osher, (1981) One-sided difference approximations for nonlinear conservation laws, Math. Comp., 36, pp. 321-353.

8. C.A.J. Fletcher & K.W. Morton (1986), Oblique shock reflection by the characteristic Galerkin method, Proc. 9th Australian Fluids Mechs. Conf., Auckland.

9. A. Harten & J.M. Hyman (1983), Self-adjusting grid methods for one-dimensional hyperbolic conservation laws, J. Comp. Phys., 50, pp. 325-269.

10. R.J. LeVeque (1987), High resolution finite volume methods on arbitrary grids via wave propogation, Preprint Dept. of Mathematics, University of Washington, Seattle, USA.

11. K.W. Morton (1983), Characteristic Galerkin methods for hyperbolic problems, Proc. 5th GAMM conference on numerical methods in fluid mechanics, M. Pandolfi & R. Riva eds., F. Vieweg & Sohn, Wiesbaden, pp. 243-250.

12. K.W. Morton, A. Priestley & E. Suli (1988), Stability analysis of the Lagrange Galerkin method with non-exact integration, to appear in Modélisation Mathématique et Analyse Numérique.

13. K.W. Morton & A. Stokes (1981), Generalized Galerkin methods for hyperbolic problems, in Proc. MAFELAP 1981, J.R. Whiteman, ed., Academic Press, London, pp. 421-431.

14. K.W. Morton & P.K. Sweby (1987), A comparison of flux-limited difference methods and characteristic Galerkin methods for shock modelling, J. Comput. Phys., 73, p.203.

15. S. Osher & F. Solomon (1982), Upwind difference schemes for hyperbolic systems of conservation laws, Math., Comp. 38, pp. 339-374.

16. S. Osher (1980), Approximation par eléments finis avee décentrage par les lois de conservation hyperboliques nonlinéaires multidimensionelles, C.R. Acad. Sci. Paris, Sér.A, v.290, pp. 819-821.

17. O. Pironneau (1982), On the transport-diffusion algorithm and its application to the Navier-Stokes equations, Numer. Math., 38, pp. 302-322.

18. P.L. Roe (1981), Approximate Riemann solvers, parameter vectors and difference schemes, J. Comp. Phys., 43, pp. 357-372.

19. P.L. Roe (1985), Some contributions to the modelling of discontinuous flow, SIAM/AMS lectures in Appl. Math., vol.22, pp. 163-193.

20. G.A. Sod (1978), A survey of several finite difference methods for systems of nonlinear conservation laws, J. Comp. Phys., 27, pp. 7-31.

21. J.L. Steger & R.F. Warming (1981), Flux-vector splitting of the inviscid gasdynamic equations with application to finite difference methods, J. Comp. Phys. 40, pp. 263-293.

22. B. Van Leer (1982), Flux-vector splitting for the Euler equations, Lecture Notes in physics 170, pp. 507-512.

23. P.R. Woodward & P. Collela (1984), The numerical simulation of two dimensional fluid flow with strong shocks, J. Comp. Phys. 54, pp. 115-173.

NUMERICAL SIMULATIONS OF COMPRESSIBLE HYDRODYNAMIC INSTABILTITIES WITH HIGH RESOLUTION SCHEMES

C.-D. Munz
Kernforschungszentrum Karlsruhe, Institut für Neutronenphysik und Reaktortechnik, Postfach 36 40, D-7500 Karlsruhe, FRG

L. Schmidt
Universität Karlsruhe, Institut für Angewandte Mathematik, Englerstraße 2, D-7500 Karlsruhe, FRG

Dedicated to Professor Martensen on the occasion of his 60th birthday

SUMMARY

The large scale motion of two-dimensional interfacial instabilities - namely the Kelvin-Helmholtz instability and the instability of a transonic jet - is examined. The numerical calculations are based on the direct simulation of the instabilities. The two-dimensional Euler equations are solved by a high resolution scheme. The movement of the interfaces is visualized by a marker particle algorithm. The interfaces are advected in a Lagrangean fashion according to the Eulerian flow field. It is shown that the numerical dissipation has a stabilizing effect similar to physical viscosity.

INTRODUCTION

Interface instabilities arise in a wide variety of physical contexts. In the present investigation we will study the instability of interfaces separating two domains of the same compressible fluid moving at different velocities, namely the Kelvin-Helmholtz instability and the instability of a jet. Our calculations are based on the direct simulation of these instabilities by the numerical solution of the equations of two-dimensional compressible fluid flow, usually called Euler equations. There are two different formulations of these equations. Numerical methods based on the Lagrangean formulation use a computational mesh traveling with the fluid. Hence these methods seem to be ideal for solving problems which involve interfaces between two fluids. However, Lagrangean calculations can typically be carried out for short time spans only. Then severe mesh distortion or mesh tangling will occur and rezoning must be performed in which all computational quantities are transferred to a new computational mesh. Because this procedure calls for much computational effort, the Lagrangean methods are not favourable for dealing with large scale computations. On the other hand, Eulerian methods, in which the mesh is fixed, are ideal for treating flows with large deformations. But interfaces are smeared out over some grid zones and the movement of the interfaces can hardly be seen.

We use a combined method: The flow field is calculated by an Eulerian method, while the interfaces are moved in a Lagrangean fashion according to this flow field. This means that we discretize the interface and within each time step calculate the new position of the dicsretized interface from the flow field. This may also be considered as a marker particle algorithm which is used only to visualize the movement of the interface. The Euler equations are solved by a so-called high resolution scheme. An efficient implementation on a vector computer permits to perform large scale computations on fine grids.

EULER EQUATIONS AND MUSCL-TYPE SCHEMES

We consider the two-dimensional equations of compressible fluid mechanics without thermal conduction and viscosity, written in the conservation form

$$U_t + f(U)_x + g(U)_y = 0 \tag{1}$$

with

$$U = \begin{bmatrix} \rho \\ \rho u \\ \rho v \\ e \end{bmatrix} , \quad f(U) = \begin{bmatrix} \rho u \\ \rho u^2+p \\ \rho uv \\ u(e+p) \end{bmatrix} , \quad g(U) = \begin{bmatrix} \rho v \\ \rho uv \\ \rho v^2+p \\ v(e+p) \end{bmatrix} \tag{2}$$

.

As usual, ρ denotes the density, u and v denote the velocity components in x and y direction, respectively, p denotes the pressure and e denotes the total energy per unit volume. The pressure is functionally related to the other variables via the equation of state of an ideal gas.

The numerical method, considered here, is based on dimensional splitting, also termed method of fractional step [13]. According to this method the two-dimensional Euler equations (1), (2) are split into two one-dimensional problems

$$U_t + f(U)_x = 0, \quad U_t + g(U)_y = 0 \quad . \tag{3}$$

These problems are then solved successively in each time step. In our calculations we use the two-cycle splitting of Strang [13] in which after each xy step the order is reversed for the following time interval: xy-yx, and which is of second-order accuracy as regards the time t. The systems (3) resemble in structure the one-dimensional Euler equations an numerical methods for these equations can be conveniently transferred.

We will restrict ourselves to describe the one-dimensional numerical method for the first equation of (3). A MUSCL-type scheme is usually formulated as a two step method with two main building blocks: a non-oscillatory interpolation and an upwind scheme (see [8]). In the first step, by means of interpolation, a piecewise linear representation of the approximate solution is calculated from the values U_i^n where U_i^n denotes an integral approximative value of the solution in the ith grid zone at time t_n. This representation defines boundary values in each grid zone - U_{i+}^n on the right and U_{i-}^n on the left side:

$$U_{i\pm}^n = U_i^n \pm \frac{\Delta x}{2} S_i^n \quad . \tag{4}$$

The value S_i^n stands for the slope in the ith grid zone, as usual Δx denotes the space increment, Δt the time increment. In order to obtain second-order accuracy with respect to time a midpoint rule is used: the boundary values are advanced to $t_{n+1/2}$

$$U_{i\pm}^{n+1/2} = U_{i\pm}^{n} - \frac{\lambda}{2}\,(F(U_{i+}^{n}) - F(U_{i-}^{n})), \quad \lambda = \frac{\Delta t}{\Delta x} \quad . \tag{5}$$

In the second step an approximative integral value at the next time level is calculated by

$$U_i^{n+1} = U_i^n - \lambda(h(U_{i+}^{n+1/2}, U_{(i+1)-}^{n+1/2}) - h(U_{(i-1)+}^{n+1/2}, U_{i-}^{n+1/2})) \quad . \tag{6}$$

where h is the flux of an upwind scheme. Any upwind method as reviewed in [5], [6] can be used for this purpose. The vector of slopes must satisfy a number of conditions. A first necessary condition for second-order accuracy in space says that the slope is a first order approximation of $U_x(x_i, t_n)$. In order to avoid spurious oscillations near strong gradients the piecewise linear representation must satisfy some monotonicity constraints. Within the MUSCL-scheme of van Leer [9] or the PPM- or PLM-method of Colella and Woodward [3] and Colella and Glaz [2] the slopes are calculated in terms of the primitive variables ρ, u, v, p. We calculate the slopes in terms of characteristic variables. This method relies on the local linearization technique of Roe [12].

In each grid zone an average value $\bar{U}_i$ is determined, e.g. $\bar{U}_i=(U_{i+1} + 2U_i + U_{i-1})/4$. Here and in the following studies the time index n is omitted as long as no misunderstandings can arise. The vector r_i^k denotes the kth right eigenvector of the Jacobi matrix evaluated at the average value. The difference quotients are then expanded in terms of this system of eigenvectors

$$\frac{1}{\Delta x}(U_{i+1}-U_i) = \sum_{k=1}^{4} \alpha_i^k r_i^k \, , \quad \frac{1}{\Delta x}(U_i-U_{i-1}) = \sum_{k=1}^{4} \beta_i^k r_i^k \quad . \tag{7}$$

The coefficients α^k, β^k measure the change of the difference quotients in direction of the kth eigenvector. A vector of slopes is obtained by

$$S_i = \sum_{k=1}^{4} s(\alpha_i^k, \beta_i^k)\, r_i^k \tag{8}$$

where s = s(a,b) denotes a slope calculation given by the scalar theory. For a scalar conservation law various suitable functions s(a,b) have been indicated and analyzed (see [10]). We use in the following slope calculations based on the class of slopes

$$S_l(a,b) = \mathrm{sign}(a)\max\,\{|\mathrm{minmod}(la,b)|, \; |\mathrm{minmod}(a,lb)|\} \tag{9}$$

with $1 \leq l \leq 2$ and

$$\text{minmod}(a,b) = \begin{cases} a & \text{for } |a| \le |b|,\ ab > 0 \\ b & \text{for } |a| > |b|,\ ab > 0 \\ 0 & \text{for } ab \le 0 \end{cases} \quad (10)$$

proposed by Sweby in terms of schemes using flux limiters (see [10]).

The main advantage of the slope calculation in terms of characteristic variables is that different slope calculations may be applied to the genuinely nonlinear characteristic fields and to the linearly degenerate fields. Hence, the numerical damping of contact discontinuities which is a severe problem for large scale computations may be strongly reduced or prevented by using a very compressive slope in the linearly degenerate fields. Such a compressive slope is the member for l=2 of the class (9) proposed by Roe and called superbee-function. The slopes (9) become more compressive with increasing l. Very compressive slopes should not be applied to the genuinely nonlinear fields. Because they may be over-compressive, they may compress each monotone profile into a discontinuity and introduce at centered rarefaction waves non-physical discontinuities. If the slopes are calculated in terms of primitive variables, the different waves cannot be treated in a different fashion. Hence, less compressive slopes must be used which introduce stronger numerical dissipation at contacts or a correction mechanism must be added which switches to a less compressive slope near sonic points.

VISUALIZATION OF INTERFACES

At the beginning of a calculation the surfaces between the fluids are discretized and replaced by a number of points. We will term these points marker or tracer particles. In each time step, at first the new flow field is calculated by the Eulerian MUSCL-type scheme. Afterwards the massless marker particles are advected in a Lagrangean fashion according to the local flow field. The movement of the interface can then be visualized by graphic display of these marker particles. The particles are overlaid on the fixed computational grid and are advected without any collisional effect between them. A quite similar technique is commonly applied in experiments using smoke or ink as marker particles.

At time t_n the kth marker particle is located at a point (x_k^n, y_k^n) of the computational domain and possesses the velocity (u_k^n, v_k^n). After computation of the flow field at the next time level by the Eulerian method the new location of the marker particle is calculated by

$$x_k^{n+1} = x_k^n + \frac{\Delta t}{2}(u_k^{n+1} + u_k^n) \ , \quad y_k^{n+1} = y_k^n + \frac{\Delta t}{2}(v_k^{n+1} + v_k^n) \quad (11)$$

The velocity of the marker particle is determined by bilinear area weighting interpolation. At first the location of the particle relative to the computational grid is calculated. For a uniform grid with constant space increments this is given by

$$i: = \text{int}\left(\frac{x_k^n - x_0}{\Delta x}\right) \ , \quad j: = \text{int}\left(\frac{y_k^n - y_0}{\Delta y}\right) \quad (12)$$

where (x_0, y_0) denotes the left lower corner of the computational domain. Next we calculate the areas $A_1,...,A_4$ in reference to Fig. 1. The x-component of the velocity of the k-th marker particle at time t_{n+1} is then determined from the flow field by the formula

$$u_k^{n+1} = (A_1 u_{i,j}^{n+1} + A_2 u_{i+1,j}^{n+1} + A_3 u_{i,j+1}^{n+1} + A_4 u_{i+1,j+1}^{n+1})/\Delta x \Delta y, \quad (13)$$

the y-component is determined in an analogous way. At the boundary some modifications may be introduced according to the physical boundary conditions. E.g., in the case of periodic boundary conditions, a particle which leaves the computational domain should reappear at the opposite boundary.

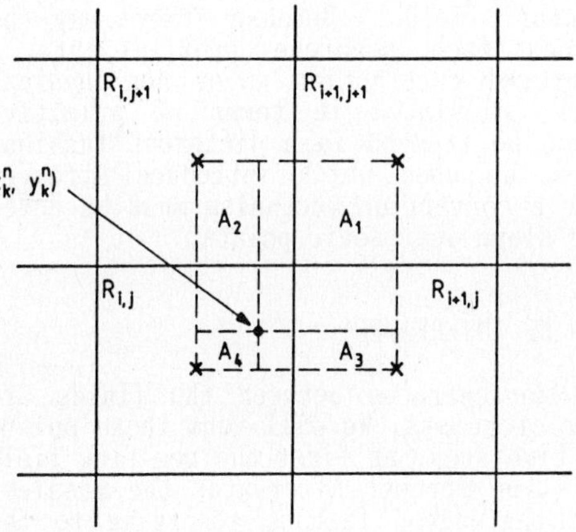

Fig. 1: Area-weighting interpolation

INTERFACIAL INSTABILITIES AND NUMERICAL RESULTS

The MUSCL-type schemes and the Lagrangean tracking of interfaces described above are applied to two-dimensional compressible interfacial instabilities. We will study interfacial instabilities which are purely inertial phenomena. We adopted two basic examples of hydrodynamic instabilities: the Kelvin-Helmholtz instability and the instability of a jet. A review of two-dimensional instabilities and their mathematical description in the incompressible case has been given by Birkhoff [1]. The initial data used in our calculations are sketched in Fig.2. In the first diagramm the fluid flows to the left in the lower half. In the second diagramm the fluid flow to the right is separated by a small band of fluid flow to the left. In both cases density and pressure are uniform in the whole domain. It is well-known that these shear layers are unstable for inviscid fluid flow in the sense that small initial perturbations will rapidly increase. Physical viscosity has a stabilizing effect.

We introduced sinusoidal perturbations of mode 2 of the initial data. In the first case the shear layer S is sinusoidally perturbed

$$S: \quad y = 0.025 \sin (4\pi x) \;, \; x \in [-0.5 \;, \; 0.5] \quad . \quad (14)$$

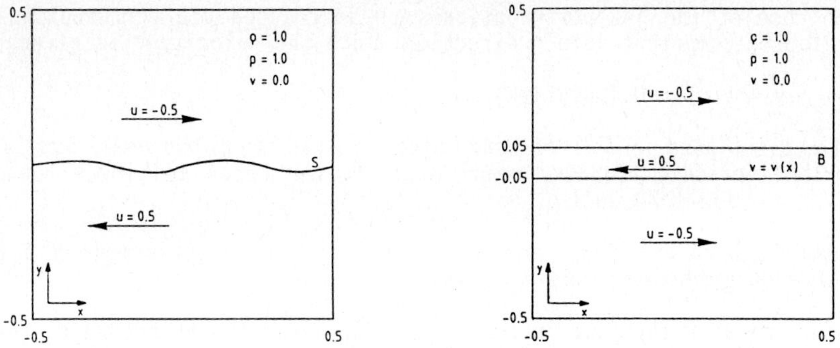

Fig. 2: Initial data of a Kelvin-Helmholtz instability and a jet

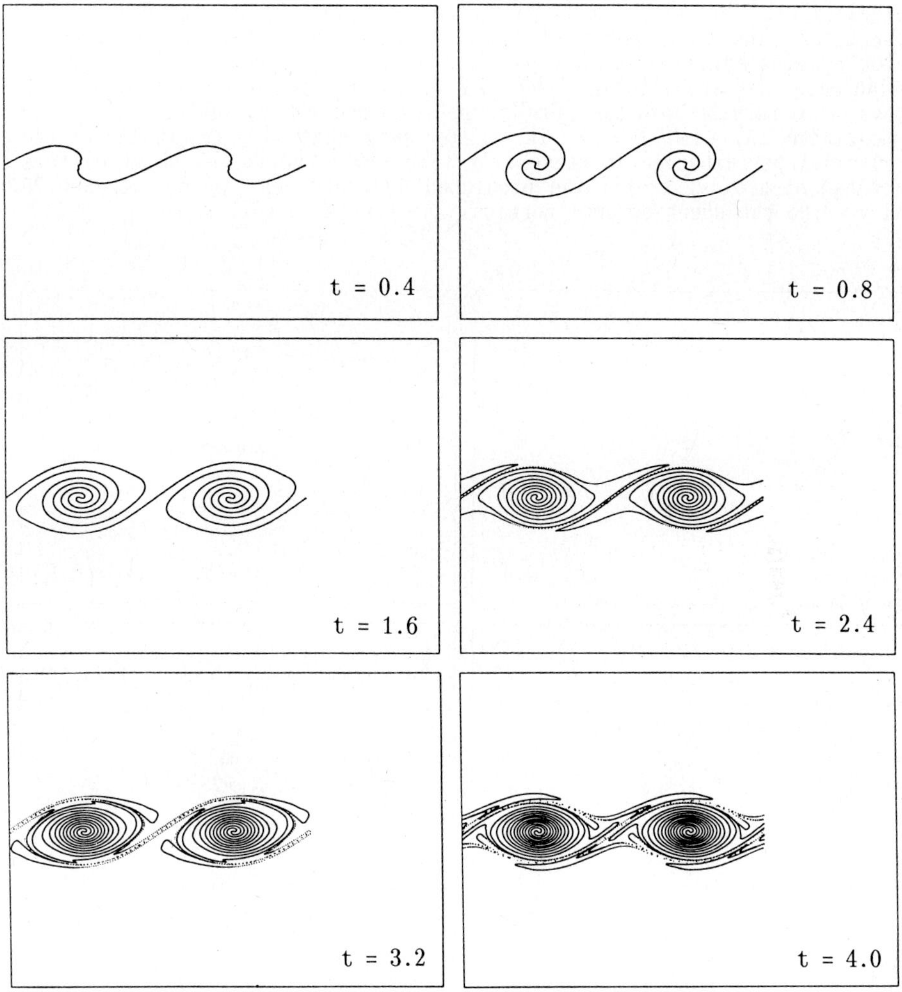

Fig. 3: Simulation of a Kelvin-Helmholtz instability

In the case of the jet perturbations are introduced via perturbations of the velocity component into y-direction. In B the velocity v is given by

$$v = v(x) = 0.1 \sin(4\pi x) \quad . \tag{15}$$

At the right-hand and left-hand side of the computational domain we prescribed periodic boundary conditions. At the upper and lower boundary those of a reflecting wall.

The numerical calculations presented here have been performed on a grid with 200x200 grid zones corresponding to step sizes $\Delta x=0.005$, $\Delta y=0.005$.

At each time step the time increment is adaptively chosen according to the Courant-Friedrichs-Lewy condition. In the second step (6) of the MUSCL-type scheme we used the simplest Godunov-type scheme theoretically investigated in [6]. Einfeldt [5] has shown how to use it for practical calculations. The marker particles (we used 40.000) are placed on the interfaces. The development of the interfaces is shown in the figures below by displaying the marker particle field at different times.Figs. 3, 4 indicate the stabilizing effect of the numerical dissipation. Fig. 3 shows the results of the MUSCL-type scheme (4)-(6) using the slope calculation (9) with l = 1. The plots show that the amplitude of the sinusoidal perturbation increases. At time t=0.4 the vortex sheet differs from the sinusoidal profile as predicted by the linear theory. At x=0.25 and x=-0.25 the sheet becomes vertical and starts to roll up.

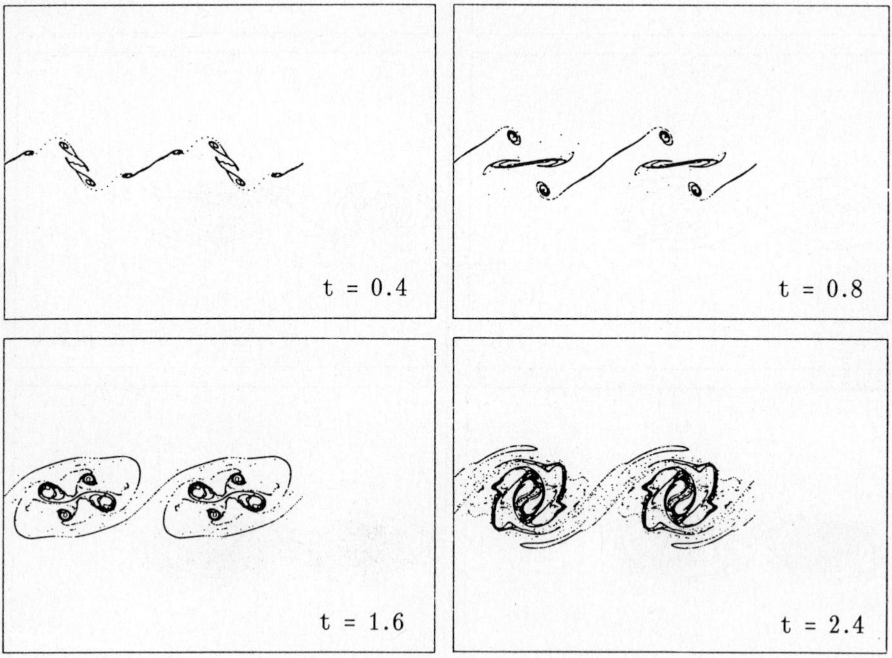

Fig. 4: Simulation of a Kelvin-Helmholtz instability

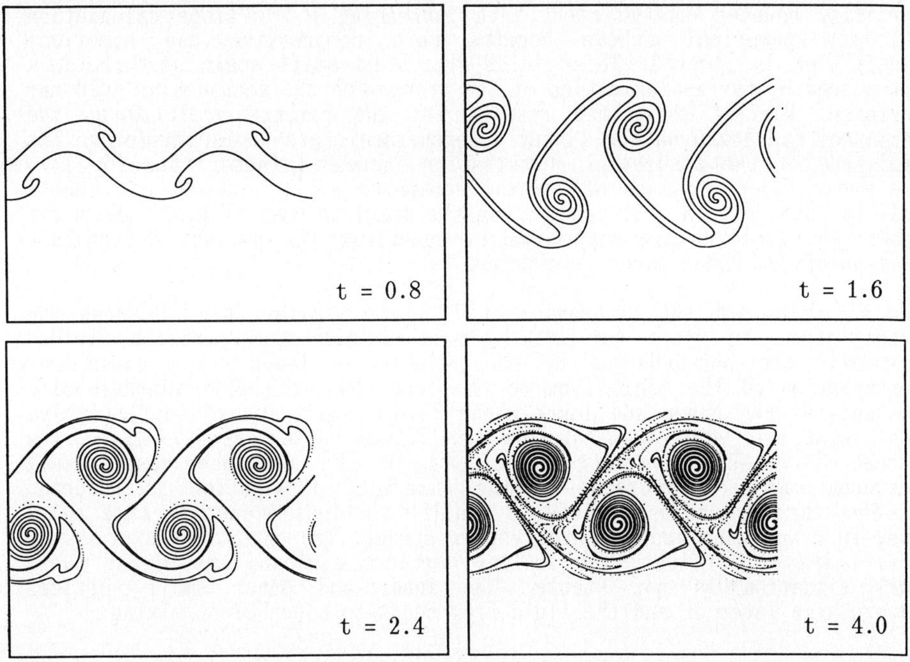

Fig. 5: Simulation of a transonic jet

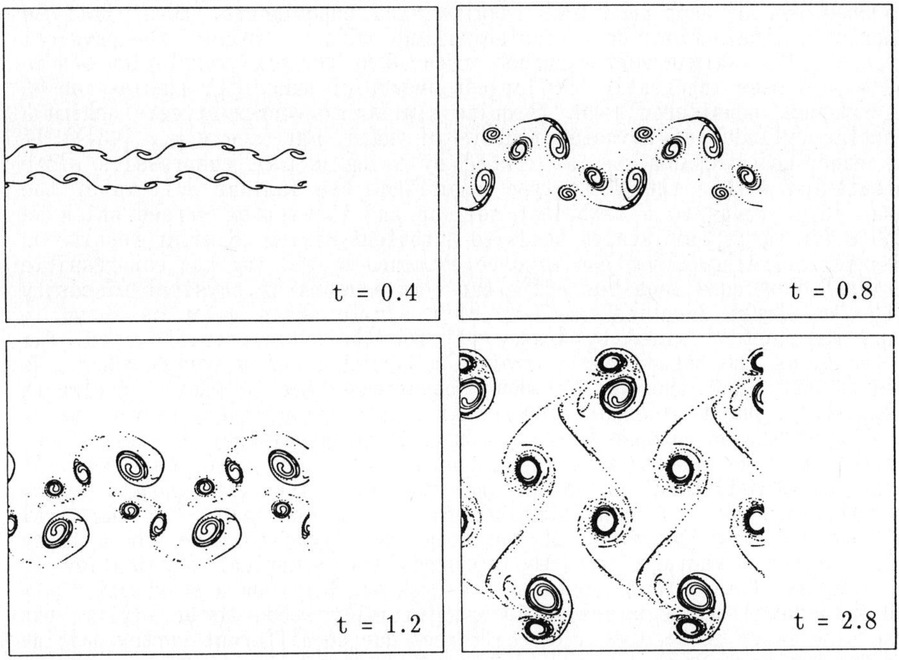

Fig. 6: Simulation of a transonic jet

This roll-up into a pair of spirals becomes obvious in the next plots. The vorticity becomes concentrated. With increasing l for slope calculation (9) the numerical scheme becomes more compressive and numerical dissipation is reduced. This yields that also small scale perturbations introduced by the approximation of the problem on the rectangular grid may increase. Fig. 4 shows the results for the parameters l=1.6 on the nonlinear and l=2.0 on the linearly degenerate characteristic fields. The amplitude of the sinusoidal perturbation increases faster than for l=1. The shear layer consists of several separated small vortices. No smooth roll-up into spirals will occur as in the previous case (Fig. 3) where the small perturbations are suppressed by numerical dissipation. A turbulent roll-up of the shear layer is obtained.

The situation for the transonic jet is quite similar. Fig. 5 shows the result when the slope for l=1 is used on all characteristic fields. Initially the perturbation of the velocity v leads to a sinusoidal perturbation of the fluid flow to the left. Due to the Kelvin-Helmholtz instability the upper and lower shear layers start to roll-up smoothly. Four asymmetric vortices occur forming a Kármán vortex street. This vortex street is stable for all times. As in the previous case, small perturbations will increase if the numerical dissipation is reduced. Besides the four main vortices, 8 smaller vortices occur. At time t=1.2 they form an inner and an outer vortex street. The outer one consists of four pairs of vortices. The smaller vortices surround the bigger ones. This situation is not stable. The inner and outer small vortices interchange later on and the fluid flow tends to a turbulent mixing.

CONCLUSIONS AND REMARKS

A comparison of our numerical results with experiments shows that the numerical dissipation has a stabilizing effect similar to physical viscosity. The Kármán vortex street produced by the high resolution scheme which possesses inherently the largest amount of numerical dissipation of all schemes, considered here, is quite similar to vortex streets behind a circular cylinder at Reynolds numbers of about 100 (see e.g., [43]). If the numerical dissipation is reduced by using a more compressive slope calculation within the MUSCL-type algorithm, the laminar roll-up of the shear layer tends to a turbulent roll-up and the vortex street which is stable for large time scales tends to turbulent mixing. Similar results as by numerical dissipation can also be obtained by solving the compressible Navier Stokes equations for different coefficients of physical viscosity [11]. An almost identical roll-up of a single shear layer as shown in Fig.2 has been obtained by Krasny [7] for the incompressible case. His calculations are based on the evolution equations of a vortex sheet. He pointed out that the vortex sheet equations have a short wavelength instability. Due to roundoff errors this instability restricts the number of discretization points for a given machine precision. In our direct simulation this short wavelength instability can also be seen. If numerical dissipation is small and the grid is fine enough, small perturbations due to approximation errors can increase. The numerical solution does not converge, if the step sizes tend to zero. On a finer grid smaller structures can be captured, the numerical dissipation is smaller, small additional vortices are created. E.g., on a grid with 500 x 500 grid zones the numerical solution initially seems to be similar but some time later it becomes quite different due to different vortex pairing and transition to turbulence.

The MUSCL-type algorithm used for our calculations has been fully optimized for the vector computers Cyber 205 and Fujitsu VP50. This could be done in a straightforward manner, because it is an explicit algorithm and contains no recursive elements. Some difficulties arise from the fact that for efficient vectorization the data should be stored contiguously within long vectors rather than two-dimensional arrays. Within the x-step of the splitting algorithm it is easy to use such one-dimensional arrays for the parts consuming most computer time, because the first index of the two-dimensional variables can be chosen to be the index of the inner loops. In the y-step this situation changes. Hence, we transpose all physical quantities at the beginning of the y-step. Thus all calculations can be performed in the same way (for the Strang-type splitting only one transposition of the physical variables per time step is necessary). By this technique we obtained a speed-up of 6-10 in comparison to Siemens 7890. The typical computer times for the calculations on a grid with 200 x 200 zones as presented above are 10 - 20 minutes, depending on the algorithm used for slope calculation. Within this time about 2000 time steps can been performed.

REFERENCES

[1] BIRKHOFF, G.: Helmholtz and Taylor instability, in Hydrodynamic Instability, G. Birkhoff, R.Bellmann, C.C.Lin (Eds.), AMS 1962.

[2] COLELLA, P., GLAZ, H.M.: Efficient solution algorithms for the Riemann problem for real gases, J.Comput.Phys 59 (1985), 264-289.

[3] COLELLA, P., WOODWARD P.R.: The piecewise parabolic method (PPM) for gas dynamical simulation, J.Comput.Phys. 54 (1984) 174-201.

[4] DYKE, M.van: An Album of Fluid Motion, Parabolic Press, Stanford, California 1982.

[5] EINFELDT, B.: On Godunov-type methods for gas dynamics, SIAM J.Numer. Anal. 25 (1988).

[6] HARTEN, A., LAX, P.D., LEER, B.van: On upstream differencing and Godunov-type schemes for hyperbolic conservation laws, Comm.Pure Appl.Math. 30 (1977), 611-638.

[7] KRASNY, R.: Desingularization of periodic vortex sheet roll-up, to appear in J.Comput.Phys.

[8] LEER, B.van: On the relation between the upwind-differencing schemes of Godunov, Engquist-Osher and Roe, SIAM J.Sci. Stat.Comput. 5 (1984), 1-21.

[9] LEER, B.van: Towards the ultimate conservative difference scheme V. A second-order sequel to Godunov's method, J.Comput.Phys.32 (1979) 101-136.

[10] MUNZ, C.-D.: On the numerical dissipation of high resolution schemes for hyperbolic conservation laws, to appear in J.Comput.Phys. 1988.

[11] MUNZ, C.-D., SCHMIDT, L.: Direkte numerische Simulation von Kelvin-Helmholtz Instabilitäten in kompressiblen Strömungen, to appear in ZAMM.

[12] ROE, P.L.: Approximate Riemann solvers, parameter vectors and difference schemes, J.Comput.Phys. 43 (1981), 357-372.

[13] STRANG, G.: On the construction and comparison of difference schemes, SIAM J.Numer.Anal.5 (1986), 506-517.

ON THE "FLUX-DIFFERENCE SPLITTING" FORMULATION

Maurizio Pandolfi
Dipartimento di Ingegneria Aeronautica e Spaziale
Politecnico di Torino, Torino, Italy

INTRODUCTION

The present paper refers to the numerical prediction of the inviscid compressible flows described by the Euler equations. We will focus our attention on unsteady flows (time marching), but any concept can be easily translated in terms of steady supersonic flows (space marching).

The Euler equations are obtained directly from the basic laws of the mass flow conservation, equilibrium of forces and the first principle of the thermodynamics. However, since they are hyperbolic, they also describe propagation of waves.

The hyperbolic character of the Euler equations is best revealed by a proper rearrangement of them, which I like to call the "formulation". There are different formulations, all related to the well-known method of characteristics. I intend the formulation as the step in which the wave-like nature of the problem is emphasized, by acknowledging that the evolution of the flow, at any grid point, is determined by the merging of signals propagated along characteristic rays. Since the phenomenon appears governed by advection equations (those which describe the convection of signals), these approaches are generally known in the literature as "upwind" methods.

Different upwind formulations have been proposed. Some are based on the quasi-linear form of the governing equations (lambda formulation, split coefficient matrix method,...), others on their divergence form (flux vector splitting, flux difference splitting,..). Non-linearity creates differences among such formulations. Indeed, they would coincide for a linearized version of the Euler equations. The main feature, common to all of them, is the attempt to interpret the usual thermodynamical properties and the velocity field in terms of waves or signals, each of them travelling along well definite paths. We note that a formulation can only be thought of for a system of equations. In fact, in a scalar problem, the only governing equation reveals immediately its hyperbolic character.

The present paper is addressed to the "flux difference splitting" formulation (FDS) and intends to review some of the forms under which it has been proposed in the literature. To the inexperienced reader, such forms may appear rather different and unrelated. Our aim is to look at them from a single standpoint and to put their common background into evidence.

In the FDS formulation, the discrete data given on a set of points is used to create a piecewise continuous distribution of values. In its simplest form, such a distribution is a sequence of constant values. Therefore a discontinuity of the flow properties occurs at the middle of each interval. Godunov [1] proposed to look at the evolution of this discontinuity along the hyperbolic coordinate, time. Its collapse generates three families of waves. A careful analysis of this system of waves provides information on how to interpret the difference of certain flow properties over each interval, in terms of propagating signals. In particular, the difference of the flux, needed for the numerical integration of the Euler equations in their original form, is split into terms contributing to the flow evolution at points located at the left and right ends of the interval.

The evolution of the discontinuity is described by the solution of a Riemann problem (RP). Subsequently, the difference of the flux is split on the basis of the resulting pattern of the waves. Finally, the terms obtained by the splitting are introduced into a numerical approximation. The latter can be characterized by the kind of assumed discretization (finite differences, finite volumes, finite elements..) and by the numerical scheme (explicit, implicit, prescribed order of accuracy...).

In his original presentation, Godunov pointed out two shortcomings in the procedure that he suggested. The first was the computational effort needed to solve the RP exactly, including tedious and time consuming numerical iterations to account for possible shocks. The second was the unsatisfactory level of accuracy in the results, due to the use of a first-order upwind scheme.

It is clear now that the practical failure of Godunov's otherwise brilliant idea was due to the inconsistency in using a first-order scheme of low accuracy after spending time in searching for an exact solution of the RP.

Consequently, efforts to circumvent the difficulty were attempted, along two different lines : (i) the search for approximate but efficient solvers of the RP, and (ii) more sophisticated interpretations of the RP, to reach satisfactory accuracy level.

We begin by reviewing Godunov's basic suggestion. Then we focus our attention on some approximate and efficient solvers. Finally we look at two sophisticated interpretations of the initial data, aimed at improving accuracy.

Some of these procedures were originally suggested with reference to the Lagrangian point of view, followed by a proper Eulerian remapping, whereas others were proposed in the Eulerian mode. Our present review is always based on the latter.

GODUNOV'S BASIC SUGGESTION

We consider the Euler equations which describe the one-dimensional unsteady flow, written in divergence form. With the usual notations, we have :

$$W_T + f_X = 0 \tag{1}$$

where:

$$W = \begin{vmatrix} \rho \\ \rho u \\ e \end{vmatrix} \qquad f = \begin{vmatrix} \rho u \\ p + \rho u^2 \\ u(p+e) \end{vmatrix}.$$

The initial data (time t_o, step K) are given at grid nodes, in particular at the points (N,N+1), which delimite the interval of length $D_N X = X_{N+1} - X_N$. With reference to Fig. 1, we consider two constant values (f_N, f_{N+1}), separated by a discontinuity at the middle point, $X_{N+1/2}$. The evolution of the discontinuity in time is described by the governing equations (Eq. 1). We expect then the generation of a system of waves, the two so-called "acoustic" waves (1,3) and the entropy wave (2). In general these waves propagate on the two sides of the interval, depending on the speeds $(u \mp a)$ for waves (1) and (3), and the speed (u) for wave (2). Two uniform regions (c,d) coexist with the initial ones ($a \equiv N$, $b \equiv N+1$).

Now we consider the difference of the flux over the interval $(D_N f)$, as split into three terms:

$$D_N f = f_{N+1} - f_N = (f_b - f_d) + (f_d - f_c) + (f_c - f_a). \tag{2}$$

Each of them is related to a wave, respectively $(f_b - f_d)$ to (3), $(f_d - f_c)$ to (2) and $(f_c - f_a)$ to (1). We define:

$$D_N f = \overleftarrow{D_N} f + \overrightarrow{D_N} f \tag{3}$$

where $\overleftarrow{D_N} f$ is built up using those terms (from Eq. 2) that belong to waves propagating to the left (negative speed) and $\overrightarrow{D_N} f$ uses terms belonging to waves moving to the right (positive speed). For the pattern of Fig. 1 :

$$\overleftarrow{D_N} f = f_c - f_a \qquad \overrightarrow{D_N} f = (f_d + f_c) + (f_b - f_d). \tag{4}$$

Once such a splitting has been done over all the intervals, we proceed to the numerical integration of Eq. 1. By confining our attention to the plain first-order difference algorithm, we have (Fig. 2):

$$\begin{aligned} W_N^{K+1} &= W_N^K + W_T \cdot DT = W_N^K - f_X \cdot DT = \\ &= W_N^K - (\overrightarrow{D_{N-1}} f + \overleftarrow{D_N} f) \cdot DT/DX \,. \end{aligned} \tag{5}$$

EXACT AND APPROXIMATE SOLUTIONS OF THE RIEMANN PROBLEM

As described above, Godunov's procedure is very simple. Unfortunately, the computational penalty to be paid to split Eq.2 according to the exact solution of the RP, is too heavy.

This is due to the non-linearity of the Euler equations (Eq.1), including the third one (energy), that accounts for variations of entropy. One or both of the "acoustic" waves (1,3) can be shocks with finite strength. In this case, the Rankine-Hugoniot equations can only be solved iteratively.

The problem has been circumvented by introducing "approximate solvers" of the RP, which, in fact, are approximate models of the RP, to be solved exactly.

Two have been the main suggestion proposed in the literature [2,3]. They have been developed indipendentely, almost at the same time. Both are widely used. Here I review these two approximate solvers and a third one developed later [5].

THE APPROXIMATE SOLVER PROPOSED BY ROE

The evolution of a discontinuity is described by the Euler equations (Eq.1). These can be written in the quasi-linear form :

$$W_T + A\, W_X = 0 \qquad (6)$$

where the matrix A depends on the variable w. Roe has proposed to linearize the physics described by Eq.6, by replacing the matrix A with a constant matrix $\tilde{A}$. The latter is determined on the basis of the flow properties at points N,N+1, which define the Riemann problem as detailed in [2].

The linear equations, which describe the evolution of the discontinuity are:

$$W_T + \tilde{A}\, W_X = 0 .$$

The constant matrix $\tilde{A}$ is constructed with proper average values $\tilde{u}$, $\tilde{H}$, $\tilde{a}$ as reported in [2]. The eigenvalues of $\tilde{A}$ (slopes of the characteristics) are:

$$\tilde{\lambda}_1 = \tilde{u} - \tilde{a} \; ; \; \tilde{\lambda}_2 = \tilde{u} \; ; \; \tilde{\lambda}_3 = \tilde{u} + \tilde{a} .$$

The corresponding right eigenvectors are:

$$r_1 = \begin{vmatrix} 1 \\ \tilde{u}-\tilde{a} \\ \tilde{H}-\tilde{u}\tilde{a} \end{vmatrix} \; ; \; r_2 = \begin{vmatrix} 1 \\ \tilde{u} \\ \tilde{u}^2/2 \end{vmatrix} \; ; \; r_3 = \begin{vmatrix} 1 \\ \tilde{u}+\tilde{a} \\ \tilde{H}+\tilde{u}\tilde{a} \end{vmatrix} .$$

The prescribed difference $D_N\, w = w_{N+1} - w_N$ can be written as:

$$D_N W = \alpha_1 r_1 + \alpha_2 r_2 + \alpha_3 r_3 .$$

Since $D_N w$ is known, we can evaluate the values of α_1, α_2, and α_3. Finally we have:

$$D_N f = \tilde{\lambda}_1 \alpha_1 r_1 + \tilde{\lambda}_2 \alpha_2 r_2 + \tilde{\lambda}_3 \alpha_3 r_3 .$$

The prescribed difference $D_N f$ appears as split into three terms, each associated to a corresponding eigenvalue. The latter represents the speed of the wave in the linear problem, constant over the (x,t) domain. By looking at Fig.3, it is easy to recognize that :

$$f_{c'} - f_a = \tilde{\lambda}_1 \alpha_1 r_1 \; ; \; f_{d'} - f_{c'} = \tilde{\lambda}_2 \alpha_2 r_2 \; ; \; f_b - f_{d'} = \tilde{\lambda}_3 \alpha_3 r_3$$

where the regions $(\tilde{c},\tilde{d})$ approximate the exact ones (c,d). Depending on the sign of the wave speed, each of these terms contribute to the split parts $(\overleftarrow{D_N} f, \overrightarrow{D_N} f)$ defined in Eq.3.

The most important feature due to the linearization is that the acoustic waves are now always concentrated on single lines of discontinuity. This picture is close to the correct one for "compression" waves, which, in the exact solution, are shock waves. On the contrary the situation is quite different for "expansion" waves, which, in the exact solution, are described by a diverging fan of characteristic lines, centered on the initial location of the discontinuity.

Unfortunately, the method fails in the following case. Assume that the initial RP is represented by an "expansion" shock at rest. This configuration is stable according to the conservation laws, Eq.1. On the contrary, the second principle of thermodynamics (not included in Eq.1) makes such a discontinuity collapse into an expansion fan, this being the only acceptable solution.

In the numerical procedure suggested by Godunov, the expansion fan determined by the exact solution is split into two terms, each of them contributing to the opposite parts appearing in Eq.3. Such contributions, expressed in terms of the flux, are equal and opposite. Consequently, the numerical result is consistent with the correct collapse of the expansion shock.

In the linearized solver, however, the expansion fan is approximated by a single vertical line and the numerical collapse of the expansion shock does not occur. Some artificial remedy can be found, as reported in [2], in order to satisfy the "entropy conditions". On the contrary, a steady shock is perfectly described by the numerical procedure.

Once the splitting has been done over each interval, we can proceed to the integration of the Euler equations. By using the first-order scheme, we find again Eq.5.

THE APPROXIMATE SOLVER PROPOSED BY OSHER

An alternate approximate solver has been proposed in [3]. The following presentation and interpretation of this procedure looks rather different from the original one, but it can provide a better understanding of this methodology in the framework of the FDS formulation.

Instead of solving a RP, as in the previous approaches, we can look at the system of three waves merging at $X_{N+1/2}$ at time t_o, thus generating the discontinuity defined by the initial data. The picture of Fig.4 shows these waves (1",2",3"), which separate the regions (c",d"). Let us suppose that we know how to predict these waves. At this point we may assume that the strenght of a wave is not largely affected by the crossing of other waves. This would be true in a linear problem, and also in the non linear homentropic problem, where the strength of a wave is given by the difference of the appropriate Riemann invariants across the wave. Therefore it would follows that:

$$f_b - f_{d'} \simeq f_{c''} - f_a \;;\quad f_{d'} - f_{c'} \simeq f_{d''} - f_{c''} \;;\quad f_{c'} - f_a \simeq f_b - f_{d''}.$$

Furthermore, we assume that the new waves (1',2',3'), coming out from the discontinuity, propagate in the same directions as the old ones (1",2",3"). Once the pattern of the new waves is known, we can soon proceed to the splitting of $D_N f$, as required by Eq.3.

We have now to predict the pattern of the waves (1",2",3") and the corresponding regions (c",d"). Following [3], we suppose that the acoustic waves can be considered as isentropic. This is certainly correct if these waves are expansions, but only approximate in the case of shocks, owing to the entropy generated through them. The only entropy variation is accounted by wave (2") and corresponds to the initial discontinuity of the entropy as provided in the initial data. It is worthwhile to remark that, due to this assumption, the RP retain its original non-linearity.

The solution of the RP requires the evaluation of six unknowns, two thermodynamical properties and the velocity in the regions (c",d"). They will be computed by matching the regions (a,c",d",b) with the conditions which hold over the waves (1",2",3"). Through the acoustic waves (1",3"), we conserve the entropy (basic assumption) and the Riemann invariant (either $2/(\gamma-1)\cdot a+u$ or $2/(\gamma-1)\cdot a-u$) which remains constant along the characteristics (u+a or u-a) running across the wave (from d" to b or from c" to a). On the contact surface (2"), we impose the continuity of pressure and velocity, as usual. Such conditions, formulated as follows:

$$\frac{2}{\gamma-1} a_{c''} - u_{c''} = \frac{2}{\gamma-1} a_a - u_a \quad ; \qquad S_{c''} = S_a$$

$$u_{c''} = u_{d''} \;;\quad p_{c''} = p_{d''} \Rightarrow a_{c''} = a_{d''} \exp\left(\frac{S_{c''} - S_{d''}}{2\gamma}\right)$$

$$\frac{2}{\gamma-1} a_{d''} + u_{d''} = \frac{2}{\gamma-1} a_b + u_b \quad ; \qquad S_{d''} = S_b$$

are easily interpreted in a Fortran code. Once the flow properties in the regions (c", d") are known, we can evaluate the fluxes $f_{c''}$ and $f_{d''}$ and find the splitting required by Eq.2. In the next step, we evaluate the directions of propagation of the waves (1",2",3"), so that we can determine to which parts of Eq.3 ($\overleftarrow{D}_{\sim} f$, $\overrightarrow{D}_{\sim} f$) the split terms contribute.

We note that the acoustic waves can be expansion or compression waves. Owing to the assumption of isentropy, they are diverging or converging fans, respectively. Just in opposition to the previous approximate solver, the expansions are now described correctly, whereas the shocks are described by converging fans, instead of single lines of discontinuity.

A very interesting feature of this approximate solver appears in the case of a sonic transition (vertical characteristic) imbedded inside an acoustic fan. The case is not an academic one: to the contrary it occurs any time we have a steady shock and prevents the formation of an expansion shock. In these cases, in the spirit of Godunov's idea, the content of the fan will be split into two additional terms which contribute to the opposite parts appearing in Eq.3. The reader interested on this point (one of the most important, since we are interested in predicting flows with shocks) can find a wider and more detailed presentation in [4].

Once these splittings have been operated and all the terms of Eqs.2,3,4 are found, we can proceed to the integration (for example, see Eq.5 for the first-order scheme).

A THIRD APPROXIMATE SOLVER

A third solver, somehow located between the previous ones, has been proposed later in [5]. It follows basic Godunov's idea of predicting the evolution of the initial discontinuity directly (as in Roe's solver), but it assumes that the acoustic waves are isentropic (as in Osher's solver). From a practical point of view, the procedure follows the solution proposed in [3], but the order in which the waves are crossed in going from region a to region b has been reversed [5].

We think worthwhile to note that the assumption used in the previous approach [3] and, later on, in the present one [5], is not new in the prediction of unsteady flows and has been suggested, many years ago, in a quite different scientific and technical context. For example, in a classical textbook written before the computer era [6], an iterative procedure is suggested, to solve the problem of interacting shocks. In the tentative initial configuration, the compression waves are assumed to be isentropic. It is also shown that, most of the times, the initial and final configurations are not very different.

In conclusion, the approximation introduced by the solver affects only the relative percentage of the terms in which the difference of the flux has been split. It does not affect the

Euler equations which will be integrated in the final step. With reference to this specific point, let us quote from [2] : "....the expense of producing an accurate solution of the Riemann problem would only be justified if the abundance of the information, which is thereby made available, could be put to some rather sophisticated use". In other words, it pays to spend time in obtaining the exact solution ONLY if in the following numerical work we are capable to profit of its results with a suitable accurate scheme. For the first-order scheme (Eq.5), and also for second order schemes, the experience has proved that the above approximate solvers provide solutions sufficiently accurate to avoid any penalty in the following numerical scheme.

HOW TO IMPROVE THE ACCURACY

The original procedure proposed by Godunov is based on a first order scheme, as shown in Eq.5. To improve the accuracy, we can use the ingredients provided by the solution of the RP (exact or approximate), that is the split terms of the difference of the flux, and incorporate them in a more sophisticated numerical scheme.

A different approach has been followed in two very accurate procedures, known as the "Piecewise Parabolic Method" (PPM) and the MUSCL method, respectively. Here the accuracy is improved in the step related to the formulation rather than in the following numerical scheme. The interpolation of the initial data is, indeed, more sophisticated than in the piecewise constant value distribution. The definition of the RP follows, but little care is taken to speed up its solution. Once the splittings are obtained, the updating of the solution follows a numerical scheme, similar to the one shown in Eq.5. The accuracy of the results, however, is much better than in the first order scheme which follows Godunov´s interpretation, due to the previous careful work in interpreting the initial data and in defining appropriate RPs.

THE PIECEWISE PARABOLIC METHOD

The procedure is reported in detail in [7]. Here we point out some of the main features.

We distinguish two steps. The first one is dedicated to the interpretation of the initial data as parabolic arcs in each cell. The second step regards the definition of a suitable RP, on the basis of the previous interpretation.

We start by reviewing the first step. With reference to Fig.5, we look for the arc of parabola which describes a general flow property φ over the cell ($X_{N+1/2}-X_{N-1/2}$) about point N. The coefficients needed to determine such an arc are evaluated by requiring that the average of this distribution equals the prescribed value φ_N and by imposing the values at $X_{N+1/2}$ and $X_{N-1/2}$.

Such values are obtained through a particular interpolation procedure. For example, in order to get the value $\varphi_{N+1/2}$ at $X_{N+1/2}$ (see Fig.6), first we evaluate the integral $\phi = \int \varphi \, dX$ numerically on the basis of the initial values of φ_N, provided at the symmetric points N-1, N, N+1, N+2. Once the values of ϕ are found at the five locations $X_{N+J+1/2}$ ($J = \mp 2, \mp 1, 0$), we look for the quartic polinomial ($\overline{\phi}$) passing through these points and we evaluate its analytic derivative $\overline{\varphi} = d\overline{\phi}/dX$. Then we compute $\overline{\varphi}$ at N+1/2. This represents the value $\varphi_{R,N}$ which delimits the arc of parabola about the point N, on the right-hand side. A similar prediction is done for the left-hand side, that is for $\varphi_{L,N}$. Since these values are also needed for the neighboring arcs, we have :

$$\varphi_{L,N} = \varphi_{R,N-1} = \varphi_{N-1/2}$$

$$\varphi_{R,N} = \varphi_{L,N+1} = \varphi_{N+1/2}$$

and so on for any other interval. At this stage the initial distribution looks as a sequence of arcs of parabola, continuously connected to each other.

Now we examine each arc and we modify those which do not look monotonic. It is at this point that we may introduce (and we certainly do it in the proximity of a captured shock) a jump at the middle of some interval ($X_{N+1/2}$), since the two arcs merging here can be modified and then $\varphi_{R,N}$ differs from $\varphi_{L,N+1}$. Further readjustments can still be introduced to achieve more definite and sharp transitions in the final numerical results through captured discontinuities, particularly in the case of contact surfaces. The distribution over a general interval (X_N, X_{N+1}) looks now as in Fig.7.

At this moment, we proceed into the second step. The picture of Fig.7 does not look like the RP configuration of Fig.1. Indeed, the relatively small discontinuity is closed by two non uniform flow regions. Therefore we try to define an "appropriate" RP, that is, to determine the two levels of the flow properties which define a RP as close as possible, to the configuration of Fig.7 in terms of wave propagation. Such values are called $\varphi_{N+1/2,L}$ and $\varphi_{N+1/2,R}$ and the RP is defined as shown in Fig.8. The equivalence of this RP with the non uniform regions configuration of Fig.7 is required to hold for the time interval (DT) determined by the CFL rule. It is specifically in the search of this appropriate RP that the wave-like nature of the phenomena is introduced in the procedure, since, up to now, no reference was made to it.

It is convenient to recall the definition of the Riemann variables, expressed in terms of pressure, density and velocity. Their differential form is given by :

$$dR_1 = \frac{dp}{a\rho} - du \qquad dR_2 = d\left(\frac{1}{\rho}\right) + \frac{dp}{\gamma p \rho} \qquad dR_3 = \frac{dp}{a\rho} + du .$$

The first and the third variables represent the "acoustic" signals, whereas the second one is related to the entropy. We can define their finite forms approximately :

$$R_1 = \frac{p}{a_N \rho_N} - u \; ; \; R_2 = \frac{1}{\rho} + \frac{p}{\gamma p_N \rho_N} \; ; \; R_3 = \frac{p}{a_N \rho_N} + u .$$

The variables (R_1, R_2, R_3) are certainly much more significant than the original ones (p, ρ, u) to emphasize the propagation of waves.

Let us now focus our attention on the search of $\varphi_{N+1/2,L}$, the left hand side value of the RP, and consider the distribution of φ over the cell about point N (see Fig.9.1). First we evaluate in which directions propagate the waves which belong to this cell. Therefore, we look at the sign of the speeds of the characteristics at point N ($u \mp a$,u). Let us suppose that ($u + a$) is positive, whereas (u) and ($u - a$) are negative. In this case we expect that the R_3 signal only flows toward the $X_{N+1/2}$ (R_2 and R_1 propagate in the opposite direction). Note that during the time interval DT, only the section ($X_{N+1/2} - (u_N + a_N)$*DT) of the initial distribution contributes to the evolution at $X_{N+1/2}$ (see Fig.9.2). We compute the averages, p_3, ρ_3 , and u_3 over this section and the average of the third signal R_3 there:

$$R_{33} = \frac{p_3}{a_N \rho_N} + u_3 .$$

This is the only wave impinging on $X_{N+1/2}$, coming from the left. Therefore, the left hand side values of the RP of Fig.8 must provide :

$$(R_3)_{N+1/2,L} = R_{33}$$

which is satisfied by assuming:

$$p_{N+1/2,L} = p_3 \; ; \; \rho_{N+1/2,L} = \rho_3 \; ; \; u_{N+1/2,L} = u_3 .$$

If (u_N) is also positive, but ($u_N - a_N$) still negative, we compute the average values (p_2, ρ_2) over the section ($X_{N+1/2} - u_N$*DT) which are related to the average value of the entropy carried on $X_{N+1/2}$ during the interval, DT. These values define :

$$R_{22} = \frac{1}{\rho_2} + \frac{p_2}{\gamma p_N \rho_N} .$$

In this case the left hand side values of the RP (Fig.8) must provide :

$$(R_3)_{N+1/2,L} = R_{33} \; ; \; (R_2)_{N+1/2,L} = R_{22}$$

which are satisfied by assuming:

$$p_{N+1/2,L} = p_3 \quad ; \quad u_{N+1/2,L} = u_3$$

$$\left(\frac{1}{\rho}\right)_{N+1/2,L} = \left(\frac{1}{\rho_2}\right) + (p_2 - p_{N+1/2,L})/(\gamma p_N \rho_N) .$$

Finally, for a supersonic flow at point N ($u_N/a_N > 1$), we also evaluate a third average (p_1, ρ_1, u_1) over the segment ($X_{N+1/2} - (u_N - a_N)$*DT) (see fig.9.4). With these values we compute:

$$R_{11} = \frac{p_1}{a_N \rho_N} - u_1 .$$

Now the left hand side values of the RP must provide three conditions:

$$(R_3)_{N+1/2,L} = R_{33} \; ; (R_2)_{N+1/2,L} = R_{22} \; ; (R_1)_{N+1/2,L} = R_{11}$$

which are satisfied by:

$$p_{N+1/2,L} = \frac{1}{2}\left[p_3 + p_1 + a_N \rho_N (u_3 - u_1)\right]$$

$$u_{N+1/2,L} = \frac{1}{2}\left[(p_3 - p_1)/(a_N \rho_N) + u_3 + u_1 \right]$$

$$(1/\rho)_{N+1/2,L} = (1/\rho_2) + (p_2 - p_{N+1/2,L})/(\gamma p_N \rho_N).$$

Similar operations must be repeated at the right hand side of the next interval, which extends from $X_{N+1/2}$ to $X_{N+3/2}$. The attention is here focused on the waves that propagate leftward. Therefore, we will only consider contributions from the waves with negative speed. The averaging process is the mirror image of the one just considered; a set of values:

$$p_{N+1/2,R} \; ; \quad \rho_{N+1/2,R} \; ; \quad u_{N+1/2,R}$$

is obtained which define the RP of Fig.8 on its right hand side.

The appropriate RP is now completely defined and the full picture of Fig.8 is determined quantitatively. Let us note that the averages obtained for each wave on the pertinent section are crucial for predicting the advection of each signal accurately and, at the same time respecting the domains of dependence.

Two further steps are still needed to complete the procedure. First, we must solve the RPs. As we have mentioned previously, no particular care is paid to speed up this step. Then, we must update the solution in time according to the scheme of Eq.5. The numerical scheme seems to have only a first-order accuracy, but the accuracy is much higher, because of the previous sophisticated procedure followed in determining the RPs.

THE "MUSCL" METHOD

Similarly to the PPM, this approach emphasizes the interpretation of the initial data and uses a simple algorithm in updating the solution in time [8].

The initial data are not only the values of the flow properties φ_N at each grid point N, as in the PPM. In addition a further variable $(\mathfrak{D}_N\varphi)$ is given, to be updated at each computational step by an integration procedure, just as φ_N. We interpret these initial data $(\varphi_N, \mathfrak{D}_N\varphi)$ by means of a piecewise linear distribution over a cell, as shown in Fig.10. Its end values at $X_{N-1/2}$ and $X_{N+1/2}$ equal

$$\varphi_{L,N} = \varphi_N - \mathcal{D}_N\varphi/2 \qquad \varphi_{R,N} = \varphi_N + \mathcal{D}_N\varphi/2$$

respectively. In the middle of the interval (X_N, X_{N+1}), we find the discontinuity

$$\varphi_{N+1/2,R} - \varphi_{N+1/2,L} \quad \text{equals} \quad \varphi_{L,N+1} - \varphi_{R,N}$$

as we can see by comparing Fig. 10 and Fig. 11. The discontinuity separates two regions with flow properties varying with X linearly and the initial configuration looks like a sequence of neighboring segments. This is not the end of the initial step. In regions of strong gradients, indeed, the slopes of the segments must be readjusted not to generate numerical oscillations in the integration step which follows.

The most peculiar feature of the MUSCL method consists in working out the initial value problem directly from the distribution shown in Fig. 11. In the following analysis, we can distinguish between the initial, instantaneous breakdown of the discontinuity and the analysis of the propagation of waves proceeding from non-uniform regions.

The former is worked out as a RP. An almost exact solution is developed. Let $\varphi^*_{N+1/2}$ be the flow properties at $X_{N+1/2}$, immediately after the collapse of the discontinuity. The tools for predicting the second phase (smooth interaction of waves) are the classical compatibility equations along characteristics. With reference to the (u-a), (u), (u+a) waves, they are written as:

$$p^*_T - (a\rho)\,u^*_T = -(u-a)\left[p^*_X - (a\rho)\,u^*_X\right]$$

$$p^*_T - \gamma\frac{p}{\rho}\,\rho^*_T = -u\left[p^*_X - \gamma\frac{p}{\rho}\,\rho^*_X\right]$$

$$p^*_T + (a\rho)\,u^*_T = (u+a)\left[p^*_X + (a\rho)\,u^*_X\right]$$

respectively. These equations hold in the non-uniform region following the breakdown of the discontinuity, at $X_{N+1/2}$. The unknowns are the time derivatives. The brackets containing the space derivatives do not change across suitable chosen characteristics; therefore such derivatives can be computed using the initial distribution of values. For example, in the case of Fig. 12, with (u+a) positive and (u) and (u-a) negative, it would result:

$$\left[p^*_X - (a\rho)\,u^*_X\right] = \left[p_X - (a\rho)\,u_X\right]_b$$

$$\left[p^*_X - \gamma\frac{p}{\rho}\,\rho^*_X\right] = \left[p_X - \gamma\frac{p}{\rho}\,\rho_X\right]_a$$

$$\left[p^*_X + (a\rho)\,u^*_X\right] = \left[p_X + (a\rho)\,u_X\right]_a$$

where the space derivatives on the right hand sides are computed from the initial data.

We are now ready to update in time both the properties φ_N and $\mathcal{D}_N\varphi$. By quoting [8], we point out that "the slopes ($\mathcal{D}_N\varphi$) are independent of the average values (φ_N); they cannot be derived from the latter, must be stored separately" and also updated in time separately.

First we evaluate the flow properties at $X_{N+1/2}$, at the time T+DT/2 (step K+1/2) and at the final time, T+DT (step K+1), by the simple algorithm:

$$\varphi_{N+1/2}^{K+1/2} = \varphi_{N+1/2}^{*} + (\varphi_{N+1/2}^{*})_T \cdot DT/2$$
$$\varphi_{N+1/2}^{K+1} = \varphi_{N+1/2}^{*} + (\varphi_{N+1/2}^{*})_T \cdot DT . \tag{7}$$

We note that this integration is carried out on the "non conservative" primitive variables (p, ρ ,u), which appear in the compatibility equations. Then we compute the difference of the flux:

$$Df = f(\varphi_{N+1/2}^{K+1/2}) - f(\varphi_{N-1/2}^{K+1/2}).$$

According to the previous Eqs.2,3,4 and to Fig.2, we recognize that:

$$Df = \overrightarrow{D_N} f + \overleftarrow{D_{N+1}} f .$$

We note that such a splitting refers to the intermediate time (T+DT/2). This improves the accuracy in time.

So far, we did not integrated the Euler equation (Eq.1) yet; we have only prepared the necessary ingredients. As for the PPM, we go to the plain scheme of Eq.5, which again seems to provide first-order accuracy, but only apparently . From the new values of the "conservative" variables w_N^{K+1} , we decode the values of the primitives ones φ_N^{K+1} , needed as initial data in the following step.

As pointed out previously, we also have to update the slopes ($\mathfrak{D}_N^{K+1}\varphi$) in time. They are computed as follows :

$$\mathfrak{D}_N^{K+1} \varphi = \varphi_{N+1/2}^{K+1} - \varphi_{N-1/2}^{K+1}$$

where the terms in the right-hand side are the ones obtained in Eq.7. We have worked out two independent integrations on φ_N and $\mathfrak{D}_N \varphi$. As pointed out in [8], "this approach potentially has the effect of a mesh refinement of a factor of two".

CONCLUDING REMARKS

In closing this short review, I would like to add some final remarks.

First, let us recall that the same ideas can be used for steady supersonic flows, to the expense of some formal complication.

A second comment refers to the extension to multidimensional flows. We know that the 1D hyperbolic problem (1D unsteady or 2D steady supersonic flows) is a very well defined problem, dealing with three compatibility equations, three signals, and three unknowns. In the multidimensional case the situation is not so clear. The number of the unknowns is increased by one or two additional components of the velocity, whilst we find a single or double infinity of available compatibility equations, characteristic rays, and signals. Some criteria have to be devised in order to select the more significant information. These are related to the choice in defining the grid and the velocity components. We would like to point out that these problems are common to any upwind formulation (whether the quasi linear or the conservative form of the governing equations is used), as well as to the original method of characteristics.

A third remark is of practical nature. Anyone of the above mentioned approximate solvers has been widely incorporated in codes for practical applications (nontrivial geometries and curvilinear grids). On the contrary, the last two accurate procedures, which require a rather heavy work in the interpretation of the initial data, have been used, to our knowledge, only for the understanding of basic physical phenomena and confined to regular Cartesian grids.

ACKNOWLEDGMENT

I would like to extend my warmest thanks to Prof. G. Moretti for his comments and suggestions during the final revision of this paper.

REFERENCES

[1] S.K.Godunov,"A finite difference method for the numerical computation of discontinuous solutions of the equations of fluid dynamics", Math. Sb., Vol 47, 1959.

[2] P.L.Roe, "Approximate Riemann solvers, parameters vectors and difference schemes", Journal of Computational Physics Vol.43, 1981.

[3] S.Osher and F.Solomon, "Upwind difference schemes for hyperbolic systems of conservation laws", Mathematics of Computation, Vol.38,1982.

[4] M.Pandolfi,"Upwind formulations for the Euler Equations", Lecture Series 1987/04 on Computational Fluid Dynamics, Von Karman Institute for Fluid Dynamics, March 1987.

[5] M.Pandolfi, "A contribution to the numerical prediction of Unsteady flows", AIAA J. Vol.22, 1984.

[6] G.Rudinger,"Nonsteady duct flow : wave diagram analysis", Dover Publ. 1969.

[7] P.Colella and P.R.Woodward, "The piecewise parabolic method for gas-dynamical simulations", Journal of Computational Physics, Vol.54, 1984.

[8] B.Van Leer, "Towards the ultimate conservative difference scheme. V. A second order sequel to Godunov´s method", Journal of Computational Physics, Vol.32, 1979.

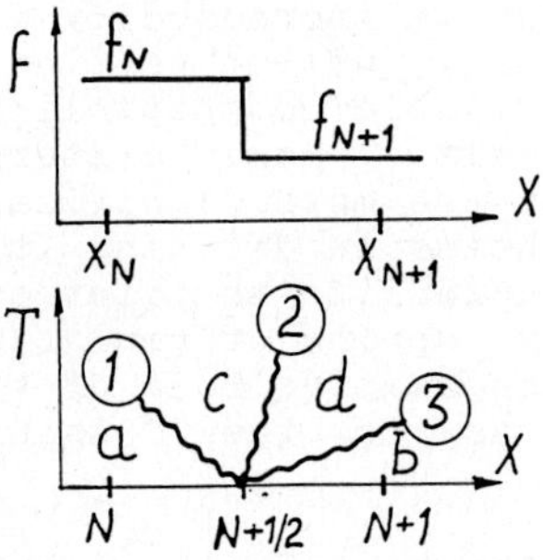

Fig.1

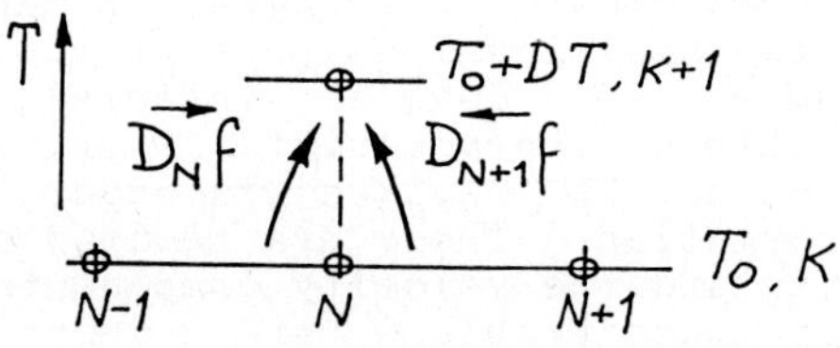

Fig.2

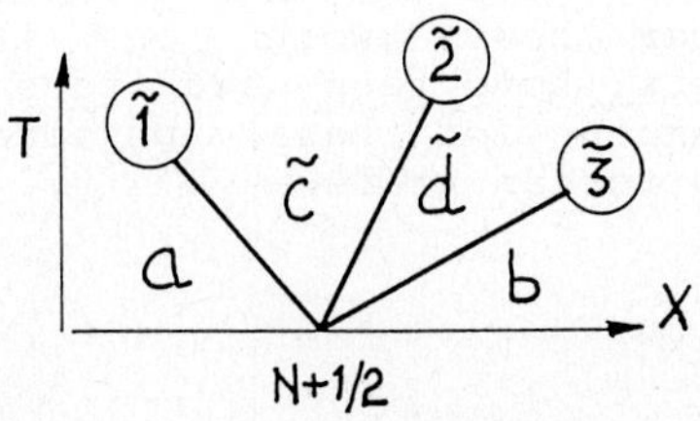

Fig.3

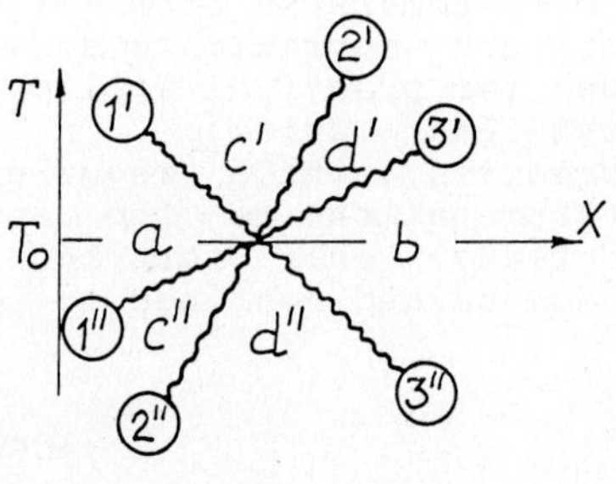

Fig.4

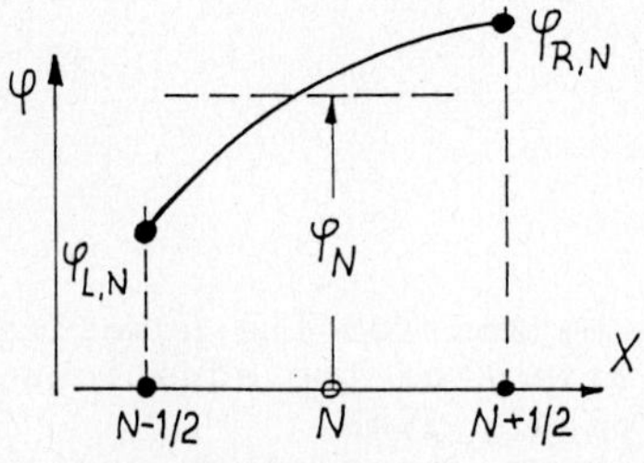

Fig.5

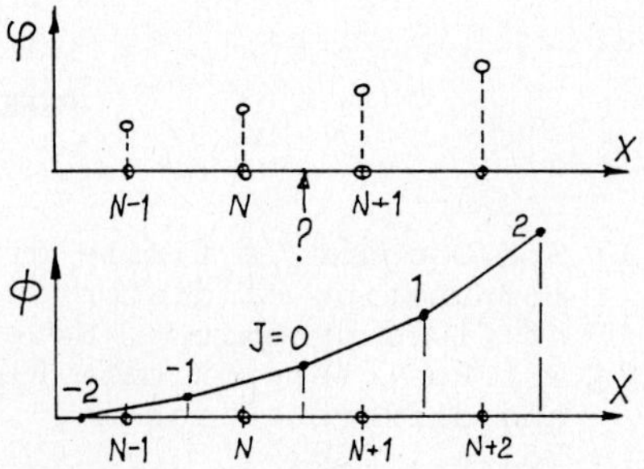

Fig.6

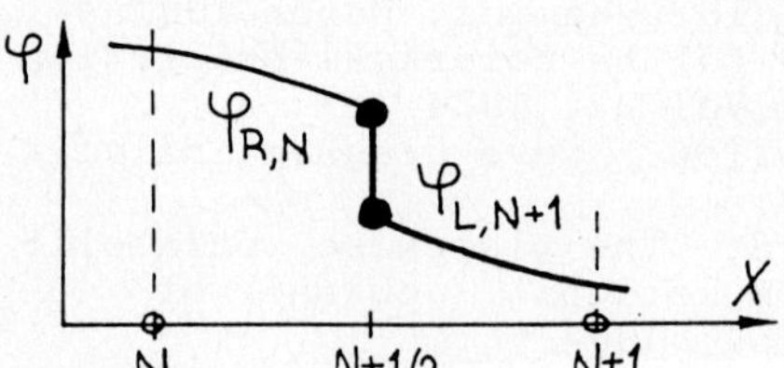

Fig.7

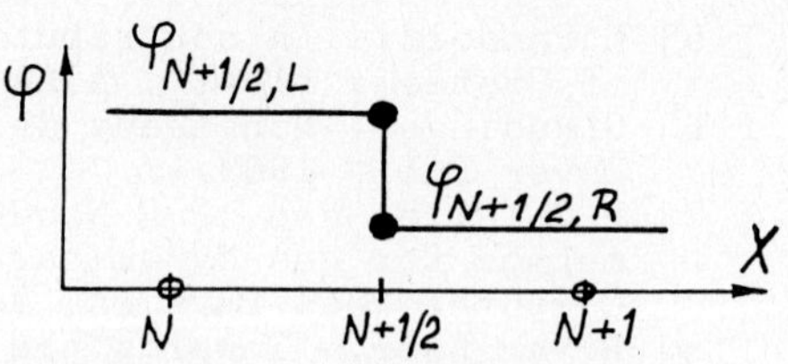

Fig.8

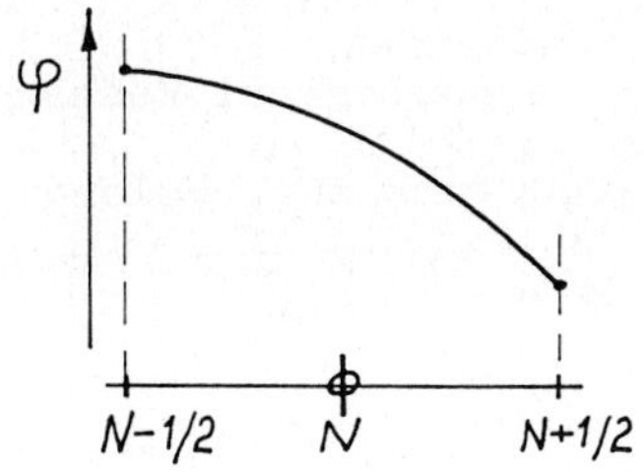

Fig. 9.1

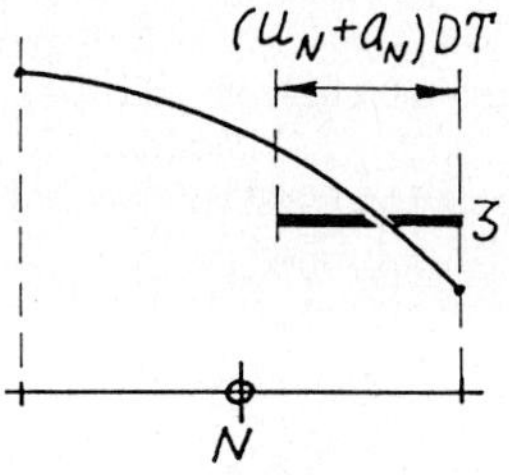

Fig. 9.2

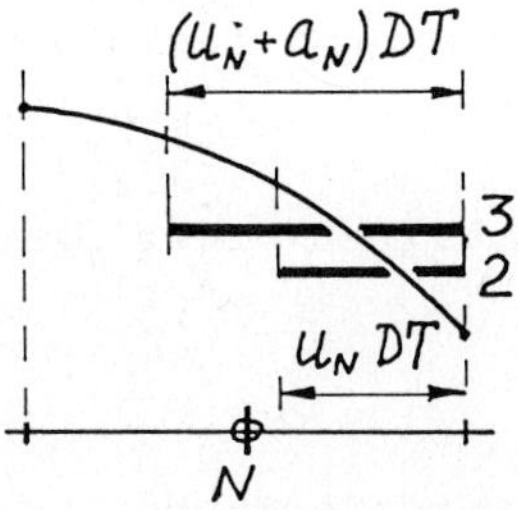

Fig. 9.3

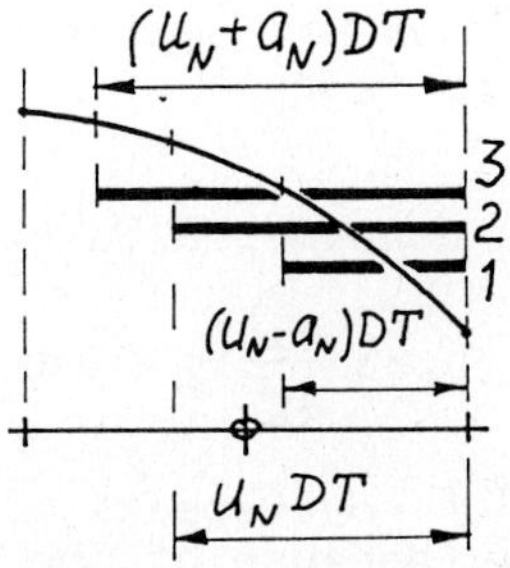

Fig. 9.4

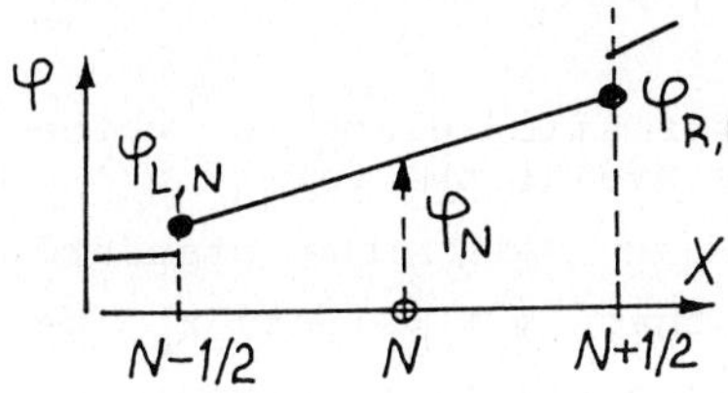

Fig. 10

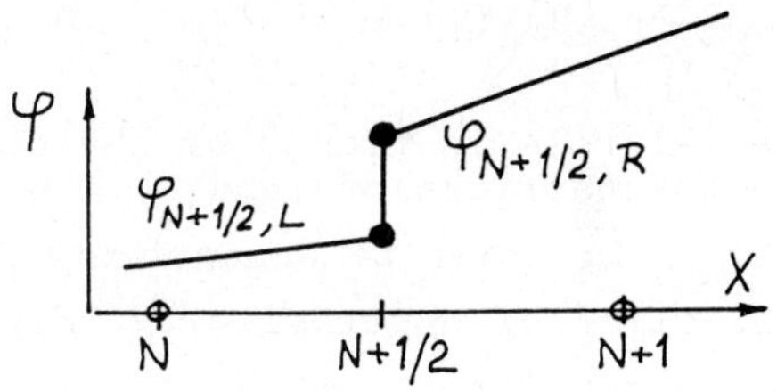

Fig. 11

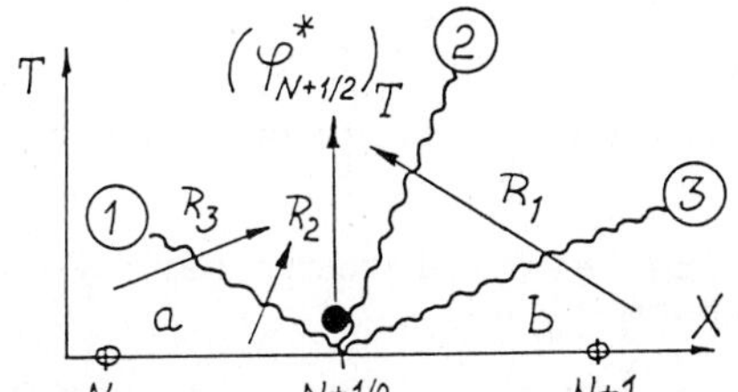

Fig. 12

ON OVERDETERMINED HYPERBOLIC SYSTEMS

Zbigniew Peradzyński
Institute of Fundamental Technological Research
Polish Academy of Sciences
Świętokrzyska 21, 00-049 Warszawa, Poland

Summary

The notion of hyperbolicity is generalized for overdetermined quasilinear systems. We define the system as hyperbolic if it is compatible and if its jet manifold is generated by the characteristic elements. The criterions of compatibility are also formulated.

Let us consider a quasilinear first order system of PDE's

$$\sum_{\nu, j} A_j^{s\nu}(u,x) \frac{\partial u^j}{\partial x^\nu} = f^s(u,x), \qquad \begin{array}{l} s = 1,\dots,m, \\ j = 1,\dots,l, \\ \nu = 1,\dots,n. \end{array} \tag{1}$$

When $m > l$ one speaks of an overdetermined system. Maxwell equations (because of the constrains div E = 0, div B = 0) or vorticity equations in hydrodynamics can be viewed as prototypes of such system.

For any u_o, x_o we define a linear space $\mathfrak{I}(u_o,x_o)$ of integral elements of the homogeneous system as composed of all $l \times n$ matrices (p_ν^j) satisfying

$$\sum_{\nu, j} A_j^{s\nu}(u_o,x_o)\, p_\nu^j = 0. \tag{2}$$

Special role is played by the characteristic elements which are the matrices of rank 1, i.e. they are of the form $p_\nu^j = \mathbf{X}^j \lambda_\nu$. In order to demonstrate it,let us recall the standard definition of hyperbolicity. Suppose that $m = l$. Let x_1 be the chosen direction. We say that system (1) is hyperbolic at (u_o,x_o) in the direction x_1 if and only if:

1° For any $\lambda = (0,\lambda_2,\dots,\lambda_n)$, $\lambda_\alpha \in \mathbb{R}$, the characteristic polynomial

$$W(\lambda) = \det\| A_j^{s1}(u_o,x_o)\lambda_1 + A_j^{s\alpha}(u_o,x_o)\lambda_\alpha \|, \qquad \alpha = 2,\dots,n,$$

has l (including multiplicities) real roots for λ_1, say $\lambda_1^1, \lambda_1^2, \dots, \lambda_1^l$.

2° The corresponding eigenvectors, say $\underset{1}{\mathbf{X}},\dots,\underset{l}{\mathbf{X}}$ span an l-dimensional space, i.e.

$$\left[A^{\cdot 1} \lambda_1^i\right] \underset{i}{\mathbf{X}} + \left[A^{\cdot \alpha} \lambda_\alpha\right] \underset{i}{\mathbf{X}} = 0, \qquad \alpha = 2,\dots,n, \quad i = 1,\dots,l,$$

$$\dim\left\{ \underset{1}{X}, \ldots, \underset{l}{X} \right\} = l .$$

As follows from this definition, for any $(\lambda_2, \ldots, \lambda_n) \in \mathbb{R}^{n-1}$ we have l different characteristic elements of the form $p = \left[X^j \lambda_\nu\right]$ where $\lambda = (\lambda^i_1, \lambda_2, \ldots, \lambda_n)$ and X is a corresponding eigenvector. Matrices p satisfy Eqs.(2).

Any system of partial differential equations can be converted into a Pfaffian system. For example, system (1) can be written as a system of one-forms

$$du - (A^{\cdot 1})^{-1} f \, dx^1 = - (A^{\cdot 1})^{-1} A^{\cdot \alpha} \, p_\alpha \, dx^1 + p_\alpha \, dx^\alpha \qquad (3)$$

in the space of $\{u, x, p_\alpha\}$ of dimension $l + n + (n-1)l$, $\alpha = 2, \ldots, n$, $p_\alpha = (p_\alpha{}^j)$, $p_\alpha{}^j \in \mathbb{R}$.

It is interesting to note that if system (1) is hyperbolic then the space $\mathfrak{I}(u_o, x_o)$ is generated (as a linear space) by the characteristic elements. In other words there exists a following representation of (3)

$$du - \tilde{f} \, dt = \xi_1 \underset{1}{X} \otimes \lambda^1 + \xi_2 \underset{2}{X} \otimes \lambda^2 + \cdots + \xi_n \underset{n}{X} \otimes \lambda^n \qquad (4)$$

where the matrices of $\underset{\alpha}{X} \otimes \lambda^\alpha$ satisfy Eqs.(2), and λ are treated as differential forms $\lambda = \lambda_\nu dx^\nu$, $\tilde{f} = (A^{\cdot 1})^{-1} f$ and $k = (n-1) \times l$.

The above property will be used in order to generalize the notion of hyperbolicity for an overdetermined system. However, this property is insufficient to make a good definition. If $k < (n-1) \times l$, the system is overdetermined and, in general, it may be incompatible. By adding compatibility conditions representation (4) can loose its validity. Indeed, these conditions can lead to some constraints among the variables u, ξ, x. If, for example, $\xi_1 = \xi_2$ is such a constraint then the right hand side of (4) is no more a linear combination of elements of rank 1 (characteristic elements) with "free coefficients" $\xi_1, \ldots \xi_k$. To avoid such complications we will require the system be compatible.

The most general notion of compatibility for C^∞ systems is expressed by the notion of "formal integrability" [6,7,9]. Janet [2] (see also [8]) used the notion of passive systems whereas Cartan [1,3] by introducing the notion of involutivity was, in addition, able to formulate an algebraic criterion allowing to verify in a finite number of steps whether the system is involutive (hence, also formally integrable) or contradictory. He also proved that after a finite number of prolongations any system becomes involutive or contradictory. There also exists a criterion of formal integrability which is expressed by the 2-acyclicity of the symbol. However, this criterion is rather cumbersome in application and therefore we are using here a less complicated notion of involutivity.

In order to give a very brief review of these ideas let us consider the following first order system

$$F^s(u,Du,x) = 0, \qquad s = 1,\dots m. \tag{5}$$

Adding to Eqs.(5) their first differential consequences

$$\partial^{\#}_{x^\nu} F^s(u,Du,x) = 0,$$

where $\partial^{\#}_{x^\nu} F^s(u,p,x) = \frac{\partial F^s}{\partial x^\nu} + \frac{\partial F^s}{\partial u^\sigma}\frac{\partial u^\sigma}{\partial x^\nu} + \frac{\partial F^s}{\partial p^j_\mu}\frac{\partial}{\partial x^\nu} p^j_\mu$ one obtains what is called the prolonged system, denoted by $p(F)$. Similarly, further prolongations can be obtained according to the formula $p^r(F) = p(p^{r-1}(F))$, $r = 2,3,\dots$. Equations (5) can be solved for some of the derivatives of u to obtain

$$u_\nu^{\,j} = \phi_\nu^{\,j}(u, \{u_{\nu'}^{\,j'}\}, x), \tag{6}$$

$(j,\nu) \in \Pi \subset \{1,..,l\} \times \{1,..,n\}$,

$(j',\nu') \in \{1,..,l\} \times \{1,..,n\} \setminus \Pi$,

Now we define all derivatives on the left hand side as the principal derivatives associated with representation (6), whereas the derivatives on the right hand side are called parametric. This generates a splitting of all higher order derivatives. By definition, any derivative of a principal derivative is principal and those which are not principal are said to be parametric. Representation (6) is said to be passive [2] if for any $r = 1,2,\dots$ the prolonged system $p^r(6)$ can be formally written as

$$P^r = \Phi^r(u,x,Q^r), \tag{7}$$

where P^r and Q^r are sets of principal and parametric derivatives, respectively, of order less or equal r. In other words, the prolongation does not restrict the freedom of parametric derivatives. It could happen that at certain stage one would obtain relations involving parametric derivatives only.

Passive systems are formally integrable: the formal solutions are Taylor series. Freedom of the general solution is given by its parametric derivatives. The above definition of a passive system is not very constructive since it requires an infinite number of prolongations be checked. The situation improves, however, when system (8) is of a special form

$$
\begin{aligned}
u^{I_1}_{,x_1} &= \Phi_1^{I_1}(x,u,u^{\bar I_1}_{,x_1}),\\
u^{I_2}_{,x_2} &= \Phi_2^{I_2}(x,u,u^{\bar I_1}_{,x_1},u^{\bar I_2}_{,x_2}),\\
&\dots\dots\dots\dots\dots\dots\\
&\dots\dots\dots\dots\dots\dots\\
u^{I_n}_{,x_n} &= \Phi_n^{I_n}(x,u,u^{\bar I_1}_{,x_1},\dots, u^{\bar I_n}_{,x_n}),
\end{aligned}
\tag{8}
$$

i.e. when the subsets $I_\alpha \subset I = \{1,2,\cdots,l\}$ are ordered in the following way $I_1 \subset I_2 \subset \cdots \subset I_n$ and where $\bar{I}_k = I \setminus I_k$. $u^{I_1}_{,x_1}, u^{I_2}_{,x_2}, \cdots, u^{I_n}_{,x_n}$ and any of their derivatives are then principal derivatives. The remaining derivatives e.g. $u^{\bar{I}_1}_{,x_1}$, $u^{\bar{I}_2}_{,x_2}, \cdots, u^{\bar{I}_n}_{,x_n}, u^{\bar{I}_1}_{,x_1x_1}, u^{\bar{I}_1}_{,x_1x_2}, \ldots$. are parametric. In this case it is possible to prove that if the first prolongation does not restrict the freedom of the parametric derivatives then it is so in all higher prolongations [3]. There also exists an algebraic criterion which, admitting linear transformations of variables x and u, answers the question of local existence of this well ordered form (8). This criterion concerns the symbol of Eqs.(5), i.e. the part of the linearized system which is homogeneous in Du. By linearizing (5) we obtain

$$A^{s\nu}_j(u_0,x_0)\, u^j_{x_\nu} = 0.$$

Then we define $\mathfrak{I}_r = \{ p^j_\nu \,;\, A^{s\nu}_j p^j_\nu = 0,\ p^j_1 = \cdots = p^j_r = 0 \}$ for $r = 0,1,\cdots,n-1$. Similarly, we define the prolongation of the space $\mathfrak{I}_r$ as the set of solutions to the homogeneous, second order, part of the prolonged system:

$$p(\mathfrak{I}_r) = \{(p^j_{\mu\nu});\ A^{s\nu}_j p^j_{\mu\nu} = 0,\ p^j_{\mu\nu} = p^j_{\nu\mu},\ p^j_{\sigma\mu} = 0 \ \text{ for } \sigma \le r\}.$$

There exists a coordinate dependent projection

$$\delta_{r+1} : p(\mathfrak{I}_r) \longrightarrow \mathfrak{I}_r \tag{9}$$

defined by

$$p(\mathfrak{I}_r) \ni (p^j_{\mu\nu}) \xrightarrow{\ \delta_{r+1}\ } (p^j_{r+1\ \nu}) \in \mathfrak{I}_r.$$

Definition. If there exist systems of coordinates for x *and* u *such that mappings* (9) *are surjective then one says that system* (5) *has an involutive symbol.*

One can prove that the set of such coordinate systems is open. It follows [1,3] that

Theorem. If system (5) *has an involutive symbol then it can be represented in the form of* (8). *Moreover, if the prolonged system can be written as*

$$P^2 = \phi(Q^2)$$

i.e. if the freedom of parametric derivatives up to the second order is not restricted, then the system is involutive and, hence, formally integrable.

Having sketched the notion of involutive systems let us

define the notion of generalized hyperbolicity. Let us confine our attention to the homogeneous systems.

Definition. A first order homogeneous system $A_j^{s\nu}(u)\, u^j_{x_\nu} = 0$ *is hyperbolic if and only if*

1°. *it can be represented as*

$$du = \xi_1 \underset{1}{X} \otimes \lambda^1 + \cdots + \xi_k \underset{k}{X} \otimes \lambda^k ,$$

i.e. it is generated by its characteristic elements,

2°. *it is involutive.*

In order to see how broad is the class of generalized hyperbolic systems and what kind of patologies does it contain let us take

$$\operatorname{rot} \vec{v} = 0$$

as an example. Let $\lambda^\alpha = (\lambda^\alpha_\nu)$ be a triple of independent vectors in $\mathbb{R}^3$. Then $\operatorname{rot} \vec{v} = 0$ is equivalent to

$$d\vec{v} = \xi_1 \lambda^1 \otimes \lambda^1 + \xi_2 \lambda^2 \otimes \lambda^2 + \xi_3 \lambda^3 \otimes \lambda^3 .$$

The system is involutive and generated by its characteristic elements therefore, according to our definition, it is hyperbolic. However, it is not hyperbolic in the usual sense. Note, that every vector is a characteristic vector!

In order to avoid these patologies one can consider systems obtained from the usual-sense well-determined hyperbolic systems by adding certain constraints (e.g. Maxwell equations). Most natural class of such systems is composed of systems generated by "full branches" of the characteristic cone. Suppose that the characteristic determinant $W(\lambda)$ can be factorized as follows

$$W(\lambda) = W_1(\lambda)\, W_2(\lambda).$$

It is possible to define a system of partial differential equations associated with the branch $W_1(\lambda)$ in the following way

$$du \in \{\, \mathbf{X} \otimes \lambda \ ;\ A_j^{s\nu}(u)\, \mathbf{X}^j \lambda_\nu = 0,\ W_1(\lambda) = 0 \,\}, \tag{10}$$

i.e. du is a linear combination of characteristic elements associated with $W_1(\lambda) = 0$.

Let us formulate a criterion concerning the question whether the symbol of a hyperbolic system is involutive.

Theorem. Suppose that for every $r = 0,1,\ldots,n-1$ *the space* $\mathfrak{I}_r(x_0,u_0)$ *is generated by the characteristic elements* $\mathfrak{I}_r = \operatorname{span}\{ \underset{1}{X} \otimes \lambda^1, \cdots, \underset{k}{X} \otimes \lambda^k \}$ *such that* $\lambda^s_{r+1} \neq 0$ *for* $s = 1,\ldots,k$. *Then the symbol of the system is involutive.*

To prove it let us notice that the elements $\{\underset{s}{X} \otimes \lambda^s \otimes \lambda^s\}$, $s = 1,\ldots,k$, belong to $p(\mathfrak{I}_r)$. We have $\delta_{r+1}(\mathbf{X} \otimes \lambda \otimes \lambda) = \lambda_{r+1} \mathbf{X} \otimes \lambda$. Therefore the mappings $\delta_{r+1} p(\mathfrak{I}_r) \longrightarrow \mathfrak{I}_r$ are surjective, which proves the theorem.

How to check the compatibility of the system

$$du = \sum_{1}^{k} \xi_\alpha \underset{\alpha}{\mathbf{X}} \otimes \lambda^\alpha \tag{11}$$

provided its symbol is already involutive? Suppose $\underset{\alpha}{\mathbf{X}}$, $\underset{\alpha}{\lambda}$ are functions of u only. Taking the exterior derivative of (11) we have

$$\sum_\alpha \underset{\alpha}{\mathbf{X}} \otimes (d\xi_\alpha \wedge \lambda^\alpha + \xi_\alpha\, d\lambda^\alpha) + \frac{1}{2}\sum_{\alpha,\beta} \xi_\alpha \xi_\beta\, [\underset{\alpha}{\mathbf{X}},\underset{\beta}{\mathbf{X}}] \otimes (\lambda^\alpha \wedge \lambda^\beta) = 0,$$

where $d\lambda^\alpha = -\sum_\beta \xi_\beta \underset{\beta}{\mathbf{X}}^j \frac{\partial}{\partial u^j} \lambda^\alpha \wedge \lambda^\beta$, $[\mathbf{X},\mathbf{Y}]$ is the commutator of the vector fields $\mathbf{X}$, $\mathbf{Y}$:

$$[\mathbf{X},\mathbf{Y}] = \mathbf{X}^j \frac{\partial}{\partial u^j} \mathbf{Y} - \mathbf{Y}^j \frac{\partial}{\partial u^j} \mathbf{X} .$$

Equation obtained by differentiating (11) should have algebraic solutions for $(d\xi_\alpha)$ at any (ξ,u,x). This can be rather easily checked if (11) does not contain too many modes (i.e. characteristic elements). For example, let $\underset{1}{\mathbf{X}},\ldots,\underset{k}{\mathbf{X}}$ and $\lambda^1,\ldots,\lambda^k$ be independent at every u. Then one gets the following compatibility conditions for (11)

a) $[\underset{\alpha}{\mathbf{X}},\underset{\beta}{\mathbf{X}}] \in \text{span}\, \{\underset{\alpha}{\mathbf{X}},\underset{\beta}{\mathbf{X}}\}$,

b) $\underset{\beta}{\mathbf{X}}^j \frac{\partial}{\partial u^j} \lambda^\alpha \in \text{span}\, \{\lambda^\alpha,\lambda^\beta\}$, $\quad \alpha \neq \beta$.

Thus, the variation of wave-vector λ^α under the influence of wave "β" must be a linear combination of $\lambda^\alpha,\lambda^\beta$. In this case the solutions can be interpreted as interacting Riemann waves [4,5].

For a high number of modes rather than with the Pfaffian forms it is easier to deal with partial differential equations.

The compatibility or incompatibility of system (10) can be interpreted in physical terms. System (10) involves only such modes which satisfy $W_1(\lambda) = 0$. Therefore if the system is compatible then the nonlinear interaction of these modes does not generate modes from other branches. Particularly, if the initial condition consists only of modes from $W_1(\lambda) = 0$, then this is also true for the solution for $t > 0$ as long as it stays continuous. Incompatibility, on the contrary, leads to the production of modes from other branches of the dispersion relation.

As an illustration let us consider the system of Euler equations for an ideal gas

$$\begin{aligned} &\frac{\partial \vec{v}}{\partial t} + \vec{v}\cdot\nabla\, \vec{v} + \frac{1}{\rho}\nabla\, p(\rho) = 0, \\ &\frac{\partial \rho}{\partial t} + \operatorname{div}(\rho\vec{v}) = 0. \end{aligned} \tag{12}$$

In this case $W(\lambda)$ has two branches [4], the linear branch

$W_1(\lambda) = \lambda_0 + \vec{v}\cdot\vec{\lambda} = 0$ of multiplicity 2 and the quadratic branch $W_2(\lambda)$. The system associated with $W_2(\lambda)$ is compatible and it defines irrotational flows. The system associated with $W_1(\lambda)$ is

$$\frac{\partial}{\partial t}\vec{v} + \vec{v}\cdot\nabla\vec{v} = 0,$$
$$\operatorname{div}\vec{v} = 0, \qquad \nabla p = 0. \tag{13}$$

It defines isobaric flows and it is incompatible what can be checked by taking the divergence of the first equation. By prolongations we obtain additional constraints for the first order derivatives $\nabla\vec{v}$ which are

$$\operatorname{Tr}(\nabla\vec{v})^2 = 0, \qquad \operatorname{Tr}(\nabla\vec{v})^3 = 0. \tag{14}$$

Note that $\operatorname{div}\vec{v} = \operatorname{Tr}(\nabla\vec{v})$. The full set of equations, i.e. (13) and (14) is involutive. Thus, if the initial conditions satisfy $p = \text{const}$ and $\operatorname{Tr}(\nabla\vec{v})^r = 0$, where $r = 1,2,3$, then there exists a solution to Eqs.(13). If, however, the initial conditions satisfy only $\operatorname{div}\vec{v} = 0$ and $\nabla p = 0$ then, in general, the solution of Eqs.(12) does not satisfy Eqs.(13). This is so because this solution may, in addition, involve the sound modes although they were not present in the initial conditions.

References

[1]. Cartan, E. Les systèmes différentielles extérieures et leurs applications scientifiques, Hermann, 1946.

[2]. Janet, M. Leçons sur les systèmes d'equations aux dérivées partielles, Cahiers scientifiques, fax. IV, Gauthiers-Villars, Paris 1929.

[3]. Kuranishi, M. Lecture on involutive systems of partial differential equations, Sao Paulo, 1967.

[4]. Peradzyński, Z. Geometry of nonlinear interactions in partial differential equations (in Polish), Institute of Fundamental Technological Research Reports, Warszawa 1981.

[5]. Peradzyński, Z. Geometry of interactions of Riemann waves, [in] Advances in Nonlinear Waves, ed. L. Debnath, Pitman, 1984.

[6]. Pommaret, J.F. Systems of partial differential equations and Lie pseudogroups, Gordon & Breach, New York, 1978.

[7]. Quillen, D.G. Formal properties of over-determined systems of partial differential equations, Thesis, Harvard University, 1964.

[8]. Riquier, C.H. Les systèmes d'equations aux dérivées partielles, Gauthier-Villars, Paris, 1910.

[9]. Spencer, D.C. Overdetermined systems of linear partial differential equations, Ann. of Math., 86, (1962).

Runge-Kutta Split-Matrix Method for the Simulation of Real Gas Hypersonic Flows

M. Pfitzner

Messerschmitt-Boelkow-Blohm GmbH
Ottobrunn , FRG

SUMMARY

A class of three step explicit Runge-Kutta type time stepping schemes for use in conjunction with second order upwind and third order upwind-biased space discretisations of the quasi-conservatively formulated Euler equations is studied. The fractional time steps are optimized to yield a second order accurate method with a maximal region of (linear) stability. The method is incorporated in a 3-D Euler code for the simulation of ideal gas and equilibrium real gas flows. A pseudo space marching method is presented to deal with the supersonic part of the flow field. It uses the time stepping scheme as relaxation procedure. The method is much more efficient than pure time relaxation and converges to the same steady state. The applications shown include 2-D and 3-D simulations of ideal and real gas flows with fitted bow shock and captured embedded shocks.

INTRODUCTION

The development of new space transport systems and of hypersonic aircraft requires the simulation of high speed air flows about realistic configurations. New concepts of propulsion systems have to be studied and here the numerical simulation of the flowfield can lead to new insights. The construction of these configurations requires an accurate prediction of flow about three-dimensional bodies at Mach numbers ranging from $0 <= M_\infty <= 30$.

The high temperatures occuring in hypersonic flow fields ($M_\infty > 4$) cause the excitation of vibrations of air molecules and at higher temperatures the dissociation of oxygen and nitrogen and the creation of oxides of nitrogen. Air then cannot be treated as an ideal gas. At low enough altitude along the shuttle reentry trajectory ($H < 50$ km), where the density of air is high enough for recombination reactions to take place, the real gas effects can be taken into account by introducing a general equation of state into the fluid dynamics equations. This implies that the gas is assumed to be in local vibrational and chemical equilibrium. Curve fit routines can be used to represent numerically the equation of state. The additional amount of computational work is then approximately 30% more than in the ideal gas case.

At those altitudes, where the equilibrium real gas assumption is valid, the Reynolds number is high enough to make the division of the flow field into an inviscid part with a sharp bow shock and a boundary layer meaningful. At higher altitudes along the shuttle reentry trajectory, where nonequilibrium chemistry has to be taken into account, the Reynolds number drops and a full nonequilibrium Navier-Stokes simulation has to be done, which is at least two orders of magnitude more expensive than an equilibrium real gas Euler - boundary layer analysis.

The relatively simple geometry of reentry bodies and the very strong hypersonic bow shock suggests the use of a bow shock fitting procedure. The algorithm must, however, be able to capture embedded shocks correctly.

The time relaxation procedure must be able to deal with the very strong transients occuring during the integration process towards the steady state. As a result, some of the integration procedures developed for the subsonic and transsonic flow regimes are not applicable or have to be modified for use in the hypersonic regime. An accurate time integration procedure avoids unphysical transients and is therefore a good candidate. Low memory requirements, good vectorizablility and the applicability of the shock fitting algorithm are also important. An iterative Runge-Kutta type time stepping scheme meets all these requirements.

SPACE DISCRETIZATION METHOD

We integrate the instationary Euler equations in quasi-conservative form in generalized coordinates:

$$Q_\tau + A^+Q_\xi{}^+ + A^-Q_\xi{}^- + B^+Q_\eta{}^+ + B^-Q_\eta{}^- + C^+Q_\zeta{}^+ + C^-Q_\zeta{}^- = 0 \tag{1}$$

where $Q = (\rho,\rho v_x,\rho v_y,\rho v_z,e)^T$ are the conservative variables.

The matrices $A^\pm,B^\pm,C^\pm$ [4] are split according to the sign of their eigenvalues [1-3]. The space derivatives of Q are calculated with a third order accurate upwind-biased formula [3]:

$$Q^+_\xi\Big|_m = \frac{1}{6\Delta\xi}(Q_{m-2} - 6Q_{m-1} + 3Q_m + 2Q_{m+1})$$
$$Q^-_\xi\Big|_m = -\frac{1}{6\Delta\xi}(Q_{m+2} - 6Q_{m+1} + 3Q_m + 2Q_{m-1}) \tag{2}$$

and a MacCormack-type artificial diffusion term prevents wiggles near shocks.

The bow shock can be captured or fitted , whereas imbedded shocks are always captured. Owing to the simple geometry of reentry vehicles and the strong bow shocks encountered there, the shock fitting option is preferred in this case. For hypersonic flows, where real gas effects are important, the pressure of the gas is a general function of the density and the internal energy. The flux matrices A,B,C can be calculated with their eigenvectors and eigenvalues for an arbitrary pressure function $p(\rho,\varepsilon)$ and the shock fitting algorithm can be generalized for this case [5,10].

TIME STEPPING SCHEME

A systematic study of three step time stepping schemes in conjunction with the third order upwind-biased and the second order upwind space discretizations [2,3] has been conducted. The iterative Runge-Kutta scheme for the stepping from time level (n) to time level (n+1) is of the general form [6]:

$$Q^{(1)} = Q^n - a_1\,\Delta t\,P(Q^n)$$
$$Q^{(2)} = Q^n - a_2\,\Delta t\,P(Q^{(1)}) \tag{3}$$
$$Q^{n+1} = Q^n - a_3\,\Delta t\,P(Q^{(2)})$$

where the operator P contains the space derivatives. The coefficients a_1, a_2 and a_3 represent the normalized fractional time steps.

An advantage of the iterative Runge-Kutta schemes is that only one intermediate set of flow variables has to be stored in addition to the time levels Q^n and $Q^{(n+1)}$. In contrast to classical Runge-Kutta methods [7], however, it is not possible to increase to formal order of time accuracy beyond second order.

For first order accuracy $a_3 = 1$ is required. Second order accuracy is achieved by setting $a_2 = 1/2$ regardless of the value of a_1. The coeffient a_1 is used to maximize the region of (linear) stability of the algorithm. Setting $a_1 = 0$ formally yields a two step Runge-Kutta method of second order accuracy, whereas with $a_2 = a_1 = 0$ the first order algorithm we used earlier [2,3,5] is recovered.

To get an estimate for the optimal choice of the parameter a_1 we carried out a linear 1-D von Neumann stability analysis of the linear advection equation:

$$f_t + u\, f_x = 0\,. \tag{4}$$

The analysis yields the linear amplification factor g, which depends on the CFL number $u\,\Delta t/\Delta x$ and on the grid wave number $k\,\Delta x$. Since the resulting expressions for g are too complicated to be discussed analytically in general, the square of the modulus of g was plotted as a function of the CFL number and of $s = \sin^2(k\,\Delta x)$. Fig. 1 shows a 3-D view of $|g|^2$ for the one step algorithm ($a_2 = a_1 = 0$) with the second order upwind and the third order upwind-biased space discretizations. The white square hiding part of the function represents the plane $|g|^2 = 1$, i.e. the stability limit. The plots reveal that the one step algorithm is unconditionally unstable for both discretizations, but the region of instability vanishes quadratically and linearly, respectively, as the CFL number approaches zero. In practice, the one step algorithm has been sucessfully used in 2-D and 3-D Euler simulations with CFL numbers on the order of 0.2.

Fig. 2 shows $|g|^2$ for the second order two step scheme ($a_3 = 1$, $a_2 = 1/2$, $a_1 = 0$). In contrast to the central space diffencing scheme, which is linearly unconditionally unstable with this time stepping, the upwind and upwind-biased space discretizations are stabilized up to CFL = 1/2 and CFL = $(2/3)^{(1/3)}$, respectively. The three step scheme with $a_1 > 0$ can extend the stability region of the algorithm. The optimal stability region for the upwind biased discretization is achieved for $0.25 < a_1 < 0.3$. In this region the stability limit is extended beyond CFL= 1.75. In contrast, the stability limit of the second order upwind algorithm cannot even be pushed up to CFL=1 by any choice of a_1. Fig. 3 shows $|g|^2$ for the upwind and upwind-biased discretizations with $a_1 = 1/4$. The central differencing scheme is unconditionally unstable below $a_1 = 1/4$ and reaches a maximal stability region of CFL=2 at $a_1 = 1/2$.

The three step Runge-Kutta time stepping with $a_3 = 1$, $a_2 = 1/2$ and $a_1 = 1/4$ has been incorporated in the shock-fitting 2-D and 3-D ideal gas and real gas Euler codes. A CFL number of 1.25 has been found to be adequate in practical computations.

PSEUDO SPACE MARCHING METHOD

For hypersonic flows the one-dimensional Mach number in the coordinate direction of the main stream is mainly greater than one. The Euler equations (1) then are hyperbolic in that coordinate. In recompression regions, however, subsonic pockets my appear. For the simulation of flows of that type we developed a marching strategy using the instationary Runge-Kutta code as basis.

A zone of coordinate planes is defined in which the instationary equations are iterated toward convergence. The inflow boundary is supersonic and the variables there are prescribed. At the outflow boundary zeroth order extrapolation is used. Fig. 4 shows the computational grid in the symmetry plane and indicates the marching zone.

The zone is marched downstream over the body. If local subsonic pockets appear in the flowfield, the width of the zone is adapted such that the subsonic part of the field is completely covered by the zone. After convergence has been achieved in the adapted zone the width of the zone is reset to its original value and the marching continues from the supersonic downstream boundary of the adapted zone.

The blunt body flow may be simulated using the same code with a fixed zone width. The marching strategy saves much CPU time since it generates excellent starting values for the instationary relaxation scheme. It does not break down if the flow becomes locally subsonic like pure space-marching schemes do. The pseudo space marching method is, however, slightly less efficient than pure space marching. A similar idea has been proposed in [9].

APPLICATIONS

Several simulations of ideal gas and equilibrium real gas hypersonic flows have been performed with the third order upwind-biased space discretization scheme in conjunction with the three step Runge-Kutta time marching algorithm with ($a_3 = 1$, $a_2 = 1/2$, $a_1 = 1/4$) and CFL=1.25. Bow shock fitting was applied and the pseudo space marching method was used where possible.

Fig. 5 shows a comparison of lines of constant temperature in flow about a hemisphere-cylinder at $M_\infty=11$, $\alpha=0$, H=50 km ($T_\infty=271$ K) for ideal gas ($\gamma=1.4$) and equilibrium real gas. A local time step with CFL=1.5 was used and the calculation was converged after 500 time steps. The standoff distance of the shock is smaller for the real gas case and the temperature near the stagnation point is much lower. The wiggle in one of the real gas temperature isolines is a result of inaccuracies of the curve-fit routine used for the calculation of the real gas temperatures [8]. It is not visible in the flow variables. Fig. 6 contains a comparison of the ideal gas and real gas static temperature distributions on the windward side of a HERMES like body for $M_\infty = 10$, $\alpha = 30°$, H = 100 km. Equal temperature increments of $\Delta T = 100$ are used for the isolines. In the real gas calculation the stagnation point temperature is 3160 K compared to 5840 K in the ideal gas case. The qualitative form of the isolines is similar, but in the real gas case the gradients are smaller and they appear smoother than in the ideal gas case. Fig. 7 contains the Mach number isolines and the computational grid in a cross-sections halfway down the body of the real gas flowfield. The imbedded crossflow shocks are clearly seen.

In Fig.8 the ideal gas and real gas temperature,Mach number and pressure distributions on the body in the plane of symmetry are compared. On the leeside the location of the canopy shock can be clearly discerned in all curves. Only every second grid point is shown. Taking this into account, one can see that 2 grid points are inside the captured canopy shock in both the ideal gas and the real gas cases. No overshoots or undershoots occur in the solution. Whereas the real gas temperature on the windward side of the vehicle is much lower compared to the ideal gas case, on the leeward symmetry line a crossover occurs and the real gas temperature lies partly above the ideal gas result.

The reason for the lower stagnation point temperature of the real gas flow is the energy swallowed by the vibrational excitation and the dissociation reaction. The temperature crossover on the lee side can occur because in the real gas case the flow velocity is smaller than in the ideal gas case.The difference to ideal gas kinetic energy is in the real gas case divided between internal degrees of freedom and temperature. The pressure distributions of the ideal and real gas simulations are virtually identical since the pressure is a kinematically dominated variable.

CONCLUSIONS

A systematic study was conducted to find the optimal choice of fractional time steps of a three step Runge-Kutta time stepping scheme in conjunction with second order upwind and a third order upwind-biased space discretisations. A pseudo space marching method is introduced, which is applied to the supersonic part of the flow field and which uses the time stepping scheme as relaxation method. The algorithm has been implemented in a 3-D Euler code capable of ideal gas and equilibrium real gas shock capturing and shock fitting. The applicability of the code to flow simulation about real 3-D configurations at hypersonic free stream Mach numbers is demonstrated.

REFERENCES

[1]: Chakravarthy S.R., Anderson D.A., Salas M.D.: The Split-Coefficient Method for Hyperbolic Systems of Gas Dynamic Equations, AIAA paper 80-0268 (1980).

[2]: Weiland C.: A Split Matrix Method for the Integration of the Quasi-Conservative Euler Equations, Notes on Numerical Fluid Mechanics, Vol.13, Vieweg 1986.

[3]: Weiland C. and Pfitzner M.: 3-D and 2-D Solutions of the Quasi-Conservative Euler Equations, Lectures Notes in Physics, Vol. 264, pp. 654-659, Springer Verlag (1986).

[4]: Whitfield D.L. and Janus J.M.: Three-Dimensional Unsteady Euler Equations Solution Using Flux Vector Splitting, AIAA paper 84-1552 (1984).

[5]: Pfitzner M. and Weiland C.: 3-D Euler Solutions for Hypersonic Mach Numbers, AGARD-CP 428, Paper No. 22, Bristol - U.K. (1987).

[6]: Turkel E. and Van Leer B.: Flux Vector Splitting and Runge-Kutta Methods, Lecture Notes in Physics, Vol. 218, pp. 566-570, Springer Verlag (1985).

[7]: Jameson A., Schmidt W. and Turkel E.: Numerical Solutions of the Euler Equations by Finite Volume Methods Using Runge-Kutta Time-Stepping Schemes, AIAA paper 81-1259 (1981).

[8]: Tannehill J.C. and Mugge P.H.: Improved Curve Fits for the Thermodynamic Properties of Equilibrium Air Suitable for Numerical Computation Using Time-dependent or Shock-capturing Methods, NASA CR-2470 (1974).

[9]: Chakravarthy S.R., Szema K.Y.: Euler Solver for Three-Dimensional Supersonic Flows with Subsonic Pockets, J.Aircraft **24**, 73-83 (1987).

[10]: Weiland C., Pfitzner M., Hartmann G.: Euler Solvers for Hypersonic Aerothermodynamic Problems, 7. GAMM conference, Louvain le Neuve 1987, Notes on Numerical Fluid Mechanics, Vieweg (1988).

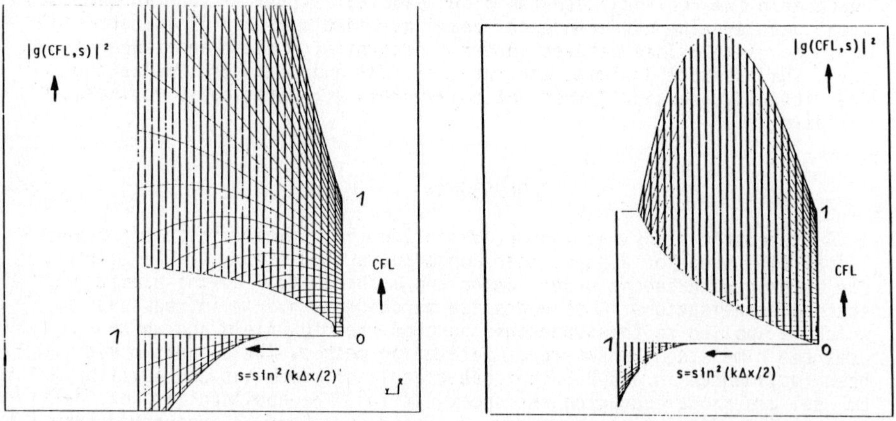

Second order upwind　　　　Third order upwind-biased

Fig.1 : 3-D plot of the square of the linear amplification factor $|g|^2$ for the one step first order time stepping

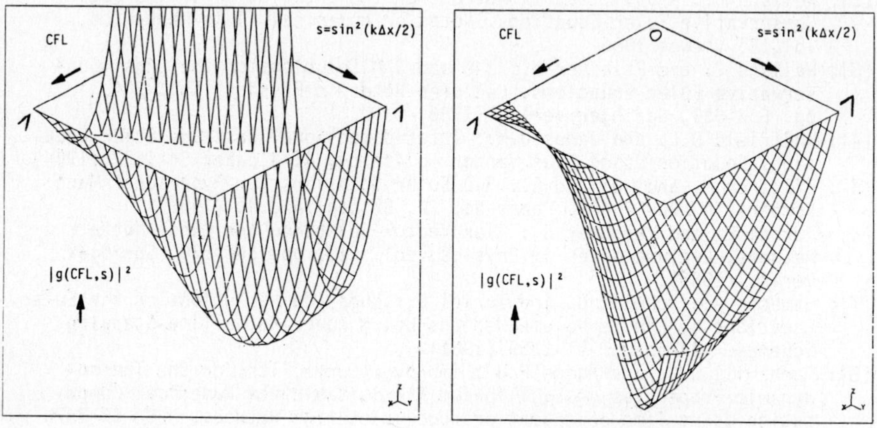

Second order upwind　　　　Third order upwind-biased

Fig.2 : 3-D plot of the square of the linear amplification factor $|g|^2$ for the two step second order Runge-Kutta time stepping

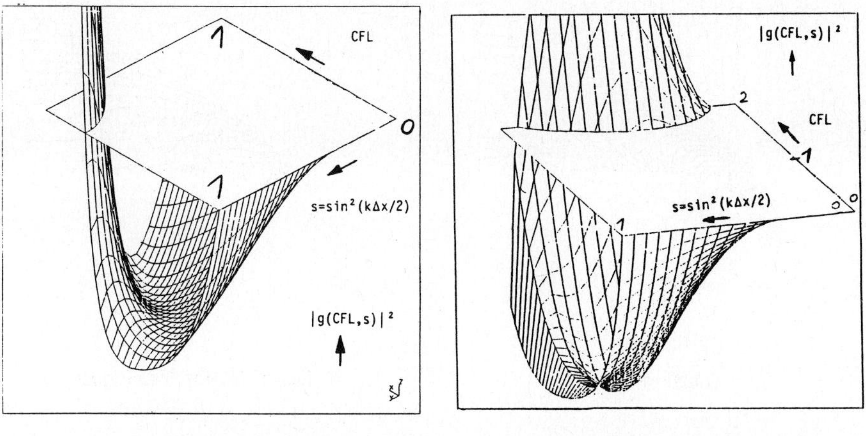

Second order upwind Third order upwind-biased

Fig.3 : 3-D plot of the square of the linear amplification factor $|g|^2$ for the three step second order Runge-Kutta time stepping (a_1 = 1/4)

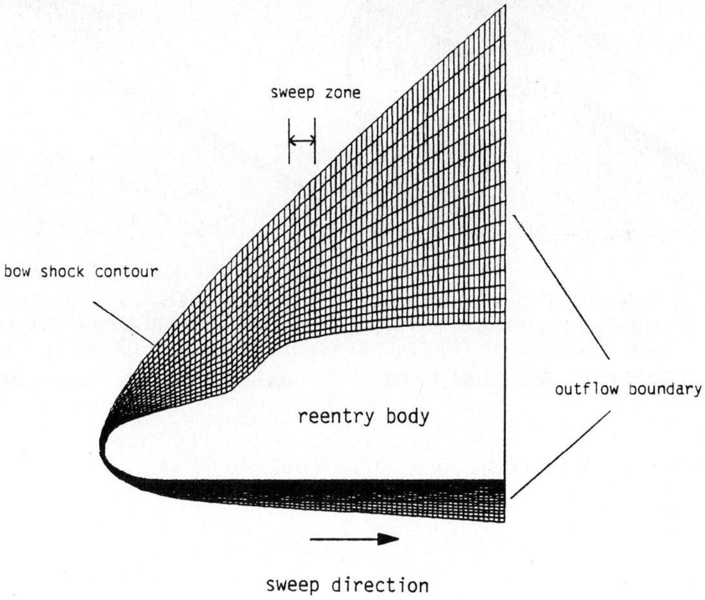

Fig.4 : Computational grid in plane of symmetry with Pseudo space marching zone

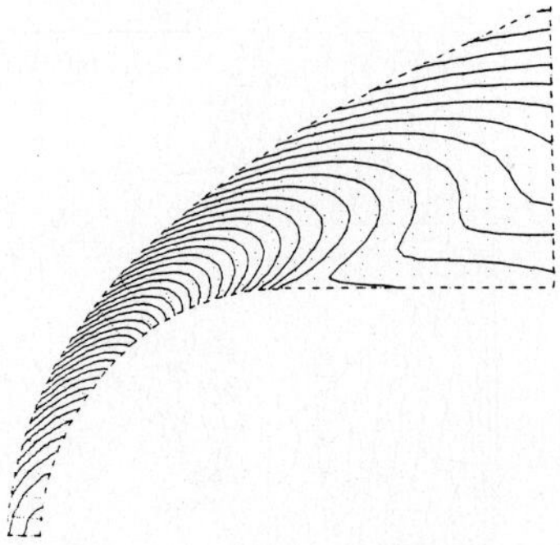

Ideal gas

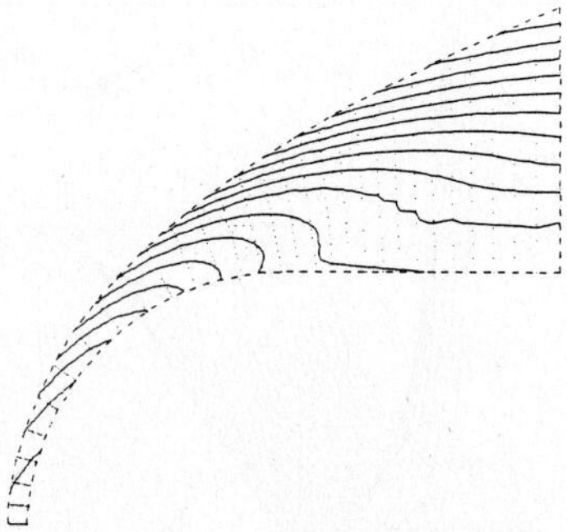

Equilibrium real gas

Fig.5 : Comparison of static temperature isolines $t=T/T_\infty$ of flow about a hemisphere-cylinder at $M_\infty = 11$, $\alpha = 0$, $H = 50$ km for ideal gas and real gas ($T_\infty = 271$ K, $\Delta t = 0.5$)

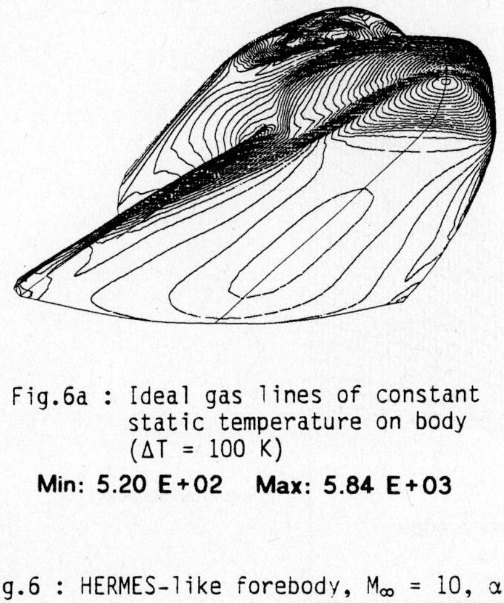

Fig.6a : Ideal gas lines of constant static temperature on body ($\Delta T = 100$ K)

Min: 5.20 E+02 Max: 5.84 E+03

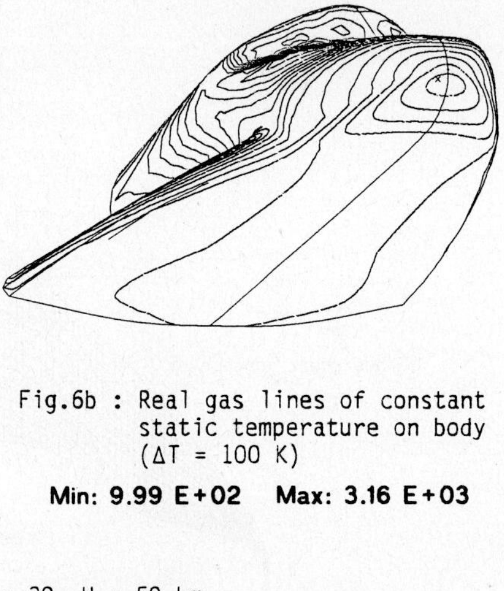

Fig.6b : Real gas lines of constant static temperature on body ($\Delta T = 100$ K)

Min: 9.99 E+02 Max: 3.16 E+03

Fig.6 : HERMES-like forebody, $M_\infty = 10$, $\alpha = 30$, $H = 50$ km

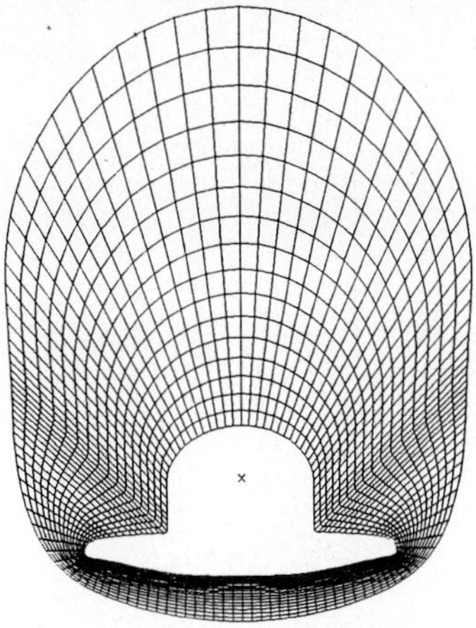

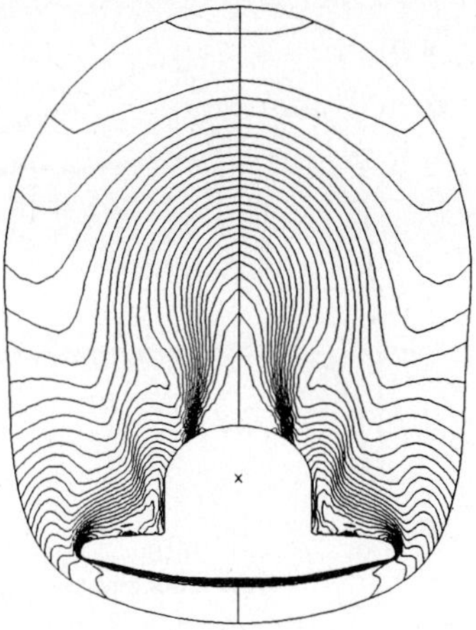

Fig.7a : Computational grid

Fig.7b : Real gas lines of constant Mach number ($\Delta M = 0.2$)

Min: 1.57 **Max: 8.02**

Fig.7 : HERMES-like forebody, $M_\infty = 10$, $\alpha = 30$, H = 50 km Cross sectional view

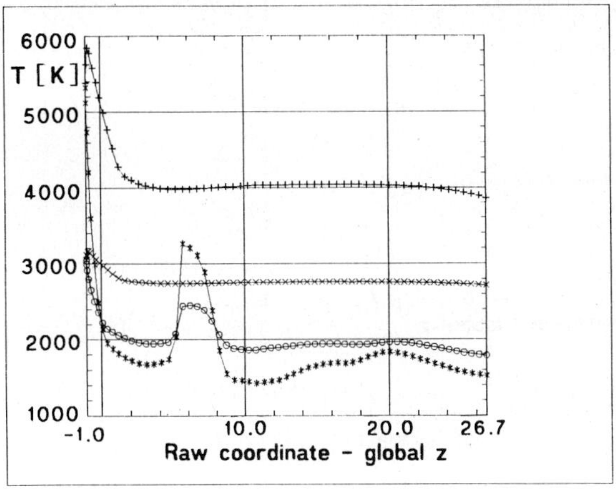

Fig.8a : Static temperature distribution on body in plane of symmetry

Fig.8 : HERMES-like forebody, $M_\infty = 10$, $\alpha = 30$, H = 50 km

+ : ideal gas luv side, * : ideal gas lee side
x : real gas luv side, o : real gas lee side

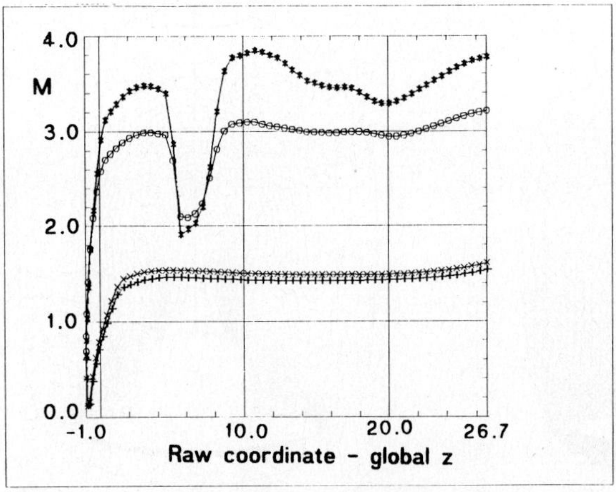

Fig.8b : Mach number distribution
on body in plane of symmetry

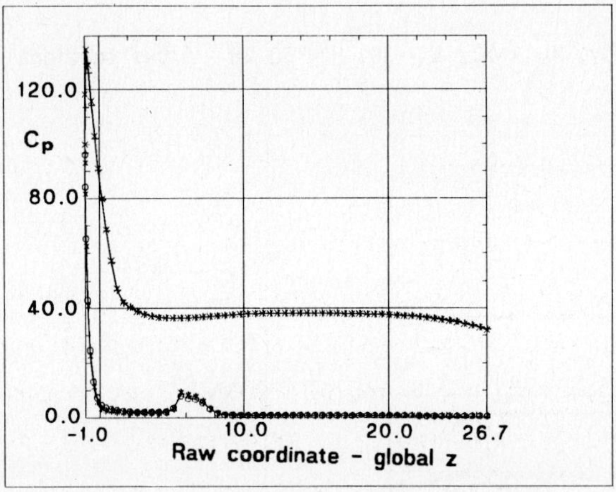

Fig.8c : Static pressure distribution
on body in plane of symmetry

Fig.8 : HERMES-like forebody, $M_\infty = 10$, $\alpha = 30$, H = 50 km

+ : ideal gas luv side, * : ideal gas lee side
x : real gas luv side, o : real gas lee side

ON SOME VISCOELASTIC STRONGLY DAMPED NONLINEAR WAVE EQUATIONS

A. Pham Ngoc Dinh
Dèpartement de Mathèmatiques
Université d'Orléans
45067 - ORLEANS CEDEX 2 - FRANCE

Dang Dinh Ang
Department of Mathematics
Dai Hoc Tong Hop
HO CHI MINH City - VIET NAM

1. INTRODUCTION

We study the problem of existence, uniqueness and asymptotic behaviour for $t \to \infty$ of (weak or strong) solutions of equation in the form

$$u_{tt} - \lambda \Delta u_t - \sum_{i=1}^{N} \partial/\partial x_i \; \sigma_i \; (u_{x_i}) + f(u, u_t) = 0 \, , (x,t) \in \Omega \times (0,T) \tag{1}$$

$$u = 0 \text{ on } \partial\Omega \tag{2}$$

$$u(x,0) = \tilde{u}_0(x) \qquad , \quad u_t(x,0) = \tilde{u}_1(x) \tag{3}$$

where $\lambda \geqslant 0$, $u_t = \partial u / \partial t$ and $u_{x_i} = \partial u / \partial x_i$.

In (1) Ω is a bounded domain in $\mathbb{R}^n$ with a sufficiently smooth boundary $\partial\Omega$, σ_i $(i = 1,\dots, N)$ are continuous functions satisfying certain monotonic and other conditions to be specified later. Equations of the type (1) with $f = 0$ and $\lambda > 0$, were given the first systematic treatment by Greenberg, MacCamy and Mizel [8] in the case of space dimension $N = 1$. They were proposed by the authors (loc. cit.) as the field equation governing the longitudinal motion of a viscoelastic bar obeying the nonlinear Voight model. For instance if we denote by x the position of a cross-section in the homogeneous rest configuration of the bar, by $u(x,t)$ the displacement at time t of the section from its rest position, by $\tau(x,t)$ the stress on the section at time t, then the equation of motion becomes (if the density is one)

$$u_{tt} = \tau_x \quad , (x,t) \in (0,1) \times (0,\infty) \, . \tag{4}$$

If a nonlinear dependence of the strain on u_x is allowed, then (4) yields the nonlinear wave equation

$$u_{tt} = \sigma'(u_x)\, u_{xx} \,. \tag{5}$$

In [9] MacCamy and Mizel assumed that σ satisfies the physically plausible conditions $\sigma(0) = 0$ and $\sigma'(\xi) > 0$, $\xi \in (-\infty, \infty)$; moreover σ' is taken to be monotone decreasing in $|\xi|$. In their paper Greenberg, MacCamy and Mizel (loc. cit.) assume the material to be a non linear Kelvin solid, that is, they assume for the strees a relation of the following form

$$\tau = \sigma(u_x) + \lambda u_{xt} \tag{6}$$

where λ -positive constant - is the viscosity coefficient. Since the appearance of the Greenberg - MacCamy - Mizel work, there has been a rather impressive literature on equations of the type (1) above, e.g., Caughey-Ellison [1], Defermos [4], Clements [2], [3], Webb [16], Yamada [17], to name but a few.

The case $\lambda = 0$, $N = 1$ and $f(u, u_t) = |u_t|^\alpha \operatorname{sgn}(u_t)$, $0 < \alpha < 1$ with nonhomogeneous boundary conditions corresponds to the motion of a linearly elastic rod in a nonlinearly viscous medium. For instance if we consider mud as the surrounding medium then α is equal to 0.1.

In section 2 we give some results with regard to equation (1), first with $\lambda > 0$ and $f(u_t) = |u_t|^\alpha \operatorname{sgn}(u_t)$, $0 < \alpha < 1$. For $N = 1$, $\tilde{u}_0$ in $H_0^1(\Omega) \cap H^2(\Omega)$, $\tilde{u}_1$ in $L^2(\Omega)$, σ in $C^1(\mathbb{R}, \mathbb{R})$ with $\sigma' > 0$ and locally Hölder continuous [5], there exists a unique strong solution $u(t)$ of the initial and boundary value problem (i.b.v. problem) (1) – (3) i.e. $t \to u(t)$ is continuous on $t \geqslant 0$ to $H_0^1(\Omega) \cap H^2(\Omega)$ and twice continuously differentiable on $t > 0$ to $L^2(\Omega)$.

In the case $N = 1$, $\lambda = 0$, $\sigma(x) = x$ and $f(u_t) = |u_t|^\alpha \operatorname{sgn}(u_t)$, $0 < \alpha < 1$, we study the equation (1) with a nonhomogeneous condition namely

$$u_x(0,t) = g(t), \qquad u(1,t) = 0 \tag{7}$$

and we prove [7] that the i.b.v. problem (1), (3) and (7) has a unique global solution on $(0,\infty)$

For the equation of the form

$$u_{tt} - \Delta u = \varepsilon f(t,u,u_t), \qquad (x,t) \in (0,1) \times (0,T) \tag{8}$$

an asymptotic expansion of order 2 in $\varepsilon(\varepsilon > 0)$ is obtained [15], for ε sufficiently small and $f \in C^1([0,\infty) \times \mathbb{R}^2)$.

In section 3 we consider the problem with $\lambda > 0$, the function f being a function of u only: $f = f(u)$, then for $\tilde{u}_0$ in $H^1_0(\Omega)$, $\tilde{u}_1$ in $L^2(\Omega)$ and a certain local Lipschitzian condition on f, a local existence and uniqueness theorem is obtained. If we strengthen the above hypotheses and assume that $1 \leqslant N \leqslant 3$, $\tilde{u}_0 \in H_0^1(\Omega) \cap H^2(\Omega)$ and $\tilde{u}_1 \in L^2(\Omega)$, with $f \in C^1(\mathbb{R}, \mathbb{R})$, $f' \geqslant -c$, $f(0) = 0$, then, the unique solution $u(t)$ exists for all $t \geqslant 0$, with the property that $\Delta u(t)$ and $u_t(t)$ decay exponentially to 0 as $t \to \infty$ [6]; this property generalizes a result of Webb [16].

2. SOME RESULTS ON EQUATIONS IN THE FORM (1)

Let $L^2 = L^2(\Omega)$, $H_0^1 = H_0^1(\Omega)$, $H^2 = H^2(\Omega)$.

A - **Problem 1**

Consider the equation

$$u_{tt} - \Delta u_t - \sum_{i=1}^{N} \partial/\partial x_i \left(\sigma_i(u_{x_i})\right) + |u_t|^{\alpha} \operatorname{sgn}(u_t) = 0 \quad (0 < \alpha < 1) \tag{9}$$

with the initial and boundary conditions (2) and (3).

Theorem 1 (weak solution). Let σ_i, $i = 1, \ldots, N$ be real-valued functions satisfying:

$$\sigma_i \text{ in } C(\mathbb{R}, \mathbb{R}), \quad \sigma_i(0) = 0$$

each $\tilde{\sigma}_i : L^2 \to L^2$ where $\tilde{\sigma}_i(f) = \sigma_i \circ f$ for f in L^2, takes bounded sets into bounded sets and is locally Lipschitzian.

Let $\tilde{u}_0$ in H_0^1 and $\tilde{u}_1$ in L^2. Then, for each $T > 0$, the i.b.v. problem (9), (2) and (3) admits a unique weak solution $u(t)$ on $(0,T)$ with the following properties:

u in $L^\infty(0,T; H_0^1)$ and u_t in $L^\infty(0,T; L^2) \cap L^2(0,T; H_0^1)$

$u(t)$ locally Hölder continuous on $[0,T)$ to H_0^1.

– If we consider the problem of global existence of strong solutions of (9), (2) and (3), we shall have to strengthen conditions on the initial data and on the σ_i's. The role of the space dimension is important and we shall limit ourselves to $N = 1$. Then, we have the following

Theorem 2 (strong solution). Let $N = 1$ and let

$\tilde{u}_0$ in $H_0^1 \cap H^2$, $\tilde{u}_1$ in L^2

σ_i in $C^1(\mathbb{R},\mathbb{R})$, $\sigma_i' > 0$, $\sigma_i(0) = 0$ and σ'_i locally Hölder continuous.

Then, there exists a unique solution of the i.b.v. problem (9), (2) and (3) with the following properties:

$t \to u(t)$ is continuous on $t \geqslant 0$ to $H_0^1 \cap H^2$

continuously differentiable on $t > 0$ to $H_0^1 \cap H^2$

continuously differentiable on $t \geqslant 0$ to L^2

twice continuously differentiable on $t > 0$ to L^2.

The idea of the proof is as follows. We take the weak solution $w(t)$ of (9), (2) and (3) which exists as per Theorem I and then, using the analytic theory of semi-group and the uniqueness of the solution, we prove that $w(t)$ is in fact the strong solution of the theorem by considering the first order differential equation with initial contition:

$$u_t = \Delta u + \tilde{u}_1 - \Delta\tilde{u}_0 + G(w) + F(w) \quad , u(0) = \tilde{u}_0 \tag{10}$$

where

$$G(w(t)) = \int_0^t Aw(s)ds \quad , \quad F(w(t)) = -\int_0^t f(w_t(s))ds \quad , \tag{11}$$

A being the nonlinear operator : $H_0^1(\Omega) \to H^{-1}(\Omega)$ defined by $Au: -(\sigma(u_x))_x$ and $f(u_t)$ the function $|u_t|^\alpha \operatorname{sgn}(u_t)$.

B - **problem 2**

$$u_{tt} - \Delta u + |u_t|^\alpha \operatorname{sgn}(u_t) = 0 . \tag{12}$$

To the equation (12) are associated nonhomogeous mixed conditions (7) and initial conditions (3). Here we shall make the following assumptions :

$$\tilde{u}_0 \in H^1(\Omega) \quad , \tilde{u}_1 \in L^2(\Omega) \quad , \Omega = (0,1)$$
$$g(t) \quad , g'(t) \in L^2(0,T) \quad ; g(0) \text{ exists} . \tag{13}$$

Let $V = \{v \in H^1(\Omega) \text{ such that } v(1) = 0\}$. We use here the Galerkin method associated to a Volterra nonlinear integral inequation namely :

$$\sigma_n(t) \leqslant D_1(t) + D_2(t) \int_0^t \sigma_n^\alpha(s)\,ds \tag{14}$$

with $\sigma_n(t) = \|u^{(n)}(t)\|_V^2 + \|u_t^{(n)}(t)\|^2 + \int_0^t |u_t^{(n)}(0,s)|^2 ds$

$u^{(n)}(t)$ being the approximate solution on a basis $(v_1, \ldots, v_n)$ of V and $D_1(t)$ and $D_2(t)$ two positive continuous functions. Here and elsewhere $\|.\|$ stands for the L^2-norm.

(14) allows one to get the required a priori estimates. To pass to the limit we shall use the fact that the function $|x|^\alpha$ sgn (x) generates a monotone operator and rely on the following lemma:

Lemma I. Let u be the solution of the following problem

$$u_{tt} - \Delta u + X = 0$$
$$u_x(0,t) = g(t) \quad , \quad u(1,t) = 0$$
$$u(0) = \tilde{u}o \quad , \quad u_t(0) = \tilde{u}_1$$

$u \in L^\infty(0,T;V)$ and $u_t \in L^\infty(0,T;L^2)$, then, we have

$$1/2\; a(u_0, u_0) + 1/2\, \|u_1\|^2 - \int_0^s < X, u_t > d\theta - \int_0^s g(\theta)\, u_t(0,\theta) d\theta$$
$$\leqslant 1/2\; a(u(s), u(s)) + 1/2\, \| u_t(s) \|^2 \quad \text{a.e. } s \in (0,T)$$

with $a(u,v) = < \partial u / \partial x \, , \, \partial v / \partial x >$, $< \, , \, >$ denoting the scalar product in $L^2(\Omega)$.

The solution is global since the Volterra nonlinear integral equation associated to (14) has a continuous solution $\forall\, t \in [0,T]$ for each $T > 0$. Finally we have the following result.

Theorem 3. For each $T > 0$, the i.b.v. problem (12), (3) and (7) under the assumptions (13) has a unique solution $u \in L^\infty(0,T;V)$ such that $u_t \in L^\infty(0,T;L^2)$.

C- **Problem 3** ,

$$u_{tt} - \Delta u = \varepsilon f(t,u,u_t) \quad , (x,t) \in (0,1) \times (0,T). \qquad (15)$$

To the hyperbolic equation are associated the conditions (2) and (3). Let the functions $\hat{u}_0$ and $\hat{u}_1$ be defined by

$$\begin{aligned} L\hat{u}_0 &= 0 \quad , (x,t) \in (0,1) \times (0,T) \\ \hat{u}_0(0,t) &= \hat{u}_0(1,t) = 0 \\ \hat{u}_0(x,0) &= \tilde{u}_0 \quad , \quad \dot{\hat{u}}_0(x,0) = \tilde{u}_1 \end{aligned} \qquad (16)$$

$$\begin{aligned} L\hat{u}_1 &= f(t, \hat{u}_0, \hat{u}_0) \quad , \quad (x,t) \in (0,1) \times (0,T) \\ \hat{u}_1(0,t) &= \hat{u}_1(1,t) = 0 \\ \hat{u}_1(x,0) &= \dot{\hat{u}}_1(x,0) = 0 \end{aligned} \qquad (17)$$

with $Lu = u_{tt} - \Delta u$ and $\dot{u} = u_t$.

Under the assumptions

$$\tilde{u}_0 \in H_0^1 \cap H^2 \ , \quad \tilde{u}_1 \in H_0^1 \qquad (18)$$

$$f(t,0,0) = 0 \qquad , \quad f \in C^1\,([\,0,\infty) \times \mathbb{R}^2\,)\ , \qquad (19)$$

we have the following theorem

Theorem 4. The unique solution $u_\varepsilon\ (t)$ of the i.b.v. problem (15), (2) and (3) has the asymptotic expansion, for $\varepsilon > 0$ "small":

$$\| u_\varepsilon - \textstyle\sum'^{1}_{r=0}\ \varepsilon^r\ \hat{u}_r \|_{L^\infty(0,T;\ H_0^1)} + \| \dot{u}_\varepsilon - \textstyle\sum'^{1}_{r=0}\ \varepsilon^r\ \dot{\hat{u}}_r \|_{L^\infty(0,T;\ L^2)} \leqslant C\varepsilon^2\ ,$$

the functions $\hat{u}_r\ (x,t)$ being defined by (16) and (17).

3. ASYMPTOTIC BEHAVIOUR FOR A DAMPED WAVE EQUATION

The following equation is considered

$$u_{tt} \ - \ \Delta u_t \ - \ \Delta u \ + \ f(u) = 0\ . \qquad (20)$$

We shall associate to (20) the initial and boundary values (2) and (3).

- First we shall make the following assumptions

$$f:\ H_0^1 \ \rightarrow \ H^{-1} \ \text{satisfies :}$$

for each bounded subset B of H_0^1, there exists $K_B > 0$ such that

$$\| f(y) \ - \ f(z) \|_{H^{-1}} \ \leqslant \ K_B \ \| \nabla y - \nabla z \| \quad \forall\ y\ ,\ \ z \in B. \qquad (21)$$

Then, we have the following

Theorem 5. Suppose f satisfies (21) and let $\tilde{u}_0 \in H_0^1$, $\tilde{u}_1 \in L^2$. Then, there exists a $T > 0$ such that the i.b.v. problem (20), (2) and (3) admits a unique solution such that $u \in C(0,T;\ H_0^1\ )$ and $u_t \in C(0,T;L^2) \cap L^2(0,T;\ H_0^1\ )$.

Furthermore, $u(t)$ is the limit of the sequence $\{\,u_n(t)\,\}$ of solutions of the following i.b.v. problems:

$$\begin{array}{l} u''_n \ - \Delta\, u'_n \ - \ \Delta u_n \ = \ -f(u_{n-1})\ ,\ n \geqslant 1\ ,\ \ u_0 = 0 \\ u_n \ = \ 0 \text{ on } \partial\Omega \ \ ;\ u_n(0) \ = \tilde{u}_0 \ \ ,\ u'_n(0) \ = \ \tilde{u} \end{array} \qquad (22)$$

$(u'' = u_{tt}\ ,\quad u' = u_t)$.

The sequence $\{u_n\}$ converges uniformly to u in $C(0, T;\ H_0^1)$ and the sequence $\{u'_n\}$ converges to u' in $L^2(0, T;\ H_0^1)$ and uniformly in $C(0, T;\ L^2)$.

– Second we shall consider the problem of global existence and asymptotoc behaviour for $t \to \infty$. To this end, we shall limit ourselves in what follows, to the case $1 \leqslant N \leqslant 3$, and furthermore , we shall restrict some what the hypotheses on f and the initial data. Thus we shall consider the following conditions on f :

$$f \in C^1(\mathbb{R}, \mathbb{R})\ ,\qquad f(0) = 0 \tag{23}$$

$$(f(u) + \varepsilon u)u \geqslant 0 \quad \text{for all } |u| \geqslant a\ , \tag{24}$$

with $0 < \varepsilon < 1$ satisfying $\varepsilon\alpha^2 < 1$ where $\alpha > 0$ is such that

$$\|u\| \leqslant \alpha\ \|\nabla u\| \text{ and } \|\nabla u\| \leqslant \alpha\ \|\Delta u\| \quad \forall\ u \in H_0^1 \cap H^2 \tag{25}$$

$$f' \geqslant -c\ ,\quad c > 0\ . \tag{26}$$

Then, we have the following

Proposition I. Let $\tilde{u}_0 \in H_0^1 \cap H^2$ and $\tilde{u}_1 \in L^2$ and let f satisfies (23) - (26). Then, there is a unique solution u(t) of the i.b.v. problem (20), (2) and (3) defined on $[0, \infty)$. Moreover the quantity

$$\|\Delta u(t)\|^2 + \|u_t(t)\|^2 + \int_0^t \|\nabla u_t(s)\|^2\, ds$$

is bounded on compact subsets of $[0, \infty)$.

The main result for the problem of asymptotoc behaviour for $t \to \infty$ can be expressed as follows :

Theorem 6. Let $\tilde{u}_0 \in H_0^1 \cap H^2$ and $u_1 \in L^2$.

Let (23) and (26) hold. Then, the solution u(t), which exists for all $t \geqslant 0$ as per Proposition I , decays exponentially to 0 as $t \to \infty$ in the following sense: there exists an $M > 0$ and $\gamma > 0$ such that

$$\|\Delta u(t)\|^2 + \|u_t(t)\|^2 \leqslant Me^{-\gamma t} \quad \text{for all } t \geqslant 0\ .$$

– For the proof of the Theorem 6 , first assume the constant c in assumption (26) satisfying the following conditions

$$0 < c < 1/2 \quad ,\quad c\alpha^2 < 1 \quad (\alpha \text{ being as in (25) , (27)) . We write} \tag{27}$$

$f(u) = g(u) - cu$, then $g'(u) \geqslant 0$ and hence f satisfies (24) and thus by proposition I the solution u(t) exists on $[0, \infty)$. Then, we show that

$$\| \Delta u(t) \|^2 + \| u_t(t) \|^2 \leqslant M \ , \quad \text{for all} \quad t \geqslant 0 \qquad (28)$$

where M is a constant.

Finally taking the inner product of (20) first with $u_t(t)\, e^{\gamma t}$ and then with $-\beta \Delta u(t) e^{\gamma t}$ and integrating with respect to the time variable from 0 to t, we find, taking (27) and (28) into account, a suitable choice for β and γ, choice which implies that there exists an $M > 0$ such that

$$\| \Delta u(t) \|^2 + \| u_t(t) \|^2 \leqslant M e^{-\gamma t} \ , \qquad \text{for all} \quad t \geqslant 0 \ . \qquad (29)$$

The restriction on c $(0 < c < 1/2)$ cazn be removed by a scaling argument that is to say $\xi = \mu x$, $\tau = \mu t$ with $\mu > 1/2$.

- On the other hand, on the question of global bound with weaker hypotheses on f and on initialdata $\tilde{u}_0$ we have the following.

Proposition 2. If $f \in C^1(\mathbb{R}, \mathbb{R})$, satisfies :

$\lim_{|x| \to \infty} f(x) \,.\, x \geqslant 0$, then under the sole condition $\tilde{u}_0 \in H_0^1$ and $u_1 \in L^2$, there exists a global bound on $\| u_t(t) \|$ and $\| \nabla u(t) \|$ for all $t \geqslant 0$.

4. NUMERICAL APPLICATIONS BASED ON THE TAU METHOD

The linear recursive schemes developed in sections 2 and 3 enables us to use the Tau method of Ortiz [10] i.e. a perturbation technique based on the ideas of best uniform approximation by polynomials. The approximate solution obtained with this technique is a polynomial which satisfies the given partial differential equation, but for a small pertubation term in the right hand side ; the supplementary (initial, boundary or mixed) conditions are satisfied exactly, provided

they are of polynomial form. Given a linear partial differential equation with polynomial coefficients.

$$Lu = f(t,x) \qquad (30)$$

we attempt to solve a slightly perturbed form of the original problem, defined by the so-called Tau problem.

$$Lu_{rs} = f(t,x) + \tau H_{rs}(t,x) \qquad (31)$$

where H_{rs} (t,x) is the product of best uniform approximations of zero, of degrees r and s respectively, on a given domain D. The parameter τ is chosen for $u_{rs}(t,x)$ to be a bivariate polynomial which satisfies the boundary and initial conditions satisfied by u.

The construction of the bivariate Tau approximation depends on two matrices of extrelemy simple structure: only elements on one line, parallel to the main diagonal, are different from zero. They lead to an algebraic problem for the coefficients of the Tau approximation with an almost block diagonal matrix. The approximate solution can be constructed in any bivariate polynomial basis. In the exemple given here we have chosen it to be the Chebyshev product basis. Computational procedures for the numerical treatment of partial differential equations with polynomial coefficients have been discussed by Ortiz and Samara [12]. In [11] we have discussed the numerical solution of semi-linear hyperbolic problem of the following type:

$$u_{tt} - \Delta u = u^2 + F(t,x) \tag{32}$$

with the initial and boundary conditions (2) and (3).

In (32) $F \in L^2(Q)$, $Q = (0,1) \times (0,1)$

To solve numerically (32) we use the linear recursive schemes defined by

$$u''_{n+1} - \Delta u_{n+1} - 2u_n u_{n+1} = -u_n^2 + F , \tag{33}$$

the u_{n+1} satisfying the conditions given in (2) and (3), but for the fact that functions have been replaced by tight polynomial approximations, to be able to use the Tau method in the numerical approximation of problem (33). By using this technique we effectively produce numerical solutions of a high accuracy. Sufficient conditions for the quadratic convergence of the equation in the form

$$u_{tt} - \Delta u = f(t,u) \tag{34}$$

are given in this paper.

Some other results on the applications of the Tau method to the numerical solution of nonlinear PDEs are reported in Ref. [13] and [14].

REFERENCES

[1] **T.K. CAUGHEY and J. ELLISON** - Existence, uniqueness and stability of solutions of a class of non linear partial differential equations, J. Math. Anal. Appl., 51 (1975), pp. 1-32.

[2] **J. CLEMENTS** - On the existence and uniqueness of solutions of the equation $u_{tt} - \partial/\partial x_i\ (\sigma_i\ (u_{x_i})) - \Delta_N u_t = f$, Canad. Math., Bull., 2 (1975), pp. 181-187.

[3] **J. CLEMENTS** - Existence theorem for a quasilinear evolution equation, SIAM J. Appl. Math., 26 (1974), pp. 745-752.

[4] **C.M. DAFERMOS** - The mixed initial boundary value problem for the equations of nonlinear one-dimensional viscoelasticity, J. Diff. Equations, 6 (1969), pp. 71-86.

[5] **DANG DINH ANG and A. PHAM NGOC DINH** - Strong solutions of a quasi-linear wave equation with nonlinear damping, SIAM J. Math. Anal., 19 (1988).

[6] **DANG DINH ANG and A. PHAM NGOC DINH** - On the strongly damped wave equation: $u_{tt} - \Delta u - \Delta u_t + f(u) = 0$. To appear in SIAM J. Math. Anal.

[7] **DANG DING ANG and A. PHAM NGOC DINH** - Mixed problem for some semi-linear wave equation with a nonhomogeneous condition. To appear in NonLinear Analysis T.M.A. (1988).

[8] **J.M. GREENBERG, R.C. MACCAMY and V.I. MIZEL** - On the existence, uniqueness and stability of solutions of the equation: $\sigma'(u_x)\, u_{xx} + \lambda\, u_{xtx} = \rho_0\, u_{tt}$ J. Math. Mech., 17 (1968), pp. 707-728.

[9] **R. C. MACCAMY and V. I. MIZEL** - Existence and nonexistence in the large of solutions to quasilinear wave equations, Arch. Ration. Mech. Anal., 25 (1967), pp. 299-320.

[10] **E. L. ORTIZ** - The Tau method. SIAM J. Numer. Anal., 6 (1969), pp. 480-492.

[11] **E.L. ORTIZ and A. PHAM NGOC DINH** - Linear recursive schemes associated with some nonlinear partial differential equations in one dimension and the Tau method, SIAM J. Math., Anal., 18 (1987), pp. 452-464.

[12] **E.L. ORTIZ and H. SAMARA** - Numerical solution of partial differential equations with variable coefficients with an operational approach to the Tau method, Comput. Math. Applic., 10 (1984), pp. 5-13.

[13] **E. L. ORTIZ and K.S. PUN** - A bidimensional Tau-elements method for the numerical solution of nonlinear partial differential equations with an application to Burger's equation, Comput. Math. Applic., 12B (1986), pp. 1225-1240.

[14] **E. L. ORTIZ and K. S. PUN** - Numerical solution of nonlinear partial differential equations, with the Tau method, J. Comput. Appl. Math., 12/13 (1985), pp. 511-516.

[15] **A. PHAM NGOC DINH and NGUYEN THANH LONG** - Linear approximation and asymptotic expansion associated to the nonlinear wave equation in 1-dim, Demonstratio Mathematica, 19 (1986), pp. 45-63.

|16] **G. F. WEBB** - Existence and asymptotic behaviour for a strongly damped nonlinear wave equation, Canad., J. Math., 32 (1980), pp. 631-643.

[17] **Y. YAMADA** -Some remarks on the equation $y_{tt} - \sigma(y_x)\, y_{xx} - y_{xtx} = f$, Osaka J. Math., 17 (1980) , pp. 303-323.

TVD SCHEMES TO COMPUTE COMPRESSIBLE VISCOUS FLOWS ON UNSTRUCTURED MESHES

Philippe ROSTAND * and Bruno STOUFFLET **

* INRIA-Menusin Domaine de Voluceau Rocquencourt BP 105 78153 Le Chesnay Cedex FRANCE
** AMD-BA 78 Quai Marcel Dassault 92214 Saint Cloud FRANCE

Abstract

Abstract : Finite volume TVD schemes derived for the Euler equations are extended to the Navier-Stokes system. The numerical diffusion introduced in the approximation of the convective part is chosen through a total variation analysis taking in account the physical diffusion. Two dimensional numerical simulations are presented, using an algorithm to solve cheaply the steady equations..

I. INTRODUCTION

Finite volume schemes based on approximate Riemann solver to solve conservation laws, also called, TVD (Total Variation Diminushing) schemes, have received considerable attention in the last twenty years, (see, among others, Harten [9], Van Leer [28], Yee [30]), and can be said to have reached a satisfactory degree of achievement .

These schemes were successfully extended to multidimensional problems, by reducing the equations to one dimension, through the finite volume formulation, and applying the one dimensional techniques. This can be done on unstructured meshes (Baba and Tabata [3], Dervieux [6], Stoufflet-Fezoui [23]).

On the other hand, several research teams have studied such algorithms on structured meshes for solving the Navier-Stokes equations. Some 3D codes on upwind schemes have been developped by Mac Cormack [4], Hänel [22] and Chakravarthy [5], among others.

A particular class of very efficient schemes is that obtained by the combination of a monotone flux formula, and of a second order extension through monotony preserving interpolation, christened Monotonic Upwind Schemes for Conservation Laws (MUSCL) by Van Leer ([29]). These schemes have been derived in order to introduce a "numerical viscosity", which will provide automatic inforcement of the entropy condition, and to provide second order accuracy, at least in regions of smoothness.

To solve advection dominated nonlinear parabolic, or incompletely parabolic equations as the compressible Navier-Stokes system, it is necessary to use an approximation which will preserve the entropy condition, but which will also provide sufficient accuracy in viscosity dependant zones, as boundary layers or wakes. In other words, one must make sure that no more diffusion than needed is added.

A model equation for compressible viscous gas dynamics is given by

$$\begin{cases} \dfrac{\partial u}{\partial t} + \dfrac{\partial f(u)}{\partial x} - \epsilon \dfrac{\partial^2 u}{\partial x^2} = 0 \\ u = u(x,t) \in R \end{cases} \tag{1.1}$$

where f is a regular function, convex or not.

In our framework, a numerical scheme to solve (1.1) is made of

. an approximation of the convection term $\partial f / \partial x$, combining

- a numerical flux function h = h(u,v) with h(u,u) = f(u)
- a MUSCL-like interpolation formula

. an approximation of the diffusion term

. an approximation of the time derivative.

In part II, we will outline the general framework of upwind TVD schemes for multidimensional gas-dynamics equations on triangular (tetrahedral) meshes.

The numerical formulation relying on the approximate Riemann solver proposed by Osher and Chakravarty [13] and a multidimensional MUSCL like interpolation will be presented. Van Leer, Thomas, Roe and Newsome in [27] have analyzed in one dimension the influence of the choice of the upwind flux formula for the convection part in terms of accuracy, and showed that some flux-vector splitting gave a bad representation of the boundary layer.This will not be our topic; we will only give our arguments in favour of Osher's scheme.

The interpolation, through which second order, or even third order accuracy for one dimensional problems, is reached, is also an important feature of the scheme. In part III, for the one dimensional scalar viscous conservation law (1.1), we will try to derive conditions on the interpolation which will insure some kind of monotony property, and still allow sufficient accuracy.

In [20],the extension of the one dimensional scalar conclusions of Part III to multidimensional systems will be detailed, together with the different possible time discretizations, including linearly implicit methods, and accelerators for the steady case. Finally, numerical results will be presented and discussed in part IV.

II. GENERAL FRAMEWORK OF THE FINITE VOLUME GALERKIN (FVG) APPROXIMATION

Let Ω an open set of R^N (N=2 or 3) and let $\Gamma = \partial\Omega$ be its boundary presumed to be smooth, τ_h a triangulation of Ω. Let $W = (\rho, \rho u, E)$ be the vector of conserved quantities ; we write the Navier-Stokes system in conservation form:

$$\frac{\partial W}{\partial t} + \nabla.F(W) = \nabla.N(W) \tag{2.1}$$

where F and N denote respectively the convective flux term and viscous term.

The formulation can be found in [24]. The space V_h is defined as follows :

$V_h = \{v_h \in C^0(\Omega)\ ;\ v_h \text{ is linear on each triangle }\}$ For each vertex $S_i \in \tau_h$ the cell C_i is defined as the union of the subtriangles having S_i as a vertex and resulting from the subdivision of each triangle of T_h by means of the medians (Fig. 1).

$$\begin{cases} \text{Find } W_h \in (V_h)^m \text{ such that } \forall S_i \in \tau_h, \\ \displaystyle\int_{C_i} \frac{\partial W_h}{\partial t} dx + \int_{\partial C_i} F(W_h).\nu_i d\sigma + \int_{\partial C_i \cap \Gamma} F(W_h).n d\sigma = R.H.S\ . \end{cases} \tag{2.2}$$

The numerical integration of the viscous terms of the RHS is carried out in a centered way.

The scheme will be completely defined if we precise now which approximation is used to compute the left hand-side integral in (2.2). For this, the boundary ∂C of the cell C_i is splitted in bisegments ∂S_{ij}, joining the middle point of the segment S_iS_j to the centroids of the triangle having S_i and S_j as common vertices (Fig. 1).

Let us introduce the following notations :

$$F_{ij}(U) = F(U).\int_{\partial S_{ij}} \nu_i d\sigma \qquad \text{and} \qquad P_{ij}(U) = \nabla.F(U).\int_{\partial Sij} \nu_i d\sigma\ .$$

Upwinding is introduced in the computation of the convection term through the numerical flux function Φ of a first-order accurate upwind scheme by :

$$\int_{\partial S_{ij}} F(W_h).\nu_i d\sigma = H_{ij}^{(1)} = \Phi_{F_{ij}}(W_i, W_j)$$

where $W_i = W_h(S_i)$ and $W_j = W_h(S_j)$.

The numerical flux function used in this study is Osher's. The numerical integration with the upwind scheme, as described previously, leads to approximations which are only first-order accurate. We present a second-order accurate MUSCL-like extension without changing the approximation space:

$$\begin{cases} \text{Find } W_h \ \in (V_h)^m \\ \displaystyle\int_{C_i} \frac{\partial W_h}{\partial t} dx + \sum_{j \in \kappa(i)} H_{ij}^{(2)} + \int_{\partial C_i \cap \Gamma_b} F(W_h).n d\sigma + \int_{\partial C_i \cap \Gamma_\infty} F(W_h).n d\sigma = R.H.S. \end{cases} \tag{2.3}$$

where

$$H_{ij}^{(2)} = \Phi_{F_{ij}}(W_{ij}, W_{ji}).$$

The arguments W_{ij} and W_{ji} are values at the interface ∂S_{ij} interpolated using upwinded gradients as described below.

We define the dowstream and upstream triangles T_{ij} and T_{ji} for each segment S_iS_j as shown in Fig. 2.

fig. 1

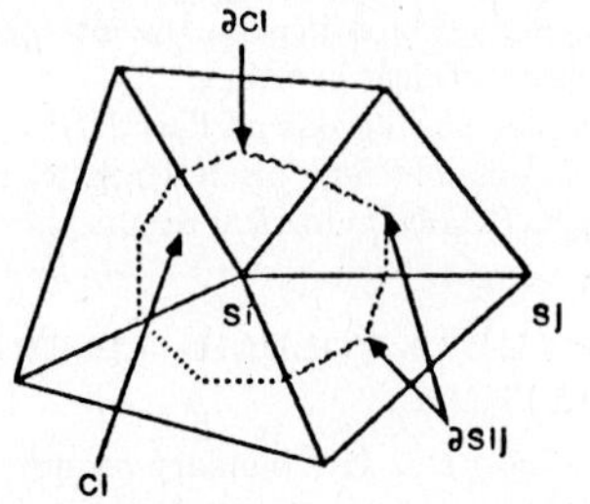

fig. 2

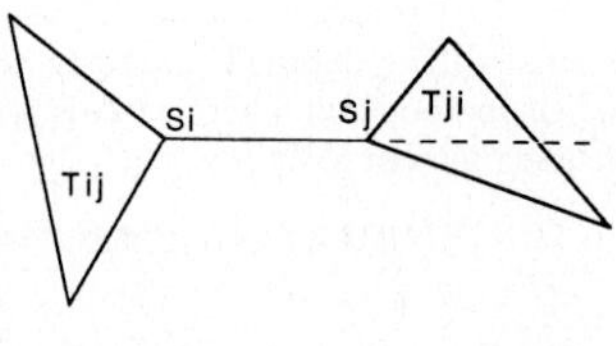

Let the centered gradient be :

$$\nabla W_{ij} = \nabla W|_{T^1_{ij}} = \nabla W|_{T^2_{ij}} \ .$$

The values at interface needed to compute the flux $H^{(2)}_{ij}$ are now given by :

$$\left\{ \begin{array}{l} W_{ij} = W_i + L_{ij} \ \left(\frac{1-\kappa}{4} \nabla W|_{T_{ij}} + \frac{1+\kappa}{4} \nabla W_{ij} \right) .S_iS_j \\ W_{ji} = W_j - L_{ji} \ \left(\frac{1-\kappa}{4} \nabla W|_{T_{ji}} + \frac{1+\kappa}{4} \nabla W_{ij} \right) .S_iS_j \end{array} \right. \tag{2.4}$$

where the parameter κ can be chosen to select the degree of upwinding in the interpolation and L_{ij} and L_{ji} are the limiting matrices, which are introduced to reduce numerical oscillations of the solution and to provide some kind of monotonicity property.

A good procedure in term of accuracy is to use limiters on characteristic variables. For this, we compute these variables by the transformation taken at midpoint of the segment. If we denote by Π_{ij} the transformation matrix corresponding to $P_{ij}(W(\frac{S_i+S_j}{2}))$, the values at interface are now given by :

$$\left\{ \begin{array}{l} W_{ij} = W_i + \Pi_{ij} Lc_{ij} \Pi_{ij}^{-1} \ \left(\frac{1-\kappa}{4} \nabla W|_{T_{ij}} + \frac{1+\kappa}{4} \nabla W_{ij} \right) .S_iS_j \\ W_{ji} = W_j - \Pi_{ij} Lc_{ji} \Pi_{ij}^{-1} \ \left(\frac{1-\kappa}{4} \nabla W|_{T_{ij}} + \frac{1+\kappa}{4} \nabla W_{ji} \right) .S_iS_j \end{array} \right. \tag{2.5}$$

where Lc_{ij} and Lc_{ji} are diagonal matrices.

III. ONE DIMENSIONAL SCALAR ANALYSIS

We first consider the one-dimensional scalar convection-diffusion law (1.1) . We define a regular mesh and apply the finite volume scheme defined in II to (1.1). We denote by h the mesh size, by x_j=jΔx, j$\in$Z, u_j=$u(x_j)$.

The equation corresponding to j is given by:

$$\left\{ \begin{array}{l} \frac{du_j}{dt} + \frac{1}{h}[\ \Phi(\ u_j + \frac{r_j}{4}((1-\kappa)(u_j - u_{j-1}) + (1+\kappa)(u_{j+1} - u_j)) \ , \\ \qquad u_{j+1} - \frac{l_{j+1}}{4}((1-\kappa)(u_{j+2} - u_{j+1}) + (1+\kappa)(u_{j+1} - u_j)) \ ) \\ \quad -\Phi(\ u_{j-1} + \frac{r_{j-1}}{4}((1-\kappa)(u_{j-1} - u_{j-2}) + (1+\kappa)(u_j - u_{j-1})) \ , \\ \qquad u_j - \frac{l_j}{4}((1-\kappa)(u_{j+1} - u_j) + (1+\kappa)(u_j - u_{j-1})) \ ) \ ] \\ -\frac{\epsilon}{h^2}(u_{j+1} - 2u_j + u_{j-1}) \qquad = 0 \ . \end{array} \right. \tag{3.1}$$

For simplicity, we have denoted by r_i (resp. l_i) the right limiter $L_{i,i+1}$ (resp. the left limiter $L_{i,i-1}$). The numerical flux function Φ satisfies $\Phi(u,u)$ = f(u); we will suppose it to be monotone, i.e. $\Phi(v,w)$ is a non increasing function of w and a non decreasing function of v.

The scheme (3.1) is defined by two parameters : the right and left slope limiters and the parameter κ defining the extrapolation. Note that the slope limiters r_i and l_i are different. However, if one wants to have the same treatment of forward and backward discontinuities, r_i and l_i must be linked, see equation (3.8).

The purpose of slope limiting is to provide some kind of monotony or TVD property [9] ; it has received considerable attention in the recent years [25], [14], [29], [26]. However, all these studies are designed for the inviscid ($\epsilon = 0$) equation. Our point is to derive a condition for total variation diminushing taking in account the physical diffusion term $\epsilon\partial^2 u/\partial x^2$.

Following Osher [12], Osher and Chakravarty [13], Harten [9], Sanders [21], we will write our five point scheme as

$$\frac{du_j}{dt} - C_{j+\frac{1}{2}}\Delta^+ u_j + D_{j-\frac{1}{2}}\Delta^- u_j = 0 \tag{3.2}$$

where

$$\Delta^+ u_j = u_{j+1} - u_j \tag{3.3.a}$$

$$\Delta^- u_j = u_j - u_{j-1} \tag{3.3.b}$$

$$C_{j+\frac{1}{2}} = C_{j+\frac{1}{2}}(u_{j+2}, u_{j+1}, u_j, u_{j-1})$$

$$D_{j-\frac{1}{2}} = D_{j-\frac{1}{2}}(u_{j+1}, u_j, u_{j-1}, u_{j-2}) \;.$$

The following result, due to Osher, Chakavarty and Sanders, will be our starting point.

Theorem 1 :

$$\text{If} \quad \begin{cases} C_{j+\frac{1}{2}} \geq 0 \\ D_{j-\frac{1}{2}} \geq 0 \end{cases} \quad \text{then} \quad \frac{d}{dt}TV(u) \leq 0$$

where TV(u) is the total variation of u.

For a Lipschitz continuous flux ϕ, we will define a local Peclet number by $\nu_i = a_i h/2\epsilon$, where a_i is the Lipschitz norm of ϕ in a neighborhood I_i of u_i of chosen size, i.e.

$$a_i = \text{ Max } \{\sup_{x\in I_i}(\sup_{\substack{y,z\in I_i \\ y\neq z}} \frac{\phi(y,x) - \phi(z,x)}{y-z}) \;,\; \sup_{x\in I_i}(\sup_{\substack{y,z\in I_i \\ y\neq z}} \frac{\phi(x,y) - \phi(x,z)}{y-z})\}$$

which is non negative because of the monotony of Φ. We denote $\delta_j = \Delta^+ u_j/\Delta^- u_j$ and we take $I_j = (u_j - k\Delta^- u_j, u_j + k\Delta^+ u_j)$ where k is a chosen positive parameter. As usual, we take r_j and l_j depending of δ_j only. We have the following result:

Theorem 2. Let $\theta \in [0,1]$.

If When $\delta_j \geq 0$ or $\delta_j \leq \dfrac{\kappa - 1}{1+\kappa}$ $\quad 0 \leq r_j \leq \dfrac{4 \;\inf(k, \theta(1 + 1/2\nu_j))\,\delta_j}{(1-\kappa) + (1+\kappa)\delta_j}$ (3.4)

When $\delta_j \leq \dfrac{\kappa - 1}{1+\kappa}$ $\quad 0 \leq r_j \leq \dfrac{4 \;\inf((k-1), (1-\theta)(1 + 1/2\nu_{j-1}))}{(\kappa - 1) - (1+\kappa)\delta_j}$ (3.5)

and When $\delta_j \geq \dfrac{1+\kappa}{\kappa - 1}$ $\quad 0 \leq l_j \leq \dfrac{4 \;\inf(k, \theta(1 + 1/2\nu_{j+1}))}{(1-\kappa)\delta_j + (1+\kappa)}$ (3.6)

When $\dfrac{1+\kappa}{\kappa - 1} \leq \delta_j \leq 0$ $\quad 0 \leq l_j \leq \dfrac{4 \;\inf((k-1), (1-\theta)(1 + 1/2\nu_j))\,\delta_j}{(\kappa - 1)\delta_j - (1+\kappa)}$ (3.7)

then $C_{j+\frac{1}{2}}$ and $D_{j+\frac{1}{2}}$ are positive numbers; the scheme defined in (3.1) is T.V.D.

The proof can be found in Rostand Stoufflet [20].

Note that by using $\theta < 1$, one can avoid the necessity to put the limiter to zero at extremas.

To treat forward and backward discontinuities in the same way, it is necessary to have the symetry condition :

$$\frac{r_i}{l_i} = \frac{1 + \kappa + (1-\kappa)\delta_i}{(1-\kappa) + (1+\kappa)\delta_i} \;. \tag{3.8}$$

If we suppose the unlimited scheme to be second order (resp. third order), we must have to keep the second order except at critical points,

$$r(x) = 1 + 0(|x-1|) \quad , \quad l(x) = 1 + 0(|x-1|) \tag{3.9a}$$

resp.

$$r(x) = 1 + 0(|x-1|^2) \quad , \quad l(x) = 1 + 0(|x-1|^2)\,. \tag{3.9b}$$

It is easy to check that the classical limiters (Van Albada's[26],Van Leer's[29], Superbee [25]) all verify the hypothesis of theorem 2, whatever the value of ν, so that they all are unnecessarily diffusive in the viscous case.

Since we have now found the constraints that the slope limiters must verify for the scheme to the T.V.D., we will investigate the influence of κ.

Our study will be based on truncature error, and comparison with exact solution, so we will use a scalar steady convection diffusion equation:

$$au_{,x} - \epsilon u_{,xx} = 0 \quad , \quad u(0) = 0 \quad , \quad u(1) = 1 \quad , \quad a > 0, \epsilon > 0\,. \tag{3.10}$$

In the unlimited case, equation (3.1) resumes to:

$$\frac{a}{h}\left[\frac{1+\kappa}{4}(u_{i+1}-u_i) + \frac{1-2\kappa}{4}(u_i - u_{i-1}) - \frac{1-\kappa}{4}(u_{i-1}-u_{i-2})\right] - \frac{\epsilon}{h^2}(u_{i+1} - 2u_i + u_{i-1}) = 0 \tag{3.11}$$

Assuming $(u_i)_i$ to be the interpolation of a regular function, with $u_i = u(x_i)$, (3.11) is equivalent to

$$au'_j - \epsilon u''_j + h^2\left(a\frac{3\kappa-1}{12}u'''_j - \frac{\epsilon}{12}u''''_j\right) + 0(h^3) = 0 \tag{3.12}$$

taking the second derivative of (3.10), we obtain : $au'''_j = \epsilon u''''_j$ so that, from (3.12), we have third order if $\kappa = 2/3$. It appears that although the approximation of u''_j is only second order, third order can be obtained in this linear scalar steady case because the errors due to convection and diffusion eliminate each other. The exact solution of (3.10) is :

$$u(x) = \frac{\exp(ax/\epsilon) - 1}{\exp(a/\epsilon) - 1} \tag{3.13}$$

replacing u_i by $u(x_i)$= u(ih) in (3.11) we obtain that, to have nodally exact results, we must have :

$$\left(\frac{1}{2} + \frac{\kappa-1}{4}(1-\exp(-\frac{ah}{\epsilon})\right) - \left(\frac{\epsilon}{ah} - \frac{1}{\exp(ah/\epsilon)-1}\right) = 0 \tag{3.14}$$

if we denote by $\nu = ah/2\epsilon$ the Peclet number, equation (3.14) gives :

$$\kappa = 1 - \frac{2(\coth\nu - 1/\nu)}{1-\exp(-2\nu)} \tag{3.15}$$

where coth is the hyperbolic cotangent. We have $\kappa(\nu) \longrightarrow \frac{2}{3}$ when $\nu \longrightarrow 0$ and $\kappa(\nu) \longrightarrow -1$ when $\nu \longrightarrow \infty$. We find again that when the mesh size tends to zero, the highest order approximation is obtained with $\kappa = 2/3$.

From this study, we conclude that in a boundary layer situation like that defined by (3.10), the best result are obtained with κ given by (3.15), at least for small values of the Peclet number. (If ν is bigger than say 2 or 3, there is no possibility to calculate correctly the boundary layer anyway).

Similar results were obtained by Hughes and Mallet [10] ; they studied the same equation (3.10) and concluded that only a fraction of the inviscid numerical diffusion had to be applied in the viscous case, depending on the Peclet number, in a way very much alike to (3.15).

We now propose the following schemes for the nonlinear equation (1.1). We use (3.1) where κ is calculated from (3.15), at least for small values of ν (big values of ν will yield $\kappa \approx$ -1, so we can limit the lower value of κ to 0, or even to 1/3) ; in (3.15), we use an average Peclet number $\nu = f'(u)h/2\epsilon$, where u is the midpoint of the considered interval (i.e. $u_{i+\frac{1}{2}}$ to calculate $\theta_{i+\frac{1}{2}}$).

We limit the extrapolation, using the value of κ obtained in the preceding step, and one of the following formulas :

- "extended superbee":

$$r_j(\delta_j) = \begin{cases} \max\left(0, \min\left(1, \dfrac{4s\delta_j}{(1-\kappa)+(1+\kappa)\delta_j}\right)\right) & \text{if } \delta_j \leq 1 \\ \dfrac{[\min(((1+\kappa)+(1-\kappa)\delta_i), 4s)]}{(1-\kappa)+(1+\kappa)\delta_i} & \text{if } \delta_j \geq 1 \end{cases} \tag{3.16a}$$

$$l_j(\delta_j) = r_j(\delta_j)\frac{1-\kappa+(1+\kappa)\delta j}{1+\kappa+(1-\kappa)\delta j} \tag{3.16b}$$

$$\text{where} \quad s = \inf\left(k, 1+\frac{1}{2\nu_j}\right). \tag{3.17}$$

The parameter k defines ν_j ; usually $k = 1$ to avoid a too complicated evaluation of ν_j. This is the most compressive limiter that will match the TVD conditions, it has the disadvantage that it is not equal to 1 in a neighborhood of $\delta_i = 1$, impeaching third order accuracy. It does verify the symetry condition (3.8) .

- "third order superbee"

$$r_j(\delta_j) = \begin{cases} \max\left(0, \min\left(1, \dfrac{4s\delta_j}{(1-\kappa)+(1+\kappa)\delta_j}\right)\right) & \text{if } \delta_j \leq 1 \\ 1 & \text{if } 1 \leq \delta_j \leq \dfrac{4s-(1-\kappa)}{1+\kappa} \\ \dfrac{4s}{(1-\kappa)+(1+\kappa)\delta_i} & \text{if } \delta_j \geq \dfrac{4s-(1-\kappa)}{1+\kappa} \end{cases} \tag{3.18.a}$$

$$l_j(\delta_j) = \begin{cases} 1 & \text{if } 1 \geq \delta_j \geq \dfrac{1+\kappa}{4s-(1-\kappa)} \\ r_j(\delta_j)\dfrac{1-\kappa+(1+\kappa)\delta j}{1+\kappa+(1-\kappa)\delta j} & \text{if not} \end{cases} \tag{3.18.b}$$

where s is given by (3.17) .

This limiter preserves third order, but doesn't verify the symetry condition (3.8) for r_j close to 1. Just as the superbee limiter, the ϕ limiters can be extended to an "extended ϕ limiter" and "extended third order ϕ limiter". $\phi = 2$ gives back the superbee. ϕ limiters with ϕ just under 2 are useful for systems, as we will see; ϕ is then a kind of security factor.

-"κ limiter"

It depends on the viscosity through the value of κ only. It is an average of Van Leer's and Van Albada's limiters, derived to be more compressive for high values of κ than for low ones.

$$r_j(\delta_j) = \begin{cases} 0 & \text{if } \delta_j \leq 0 \\ \max\left(\dfrac{4\delta_j}{(1-\kappa)+4\delta_j+(1+\kappa)\delta_j^2}, \dfrac{4\delta_j}{(1-\kappa)(1+\delta_j)+(1+\kappa)(1+\delta_j^2)}\right) & \text{if not} \end{cases} \tag{3.19.a}$$

$$l_j(\delta_j) = \frac{(1-\kappa)\delta_j+1+\kappa}{1-\kappa+(1+\kappa)\delta_j}\, r_j(\delta_j). \tag{3.19b}$$

This limiter is less compressive than the two preceeding one, but has smoother variations.

IV. NUMERICAL RESULTS

We first compared the results of the unlimited scheme ($r_j = l_j = 1$) for different values of κ. A transonic flow at a Mach number of $M_\infty = 0.85$ and a constant Reynolds number of 500 was computed on an undermeshed grid (3114 nodes), for κ = -1 and κ = 2/3. On figure 3, the iso-mach lines are compared. The fully upwind scheme (3b) yields more numerical viscosity as can be seen in the wake, while the third order scheme (κ = 2/3, 9a) allows spurious oscillations behind the schock, although very weak, but gives a better result in the wake and, though less obviously, in the boundary layer. This confirms our statement that different values of κ are needed in different zones, depending on the local Peclet number.

The history of convergence in the case of $\kappa = 2/3$ is presented on fig. 4 ; (4 a) shows the logarithm of the residual versus the number of time steps for the explicit scheme, (4 b) is the same for the implicit scheme (details of the algorihm can be found in[20], (4 c) is a comparison of the two schemes in terms of CPU time on CRAY II,using a vectorized program [2]. Machine-precision convergence is achieved in 150 time steps or 120 seconds by the implicit scheme, while the explicit scheme takes 1200 time steps or 240 seconds to reach a residual superior to 0.001. For a steady calculation, the use of the implicit scheme divides the needed CPU time by more than 10, although the awaited Newton-like quadratic convergence is not achieved.

The Peclet dependent κ,together with the unlimited scheme was used to compute the same flow on an adapted mesh, still rather coarse (fig. 5, 2970 nodes), and on a thinner mesh (fig. 6, 5712 nodes). The mesh, the iso-mach lines and the pressure coefficients on the body are compared. It can be seen that a good solution is obtained on the smaller grid, although quite coarse.

To compare the limiters, a hypersonic flow over a flat plate was computed. The Mach number is $M_\infty = 10$, the Reynolds number $Re/m = 5.10^5$, the length of the plate is 2, the temperature at the inflow is 83,5K the temperature at the wall is 525 K ; Sutherland's law is used for viscosity. A first computation was made with the κ limiter, and the Peclet dependent κ. The mesh (fig. 7 a, partial view), speed vectors (fig. 7 b), pressure coefficient (fig. 7 c) and skin friction coefficient compared with laminar boundary layer theory results (fig. 7 d) are presented.

There are about 15 nodes in the boundary layer. The shock at x = 0 is captured ; no oscillation is seen. The agreement with theory is excellent ; the same flow was computed with the Van Albada limiter and $\kappa = 1/3$, giving extremely similar results (not shown), so we consider this result as a reference, and used a coarser grid. The same flow is computed on a mesh with 8 nodes in the boundary layer, using $\kappa = 1/3$ and the Van Albada limiter for the non linear fields, and the viscosity dependant κ with the κ-limiter for the contacts. The κ limiter was experimented to be too compressive for use on the nonlinear waves. Figure 8 shows the speed vectors. Figure 9 is a comparison of skin friction obtained on the preceeding grid (full), using $\kappa = 1/3$, Van Albada limter for the four fields (long broken), using for the contacts the "third order Φ limiter", with $\Phi = 1.6$ and the Peclet dependent κ (short broken), or the κ limiter and Peclet dependent κ (broken doted). Figure 10 shows the skin friction on the second half of the plate, for schemes which all use $\kappa = 1/3$, Van Albada limiter for the nonlinear fields, and the Peclet dependent κ for the contact discontinuities. It is seen that the "extended superbee", or "extended Φ limiters", are too compressive, even for a contact, but that both the "κ limiter" and "third order extended Φ limiter" with $\Phi \simeq 1.6$ give results in agreement within one or two percents with our reference result, at least on the second half of the plate. This is an improvement over the $\kappa = 1/3$, Van Albada limiter, which gives more than 10 % error.

Another hypersonic computation was performed around an ellipse: the Mach number is $M_\infty = 8$, the Reynolds number is constant and equal to $Re_\infty/m = 1000$, the angle of attack is $\alpha = 40^o$. The mesh and iso-mach lines are shown (fig. 11). There is no over shoot at the shock and with the implicit scheme, the after body flow can be computed without any special treatment ; this cannot be achieved with an explicit code. The mesh was adapted by an automatic local mesh refinement algorithm, due to C. Pouletty [16] and B. Palmerio [15].

An easier calculation was performed to compare the convergence of the explicit and implicit codes; fig. 12 shows the convergence history for a flow around an ellipse at a Mach number of 4, for the explicit and implicit scheme.The mesh has 1378 nodes (not shown).It is seen that the implicit scheme allows schock capturing with a courant number C=100.

V. CONCLUSION

A numerical scheme to solve the compressible Navier-Stokes equations, based on a "TVD" finite volume formulation, has been obtained, by extending a method first derived for perfect gas. We have obtained a condition on the limiters for the scheme to be TVD, taking in account the physical viscosity. Different limiters have been proposed which match this condition, and compared from a numerical point of view. The upwinding also depends on the local amount of physical viscosity. It has been shown that laminar boundary layers can be calculated with our scheme, on a triangular mesh, with less than ten nodes in the boundary layer.

An efficient algorithm has been proposed for the steady case, which allows cheap computation of very stiff problems, as hypersonic flows on geometries with rear body. Really unsteady flow remain a challenge because of their computational cost.

Aknowledgment

P.Rostand is supported by DRET contract n_o 03 40 79 01.

The computations where made on the CRAY II of C.C.V.R.

References

[1] F.Angrand,"Numerical Simulations of Compressible Navier Stokes Flows", GAMM Workshop,M.O.Bristeau,R.Glowinski, J.Periaux,H.Viviand (eds),Friedr. Vieweg und Sohn,1987,p 69-85.

[2] F.Angrand,J.Erhel,"Vectorized F.E.M. codes for compressible flows", Proc. of the 6th International Symposium on Finite Element Methods in flow problems,Antibes,Juin 1896.

[3] K.Baba,M.Tabata,"On a Conservative Upwind Finite Element Scheme for Convection Diffusion Equations",RAIRO Numerical Analysis,15 n1,1981.

[4] G.V.Candler,R.W.Mac Cormack, "Hypersonic Flow past 3D Configurations", AIAA 87-0480,Reno,January 12-16,1987.

[5] S.Chakravarty,"High Resolution Upwind Formulations for the Navier-Stokes Equations",Von Karman Institute Lecture Series 1988-05,March 7-11, 1988.

[6] A.Dervieux,"Steady Euler Simulations using Unstructured Meshes", Von Karman Institute Lecture Series,1983.

[7] B. Engquist and S. Osher, "Stable and entropy satisfying approximations for transonic flow calculations", Math. Comp. V. 34, 1980, pp. 45-75.

[8] B. Engquist and S. Osher,"One sided difference approximations for nonlinear conservation laws", Math. Comp. V. 36, 1981, pp. 321-352.

[9] A. Harten, "High resolution schemes for hyperbolic conservation laws", JCP, Vol. 49, pp. 357-393, 1983.

[10] T.J.R.Hughes,M.Mallet,"A New Finite Element Formulation for Computational Fluid Dynamics:III The Generalized Streamline Operator for Multidimensional Advective-Diffusive Systems", Computer Methods in Applied Mechanics and Engineering,n_o 58,1986,p 305-328.

[11] M.H.Lallemand,"Schemas Decentres Multigrilles pour la Resolution des equations d'Euler en elements finis",Thesis,Universite de Provence,Mars 1986.

[12] S. Osher, "Convergence of generalized MUSCL schemes", ICASE-NASA contractor report 172-306, Feb. 1984.

[13] S.Osher,S.Chakravarty,"Upwind difference schemes for the hyperbolic systems of conservation laws", Mathematics of Computation,April 1982.

[14] S.Osher,S.Chakravarty,"High resolution schemes and the entropy condition", SIAM J. Num. An. Vol. 21, n° 5, October 1985.

[15] B.Palmerio,"Self Adaptive F.E.M. Algorithms for the Euler Equations", INRIA report 338,1985.

[16] C.Pouletty,"Generation et Optimisation de Maillages en Elements Finis,Application a la resolution des equations de la mecanique des fluides",These de Docteur Ingenieur,Ecole Centrale,Dec 1985.

[17] P.L. Roe, "Some contributions to the modelling of discontinuous flows", Proc. AMS/SIAM Seminar, San Diego, 1988.

[18] P.L. Roe, "Finite volume methods for the compressible Navier-Stokes equations", Proceedings of "Numerical methods on Laminar and Turbulent flow", Montreal, 1987.

[19] P.Rostand,Thesis,Univesite Paris VI,in preparation.

[20] P.Rostand,B.Stoufflet, "Finite volume Galerkin Methods for Compressible Viscous Flows", INRIA report,to appear.

[21] R. Sanders, "On convergence of monotone finite difference schemes with variable spatial differencing", Math. Comp. 40, (1983), pp. 91-106.

[22] R.Schwane,D.Hanel,"Computation of Viscous Supersonic Flows around Blunt Bodies,7th GAMM Conference on Numerical Methods in Fluid Mechanics,Louvain la Neuve,1987.

[23] B. Stoufflet,L.Fezoui,"A Class of Implicit Upwind Schemes for Euler Simulations with Unstructured Meshes",to appear.

[24] B. Stoufflet,J.Periaux,L.Fezoui,A.Dervieux,"Numerical Simulations of 3-D Hypersonic Euler Flows Around Space Vehicles Using Adapted Finite Elements",AIAA paper 87 0560 Reno 1987.

[25] P.K. Sweby, "High resolution schemes using flux limiters for hyperbolic conservation laws", SIAM J. of Num. An., Vol. 21, n° 5, October 1984.

[26] G.D. Van Albada, B. Van Leer, W.W. Roberts, Jr. (1982), Astron. Astrop. 108, 1976.

[27] B.Van Leer,J.L.Thomas,P.L.Roe,R.W.Newsome,"A Comparison of Numerical Flux Formulas for the Euler and Navier Stokes Equations",proceedings of the AIAA Honolulu meeting,1987,p 36-41.
[28] B.Van Leer,"Computational Methods for Ideal Compressible Flow",Von Karman Institute for fluid dynamics,lecture series 1983-04,Computational Fluid Dynamics, March 7-11 1983.
[29] B. Van Leer, "Towards the ultimate conservative difference scheme, II. Monotonicity and conservation combined in a second order scheme", JCP, 14 (1974), pp. 361-370.
[30] H.C.Yee,"Upwind and Symetric Shock-Capturing Schemes",NASA TM 89464,May 1987.

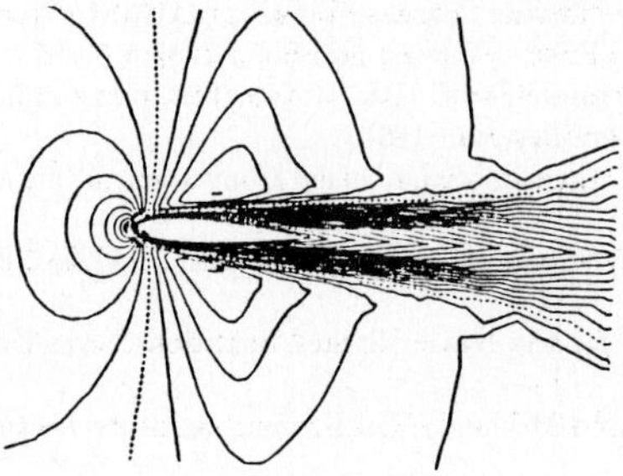

fig. 3a : $\kappa = \frac{2}{3}$

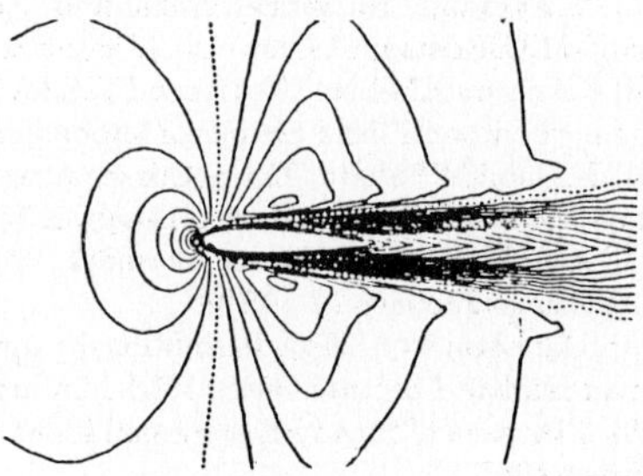

fig. 3b : $\kappa = -1$

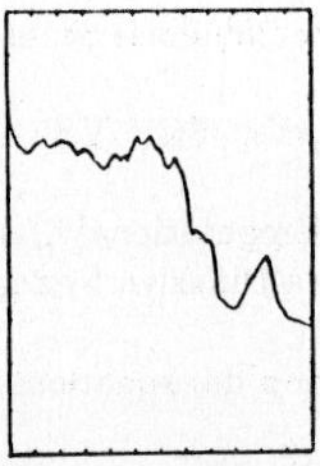

fig. 4a

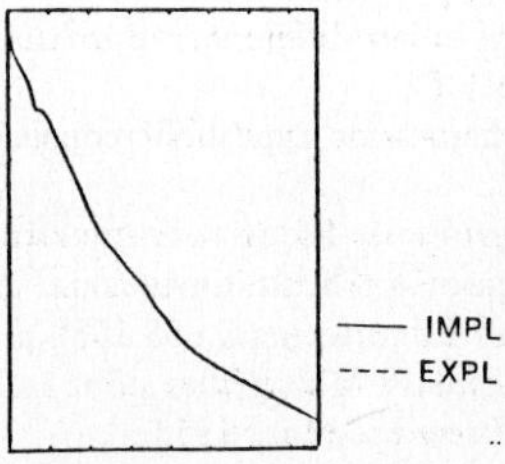

fig. 4b

fig. 4c

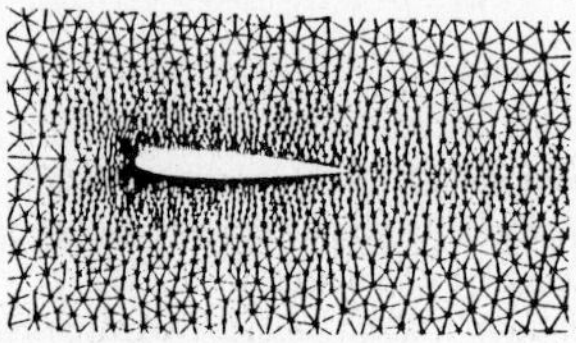

fig. 5a : 2970 Nodes,

5764 Elements (partial view)

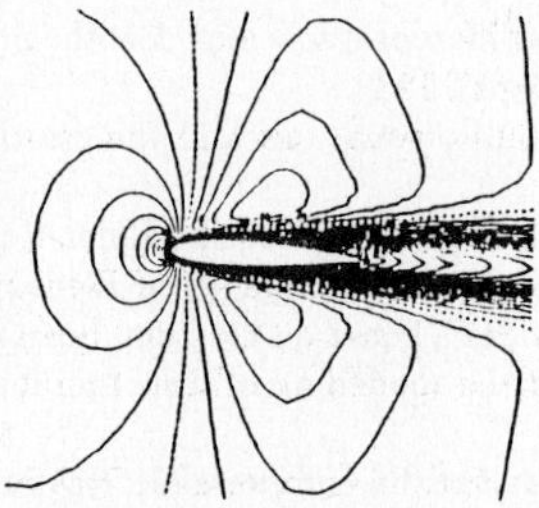

fig. 5b : iso Mach lines

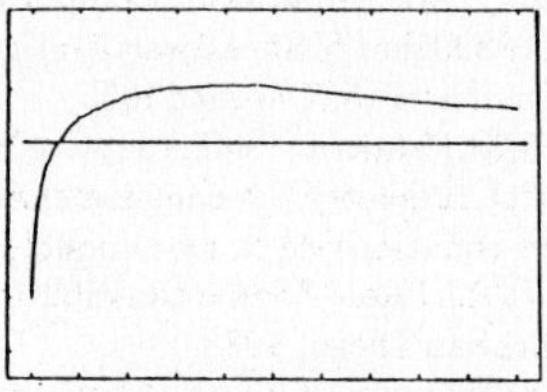

fig. 5c : pressure coefficient

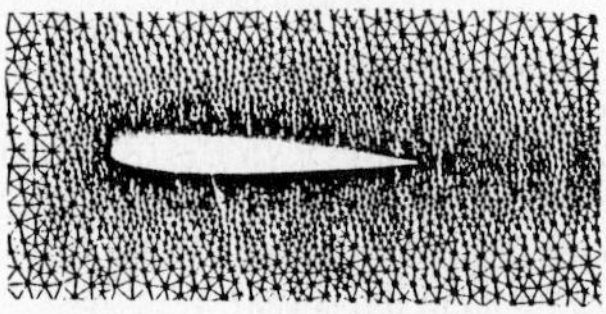

fig. 6a : 5712 Nodes,

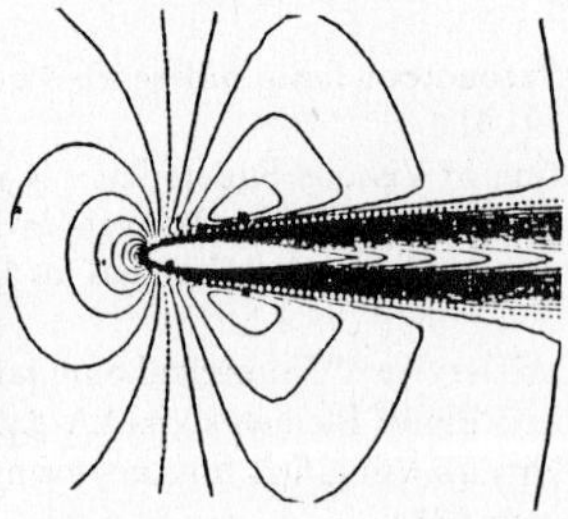

fig. 6b : iso Mach lines

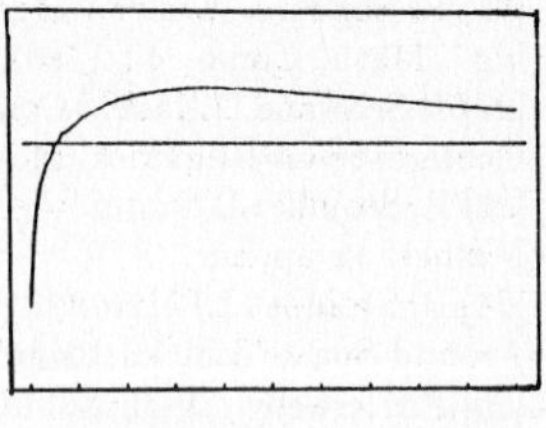

fig. 6c : pressure coefficient

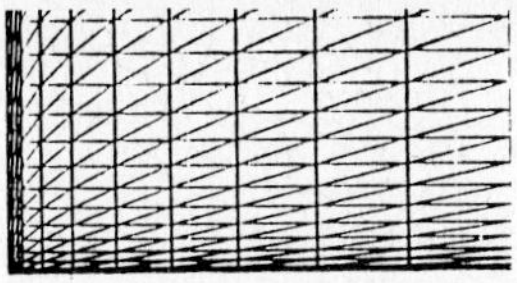

fig. 7a : mesh (partial view)

fig. 7b : speed vectors

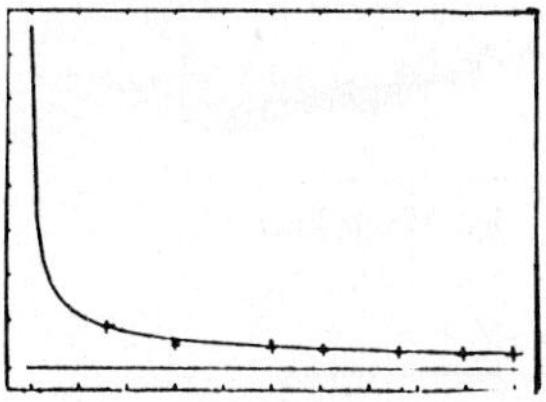

fig. 7c : skin friction coefficient

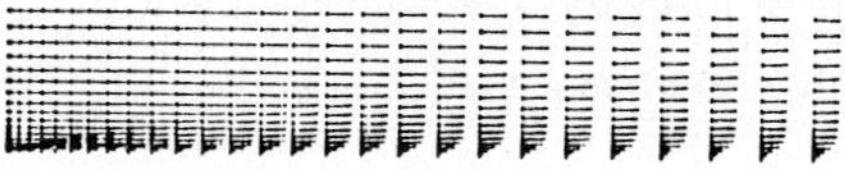

fig. 8 : speed vectors

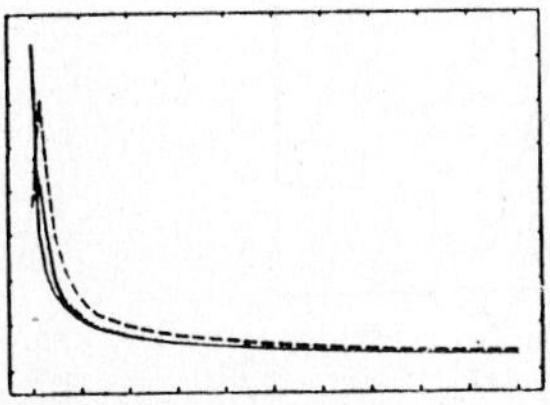

fig. 9

comparison of skin friction for different limiters. Long broken: Van Albada limiter, short broken: third order extended ϕ limiter, broken dotted: κ limiter, full: reference result.

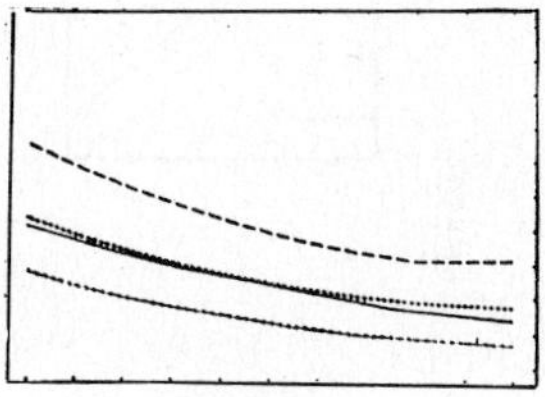

fig. 10a

comparison of skin friction for different limiters, enlargement. Long broken: Van Albada limiter, short broken: κ limiter, full: reference result, broken dotted: third order extended superbee .

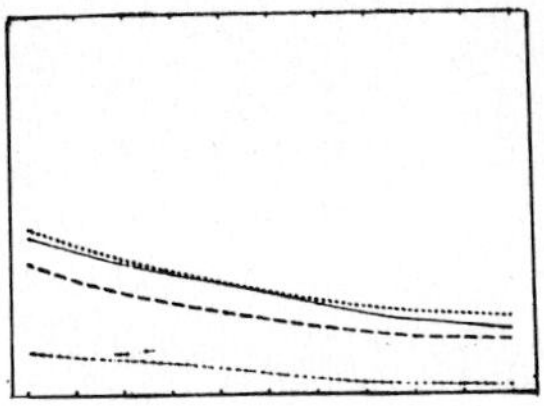

fig. 10b

comparison of skin friction for different limiters, enlargement. Long broken: extended ϕ limiter, short broken: κ limiter, full: reference result, broken dotted: extended superbee .

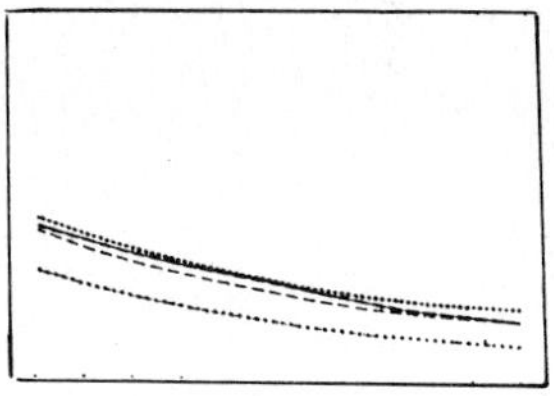

fig. 10c

comparison of skin friction for different limiters, enlargement. Long broken: extended third order ϕ limiter, short broken: κ limiter, full: reference result, broken dotted: extended third order superbee .

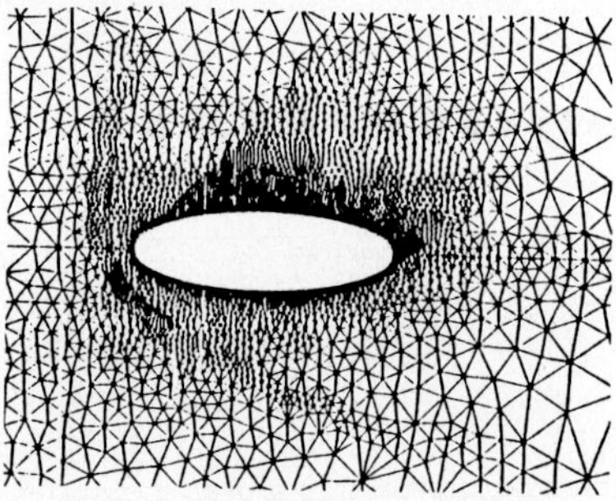

fig. 11a : 5172 Nodes, 9986 Elements (partial view)

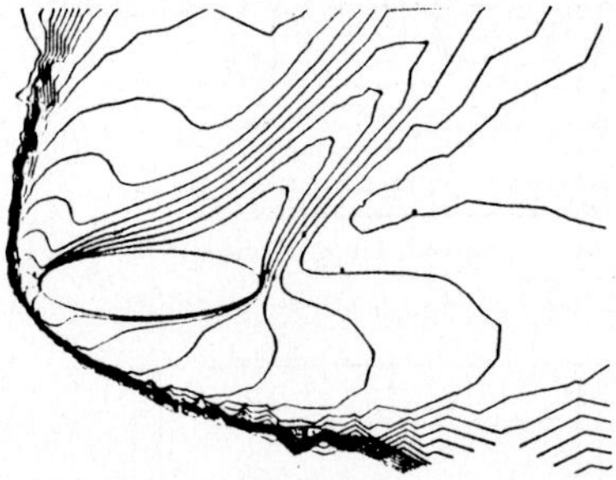

fig. 11b : iso Mach lines

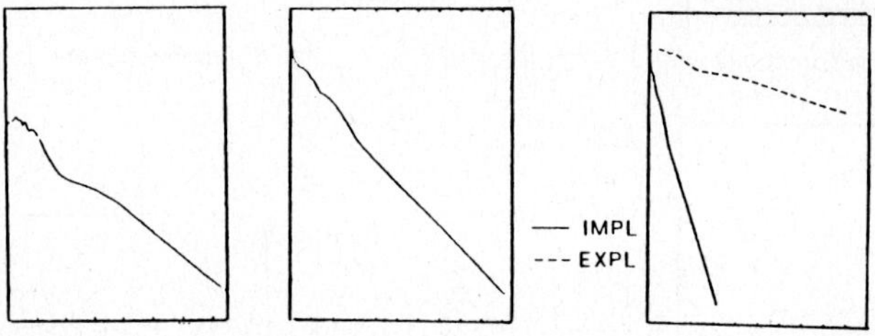

fig. 12

NONSTATIONARY SHOCK WAVE GENERATION IN DROPLET VAPOUR MIXTURES

P. Schick[1)] and K. Hornung

Universität der Bundeswehr München, Fakultät LRT

D-8014 Neubiberg, W.-Germany

SUMMARY

If a piston is accelerated into a tube filled with gas, shock waves will be generated. A mixture of droplets and their own vapour will show the same behaviour, if the fraction of the liquid phase is small. But when the fraction of the droplets becomes greater, a significant shock damping may occur. This process has been investigated by a method of characteristics, simultaneously taking into account mass-, momentum- and energy transfer between the phases. The differential equations can be formulated in a well posed way. Local linear stability is present as well as convergence. The method uses two grids, one for the gas phase and one for the droplet phase. CPU-time and storage is saved by controlling the number of characteristics to be calculated. Example calculations show that the relaxation zone consists of three distinct parts as predicted by earlier stationary investigations: first a fast condensation zone, then a friction zone and finally a heat conduction zone.

INTRODUCTION

A mixture of homogeneously dispersed droplets together with their own vapour and no additional carrier gas is considered. The flow of such fluids has many applications. For example in steam turbines, reactor cooling, heat pipes, cryo-techniques etc. One most important feature of such fluids is that they can transport a large amount of latent energy by varying the liquid/gas mass ratio. The consideration of the phase change, vaporization and condensation, is essential.

In this paper some fundamental aspects of the fluid being at thermodynamic nonequilibrium are discussed. This situation is of practical importance since often the time scales of nonstationary effects and of transport processes happen to be in the same order of magnitude.

As a numerical example the shock wave generation by an accelerated piston is selected because it is a classical fundamental problem. Moreover shock waves form an excellent tool for studying nonequilibrium phenomena at very defined conditions, f.e. in a shock tube. Finally wet vapour shock waves may form in practical flow situations as well, f.e. in accident situations.

Previous work is on the two phase piston-cylinder problem with solid particles in gases or on droplets in a carrier gas (see f.e. Refs. [1],[2]). There is also a stationary analysis of the nonequilibrium wet vapour shock wave [3]. The present analysis includes the full nonstationary treatment under the conditions of general phase nonequilibrium up to high liquid mass fractions. An example will be discussed for water as the working substance.

CONSERVATION- AND TRANSPORT EQUATIONS

A number of model assumptions is necessary to arrive at a tractable set of differential equations. One of them being the treatment of the fluid system as two interacting continua, characterized by the volume fractions ε_l and ε_g of liquid and gas. These give volume averaged densities if multiplied by the individual real densities ρ_l and ρ_g. n_p is the inverse volume of one droplet. The number of droplets per unit volume of the flow field is $\varepsilon_l n_p$.

Transport of mass, momentum and energy between the droplets and the surrounding vapour are considered. The transport within the single phases like gas phase viscosity or heat conduction within the droplet is neglected. This is possible, because there are no boundary layers in the flow field and the surface to volume ratio of micron sized droplets is large. It is important to consider droplet-droplet interactions if one deals with high liquid mass loadings ($1-\varphi > 50\%$, φ = gas mass fraction). Moreover a consideration of the nonuniform pressure distribution around the droplet is essential. The flow is one-dimensional in space. With these assumptions the conservation equations are written in flux form as

droplet number

$$\partial/\partial t\,(n_p\varepsilon_l) + \partial/\partial x\,(n_p\varepsilon_l v_l) = 0\ , \tag{1}$$

mass

$$\partial/\partial t\,(\varepsilon_g\rho_g) + \partial/\partial x\,(\varepsilon_g\rho_g v_g) = (n_p\varepsilon_l)\,I\ , \tag{2}$$

$$\partial/\partial t\,(\varepsilon_l\rho_l) + \partial/\partial x\,(\varepsilon_l\rho_l v_l) = -\,(n_p\varepsilon_l)\,I\ , \tag{3}$$

momentum

$$\partial/\partial t\,(\varepsilon_g\rho_g v_g) + \partial/\partial x\,(\varepsilon_g\rho_g v_g v_g) = -\,\partial/\partial x\,(\varepsilon_g p) - (p-\zeta\rho_g(v_l-v_g)^2)\,\partial/\partial x\,(\varepsilon_l) + (n_p\varepsilon_l)\,(F_D+I\,v_l), \tag{4}$$

$$\partial/\partial t\,(\varepsilon_l\rho_l v_l) + \partial/\partial x\,(\varepsilon_l\rho_l v_l v_l) = -\,\partial/\partial x\,(\varepsilon_l p) + (p-\zeta\rho_g(v_l-v_g)^2)\,\partial/\partial x\,(\varepsilon_l) - (n_p\varepsilon_l)\,(F_D+I\,v_l), \tag{5}$$

energy

$$\partial/\partial t\,(\varepsilon_g\rho_g(v_g^2/2+u_g)) + \partial/\partial x\,(\varepsilon_g\rho_g v_g(v_g^2/2+u_g)) = -\,\partial/\partial x\,(\varepsilon_g p v_g) - v_l(p-\zeta\rho_g(v_l-v_g)^2)\,\partial/\partial x\,(\varepsilon_l) + (n_p\varepsilon_l)\,(Q+F_D v_l+I\,(v_l^2/2+h_g(T_l)))\ , \tag{6}$$

$$\partial/\partial t\,(\varepsilon_l\rho_l(v_l^2/2+u_l)) + \partial/\partial x\,(\varepsilon_l\rho_l v_l(v_l^2/2+u_l)) = -\,\partial/\partial x\,(\varepsilon_l p v_l) + v_l(p-\zeta\rho_g(v_l-v_g)^2)\,\partial/\partial x\,(\varepsilon_l) - (n_p\varepsilon_l)\,(Q+F_D v_l+I\,(v_l^2/2+h_g(T_l)))\ . \tag{7}$$

Here v_l, v_g are the liquid and gas phase velocities, u_l, u_g the internal energies, $h_g(T_l)$ is the gas enthalpy at the temperature T_l of the liquid and ς is defined below (Equ.(24)). There is one extra equation for the number of droplets. The zero on its right hand side means that no new droplets are formed by nucleation, and droplets do not vanish completely upon evaporation. The mass equations contain source terms I for net vaporization. The pressure forces in the momentum equations contain a term which is multiplied by the change of the liquid volume fraction ($\partial\varepsilon_l/\partial x$). Although being small, this term is essential for the numerical stability and for the hyperbolic nature of the system of equations. Physically it contains the nonuniform pressure distribution around the droplet mentioned before, combined with the effects of cutting droplets by the boundaries of the volume element according to Stuhmiller [4]. Various source terms in the momentum-

and energy equations depend on the net evaporation rate I. F_D is the drag force and Q is the heat flux by conduction.

After differentiation and inserting mass- into momentum- as well as momentum- into energy equations, by furthermore assuming ideal gas with constant specific heat for the gas phase, the conservation equations are written in a characteristic form

gas

$$(\partial/\partial t + v_g\, \partial/\partial x)(\varepsilon_g p) - c^2 (\partial/\partial t + v_g\, \partial/\partial x)(\varepsilon_g \rho_g) = (\gamma-1)(R_e - (c^2/\gamma) R_m), \tag{8}$$

$$(\partial/\partial t + (v_g + c)\, \partial/\partial x)(\varepsilon_g p) + c(\varepsilon_g \rho_g)(\partial/\partial t + (v_g + c)\, \partial/\partial x)(v_g) = (\gamma-1) R_e + (c^2/\gamma) R_m + c R_i, \tag{9}$$

$$(\partial/\partial t + (v_g - c)\, \partial/\partial x)(\varepsilon_g p) - c(\varepsilon_g \rho_g)(\partial/\partial t + (v_g - c)\, \partial/\partial x)(v_g) = (\gamma-1) R_e + (c^2/\gamma) R_m - c R_i, \tag{10}$$

where

$$R_m = (n_p \varepsilon_l) I,$$

$$R_i = (p - \zeta \rho_g (v_l - v_g)^2)\, \partial/\partial x (\varepsilon_l) + (n_p \varepsilon_l)(F_D + I(v_l - v_g)),$$

$$R_e = -(v_l - v_g)(p - \zeta \rho_g (v_l - v_g)^2)\, \partial/\partial x (\varepsilon_l) + (n_p \varepsilon_l)(Q + F_D (v_l - v_g) + I((v_l - v_g)^2/2 + h_g(T_l) - u_g)),$$

droplets

$$(\partial/\partial t + v_l\, \partial/\partial x)(n_p \varepsilon_l) = -(n_p \varepsilon_l)\, \partial/\partial x (v_l), \tag{11}$$

$$(\partial/\partial t + v_l\, \partial/\partial x)(\varepsilon_l \rho_l) = -(\varepsilon_l \rho_l)\, \partial/\partial x (v_l) - (n_p \varepsilon_l) I, \tag{12}$$

or:

$$(\partial/\partial t + v_l\, \partial/\partial x)((\varepsilon_l \rho_l)/(n_p \varepsilon_l)) = -I,$$

$$(\varepsilon_l \rho_l)(\partial/\partial t + v_l\, \partial/\partial x)(v_l) = -\partial/\partial x (\varepsilon_l p) + (p - \zeta \rho_g (v_l - v_g)^2)\, \partial/\partial x (\varepsilon_l) - (n_p \varepsilon_l) F_D, \tag{13}$$

$$(\varepsilon_l \rho_l)(\partial/\partial t + v_l\, \partial/\partial x)(u_l) - (\varepsilon_l p)/(n_p \varepsilon_l)(\partial/\partial t + v_l\, \partial/\partial x)(n_p \varepsilon_l) = -(n_p \varepsilon_l)(Q + I h_v), \tag{14}$$

where c is the gas phase velocity of sound. Note that the right hand sides contain spacial derivatives. These form small correction terms for the gas but they are essential for the droplets. This will be discussed below.

Besides using ideal gas law, the constitutive equations contain the temperature dependence of the gas phase viscosity η, thermal conductivity λ, heat of vaporization h_v, liquid density ρ_l, and vapour pressure p_s. From these the interphase transport equations follow as

$$I = \pi d^2 \rho_g \beta_j \alpha_c c (p_s(T_l)/p - 1) \quad ([3],[5]), \tag{15}$$

$$\beta_j = 1/(1 + Kn((4/3 Kn + 0.71)/(Kn + 1) - 4/3 + 4/(3\alpha_j))) \quad ([5]), \tag{16}$$

$$\alpha_j = 1, \tag{17}$$

$$\alpha_c = 0.26 \quad ([3]), \tag{18}$$

$$F_D = \pi d^2 c_D \varepsilon_g^{-2.7} \rho_g/2\, (v_l - v_g)|v_l - v_g| \beta_f, \tag{19}$$

$$c_D = \text{const.} \quad ([6]),$$

$$\varepsilon_g^{-2.7}\text{: droplet interaction} \quad ([4]), \tag{20}$$

$$\beta_f = 1/(1/(1 + 0.42 Kn) + 1.67 Kn/\alpha_f) \quad ([7]), \tag{21}$$

$$\alpha_f = 1, \tag{22}$$

$(p-\zeta\rho_g(v_l-v_g)^2)\ \partial/\partial x\ (\varepsilon_l)$ ([4]), (23)

$\zeta = 0.37 c_D \varepsilon_g^{-2.7}$ ([8],[4]) (24)

(nonsymmetric pressure distribution around the droplet),

$Q = (\pi d)\ Nu\ \lambda\ (T_l-T_g)\ \beta_q$, (25)

$\beta_q = 1/(1/(1+Kn)+3.75Kn/\alpha_q)$ ([7]) , (26)

$\alpha_q = 1$. (27)

The net mass transport is driven by the difference of vapour pressure and gas pressure. It is not limited by diffusion in the gas phase like in the case of an additional carrier gas. A Knudsen correction accounts for small droplets (Kn = Knudsen number). For the drag force a constant drag coefficient is used as suggested by recent shock tube data. There is a correction for droplet-droplet interaction, Knudsen- and nonuniform pressure corrections. Heat conduction uses standard Nusselt formulation and a Knudsen correction.

STABILITY ANALYSIS

As mentioned before, there are derivatives remaining on the right hand sides of the characteristic equations. So the system can not be of strict hyperbolic nature. Hence one can only investigate local linear stability from which convergence can be expected through Lax's equivalence theorem and by experience. This is done by applying a method described by Ramshaw and Trapp [9] . The set of equations is written in matrix form. The coefficients are assumed being locally constant, including incompressibility of the gas phase.

Scheme of the set of equations $\mathbf{A}(\partial \mathbf{v}/\partial t) + \mathbf{B}(\partial \mathbf{v}/\partial x) + \mathbf{c} = 0$ (**A**,**B**,**c** are constant) , (28)

Eigenvalues $|\mathbf{B}-\lambda_{ei}\mathbf{A}| = 0,\ i = 1,...7$, (29)

$\lambda_{e1},...\lambda_{e5}$: v_l, v_l, v_l, v_g, v_g (all real) ,

$\lambda_{e6},.\lambda_{e7}$ are real, if $\varepsilon_l\rho_g\zeta + \varepsilon_g\rho_g\zeta \geq \varepsilon_g\rho_g\varepsilon_l$ (30)

or $0.37 c_D \varepsilon_g^{-2.7} \geq \varepsilon_l$. (31)

By neglecting the dynamic pressure corrections we get, as usual, real eigenvalues for the gas characteristics, but not so for the droplets. If, on the other hand, these corrections are included, the gas characteristics still remain real and positive and the droplet eigenvalues are real if the above condition for ε_l is fulfilled. In the examples discussed, this is the case. Thus the variable pressure distribution as well as the particle-particle interaction ensure local stability.

In the following a first order method of characteristics is defined. For the gas the three characteristic directions $v+c$, $v-c$ and v are used. The nonisentropic changes are transported along these directions. The difference scheme for the gas phase is

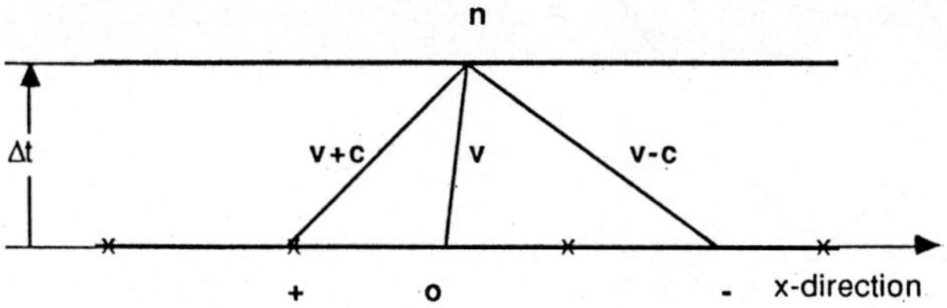

Fig. 1 Gas phase characteristics.

For constant time steps the $v+c$ - characteristics are computed to get the new spacial coordinates. The base points of the remaining two characteristics are interpolated.

$$(v_g)_n = (((\gamma-1)R_e + (c^2/\gamma)R_m) + c\,R_i)_+\Delta t - ((\gamma-1)R_e + (c^2/\gamma)R_m) - c\,R_i)_-\Delta t \qquad (32)$$
$$(v_g\,c\,\varepsilon_g\rho_g)_+ + (v_g\,c\,\varepsilon_g\rho_g)_-)/((c\,\varepsilon_g\rho_g)_+ + (c\,\varepsilon_g\rho_g)_-),$$

$$(\varepsilon_g p)_n = (\varepsilon_g p)_+ - (c\,\varepsilon_g\rho_g)_+ ((v_g)_n - (v_g)_+) + ((\gamma-1)R_e + (c^2/\gamma)R_m) + c\,R_i)_+\Delta t\ , \qquad (33)$$

$$(\varepsilon_g\rho_g)_n = (\varepsilon_g\rho_g)_o + ((\varepsilon_g p)_n - (\varepsilon_g p)_o)/c^2{}_o - ((\gamma-1)(R_e - (c^2/\gamma)R_m))_o\Delta t/c^2{}_o\ , \qquad (34)$$

$$R_m = (n_p\varepsilon_l)\,I\ , \qquad (35)$$

$$R_i = (p-\zeta\rho_g(v_l-v_g)^2)\,\Delta\varepsilon_l/\Delta x + (n_p\varepsilon_l)\,(F_D + I\,(v_l-v_g))\ , \qquad (36)$$

$$R_e = -(v_l-v_g)\,(p-\zeta\rho_g(v_l-v_g)^2)\,\Delta\varepsilon_l/\Delta x + (n_p\varepsilon_l)\,(Q+F_D\,(v_l-v_g) + I\,((v_l-v_g)^2/2 + h_g(T_l) - u_g))\ . \qquad (37)$$

For the droplets there is only one characteristic direction given by the velocity v_l

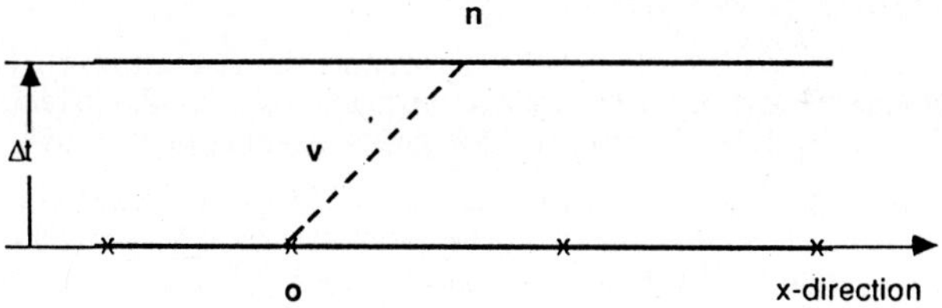

Fig. 2 Droplet characteristic.

$$(n_p\varepsilon_l)_n = (n_p\varepsilon_l)_o - ((n_p\varepsilon_l)\,\Delta v_l/\Delta x)_o\Delta t\ , \qquad (38)$$

$$(\varepsilon_l\rho_l)_n = (\varepsilon_l\rho_l)_o - ((\varepsilon_l\rho_l)\,\Delta v_l/\Delta x)_o\Delta t - ((n_p\varepsilon_l)\,I)_o\Delta t\ , \qquad (39)$$

or:

$$((\varepsilon_l\rho_l)/(n_p\varepsilon_l))_n = (\varepsilon_l\rho_l)_o/(n_p\varepsilon_l)_o - I_o\Delta t\ ,$$

$$(v_l)_n = (v_l)_o - ((\Delta(\varepsilon_l p)/\Delta x + (p-\zeta\rho_g(v_l-v_g)^2)\,\Delta\varepsilon_l/\Delta x + (n_p\varepsilon_l)\,F_D)_o/(\varepsilon_l\rho_l)_o)\Delta t\ , \qquad (40)$$

$$(u_l)_n = (u_l)_o + ((\varepsilon_l p)_o/(\varepsilon_l\rho_l)_o\,((n_p\varepsilon_l)_n/(n_p\varepsilon_l)_o - 1) + ((n_p\varepsilon_l)/(\varepsilon_l\rho_l)\,(Q + I\,h_v))_o\Delta t\ . \qquad (41)$$

The left side boundary is given by an accelerated piston within a certain time (40 µsec in the example discussed below). Then the piston

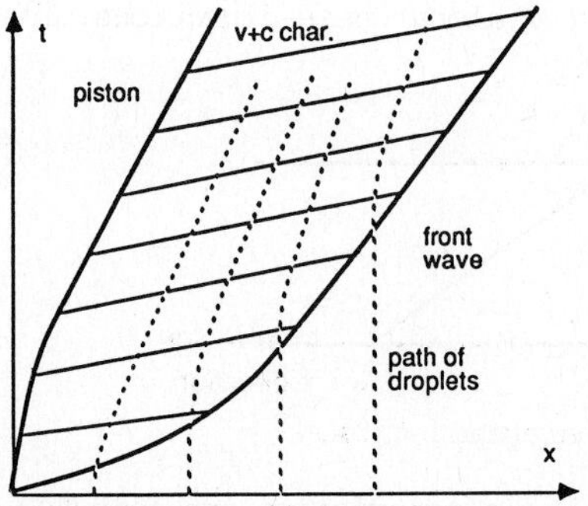

Fig. 3 Numerical grids (without the lines t= const.).

velocity remains constant (in the example below its Mach number is Ma_p=1.25 using the upstream velocity of sound). The boundary condition is nonisentropic. The piston is assumed to form a thermally insulating wall. Droplets which hit the piston get trapped. A v-c - characteristic gives information about the pressure in front of the piston.

The right side boundary consists of a Mach wave until a shock wave is generated by the intersection of two v+c - characteristics. Here the first v+c - characteristic behind the shock wave is combined with the changes along the direction of the shock front, by using a differentiated form of the gas phase jump conditions [10]. The two families of curves shown in figure 3 form two independent grids together with the lines t=const.

During the course of the calculations the number of characteristics increases with each time step. In order to arrive at reasonable storage requirements and computing time, the number of characteristics is reduced at a later stage if the gradients along the characteristic lines become sufficiently small. A change larger than 0.02 of the value at the base point should not be used because of stability problems occuring. Convergence is proved in the usual way using calculations with reduced time steps. The order of convergence is between 0.5 and 1 .

For the initial condition a tube is assumed, filled with water vapour at 20 mbar at room temperature (phase equilibrium). The diameter of the droplets is 70 µm . The initial liquid mass fraction is 65% (φ_o= 0.35).

Example calculations show profiles for different times up to 4 msec . In the pressure profiles the intial jump across the vapour shock front can be seen, which is then followed by an additional strong increase due to the acceleration of the droplets and conversion of their kinetic energy into thermal energy of the gas phase (Fig. 4). The temperature profiles (Fig. 5) also show the initial shock and the following dissipation of the droplet`s kinetic energy. The dotted line is the droplet`s temperature. As predicted by earlier stationary studies by Marble [3], it first rises behind the shock front by condensation until the droplet`s vapour pressure has reached the pressure of the surrounding gas. From this moment on any transfer of heat from the gas to the droplet leads to a slight evaporation, thus stabilizing the droplet`s temperature by vaporization cooling. The parameter $1-p_s/p$ indicates the direction of the phase change (positive: condensation, negative: evaporation). Temperature equilibrium is not reached for the present example.

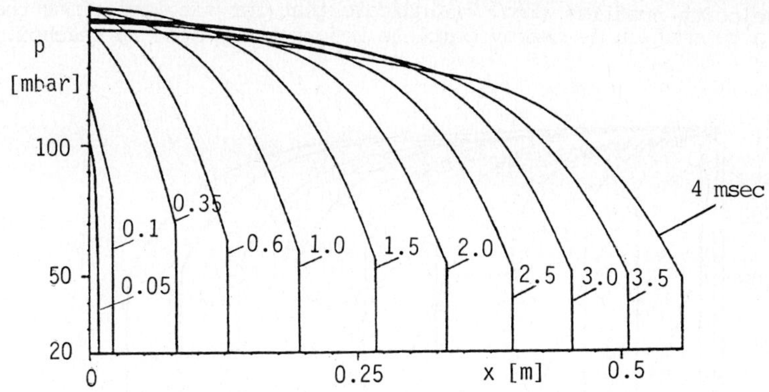

Fig. 4 Pressure profiles for the example H_2O, T_o = 293K, Ma_p = 1.25, φ_o = 0.35, r_o = 35 µm, c_D = 70 (x is the distance from the moving piston).

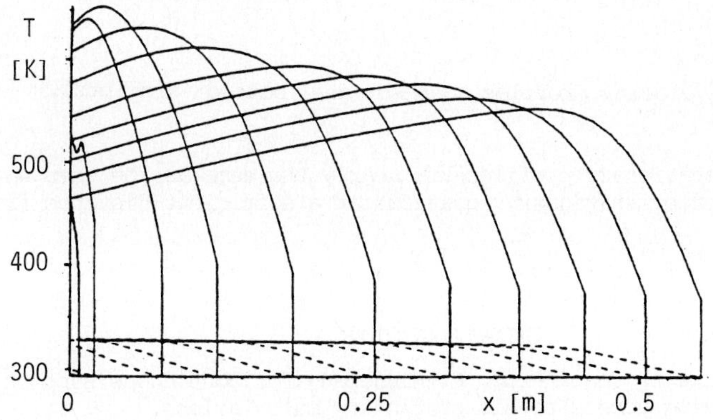

Fig. 5 Temperature profiles (dashed: gas, dotted: droplets).

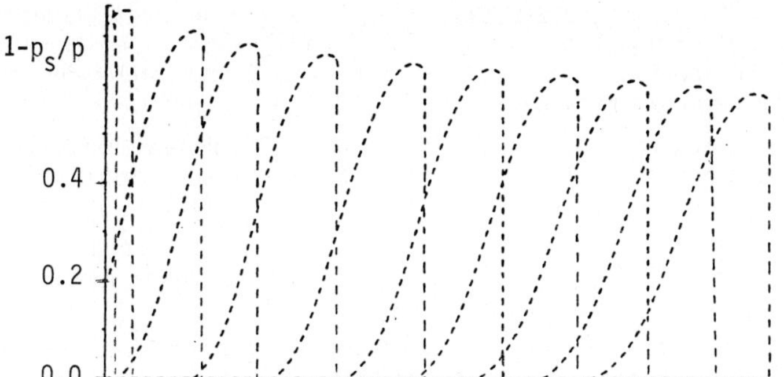

Fig. 6 Profiles of the parameter $1-p_s/p$.

The velocity profiles (Fig. 7) indicate that for the late times the droplet`s acceleration is nearly complete before the piston is reached.

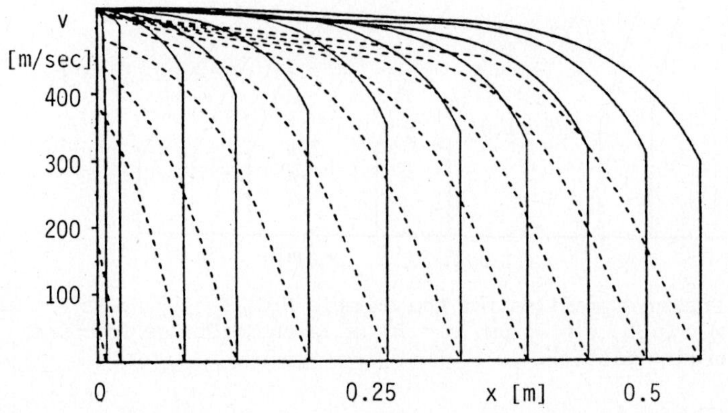

Fig. 7 Velocity profiles (dashed: gas, dotted: droplets).

The quasi-stationary profiles are nearly the same as the ones which can be calculated by stationary equations at a sufficient distance from the piston.

ACKNOWLEDGEMENTS

We would like to thank Prof. F. Hindelang for continuous support and P. Luger for giving access to his stationary calculations.

REFERENCES

[1] Marconi F., Rudman S., Calia V., "One dimensional unsteady two-phase flows with shock waves", AIAA paper 80-1448 (1980).

[2] Ivandeev A.I., Kutushev A.G., Nigmatulin R.I., "Numerical investigation of the expansion of a cloud of dispersed particles or drops under the influence of an explosion", Fluid Dynamics No. 1 (1982), pp. 68-74.

[3] Marble F.E., "Some gasdynamic problems in the flow of condensing vapors", Astron. Acta 14 (1969), pp. 585-614; "Dynamics of dusty gases", Ann. Rev. of Fluid Mech. 2 (1970), pp. 397-448.

[4] Stuhmiller J.H., "The influence of interfacial pressure forces on the character of two-phase flow model equations", Int. J. of Multiphase Flow 3 (1977), pp. 551-560.

[5] Roth P., Fischer R., "An experimental shock wave study of aerosol droplet evaporation in the transition regime", Phys. Fluids 28, 6 (1985), pp. 1665-1672 .

[6] Luger P., Hindelang F., Hornung K., "Time-resolved laser doppler anemometry applied to shock waves in wet steam", Proc. 16th Int. Symp. on Shock Tubes and Waves, Aachen 1987, Ed. H. Grönig, VCH Verlag, Weinheim (1988), pp. 839-845.

[7] Reichelt B., Roth P., Wang L., "Stoßwellen in Gas-Partikel - Gemischen mit Massenaustausch zwischen den Phasen", Wärme- und Stoffübertragung 19 (1985), pp. 101-111.

[8] Schlichting H., Boundary Layer Theory, Mc Graw Hill, N.Y. (1959).

[9] Ramshaw J.D., Trapp J.A., "Characteristics, stability and short wavelength phenomena in two-phase flow equation systems", Nucl. Sci. and Engineering 66 (1978), pp. 93-102.

[10] Sardei F., private communication (1986).

1) Present adress: BMW AG, AQ 32, D-8000 München 40, W.-Germany.

NONLINEAR RESONANCE PHENOMENA FOR THE EULER - EQUATIONS COUPLED WITH CHEMICAL REACTION - KINETICS

St.U. Schöffel

Mechanische Verfahrenstechnik und Strömungsmechanik

Fachbereich Maschinenwesen, Universität Kaiserslautern

Erwin-Schrödinger-Straße, D-6750 Kaiserslautern, FRG

SUMMARY

The investigation of detonation-dynamics and its cellular structure for gaseous, explosive mixtures is of importance in order to find safety-relevant criteria for the onset, respectively failure of detonative combustion. It was shown by the author in [1] by comparing characteristic time- and length-scales that a hyperbolic mathematical model, which ignores dissipative transport-processes is appropriate, in order to describe the cellular detonation-structure. The detonation-dynamics may therefore be investigated by the inviscid, non-conductive and non-diffusive model of the gasdynamic EULER-equations together with kinetic balance-equations for the (adiabatic) induction-reaction and the exothermic reaction. Latter describes the rate for release of chemically bounded heat. A steady solution of the model-equations constitutes the ZEL´DOVICH-DÖRING-V.NEUMANN (ZDN) - equilibrium structure. In this paper stability-criteria for the onset of transverse waves and unsteady nonequilibrium-processes are investigated numerically starting from ZDN - initial data for global reaction-kinetics.

The equations are solved by a two-step predictor-corrector MACCORMACK-scheme, which is monotonized by adding nonlinear flux-limiters. In the case of a linear transport-equation V.NEUMANN stability-analysis may be performed. The stability-bound for the COURANT-number is determined by neglecting the nonlinear source-terms. Exploiting the stability-properties of the applied scheme a spontaneous establishment of many experimentally observed flow-phenomena, e.g. Mach-stems, transverse waves and oscillatory behavior can be obtained. The calculations are performed for two-dimensional channel-geometry in a shock-fixed, detonation-front oriented frame of reference (GALILEI-transformation).

PREFACE

The author is grateful to Prof. Dr. Fritz Ebert for stimulating his research in the attractive field of gasdynamics of combustion and reactive flow-phenomena. This paper is devoted to Prof. Ebert on the occasion of his 50th birthday in November 88.

INTRODUCTION

Combustion phenomena occurring in jet propulsion-systems, particularly in closed combustion-chambers and afterburners are known to be associated with self-excited nonlinear oscillations (see for instance the well-known phenomenon of the singing flame). An intimately strong coupling between the heat released due to exothermic chemical reactions and the created pressure-waves occurs for detonative combustion.

In spite of rich experimental evidence the transition phenomenon of a high-speed deflagration to a shock-induced detonation still lacks a profound theoretical explanation. An appropriate mathematical model for studying the transition phenomenon must consider the different transport mechanisms for deflagrative, respectively detonative combustion. Whereas a pure flame-front propagation (i.e. a deflagration) requires molecular transport-processes (heat conduction and species diffusion), a detonation-process is sustained by self-ignition caused by adiabatic compression due to a shock-wave with sufficiently high mach-number.

Since the beginning of the 80s of this century large-scale experiments connected with light-water reactor safety have been world-wide performed. The experimental results provide evidence that the detonation-limits are geometry-dependent and no pure material-properties. They indicate, however, that the detonation-limits may be scaled by the detonation cell-size. Latter constitutes the most important dynamic detonation-parameter (J. Lee [2]), since it is a reference-value for the other parameters like the critical initiation-energy or the critical transmission-distance from a confined into an unconfined configuration.

Self-sustained detonation propagation is known to be connected with transverse wave-phenomena (cellular detonation-structure). In a dynamic detonation-process the transverse pressure-waves are reflected shock-waves of a single Mach-reflexion of the precursor detonation-front. For planar channel-geometry the right- and left-running family of transverse waves must be distinguished. Additionally in rectangular, spatial geometry the family of slapping and galloping transverse waves may be discerned (see W. Fickett & W.C. Davis for further reference [3]). The formation (birth) of a detonation-cell is caused by collision, resp. focussing of the transverse waves, resulting in a Mach-stem near the focus (B. Sturtevant & V.A. Kulkarny [4]). The whole life (opening and closure) of a detonation-cell can only be described by the mechanism of double Mach-reflexion (see fig.2 on the next page). At the apex of the detonation-cell Mach-stem and primary (incident) shock interchange. The transverse waves move approximately with the equilibrium sound-velocity c_{eq} of the burnt gases.

MATHEMATICAL MODEL

The time-dependent thermo-fluid-dynamic, differential balance-equations governing the cellular detonation-dynamics may be easily derived from their integral form. Fig.1 shows a control-volume $\mathcal{G}$ with capacity V. The normal unit-vector $\underset{\sim}{n}^{o}$ is oriented in the same direction like the surface-element vector $d\underset{\sim}{A} = dA\, \underset{\sim}{n}^{o}$ of the control volume´s envelope $\partial\mathcal{G}$. $\underset{\sim}{s}$ stands for the vector of internal (or external) sources of matter, momentum, resp. energy. The integral form of the balance-equation, equation (1) reads:

$$\int_{(\mathcal{G})} \underset{\sim}{U}\, dV \Big|_{t_o}^{t} + \int_{t_o}^{t} \oint_{(\partial\mathcal{G})} \tilde{F}(\underset{\sim}{U}) \cdot \underset{\sim}{n}^{o}\, dA\, dt = \int_{(\mathcal{G})} \rho \underset{\sim}{s}\, dV \Big|_{t_o}^{t},$$

where the temporal integration-limits are the initial-time t_o and the time of immediate interest t.

$$\underset{\sim}{U} = [\rho, \rho \underset{\sim}{v}, \rho e_o, \rho r]^T \qquad (1.1)$$

denotes the vector of transport-variables in the volume-element dV,

$$\underset{\sim}{s} = [\underset{\sim}{0}, w_r q, w_r]^T\, \mathcal{H}(t - \tau_I) \qquad (1.2)$$

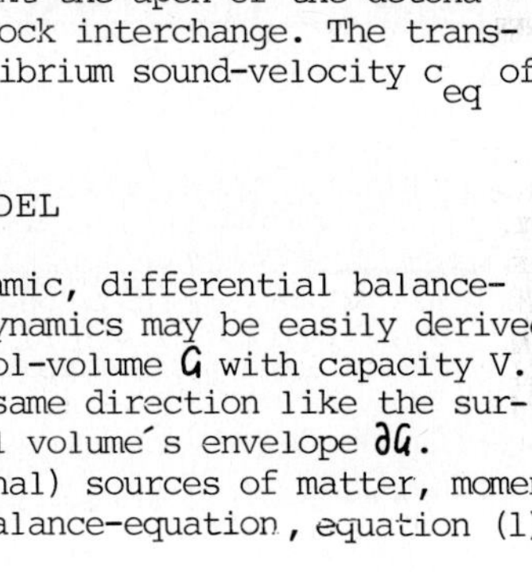

Fig. 1

a source-vector containing the exothermic heat-release $w_r q$ per time-unit

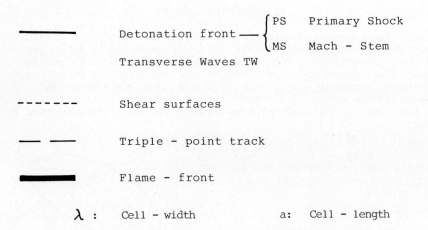

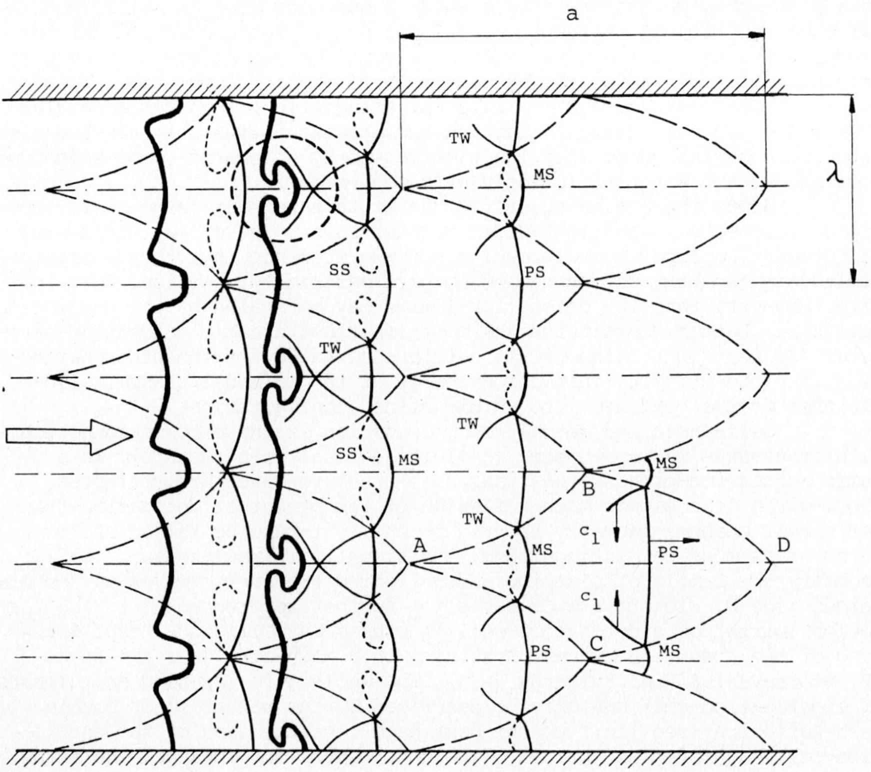

Detonation front — {PS Primary Shock / MS Mach - Stem}

Transverse Waves TW

Shear surfaces

Triple - point track

Flame - front

λ : Cell - width   a: Cell - length

Fig. 2: Schematic sketch of dynamic phenomena occurring during a detonation propagation-process due to the temporal change of the heat release-parameter, the reaction rate - velocity

$$w_r = Dr/Dt = w_r(r,p,T) \qquad (1.3),$$

multiplied with the net-heat production per unit mass of the gas-mixture q. In equation (1.1) ρ stands for the density of the mixture, $\underset{\sim}{v}$ its fluid-velocity, p the thermodynamic pressure and ρe_o the specific total energy, which is composed of the specific internal energy $\rho\varepsilon$ and the kinetic energy per unit-volume $\rho \underset{\sim}{v}^2/2$, i.e.:

$$e_o = \varepsilon + \underset{\sim}{v}^2/2 \qquad (1.4).$$

For a polytropic gas follows from the thermal equation of state

$$p = p(\rho, T) = \rho RT \qquad (1.5)$$

that $\varepsilon(T,\rho) = \varepsilon(T) = c_v T = RT/(\kappa - 1) = \dfrac{p}{\rho(\kappa - 1)}$ (1.6),

where R denotes the (universal) gas-constant, $c_v = (\partial\varepsilon/\partial T)_\rho$ the constant-volume heat-capacity per mass-unit and the isentropic exponent

$$\kappa = 1 + R/c_v \qquad (1.7).$$

According to a simple <u>two-substance model</u> the rate-parameter r in (1.1) signifies the mass-concentration of burnt gas. Therefore r = 0 corresponds to fresh (unburnt) gas and r = 1 to complete heat-release. The approach of a value r = 1 is unrealistic, since endothermic, backwards-directed reactions ensure the formation of a dynamic thermochemical equilibrium (lower index eq) with vanishing reaction-rate $w_r \mapsto 0$ together with $r = r_{eq}$ at the end of the reaction - zone.

Usually the exothermic rate-velocity w_r in (1.3) obeys simple ARRHENIUS - kinetics and writes:

$$w_r = p^{n-1} k_2 \left[r^m \exp\left(\frac{-E_2}{RT}\right) - (1-r)^m \exp\left(-\frac{E_2+q}{RT}\right)\right] \qquad (1.8),$$

where the exponents n and m denote the order of the reaction and its concentration - influence. E_2 stands for the activation-energy of the exothermic reaction. Before any heat will be set free, a (non-exothermic) induction - reaction takes place. The induction-delay is taken into account in the source-term (1.2) by means of a temporal shift applying the HEAVISIDE - distribution

$$\mathcal{H}(t - \tau_I) = \begin{cases} 1 & \text{for } t \geqq \tau_I \\ 0 & // \ t < \tau_I \end{cases} \qquad (1.9)$$

with the induction-delay time

$$\tau_I = k_1 \rho^\alpha \exp\left(-\frac{E_1}{RT}\right) \quad \text{and} \quad -1 < \alpha < -.5 \qquad (1.9.1).$$

The activation-energy E_1 of the induction-reaction together with E_2 and the pre-constants k_1, resp. k_2 in equations (1.8), resp. (1.9.1) can be obtained from laboratory - experiments (see for instance Korobeinikov et. al.´s data in [5]).

The scalar (dot) product of the flux - dyadic $\tilde{F}(\underset{\sim}{U})$ (second - order tensor) with the normal unit-vector $\underset{\sim}{n}^o$ appearing in equation 1 at the beginning of this section results in a vector $\underset{\sim}{f}$ of convective fluxes of the transport - variable $\underset{\sim}{U}$ according to the form:

$$\underset{\sim}{f}(\underset{\sim}{U}) = \rho\left[\underset{\sim}{v}\cdot\underset{\sim}{n}^o,\ \underset{\sim}{v}(\underset{\sim}{v}\cdot\underset{\sim}{n}^o)+p/\rho\,\underset{\sim}{n}^o,\ (e_o+p/\rho)\,\underset{\sim}{v}\cdot\underset{\sim}{n}^o,\ r\,\underset{\sim}{v}\cdot\underset{\sim}{n}^o\right] = \tilde{F}(\underset{\sim}{U})\cdot\underset{\sim}{n}^o \qquad (2)$$

The integral-form (1) of the transport-equations is the starting-point for a numerical scheme based on the finite-volume method. By means of GAUSS - GREEN´s integral-law it follows directly from (1) that

$$\int_{(G)} \left(\underset{\sim}{U} \Big|_{t_o}^{t} + \int_{t_o}^{t} \nabla \cdot \tilde{F} \right) dV\, dt = \int_{t_o}^{t} \iint_{(G)} \rho \underset{\sim}{s}\, dV\, dt \qquad (3),$$

where ∇ denotes the Nabla (- or gradient) operator, which reads for two-dimensional, cartesian x,y - geometry with orthonormal basis - vectors $(\underset{\sim}{e}_x, \underset{\sim}{e}_y)$: $\nabla = \partial/\partial x\, \underset{\sim}{e}_x + \partial/\partial y\, \underset{\sim}{e}_y$. The flux - dyadic $\tilde{F}$ has in this coordinate - system the form:

$$\tilde{F} = \underset{\sim}{F}_x ; \underset{\sim}{e}_x + \underset{\sim}{F}_y ; \underset{\sim}{e}_y \qquad (3.1).$$

The differential-form of the balance-equations cast in <u>conservation</u> (divergence or flux) - form equivalent to (3) now reads:

$$\partial \underset{\sim}{U}/\partial t + \partial \underset{\sim}{F}_x/\partial x + \partial \underset{\sim}{F}_y/\partial y = \rho \underset{\sim}{s} \qquad (4)$$

with $\underset{\sim}{U} = [\rho, \rho u, \rho v, \rho e_o, \rho r]^T$ (4.1) and the source-term

$\underset{\sim}{s} = [\underset{\sim}{\sigma}, q, 1]^T \; \rho\, w_r\, \varkappa(t - \tau_I)$ according to equation (1.2).

The semicolon in (3.1) denotes a dyadic (tensor) - product of the vector-fluxes in the x-, resp. y-direction with the basis-vectors

$$\begin{aligned} \underset{\sim}{F}_x &= \tilde{F} \cdot \underset{\sim}{e}_x = [\rho u, \rho u^2 + p, \rho uv, u(\rho e_o + p), \rho r]^T \\ \underset{\sim}{F}_y &= \tilde{F} \cdot \underset{\sim}{e}_y = [\rho v, \rho vu, \rho v^2 + p, v(\rho e_o + p), \rho r]^T \end{aligned} \qquad (4.2)$$

In (4.1), resp. (4.2) u and v are the components of the fluid-velocity $\underset{\sim}{v} = u\, \underset{\sim}{e}_x + v\, \underset{\sim}{e}_y$ (4.3).

The system (4) is of hyperbolic type, since the JACOBIAN - matrices

$J_x = \partial \underset{\sim}{F}_x/\partial \underset{\sim}{U}$, $J_y = \partial \underset{\sim}{F}_y/\partial \underset{\sim}{U}$ (5) have real Eigen - values (characteristics) λ_x, resp. λ_y with

$$\underset{\sim}{\lambda}_x = [u, u, u, u-c, u+c]^T \quad \& \quad \underset{\sim}{\lambda}_y = [v, v, v, v-c, v+c]^T \quad (5.1).$$

The conservation-form of a scalar hyperbolic transport-equation allows the application of the so-called ´shock-capturing´- technique. According to this concept a consistently formulated numerical scheme converges in a weak (or integral) sense to the jump-conditions (see P.Lax & B.Wendroff [6]). Latter hold across gasdynamic-discontinuities (particularly shocks) and are called RANKINE - HUGONIOT - relations.

The square of the velocity of sound c in (5.1) with $c^2 = (\partial p/\partial \rho)_s = \kappa (\partial p/\partial \rho)_T$ as the (isentropic) propagation-velocity of weak pressure-waves without change of local specific entropy s can be written for a polytropic gas in the form: $c^2 = \kappa p/\rho = \kappa R T$ (5.2)

NUMERICAL METHOD

The system (4) is solved by means of a two-step predictor-corrector MacCormack - scheme. In the case of a linear transport-equation the MacCormack - algorithm is equivalent to the Lax-Wendroff - scheme described in [6]. A.Harten introduced in [7] the notion of so-called total-variation-diminishing (briefly TVD-) schemes, which combine the smoothing of the high-frequency-oscillations with second-order accuracy and high-resolution. S.F. Davis showed in [8] by applying flux-limiters in a form proposed by P.K. Sweby in [9] that the spurious oscillations (wiggles) inherent of the Lax-Wendroff - method may be avoided. The critical amounts of artificial viscosity needed to satisfy the TVD-property for schemes of Lax-Wendroff, resp. MacCormack-type are determined in Davis´ paper [8].

The inhomogeneous term $\rho \underset{\sim}{s}$ occurring due to the exothermic reaction-rates is considered adopting a strategy proposed by H.A.Dwyer et.al. in [10]. The complete algorithm for the solution of a scalar hyperbolic transport-equation of the form

$$\partial U/\partial t + \partial F(U)/\partial x - h(U) = 0 \qquad (6)$$

reads:

$$\mathcal{L}_{pred}: \; U_i^* = U_i^n - \Delta t/\Delta x \; (F_{i+1}^n - F_i^n) + h_{i+1/2}^n \, \Delta t \qquad (7.1)$$

$$\mathcal{L}_{corr}: \; U_i^{n+1} = [(U_i^n + U_i^*) - \Delta t/\Delta x \; (F_i^* - F_{i-1}^*) + h_{i-1/2}^* \, \Delta t]/2$$

$$+ 1/8 [\Theta_{i+1/2}^n \; (U_{i+1}^n - U_i^n) - \Theta_{i-1/2}^n \; (U_i^n - U_{i-1}^n)] \qquad (7.2)$$

$$\text{with } h_{i \pm 1/2} = (h_i + h_{i \pm 1})/2 \qquad (7.3)$$

The subscript (i) in (7.1) and (7.2) denotes the discretized x-position, the superscripts n, * and n+1 stand for the old, predicted and corrected (new) time-level. The chosen mesh-size is Δx and the time-step Δt. The parameter Θ is a switch, which is connected with Sweby's 'flux-limiter' - function $\varphi(\nu)$ in the paper [9] according to following equation:

$$\Theta_{i+1/2} = 8 \; \vartheta \; (1 - \vartheta) \; [1 - \varphi(\nu_i^+)] \qquad (7.4),$$

where $\vartheta = \lambda \Delta t / \Delta x$ (7.5) denotes the COURANT - number as the ratio of the physical (characteristic) wave-velocities $\lambda(U) = \partial F(U)/\partial U$ and the numerical signal-velocity $\Delta x/\Delta t$.
The parameter $\nu_i^+ = (U_i - U_{i-1})/(U_{i+1} - U_i)$ in (7.4) stands for the ratio of discrete, neighbour-gradients of the transport-variable U in (6).

$\mathcal{L}_{pred}$ in (7.1) is the solution-operator of the scheme's predictor-step, $\mathcal{L}_{corr}$ in (7.2) the operator for the corrector-step. Hence the value of the solution determined at the new time-level (n+1) for the one-dimensional equation (6) reads: $U^{n+1} = \mathcal{L}_x(\Delta t)\, U^n$ with $\mathcal{L}_x = \mathcal{L}_{corr}\mathcal{L}_{pred}$ (8), where U^n denotes the solution at time t_n and U^{n+1} at time $t_{n+1} = t_n + \Delta t$. In the case of a two-dimensional transport-equation (here equation (4)) the solution is advanced to the new time-level t_{n+1} according to : $U^{n+1} = \mathcal{L}_x(\Delta t/2)\, \mathcal{L}_y(\Delta t/2)\, \mathcal{L}_y(\Delta t/2)\, \mathcal{L}_x(\Delta t/2)\cdot U^n$, where the $\mathcal{L}_y(\Delta t/2)$ - operators may be combined resulting in:

$$U^{n+1} = \mathcal{L}_x(\Delta t/2)\; \mathcal{L}_y(\Delta t)\; \mathcal{L}_x(\Delta t/2)\cdot U^n \qquad (9).$$

The so-called STRANG - type operator-splitting in (9) ensures that the order of the method (here: second-order accuracy) may be preserved in the multi-dimensional case.
Figure 3 on the next page shows the computational domain and the boundaries for assumed planar channel-geometry. A numerical solution of the hyperbolic system (4) requires an appropriate set of initial- and boundary-conditions. In the finite-volume method applied here boundary-values are prescribed in the image-cell region hatched in figure 3.

INITIAL - CONDITION

The initial-data are steady, planar solutions of the balance-equations (4). They form the well-known, stationary ZELDOVICH-DÖRING-V.NEUMANN (ZDN)- detonation structure, which is composed of a shock-wave separated from a finite-size reaction-zone of exothermic heat-release by a distance resulting from the induction-delay. The ZDN-model is of great significance for predicting dynamic detonation-parameters like the cell-size by applying the empirically confirmed assumption that fixed ratios exist between these parameters and characteristic chemical length-scales.

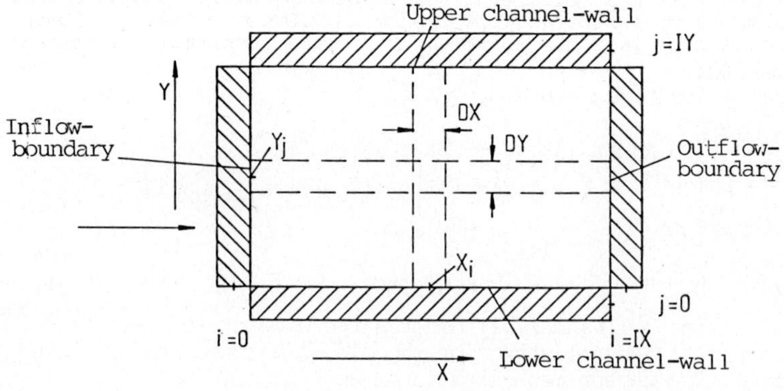

Fig. 3: Computational domain and boundaries

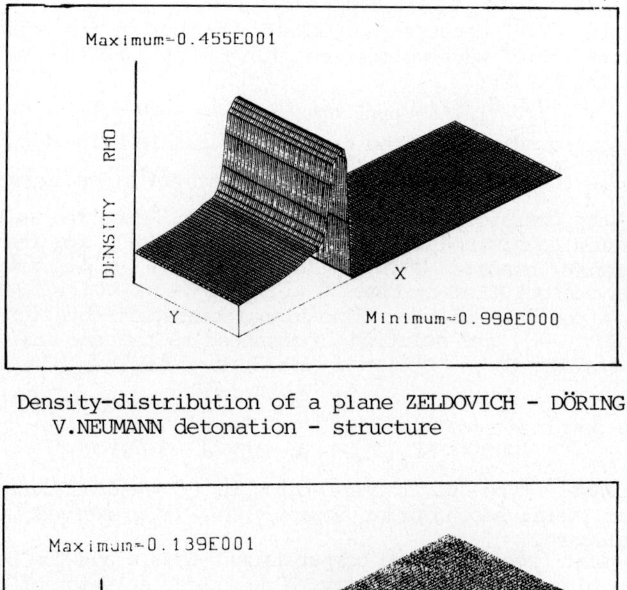

Fig.4: Density-distribution of a plane ZELDOVICH - DÖRING - V.NEUMANN detonation - structure

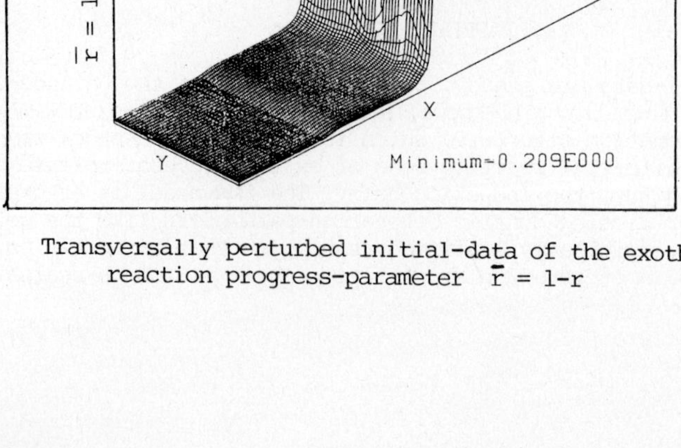

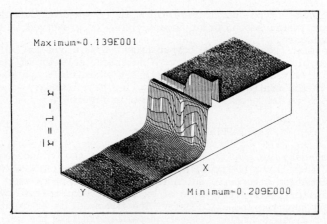

Fig.5: Transversally perturbed initial-data of the exothermic reaction progress-parameter $\bar{r} = 1-r$

K.I. Shchelkin & Ya.K. Troshin were the first, who presented in [11] empirical values of a scaling-factor, in order to estimate the cell-size for any gaseous mixture from a chemical length-scale of the ZDN-structure. For the global chemistry according to Korobeinikov et.al. ([5]) for an Argon (,resp. Helium) - diluted, stoichiometric oxygen-hydrogen mixture applied here the ZDN initial-data are determined by means of adaptive ROMBERG - integration of the equation of motion of a fluid-element in a reference-system, which moves with the average (so-called) CHAPMAN-JOUGUET (CJ)-detonation-velocity D_{CJ}. From the integral, steady balance-equations (reaction-front equations) follows that there exists a minimum propagation-velocity, the so-called CJ-velocity, below which only deflagrative combustion is feasible. The experimentally well-known fact of preference of the minimum-velocity to larger velocities and corresponding thermodynamic variables is supported theoretically by a stability-criterion given by E. Jouguet at the beginning of this century (see chapter 3.8 in the book of F. Bartlmä [12]).

The equation of motion for a plane ZDN-structure reads in a GALILEI-transformed (detonation front-fixed) frame of reference $Dx/Dt = D_{CJ}-u(r)$ (10), where x denotes the distance from the precursor shock-front and u the fluid-velocity in the laboratory-system. It follows from the chain-rule of differentiation that

$$x(r) = \int_0^r [D_{CJ} - u(w)] / w_w \, dw + l_I \qquad (10.1)$$

with the exothermic reaction-velocity $w_r = Dr/Dt$ and the induction reaction-length l_I. For the induction reaction-length holds the equation:

$$l_I = \int_0^r (D_{CJ} - u)\, dt = (D_{CJ} - u_{VN})\, \tau_I \qquad (10.2),$$

where τ_I obeys simple ARRHENIUS - kinetics (see [5]) according to equation (1.9.1). The transport-variables in the induction-zone of a plane ZDN-structure are assumed to be constant and equal to the so-called V.NEUMANN (post shock)- state (subscript VN). For the detonation mach-number $Ma_{CJ} = D_{CJ}\, c_o$ with the sound-velocity $c_o = (\kappa RT_o)^{1/2}$ of the unburnt (fresh) gas follows from the classical CHAPMAN-JOUGUET (CJ)-theory the relationship:

$$Ma_{CJ} = \left(1 + \frac{\kappa^2 - 1}{2\kappa} \frac{q\, r_{eq}}{R\, T_o}\right)^{1/2} + \left(\frac{\kappa^2 - 1}{2\kappa} \frac{q\, r_{eq}}{R\, T_o}\right)^{1/2} \qquad (11),$$

where r_{eq} stands for the exothermic reaction-rate parameter according to thermo - chemical equilibrium.

Experiments show that the dynamic detonation-process, although multidimensional and highly unsteady in average, proves to be stationary and one - dimensional in average. Algebraic terms for the velocity u, pressure p, density ρ and temperature T in the exothermic reaction-zone are an immediate outcome of the integral reaction-front equations:

$$u(r)/c_o = \Phi / [(\kappa+1)\, Ma_{CJ}], \quad p(r)/p_o = 1 + \frac{\kappa}{\kappa+1} (Ma_{CJ}^2 - 1)$$

$$\rho_o/\rho(r) = 1 - \Phi/[(\kappa+1)\, Ma_{CJ}^2] \quad \text{and} \quad T/T_o = p/p_o\, \rho_o/\rho \qquad (12),$$

where $\Phi = (Ma_{CJ}^2 - 1) + [(Ma_{CJ}^2 - 1)^2 - 2(\kappa+1) Ma_{CJ}^2\, qr/(c_p T_o)]^{1/2}$

with $c_p = R\kappa/(\kappa-1)$ as the constant pressure heat-capacity per unit-volume.

In order to ascertain the ZDN-structure the inverse-function of x(r) in (10.1) is determined, which can easily be done numerically by simple NEWTON-iteration $r_{n+1} = r_n - [x_n(r_n) - x_i]/ f(r)$ with $f(r) = (dx/dr)_n$ for a monotonous function x(r) with given discretisation x_i (i=1,...,N). Fig.4 shows the calculated three-dimensional density-distribution of the plane ZDN-structure for the assumed planar channel-geometry.

In contrast to the significance of the ZDN-model for predicting dynamic detonation-parameters stands the fact that all previous numerical simulations of detonation-dynamics avoided to attack the question of hydrodynamic stability of the ZDN-structure. Previous numerical studies of the cellular detonation-dynamics by Oran et.al. (1982) [13], Markov (1981) [14] and Taki & Fujiwara (1982) [15]were restricted to arbitrary, blast-wave perturbed initial data. Furthermore unrealistic, large-scale (transverse) perturbations of the reaction-progress parameter r with a wave-length of the order of the combustion channel-width were applied in order to establish the complex wave-structure. Fig. 5 shows such perturbed initial-data of the parameter $\overline{r} = 1 - r$ applied by former investigators for obtaining transverse wave-phenomena in a laboratory reference-system.
However, in a reference-system fixed to the detonation-front the applied numerical scheme (7) yields a spontaneous establishment of the transverse wave-structure (see also the author's paper [16]).

BOUNDARY - CONDITIONS

According to the theory of characteristics the boundary-conditions, which must be prescribed at the left, resp. right side of the computational mesh depend on the underlying mathematical frame of reference. In the subsequent figure 6 the characteristic planes in an (x,y,t)-coordinate-system are drawn either in case of a laboratory frame of reference (fig.6a) or for a precursor-shock oriented (so-called GALILEI-transformed) system (fig.6b). The implications and advantages of the latter, namely the shock reference-system,are discussed in [16] (p.782).

The boundary-values required for the transport-variables at the image-cells of the solid channel-walls are corresponding to the underlying in-

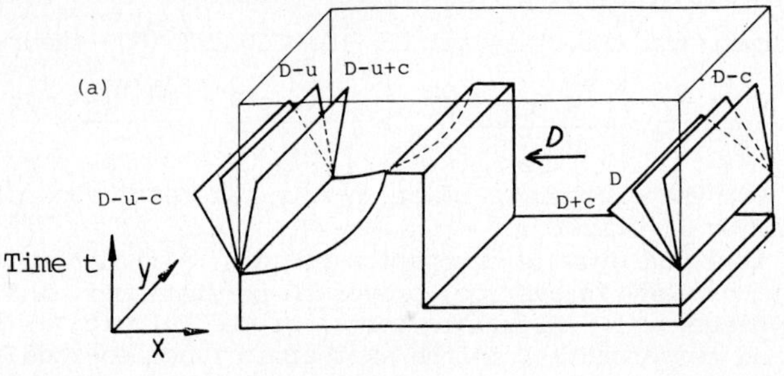

Fig. 6a: Characteristic planes in a shock-reference system

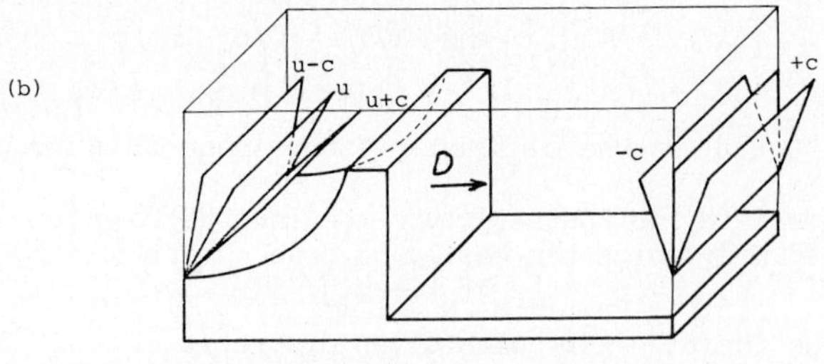

Fig. 6b: Characteristic planes in a laboratory reference -system

cid wave - model non-reflecting (symmetric) slip-conditions for the velocity u tangential to the wall and reflecting (antimetric) conditions for the normal-velocity v. The other variables (density ρ, internal energy ε and reaction-progress parameter r) are also subject to symmetric or absorbing conditions.

V. NEUMANN STABILITY - ANALYSIS

An essential concern of the author was to allow only the growth of the physically relevant wave-numbers $k = 2\pi/\lambda$ of the transverse wave - structure. Experimentally these wave-numbers are known to be harmonics, which fulfil the standing wave-condition $k = \bar{n}\pi / W$ (13) with the mode-number $\bar{n}$ and the channel-width W . Since a plane ZDN-detonation-front breaks up on the smallest scales forming a caustics (or singularity), a rise of the highest resolvable frequencies must be admitted. The NYQUIST - frequency with $k\Delta x = \pi$ for a mesh-size Δx and a wave-length $\lambda = 2\Delta x$ is the largest frequency that can be identified in any finite difference (or volume)-method.

In order to allow an exponential increase of the state-variables $\underset{\sim}{U}$ caused by the source-term $\rho\underset{\sim}{s}$ appearing in equation (4), the linear stability-analysis should supply an amplification-factor α, which obeys the inequality

$$|\alpha| = |\hat{U}_i^{n+1} / \hat{U}_i^n| \leqq 1 + \underline{\mathcal{O}(\Delta t)} \qquad (14)$$

with the amplitudes $\hat{U}_i$ according to the Ansatz: $U_i = \hat{U}_i \exp(ijk\Delta x)$, where $j^2 = -1$ and $\mathcal{O}$ in eq. (14) denotes the LANDAU - symbol. Concerning the foundation of relation (14) see chapter III-F-5 in P.Roache´s book [17]. The complex amplification-factor α obtained by means of linear V.NEUMANN-stability-analysis for the Lax-Wendroff-method is an ellipse with the center $(1 - \vartheta^2)$ and the half-axes ϑ^2, resp. ϑ, where again ϑ denotes the COURANT-number according to eq. (7.5). It is a well-known fact that the classical COURANT-FRIEDRICHS-LEWY (CFL) - criterion with $|\vartheta| \leqq 1$ (15) for the numerical stability of a scalar, hyperbolic transport-equation is an immediate consequence of the V.NEUMANN stability-criterion $|\alpha| \leqq 1$ (15.1).
The additional terms of numerical viscosity added to the mere Lax-Wendroff ($\mathcal{LW}$) - scheme according to Davis´ paper 8 lead in the region of steep gradients to a restriction of the classical CFL-criterion $|\vartheta| \leqq \beta < 1$ (16). In the induction-zone behind the shock-wave the ´viscosity-switch´ Θ was chosen equal to one. Then β in (16) becomes $\sqrt{3}/2$, and the complex amplification-factor α transforms for $\Theta \equiv 1$ into an ellipse with center 0 and the half-axes 1, resp. ϑ . The amplification-factors for the different methods are shown in figure 7 on the next page. In fig. 7 the legends Re and Im at the coordinate-axis denote the real part, resp. imaginary part of the amplification-factor in the complex plane.
In the code Davis´ TVD - Lax-Wendroff-method was tested with $\Theta \equiv 1$ and a COURANT-number $\vartheta = 0.95\sqrt{3}/2$ in the induction-zone, which is according to (16) close to the linear stability-limits for high-frequency oscillations with wave-number $k\Delta x = \pi$.
The numerical results show that the energy of the spontaneously developing small-scale structures is aliased to the large-scale, transverse wave - structure with $k\Delta x \mapsto 0$. Calculations performed with accuracy Real*16 instead of Real*8 showed that the transverse wave-structure evolves not due to the machine-error but by the nonlinearity inherent to the balance-equations.

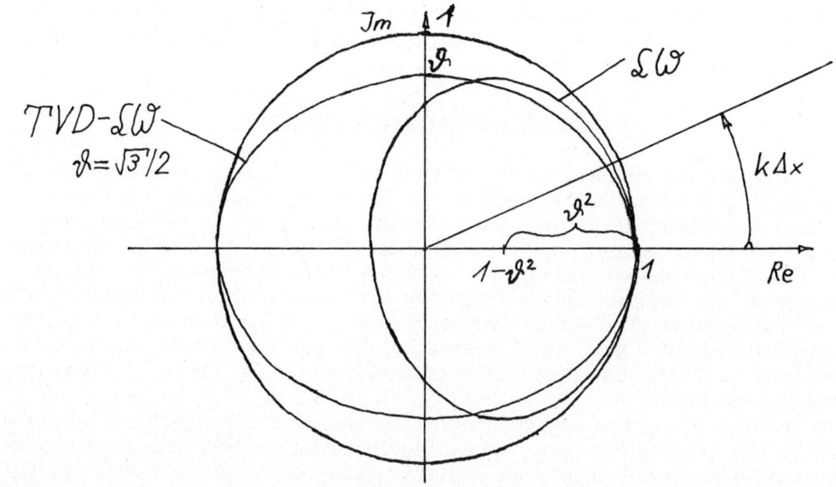

Fig.7: Complex amplification-factors for mere Lax-Wendroff- and TVD - Lax-Wendroff-method

RESULTS

By means of the method described above a heuristic stability-criterion for the onset of transverse wave-phenomena given by Shchelkin in [11], chapter 1/3 could be reaffirmed successfully. The reason for the excitation of transverse pressure-waves in a plane ZDN-structure is a highly nonlinear dependence of the ignition-time from temperature according to equation (1.9.1). The instability-criterion given by Shchelkin reads:

$$\partial \tau_I / \partial T \cdot (T - T_{VN}) \big|_u \geqq \tau_I \qquad (15),$$

where u denotes the state of the unburnt gas expanding from the shocked V.NEUMANN-state (subscript VN). Assuming for an unburnt pocket in the exothermic reaction-zone an isentropic expansion from the VN-pressure-level behind the precursor shock-wave to a pressure $p_{eq} = p(r_{eq})$ according to thermochemical equilibrium, it follows from (15) that

$$E_1/RT_{VN} \; (1-T/T_{VN}) \big|_u = E_1/RT_{VN} \; \left[1 - (p_{eq}/p_{VN})^{(\kappa -1)/\kappa} \right] \geqq 1 \qquad (15.1)$$

is necessary to obtain instabilities in the plane ignition-front during detonation. The numerical simulations in two (recently three) space-coordinates provide evidence that the ZDN-data are only stable, if either the Shchelkin - criterion of instability is violated or the channel-width is below a critical value, which is experimentally known to be the critical kernel for the single head-spin-propagation (see [11], chapter 1/6).
In fig.8 the pressure-distribution for a detonation-propagation with mode-number $\overline{n}$ = 2.5 is plotted after Kend = 285, 290, 300, resp. 310 time-steps according to the CFL-stability-criterion.
Condition (15), resp. (15.1) must be fulfilled in order to obtain nonlinear resonance-phenomena between the heat-release due to chemical reactions and the transverse pressure-modes. By varying the channel-width and keeping all other system-parameters as constant also non-uniqueness of the mode-number and even chaotic cell-structures could be determined. In the latter case no resonance could be obtained.

The shape and size of the detonation-cells agree surprisingly well with experimental soot-tracks obtained by R.A. Strehlow et.al. (see [1] and [16] for further reference).

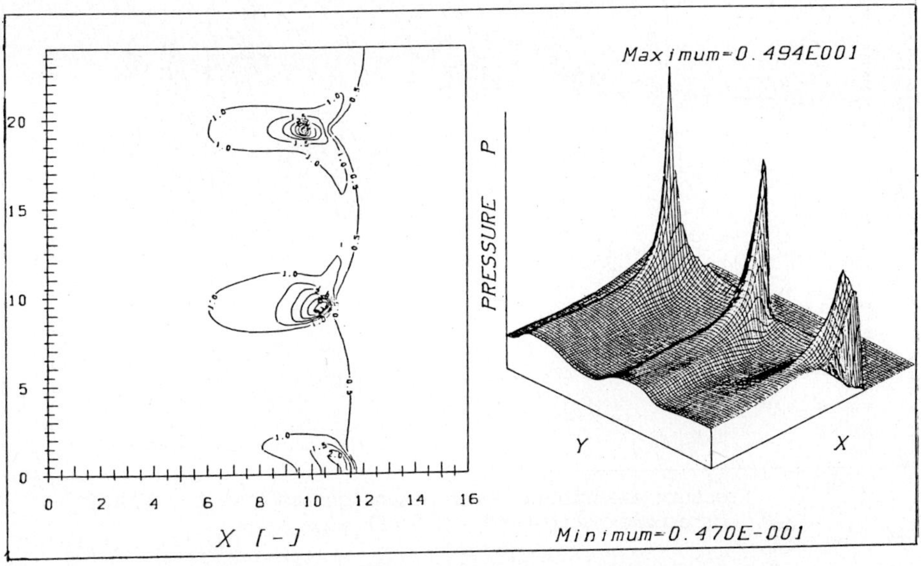

Fig. 8/1: Pressure relief and iso-lines for dimensionless time t = 31.049 (corresponding to Kend = 285 time-steps according to the CFL - stability - criterion)

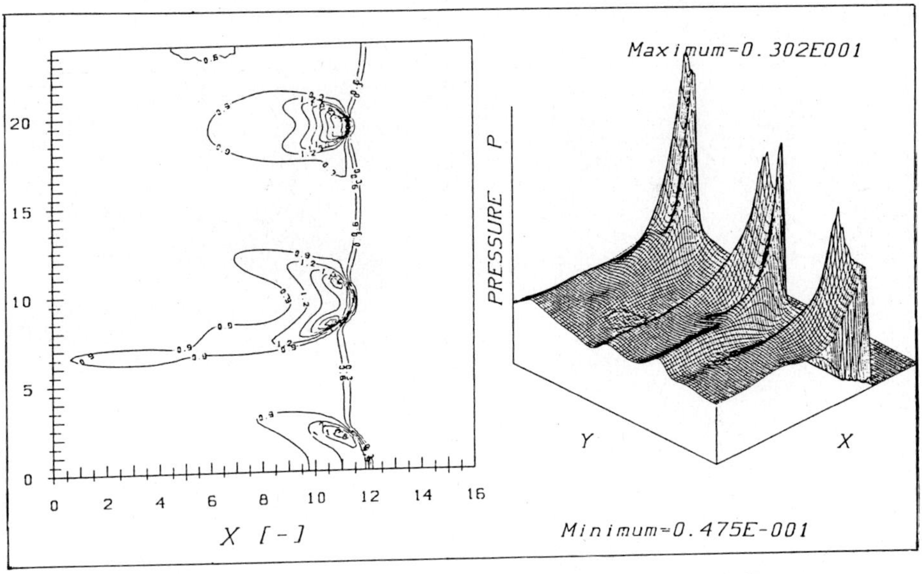

Fig. 8/2: Pressure distribution for t = 31.583 (corresponding to Kend = 290 CFL time steps)

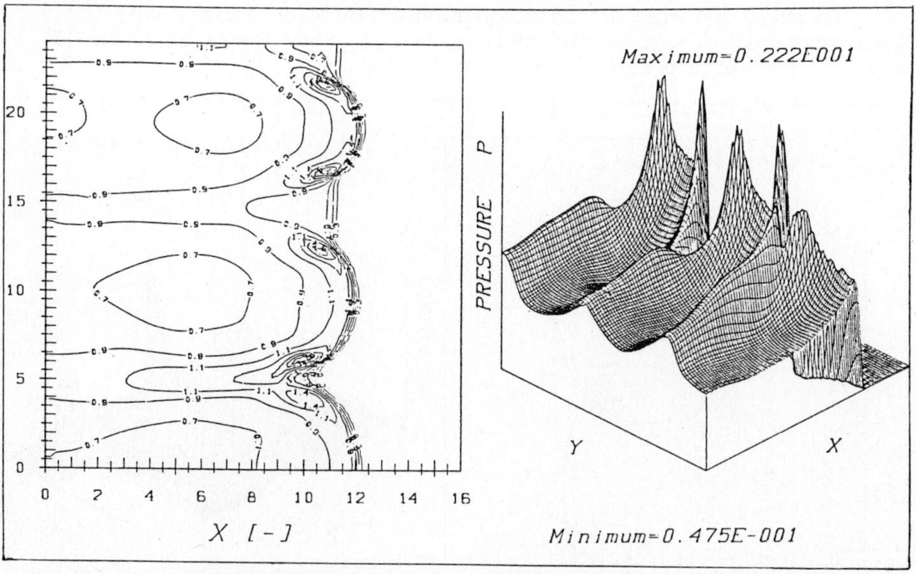

<u>Fig. 8/3</u>: Pressure distribution for dimensionless time t = 32.642 (corresponding to Kend = 300 CFL time steps)

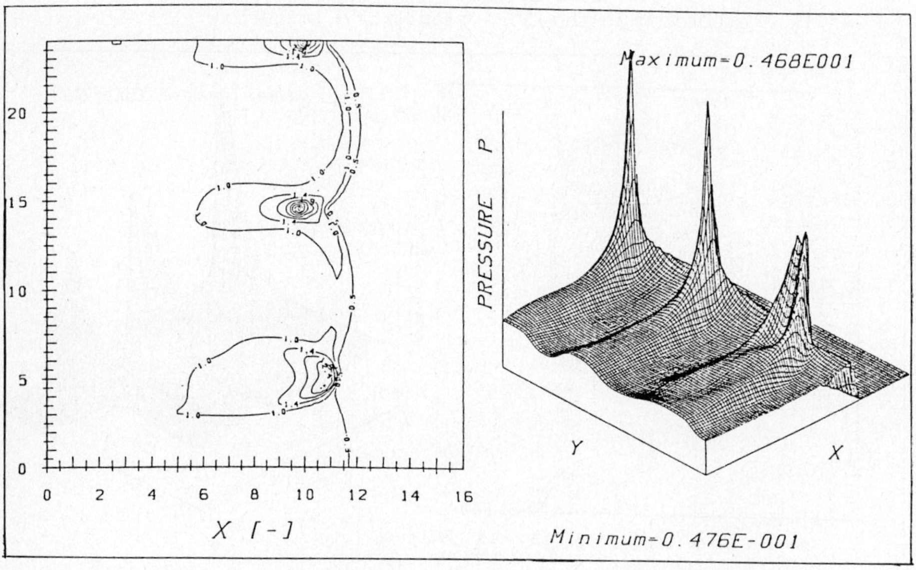

<u>Fig. 8/4</u>: Pressure distribution for t = 33.687 (corresponding to Kend = 310 CFL time steps)

CONCLUSION

Besides the influence of the channel-width W on the resonance-frequency and cell-size already discussed in [1] the author tries to investigate the effect of changes of the reaction-kinetic parameters on the regularity of the cell-structure. Moreover the code has been extended to three space-coordinates in order to detect other secondary flows associated with the transverse wave-structure.

ACKNOWLEDGEMENT

The Deutsche Forschungsgemeinschaft is acknowledged for sponsoring the work in the Priority Research Program ´Finite Approximations in Fluid Mechanics´.

REFERENCES

[1] SCHÖFFEL, St., EBERT, F.: "Numerical analyses concerning the spatial dynamics of an initially plane gaseous ZDN-detonation", 11th. Internat.Conf. on Dynamics of Explosions and Reactive Systems, Warsaw (Poland), Aug.3-7 (1987), p.35, appears in AIAA-Progress in Astronautics & Aeronautics, ed.: A.L. KUHL.

[2] LEE, J.H.: "Dynamic parameters of gaseous detonations", Ann.Rev. of Fluid Mechanics 16, p.311-336 (1984).

[3] FICKETT, W. & DAVIS, W.C.: "Detonation", Univers. of California Press, Berkeley (1979).

[4] STURTEVANT, B., KULKARNY, V.A.: "The focusing of weak shock waves", Journal of Fluid Mechanics, vol.173, part 4, pp. 651-671 (1976).

[5] KOROBEINIKOV, V.P., LEVIN, V.A., MARKOV, V.V. & CHERNYI, G.G.: "Propagation of blast-wave in a combustible gas", Astronautica Acta, vol. 17, pp. 529-537 (1972).

[6] LAX, P.D., WENDROFF, B.: "Systems of conservation laws", Comm. Pure Appl. Math. 13 (1960), p.217-237.

[7] HARTEN, A.: "High-resolution schemes for hyperbolic conservation-laws" Journal Computational Physics, vol.49, pp.357-393 (1983).

[8] DAVIS, S.F.: "TVD finite-difference schemes and artificial viscosity", ICASE-report N 84-20, NASA contractor report 172373 (1984).

[9] SWEBY, P.K.: "High-resolution schemes for hyperbolic conservation laws", SIAM J. Numerical Analysis 21 (1984).

[10] DWYER, H.A., ALLEN, R., WARD, M., KARNOPP, D. & MARGOLIS, D.: "Shock-Capturing Finite-Difference Methods for Unsteady Gas Transfer", AIAA - paper 74-251 (1974).

[11] SHCHELKIN, K.I. & TROSHIN, Ya.K.: "Gazodinamika Goreniya (Gasdynamics of Combustion)", Baltimore, Mono Book Corporation (1965).

[12] BARTLMÄ, F.:"Gasdynamik der Verbrennung", Springer 1975.

[13] ORAN, E.S., BORIS, J.P., YOUNG, T. et.al.: "Numerical Simulations of detonations in hydrogen-air and methane-air mixtures", 18th. Symposium (Int.) on Combustion (1982), pp. 1641-1649, The Comb.Institute, Pittsburgh, Pa.

[14] MARKOV, V.V.: Dokl. AN SSSR 258, pp. 314-317 (1981).

[15] TAKI, S. & FUIJIWARA, T.: "Numerical simulation of triple-shock behavior of gaseous detonations", 18th. Symp. (Int.) on Combustion (1982), pp. 1671-1681, The Combustion Institute, Pittsburgh, Pa.

[16] SCHÖFFEL, St., EBERT, F.: "A Numerical Investigation of the Reestablishment of a Quenched Gaseous Detonation in a Galilei-Transformed System", Proc. of the 16th. Internat. Sympos. on Shock Tubes and Waves, Aachen, July 26-31 (1987), VCH - Verlagsgesells., ed.: H. GRÖNIG

[17] ROACHE, P.: "Computational Fluid-Dynamics", Hermosa Publishers, Albuquerque, New Mex. (1982).

THE DESIGN OF ALGORITHMS FOR HYPERSURFACES MOWING WITH CURVATURE-DEPENDENT SPEED

James Sethian [1]
Department of Mathematics
University of California
Berkeley, California 94720

Stanley J. Osher [2]
Department of Mathematics
University of California
Los Angeles, California 90024

SUMMARY

The need to follow fronts moving with curvature-dependent speed arises in the modeling of a wide class of physical phenomena, such as crystal growth, flame propagation and secondary oil recovery. In this paper, we show how to design numerical algorithms to follow a closed, non-intersecting hypersurface propagating along its normal vector field with curvature-dependent speed. The essential idea is an Eulerian formulation of the equations of motion into a Hamilton-Jacobi equation with parabolic right-hand side. This is in contrast to marker particle methods, which are rely on Lagrangian discretizations of a moving parameterized front, and suffer from instabilities, excessively small time step requirements, and difficulty in handling topological changes in the propagating front. In our new Eulerian setting, the numerical algorithms for conservation laws of hyperbolic systems may be used to solve for the propagating front. In this form, the entropy-satisfying algorithms naturally handle singularities in the propagating front, as well as complicated topological changes such as merging and breaking. We demonstrate the versatility of these new algorithms by computing the solutions of a wide variety of surface motion problems in two and three dimensions showing sharpening, breaking and merging.

[1] Supported in part by the Applied Mathematics Subprogram of the Office of Energy Research under contract DE-ACO3-76SF00098, NSF under the National Science Foundation Mathematical Sciences Program, and the Sloan Foundation

[2] Supported in part by NSF Grant No. DMS85-03294, ARO Grant No. DAAG29-85-K-0190, DARPA Grant in the ACMP Program, ONR Grant N00014-86-K-0691, NASA Langley Grant NAG1-270.

1. INTRODUCTION: EQUATIONS OF MOTION

We wish to follow the evolution of an initial surface $\gamma(0)$ propagating along its gradient field with speed $F(K)$ a given function of the curvature K (either mean or Gaussian). The key idea, as derived in [5], is to view the evolving front $\gamma(t)$ as the level set of a higher-dimensional function ϕ. To be more precise, let the initial surface $\gamma(0)$ be a closed, non-intersecting hypersurface of dimension $N-1$. We construct the function ϕ by letting $\phi(\bar{x},0)=(1\pm d)$ $\bar{x}\in R^N$ where g is the distance from $\bar{x}$ to $\gamma(0)$, with the plus (minus) sign chosen if $\bar{x}$ is inside (outside) $\gamma(0)$. Then, at t=0, the level set $\left\{ \bar{x} \mid \phi(\bar{x},0)=1 \right\}$ gives $\gamma(0)$. We now require a time-dependent differential equation for ϕ corresponding to the evolution of $\gamma(t)$. If the family of level sets ϕ=C, where C is a constant, flow such that each level surface propagates with speed given by $F(K)$, then it can be shown (see [5]), that

$$\phi_t = F(K)\,\nabla\phi \tag{1}$$

$$\phi(\bar{x},0) = \text{given} \ .$$

Equation (1) specifies the complete initial value partial differential equation. Note that

1) ϕ is a function in $R^N \mathrm{x}[0,\infty)\to R$, thus we have added an extra dimension to the problem.

2) At any time t, the position of the front $\gamma(t)$ is just the level set $\left\{ \bar{x} \mid \phi(\bar{x},t)=1 \right\} = \gamma(t)$.

Eqn. (1) is an Eulerian formulation of the front propagation problem. The level surface ϕ=1 may change topology as it moves, either breaking into multiple parts or fusing together. For any fixed t, slicing ϕ by the level plane at height 1 retrieves the position of the front.

2. HAMILTON-JACOBI EQUATIONS: THE ROLE OF CURVATURE AS VISCOSITY

To see the effect of curvature on a propagating front, consider a propagating closed curve in R^2 and special speed function $F(K)$=$1-\varepsilon K$. Using the expression for the mean curvature in terms of ϕ, we substitute into Eqn. (1) to produce

$$\phi_t - H(\nabla\phi) = \varepsilon\left[\frac{\phi_{xx}\phi_y^2 - 2\phi_x\phi_y\phi_{xy} + \phi_{yy}\phi_x^2}{(\phi_x^2+\phi_y^2)^{3/2}} \right] \tag{2}$$

where $H(\nabla\phi)=(\phi_x^2+\phi_y^2)^{1/2}$. Eqn. (2) is a Hamilton-Jacobi equation with parabolic right-hand-side, which has a type of "viscosity" solution discussed in [1]. Thus, the role of curvature (εK) is to smooth propagating fronts so that sharp corners do not develop. In the limit as $\varepsilon\to 0$

(curvature term vanishes and $F(K)=1$), corners develop, and a weak solution is obtained from an appropriate entropy condition (see [6,7,8]). Thus the role of curvature in this Hamilton-Jacobi formulation for propagating fronts is identical to the the role of viscosity in hyperbolic conservation laws: it inhibits the formation of corners, that is, shocks in the tangent vector.

3. NUMERICAL ALGORITHMS BASED ON HYPERBOLIC CONSERVATION LAWS

Our goal is to approximate the solution to the initial value problem given in Eqn. (1). In [5], a class of non-oscillatory, upwind, entropy-satisfying algorithms of arbitrary order were given to solve this equation, based in part on ideas in [3,4]. The central idea behind these algorithms is to exploit the conservation form of theses schemes directly into the initial value Hamilton-Jacobi equation. As a motivation to understand the scheme, consider the initial value Hamilton-Jacobi equation

$$\psi_t - F(K)(1+\psi_x^2)^{1/2} = 0 \tag{3}$$

where $x \in R$ and $\psi: R \times [0,\infty) \rightarrow R$. This is a simplified version of Eqn. (1), and applies when the propagating curve $\gamma(0)$ can be written as a function $\psi(x,t)$ for all time. Furthermore, in the simple case $F(K)=1$, we have

$$\psi_t - (1+\psi_x^2)^{1/2} = 0 \ . \tag{4}$$

Eqn. (4) is a Hamilton-Jacobi equation. If we differentiate with respect to t, and let $u = \psi_x$, we have

$$u_t + [G(u)]_x = 0 \tag{5}$$

where $G(u) = -(1+u^2)^{1/2}$. Eqn. (5) is hyperbolic conservation law which may be solved by a variety of methods. The key lies in an adequate numerical flux function $g_{j+1/2} = g(u_{j-p+1},.......,u_{j+q+1})$ which approximates the flux $G(u)$. Rather than differentiate the numerical flux function to achieve an approximation to Eqn. (5), we work directly with Eqn. (4) and write

$$\psi_j^{n+1} = \psi_j^n - \Delta t g \ . \tag{6}$$

A wide class of flux functions are described in [5], leading to a collection of upwind, non-oscillatory, entropy-satisfying algorithms in several space dimensions for the original Hamilton-Jacobi initial value problem (Eqn. 1). The upwind nature of these schemes is crucial in the formulation of far-field boundary conditions. Finally, parabolic right-hand-sides (resulting from the curvature component of $F(K)$) are approximated by straight-forward central differences.

4. EXAMPLES

A. Level Curve, Burning out, Development of Corners

We consider a seven-pointed star

$$\gamma(s) = (.1+(.065)\sin(7\cdot 2\pi s)(\cos(2\pi s),\sin(2\pi s))$$

$$s \in [0,1]$$

as the initial curve and solve the initial equations with speed function $F(K)=1$. The computational domain is a square centered at the origin of side length 1/2. We use 300 mesh points per side and a time step $\Delta t=.0005$. At any time $n\Delta t$, the front is plotted by passing the discrete grid function ϕ_{ij}^n to a standard contour plotter and asking for the contour $\phi=1$. The initial curve corresponds to the boundary of the shaded region, and the position of the front at various times is shown in Fig. 1. The smooth initial curve develops sharp corners which then open up as the front burns, asymptotically approaching a circle.

B. Level Curve, Motion Under Curvature

We consider the initial wound spiral

$$\gamma(s) = (.1e^{(-10y(s))} - (.1-x(s))/20)(\cos(a(s)), \sin(a(s)))$$

where $a(s) = 25\tan^{-1}(10y(s))$ and

$$x(s) = (.1)\cos(2\pi s)+.1 \quad y(s) = (.05)\sin(2\pi s)+.1 \quad s \in [0,1].$$

and let $F(K)=-K$, corresponding to a front moving in with speed equal to its curvature. It has recently been shown (see [2]), that any non-intersecting curve must collapse smoothly to a circle under this motion. With $N_{point}=200$ and $\Delta t=.0001$, Figure 2 shows the unwrapping of the spiral from $t=0$ to $t=0.65$. In Figures 2a-d we show the collapse to a circle and eventual disappearance at $t=.295$ (The surface vanishes when $\phi_{ij}^n < 1$ for all ij.)

C. Level Surface, Torus, $F(K) = 1-\varepsilon K$

We evolve the toroidal initial surface, described by the set of all points (x,y,z) satisfying

$$z^2 = (R_0)^2-((x^2+y^2)^{1/2}-R_1)^2$$

where $R_0=.5$ and $R_1=.05$. This is a torus with main radius .5 and smaller radius .05. The computational domain is a rectangular parallelpiped with lower left corner $(-1,-1,-.8)$ and upper right corner (1.,1.,.8). We evolve the surface with $F(K)=1-\varepsilon K$, ε=.001, Δt=.01, and N_{point}=90 points per x and y side of the domain and the correct number in the z direction so that the mesh is uniform. Physically, we might think of this problem as the boundary of a torus separating products on the inside from reactants outside, with the burning interface propagating with speed $K=1-\varepsilon K$. Here, K is the mean curvature. In Figure 3, we plot the surface at various times. First, the torus burns smoothly (and reversibly) until the main radius collapses to zero. At that time (T=0.3), the genus goes from 1 to 0, characteristics collide, and the entropy condition is automatically invoked. The surface then looks like a sphere with deep inward spikes at the top and bottom. These spikes open up as the surface moves, and the surface approaches the asymptotic spheroidal shape. When the expanding torus hits the boundaries of the computational domain, the level surface ψ=1 is clipped by the edges of the box. In the final frame (T=0.8), the edge of the box slices off the top of the front, revealing the smoothed inward spike.

REFERENCES

1) Crandall, M.G., and Lions, P.L., Math. Comp., **43,** 1, (1984).

2) Grayson, M. "The heat equation shrinks embedded plane curves to round points, J. Diff. Geom., to appear.

3) Harten, A., Engquist, B., Osher, S., Chakravarthy, S., Uniformly high order accurate essentially non-oscillatory schemes III, J. Comput. Phys., to appear.

4) Osher, S., Siam J. Num. Anal., **21,** 217, (1984).

5) Osher, S., and Sethian, J.A., Front Propagating with Curvature-Dependent Speed: Algorithms Based on Hamilton-Jacobi Formulations, to appear, J. Comp. Phys., 1988.

6) Sethian, J.A.: An analysis of flame propagation. Ph.D. Dissertation, University of California, Berkeley, California, June 1982; CPAM Rep. 79.

7) Sethian, J.A.: Curvature and the Evolution of Fronts. Comm. Math. Phys, **101,** 487, (1985).

8) Sethian, J.A.: Numerical Methods for Propagating Fronts, in *"Variational Methods for Free Surface Interfaces"*, edited by P. Concus and R. Finn, (Springer-Verlag, New York, 1987).

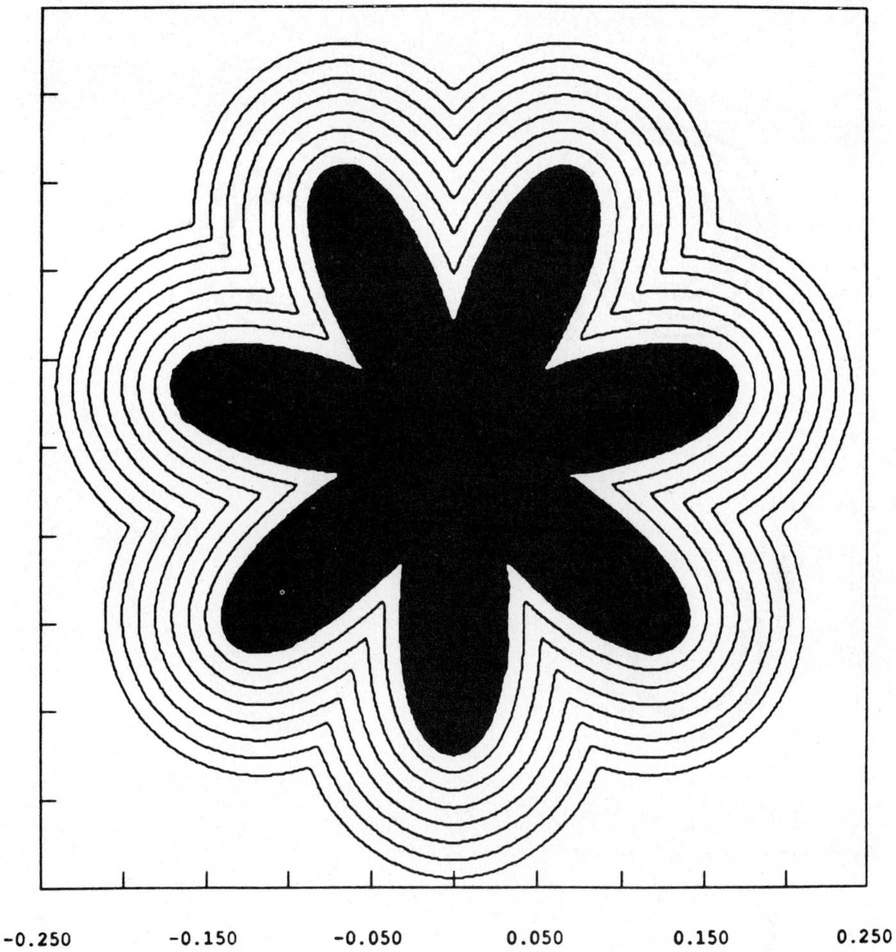

FIGURE 1: EXPANDING STAR-SHAPED INITIAL REGION

$F(K) = 1,$

T=0.0, 0.07 (0.1)

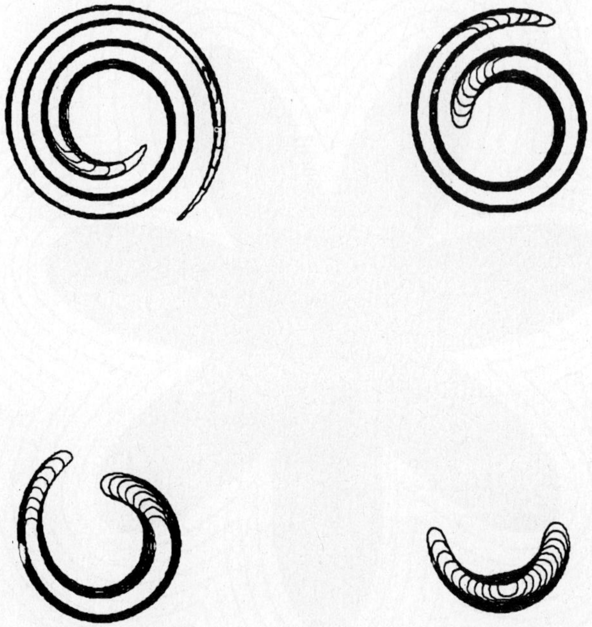

FIGURE 2: SPIRAL COLLAPSING UNDER ITS OWN CURVATURE

$$F(K) = -K$$

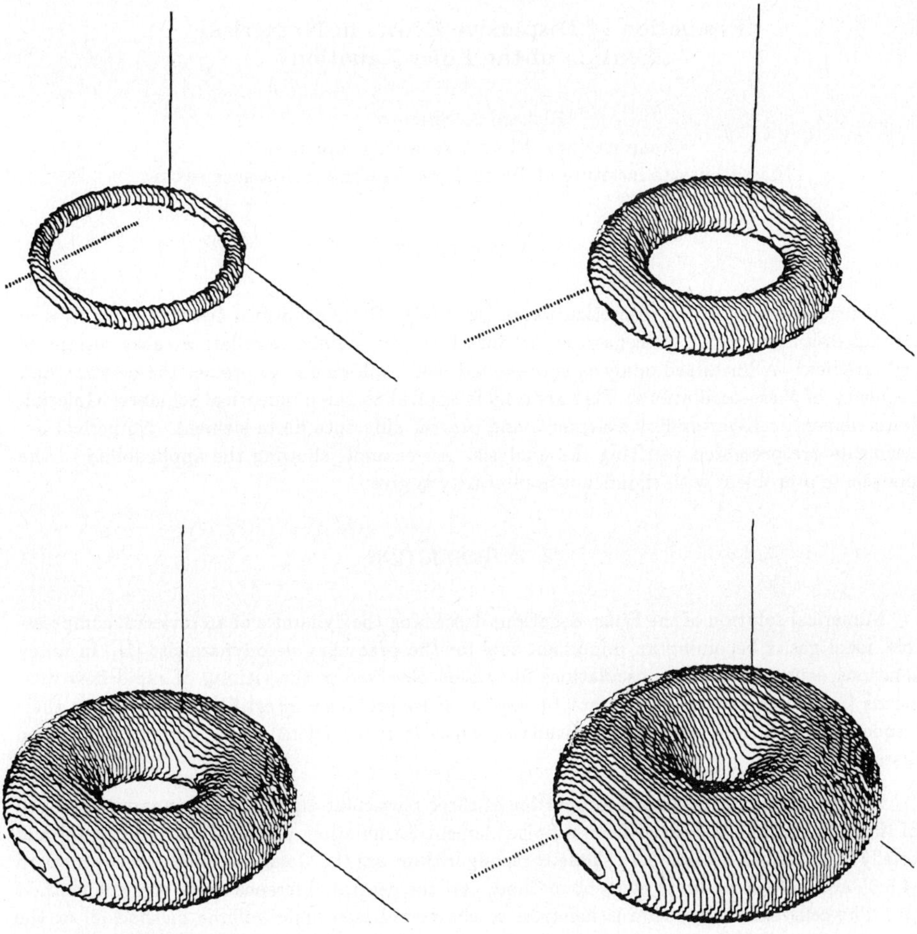

FIGURE 3: EXPANDING TORUS

$F(K) = 1 - \varepsilon K$, $\varepsilon = .01$

T=0.0, 0.1, 0.2, 0.3

Prediction of Dispersive Errors in Numerical Solution of the Euler Equations

Richard A. Shapiro
Computational Fluid Dynamics Laboratory
Massachusetts Institute of Technology, Cambridge, Massachusetts

1. ABSTRACT

Dispersive errors in the discretization of the steady Euler equations describing the flow of a compressible, inviscid, ideal gas can produce low wave number oscillations near regions of high gradient. A linearized analysis is presented which allows one to predict the location and frequency of these oscillations. This analysis is applied to three numerical schemes: Galerkin finite element, cell-vertex finite element, and central difference finite element. Numerical experiments are presented verifying the analysis. An example showing the applicability of the analysis to a problem with significant nonlinearity is given.

2. INTRODUCTION

Numerical solution of the Euler equations describing the dynamics of an inviscid, compressible, ideal gas is becoming an important tool for the practicing aerodynamicist [1]. In many solutions, low wave number oscillations have been observed in the vicinity of rapid flow variations [2]. These oscillations cannot be explained by problems in artificial viscosity, as their frequency is very low, and the amplitude is relatively independent of the amount of artificial dissipation used.

In this paper, the dispersive properties of three particular algorithms are examined. Each of these algorithms is derived from a finite element formulation, discussed in detail in [3] and briefly discussed in Section 3.1. These three algorithms are the Galerkin finite element method [4,5,6], the cell-vertex finite element method, and the central difference finite element method [7]. The cell-vertex algorithm is identical to the node-based finite volume method [8] or the first-order step in Ni's method [9]. On grids with parallelogram elements, the central difference method is equivalent to Jameson's cell-centered finite volume method [10].

The approach taken is to analyze the dispersive properties of the linearized, steady Euler equations on a regular mesh, using the spatial derivative operator for each of three methods discussed below (Galerkin, cell-vertex, central difference). This analysis is applied to a model problem, and the prediction of the frequency and location of the dispersive oscillations is demonstrated. Finally, the analytic theory is validated by comparison with numerical experiments.

3. SOLUTION ALGORITHM

In this study, the two-dimensional Euler equations describing the flow of an inviscid, compressible fluid are considered. To allow the capture of shocks and other discontinuous phenomena (such as slip lines), the Euler equations are written in conservative vector form as

$$\frac{\partial}{\partial t}\begin{bmatrix}\rho\\ \rho u\\ \rho v\\ \rho e\end{bmatrix}+\frac{\partial}{\partial x}\begin{bmatrix}\rho u\\ \rho u^2+p\\ \rho uv\\ \rho uh\end{bmatrix}+\frac{\partial}{\partial y}\begin{bmatrix}\rho v\\ \rho uv\\ \rho v^2+p\\ \rho vh\end{bmatrix}=0 \tag{1}$$

or

$$\frac{\partial \vec{U}}{\partial t}+\frac{\partial \vec{F}}{\partial x}+\frac{\partial \vec{G}}{\partial y}=0 \tag{2}$$

where e is total energy, p is pressure, ρ is density, u and v are the x and y flow velocities, $\vec{U}$ is a vector of state variables, $\vec{F}$ and $\vec{G}$ are flux vectors in the x and y directions, and h is the total enthalpy, given by the thermodynamic relation

$$h=e+\frac{p}{\rho}. \tag{3}$$

In addition, one requires the equation of state

$$\frac{p}{\rho}=(\gamma-1)\left[e-\frac{1}{2}(u^2+v^2)\right] \tag{4}$$

where the specific heat ratio γ is taken as a constant (1.4) for all calculations reported.

3.1 Spatial discretization

The finite element approach to discretizing these equations divides the domain into elements determined by some number of nodes (4 in this report).

Within each element the state vector $U^{(e)}$ and flux vectors $F^{(e)}$ and $G^{(e)}$ are written

$$U^{(e)} = \sum N_i^{(e)}U_i^{(e)} \tag{5}$$
$$F^{(e)} = \sum N_i^{(e)}F_i^{(e)} \tag{6}$$
$$G^{(e)} = \sum N_i^{(e)}G_i^{(e)} \tag{7}$$

where $U_i^{(e)}$, $F_i^{(e)}$, and $G_i^{(e)}$ are the nodal values of the state vector in element e and the $N_i^{(e)}$ are a set of bilinear interpolation functions on that element. These interpolation functions are expressed in terms of local coordinates (ξ,η), which are related to (x,y) by an isoparametric transformation. Thus, inherent in the formulation that follows are some transformational metrics, which are not shown for clarity.

These expressions can be differentiated to obtain an expression for the derivative in each element in terms of the nodal values (shown here for the state vectors)

$$\frac{\partial U^{(e)}}{\partial x_j}=\sum_{i=1}^{4}\frac{\partial N_i^{(e)}}{\partial x_j}U_i^{(e)}. \tag{8}$$

The flux vector derivatives are calculated the same way.

The expression for the derivatives is substituted into equation (2) and summed over all elements to obtain

$$\frac{\partial \vec{U}_i}{\partial t}=-\frac{\partial \vec{N}}{\partial x}\vec{F}_i-\frac{\partial \vec{N}}{\partial y}\vec{G}_i \tag{9}$$

where $\vec{N}$ is now a global vector of interpolation functions, determined by summing the interpolation functions for each element.

The next step can be thought of as a projection onto the space spanned by some other functions N', called test functions, such that the error in the discretization is orthogonal to the space spanned by the test functions (for more detail on the mathematics involved see [11]). To do this, multiply Eq. (9) by $\vec{N}'$ and integrate over the entire domain. This results in the semi-discrete equation

$$M\frac{d\vec{U}_i}{dt} = -\iint (\vec{N}'^T\frac{\partial \vec{N}}{\partial x}\vec{F}_i + \vec{N}'^T\frac{\partial \vec{N}}{\partial y}\vec{G}_i)\, dV \tag{10}$$

$$M = \iint \vec{N}'^T \vec{N} dV \tag{11}$$

or

$$M\frac{d\vec{U}_i}{dt} = -R_x\vec{F}_i - R_y\vec{G}_i \tag{12}$$

where M is the consistent mass matrix and R_x and R_y are residual matrices. The mass matrix M is sparse, symmetric, and positive definite, but not structured, so it is replaced by a lumped (diagonal) mass matrix M_L in which each diagonal entry is the sum of all the elements in the corresponding row of M. This allows Eq. 12 to be solved explicitly. The lumping does not change the steady-state solution, but does modify the time behavior of the algorithm. Finally, this set of ODE's is integrated in time to obtain a steady solution. The details of the time integration, artificial viscosity formulation, and boundary conditions are not critical to the understanding of dispersion, and can be found in [3].

3.2 Choice of Test Functions

Various choices for N' are possible, each giving rise to a particular discretization. If one choses each $N_i'^{(e)}$ to be the corresponding $N_i^{(e)}$, one obtains the Galerkin finite element approximation. If one chooses each $N_i'^{(e)}$ to be a constant, the "cell-vertex" approximation [8] results. This approximation is identical to a node-based finite volume method. Finally, if the $N_i'^{(e)}$ are chosen as

$$N'^T = \left[\begin{array}{cccc} \frac{(1-3\xi)(1-3\eta)}{4} & \frac{(1+3\xi)(1-3\eta)}{4} & \frac{(1+3\xi)(1+3\eta)}{4} & \frac{(1-3\xi)(1+3\eta)}{4} \end{array} \right] \tag{13}$$

one obtains the central difference or collocation approximation [7]. On a mesh of parallelograms, this is identical to a cell-based finite volume method.

4. LINEARIZATION OF THE EQUATIONS

This section describes the linearizations of the Euler equations. The 2-D Euler equations (Eq. 1) can be rewritten

$$\frac{\partial U}{\partial t} + A\frac{\partial U}{\partial x} + B\frac{\partial U}{\partial y} = 0 \tag{14}$$

where

$$U = \begin{bmatrix} \rho \\ u \\ v \\ p \end{bmatrix} \quad A = \begin{bmatrix} u & \rho & 0 & 0 \\ 0 & u & 0 & \frac{1}{\rho} \\ 0 & 0 & u & 0 \\ 0 & \gamma p & 0 & u \end{bmatrix} \quad B = \begin{bmatrix} v & 0 & \rho & 0 \\ 0 & v & 0 & 0 \\ 0 & 0 & v & \frac{1}{\rho} \\ 0 & 0 & \gamma p & v \end{bmatrix}. \tag{15}$$

The equations are linearized by "freezing" the A and B matrices. In the steady state, the time derivative vanishes, so we can write the linearized Euler equations in operator form as

$$(As_x + Bs_y)U = 0 \tag{16}$$

where s_x and s_y are the x and y derivative operators. If we desire non-trivial solutions to this equation, the operator matrix $(As_x + Bs_y)$ must have zero determinant. This is the statement that

$$\begin{vmatrix} us_x + vs_y & \rho s_x & \rho s_y & 0 \\ 0 & us_x + vs_y & 0 & \frac{s_x}{\rho} \\ 0 & 0 & us_x + vs_y & \frac{s_y}{\rho} \\ 0 & \gamma p s_x & \gamma p s_y & us_x + vs_y \end{vmatrix} = 0. \tag{17}$$

Define

$$r = \frac{s_x}{\sqrt{{s_x}^2 + {s_y}^2}} \tag{18}$$

$$s = \frac{s_y}{\sqrt{{s_x}^2 + {s_y}^2}} \tag{19}$$

and Eq. (17) can be expanded to

$$(ru + sv)^2 \left[a^2(r^2 + s^2) - (ru + sv)^2\right] = 0 \tag{20}$$

where a is the speed of sound. This has solutions

$$ru + sv = \begin{cases} 0 \\ \pm a \end{cases}. \tag{21}$$

Now let

$$\vec{s} = \begin{bmatrix} r \\ s \end{bmatrix} \qquad \vec{u} = \begin{bmatrix} u \\ v \end{bmatrix}$$

so that Equation (21) becomes

$$\vec{s} \cdot \vec{u} = \begin{cases} 0 \\ \pm a \end{cases}. \tag{22}$$

Since $\vec{s}$ has unit norm, the non-zero solution will exist only if the flow is supersonic. So far, no restrictions have been placed on the derivative operators s_x and s_y. The analysis above applies to the exact derivative operators as well as any of the discrete operators. The next section introduces the discrete equations and their solution.

5. FOURIER ANALYSIS OF THE LINEARIZED EQUATIONS

This section introduces the spatial discretizations of the equations into the linear model, and discusses the consequences of the truncation error in the approximations. Many of the ideas used here can be found in [12], but those analyses were performed for a scalar problem involving only one spatial direction and time.

Table 1: Spatial Derivative Operators for Various Methods

Method	s_x/i	$AR s_y/i$
Exact Derivative	ϕ	θ
Galerkin	$\frac{1}{3}\sin\phi(2+\cos\theta)$	$\frac{1}{3}\sin\theta(2+\cos\phi)$
Cell-Vertex	$\frac{1}{2}\sin\phi(1+\cos\theta)$	$\frac{1}{2}\sin\theta(1+\cos\phi)$
Central Difference	$\sin\phi$	$\sin\theta$

For purposes of analysis, assume that the equations are discretized on a Cartesian $N_x \times N_y$ mesh with grid spacings in the x and y directions of Δx and Δy. Let $x = j\Delta x$ and $y = k\Delta y$, then assume the state vector is of the form

$$U(j\Delta x, k\Delta y) = \sum_{m=0}^{N_x-1}\sum_{n=0}^{N_y-1} \exp i(j\phi_m + k\theta_n)U'_{mn} \tag{23}$$

where ϕ_m and θ_n are spatial frequencies in the x and y directions and U'_{mn} is some eigenvector. The spatial frequencies are related to m and n by the relations

$$\phi_m = \frac{2\pi m}{N_x} \tag{24}$$

$$\theta_n = \frac{2\pi n}{N_y}. \tag{25}$$

Now consider a model problem in which $\Delta x = 1$, $\Delta y = AR$ and $v \ll u$, and $u = Ma$. Then Eq. (21) has the solution

$$\frac{s_x}{\sqrt{{s_x}^2 + {s_y}^2}} = \pm\frac{1}{M}. \tag{26}$$

For a particular choice of spatial discretization, there is a particular dispersive character for a given Mach number M. Table 5 shows s_x and s_y for the Galerkin, cell-vertex and central difference methods, as well as the exact spatial derivative, assuming that ϕ and θ are continuous rather than discrete. Now introduce $s_1 = s_x$ and $s_2 = AR s_y$, square Eq. (26) and solve for s_1/s_2 to obtain

$$\frac{s_1}{s_2} = AR\sqrt{M^2-1}. \tag{27}$$

This representation of the dispersion relation has the properties that s_1 and s_2 are functions only of the non-dimensional spatial frequencies ϕ and θ, and all the problem and grid dependent terms are contained in the quantity $AR\sqrt{M^2-1}$, which will be called κ. Problems with similar values of κ should have similar dispersive behavior.

One can obtain useful information from these plots of θ vs. ϕ. The slope of a curve on which κ is constant is the spatial "group velocity", or the *angle* at which waves propagate. Waves with large spatial group velocity (the angle on the θ/ϕ plot is close to vertical) will travel at a shallow spatial angle (the wave will move a long way in x for a little change in y). This allows one to predict where the dispersed waves will appear. For the exact spatial difference operator, the curves of constant κ are straight lines, indicating that all frequencies travel at the same angle. Moreover, for $\mathcal{R} = 1$, the waves have angle $\tan^{-1}(1/\sqrt{M^2 - 1}) = \sin^{-1}(1/M)$, which is just the Mach angle, as expected.

Figure 2 shows the contours for the Galerkin method. Note that the curves are multiple-valued. In practice, the high-frequency branch is of no consequence due to the presence of artificial damping in the solution scheme. Note also that the curves depart from the exact Euler solution much later than all the other methods. This is due to the fact that on a uniform, Cartesian mesh, the Galerkin method is fourth-order accurate for the linearized Euler equations [13].

Figure 3 shows the dispersion plot for the central difference method. Note that the character of the diagram is similar to the Galerkin plot. One would expect the dispersive behavior to be similar to the Galerkin dispersive behavior, and to some extent, this is the case.

Figure 4 shows the dispersion curves for the cell-vertex scheme. Note that the curves are single-valued. Also note that the curvature is opposite the curvature for the Galerkin and central difference methods. For a particular choice of κ, the dispersion curve for the cell-vertex method will lie on the opposite side of the exact dispersion line than the curves for the Galerkin and central difference methods. This implies that the oscillations due to dispersion at a feature (a shock, for example) should appear on the opposite side (ahead or behind) of the feature compared to the Galerkin and central difference oscillations.

An important application of these curves is the prediction of oscillations due to discontinuities such as shocks. In some problems, oscillations before or after a shock can cause the solution algorithm to diverge. For example, in a strong expansion, a post-expansion oscillation may drive the pressure negative, while a pre-expansion oscillation may not be harmful. The dispersion curves allow one to predict the location of these oscillations and choose a solution algorithm which will place them in a safe place. The location of oscillations may be predicted by the following rule: If the θ vs. ϕ curve is concave up, the oscillations will be behind the feature (they travel faster than the exact solution), and if the curve is concave down, the oscillations will be ahead of the feature. For the cell-vertex method, this means that one will see pre-feature oscillations for $\kappa > 1$ and post-feature oscillations for $\kappa < 1$. For the Galerkin and central difference methods, this is reversed: $\kappa < 1$ implies pre-feature oscillations, and $\kappa > 1$ implies post-feature oscillations.

6. NUMERICAL VERIFICATION

Flows over a 1/2 degree wedge in a channel were used to verify the dispersive properties numerically. Figure 1 shows the geometry and flow topology for a typical problem. All the calculations were performed on 50x20 grids, and result in similar flow topologies.

The first set of experiments demonstrates the validity of the similarity parameter κ. Three numerical test cases were run: Mach 2 flow with $\mathcal{R} = 1$; Mach 1.323 flow with $\mathcal{R} = 2$ and Mach 3.606 flow with $\mathcal{R} = 1/2$. A quick examination of the flow geometry gives the physical significance of κ as the ratio of the number of x grid lines crossed by the feature per y grid line crossed. In the Mach 3.606 flow, the shock lies at a much shallower angle, so that for a smaller Δy the same crossing ratio is obtained. A similar argument holds for the Mach 1.323 flow. Figure 6 shows the Mach number at mid-channel for the central difference method, scaled by the free stream Mach number for the different Mach numbers above. The central difference method is used here because it exhibits the most oscillation with the greatest amplitude. The exact Mach number ratios (M/M_∞) for these shocks are 0.991 for $M_\infty = 2$ and $M_\infty = 3.606$ and 0.986 for $M_\infty = 1.323$. These compare well with the actual data, and explain why two of the curves lie on top of each other. Note that the frequencies of the oscillation are nearly identical. Also note that the frequency changes slightly as one moves further downstream of the shock. This is as predicted by the dispersion curve. As one moves downstream the spatial group velocity increases, meaning ϕ increases slightly. The wavelength predicted by the dispersion relation at $(x, y) = (1.5, 0.5)$ should be about 10.5 points, and the measured wavelength (crest-to-crest) is either 10 or 11 points, depending on where one defines the crest.

The next set of data shows the use of the dispersion curves in predicting the location of the oscillations. Figure 5 shows the dispersion curves for all three numerical methods and the exact spatial derivatives on a single plot for $\kappa = \sqrt{3}$. Here it is apparent that the Galerkin curve stays much closer to the exact curve. Numerical examples were computed for the Mach 2 case above using all three methods. In following figures Mach number at mid-channel is plotted. Figure 7 shows the plot for the Galerkin method, Figure 8 for the central difference method and Figure 9 for the cell-vertex method. Note that both the Galerkin and central difference methods exhibit post-shock oscillation, while the cell-vertex exhibits pre-shock oscillation. Also note that the frequency of the Galerkin oscillations is much higher, and with a lower amplitude than the central difference approximation. This is expected since the Galerkin method group velocity errors occur at higher spatial frequencies, (see Fig. 5). As an interesting aside, note that in Fig. 9 the pre-shock oscillations from the reflected shock are visible at the right side of the plot. These examples verify the use of the dispersion curves to predict the location of dispersive phenomena.

7. CONCLUSIONS

The primary conclusion of this study is that the low frequency oscillations sometimes seen near shocks are due to dispersion in the numerical scheme. The linearized analysis presented gives one a method for predicting the location and frequency of these oscillations. The linear analysis is effective in predicting the location of oscillations, even for problems with significant nonlinearity. The central difference finite element method is shown to be inferior to the Galerkin and cell-vertex methods due to its poor dispersive behavior. For the practical analyst, either Galerkin or cell-vertex provides adequate performance.

8. ACKNOWLEDGEMENTS

This work was supported by the Air Force Office of Scientific Research under contract number AFOSR-87-0218, Dr. James Wilson, Technical Monitor, and by the Fannie and John Hertz Foundation. The author wishes to acknowledge the help of Prof. Earll Murman in preparing the report, and of Ellen Mandigo in proofreading the manuscript.

REFERENCES

[1] A. Jameson, "Successes and Challenges in Computational Aerodynamics," AIAA Paper 87-1184, 1987.

[2] R. Haimes, Personal communication.

[3] R. Shapiro and E. Murman, "Adaptive Finite Element Methods for the Euler Equations," AIAA Paper 88-0034, January 1988.

[4] R. Löhner, K. Morgan, J. Peraire, and O. C. Zienkiewicz, "Finite Element Methods for High Speed Flows," AIAA Paper 85-1531, 1985.

[5] K. Bey, E. Thornton, P. Dechaumphai, and R. Ramakrishnan, "A New Finite Element Approach for Prediction of Aerothermal Loads–Progress in Inviscid Flow Computations," AIAA Paper 85-1533, 1985.

[6] R. Shapiro and E. Murman, "Cartesian Grid Finite Element Solutions to the Euler Equations," AIAA Paper 87-0559, January 1987.

[7] R. J. Prozan, L. W. Spradley, P. G. Anderson, and M. L. Pearson, "The General Interpolants Method: A Procedure for Generating Numerical Analogs of the Conservation Laws," In *Proceedings of the AIAA Third Computational Fluid Dynamics Conference*, 1977, pp. 106–115.

[8] M. G. Hall, *Cell Vertex Schemes for the Solution of the Euler Equations*, Technical Memo Aero 2029, Royal Aircraft Establishment, March 1985.

[9] R. H. Ni, "A Multiple-Grid Scheme for Solving the Euler Equations," *AIAA Journal*, Vol. 20, No. 11, November 1982, pp. 1565–1571.

[10] A. Jameson, W. Schmidt, and E. Turkel, "Numerical Solutions of the Euler Equations by a Finite Volume Method Using Runge-Kutta Time Stepping Schemes," AIAA Paper 81-1259, June 1981.

[11] G. Strang, *Introduction to Applied Mathematics*, Wellesley-Cambridge Press, Wellesley, Massachusetts, 1986.

[12] R. Vichnevetsky and J. B. Bowles, *Fourier Analysis of Numerical Approximations of Hyperbolic Equations*, SIAM, 1982.

[13] S. Abarbanel and A. Kumar, *Compact Higher-Order Schemes for the Euler Equations*, ICASE Report 88-13, ICASE, February 1988.

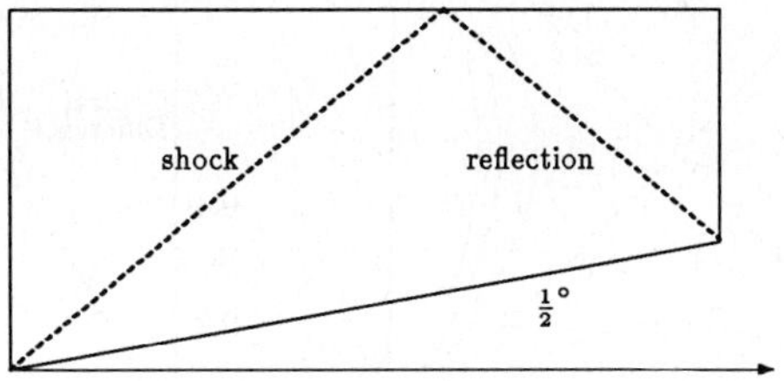

Figure 1: Geometry for Wedge Numerical Test Cases

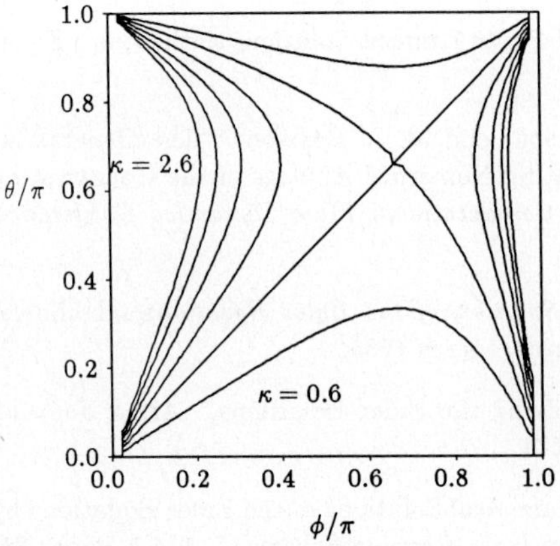

Figure 2: Lines of Constant κ for Galerkin Method

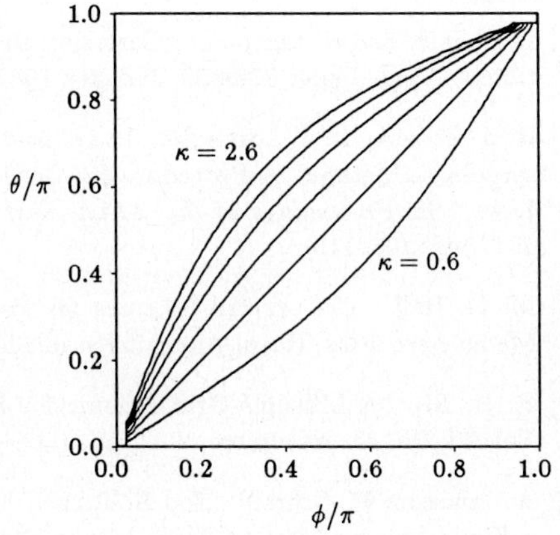

Figure 4: Lines of Constant κ for Cell-Vertex Method

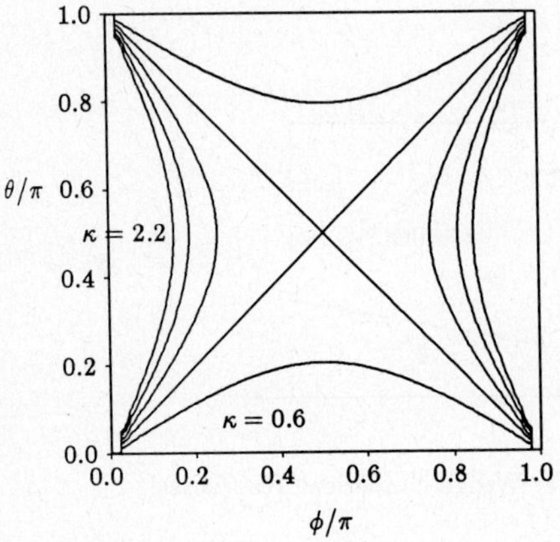

Figure 3: Lines of Constant κ for Central Difference Method

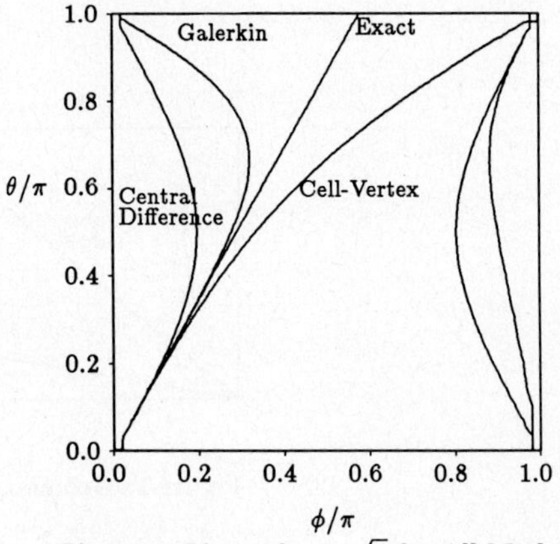

Figure 5: Lines of $\kappa = \sqrt{3}$ for All Methods

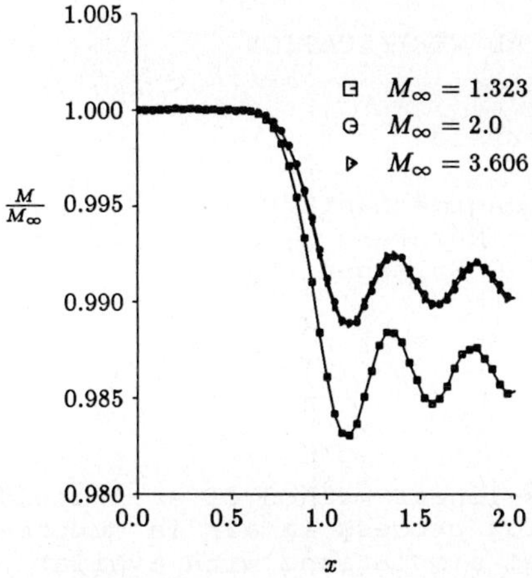

Figure 6: M/M_∞ for $\kappa = \sqrt{3}$, Central Difference Method

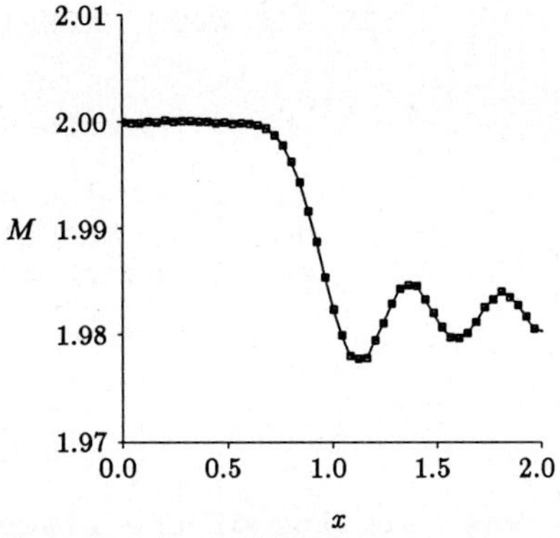

Figure 8: Mid-channel Mach Number, $M_\infty = 2$, Central Difference Method

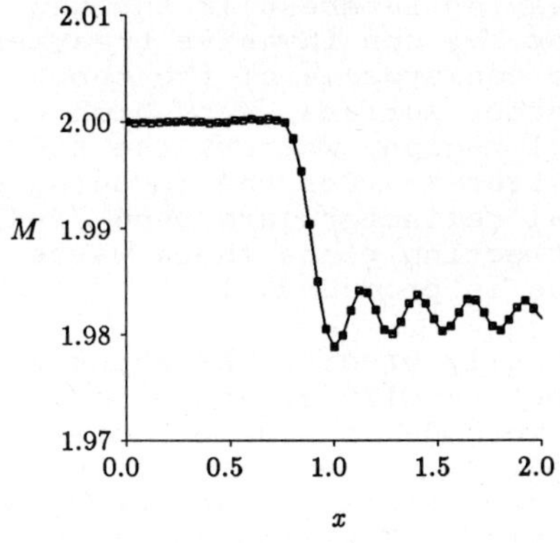

Figure 7: Mid-channel Mach Number, $M_\infty = 2$, Galerkin Method

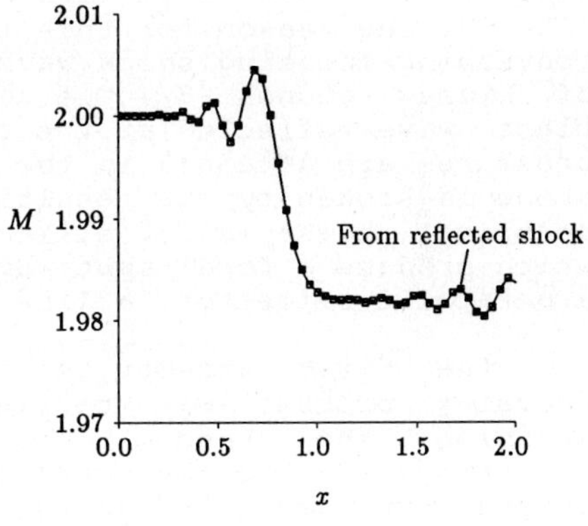

Figure 9: Mid-channel Mach Number, $M_\infty = 2$, Cell-Vertex Method

NUMERICAL PREDICTION OF SHOCK WAVE FOCUSING PHENOMENA IN AIR WITH EXPERIMENTAL VERIFICATION

Martin Sommerfeld

Lehrstuhl für Strömungsmechanik
Universität Erlangen-Nürnberg
Egerlandstr. 13, 8520 Erlangen, FRG

SUMMARY

The ability of the piecewise-linear method of accurately predicting the shock wave focusing process in air is demonstrated by comparing the numerical simulations with available experimental results. After proving the grid-independence of the numerical results comparisons are introduced for the shock wave pattern evolving during the focusing process and the pressure histories at different location in the flow field. The agreement between experiment and calculation is found to be very good for the considered weak and strong shock cases.

INTRODUCTION

In recent years shock wave focusing by means of different types of concave reflectors has received increasing attention /1 - 7/. The reason for this increasing interest is the use of converging focusing shock waves for the non invasive treatment of kidney stones /8/. Due to the convergence of the shock or blast wave reflected at the reflector surface, very high peak pressures are attained in the focal region, whereby the kidney stone is broken by the resulting stresses. For the focusing of spherical blast waves, ellipsoidal reflectors are used /2-4/, which produce a focal spot. By reflecting plane shock waves at a parabolic reflector, a line focus is produced /1/.

The first attempt to numerically predict the shock wave focusing process was carried out by Olivier and Grönig /5/ applying the Random Choice methode based on operator splitting. Due to the operator splitting however, this method looses the great advantage of representing shock waves and other discontinuities within one mesh and strong oscillations are found behind shock fronts. The agreement of the pressure histories with the experimental results of Sturtevant and Kulkarney /1/ was reasonable inspite of the above mentioned oscillations in the calculated pressure traces.

Recently, a second order extension of Goudunov's method, called piecewise-linear method /9/, was applied to the numerical prediction of the shock wave focusing process in air /6/.

Since the experiments conducted by Nishida et al. /6/ gave rather low pressure amplifications in the focal region, the agreement with the numerical calculations is very poor. The comparison of the shadowgraphs with numerically simulated shadowgraphs, however showed quite good qualitative agreement /6/.

Numerical simulations of shock wave focusing in water for plane and axisymmetric configurations were recently performed by introducing an equation of state for water, namely the Tait equation /7/. The calculations which are also based on the piecewise-linear method, showed good quantitative agreement in the pressure histories when compared to experimental results obtained with an ellipsoidal reflector /3/. In the region close to the geometrical focus however, the calculated peak pressure is about four times smaller than the measured value. A test of the dependence of the maximum peak pressure at the focus on the mesh size showed that by decreasing the mesh size the peak pressure still was increasing further /7/. This indicates that for the numerical simulations of the shock wave focusing in water, a very fine mesh is necessary in order to obtain a grid-independent result.

Besides the importance of testing numerical schemes against experimental results, the above mentioned problem was a reason for performing further numerical studies on the shock focusing in air and testing the grid dependence of the results. As a basis of the numerical simulations the experiments of Sturtevant and Kulkarny /1/ were chosen.

BASIC EQUATIONS AND NUMERICAL SCHEME

The basic equations are those for a compressible inviscid, non-heat conducting fluid, namely the Euler equations for two dimensions written in conservation form.

$$U_t + F_x + G_r = I$$

$$U = \begin{bmatrix} \varrho \\ \varrho u \\ \varrho v \\ \varrho E \end{bmatrix}, \quad F = \begin{bmatrix} \varrho u \\ \varrho u^2 + p \\ \varrho u v \\ u(\varrho E + p') \end{bmatrix},$$

$$G = \begin{bmatrix} \varrho v \\ \varrho v u \\ \varrho v^2 + p \\ v(\varrho E + p) \end{bmatrix}, \quad I = -\frac{j}{r} \begin{bmatrix} \varrho v \\ \varrho v u \\ \varrho v^2 \\ v(\varrho E + p) \end{bmatrix}$$

were the variables ρ, u, v, p and E are the density, the velocity in x- and r-direction, the pressure and the specific total energy per unit mass which may be expressed by

$$E = \frac{P/\varrho}{\gamma - 1} + 0.5(u^2 + v^2).$$

The equation of state is given by

$$p = \varrho RT.$$

Since the above equations are written for both the two-dimensional and the axisymmetric form, we have to set the parameter j accordingly (j = 0: two-dimensional plane flow; j = 1: axisymmetric flow). For the present calculations the two-dimensional form of the equations is used and the solution to the two-dimensional equations is obtained by a second-order accurate operator splitting of the form

$$U^{n+1} = \left\{ L_x\left(\frac{\Delta t}{2}\right) L_y(\Delta t)\, L_x\left(\frac{\Delta t}{2}\right)\right\} U^n$$

which results in three one-dimensional sweeps to yield the solution at the next time level n + 1 from a given solution or initial condition at time level n. The one-dimensional sweeps are solved by the piecewise-linear method (PLM) which was proposed by Colella and Glaz /9/. The PLM basically consists of four steps which are successively carried out for each sweep:

(1) The calculation of interpolated profiles which are taken to be piecewise linear for the dependent variables by applying some monotonicity constraints to avoid physically unrealistic oscillations.
(2) The construction of time-centered right and left states (V_r, V_l) of the dependent variables at $x_{i+1/2}$ (Fig. 1) by taking into account the direction of the associated characteristics (λ^+, λ° and λ^-).
(3) The solution of the Riemann problem for the right and left states (V_r, V_l) to give a solution at n+1/2, i+1/2.
(4) Conservative differencing of the fluxes, which are calculated from the solution of the Riemann problem.

Further details of the PLM may be found in the paper of Colella and Glaz /9/.

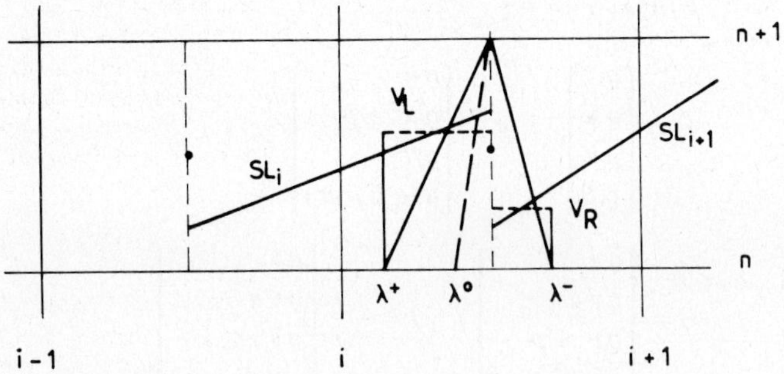

Fig. 1: Construction of the time centered right and left states

INFLUENCE OF MESH SIZE AND BOUNDARY CONDITIONS

The calculations were done according to the experiments of Sturtevant and Kulkarny, where a parabolic plane reflector was placed at the end wall of a shock tube. The reflector width was 203.2 mm and hence the geometrical focus is located 60.3 mm ahead of the reflector (Fig. 2).

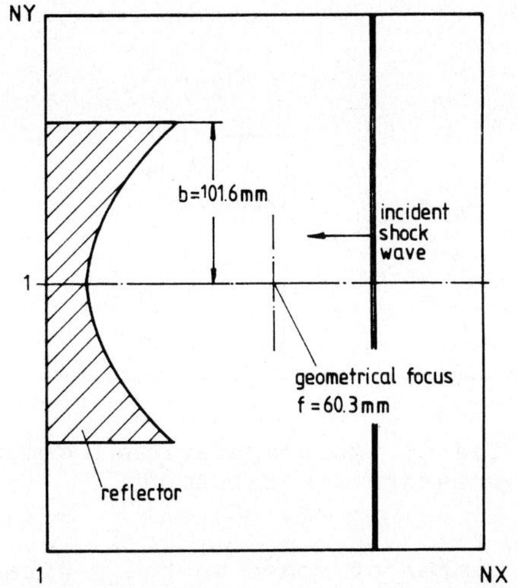

Fig. 2: Reflector configuration and dimensions

Due to the symmetry of the flow field only one half of the whole domain of Fig. 2 was calculated. The boundary conditions employed are reflection conditions at the reflector axis and the reflector surface, inflow conditions at the right hand boundary and outflow conditions at the upper boundary.

In order to prove the grid-independence of the numerical results, the shock wave focusing process was calculated by employing several different mesh sizes ranging from 1.5 mm to 0.75 mm. Smaller mesh sizes yield of course a better resolution of the discontinuities and shock waves and, by decreasing the mesh size, the calculated value of the maximum pressure at the focus approaches a limiting value (Table); which is close to the experimental one. Comparing the result obtained using the coarsest and finest mesh shows that the difference is about 7%. Further numerical tests were run to optimize the number of grid points above the reflector edge (Fig. 2), which are necessary to guarantee the appropriate prediction of the expansion eminating from the reflector edge. As a criterium to judge the proper number of meshes above the reflector edge, the maximum pressure on the reflector axis obtained throughout the focusing process is compared (Fig. 3).

Table 1: Dependence of peak pressure in mesh size

mesh size mm	peak pressure P /P1	peak pressure ratio
1.5	2.405	0.915
1.0	2.540	0.97
0.75	2.629	1.0

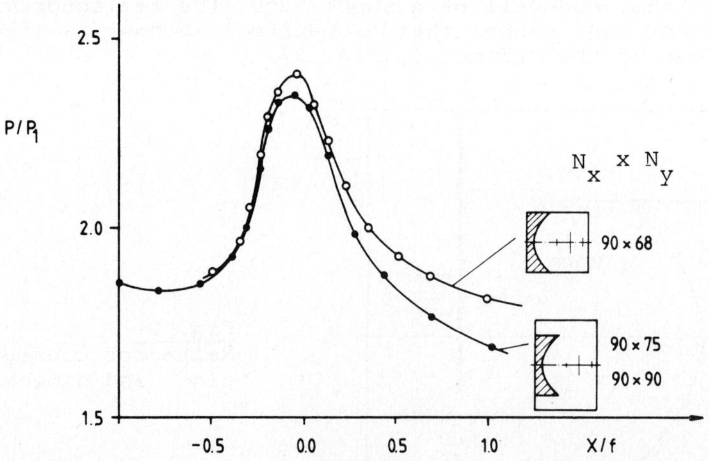

Fig. 3: Effect of the choice of the computational domain on the maximum on-axis pressure distribution

When increasing the number of nodes in the y-direction from 75 to 90, no considerable difference in the pressure distribution is observed. Since quadratic meshes are used, the mesh number of 90 x 75 in x- and y-direction is taken to be sufficient for simulating the expansion waves from the reflector edges. In order to give a good resolution of the shock waves the final calculations were conducted with a finer mesh of 180 x 150 nodes in x- and y-direction, respectively.

NUMERICAL RESULTS IN COMPARISON TO THE EXPERIMENTS

The efficiency of the shock wave focusing, namely the amplification of the pressure at the focus and the size of the focal region, are strongly dependent on the strength of the incident shock front. The pressure amplification decreases with increasing incident shock strength due to non-linear effects in the focusing process.

The numerically simulated evolution of the wave pattern during shock focusing for a relatively weak Mach number of 1.1 is shown in Fig. 4 at four different instants. In Fig. 4a, the center part of the incident shock wave is still propagating towards the reflector, while the outer parts are already reflected and defracted at the reflector edge. When the whole shock wave is reflected, the concave part of the reflected shock front converges towards the focal point whereby its strength is amplified (Fig. 4b). This converging shock is overtaken by expansion waves eminating from the edges of the reflector, whereby the outer parts of the converging shock fronts are weakend. When the intersection of the converging shock front with the head of the expansion meets the reflector axis (Fig. 4c), the maximum pressure is attained during the focusing process. This exhibits an important non-linear effect, since for a finite shock strength, this happens before the converging shock front has reached the geometrical focus. The expansion waves colliding with each other are reflected as outward propagating compression waves (Fig. 4c). At a later stage (Fig. 4d) the outer parts of the converging shock fronts are seen to be reflected from each other in a regular way, which is due to the small angle between the upper and lower part of the fronts. The outward propagating compression waves have steepend to form a shock front which follows the crossing of the outer parts of the converging shock fronts and form a three shock intersection with the reflected parts behind the crossing.

A direct comparison of the numerical predictions (density contour lines) with the shadowgraphs obtained by Sturtevant and Kulkarny /1/ for a Mach number of 1.1 is shown in Fig. 5. The times given in this figure are counted from the instant when the incident shock front hits the reflector edge and are identical for experiment and calculation. The length scale is the same for the shadowgraphs and the density contour plots. The calculated time evolution of the shock wave focusing process is found to be in very good agreement with the experiment and the evolving wave pattern at the different stages coincide very well. The resolution of the shock fronts and slip lines in the numerical simulation is comparable to that seen on the shadowgraphs. In Figs. 5c and 5d, the wave pattern after focusing are shown in more detail. In the shadowgraphs a dark elongated spot is visible in the focal region which is a region of hot gas. This region is connected by slip lines to the three shock intersection of the outward propagating, nearly spherical shock and the reflections of the outer parts of the converging shocks. All these details are also found in the numerical simulation. Since the slip lines are very weak discontinuities they appear only as small kinks in the density contour lines.

At a higher Mach number ($M_S = 1.3$) the shock wave pattern at the different stages after focusing are much different to the weak-shock case. The situation before focusing is comparable to the weak-shock case (Fig. 6a). As soon as the intersections of the head of the expansion waves with the converging shock wave meet each other at the axis of the reflector the maximum pressure is attained in the focusing process. This occurs earlier than in the weak-shock case and the location of maximum pressure is closer to the reflector surface. Since the angle between the converging shock fronts is larger than for the weak case, a Mach-type reflection is evolving having a rather plane stem shock (Figs. 6b and 6c). The three shock intersections between the outward propagating steepened compression waves, the outer part of the converging shock and the Mach stem are connected by two slip lines with the hot spot at the focus (Fig. 6c). All the details of the wave pattern seen on the shadowgraphs are found in the numerical simulations and the agreement at the different stages is very good. Also the slip lines and the rolling up of the slip line near the focus at a later stage comes out very clearly in the calculations (Fig. 6d).

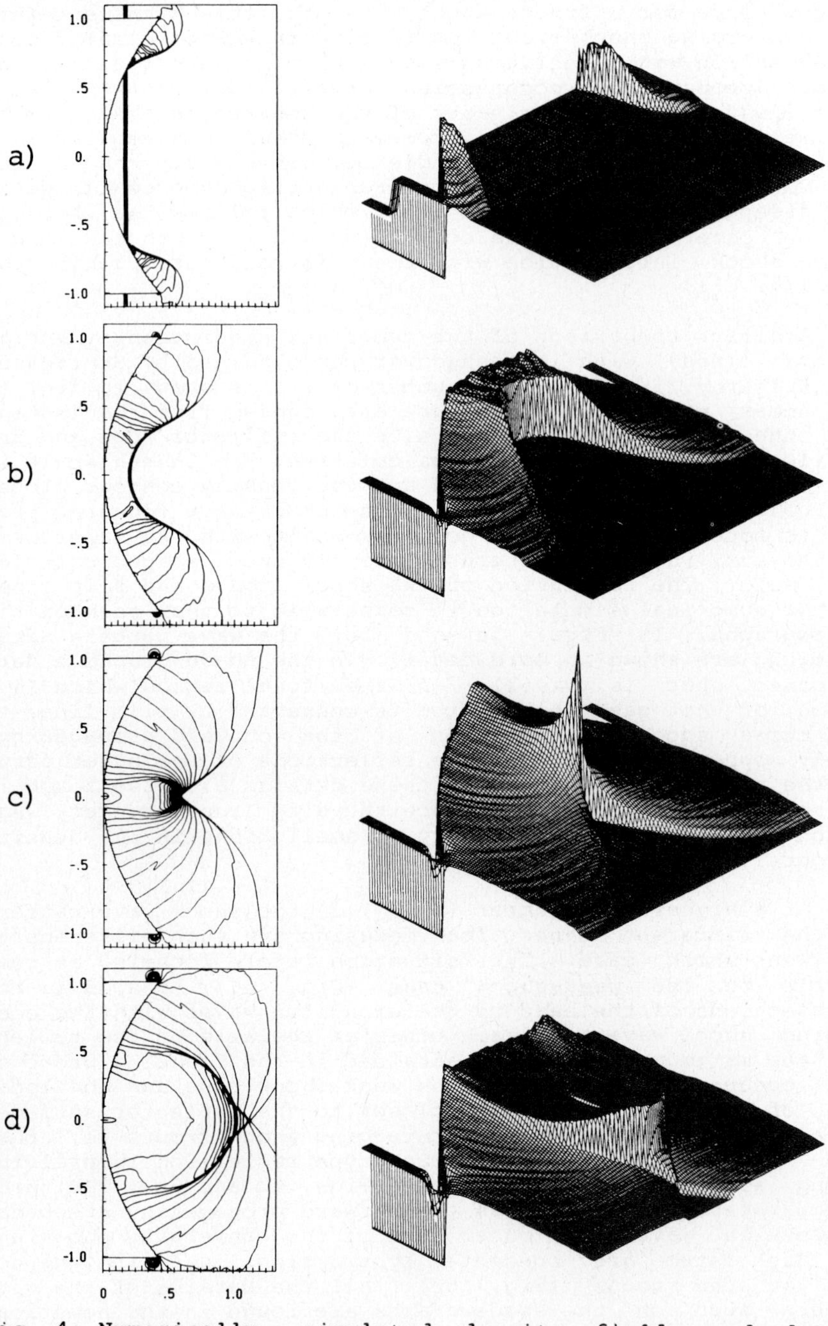

Fig. 4: Numerically simulated density fields and density contour lines for the shock wave focusing process ($M_S = 1.1$)

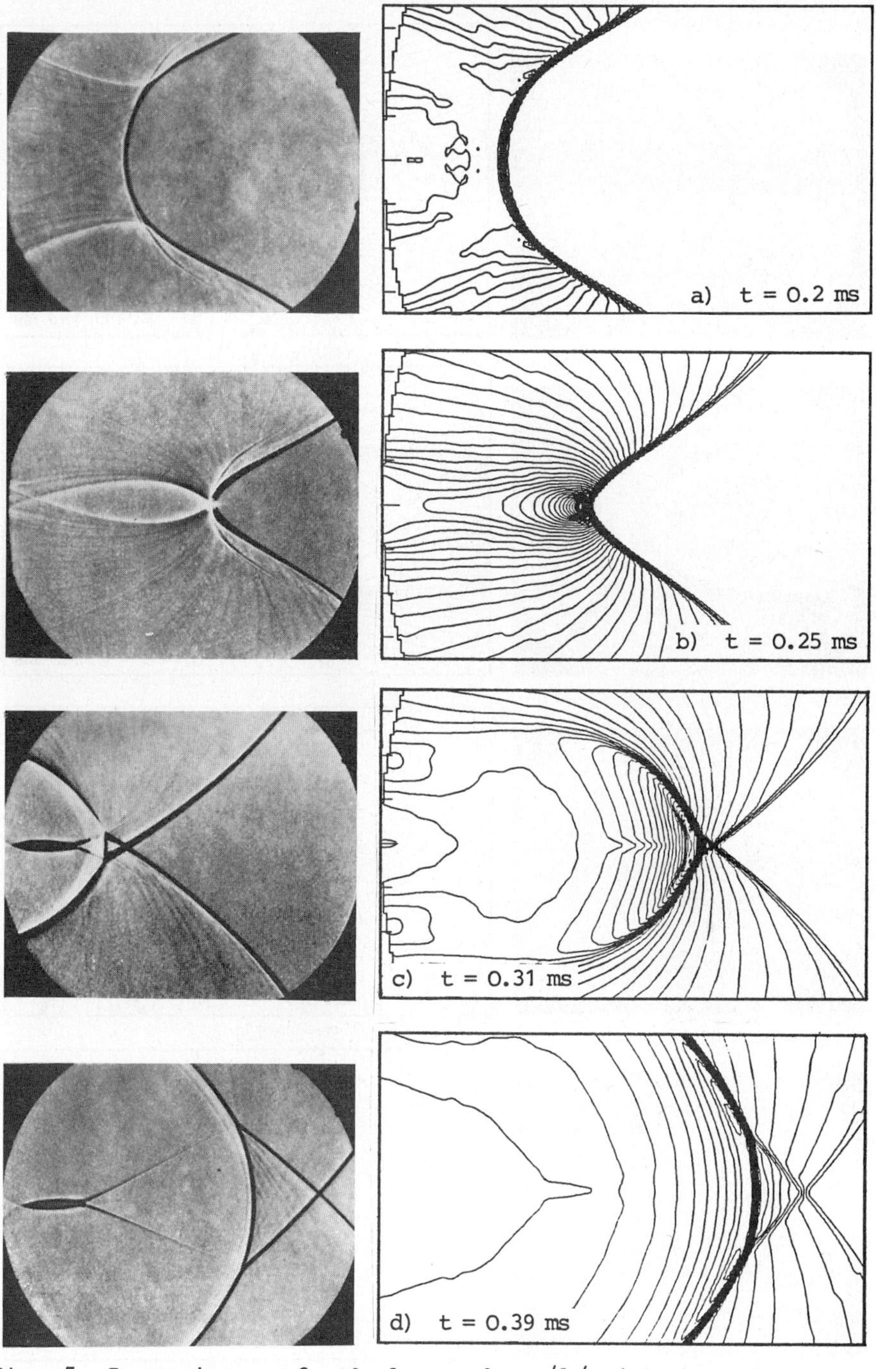

Fig. 5: Comparison of shadowgraphs /1/ with density contour lines of the numerical calculation ($M_S = 1.1$)

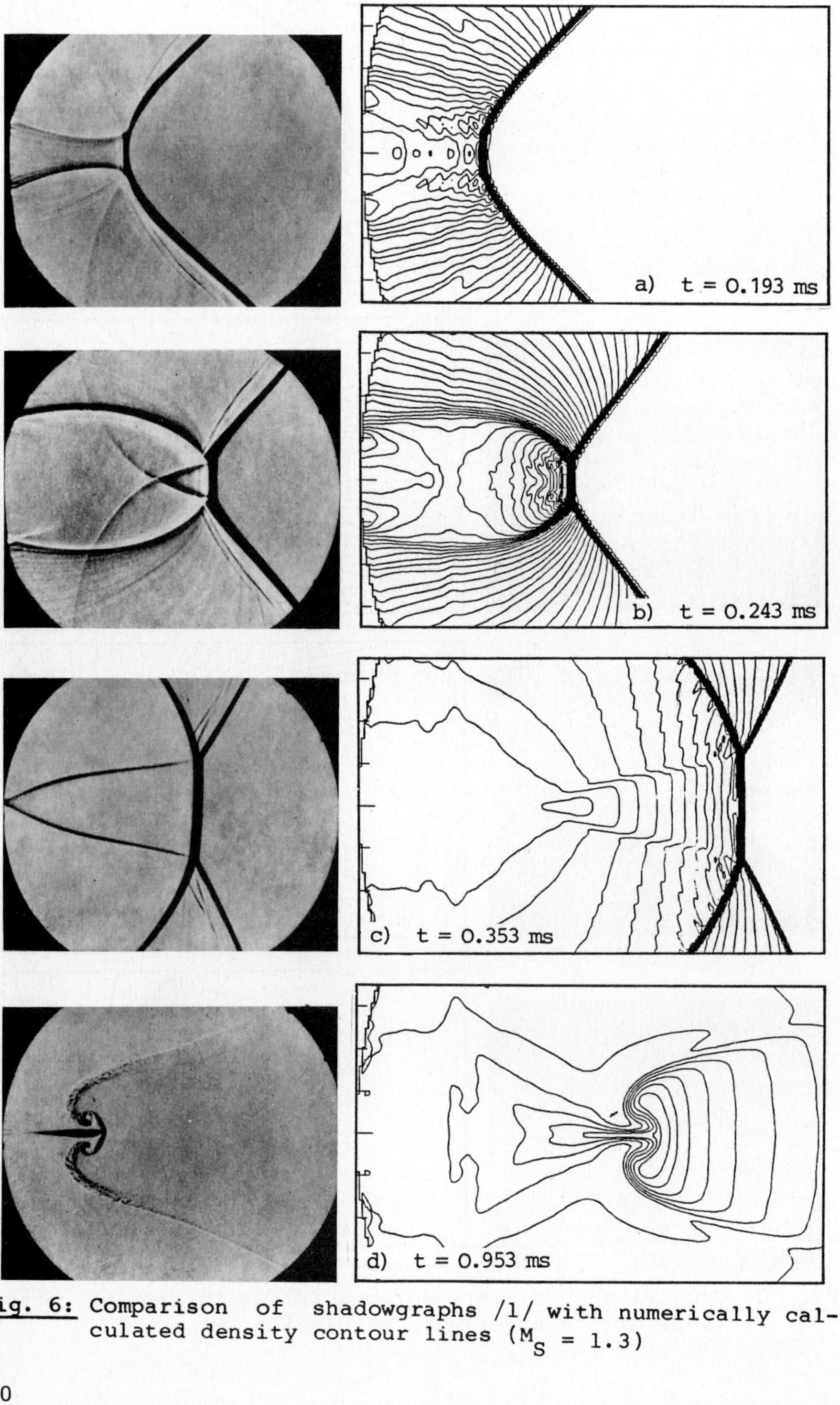

Fig. 6: Comparison of shadowgraphs /1/ with numerically calculated density contour lines ($M_S = 1.3$)

The plot of the triple point trajectories originating at the reflector edge for the weak and strong-shock case gives a more quantitative comparison between experiment and simulation (Fig. 7), and the agreement is quite good. For the weak-shock case (M_S = 1.1) however, the triple point seems to propagate slower outward in the numerical result. Furthermore, the measured pressure histories for the weak-shock case measured at two locations near the focal spot and at two locations between the reflector and the focal spot /1/ are compared with the numerical predictions (Fig. 8). The agreement is again found to be very good for both the pressure amplitudes and the arrival times of the different shock waves. At the locations between the focus and the reflector surface the pressure traces exhibit that the expansion wave originated at the reflector edge and the associated reflected compression waves propagate a bit slower in the numerical simulation.

Finally, the maximum pressure attained on the reflector axis throughout the whole focusing process is compared for both considered shock strengths, Fig. 9. This clearly reveals the non-linear effects with increasing shock strength, whereby the amplification is reduced and the location of maximum pressure is shifted away from the geometrical focus towards the reflector. The pressure is normalized with the pressure jump of the reflected converging shock as it leaves the reflector surface and the distance is normalized by the focal distance. In the calculated pressure distribution the pressure decay after focusing is not so pronounced as in the experiment which may be due to boundary layer effects. The maximum pressure location of the calculation is identical with experimentally found location for the higher shock strength and closer to the geometrical focus for the weak-shock case.

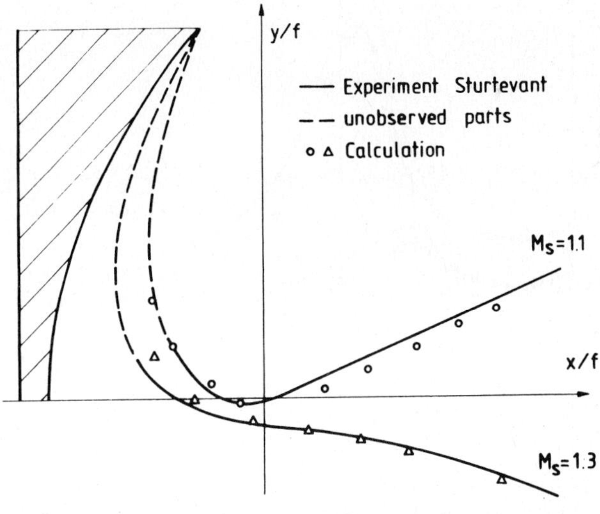

Fig. 7: Trajectories of the triple point (comparison of measurement and calculation)

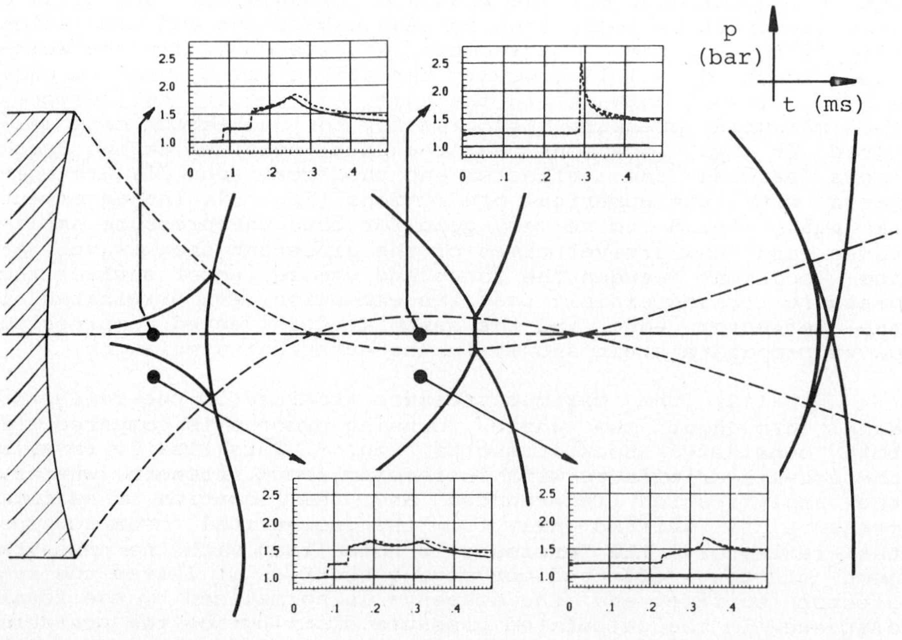

<u>Fig. 8:</u> Measured /1/ and predicted pressure histories for M_S = 1.1 (closed line: measurement; dashed line: prediction)

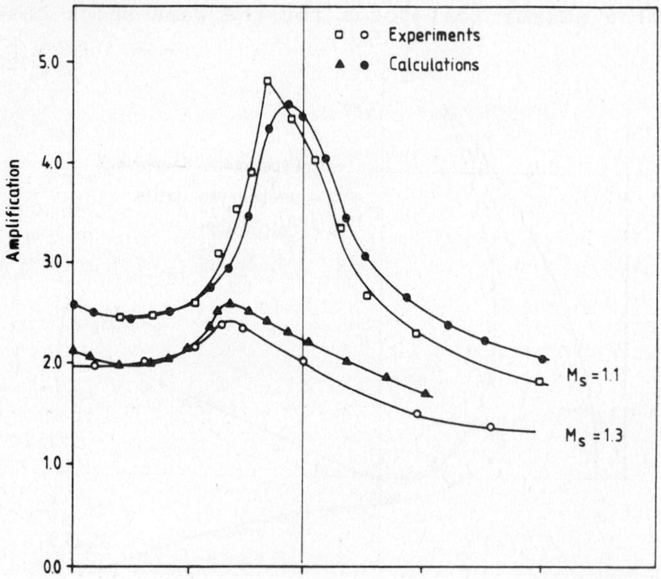

<u>Fig. 9:</u> Measured and predicted maximum on-axis pressure distribution

CONCLUSIONS

The numerical simulation of the shock wave focusing process in air by the piecewise-linear method (PLM) gave a very good agreement with the experimental results obtained by Sturtevant and Kulkarny /1/ for both the developing shock wave pattern and the pressure histories during the focussing process. The resolution of the shock fronts and slip lines by the numerical calculation for a mesh of 180 x 150 (mesh size 0.75 mm) was found to be comparable to the resolution on the experimentally obtained shadowgraphs even for very weak shock waves. The associated mesh size was found by testing the grid-independence of the numerically predicted results.

REFERENCES

/1/ Sturtevant, B. and Kulkarny, V.A.: The focusing of weak shock waves, J. Fluid Mech., 73, (1976), 651-671.

/2/ Holl, R.: Wellenfokussierung in Fluiden, Dissertation, RWTH Aachen (1982).

/3/ Müller, H.M. and Grönig, H.: Experimental investigations on shock wave focusing in water, Proc. 12th Int. Congr. on Acoustics, Vol. III, H3-3, Toronto (1986).

/4/ Takayama, K., Esashi, H. and Sanada, N.: Propagation and focusing of spherical shock waves produced by underwater microexplosions, In: Proceedings of the 14th Int. Symp. on Shock Tubes and Shock Waves (Eds. Archer, R.D., Milton, B.E), pp. 553-562, Kensington/Australia: New South Wales University Press (1983).

/5/ Olivier, H. and Grönig H.: The Random choice method applied to two-dimensional shock focusing and diffraction, J. Comp. Phys., 63, (1986), 85-106.

/6/ Nishida, M., Nakagawa, T., Saito, T. and Sommerfeld, M.: Interaction of weak shock waves reflected on concave walls, In: Proceedings of the 15th Int. Symp. on Shock Waves and Shock Tubes (Eds. Bershader, D., Hanson, R.), pp. 211-217, Stanford/CA: Stanford University Press (1985).

/7/ Sommerfeld, M. and Müller, H.M.: Experimental and numerical studies of shock wave focusing in water, Experiments in Fluids, 6, (1987), 209-216.

/8/ Chaussy, Ch., Forßmann, B., Brendel, N., Jocham, D., Eisenberger, F., Hepp, N. und Gokel, J.M.: Berührungsfreie Nierensteinzertrümmerung durch extrakorporal erzeugte, fokussierte Stoßwellen, In: Beitrag zur Urologie, Bd. 2 (Eds. Chaussy, Ch., Staehler, G.), Basel: Karger, (1980).

/9/ Colella, P. and Glaz, H.M.: Efficient solution algorithms for the Riemann problem for real gases, J. Comp. Phys., 59, 264-289, (1985).

FUNDAMENTAL ASPECTS OF NUMERICAL METHODS FOR THE PROPAGATION OF MULTI-DIMENSIONAL NONLINEAR WAVES IN SOLIDS

M. Staat, J. Ballmann
Lehr- und Forschungsgebiet Mechanik
RWTH Aachen
D - 5100 Aachen, FRG

SUMMARY

The nonlinear scalar constitutive equations of gases lead to a change in sound speed from point to point as would be found in linear inhomogeneous (and time dependent) media. The nonlinear tensor constitutive equations of solids introduce the additional local effect of solution dependent anisotropy. The speed of a wave passing through a point changes with propagation direction and its rays are inclined to the front. It is an open question wether the widely used operator splitting techniques achieve a dimensional splitting with physically reasonable results for these multi-dimensional problems.

May be this is the main reason why the theoretical and numerical investigations of multi-dimensional wave propagation in nonlinear solids are so far behind gas dynamics. We hope to promote the subject a little by a discussion of some fundamental aspects of the solution of the equations of nonlinear elastodynamics. We use methods of characteristics because they only integrate mathematically exact equations which have a direct physical interpretation.

INTRODUCTION

Many characteristic-based methods have been devised for the solution of hyperbolic problems with more than two independent variables (e.g. two- and three-dimensional wave propagation). Most of them (e.g. GODUNOV-Type-Methods and GLIMM's Random-Choice-Method) use schemes developed for one-dimensional wave propagation by various operator splitting techniques. May be this is the reason, why promising results are only known for nonlinear media with scalar constitutive equations so far.

The mechanical state variables for solids are second order tensors, and thus only physically one-dimensional problems can be modelled by scalar laws, whereas tensor constitutive equations describe the material behaviour in multi-dimensional problems. Thereby a strong coupling of the different spatial directions may result, and, if the material is nonlinear, local effects of anisotropy may occur. Such effects are probably best known from magnetohydrodynamics and from optical and mechanical waves in linear anisotropic solids [1. 2]. Due to the dependence on the solution, the situation is even more complicated for nonlinear elastic and plastic waves in solids. A scalar nonlinear constitutive equation introduces a solution dependent inhomogeneity . Nonlinear tensor constitutive equations cause the additional local effect of solution dependent anisotropy.

For a numerical treatment of these nonlinear problems. methods of nearcharacteristics have been devised. which become methods of bicharacteristics. if the local scheme uses the axes of symmetry of the local wave fronts of point disturbances [3. 4]. The complete set of PDE's describing a general nonlinear elastic solid can be solved numerically for arbitrary large deformations and large displacements. For convenience we restrict our discussion to hyperelastic materials . One can easily dispense with the lengthier treatment of CAUCHYelastic materials as included in [4]. since in a purely mechanical theory every stable passive elastic material is hyperelastic (or GREEN-elastic) [5]. Furthermore we exclude all physical situations for incompressible solids that do not permit longitudinal waves. e.g. plain strain problems but not plain stress problems.

BASIC EQUATIONS

The material points of a body are denoted by their coordinates in a possible reference configuration $\bar{B}^*$ and the actual configuration B^* in space and time by point-coordinates ξ^a and x^a $(a = 0, 1, 2, 3)$, respectively. The time-like coordinates are $\xi^o = c\tau$ and $x^o = ct$ with time $\tau = t$, and some constant speed c. Co- and contravariant basis vectors are introduced in both configurations in the usual manner. The material points with position vector $\bar{\mathbf{r}}^* = \bar{\mathbf{r}} + \xi^o \bar{\mathbf{g}}_o$ in $\bar{B}^*$ are moved by a displacement field $\mathbf{u}(\bar{\mathbf{r}}^*)$ into their position $\mathbf{r}^* = \mathbf{r} + x^o \mathbf{g}_o$ with $\mathbf{r} = \bar{\mathbf{r}} + \mathbf{u}$. The dyadic notation of the material displacement of a field $f(\bar{\mathbf{r}}^*)$ over $\bar{B}^*$ is given by

$$\bar{\nabla}^* f := \frac{1}{c}\frac{\partial f}{\partial \tau} \circ \bar{\mathbf{g}}^o + \bar{\nabla} f =: \frac{\partial f}{\partial \xi^\alpha} \circ \bar{\mathbf{g}}^\alpha, \quad (\alpha = 0, 1, 2, 3). \tag{1}$$

The local approximation of the bijective mapping in space $x^i(\xi^1, \xi^2, \xi^3; \xi^o)$, $i = 1, 2, 3$, is the deformation gradient

$$\bar{\mathbf{F}} := \bar{\nabla}\mathbf{r}. \tag{2}$$

Other useful kinematic tensors are the displacement derivative $\bar{\mathbf{H}}$. the positiv definite right CAUCHY-GREEN-tensor $\mathbf{C}$ or the GREEN-deformation tensor $\mathbf{G}$. With the purely space-like unit dyadic $\mathbf{E}$ we have

$$\bar{\mathbf{H}} := \bar{\nabla}\mathbf{u} = \bar{\mathbf{F}} - \mathbf{E}, \quad \delta\bar{\mathbf{H}} = \delta\bar{\mathbf{F}}, \tag{3a}$$

$$\mathbf{C} := \bar{\mathbf{F}}^T\bar{\mathbf{F}}, \quad \mathbf{C}^T = \mathbf{C}, \quad III := \det \mathbf{C} = \det^2 \bar{\mathbf{F}} \neq 0, \tag{3b}$$

$$\mathbf{G} := \frac{1}{2}(\mathbf{C} - \mathbf{E}) = \frac{1}{2}(\bar{\mathbf{H}} + \bar{\mathbf{H}}^T + \bar{\mathbf{H}}^T\bar{\mathbf{H}}). \tag{3c}$$

To any of these deformation tensors, there is a conjugate stress tensor which is a single valued tensor function of the deformation for elastic media. For hyperelastic materials the simultaneous invariant of the pairs of conjugate tensors is the stored energy density.
For an elastic material the stress tensor is a single valued function of **C** and thus also a deformation measure. This makes the theory of elasticity mathematically attractive, although in the nonlinear theory there is no analytical presentation for the inverse function, [6]. For an isotropic elastic solid the second PIOLA-KIRCHHOFF-stress $\bar{\bar{\sigma}}$ is an isotropic tensor function of **C**. In three space dimensions we use the following presentations for compressible or incompressible solids, respectively:

$$\bar{\bar{\sigma}}(\mathbf{G}) = \varphi'_0 \mathbf{E} + \varphi'_1 \mathbf{G} + \varphi'_2 \mathbf{G}^2 , \tag{4a}$$

$$\bar{\bar{\sigma}}(\mathbf{C}) = \psi'_0 \mathbf{E} + \psi'_1 \mathbf{C} - p\mathbf{C}^{-1} . \tag{4b}$$

Here φ'_0 . φ'_1 . φ'_2 and ψ'_0, ψ'_1 are scalar functions of the invariants of **G** and **C**, respectively. Note, for an incompressible solid only the deviatoric stress is determined from the deformation in three dimensions, since III = 1. In this case the hydrostatic pressure p can be calculated as the solution of a boundary value problem. Furthermore, the incompressible solid allows no longitudinal waves in three dimensions. We therefore exclude this special case from our discussion. For the plane stress problem of a plate one can calculate p(I, II) — where I, II are principal invariants of **C** — and thus obtain for the compressible and incompressible solid the formally similar representations

$$\bar{\bar{\sigma}}(\mathbf{G}) = \varphi_0 \mathbf{E} + \varphi_1 \mathbf{G} , \tag{4c}$$

$$\bar{\bar{\sigma}}(\mathbf{C}) = \psi_0 \mathbf{E} + \psi_1 \mathbf{C} . \tag{4d}$$

It is understood that now $\bar{\bar{\sigma}}$, **G**, **C** are tensors in two-dimensional space. The one-dimensional stress $\bar{\bar{\sigma}} := \sigma \bar{\mathbf{g}}_{\bar{1}} \circ \bar{\mathbf{g}}_{\bar{1}}$ may be calculated from a scalar law $\sigma = f(\varepsilon)$ with $\varepsilon := \bar{\mathbf{g}}_{\bar{1}} \mathbf{G} \bar{\mathbf{g}}_{\bar{1}}$. But **G** is still three-dimensional (with cylindrical symmetry). Only the trivial hydrostatic stress reduces the constitutive equation to a scalar law, where both $\bar{\bar{\sigma}}$ and **G** are spherical tensors. There are longitudinal waves in an incompressible plate, due to the variation of its thickness. Therefore we exclude plane strain for those materials. In the local balance of momentum, written in the reference configuration (multiple dots denote multiple transvection),

$$\bar{\rho}c(\bar{\nabla}^* \mathbf{v})\mathbf{E} - (\bar{\nabla}^* \bar{\sigma}) : \mathbf{E} - \bar{\rho}\mathbf{b} = \mathbf{0} , \tag{5}$$

appears the first PIOLA-KIRCHHOFF-stress $\bar{\sigma}$,

$$\bar{\sigma} := \bar{\mathbf{F}} \bar{\bar{\sigma}} , \quad \bar{\sigma}^T \neq \bar{\sigma} , \tag{6}$$

which is a single valued tensor function of $\overline{\mathbf{F}}$ by eqs. (3), (4). $\mathbf{v}$, $\bar{\rho}$, $\mathbf{b}$ are the particle velocity, the mass density in the reference configuration and the density of body force, respectively. The material is called hyperelastic if the stress function can be derived from a stored energy density U. This imposes some integrability conditions upon eq. (4). With the fourth order stiffness $\mathbf{A}(\overline{\mathbf{F}})$, the constitutive equations may be written in the form

$$\delta\bar{\sigma} = \mathbf{A} : \delta\overline{\mathbf{F}} \quad , \tag{7a}$$

where $\mathbf{A}$ can be derived from U for hyperelastic solids:

$$\delta\overline{\mathbf{F}} : \mathbf{A} : \delta\overline{\mathbf{F}} = \bar{\rho}\delta^2 U \, . \tag{7b}$$

From eqs. (3a), (5) and (7a) we have the final balance of momentum

$$\mathbf{l} = \mathbf{0} \, , \quad \mathbf{l} := \bar{\rho} c (\overline{\nabla}^* \mathbf{v}) \mathbf{E} - \mathbf{A} : \overline{\nabla}^* \overline{\mathbf{H}}^T - \bar{\rho} \mathbf{b} \, . \tag{8a}$$

In addition, RICCI's lemma on the second covariant derivative of an integrable displacement field $\mathbf{u}(\bar{\mathbf{r}}^*) = \mathbf{u}(\mathbf{r}^*)$ reduces to SCHWARZ's lemma if the RIEMANN-CHRISTOFFEL-curvature tensor vanishes everywhere. Then it holds also for the jumps on an acceleration wave front. It reads

$$\bar{\mathbf{L}}^* = \mathbf{0} \, , \quad \bar{\mathbf{L}}^* := (\overline{\nabla}^* \mathbf{v}) \mathbf{E} - c (\overline{\nabla}^* \overline{\mathbf{H}}) \bar{\mathbf{g}}_{\bar{o}} \tag{8b}$$

and

$$(\overline{\nabla}^* \overline{\mathbf{H}}) : (\bar{\mathbf{g}}_{\bar{k}} \circ \bar{\mathbf{g}}_{\bar{l}} - \bar{\mathbf{g}}_{\bar{l}} \circ \bar{\mathbf{g}}_{\bar{k}}) = 0 \, , \qquad k, l = 1, 2, 3 \tag{8c}$$

on a purely spatial and a time-like manifold, respectively. Given initial fields of displacement and velocity, $\mathbf{u}_a(\bar{\mathbf{r}})$ and $\mathbf{v}_a(\bar{\mathbf{r}})$, at time $\tau = t_a$ we have the conditions

$$\mathbf{u}(\bar{\mathbf{r}}, t_a) - \mathbf{u}_a = \mathbf{0} \, , \quad \mathbf{v}(\bar{\mathbf{r}}, t_a) - \mathbf{v}_a = \mathbf{0} \, . \tag{8d}$$

Boundary conditions shall be given for place on the part $\partial\overline{B}_v$ (points with $\bar{\mathbf{r}} = \bar{\mathbf{r}}_v$) and for traction on the part of $\partial\overline{B}_\sigma$ (points with $\bar{\mathbf{r}} = \bar{\mathbf{r}}_\sigma$) of $\partial\overline{B}$, $\partial\overline{B} = \partial\overline{B}_v \cup \partial\overline{B}_\sigma$, $\partial\overline{B}_v \cap \partial\overline{B}_\sigma = \emptyset$.

The boundary condition of place at a point $\bar{\mathbf{r}}^*_v := \bar{\mathbf{r}}_v + \xi^o \bar{\mathbf{g}}_o$ shall be

$$\mathbf{v}(\bar{\mathbf{r}}^*_v) - \mathbf{v}_b = \mathbf{0} \tag{8e}$$

with velocities $\mathbf{v}(\bar{\mathbf{r}}^*_v)$ prescribed on $\partial\overline{B}^*_v$ for all values of time. With a load vector $\mathbf{k}(\bar{\mathbf{r}}^*_\sigma)$ prescribed on $\partial\overline{B}^*_\sigma$ ($\bar{\mathbf{r}}^*_\sigma := \bar{\mathbf{r}}_\sigma + \xi^o \bar{\mathbf{g}}_o$) the typical boundary value condition of traction is the nonlinear equation

$$(\bar{\sigma}\bar{\mathbf{n}}_b)\big|_{\bar{\mathbf{r}}^* = \bar{\mathbf{r}}^*_\sigma} = \mathbf{k}\,(\det\bar{\mathbf{F}}\,\sqrt{\bar{\mathbf{n}}_b\,\mathbf{C}^{-1}\,\bar{\mathbf{n}}_b}\,)\big|_{\bar{\mathbf{r}}^* = \bar{\mathbf{r}}^*_\sigma} \quad , \tag{8f}$$

where $\bar{\mathbf{n}}_b$ is the outer normal on $\partial\bar{\mathbf{B}}$. Eq. (8f) calls for an iterative solution, [4].

A is strongly elliptic for materials that are infinitesimally stable in HADAMARD's sense in statics. Then the boundary value problem derived from eqs. (8a,b, c,e,f) is elliptic and the initial-boundary value problem eqs. (8a - f) is hyperbolic. Therefore we apply the theory of characteristics to derive exact qualitative results and to develop numerical solution schemes.

METHOD OF CHARACTERISTICS

It is well known that systems of hyperbolic equations exhibit undetermined derivatives in certain normal directions $\bar{\mathbf{n}}^*$ in space and time. These normals define singular surfaces (so-called characteristic manifolds), on which interior derivatives are continuous, but jump discontinuities of certain exterior derivatives are admitted. The conclusive equations - the characteristic condition and the so-called compatibility equations - may be derived from the general eigenvalue problem associated with the PDE's. We make the ansatz

$$\bar{\mathbf{n}}^* := -\frac{\nu}{c}\bar{\mathbf{g}}_{\bar{o}} + \bar{\mathbf{n}} \tag{9}$$

for the system (8a - c) and find the characteristic condition

$$0 = \frac{1}{\bar{\rho}^2}\nu_o^2 \det\,(\mathbf{Q}\,(\bar{\mathbf{n}}) - \bar{\rho}\nu_\varepsilon^2\,(\bar{\mathbf{n}})\,\mathbf{E})\,, \qquad \varepsilon = L,\,T_1\,,\,T_2 \tag{10}$$

where $\mathbf{Q}(\bar{\mathbf{n}})$ is the acoustic tensor, defined by

$$\mathbf{Q}\,(\bar{\mathbf{n}}) := (\mathbf{g}_\lambda \circ \bar{\mathbf{n}} : \mathbf{A} : \mathbf{g}^\mu \circ \bar{\mathbf{n}})\,\mathbf{g}^\lambda \circ \mathbf{g}_\mu \ . \tag{11}$$

The condition (10) may be taken as the equation for the time-like component ν of $\bar{\mathbf{n}}^*$, with an arbitrary choice of all components of the spatial normal $\bar{\mathbf{n}}$, $\bar{\mathbf{n}} \neq \mathbf{0}$.

Besides $\nu_o = 0$ one obtains solutions $\nu_\varepsilon(\bar{\mathbf{n}})$ of eq. (10) from the eigenvalue problem of the acoustic tensor **Q** which, in contrary to the locally isotropic wave propagation in compressible fluids, depends on the spatial direction $\bar{\mathbf{n}}$.

Therefore, for every spatial direction $\bar{\mathbf{n}}$ one has a specific eigenvalue problem. For hyperelastic materials $\mathbf{Q}(\bar{\mathbf{n}})$ is symmetric and positive definite, and thus ν_ε^2 is positive. The eigenvectors $\mathbf{q}_\varepsilon(\bar{\mathbf{n}})$ of $\mathbf{Q}\,(\bar{\mathbf{n}})$ are real and orthogonal for any normal $\bar{\mathbf{n}}$ at a point. This situation holds even for multiple eigenvalues. $\bar{\mathbf{n}}$ and $\mathbf{q}_\varepsilon(\bar{\mathbf{n}})$ are normalized. Then in three-dimensional space $\bar{\mathbf{n}}$ has two independent components. Furthermore, the ν_o, ν_ε are first order homogeneous functions of $\bar{\mathbf{n}}$. If one varies the components of $\bar{\mathbf{n}}$ as parameters at a material point, the vectors

$$\bar{\mathbf{n}}^*_o = \bar{\mathbf{n}}\ ,\quad \bar{\mathbf{n}}^*_\varepsilon = -\frac{\nu_\varepsilon}{c}\bar{\mathbf{g}}_{\bar{o}} + \bar{\mathbf{n}} \tag{12}$$

generate the normal hypercones N_o^* , N_ε^* . The according local characteristic hyperplanes envelope the local MONGE-hypercones M_o^* , M_ε^* . The lines of tangency are given by the bicharacteristics $\bar{\mathbf{m}}_o^*$, $\bar{\mathbf{m}}_\varepsilon^*$ (generators of M_o^* , M_ε^*). The local characteristic hyperplanes represent plane wave fronts and the MONGE-hypercones represent the wave fronts emanating from a point disturbance at their apex. For two space dimensions and time the situation is demonstrated in Fig. 1. It shows typical calculated wave fronts emanating from point P_9 in an isotropic plate under pure stretch.

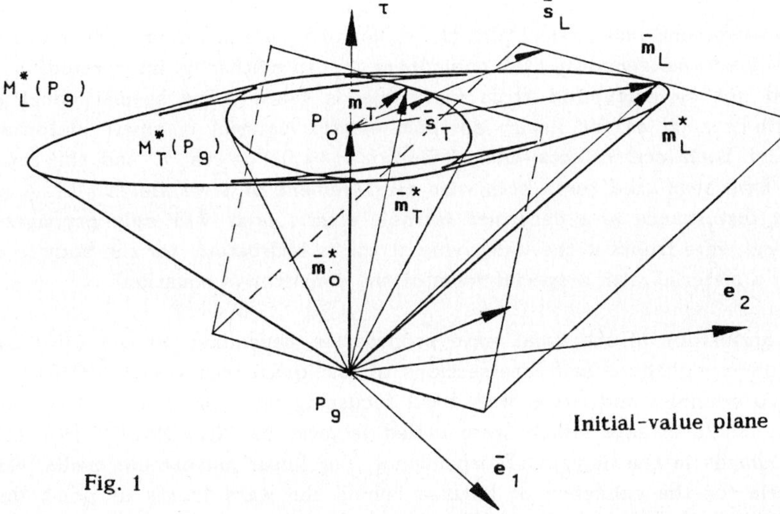

Fig. 1

The normal $\bar{\mathbf{n}}_o^*$ is purely spatial and thus defines a material singular surface. Crossing it the lowest order discontinuous derivative is purely spatial. The according MONGE-cone degenerates to the particle path line. The normals $\bar{\mathbf{n}}_\varepsilon^*$ define acceleration waves with possibly discontinuous time-like derivatives $(\bar{\nabla}\ {}^*\bar{\mathbf{H}})\bar{\mathbf{n}}_\varepsilon^*$, $(\bar{\nabla}_\varepsilon^*\ \mathbf{v})\,\bar{\mathbf{n}}_\varepsilon^*$. From the orthogonality of the eigenvectors $\mathbf{q}_\varepsilon(\bar{\mathbf{n}})$ one deduces that the jumps on different MONGE-hypercones through a point are orthogonal to each other. But these waves are purely longitudinal and transversal only on the axis of symmetry of the cones, because only there $\mathbf{q}\ (\bar{\mathbf{n}})$ is parallel to $\bar{\mathbf{n}}$. These axes are called acoustical axes, and waves propagating in these directions are called principal waves. In a deformed isotropic elastic material the eigenvectors of $\bar{\bar{\sigma}}$ and $\mathbf{C}$ or $\mathbf{G}$ are collinear with the acoustical axes $\bar{\mathbf{e}}$. Transformation into the actual configuration is via $\mathbf{e} = \bar{\mathbf{F}}\bar{\mathbf{e}}\,|\bar{\mathbf{F}}\,\bar{\mathbf{e}}\,|^{-1}$. If some initial amplitude of the jump discontinuity is known, its future magnitude can be calculated from the transport equation, [7].

Definition (11) allows **A** to be written in terms of principle wave speeds, which can be found from ultrasound wave speed measurements. Unfortunately, not all components may be measured from sound disturbances superimposed on the standard uniaxial tension test. There are two principal axes $\bar{\mathbf{e}}_1, \bar{\mathbf{e}}_2$ in two-dimensional stress given bei eqs. (4c, d). Taking care of $\nu_T(\bar{\mathbf{e}}_1) = \nu_T(\bar{\mathbf{e}}_2) =: \nu_T$ we have

$$\frac{1}{\rho}\mathbf{A} := \nu_L^2(\bar{\mathbf{e}}_1)\,\mathbf{e}_1 \circ \bar{\mathbf{e}}_1 \circ \mathbf{e}_1 \circ \bar{\mathbf{e}}_1 + \nu_L^2(\bar{\mathbf{e}}_2)\,\mathbf{e}_2 \circ \bar{\mathbf{e}}_2 \circ \mathbf{e}_2 \circ \bar{\mathbf{e}}_2$$

$$+ \nu_T^2(\mathbf{e}_2 \circ \bar{\mathbf{e}}_1 \circ \mathbf{e}_2 \circ \bar{\mathbf{e}}_1 + \mathbf{e}_1 \circ \bar{\mathbf{e}}_2 \circ \mathbf{e}_1 \circ \bar{\mathbf{e}}_2)$$

$$+ \varkappa^2(\mathbf{e}_1 \circ \bar{\mathbf{e}}_1 \circ \mathbf{e}_2 \circ \bar{\mathbf{e}}_2 + \mathbf{e}_2 \circ \bar{\mathbf{e}}_2 \circ \mathbf{e}_1 \circ \bar{\mathbf{e}}_1). \tag{13}$$

In three-dimensional stress fields the values of ν_T are different in different principal directions generally. The component $\varkappa^2$ can neither be interpreted as wave speed nor be calculated from wave speeds (except for special constitutive equations such as for linear isotropic elastic material in small deformation theory). Both local effects, the difference $\nu_L(\bar{\mathbf{e}}_1) - \nu_L(\bar{\mathbf{e}}_2)$ and the rotation of $\bar{\mathbf{e}}$ have been used for a pointwise measurement of stressfields [8]. A small point disturbance in a deformed istropic elastic body will only propagate on sherical wave fronts if the underlying stress is hydrostatic (or the body is made form a material with a special form of the constitutive equation).

The anisotropy of the local wave propagation depending on the local deformation may lead to self-intersections of the quasi-transversal MONGE-cones and to crunodes and cusps with local focussing on their conics. These phenomena result in gaps which were called lacunae by PETROWSKY [9] and lie like islands in the domain of dependence. For linear anisotropic media, various criteria for the existence of lacunae behind the wave fronts of point disturbances have been found [1, 4, 9, 10]. But no direct results are known for the nonlinear case except by arguments for the locally linearized equations [11]. Further information on anisotropic wave propagation in solids may be found in [12] for linear elastic deformation and in [13] for compressible plastic deformation.

There are infinitely many ways to describe the propagation of a plane wave. We use the bicharacteristics $\mathbf{m}_o^*$ and $\mathbf{m}_\varepsilon^*$ with the rays $\bar{\mathbf{m}}_\varepsilon$ in space,

$$\bar{\mathbf{m}}_o^* := c\,\bar{\mathbf{g}}_o \quad , \tag{14a}$$

$$\bar{\mathbf{m}}_\varepsilon^* := \bar{\mathbf{m}}_o^* + \bar{\mathbf{m}}_\varepsilon = \bar{\mathbf{m}}_o^* + \frac{\partial \nu_\varepsilon}{\partial \bar{\mathbf{n}}} = \bar{\mathbf{m}}_o^* + \frac{1}{2\bar{\rho}\nu_\varepsilon}\,\mathbf{q}_\varepsilon \circ \mathbf{q}_\varepsilon : \frac{\partial \mathbf{Q}}{\partial \bar{\mathbf{n}}}\,, \tag{14b}$$

and the near-characteristic $\bar{\mathbf{s}}_\varepsilon^*$ with the normal velocity $\bar{\mathbf{s}}_\varepsilon$

$$\bar{\mathbf{s}}_\varepsilon^* := \bar{\mathbf{m}}_o^* + \bar{\mathbf{s}}_\varepsilon = \bar{\mathbf{m}}_o^* + \nu_\varepsilon \bar{\mathbf{n}} = \bar{\mathbf{m}}_o^* + \frac{1}{\bar{\rho}\nu_\varepsilon}\,\mathbf{q}_\varepsilon \circ \mathbf{q}_\varepsilon : \mathbf{Q} \circ \bar{\mathbf{n}}\,. \tag{15}$$

Any other time-like vector in the characteristic hyperplane tangent to the MONGE cone M_ε^* is also called near-characteristic, but will not be needed here. For plane problems the space-like tangent $\bar{\mathbf{t}}$ and either $\bar{\mathbf{m}}_\varepsilon^*$ or $\bar{\mathbf{s}}_\varepsilon^*$ span the characteristic surface elements. By the vectors $\bar{\mathbf{m}}_\varepsilon^*$ and $\bar{\mathbf{s}}_\varepsilon^*$ two total time derivatives in a characteristic hyperplane can be introduced,

$$(\bar{\nabla}^* f)\,\bar{\mathbf{m}}_\varepsilon^* := \frac{\partial f}{\partial \tau} + (\bar{\nabla} f)\,\bar{\mathbf{m}} \tag{16a}$$

and the so-called δ-time derivative

$$(\bar{\nabla}^* f)\,\bar{\mathbf{s}}_\varepsilon^* := \frac{\partial f}{\partial \tau} + (\bar{\nabla} f)\,\bar{\mathbf{s}}\,, \tag{16b}$$

also known as displacement derivative. The discussion was only local so far. In finite time a point disturbance propagates along MONGE-conoids which may be twisted. For the integration of some function f on M_ε one may use the canonical HAMILTON equations for the HAMILTONian ν_ε . With $\bar{\mathbf{n}}$ normalized and ν_ε as a function of $\bar{\mathbf{n}}$, $\bar{\mathbf{r}}$, τ we derive the special form of the HAMILTON equations

$$(\bar{\nabla}^* \bar{\mathbf{n}})\,\bar{\mathbf{m}}_\varepsilon^* = (\bar{\mathbf{n}} \circ \bar{\mathbf{n}} - \mathbf{E})\,\bar{\nabla} \nu_\varepsilon\,, \tag{17a}$$

$$(\bar{\nabla}^* \bar{\mathbf{r}})\,\bar{\mathbf{m}}_\varepsilon^* = \frac{\partial \nu_\varepsilon}{\partial \bar{\mathbf{n}}}\,. \tag{17b}$$

Different proofs of these equations were given in [4. 14] . Instead of eqs. (17a,b), we get for $\bar{\mathbf{s}}_\varepsilon^*$

$$(\bar{\nabla}^* \bar{\mathbf{n}})\,\bar{\mathbf{s}}_\varepsilon^* = (\bar{\mathbf{n}} \circ \bar{\mathbf{n}} - \mathbf{E})\,\bar{\nabla} \nu_\varepsilon\,, \tag{18a}$$

$$(\bar{\nabla}^* \bar{\mathbf{r}})\,\bar{\mathbf{s}}_\varepsilon^* = \nu_\varepsilon \bar{\mathbf{n}}\,. \tag{18b}$$

Different proofs of the famous HAYES-THOMAS-formula (18a) may be found in [4,15 ,16 ,17 ,18]. It is assumed that the equation of the wave surface in space and time has continuous second derivatives in space. Thus eqs. (17a), (18a) do not hold on cusps. In [4] it is shown that no torsion of the wave front occurs in plane problems.

The purpose of the discussion of point disturbances is to replace the initial set of PDE's (8a, b) by a linearly independent set of so-called compatibility equations in which no undetermined derivatives appear. These equations only hold in characteristic surfaces. On M_0^* one gets for $\bar{\mathbf{n}}_0^* = \bar{\mathbf{n}}$

$$0 = \bar{\mathbf{L}}^* \bar{\mathbf{t}}\,, \tag{19a}$$

and on M_ε^* for $\bar{\mathbf{n}}_\varepsilon^*$

$$0 = \mathbf{q}_\varepsilon(\bar{\mathbf{n}})\,(\nu_\varepsilon(\bar{\mathbf{n}})\mathbf{1} - \mathbf{A} : (\bar{\mathbf{n}} \circ \bar{\mathbf{L}}^*)). \tag{19b}$$

In a more extended notation for plane problems,

$$0 = (\bar{\nabla}^* \bar{\mathbf{H}}) : \bar{\mathbf{t}} \circ \bar{\mathbf{m}}_0^* - (\bar{\nabla}^* \mathbf{v}) \bar{\mathbf{t}} \,, \tag{20a}$$

$$\begin{aligned} 0 = {} & (\bar{\rho} \nu_\varepsilon \mathbf{q}_\varepsilon \circ \bar{\mathbf{m}}_\varepsilon^* + \mathbf{q}_\varepsilon \circ \bar{\mathbf{n}} : \mathbf{A} \bar{\mathbf{t}} \circ \bar{\mathbf{t}} - \bar{\rho} \nu_\varepsilon (\bar{\mathbf{m}}_\varepsilon \bar{\mathbf{t}}) \, \mathbf{q}_\varepsilon \circ \bar{\mathbf{t}}) : \bar{\nabla}^* \mathbf{v} \\ & - (\mathbf{q}_\varepsilon \circ \bar{\mathbf{n}} : \mathbf{A} \circ \bar{\mathbf{m}}_\varepsilon^* + \nu_\varepsilon \mathbf{q}_\varepsilon \circ \bar{\mathbf{t}} : \mathbf{A} \circ \bar{\mathbf{t}} - (\bar{\mathbf{m}}_\varepsilon \bar{\mathbf{t}}) \mathbf{q}_\varepsilon \circ \bar{\mathbf{n}} : \mathbf{A} \circ \bar{\mathbf{t}}) : \bar{\nabla}^* \bar{\mathbf{H}} \\ & - \bar{\rho} \nu_\varepsilon \mathbf{q}_\varepsilon \mathbf{b} \,, \end{aligned} \tag{20b}$$

one can see that only interior derivatives remain. These are the cross derivatives in direction of $\bar{\mathbf{t}}$ and the derivatives in characteristic directions $\bar{\mathbf{m}}_\varepsilon^*$ which may be replaced by $\bar{\mathbf{s}}_\varepsilon^*$ in a near-characteristics method. Using $\bar{\mathbf{s}}_\varepsilon \bar{\mathbf{t}} = 0$ eq. (20b) becomes

$$\begin{aligned} 0 = {} & (\bar{\rho} \nu_\varepsilon \mathbf{q}_\varepsilon \circ \bar{\mathbf{s}}_\varepsilon^* + \mathbf{q}_\varepsilon \circ \bar{\mathbf{n}} : \mathbf{A} \bar{\mathbf{t}} \circ \bar{\mathbf{t}}) : \bar{\nabla}^* \mathbf{v} \\ & - (\mathbf{q}_\varepsilon \circ \bar{\mathbf{n}} : \mathbf{A} \circ \bar{\mathbf{s}}_\varepsilon^* + \nu_\varepsilon \mathbf{q}_\varepsilon \circ \bar{\mathbf{t}} : \mathbf{A} \circ \bar{\mathbf{t}}) : \bar{\nabla}^* \bar{\mathbf{H}} - \bar{\rho} \nu_\varepsilon \mathbf{q}_\varepsilon \mathbf{b} \,. \end{aligned} \tag{20c}$$

To this end all equations are exact and hold for arbitrary large deformation and large displacement of a general isotropic nonlinear hyperelastic body. The assumption of the existence of a stored energy density was a mere simplification of notation. Its only effect is the symmetry imposed upon **A** by eq. (7b).

NUMERICAL SOLUTION

Given the inital data at points of a suitable mesh on the inital value plane at time $\tau = t_a$, the solution at a point P_0 on the plane $\tau = t_0 + \Delta\tau$ may be computed numerically by integrating the HAMILTON equations (17 a, b) and the compatibility equations, which are given for the plane problem by equations (20a, b). We actually integrate the simplified equations (18 a, b) and (20a, c). From Fig. 1 it is clear that a near-characteristics method uses points outside the analytical domain of dependence. Therefore, we integrate in characteristic hypersurfaces which are principal waves at point P_0. For principal waves, the two integration paths are equivalent in the limit $\Delta\tau \rightarrow 0$, since $\bar{\mathbf{m}}_\varepsilon^* = \bar{\mathbf{s}}_\varepsilon^*$. The principal axes at P_0 depend on the iterative solution. Therefore, we have to rotate the local basis in each iteration step. This expense is compensated by the condensed presentation (13) of **A**. In the initial value surface all functions and the cross derivatives may be calculated from a constrained least squares approximation with constraints following from eq. (8c). If the cross derivatives at P_0 are considered as additional unknowns, eqs. (18a,b), (20a,c) contain only total time derivatives in directions $\bar{\mathbf{s}}_0^*$ and $\bar{\mathbf{s}}_\varepsilon^*$. These ODE's may be integrated by HEUN's second order method. It can be shown that the number of linearly independent difference equations derived from eqs. (20a,c) is less than the number of unknown functions including the cross derivatives at P_0. Therefore, we integrate the non characteristic balance of momentum (8a) along the path line to get two additional equations. This was suggested for problems of linear elastodynamics in [19] and for gas dynamics in [20]. On the path line all derivatives in (8a) are continuous because it is a material singular surface. Fig. 2 shows the scheme we use for an interior point of the plate. The numbers of the points in the scheme correspond to the indices written in the difference equations. Fig. 1 and Fig 2 indicate that the CFL stability condition is satisfied. Note that the iteration may be started with the first order EULER-CAUCHY-integration of (8b) along the path line to give $\mathbf{H}_0$ at P_0, since a first guess of $\mathbf{v}_0$ is not needed in elastodynamics. The algorithm may be read from Fig. 2 and the following formulas:

Initial-value plane

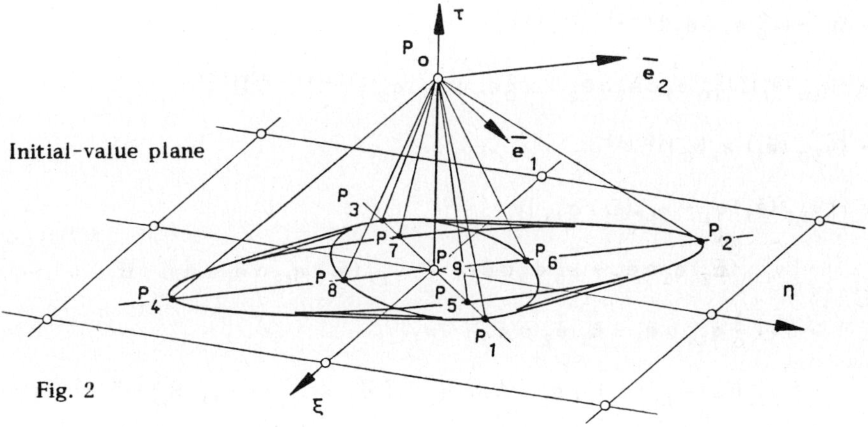

Fig. 2

Praedictor (first guess): $i = 1(1)9$ with $\nu_0 = 0$ for $i = 9$

$$\bar{\mathbf{r}}_i^{*(0)} = \bar{\mathbf{r}}_9^{*} + \Delta\tau\,[\bar{\mathbf{e}}\,\nu_{\varepsilon 0}(\bar{\mathbf{e}})]^{(0)}, \tag{21}$$

and from the compatibility equation along the path line

$$\bar{\mathbf{H}}_0^{(0)} = \bar{\mathbf{H}}_9 + \Delta\tau\bar{\nabla}\mathbf{v}_9. \tag{22a}$$

Corrector (k-th iteration step):

$$\bar{\mathbf{r}}_i^{*(k)} = \bar{\mathbf{r}}_9^{*} + \frac{\Delta\tau}{2}\,[\bar{\mathbf{e}}\,(\nu_{\varepsilon i}(\bar{\mathbf{e}}) + \nu_{\varepsilon 0}(\bar{\mathbf{e}}))]^{(k-1)}, \tag{21b}$$

$$\bar{\mathbf{H}}_0^{(k)} = \bar{\mathbf{H}}_9 + \frac{\Delta\tau}{2}\,[\bar{\nabla}\mathbf{v}_9 + \bar{\nabla}\mathbf{v}_0^{(k-1)}]. \tag{22b}$$

We derive the difference equations for principal waves arriving at point P_0 (see Fig. 2) from the compatibility equations on the quasi-longitudinal cone M_L^* and the quasi-transversal cone M_T^*. The appropriate choices of $\bar{\mathbf{n}}_{oi}$ (i = 1 (1) 4) and $\bar{\mathbf{n}}_{oj}$ (j = 5 (1) 8) are $\bar{\mathbf{e}}_1, \bar{\mathbf{e}}_2, -\bar{\mathbf{e}}_1, -\bar{\mathbf{e}}_2$. We give an example of the difference equations on each cone.

From the compatibility equation on the quasi-longitudinal cone M_L^* with $\bar{\mathbf{n}}_{Oi} := \bar{\mathbf{e}}_1$ and $\bar{\mathbf{t}}_{Oi} := \bar{\mathbf{e}}_2$ and $\mathbf{q}_{Li} = \mathbf{q}_{Li}(\bar{\mathbf{e}}_1)$ for i = 1 it follows :

$$
\begin{aligned}
&[\nu_{LO}(\bar{\mathbf{e}}_1)\,\mathbf{e}_1 + \nu_{Li}(\bar{\mathbf{e}}_1)\mathbf{q}_{Li}]^{(k-1)}\,\mathbf{v}_O^{(k)} \\
&- [\nu_{LO}^2(\bar{\mathbf{e}}_1)\,\mathbf{e}_1 \circ \bar{\mathbf{e}}_1 + \varkappa_0^2\,\mathbf{e}_2 \circ \bar{\mathbf{e}}_2 + \tfrac{1}{\rho}\,\nu_{Li}(\bar{\mathbf{e}}_1)\mathbf{q}_{Li} \circ \bar{\mathbf{e}}_1 : \mathbf{A}_i]^{(k-1)} : \bar{\mathbf{H}}_O^{(k)} \\
&- \Delta\tau\,[-(\varkappa_0^2\,\mathbf{e}_2 \circ \bar{\mathbf{e}}_2)^{(k-1)} : \bar{\nabla}\mathbf{v}_O^{(k)} \\
&+ (\nu_{LO}(\bar{\mathbf{e}}_1)(\nu_{TO}^2\,\mathbf{e}_1 \circ \bar{\mathbf{e}}_2 \circ \bar{\mathbf{e}}_2 + \varkappa_0^2\mathbf{e}_2 \circ \bar{\mathbf{e}}_1 \circ \bar{\mathbf{e}}_2))^{(k-1)} : \bar{\nabla}\bar{\mathbf{H}}_O^{(k)} \\
&+ (\nu_{LO}(\bar{\mathbf{e}}_1)\,\mathbf{e}_1\mathbf{b}_O)^{(k-1)}] \\
&= \Big\{[\nu_{LO}(\bar{\mathbf{e}}_1)\mathbf{e}_1 + \nu_{Li}(\bar{\mathbf{e}}_1)\,\mathbf{q}_{Li}]\mathbf{v}_i \\
&\qquad - [\nu_{LO}^2(\bar{\mathbf{e}}_1)\mathbf{e}_1 \circ \bar{\mathbf{e}}_1 + \varkappa_0^2\,\mathbf{e}_2 \circ \bar{\mathbf{e}}_2 + \tfrac{1}{\rho}\;\nu_{Li}(\bar{\mathbf{e}}_1)\,\mathbf{q}_{Li} \circ \bar{\mathbf{e}}_1 : \mathbf{A}_i] : \bar{\mathbf{H}}_i \\
&\qquad - \Delta\tau\,[\tfrac{1}{\rho}\mathbf{q}_{Li} \circ \bar{\mathbf{e}}_1 : \mathbf{A}_i\;\bar{\mathbf{e}}_2 \circ \bar{\mathbf{e}}_2 : \bar{\nabla}\mathbf{v}_i \\
&\qquad + \tfrac{1}{\rho}\,\nu_{Li}(\bar{\mathbf{e}}_1)\,\mathbf{q}_{Li}(\bar{\mathbf{e}}_1) \circ \bar{\mathbf{e}}_2 : \mathbf{A}_i \circ \bar{\mathbf{e}}_2 : \bar{\nabla}\bar{\mathbf{H}}_i + \nu_{Li}(\bar{\mathbf{e}}_1)\,\mathbf{q}_{Li}\,\mathbf{b}_i]\Big\}^{(k-1)}. \qquad (23)
\end{aligned}
$$

From the compatibility equation on the quasi-transversal cone M_T^* with $\bar{\mathbf{n}}_{Oj} := \bar{\mathbf{e}}_1$ and $\bar{\mathbf{t}}_O := \bar{\mathbf{e}}_2$ and $\mathbf{q}_{Tj} = \mathbf{q}_{Tj}(\bar{\mathbf{e}}_1)$ for j = 5 it follows :

$$
\begin{aligned}
&[\nu_{TO}^2\,\mathbf{e}_2 + \nu_{Tj}\,\mathbf{q}_{Tj}]^{(k-1)}\,\mathbf{v}_O^{(k)} \\
&- [\nu_{TO}^2\,\mathbf{e}_2 \circ \bar{\mathbf{e}}_1 + \varkappa_0^2\,\mathbf{e}_1 \circ \bar{\mathbf{e}}_2 + \tfrac{1}{\rho}\,\nu_{Tj}\mathbf{q}_{Tj} \circ \bar{\mathbf{e}}_1 : \mathbf{A}_j]^{(k-1)} : \bar{\mathbf{H}}_O^{(k)} \\
&- \Delta\tau\,[-(\varkappa_0^2\,\mathbf{e}_1 \circ \bar{\mathbf{e}}_2)^{(k-1)} : \bar{\nabla}\mathbf{v}_O^{(k)} \\
&+ (\nu_{TO}(\nu_{LO}^2(\bar{\mathbf{e}}_2)\,\mathbf{e}_2 \circ \bar{\mathbf{e}}_2 \circ \bar{\mathbf{e}}_2 + \varkappa_0^2\,\mathbf{e}_1 \circ \bar{\mathbf{e}}_1 \circ \bar{\mathbf{e}}_2))^{(k-1)} : \bar{\nabla}\bar{\mathbf{H}}_O^{(k)} \\
&+ (\nu_{TO}\,\mathbf{e}_2\,\mathbf{b}_O)^{(k-1)}] \\
&= \Big\{[\nu_{TO}\,\mathbf{e}_2 + \nu_{Ti}\,\mathbf{q}_{Tj}]\mathbf{v}_j \\
&\qquad - [\nu_{TO}^2\,\mathbf{e}_2 \circ \bar{\mathbf{e}}_1 + \varkappa_0^2\,\mathbf{e}_1 \circ \bar{\mathbf{e}}_2 + \tfrac{1}{\rho}\,\nu_{Tj}\,\mathbf{q}_{Tj} \circ \bar{\mathbf{e}}_1 : \mathbf{A}_j] : \bar{\mathbf{H}}_j \\
&\qquad + \Delta\tau\,[\tfrac{1}{\rho}\mathbf{q}_{Tj} \circ \bar{\mathbf{e}}_1 : \mathbf{A}_j\,\bar{\mathbf{e}}_2 \circ \bar{\mathbf{e}}_2 : \bar{\nabla}\mathbf{v}_j \\
&\qquad + \tfrac{1}{\rho}\nu_{Tj}\mathbf{q}_{Tj} \circ \bar{\mathbf{e}}_2 : \mathbf{A}_j \circ \bar{\mathbf{e}}_2 \vdots\; \bar{\nabla}\bar{\mathbf{H}}_j + \nu_{Tj}\,\mathbf{q}_{Tj}\,\mathbf{b}_j\,]]\Big\}^{(k-1)}. \qquad (24)
\end{aligned}
$$

We complete the difference equations with the balance of momentum

$$\mathbf{v}_0^{(k)} - \frac{\Delta\tau}{2}\left[(\frac{1}{\rho}\mathbf{A}_0)^{(k-1)} : (\bar{\nabla}\mathbf{H}_0^T)^{(k)} + \mathbf{b}_0\right] = \mathbf{v}_9 + \frac{\Delta\tau}{2}(\frac{1}{\rho}\mathbf{A}_9 : \bar{\nabla}\mathbf{H}_9^T + \mathbf{b}_9) . \qquad (25)$$

If we are not interested in the gradients at P_o we may eliminate them by linear combinations of the difference equations and just calculate $\mathbf{v}_0$ and $\mathbf{H}_0$. After the solutions at all points at time $\tau = t_a + \Delta\tau$ are calculated they may be used as initial data for the following time step.

On boundaries some difference equations are not needed. But only in linear problems they may be replaced by boundary values. The schemes for the nonlinear problem use the approximation of $\bar{\nabla}\mathbf{H}_0$ at P_0 [4]. They also need some iteration to solve (8f). The only reason for both disadvantages ist the nonlinearity of the boundary condition (8f).

NUMERICAL EXAMPLES

Fig. 3a shows the simple elongation strain path of a typical rubber-like material using TRELOAR' s approximation with data from [21]. The functions ψ_o' and ψ_1' in eq. (4b) are approximated by second degree polynomials in the eigenvalues of **C**. Up to point 3 lower degree polynomials would result in roughly the same curve since the geometrical nonlinearity is predominant. But this is not the case for the wave speeds [4]. Obviously only $\nu_L(\bar{\mathbf{e}}_1)$ can be deduced from the uniaxial stress-strain curve. Fig. 3b represents the related principal wave speeds. For three different homogeneous states of stresses (uniaxial pressure (1), no stresses (2), uniaxial tension (3)) Fig. 3c shows typical wave fronts of infinitesimal point disturbances in a plate, i.e. lines of intersection of the MONGE-cones with a space-like plane. The state (3) was used for a numerical test, where a small shear deformation was superimposed by an initial disturbance of the velocity component $\mathbf{ve}_2$. Both of the MONGE-cones come out nicely and the two wave types seem decoupled. The amplitudes of $\mathbf{ve}_2$ are approximately ten times higher than those of $\mathbf{ve}_1$. In another example, for a plate made of a compressible material, we found a strong coupling of the two types of waves [4].

Since at the state 1 $\nu_L(\mathbf{e}_1)$ increases steepely with growing pressure one expects pressure waves to form shocks quickly. The initial profile of the wave in Fig. 4 changes in the predicted way while moving to the right. Clearly, there is also an equalisation in the transverse direction.

ACKNOWLEDGEMENT

The present investigation was financially supported by the Deutsche Forschungsgemeinschaft.

REFERENCES

[1] PAYTON, R.G.: Elastic Wave Propagation in Transversely Isotropic Media. Martinus Nijhoff, The Hague (1983).

[2] JEFFREY, A.; TANIUTI, T.: Non-Linear Wave Propagation with Applications to Physics and Magnetohydrodynamics. Academic Press, New York (1964).

[3] BALLMANN, J.; RAATSCHEN, H.J.; STAAT, M.: High Stress Intensities in Focussing Zones of Waves. In P. LADEVESE (Ed.): Local Effects in the Analysis of Structures. Elsevier, Amsterdam (1985).

[4] STAAT, M.: Nichtlineare Wellen in elastischen Scheiben. Dr.-Ing. Thesis, RWTH-Aachen (1987).

[5] KRAWIETZ, A.: Materialtheorie, Springer, Berlin (1986).

[6] STAAT, M.; BALLMANN, J.: Zur Problematik tensorieller Verallgemeinerungen einachsiger nichtlinearer Materialgesetze. To be published in ZAMM.

[7] BRAUN, M.: Wave Propagation in Elastic Membranes. In U. NIGUL, J. ENGELBRECHT (Ed.): Nonlinear Deformation Waves. Springer, Berlin (1983).

[8] PAO, Y.-H.; SACHSE, W.; FUKUOA, H.: Acoustoelasticity and Ultrasonic Measurements of Residual Stresses. Physical Acoustics 17, Academic Press, New York (1984) Chap. 2.

[9] PETROWSKY, I.G.: On the Diffusion of Waves and the Lacunas for Hyperbolic Equations. Rec. Math. (Math. Sbornic) 17 (1945) 289-370.

[10] BURRIDGE, R.: Lacunas in Two-Dimensional Wave Propagation. Proc. Camb. Phil. Soc. 63 (1967) 819-825.

[11] PAYTON, R.G.: Two Dimensional Wave Front Shape Induced in a Homogeneously Strained Elastic Body by a Point Perturbing Body Force. Arch. Rat. Mech. 32 (1969) 311-330 and 35 (1969) 402-408.

[12] MUSGRAVE, M.J.P.: Crystal Acoustics. Holden-Day, San Francisco (1970).

[13] SAUERWEIN, H.: Anisotropic Waves in Elastoplastic Soils. Int. J. Engng. Sci. 5 (1967) 455-475.

[14] VARLEY, E.; CUMBERBATCH E.: Non-linear Theory of Wave-front Propagation. J. Inst. Maths Applics, 1 (1965) 101-112.

[15] ERINGEN, A.C.; SUHUBI, E.S.: Elastodynamics I, Academic Press, New York (1974).

[16] THOMAS, T.Y.: Plastic Flow and Fracture in Solids. Academic Press, New York (1961).

[17] TRUESDELL, C.; TOUPIN, R.A.: The Classical Field Theories. In S. FLÜGGE (Ed.): Handbuch der Physik III/1. Springer, Berlin (1960).

[18] WANG, C. C.; TRUESDELL, C.: Introduction to Rational Elasticity. Nordhoff, Leyden (1973).

[19] CLIFTON, R.J.: A Difference Method for Plane Problems in Dynamic Elasticity. Quart. of Appl. Math. 25 (1967) 97-116.

[20] BUTLER, D.S.: The Numerical Solution of Hyperbolic Systems of Partial Differential Equations in Three Independent Variables. Proc. Roy. Soc., A 255 (1966) 232-252.

[21] HAINES, D.W.; WILSON, W.D.: Strain-Energy Density Function for Rubber-like Materials, J.Mech. Phys. Solids 27 (1979) 331-343

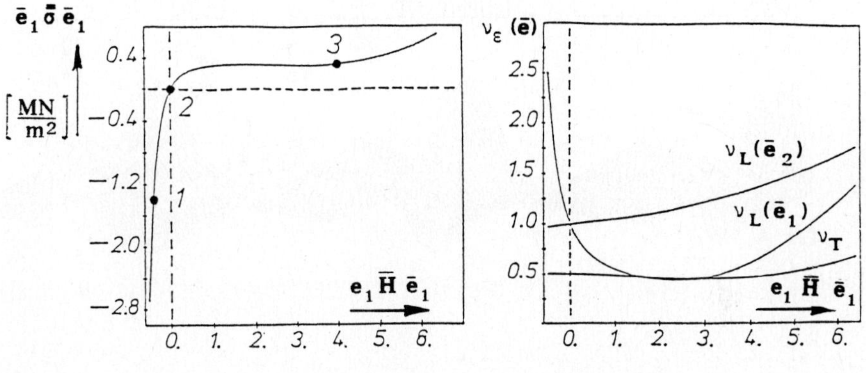

Fig. 3a

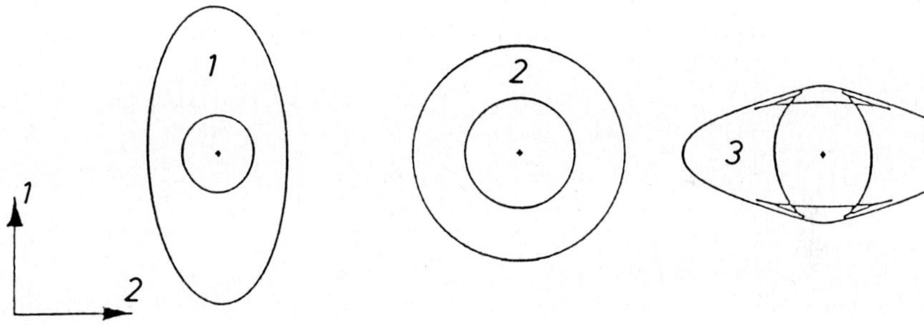

Fig. 3b

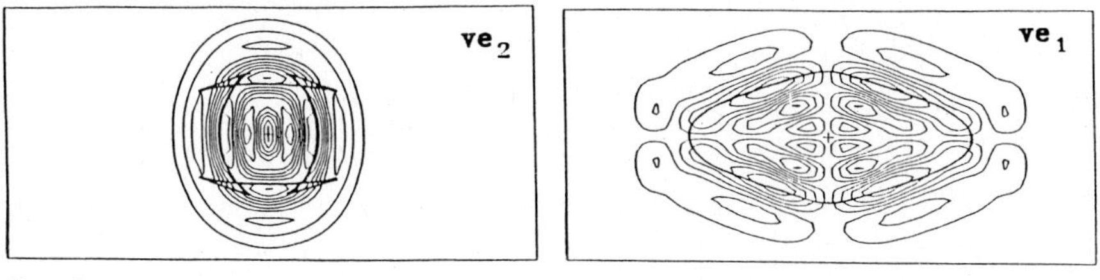

Fig. 3c

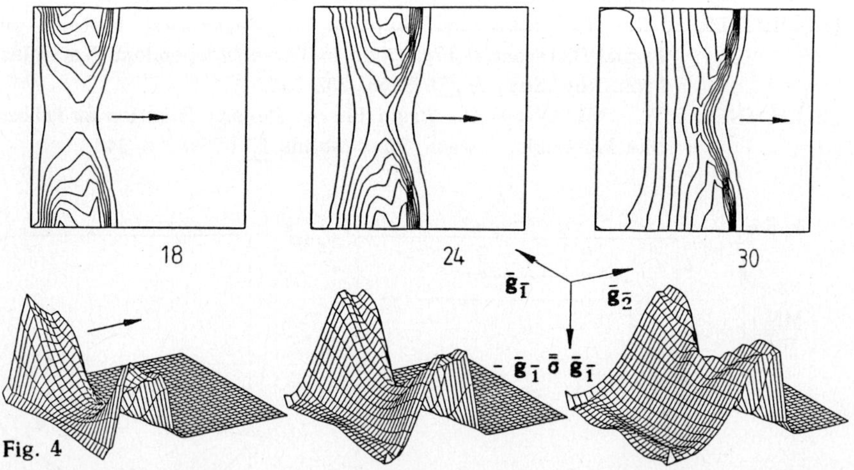

Fig. 4

ON A NONLINEAR TELEGRAPH EQUATION WITH A FREE BOUNDARY

I. Straškraba

Mathematical Institute of Czechoslovak Academy of Sciences,
Žitná 25, 115 67 Praha 1, Czechoslovakia

SUMMARY

An existence result for a nonlinear telegraph equation with a free boundary problem in one space dimension is given. The model is applied to two (separated) fluids flow in a tube.

INTRODUCTION

The present contribution concerns the following problem: For given functions f_o, f_1, u_o, u_1 to find functions $u = u(x,t)$, $\xi = \xi(t)$, $(x \in [0,\ell]$, $t \in [0,T]$, $\ell > 0$, $T > 0)$ such that

$$u_{tt} - c^2 u_{xx} + f_o(u)_t = 0 \ , \ 0 < x < \xi(t) \ , \tag{1}$$

$$u_{tt} - c^2 u_{xx} + f_1(u)_t = 0 \ , \ \xi(t) < x < \ell, \ t \in (0,T) \ , \tag{2}$$

$$u(\xi(t)-,t) = u(\xi(t)+,t) \ , \ u_x(\xi(t)-,t) = u_x(\xi(t)+,t) \ , \tag{3}$$

$$u_x(0,t) = u_x(\ell,t) = 0 \ , \ t \in [0,T] \ , \tag{4}$$

$$u(x,0) = u_o(x) \ , \ u_t(x,0) = u_1(x) \ , \ x \in [0,\ell] \tag{5}$$

$$\xi(t) = G\left(\int_0^t u(\xi(\tau),\tau)\,d\tau \right) \ , \tag{6}$$

where c is a positive constant and

$$G(y) = \begin{cases} 0, & y \leq 0 \\ y, & 0 < y < \ell \\ \ell, & y \leq \ell. \end{cases} \tag{7}$$

This problem is motivated by a model developped in [1] describing flow of two fluids separated by a free boundary in a cylindric compartment. The purpose is to find a velocity u and a pressure p (which can be recovered from u as soon as the problem (1) - (6) has been solved) as functions of a space variable x along the axis of the cylinder (the cross sectional derivatives being neglected) and a time t, as well as a free boundary $x = \xi(t)$. A particular technological application consists in using this model to describe the filling the compartment of the machine feeding a pipeline by ash-water suspension to be transported away from classical power stations. The original governing Euler equations of a compressible fluid flow with a lower order friction term (cf. [3],[7]), are simplyfied to the equations (1), (2). The nonlinear functions f_o, f_1 describe the friction of the respective fluid near the wall of the compartment.

In the next Section, after necessary preliminaries, we define a solution to (1) - (6), impose some natural assumptions on the functions f_o, f_1, u_o, u_1 formulate the existence theorem, present its proof using Faedo-Galerkin approximations, apriori estimates and a method of compactness and finally, add a few remarks.

MAIN RESULT

In the sequel we use the usual notation of function spaces. Namely, for a region $\Omega \subset R^n$, $k \geq 0$ integer or $k = \infty$ we denote by $C^k(\Omega)$ the space of functions continuous together with their derivatives up to the order k in Ω, $C^{(k)+\alpha}(\Omega)$, $0 < \alpha \leq 1$ the subspace of $C^k(\Omega)$ of functions which are Hölder continuous in Ω with the exponent α together with all their derivatives up to the order k, $H^k(\Omega)$ $(k \geq 0)$ the space of functions which are square integrable together with their (fractional order) generalized derivatives up to the order k in Ω, $L^p(\Omega)$, $1 \leq p \leq \infty$ the usual L^p-spaces, $W^{k,\infty}(\Omega) = \{u \in L^\infty(\Omega)\ ;\ D^j u \in L^\infty(\Omega),\ 1 \leq |j| \leq k\}$. If B is a Banach space, $T > 0$ then $L^\infty(0,T;B)$ denotes the space of functions $u:(0,T) \to B$ such that $\operatorname{ess\,sup}_t ||u(t)||_B < \infty$. For a region with a Lipschitz continuous boundary $\partial\Omega$ we shall make use frequently the following embedding theorems:

$$H^k(\Omega) \hookrightarrow\hookrightarrow L^p(\Omega) \quad \text{for} \quad k \leq \frac{n}{2}\ ,\ 1 \leq p < \frac{2n}{n-2k}\ ;$$

for $k > \frac{1}{2}$ the the operator of traces is continuous from

$$H^k(\Omega) \quad \text{into} \quad H^{k-\frac{1}{2}}(\partial\Omega)\ ;$$

$$H^k(\Omega) \hookrightarrow\hookrightarrow H^\ell(\Omega) \quad \text{for any} \quad k > \ell \geq 0\ ;$$

$$H^k(\Omega) \hookrightarrow\hookrightarrow C(\Omega)\ ,\ k > \frac{n}{2}\ ;$$

$$H^k(\Omega) \hookrightarrow C^{(0)+k-\frac{1}{2}}(\Omega)\,, \quad \Omega\subset R\,, \quad \frac{1}{2} < k \le 1\,.$$

These results can be found e.g. in [2], [4].

Assumptions

(i) The functions f_o, f_1 are continuously differentiable in R with locally Lipschitz continuous derivatives f_o', f_1';

(ii) there exists a constant $\alpha \in R$ such that

$$f_o'(u) \ge \alpha\,, \quad f_1'(u) \ge \alpha \quad \text{for all } u\in R\,;$$

(iii) $u_o \in H^1(0,\ell)$, $u_1 \in L^2(0,1)$.

DEFINITION. *By a weak solution of the problem* (1) - (6) *on* $[0,T]$ *we call a couple* $(u,\xi)\in L^\infty(0,T;\, H^1(0,\ell))\times W^{1,\infty}(0,\ell)$ *such that* $u_t \in L^\infty(0,T;\, L^2(0,\ell))$, $u(x,0)=u_o(x)$ *in the sense of traces and*

$$\int_0^T\int_0^\ell (c^2u_x\phi_x - u_t\phi_t)\,dxdt + \int_0^T\int_0^{\xi(t)} f_o(u)_t\phi\, dxdt + \int_0^T\int_{\xi(t)}^\ell f_1(u)_t\phi\, dxdt - \int_0^\ell u_1(x)\phi(x,0)\,dx = 0 \tag{8}$$

for all $\phi\in C^1([0,\ell]\times[0,T])$ *satisfying* $\phi(x,T) = 0$, $x\in[0,\ell]$.

THEOREM. *Let the functions* f_o, f_1, u_o, u_1 *satisfy the assumptions* (i) - (iii). *Then there exists at least one weak solution of the problem* (1) - (6).

Proof. For the construction of a solution we shall make use of Faedo-Galerkin approximations and a method of compactness. Let

$$v_o(x) = \ell^{-1/2}, \quad v_j(x) = \left(\frac{2}{\ell}\right)^{1/2}\cos\left(\frac{j\pi x}{\ell}\right), \quad j = 1,2,\ldots, \quad x\in[0,\ell]\,.$$

The functions $\{v_j\}_{j=0}^\infty$ form the orthonormal basis in $L^2(0,\ell)$ of eigenfunctions of the problem

$$-c^2v''(x) = \lambda v(x)\,, \quad x\in(0,\ell)\,,$$
$$v'(0) = 0\,, \quad v'(\ell) = 0\,.$$

The corresponding eigenvalues are given by

$$\lambda_j = c^2\pi^2\ell^{-2}j^2 \ , \ j = 0,1,2,\dots \ .$$

F i n i t e - d i m e n s i o n a l   a p p r o x i m a t i o n s
The finite-dimensional approximations are sought in the form

$$u^n(x,t) = \sum_{j=0}^{n} u_j^n(t)\ v_j(x)\ ,\ (x,t)\in\Omega \equiv [0,\ell]\times[0,T]\ ,\qquad n = 0,1,2,\dots \ . \tag{9}$$

The coefficients $u_j^n(t)$, denoted simply by $u_j(t)$, are to be chosen so that

$$\int_0^\ell [u_{tt}^n(x,t) - c^2 u_{xx}^n(x,t)]\,v_k(x)\,dx + \int_0^{\xi_n(t)} f_o(u^n(x,t))_t v_k(x)\,dx + \int_{\xi_n(t)}^{\ell} f_1(u^n(x,t))_t v_k(x)\,dx = 0,\ k = 0,1,\dots,n,\ t\in[0,T], \tag{10}$$

$$u^n(x,0) = \sum_{j=0}^{n} \int_0^\ell u_o(y)\,v_j(y)\,dy\ v_j(x) \equiv u_o^n(x)$$

$$u_t^n(x,0) = \sum_{j=0}^{n} \int_0^\ell u_1(y)\ v_j(y)\ dy\ v_j(x) \equiv u_1^n(x)\ ,\ x\in[0,\ell]\ ,$$

where the approximations $x = \xi_n(t)$ of the free boundary are given by

$$\xi_n(t) = G\Big(\int_0^t u^n(\xi_n(\tau),\tau)\,d\tau\Big)\ ,\ n = 0,1,\dots \ . \tag{11}$$

The equations (10), (11) yield the following integrodifferential system:

$$\ddot{u}_k(t) + \lambda_k u_k(t) + \int_0^{\xi_n(t)} f_o'\Big(\sum_{j=0}^{n} u_j(t)\ v_j(x)\Big) \sum_{j=0}^{n} \dot{u}_j(t)\ v_j(x)\ v_k(x)\ dx + \tag{12}$$

$$+ \int_{\xi_n(t)}^{\ell} f_1'\Big(\sum_{j=0}^{n} u_j(t)\ v_j(x)\Big) \sum_{j=0}^{n} \dot{u}_j(t)\,v_j(x)\,v_k(x)\,dx = 0\ ,$$

$$k = 0,1,\dots,n, \ t \in (0,T] ,$$

$$u_k(0) = \int_0^\ell u_0(x) v_k(x)\, dx, \quad \dot{u}_k(0) = \int_0^\ell u_1(x) v_k(x)\, dx , \tag{13}$$

$$k = 0,1,\dots,n$$

$$\xi_n(t) = G\left(\int_0^t \sum_{j=0}^n u_j(\tau) v_j(\xi_n(\tau))\, d\tau \right) , \ t \in [0,T], \ n = 0,1,\dots . \tag{14}$$

It is almost standard that this system has a maximal solution $(u_0(t),\dots,u_n(t),\xi_n(t))$ and that it is defined on $[0,T]$ if $u_j(t)$, $\dot{u}_j(t)$, $\xi_n(t)$ are bounded on $[0,T]$. The details are given in [5].

A p r i o r i e s t i m a t e

Denote

$$E_n(t) = \int_0^\ell [u_t^n(x,t)^2 + c^2 u_x^n(x,t)^2]\, dx = \sum_{j=0}^n [\dot{u}_j(t)^2 + \lambda_j u_j(t)^2] .$$

Then multiplying (12) by $\dot{u}_k(t)$ and summing up for $k = 0,1,\dots,n$ we get

$$\dot{E}_n(t) = -2 \int_0^{\xi_n(t)} f_0'(u^n(x,t)) u_t^n(x,t)^2 dx - \tag{15}$$

$$- 2 \int_{\xi_n(t)}^\ell f_1'(u_n^n(x,t)) \ u_t^n(x,t)^2 dx \le$$

$$\le - 2\alpha \int_0^\ell u_t^n(x,t)^2 dx \le 2|\alpha| \ E_n(t) ,$$

where we have used the assumption (ii). We have from (14) $|\xi_n(t)| \le \max_{\eta \in R} G(\eta) = \ell$. Using in (15) the Gromwall inequality we get the estimate ensuring the global existence for the system (12) - (14) namely, we find

$$\int_0^\ell [u_t^n(x,t)^2 + c^2 u_x^n(x,t)^2]\, dx \le \tag{16}$$

$$\leq e^{2|\alpha|T} \int_0^{\ell} [u_1(x)^2 + c^2 u_0'(x)^2] dx < \infty, \quad t \in [0,T], \quad n = 0,1,2,\ldots$$

C o n v e r g e n c e

From (16) and the above mentioned embeddings we get that given $p \in [1,\infty)$, there exists a subsequence of $\{u^n\}_{n=0}^{\infty}$ denoted again by u^n such that

$$u^n \to u \quad \text{weak * in } L^\infty(0,T;H^1(0,\ell));$$

$$u_t^n \to u_t \quad \text{weak * in } L^\infty(0,T;L^2(0,\ell));$$

$$u^n \to u \quad \text{weakly in } H^1(\Omega), \ (\Omega = (0,\ell)\times(0,T)); \tag{17}$$

$$u^n \to u \quad \text{strongly in } L^p(\Omega); \tag{18}$$

$$u^n \to u \quad \text{a.e. in } \Omega, \tag{19}$$

where $u \in L^\infty(0,T;H^1(0,\ell))$, $u_t \in L^\infty(0,T;L^2(0,\ell))$.

From (16) and $(10)_2$ it may be easily seen that

$$\sup\{||u^n(.,t)||_{H^1(0,\ell)}; \ t \in [0,T], \ n = 0,1,2,\ldots\} < \infty .$$

This together with the embedding $H^1(0,\ell) \hookrightarrow C([0,\ell])$ yields

$$\sup\{|u^n(x,t)|; \ x \in [0,\ell], \ t \in [0,T], \ n = 0,1,2,\ldots\} < \infty. \tag{20}$$

In particular we have

$$|u^n(\xi_n(t),t)| \leq k < \infty, \quad t \in [0,T], \ n = 0,1,2,\ldots . \tag{21}$$

Let $t_1, t_2 \in [0,T]$ be arbitrary. By (11), (7) and (21) we have

$$|\xi_n(t_1) - \xi_n(t_2)| \leq \left|\int_{t_2}^{t_1} u^n(\xi_n(\tau),\tau) d\tau\right| \leq K\,|t_1 - t_2|. \tag{22}$$

Hence, by the Arzelà-Ascoli theorem the choice of the subsequence can be done so that

$$\xi_n \text{ converges uniformly to a } \xi \in C([0,T]) . \tag{23}$$

By the help of (22) and (23) it is easy to show that ξ is Lipschitz continuous and $\xi \in W^{1,\infty}(0,T)$.

Further, by the embedding $H^1(0,\ell) \hookrightarrow C^{(0)+\frac{1}{2}}([0,\ell])$ and the boundedness of u^n in $L^\infty(0,T;H^1(0,\ell))$ we find

$$|u^n(\xi_n(t),t) - u^n(\xi(t),t)| \leq$$

$$\leq ||u^n(.,t)||_{C^{(0)+\frac{1}{2}}([0,\ell])} |\xi_n(t) - \xi(t)|^{1/2} \leq$$

$$\leq \text{const.}\ ||u^n(.,t)||_{H^1(0,\ell)} |\xi_n(t) - \xi(t)|^{1/2} \leq$$

$$\leq \text{const.}\ |\xi_n(t) - \xi(t)|^{1/2}$$

what together with (23) yields

$$\lim_{n\to\infty} |u^n(\xi_n(t),t) - u^n(\xi(t),t)| = 0 \quad \text{uniformly for} \quad t \in [0,T] .$$

Besides, for any $r \in [1,\infty)$ the operator of traces maps $H^1(\{0 < x < \xi(t);\ t \in (0,T)\})$ compactly into $L^r(\{x = \xi(t);\ t \in (0,T)\})$, (see [4] , Theorem 6.2). Thus we have proved that

$$u^n(\xi_n(.),.) \to u(\xi(.),.) \quad \text{strongly in } L^r(0,T) , \qquad (24)$$

$$u^n(\xi_n(t),t) \to u(\xi(t),t) \quad \text{a.e. in } (0,T) . \qquad (25)$$

L i m i t i n g p r o c e s s

Now, let $\phi \in C^1(\bar{\Omega})$, $\phi(x,T) = 0$ for $x \in [0,\ell]$.

Put

$$\phi^n(x,t) = \sum_{j=0}^{n} \left(\int_0^\ell \phi(y,t)\, v_j(y)\, dy \right) v_j(x) \quad .$$

Then

$$\phi^n \to \phi \qquad \text{strongly in } H^1(\Omega) . \qquad (26)$$

Multiply $(10)_1$ by $\int_0^\ell \phi(y,t)\, v_k(y)\, dy$, sum up over k from 0 to n , integrate over $[0,T]$ and then integrate by parts.
We get

$$\int_0^T \int_0^\ell (c^2 u^n \phi^n_x - u^n_t \phi^n_t)\, dxdt + \int_0^T \int_0^{\xi_n(t)} f_o'(u^n)\, u^n_t \phi^n\, dxdt \quad +$$

$$+ \int_0^T \int_{\xi_n(t)}^{\ell} f_1'(u^n) u_t^n \phi^n \, dxdt - \int_0^{\ell} u_1^n(x) \phi^n(x,0) \, dx = 0 .$$

From (17), (26) it is clear that

$$\lim_{n\to\infty} \int_0^T \int_0^{\ell} (c^2 u_x^n \phi_x^n - u_t^n \phi_t^n) \, dxdt = \int_0^T \int_0^{\ell} (c^2 u_x \phi_x - u_t \phi_t) \, dxdt .$$

Show that

$$\lim_{n\to\infty} \int_0^T \int_0^{\xi_n(t)} f_0'(u^n) u_t^n \phi^n \, dxdt = \int_0^T \int_0^{\xi(t)} f_0'(u) u_t \phi \, dxdt .$$

We have

$$\left| \int_0^T \int_0^{\xi_n(t)} f_0'(u^n) \, u_t^n \phi^n dxdt - \int_0^T \int_0^{\xi(t)} f_0'(u) u_t \phi dxdt \right| \le$$

$$\le \left| \int_0^T \int_{\xi(t)}^{\xi_n(t)} f_0'(u^n) u_t^n \phi^n dxdt \right| + \left| \int_0^T \int_0^{\xi(t)} f_0'(u^n) u_t^n (\phi^n - \phi) dxdt \right| +$$

$$+ \left| \int_0^T \int_0^{\xi(t)} f_0'(u^n) (u_t^n - u_t) \phi dxdt \right| + \left| \int_0^T \int_0^{\xi(t)} [f_0'(u^n) - f_0'(u)] u_t \phi dxdt \right| \equiv$$

$$\equiv I_1^n + I_2^n + I_3^n + I_4^n .$$

Clearly

$$I_1^n \le \text{const.} \int_0^T \int_{\min\{\xi(t), \xi_n(t)\}}^{\max\{\xi(t), \xi_n(t)\}} |u_t^n(x,t)| \, dxdt \to 0, \ (n \to \infty) .$$

Further, by (20), (16), (26)

$$I_2^n \le \left(\int_0^T \int_0^{\xi(t)} f_0'(u^n)^2 (u_t^n)^2 dxdt \right)^{1/2} \left(\int_0^T \int_0^{\xi(t)} (\phi^n - \phi)^2 dxdt \right)^{1/2} \le$$

$$\le \text{const.} ||\phi^n - \phi||_{L^2(\Omega)} \to 0 , \ (n \to \infty) .$$

From (19) and (20) we have $f_0'(u^n) \to f_0'(u)$ a.e. in Ω and $|f_0'(u^n)| \le$ const. in Ω. Hence by the Lebesgue theorem $f_0'(u^n)\phi \to f_0'(u)\phi$ in $L^2(\Omega)$.

Besides from (17) $u_t^n \to u_t$ weakly in $L^2(\Omega)$. Thus $I_3^n \to 0$ $(n \to \infty)$. Finally, by the Schwartz inequality and the Lebesgue theorem $I_4^n \to 0$ $(n \to \infty)$ as well. A similar reasoning with

$$\int_0^T \int_{\xi_n(t)}^{\ell} f_1'(u^n)\, u_t^n \phi^n \, dxdt$$ and the obvious relation

$$\lim_{n\to\infty} \int_0^{\ell} u_1^n \phi^n dx = \int_0^{\ell} u_1 \phi dx$$ yields (8). Since it is easy to show that $u(x,0) = u_o(x)$ in the sense of traces and by density (8) holds for all ϕ as in the definition, the couple (u,ξ) is a weak solution of (1) - (6).

REMARKS

1. The free boundary equation (6) corresponds to the original physical problem only if $\int_0^t u(\xi(\tau),\tau)d\tau$ belongs to $[0,\ell]$. In general instead of (6) a variational inequality

$$(6') \qquad (\dot{\xi}(t) - u(\xi(t),t))(\eta - \xi(t)) \geq 0, \ \eta \in [0,\ell] \ ,$$

$$\xi(t) \in [0,\ell] \ , \ t \in [0,T]$$

should be employed. The arising problem can be solved as well but for the lack of space we cannot present it here. This result will be published in [6] .

2. Although we are not able to prove the uniqueness of the solution to (1) - (6), it seems that the uniqueness can be obtained for more regular (strong) solutions, namely for $u \in H^2(\Omega)$, $\xi \in W^{1,\infty}(0,\ell)$. Unfortunately, we are not able to show the existence of a strong solution.

3. An interesting problem is how to show the existence of the solution in the case of different (but still constant) sound speeds c_o, c_1 in separate fluids. In that case the second interface condition in (3) should be replaced by an appropriate interpretation of the continuity of the pressure accross $x = \xi(t)$.

REFERENCES

[1] KOLARČÍK, V.: "Formulace úlohy plnění komory hydrosměsí" ("Formulation of the problem filling the compartment by a hydro-suspension") VÚ Sigma, Olomouc 1986.

[2] LIONS, J.L., MAGENES, E.: " Problèmes aux limites non homogènes et applications". Vol. 1, Dunod Paris 1968.

[3] LUSKIN, M.: "On the existence of global smootn solutions for a model equation for fluid flow in a pipe". J. Math. Anal. Appl. 84, 614-630 (1981).

[4] NEČAS, J.: "Les méthodes directes en théorie des équations elliptiques", Academia Prague 1967.

[5] STRAŠKRABA, I.: "An existence result for a nonlinear telegraph equation with a free boundary". Math. Inst. of Czech. Acad. Sci 1987, Preprint No. 27.

[6] STRAŠKRABA, I.: "On a telegraph equation with a free boundary". To appear in Czech. Math.J.

[7] STREETER, V.: "Fluid Mechanics". 5th ed., McGraw-Hill, New York, 1971.

"TVD" Schemes
for
Inhomogeneous Conservation Laws.

Peter K. Sweby
Department of Mathematics, University of Reading,
Whiteknights, Reading, England.

Summary.

Many schemes have been developed for the numerical solution of homogeneous conservation laws giving high resolution, oscillation free results. However the TVD criterion used in these schemes is inappropriate for inhomogeneous problems. Despite this, there have been various attempts to apply such schemes to these problems. We review here one such empirical technique which has been used successfully, although we demonstrate that its successful behaviour cannot be guaranteed. We then utilise a change of dependent variable to reduce the inhomogeneous problem to homogeneous form and thus suggest a correct way to apply TVD schemes to such a problem.

1. Introduction.

In recent years much effort has been devoted to the design of numerical schemes which give high resolution, oscillation free, solutions to systems of homogeneous conservation laws,

$$\underset{\sim}{u}_t + \underset{\sim}{f}(\underset{\sim}{u})_x = 0. \tag{1.1}$$

All such schemes are non–linear, their coefficients being data dependent (see e.g. [1],[2],[3]), and all use as a criterion to monitor oscillations the total variation (we consider here the scalar case)

$$TV(u^n) = \Sigma | u_k^n - u_{k-1}^n | \tag{1.2}$$

of the numerical solution at time $n\Delta t$. This mimics the total variation of the analytic solution

$$TV(u(\cdot,t)) = \int | u_x(\cdot,t) | \, dx \tag{1.3}$$

which has the property of being non–increasing for scalar equations of the form (1.1). Schemes are designed therefore to be Total Variation Diminishing (TVD), i.e.

$$TV(u^{n+1}) \leq TV(u^n) \tag{1.4}$$

(Harten [3]).

Solutions to inhomogeneous conservation laws

$$u_t + f(u)_x = b, \tag{1.5}$$

however, do not possess this total variation non–increasing property, indeed the right hand side of (1.5) often represents a source term which will actively increase the variation of the analytic solution.

The problem therefore is how to obtain high resolution oscillation free numerical solutions to (1.5) without inhibiting any natural growth in the variation of the solution due to the inhomogeneity of the problem.

In the next section we look at one empirical technique which performs well in many situations, but whose performance we show cannot be guaranteed. In section 3 we utilize a transformation of dependent variable to reduce (1.5) to an homogeneous equation which does possess the TVD property and in section 4 we suggest how this can be used to apply TVD schemes to inhomogeneous equations. In the final section we make concluding remarks, highlighting the need to develop better techniques for the treatment of the source terms themselves.

Throughout the paper techniques are illustrated by the use of Flux Limiter schemes [1] coupled with Roe's approximate Riemann solver [4] to extend them to systems.

2. Roe's Approach.

In [5] Roe proposed an empirical technique for applying high resolution TVD schemes to inhomogeneous problems. His approach employed the application of the TVD scheme to a modified flux derived from a Lax–Wendroff like Taylor expansion. An outline of the process is as follows:

1. Calculate $u^{n+\frac{1}{2}}$ using a low order scheme with time increment $\frac{1}{2}\Delta t$.
2. Use this to compute $b^{n+\frac{1}{2}}$, the source term evaluated at the half time step.
3. Define a modified flux $(b^{n+\frac{1}{2}} - f_x^n)$ where a suitable difference is used for the derivative.
4. Apply the TVD scheme to the modified flux to obtain u^{n+1} from u^n.

This technique has been successfully used in various situations (see [5],[6]), although Roe himself observed some slight over/under shoots due to its empirical nature. We now explore this further.

Consider the following test problem, to which an analytic solution is available [7].

$$\begin{pmatrix} \rho \\ \rho u \\ e \end{pmatrix}_t + \begin{pmatrix} \rho u \\ p+\rho u^2 \\ u(e+p) \end{pmatrix}_x = \frac{H(-x)C}{a} \begin{pmatrix} \rho \\ \rho u \\ \rho h \end{pmatrix} \tag{2.1}$$

where ρ, u, p, e, h, and a are density, velocity, pressure, energy, enthalpy and sound speed respectively, $H(\cdot)$ is the Heavyside step function and C is a constant. The situation therefore is the Euler equations with a source term in the left hand half plane.

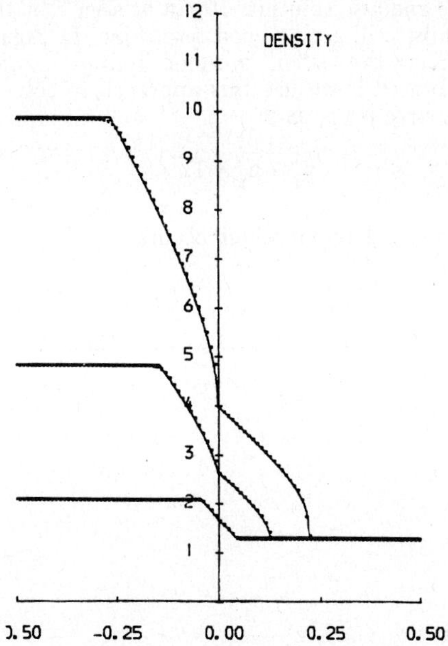

Figure 1. Modified flux and superbee limiter.

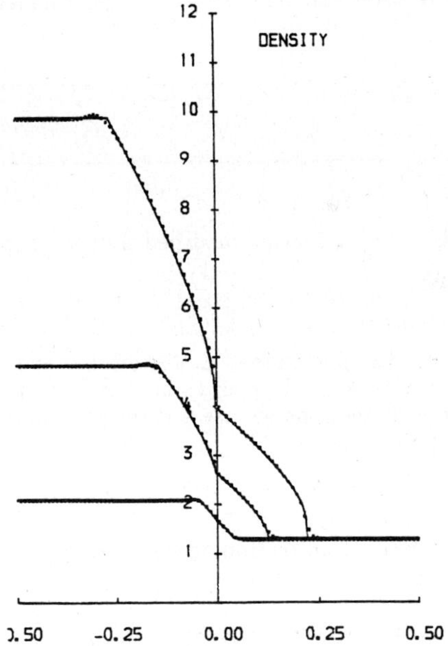

Figure 2. Modified flux and minmod limiter.

Results of the problem using the superbee flux limiter are shown in Figure 1, the solid lines denoting the analytic solution. It can be seen that the numerical solution is free from oscillations whilst still giving good resolution. In Figure 2 .however, which displays results obtained using the "safer" minmod limiter, it can clearly be seen that overshoots are occurring. Indeed if we use this approach to solve the linear advection equation with a constant source term, as in

$$u_t + u_x = 1, \tag{2.2}$$

Figure 3 shows the oscillatory behaviour which occurs.

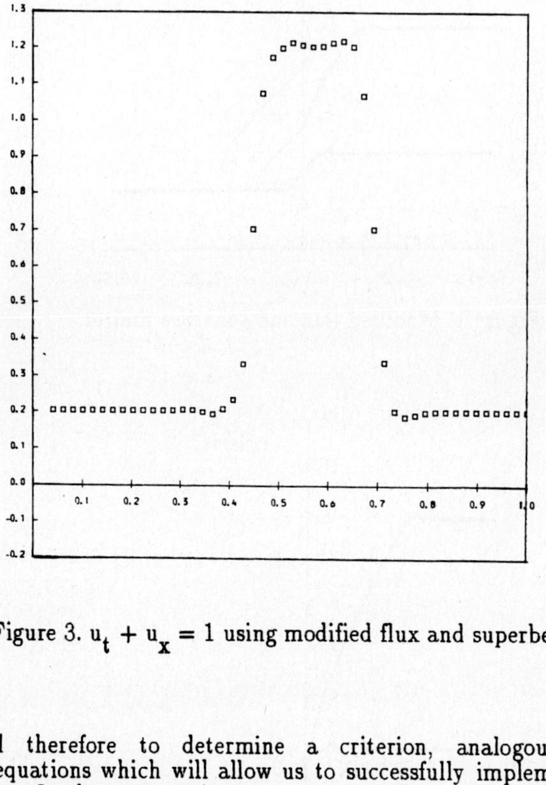

Figure 3. $u_t + u_x = 1$ using modified flux and superbee.

We need therefore to determine a criterion, analogous to TVD, for inhomogeneous equations which will allow us to successfully implement existing high resolution schemes. In the next sections we use a transformation of dependent variable to investigate such a possibility.

3. Reduction to Homogeneous Form.

Consider the scalar inhomogeneous equation

$$u_t + f(u)_x = b(x,t) \tag{3.1}$$

which can equivalently be written as

$$u_t + a(u)u_x = b(x,t) \tag{3.2}$$

where

$$a(u) = f'(u). \tag{3.3}$$

Note that we are considering here functions $b(x,t)$ for algebraic simplicity, however the following analysis can be shown to hold for the more general source $b(u,x,t)$.

The characteristics of (3.2) are given by

$$\frac{dx}{dt} = a(u) \tag{3.4}$$

along which

$$\frac{du}{dt} = b(x,t). \tag{3.5}$$

Since u is not constant along the characteristics they are no longer straight lines as in the homogeneous case. By integrating back along the characteristics we can write the solution $u(x,t)$ as

$$u(x,t) = u(x_0,0) + \int_0^t \hat{b}(x,t;\tau)d\tau \tag{3.6}$$

where x_0 is where the characteristic crosses the x axis and where

$$\hat{b}(x,t;\tau) = b\left[x - \int_\tau^t a\,ds\ ,\ \tau\right], \tag{3.7}$$

the integrations being along the characteristic.

Now define a new variable v by

$$v(x,t) = u(x,t) - \int_0^t \hat{b}(x,t;\tau)d\tau \tag{3.8}$$

and substitute into (3.1). We have, using (3.7),

$$\begin{aligned} u_t &= v_t + \frac{\partial}{\partial t}\int_0^t \hat{b}(x,t;\tau)d\tau \\ &= v_t + b(x,t) + \int_0^t \frac{\partial}{\partial t}\hat{b}(x,t;\tau)d\tau \\ &= v_t + b(x,t) - a(u)\int_0^t \partial_1 b\ d\tau \end{aligned} \tag{3.9}$$

and similarly

$$u_x = v_x + \frac{\partial}{\partial x}\int_0^t \hat{b}(x,t;\tau)d\tau$$

$$= v_x + \int_0^t \partial_1 b \, d\tau \tag{3.10}$$

and so

$$u_t + a(u)u_x = v_t + a(u)v_x + b(x,t) + \left\{a(u)-a(u)\right\}\int_0^t \partial_1 b \, d\tau \tag{3.11}$$

from which we deduce

$$v_t + a(u)v_x = 0. \tag{3.12}$$

If we label the integral in (3.8) as $c(x,t)$, and so $v = u - c$, we also note that the above analysis gives us

$$\left.\begin{aligned} c_t + a(u)c_x &= b(x,t) \\ c(x,0) &= 0 \end{aligned}\right\} \tag{3.13}$$

i.e. $c(x,t)$ is a "particular integral" of (3.1).

Notice that the homogeneous equation (3.12), which will have the same initial data as (3.1), does possess the Total Variation non–increasing property, and in the next section we will see how we can use this to indicate how to correctly apply TVD schemes to inhomogeneous problems.

4. Application of TVD Schemes.

We now use the homogeneous form obtained in the previous section to apply a TVD scheme to an inhomogeneous problem. For illustration we use the class of Flux Limiter schemes, which may be written in the homogeneous case as a first order scheme plus a limited antidiffusive term, i.e.

$$u^{n+1} = u^n - \lambda a(u^n)\Delta u^n - \mathscr{L}(u^n) \tag{4.1}$$

where $\lambda = \Delta t/\Delta x$ the mesh ratio, Δu^n is an upwind difference and $\mathscr{L}$ encompasses the limiter terms.

If we now apply such a scheme to (3.12) we have

$$v^{n+1} = v^n - \lambda a(u^n)\Delta v^n - \mathscr{L}(v^n) \tag{4.2}$$

or, putting $v = u - c$,

$$u^{n+1} = u^n - \lambda a(u^n)\Delta(u^n - c^n) - \mathcal{L}(u^n - c^n) + c^{n+1} - c^n. \tag{4.3}$$

We can also solve (3.13), viz

$$c^{n+1} = c^n - \lambda a(u^n)\Delta c^n + \mathcal{H}(c^n) + \mathcal{S}(b^n) \tag{4.4}$$

where $\mathcal{H}$ is some high order operator and $\mathcal{S}$ is the treatment of the source term.

If we now combine (4.3) and (4.4) we obtain

$$u^{n+1} = u^n - \lambda a(u^n)\Delta u^n - \mathcal{L}(u^n - c^n) + \mathcal{H}(c^n) + \mathcal{S}(b^n) \tag{4.5}$$

which still risks contamination through the $\mathcal{H}$ term – however if we reset time in (3.13) at each timestep, i.e. take $c^n \equiv 0$, we obtain

$$u^{n+1} = u^n - \lambda a(u^n)\Delta u^n - \mathcal{L}(u^n) + \mathcal{S}(b^n), \tag{4.6}$$

that is the TVD scheme is applied only to the flux of the homogeneous equation, the source term being treated separately. Figure 4 shows results, using both minmod and superbee limiters, of the above splitting with upwinded second order treatment of the source term. Alternatively, multistage splittings could be derived form (4.3) and (4.4) by application of the technique recursively to (4.4) and setting $c^{n-m} \equiv 0$.

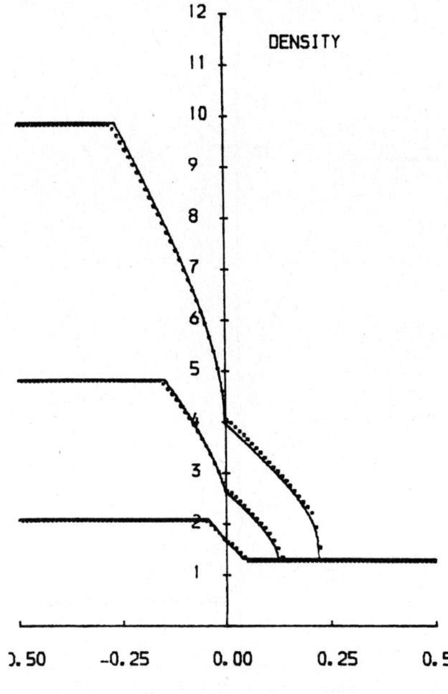

Figure 4a. Superbee applied to homogeneous part only.

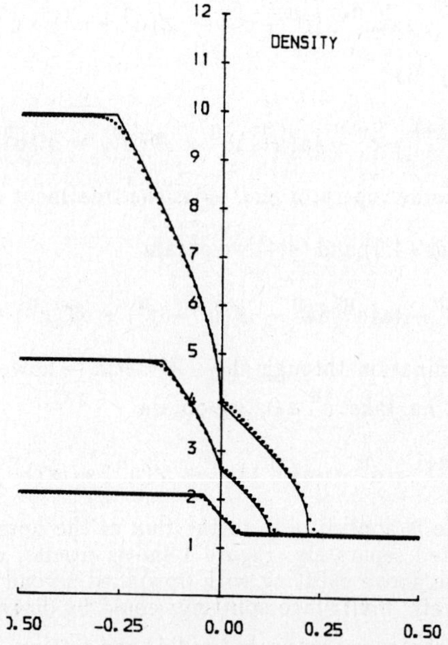

Figure 4b. Minmod applied to homogeneous part only.

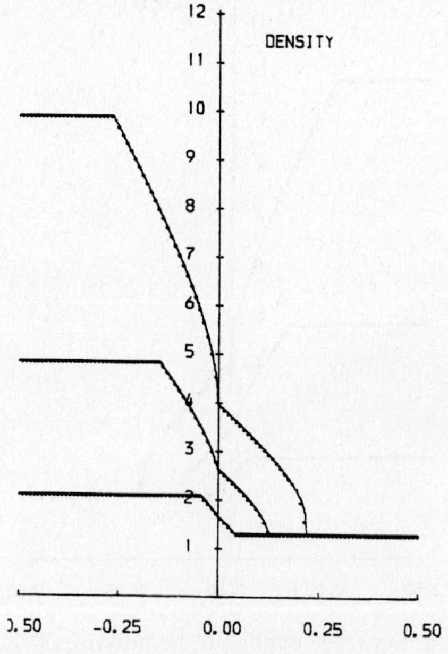

Figure 5. As Fig 4a but using pointwise application of source.

5. Concluding Remarks.

We have devised a technique for applying TVD schemes to inhomogeneous problems which is sufficient to prevent spurious oscillations whilst not adversely affecting any natural increase in variation. However there is still a large question remaining – what is the best treatment of the source term itself? Figure 5 shows results of the test problem using a pointwise application of the source term as opposed to the upwinded average used in Figure 4. For this example the pointwise appears to be the better treatment – however this is not always the case, see [6],[8].

References.

1. P.K. SWEBY, *High resolution schemes using flux limiters for hyperbolic conservation laws.* SIAM J. Numer. Anal. 21 (1984).

2. S. OSHER and P.K. SWEBY, *Recent developments in the numerical solution of nonlinear conservation laws.* The State of the Art in Numerical Analysis. Eds. A. Iserles and M.J.D. Powell, IMA Conference Series 9. OUP (1987).

3. A. HARTEN, *High resolution schemes for hyperbolic conservation laws.* J. Comput. Phys. 49, (1983)

4. P.L. ROE, *Approximate Riemann solvers, parameter vectors, and difference schemes.* J. Comput. Phys., 43, (1981).

5. P.L. ROE, *Upwind differencing schemes for hyperbolic conservation laws with source terms.* Proc. Nonlinear Hyperbolic Problems, Eds. C. Carasso, P.A. Raviart, D. Serre, Springer Lecture Notes in Mathematics 1270 (1986).

6. A. PRIESTLEY, *Roe type schemes for the 2–D shallow water equations.* University of Reading, Department of Mathematics, Numerical Analysis Report 8/87.

7. J.F. CLARKE, *Compressible flow produced by distributed sources of mass: An exact solution.* Cranfield College of Aeronautics report 8710, (1987).

8. P. GLAISTER, *Flux difference splitting for the Euler equations in one spatial co–ordinate with area variation.* Int. J. Num. Meth. Fluids, 8, (1988).

The L^1-Norm Distinguishes the Strictly Hyperbolic from a Non-Strictly Hyperbolic Theory of the Initial Value Problem For Systems of Conservation Laws

Blake Temple*

Department of Mathematics

University of California, Davis

Davis, CA 95616

Abstract

We discuss recent work of the author in which he proves that solutions to systems of two strictly hyperbolic genuinely nonlinear conservation laws are weakly stable in the global L^1-norm. We contrast this with the theory of the initial value problem for a nonstrictly hyperbolic system in which weak stability in L^1 is shown to fail. This is understood from a study of the asymptotic wave patterns to which solutions in this problem decay as $t \to +\infty$. Since solution in both cases have been shown to be stable in the total variation and sup norms, we conclude that the L^1 estimate is the first stability result in a norm that distinguishes the strictly hyperbolic from a nonstrictly hyperbolic theory of the initial value problem.

In this talk we compare the theory of the initial value problem for a 2x2 non-strictly hyperbolic system of conservation laws to the corresponding strictly hyperbolic theory. In terms of the total variation and supnorms the theories look the same. Here we demonstrate that the theories diverge at the L^1-norm. In particular, recent work of the author gives a proof of the weak stability in the global L^1-norm for systems of two strictly hyperbolic equations. In contrast to this, a study of the asymptotic wave structures in a nonstrictly hyperbolic system leads directly to the conclusion that no such stability result holds in a special nonstrictly hyperbolic problem. We first discuss the weak stability result (see "Weak Stability in the global L^1-norm for systems of conservation laws" by Blake Temple, Davis preprint), and then we discuss the asymptotic wave patterns in a simple nonstrictly hyperbolic system with an eye toward seeing how L^1 stability fails (see "The structure of asymptotic states in a singular system of conservation laws" with E. Isaacson, Davis preprint.)

We consider the initial value problem

$$u_t + F(u)_x = 0, \qquad u = (u_1, u_2), \qquad u(x,0) = u_0(x) \qquad F = (F_1, F_2)\,. \tag{C}$$

In the strictly hyperbolic case, Glimm demonstrated in his fundamental paper of 1965 [3] that solutions of (C) generated by the random choice method are stable in the supnorm and in the total variation norm. Indeed, it is stability in the total variation that gives compactness of the approximate solutions, and this resulted in the first existence theory for systems of conservation laws. (We remark that in general we have no proof of uniqueness or continuous dependence for solutions generated by this method.) We state Glimm's result precisely [25].

Theorem (Glimm 1965): Assume (C) is genuinely nonlinear and strictly hyperbolic in both characteristic fields in a neighborhood of a state $\bar{u} \in \mathbb{R}^2$. Then $\forall\, V > 0$ there exists $\delta << 1$ such that if

$$TV\,\{u_0\,(0)\} < V\,, \qquad \|u_0\,(0) - \bar{u}\|_{sup} < \delta\,,$$

then there exists a solution to (C) satisfying

*This work supported by the NSF under the grant NSF–DMS–86–13450.

$$\mathrm{TV}\,\{u(\cdot,t)\} < C\,V, \qquad \text{(TV)}$$
$$\|u(\cdot,t) - \bar{u}\|_{\sup} < C\,\delta, \qquad \text{(SUP)}$$
$$\|u(\cdot,t) - \bar{u}(\cdot,s)\|_{L^1} \le C|t-s|\,. \qquad \text{(LIP)}$$

Here C denotes a generic constant, TV denotes the total variation and $\|\ \|_{\sup}$ denotes the supnorm. Note that (LIP) implies that the data is taken on in the L^1 sense.

The author recently proved the following weak stability result in the global L^1-norm for solutions generated by Glimms method [24]:

$$\|u(\cdot,t) - \bar{u}\|_{L^1} \le G\,(t, \|u_0(\cdot) - \bar{u}\|_{L^1}) \qquad (L^1)$$

where G is an explicitly constructed smooth function satisfying $G(t,\xi) \to 0$ as $\xi \to 0$ for every fixed $t \ge 0$. Here we assume that $u_0(\pm\infty) = \bar{u}$.

We now contrast this with a corresponding existence theory for a non-strictly hyperbolic system in which (TV), (SUP) and (LIP) have been shown to hold (cf [20]), but (L^1) fails for every smooth function G satisfying $G(t,\xi) \to 0$ as $\xi \to 0$. We conclude that (L^1) gives the first stability result in a norm that distinguishes the two theories. That (L^1) fails in the next example follows directly from an understanding of the asymptotic wave structures to which solutions decay as $t \to +\infty$. This was studied in joint work with E. Isaacson, Dept. of Math., Univ. of Wyoming. Consider the 2x2 system of polymer equations:

$$s_t + f(s,c)_x = 0, \quad u = (s, cs), \quad (sc)_t + \{cf(s,c)\}_x = 0, \quad F = (f, cf). \qquad \text{(P)}$$

In general, system (P) is not strictly hyperbolic when $f(\cdot,c)$ is non-convex. E Isaacson first derived (P) from a simple two component flow problem, and he solved the corresponding nonconvex Riemann problem [4]. In [8], B. Keyfitz and H. Kranzer earlier solved the Riemann problem for a system formally equivalent to (P). In [20] the author proved a global existence theorem by Glimm's method. We state it here in order to compare it with the strictly hyperbolic case:

<u>Theorem (Te)</u>: If $u_0(\cdot)$ is initial data for (P) satisfying

$$\mathrm{TV}\,\{u_0(\cdot)\} < V < \infty, \qquad \|u_0(\cdot) - \bar{u}\|_{\sup} < \delta,$$

then there exists a global weak solution of (P) with initial data u_0 satisfying

$$\mathrm{TV}\,\{u(\cdot,t)\} < C\,V, \qquad \text{(TV)}$$
$$\|u(\cdot,t) - \bar{u}\|_{\sup} < \delta, \qquad \text{(SUP)}$$
$$\|u(\cdot,t) - u(\cdot,s)\|_{L^1} < C|t-s|\,. \qquad \text{(LIP)}$$

Here total variation is measured in the singular coordinate system of Riemann invariants, and this leads to a modified convergence proof, but formally, the results look the same as in the strictly hyperbolic case of Theorem (Glimm). In joint work with E. Isaacson, we determine the asymptotic waves that these solutions decay to as $t \to +\infty$, and this leads directly to the following result which implies that the two theories diverge on the level of the L^1-norm (cf. [5]).

For the solutions u(x,t) of (P) generated by Theorem (Te) and satisfying $u_0(\pm\infty) = \bar{u}$, the L^1-norm at time t cannot be controlled by the L^1-norm at time t = 0, through any nonlinear function; i.e.,

Theorem ($\neg L^1$): The estimate (L^1) FAILS in general for every smooth G satisfying $G(t,\xi) \to 0$ as $\xi \to 0$ for each fixed t. Specifically, there exists a sequence of solutions $u^\sigma(x,t)$ of (P), $0 \leq \sigma \leq 1$, such that $u_0^\sigma(\pm\infty) = u_L$,

$$\lim_{\sigma\to 0} \|u_0^\sigma(\cdot) - u_L\|_{L^1} = 0, \qquad (\neg L^1)$$

but

$$\lim_{\sigma\to 0} \|u^\sigma(\cdot,t) - u_L\|_{L^1} \neq 0 \qquad (\neg L^1)$$

at any $t > 0$.

In the next section we discuss the asymptotic states for solutions of (P) with an eye toward seeing ($\neg L^1$). We comment on the interesting role played by the admissible solutions of the Riemann problem in this nonstrictly hyperbolic problem. In section 3 we return to the strictly hyperbolic case, and discuss the proof of (L^1). The estimate (L^1) is a consequence of the author's decay result [22] which states that

$$\|u(\cdot,t) - \bar{u}\|_{sup} \leq F\left(\frac{t}{\|u_0(\cdot) - \bar{u}\|_{L^1}}\right),$$

where $F(\xi)$ is an explicitly constructed function satisfying $F(\xi) \to 0$ as $\xi \to +\infty$, together with the new estimate

$$\|u(\cdot,t) - \bar{u}\|_{L^1} \leq \|u_0(\cdot) - \bar{u}\|_{L^1} + C\,\delta\, t \qquad \text{(E)}$$

where δ denotes the supnorm of the initial data $u_0(\cdot)$. The details of the proof of this new estimate (E) together with a further discussion can be found in the author's paper [24].

§2 The structure of asymptotic wave patterns for (P).

We view (P) as modeling the polymer flood of an oil reservoir in one space dimension as first developed by Isaacson in [4]. By a polymer flood we mean a two component flow of immiscible fluids, oil and a mixture consisting of water together with polymer. The polymer is a thickener which moves passively with the water and which is assumed to affect the mutual flow of the two components in the porous media. Here, $s \equiv$ saturation of the aqueous phase, $c \equiv$ concentration of polymer in water, $0 \leq s \leq 1$, $0 \leq c \leq 1$, and $g(s,c) = \frac{f(s,c)}{s}$ is the particle velocity of the water. In this way (P1) represents conservation of water plus polymer, (P2) represents conservation of polymer, and $f(s,c)$ gives the fraction of the total flow associated with the aqueous component at each position x of the reservoir. The system is determined once the constitutive function $f(s,c)$ is specified. Properties of the flow are determined by quantitative properties of f, and we assume only that $f(\cdot,c)$ is S - shaped for each fixed c, and that $\frac{\partial f}{\partial c} < 0$. (See Fig. 1, cf. [4,20].) These assumptions can be justified by an argument based on Darcy's Law [4].

In this section we describe the structure of the noninteracting waves to which the solutions constructed in [20] decay as $t \to +\infty$. We then discuss the relationship between the admissible solution of a given Riemann problem (P),

$$u_0(x) = \begin{cases} u_L & \text{for } x < 0 \\ u_R & \text{for } x \geq 0, \end{cases} \qquad \text{(RP)}$$

and the asymptotic waves to which a given solution u(x,t) of (P) satisfying

$$u_0(-\infty) = u_L,\; u_0(+\infty) = u_R \qquad \text{(AS)}$$

decays as $t \to +\infty$. In this problem the admissible solution of the Riemann problem is the solution (shown in [4] to be unique) constructed from waves which satisfy the Lax characteristic criterion. Alternatively, these are the solutions which do not spontaneously introduce "extra" polymer into the flow over and above that accounted for in the states u_L and u_R. The noninteracting waves to which a general solution satisfying (AS) decays as $t \to +\infty$ represent an alternate solution of the Riemann problem (P), (RP) which in general is inadmissible by the Lax characteristic criterion. This is because the asymptotic state must account for the "extra" polymer contained in the initial data between $x = -\infty$ and $x = +\infty$. The conclusion then is that in contrast to the classical strictly hyperbolic theory, the asymptotic states do not depend on $u_L = u_0(-\infty)$ and $u_R = u_0(+\infty)$ alone, but on

$$c_{max} = \operatorname*{Sup}_x \{c_0(x)\}$$

as well. The analysis leads to the result that the solutions are not well-posed in the L^1-norm (i.e., $(\neg L^1)$ holds) even though the admissible solutions of the Riemann problem depend continuously on u_L and u_R in L^1_{loc}, and despite the fact that the solutions are Lipshitz continuous in time in the L^1-norm. Moreover, the two component flow interpretation indicates that the lack of well-posedness in one dimension may be related to fingering instabilities in higher dimensions. It also appears that well-posedness is retrieved when viscosity is not neglected. In this problem, the admissible solutions of the Riemann problem play an interesting and special role.

We first review the solution of the Riemann problem as first presented by Isaacson [4]. One can easily verify that the eigenvalues of dF (the wave speeds for system (P)) are given by

$$\lambda_s = \frac{\partial f}{\partial s}(s,c)\ , \qquad \lambda_c = \frac{f(s,c)}{s}\ ,$$

and the integral curves of the corresponding eigenvectors through a state $\bar{u}$ are given by

$$R\,(\bar{u}) = \{u: c(u) = c(\bar{u})\}\ , \qquad R\,(\bar{u}) = \{u: g(u) = g(\bar{u})\}\ .$$

Because $f(\cdot,c)$ is S - shaped, it is clear that $\lambda_s = \lambda_c$ on a curve in state space labeled T for the transition curve (see Fig. 1, 2).

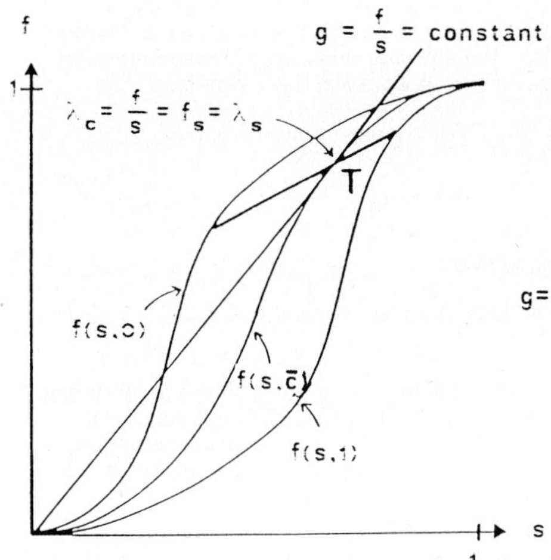

Fig. 1

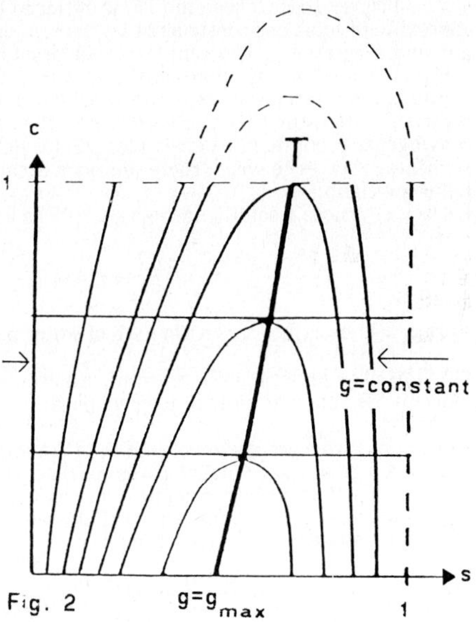

Fig. 2

For this system, the shock and rarefaction curves coincide, and the elementary waves which satisfy the Lax characteristic criterion consist of s-waves and c-waves. Here, s—waves solve the non-convex scalar conservation law which (P) reduces to when c = const., and c-waves are contact discontinuities at g = const. The Lax condition for the c-waves translates into the condition that c—waves cannot cross the Transition curve. The solution of the Riemann problem is summarized in the following theorems (see [4,8]).

Theorem (Is): For each u_L and u_R in the region $0 < s < 1$, $0 < c < 1$, there exists a unique solution to the Riemann problem (P), (RP) in the class of s-waves and c-waves. The solutions are diagramed in Figures 3 and 4. Moreover, these solutions depend continuously on u_L and u_R in L^1_{loc} at each time.

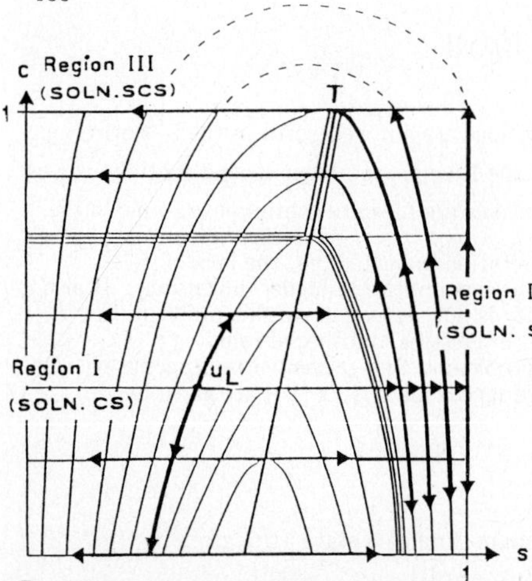

Fig. 3 Riemann problem solution for u_L left of T

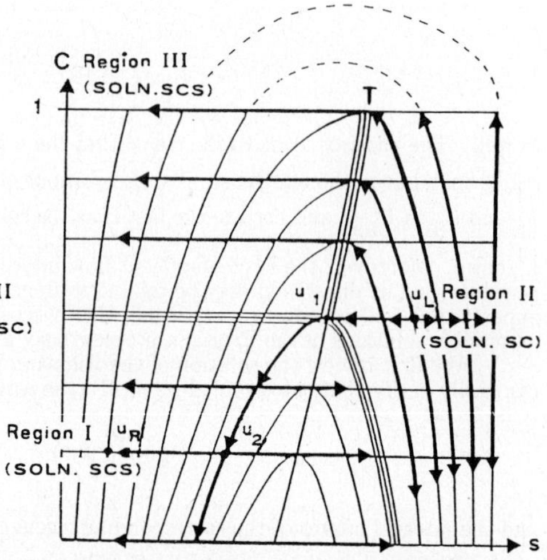

Fig. 4 Riemann problem solution for u_L right of T

The existence Theorem (Te) is obtained by extracting a convergent subsequence from approximate solutions constructed by the random choice method using the solutions of the Riemann problem generated in Theorem (Is). The proof relies on a positive non-increasing function F(t) which is defined on the approximate solutions, and which dominates the total variation of the approximate solutions at time t as measured in the singular coordinate system of Riemann invariants. Because the total variation in the conserved quantities cannot be bounded, a modified convergence proof must be given (see [20] for details). We now ask, what are the noninteracting elementary waves to which these solutions decay as $t \to +\infty$? We answer this by means of the following claim:

Let u(x,t) denote a solution generated by Theorem (Te). For a given u, let x(t) satisfy

$$\frac{dx}{dt} = g(s,c) \,, \qquad\qquad x(0) = x_0 \,,$$

so that x(t) describes a particle path of water in the solution.

CLAIM: The particle paths are continuous curves defined and nonintersecting for all $t \geq 0$, and the value of c is constant on each particle path.

We do not give a complete proof of the CLAIM, but we argue for it as follows. Since c-waves move with speed g, we argue first that the particle paths do not cross c-waves in the weak solutions. Since the particle paths are nonintersecting in smooth solutions and Lipshitz continuous across s-waves, we conclude that the particle paths are defined and nonintersecting throughout the weak solutions. Moreover, for smooth solutions,

$$\frac{dc(x(t),t)}{dt} = c_x\, x_t + c_t = c_t + g\, c_x = 0$$

because equations (P2) gives

$$\begin{aligned} 0 &= c_t\, s + c\, s_t + c\, f_x + f\, c_x \\ &= s\,(c_t + g\, c_x) + c\,(s_t + f_x) \\ &= s\,(c_t + g\, c_x)\,; \end{aligned}$$

and since c is constant across s-waves and we have argued that particle paths don't cross c-waves, we conclude that c is constant on particle paths of the weak solutions. An actual proof of this is made difficult by the fact that the claim is false for approximate solutions of the random choice method. We conclude from the claim that the total variation in c is passively transported along particle paths. Thus in particular, the value

$$\bar{c} = \sup_x c_0(x)$$

satisfies

$$\bar{c} = \sup_x u(x,t)$$

for every $t \geq 0$. We now determine the asymptotic waves through the following theorem:

Theorem (Is, Te): for each $\bar{c}$, u_L and u_R in our domain, there exists a unique set of noninteracting waves taking u_L to u_R, and taking on the value $\bar{c}$ as the maximum value of c at each time. In general, these waves correspond to an inadmissible solution of the Riemann problem. Moreover, the positive nonincreasing function F(t) used in the existence theory is minimized on these waves among all sequences of elementary waves taking u_L to u_R and taking on $\bar{c}$ as the maximum value of c. These waves are diagramed in Figures 5 – 9 according to whether u_L ties in regions A, B or C determined by the value of $\bar{c}$ (see Fig. 5).

We conclude from Theorem (Is, Te) that the solutions generated in Theorem (Te) decay to the noninteracting waves determined by $u_0(-\infty) \equiv u_L$, $u_1(+\infty) \equiv u_R$ and $\bar{c} = \operatorname{Max}_x c_0(x)$. A proof here would be complete were one to show rigorously that F(t) decreases to its minimum possible value in each solution.

In order to contrast the situation here with the classical strictly hyperbolic case, consider the example of the asymptotic state corresponding to the values $u_R \equiv u_L$ and $\bar{c} = c(\bar{u})$ diagramed in Figure 8, and corresponding to u_L in Region B. This is the region for which the structure of asymptotic states differ strikingly from the structure of asymptotic states in a strictly hyperbolic problem. For example, assume that the initial data is given by

$$u_0(x) = \begin{cases} u_L & x \leq 0 , \\ \bar{u} & 0 < x < \sigma , \\ u_L & x \geq \sigma . \end{cases}$$

The exact solution, which corresponds to the asymptotic state $u_0(-\infty) = u_L = u_0(+\infty)$, $\bar{c} = c(\bar{u})$, is drawn in Figure 10. In a strictly hyperbolic problem such a solution would decay to zero, because the admissible solution of the Riemann problem for $u_0(-\infty) = u_L = u_0(+\infty)$ is the constant solution $u \equiv u_L$ (cf [2, 11–14]). For (P), however, the solution decays to a solution containing two strong nonlinear s–waves separated by a contact discontinuity. We can now observe Theorem ($\neg L^1$) by taking the limit $\sigma \to 0$. Indeed, when $\sigma = 0$, the solution is the constant state $u \equiv u_L$, but for $\sigma > 0$ the solution at times $t > 0$ is far from the solution $u \equiv u_L$ in the L^1–norm. This occurs despite the Lipshitz continuity of the solutions in L^1. We conclude that a small amount of polymer at $x = 0$, $t = 0$ drastically alters the flow in this model.

The admissible solutions of the Riemann problem play a different role in the theory of this non–strictly hyperbolic problem than they play in the classical strictly hyperbolic theory of Lax. We explore this difference in the following comments.

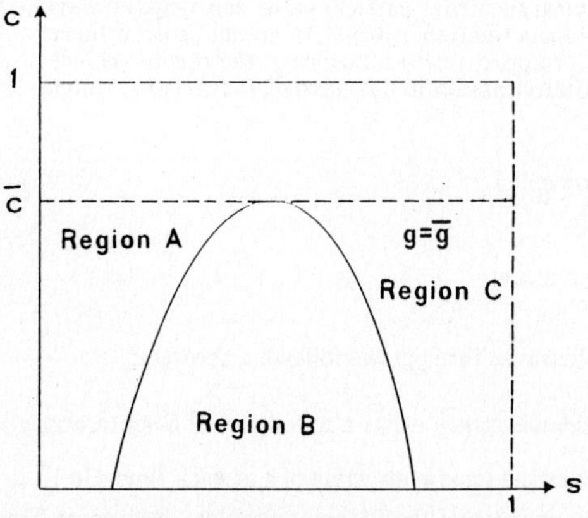

Fig. 5 The three regions for the asymptotic states corresponding to $\bar{c} = \max_x [c_O(x)]$, sc-plane

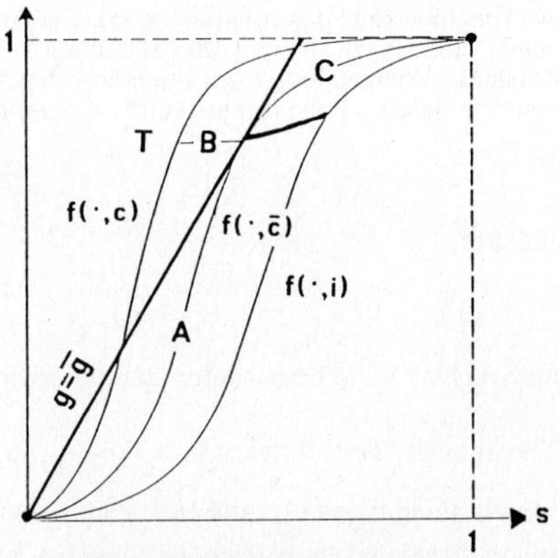

Fig. 6 The three regions for the asymptotic states corresponding to $\bar{c} = \max_x [c_O(x)]$, sf-plane

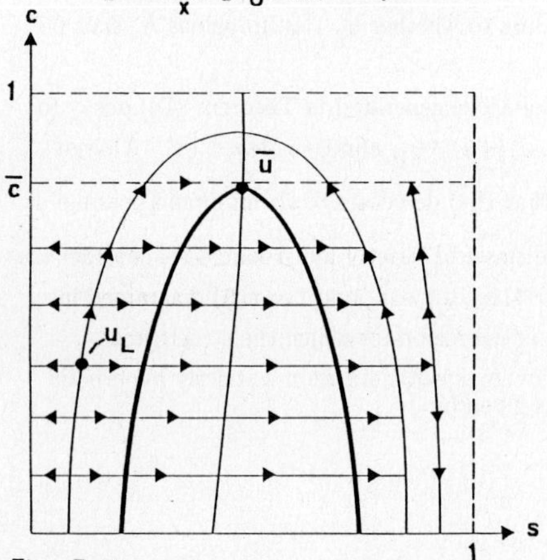

Fig. 7 The asymptotic states for $u_L \epsilon$ A

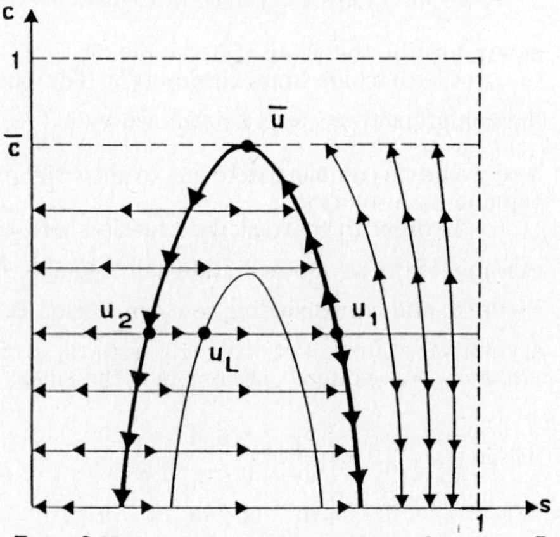

Fig. 8 The asymptotic state for $u_L \epsilon$ B

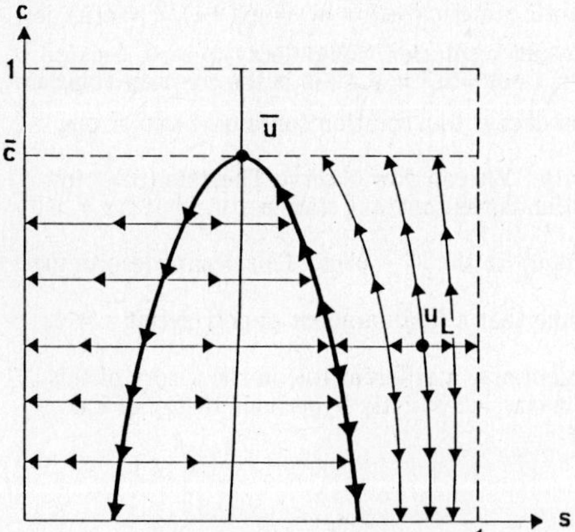

Fig. 9 The asymptotic state for $u_L \epsilon$ C

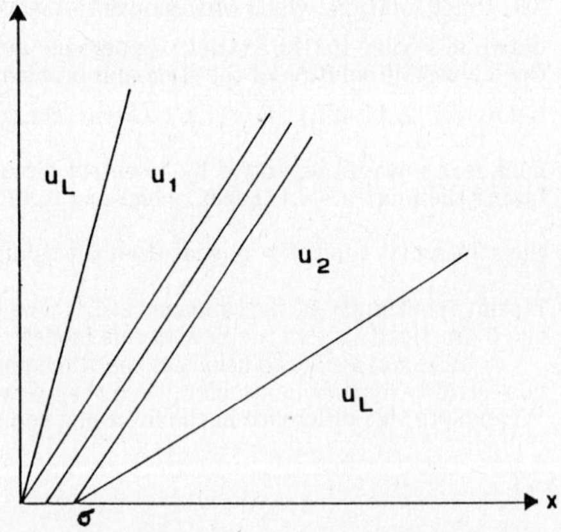

Fig. 10 The asymptotic state for u_L and $\bar{c}$ diagrammed in Fig. 8

Comments

(1) The classical strictly hyperbolic theory of conservation laws is a generalization of the theory of Euler's equations in gas dynamics. One can take the point of view that the Riemann problem is relevant because it represents the local asymptotic state in a general flow. By the scale invariance of the equations, the flow should locally look like an asymptotic state, and Glimm's theorem can be viewed as a justification of this picture; the random choice method replaces the solution locally by an asymptotic state. For system (P), the asymptotic solutions are not the admissible solution of the Riemann problem, but in fact one can speed up the convergence of the random choice method by replacing the solution of the Riemann problem by the asymptotic solution in each cell. Since the limit solution in this case conserves c–values, we expect this to generate the same solution as that generated in Theorem (P). The admissible solutions of the Riemann problem are special in that all asymptotic wave structures are concatenations of these. Thus the admissible solutions can be characterized as the only solutions of the Riemann problem which give convergence to the polymer conserving solution by the random choice method, but which require only the values of u_L and u_R in each cell, and not the further information of $\bar{c}$.

(2) From the example above, it appears that continuous dependence in L^1 is recovered when diffusion is not neglected. For example, if ϵu_{xx} is added to the right hand side of (P), then we expect the spike in Fig. 10 to diffuse away as $t \to +\infty$, and the solution to decay to the constant state $u \equiv u_L$. Moreover the rate of decay would increase as $\sigma \to 0$, so we expect continuous dependence in L^1 as $\sigma \to 0$.

(3) We believe that the weak solutions generated by Theorem (Te) are limits of the viscously perturbed equations as $\epsilon \to 0$. If this is indeed the case (we have no proof), then we can also characterize the admissible solutions of the Riemann problem as follows: Let $u^\epsilon(x,t)$ denote a solution of the initial value problem for the viscous equation

$$u_t + f(u)_x = \epsilon u_{xx}, \qquad (P^\epsilon)$$

where u and f are given in (P). Let Q_1 and Q_2 denote the asymptotic states defined by

$$Q_1 \equiv \lim_{t\to\infty} \lim_{\epsilon\to 0} u^\epsilon, \qquad Q_2 \equiv \lim_{\epsilon\to 0} \lim_{t\to\infty} u^\epsilon . \qquad (*)$$

If solutions of (P) are limits of solutions of (P^ϵ) as $\epsilon \to 0$, then Q_1 is the actual asymptotic solution determined by $u_0(-\infty) = u_L$, $u_0(+\infty) = u_R$ and $\bar{c} = \text{Max}\, c_0(x)$. However, our example indicates that the limit Q_2 should be the admissible solution of the Riemann problem $[u_L, u_R]$. In this case, the admissible solutions of the Riemann problem are special because $Q_1 \equiv Q_2$ only when the asymptotic state is the admissible solution of the Riemann problem. Thus the admissible solutions are the ones for which it is valid to interchange the limits in (*). (This comment was suggested to the author by Philip Collela of Lawrence Livermore Laboratories).

(4) In the polymer flood interpretation of (P) it is clear that the narrow "spike" in the example of Figure 10 is unstable to figuring in higher dimension. We wonder whether a lack of continuous dependence corresponds to the presence of higher dimensional instabilities in some general setting.

In conclusion, we comment that system (P) probably represents the simplest setting in which one finds a singular hyperbolic problem. It is surprising that one can give an almost complete analysis of the initial value problem in this case. We hope that this study of the Riemann problem and the structure of asymptotic states can help to shed light on the role of admissibly criteria and the non–uniqueness of Riemann problem solutions in more complicated problems in which strict hyperbolicity is lost.

References

[1] Courant, R. and Friedrichs, K. O. *Supersonic Flow and Shock Waves*, Wiley, New York, 1948.

[2] DiPerna, R., "Decay and asymptotic behavior of solutions to nonlinear hyperbolic systems of conservation laws", *Ind. Univ. Math. J.*, *24*, 1975, pp. 1047–1071.

[3] Glimm. J., "Solutions in the large for nonlinear hyperbolic systems of equations", *Comm. Pure Appl. Math.*, 18, 1965, pp. 697–715.

[4] Isaacson, E., "Global solution of a Riemann problem for a non–strictly hyperbolic system of conservation laws arising in enhanced oil recovery", *J. Comp. Phys.*, to appear.

[5] Isaacson, E. and Temple, B., "The structure of asymptotic states in a singular system of conservation laws" Davis preprint.

[6] Isaacson, E., Marchesin, D. Plohr, B., and Temple, B., "The classification of solutions of quadratic Riemann problems (I), Joint MRC, PUC/RJ Report, 1985.

[7] Isaacson, E. and Temple, B., "Examples and classification of non–strictly hyperbolic systems of conservation laws", *Abstracts of AMS*, January 1985. Presented in the Special Session on "Non–Strictly Hyperbolic Conservation Laws" at the Winter Meeting of AMS, Anaheim, January 1985.

[8] Keyfitz, B. and Kranzer, H., "A system of non–strictly hyperbolic conservation laws arising in elasticity theory", *Arch. Rat. Mech. Anal.*, 72, 1980.

[9] Lax, P. D., "Hyperbolic systems of conservation laws, II", *Comm. Pure Appl. Math.*, 19, 1957, pp. 537–566.

[10] Lax, P. D., "Shock waves and entropy", in *Contributions to Nonlinear Functional Analysis*, E. H. Zarantonello, editor, Academic Press, New York, 1971, pp. 634–643.

[11] Liu, T.–P., "Invariants and asymptotic behavior of solutions of a conservation law", preprint.

[12] Liu, T.–P., "Asymptotic behavior of solutions of general systems of nonlinear hyperbolic conservation laws", *Indiana Univ. J.*, to appear.

[13] Liu, T.–P., "Decay to N–waves of solutions of general systems of nonlinear hyperbolic conservation laws", *Comm. Pure Appl. Math.*, 30, 1977, pp. 585–610.

[14] Liu, T.–P., "Large–time behavior of solutions of initial and initial–boundary value problems of general systems of hyperbolic conservation laws", *Comm. Math. Phys.*, 1977.

[15] Schaeffer, D. G. and Shearer, M., "The classification of 2×2 systems of non–strictly hyperbolic conservation laws, with application to oil recovery", with Appendix by D. Marchesin, P. J. Paes–Leme, D. G. Schaeffer, and M. Shearer, preprint, Duke University.

[16] Serre, D., "Existence globale de solutions faibles sous une hypothese unilaterale pour un systeme hyperbolique non linaere", *Equipe d'Analyse Numerique*, Lyon, Saint–Etienne, July 1985.

[17] Serre, D., "Solutions a variation bornees pour certains systemes hyperboliques de lois de conservation", Equip d'Analyse Numerique, Lyon, Saint–Etienne, February 1985.

[18] Shearer, M., Schaeffer, D. G., Marchesin, D., and Paes–Leme, P. J., "Solution of the Riemann problem for a prototype 2×2 system of non–strictly hyperbolic conservation laws", preprint, Duke University.

[19] Smoller, J. A., *Shock Waves and Reaction Diffusion Equations*, Springer–Verlag, 1980.

[20] Temple, B., "Global solution of the Cauchy problem for a class of 2×2 non–strictly hyperbolic conservation laws", *Adv. Appl. Math.*, 3, 1982, pp. 335–375.

[21] Temple, B., "Systems of conservation laws with coinciding shock and rarefaction curves", *Contemporary Mathematics*, 17, 1983.

[22] Temple, B., "Decay with a rate for non–compactly supported solutions of conservation laws", *Trans. of Am. Math. Soc.*, Vol. 298, No. 1, November 1986.

[23] Temple, B., "Degenerate systems of conservation laws", *Contemporary Mathematics*, Vol. 60, 1987.

[24] Temple, B., "Weak stability in the global L^1–norm for systems of conservation laws".

[25] Temple, B., "Supnorm estimates in Glimm's method", MRC Technical Summary Report #2855

[26] Temple, B., "No L^1–contractive metrics for systems of conservation laws" Trans. Am. Math Soc., Vol. 288, No. 2, April 1985.

THE RIEMANN PROBLEM WITH UMBILIC LINES FOR WAVE PROPAGATION IN ISOTROPIC ELASTIC SOLIDS

T. C. T. Ting

Department of Civil Engineering, Mechanics and Metallurgy

University of Illinois at Chicago

Box 4348, Chicago, IL 60680, USA

SUMMARY

The governing equations for plane waves in isotropic elastic solids are a 6x6 system of hyperbolic conservation laws. It is shown that, for the Riemann problem, the system is equivalent to a prototype 2x2 system of hyperbolic conservation laws. We then discuss simple wave curves, shock curves, umbilic points and umbilic lines for systems which possess a potential. This is the case for hyperelastic materials which possess a complementary strain energy. For the third-order hyperelastic materials it is shown that one may have (i) a straight umbilic line, or (ii) an umbilic curve which is a hyperbola, parabola or ellipse. The case (ii) is studied in detail here.

INTRODUCTION

In a fixed rectangular coordinate system x_1, x_2, x_3, consider a plane wave propagating in the x_1-direction. Let σ, τ_1, τ_2 be, respectively, the normal and two shear stresses on the plane x_1 = constant. Likewise, let u, v_1, v_2 be the particle velocity in the x_1-, x_2-, x_3-direction, respectively. The equations of motion and the continuity of displacement can be written as a 6x6 system of hyperbolic conservation laws [1,2],

$$\left.\begin{aligned} &\underset{\sim}{U}_x - \underset{\sim}{F}(\underset{\sim}{U})_t = \underset{\sim}{0}\,, \\ &\underset{\sim}{U} = (\sigma,\tau_1,\tau_2,u,v_1,v_2)\,, \\ &\underset{\sim}{F}(\underset{\sim}{U}) = (\rho u,\rho v_1,\rho v_2,\varepsilon,\gamma_1,\gamma_2)\,. \end{aligned}\right\} \tag{1}$$

In the above, $x = x_1$, t is the time, ρ is the mass density in undeformed state, ε, γ_1, γ_2 are, respectively, the normal strain and the two shear strains. For isotropic elastic materials, σ, τ_1, τ_2 are functions of ε, γ_1, γ_2 and

$$\tau_1/\tau_2 = \gamma_1/\gamma_2\,.$$

Hence we may let

$$\left.\begin{aligned} \tau_1 &= \tau\cos\theta\ , & \tau_2 &= \tau\sin\theta\ , \\ \gamma_1 &= \gamma\cos\theta\ , & \gamma_2 &= \gamma\sin\theta\ , \end{aligned}\right\} \tag{2}$$

in which τ and γ are, respectively, the total shear stress and shear strain. If we regard (τ_1,τ_2,σ) as a rectangular coordinate system in the stress space, (τ,θ,σ) would be a cylindrical coordinate system.

EQUIVALENT TO A 2x2 SYSTEM

It is shown in [1,2] that the characteristic wave speeds of (1) are $\pm c_i$, $i = 1,2,3$. The simple waves associated with c_2 are <u>circularly polarized</u> because along the c_2 wave curve σ and τ are constant while θ is variable. On the other hand, the simple wave curves associated with c_1 and c_3 are <u>plane polarized</u>. Along the c_1 or c_3 simple wave curve θ is a constant. (See also [3]). The system is linearly degenerate with respect to c_2. Therefore, the simple wave associated with c_2 is in fact a shock wave. Thus, for the Riemann problem, it is sufficient to consider wave curves associated with c_1 and c_3 only. Without loss of generality, we consider wave curves on the radial plane $\theta = 0$.

With $\theta = 0$ in (2), (1) reduces to a 4x4 system

$$\left.\begin{aligned} &\underset{\sim}{U}_x - \underset{\sim}{F}(\underset{\sim}{U})_t = \underset{\sim}{0}\ , \\ &\underset{\sim}{U} = (\sigma,\tau,u,v)\ , \\ &\underset{\sim}{F}(\underset{\sim}{U}) = (\rho u,\rho v,\varepsilon,\gamma)\ . \end{aligned}\right\} \tag{3}$$

We consider the domain of σ, τ in which the stress-strain laws are invertable. We then consider ε, γ to be given functions of σ and τ. In fact ε is an even function of τ while γ is an odd function of τ [1]. For the Riemann problem [4-7], $\underset{\sim}{U}$ depends on one parameter $\lambda = x/t$ only and $(3)_1$ reduces to

$$(\underset{\sim}{I} + \lambda\underset{\sim}{A})\underset{\sim}{U}' = \underset{\sim}{0}\ , \tag{4}$$

$$\underset{\sim}{A} = \begin{bmatrix} \underset{\sim}{0} & \rho\underset{\sim}{I} \\ \underset{\sim}{G} & \underset{\sim}{0} \end{bmatrix}\ , \qquad \underset{\sim}{G} = \begin{bmatrix} \varepsilon_\sigma & \varepsilon_\tau \\ \gamma_\sigma & \gamma_\tau \end{bmatrix}\ , \tag{5}$$

where the prime denotes differentiation with λ, the subscripts σ and τ denote partial differentiation with these variables and $\underset{\sim}{I}$ is an identity matrix. Equation (4) tells us that $\lambda = c$ is the characteristic wave speed and $\underset{\sim}{U}'$ is the associated right eigenvector. The four equations in (4) can be reduced to two by eliminating u' and v' components of $\underset{\sim}{U}'$. We have

$$(\underset{\sim}{G} - \xi\underset{\sim}{I})\underset{\sim}{s}' = \underset{\sim}{0}\ , \tag{6}$$

$$\underset{\sim}{s} = (\sigma,\tau) , \qquad \xi = (\rho c^2)^{-1} \tag{7}$$

If $\lambda = x/t$ is the shock wave, $\underset{\sim}{U}$ is discontinuous at λ and $(3)_1$ is replaced by

$$[\underset{\sim}{U}] + \underset{\sim}{V}[\underset{\sim}{F}(\underset{\sim}{U})] = \underset{\sim}{0} ,$$

in which [•] denotes the discontinuity across the shock wave and V is the shock wave speed. Again, elimination of the [u], [v] components of $[\underset{\sim}{U}]$ leads to

$$[\underset{\sim}{p}] - \eta[\underset{\sim}{s}] = \underset{\sim}{0} , \tag{8}$$

$$\underset{\sim}{p} = (\varepsilon,\gamma) , \qquad \eta = (\rho V^2)^{-1} . \tag{9}$$

Equations (6) and (8) are identical to the Riemann problem for the 2x2 hyperbolic system [8-11]

$$\left.\begin{aligned} &\underset{\sim}{U}_t + \underset{\sim}{F}(\underset{\sim}{U})_x = \underset{\sim}{0} , \\ &\underset{\sim}{U} = \underset{\sim}{s} = (\sigma,\tau) , \qquad \underset{\sim}{p} = (\varepsilon,\gamma) . \end{aligned}\right\} \tag{10}$$

Therefore, the 4x4 system of (3) and the 2x2 system of (10) are mathematically identical. The only difference is that if c and V are, respectively, the characteristic wave speed and the shock wave speed for (3), the corresponding quantities for (10) are ξ and η.

Before we close this section, we present below the characteristic wave speeds c of (3) which is related to ξ through $(7)_2$:

$$\left.\begin{aligned} \xi_1 &= \frac{1}{2}\left\{(\varepsilon_\sigma + \gamma_\tau) - Y\right\} = (\rho c_1^2)^{-1} , \\ \xi_3 &= \frac{1}{2}\left\{(\varepsilon_\sigma + \gamma_\tau) + Y\right\} = (\rho c_3^2)^{-1} , \\ Y &= \left\{(\varepsilon_\sigma - \gamma_\tau)^2 + 4\varepsilon_\tau\gamma_\sigma\right\}^{1/2} . \end{aligned}\right\} \tag{11}$$

Assuming that c_1 and c_3 are real, we have

$$c_1 \geq c_3 \geq 0 .$$

For hyperelastic materials, there exists a complementary strain energy $W(\sigma,\tau)$ [12] such that the strains ε and γ are obtained by differentiating W with respect to σ and τ, i.e.,

$$\varepsilon = W_\sigma , \qquad \gamma = W_\tau . \tag{12}$$

The matrix $\underset{\sim}{G}$ in $(5)_2$ now becomes

$$\underset{\sim}{G} = \begin{bmatrix} W_{\sigma\sigma} & W_{\sigma\tau} \\ W_{\sigma\tau} & W_{\tau\tau} \end{bmatrix} \tag{13}$$

which is symmetric. The eigenvalues of $\underset{\sim}{G}$ are, from (6) or (11),

$$\left.\begin{aligned} \xi_1 &= \frac{1}{2}\left\{(W_{\sigma\sigma} + W_{\tau\tau}) - Y\right\} = (\rho c_1^2)^{-1} , \\ \xi_3 &= \frac{1}{2}\left\{(W_{\sigma\sigma} + W_{\tau\tau}) + Y\right\} = (\rho c_3^2)^{-1} , \\ Y &= \left\{(W_{\sigma\sigma} - W_{\tau\tau})^2 + 4W_{\sigma\tau}^2\right\}^{1/2} . \end{aligned}\right\} \tag{14}$$

Substituting $\underset{\sim}{G}$ and ξ of (13) and (14) into (6), the differential equation for simple wave curves is given by [13]

$$\frac{d\sigma}{d\tau} = \frac{(W_{\sigma\sigma} - W_{\tau\tau}) \mp Y}{2W_{\sigma\tau}} = \frac{2W_{\sigma\tau}}{-(W_{\sigma\sigma} - W_{\tau\tau}) \mp Y} , \tag{15}$$

in which the upper (or lower) sign is for the c_1 (or c_3) simple wave curve. The equation for shock wave is, from (8),

$$\eta = \frac{[W_\sigma]}{[\sigma]} = \frac{[W_\tau]}{[\tau]} . \tag{16}$$

To insure that c_1 and c_3 of (14) are real, $\underset{\sim}{G}$ must be positive definite, i.e.,

$$W_{\sigma\sigma} > 0 , \qquad W_{\sigma\sigma}W_{\tau\tau} - W_{\sigma\tau}^2 > 0 .$$

We notice from the theory of differential geometry that the eigenvectors of $\underset{\sim}{G}$ are lines of curvature of the surface $z = W(\sigma,\tau)$. By (6), this means that the simple wave curves on the (σ,τ) plane are lines of curvature of the surface $z = W(\sigma,\tau)$ [8].

At an umbilic point, $c_1 = c_3$ and, by (14), we have

$$W_{\sigma\tau} = 0 , \qquad W_{\sigma\sigma} = W_{\tau\tau} . \tag{17}$$

When (17) hold, the eigenvalue ξ of $\underset{\sim}{G}$ is $\xi = W_{\sigma\sigma} = W_{\tau\tau}$ and the eigenvector of $\underset{\sim}{G}$ is arbitrary. Therefore, at an umbilic point, the

the simple wave curve can be directed towards any direction. It follows from this argument that, if we have an umbilic line on which (17) hold, the umbilic line is itself a simple wave curve.

UMBILIC LINES FOR THIRD-ORDER HYPERELASTIC MATERIALS

If we expand W in powers of σ and τ of order up to four, noticing that the constant terms and the linear terms have no effects on simple wave curves and that W is even in τ, we have

$$W = \frac{a}{2}\sigma^2 + \frac{d}{2}\tau^2 + \frac{b}{6}\sigma^3 + \frac{e}{2}\sigma\tau^2 + \frac{1}{12}\delta_1\sigma^4 + \frac{1}{12}\delta_2\tau^4 + \frac{1}{2}\delta_3\sigma^2\tau^2 , \quad (18)$$

in which $a, d, b, e, \delta_1, \delta_2, \delta_3$ are constants. The stress-strain laws are then obtained from (12) and we see that a and d are the elastic constants for linear materials, b and e are the second order material constants while $\delta_1, \delta_2, \delta_3$ are the third order constants. The wave curves for the second order materials, i.e., for the case when $\delta_1 = \delta_2 = \delta_3 = 0$ have been discussed in [1] for special initial and boundary conditions and in [2] for arbitrary combinations of initial and boundary conditions. There is one umbilic point at which $c_1 = c_3$ and, if $e = 0$, there is one umbilic line [1]. For the third order materials given by (18), the existence of more umbilic points and umbilic lines are discussed in [2] but the characteristics of the umbilic lines and the wave curves associated with the umbilic lines are not studied.

With W given by (18), (17) lead to

$$\left.\begin{aligned} &\tau(e + 2\delta_3\sigma) = 0 , \\ &a + b\sigma + \delta_1\sigma^2 + \delta_3\tau^2 = d + e\sigma + \delta_3\sigma^2 + \delta_2\tau^2 . \end{aligned}\right\} \quad (19)$$

For the umbilic lines, both equations must be satisfied for a one parameter family of points. Equation $(19)_1$ is satisfied if (a) $\tau = 0$, (b) $\delta_3 \neq 0$ and $\sigma = -e/2\delta_3$, or (c) $\delta_3 = 0$ and $e = 0$. Consideration of $(19)_2$ shows that only (b) and (c) lead to an umbilic line. They are:

(i) $\delta_3 = \delta_2 \neq 0$, and, letting $\sigma_* = -e/2\delta_3$,

$$a + b\sigma_* + \delta_1\sigma_*^2 = d + e\sigma_* + \delta_3\sigma_*^2 .$$

We have an umbilic line at $\sigma = \sigma_*$, τ arbitrary.

(ii) $\delta_3 = 0$ and $e = 0$. We have the umbilic line given by

$$a + b\sigma + \delta_1\sigma^2 = d + \delta_2\tau^2 . \quad (20)$$

Assuming that $\delta_2 \neq 0$, this is a parabola, hyperbola, or ellipse depending on whether $\delta_1\delta_2$ is zero, positive, or negative, respectively. The case (ii) appears more interesting and is discussed below.

WAVE CURVES FOR $\delta_3 = 0$ AND $e = 0$

When $\delta_3 = 0$ and $e = 0$, we have the umbilic lines given by (20). The stress-strain laws are, in this case,

$$\varepsilon = a\sigma + \frac{b}{2}\sigma^2 + \frac{1}{3}\delta_1\sigma^3 , \tag{21}$$

$$\gamma = d\tau + \frac{1}{3}\delta_2\tau^3 . \tag{22}$$

We see that the normal stress-strain law and the shear stress-strain law are uncoupled. Thus, instead of a 2x2 system we have two 1x1 systems. The simple wave curves are therefore straight lines parallel to the σ-axis or the τ-axis. However, the wave speed associated with σ is not always c_1, the fast wave speed. Likewise the wave speed associated with τ is not always c_3, the slow wave speed.

We first examine the stress-strain law for shear given by (22) which is shown in Fig. 1. For $\delta_2 < 0$, τ_o is the point at which the wave speed becomes infinity. For $|\tau| > \tau_o$ the wave speed becomes imaginary and hence the range of validity for $\delta_2 < 0$ is limited to $|\tau| < \tau_o$.

We next examine the stress-strain law for the normal stress given by (21). We see that if we change the sign of b, σ and ε, (21) remains the same. Therefore we consider

$$b > 0$$

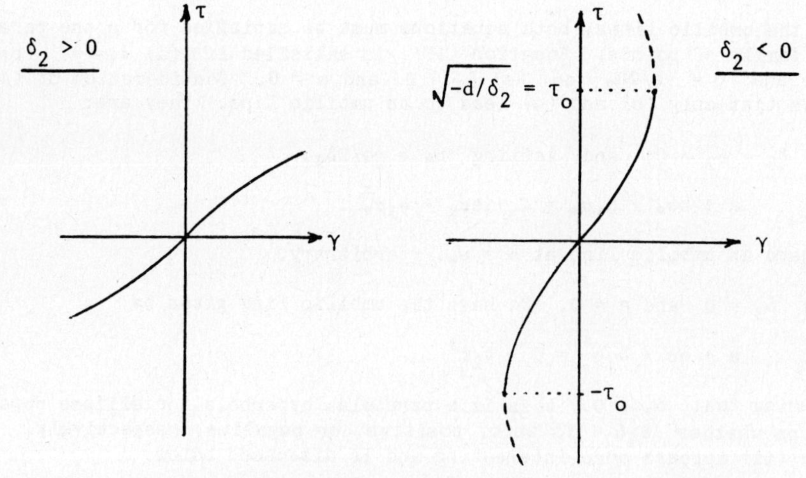

Fig. 1 Stress-strain curves for shear: $\gamma = d\tau + \frac{1}{3}\delta_2\tau^3$

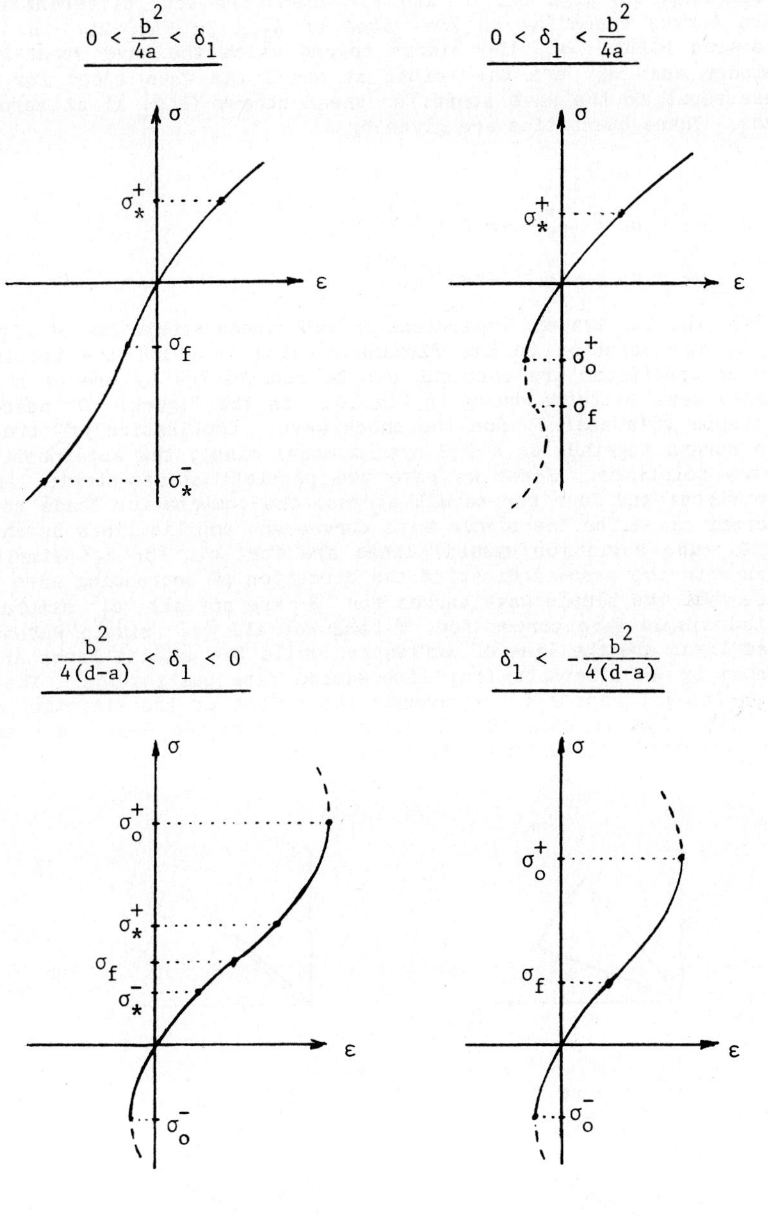

Fig. 2 Stress-strain curves for normal stress:

$$\varepsilon = a\sigma + \frac{b}{2}\sigma^2 + \frac{1}{3}\,\delta_1\sigma^3\,, \quad b > 0.$$

only because the case for $b < 0$ can be obtained from that for $b > 0$ by reversing the sign of σ and ε. There are four different stress-strain curves depending on the value of δ_1. In Fig. 2, σ_f is the reflection point, $\sigma_o^{\pm}$ the limits beyond which the wave speed becomes imaginary and $\sigma_*^{\pm}$ are the points at which the wave speed for normal stress equal to the wave speed for shear stress (Fig. 1) at zero shear stress. These quantities are given by

$$\sigma_f = -b/2\delta_1 ,$$

$$\sigma_o^{\pm} = \sigma_f \pm \{\sigma_f^2 - (a/\delta_1)\}^{1/2} ,$$

$$\sigma_*^{\pm} = \sigma_f \pm \{\sigma_f^2 + (d-a)/\delta_1\}^{1/2} .$$

For the 1x1 system, regardless of the stress-strain law of Fig. 1 or Fig. 2, the solution to the Riemann problem in which the initial and boundary conditions are constant can be represented by one of the four possible wave patterns shown in Fig. 3. In the figure, C stands for the simple wave and V for the shock wave. Combination of normal and shear stress together as a 2x2 system means simply the superposition of the two solutions. Since we have two possible stress-strain laws for shear stress and four for normal stress, the combination leads to eight different cases for the simple wave curves and umbilic lines as shown in Fig. 4. The solid (or dashed) lines are for c_1 (or c_3) simple wave curves with the arrow indicating the direction of decreasing wave speed. We see that the simple wave curves for σ are not all c_1 simple waves and the simple wave curves for τ are not all c_3 simple waves. The dotted lines are the line of inflection while the umbilic lines are represented by an alternate long line/dashed line combination. The value τ_* in Cases I^- and III^+ represents the height of the elliptic umbilic line while that in Case IV^- represents the distance from the σ-axis to the hyperbolic umbilic line. It is given by

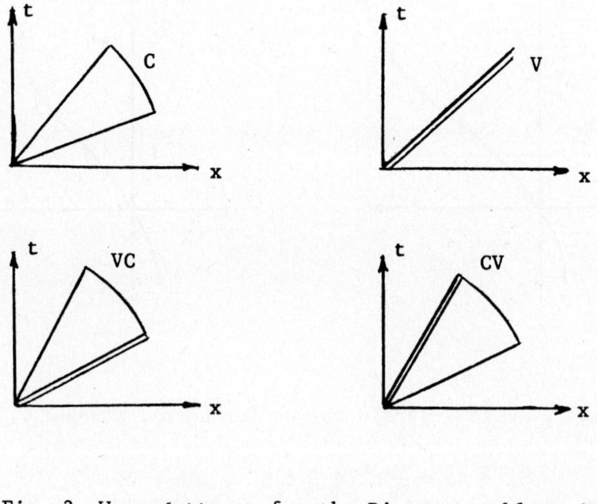

Fig. 3 Wave patterns for the Riemann problem with stress-strain law given by Fig. 1 or Fig. 2.

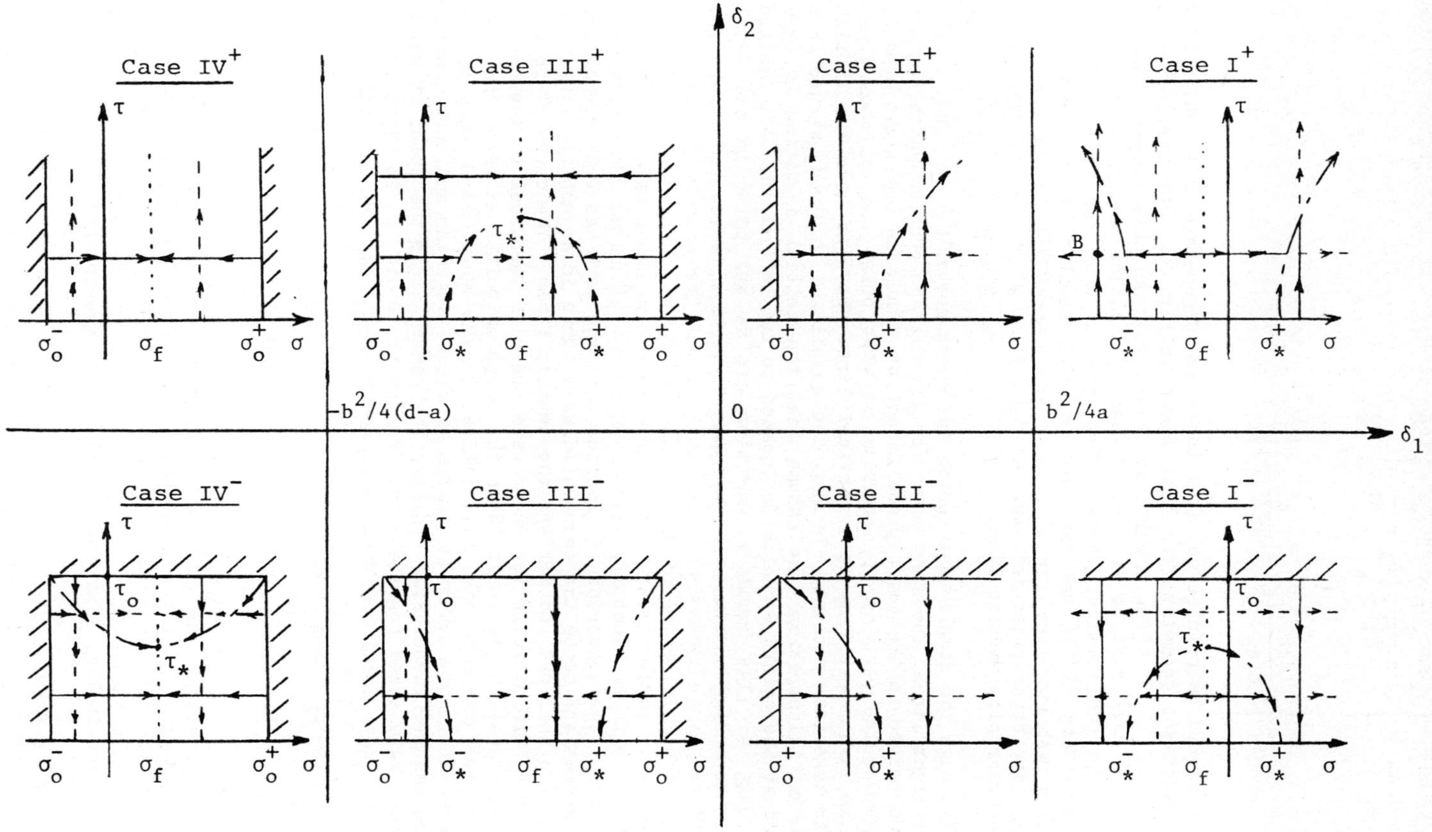

Fig. 4 Simple wave curves and umbilic lines for third-order hyperelastic materials with $\delta_3 = 0$ and $e = 0$.

$$\tau_* = \left\{\tau_o^2 + \frac{a}{\delta_1\delta_2}\left(\delta_1 - \frac{b^2}{4a}\right)\right\}^{1/2}$$

$$= \left\{-\frac{(d-a)}{\delta_1\delta_2}\left(\delta_1 + \frac{b^2}{4(d-a)}\right)\right\}^{1/2} .$$

It should be pointed out that there are degenerate cases which have quite different geometry from that shown in Fig. 4. For instance, for $\delta_2 = 0$ and $\delta_1 > b^2/4a$, which is the degenerate case between Case I^+ and Case I^- (and will be denoted by Case I^+/I^-), the umbilic line degenerates into two vertical lines. The same is true for Case III^+/III^- while for Case II^+/II^- we have one vertical umbilic line. There is a parabolic umbilic line for $\delta_1 = 0$. For Case III^-/IV^-, however, the umbilic line degenerates into two straight lines which are the asymptotes of the hyperbolic umbilic line.

Although the simple wave curves and shock curves for σ and τ are all straight lines, the wave curve for the Riemann problem which connect any two points in the (σ,τ) plane is not as simple as one might expect. The complicated part is to determine at what point one switches from, say, the horizontal line to a vertical line. We also have the umbilic line to take into account. To show the complicacy involved, we use Case I^+ for an illustration. We assume that the initial condition is prescribed at the point B which is located to the left of the left umbilic line, Fig. 4. The boundary condition can be any point in the (σ,τ) plane. Depending on where the boundary condition is located, there are 14 different wave curves as shown in Fig. 5. The associated wave patterns in the (x,t) plane are also shown in Fig. 5. As in Fig. 3, C and V stand for, respectively, the simple waves and shock waves. The subscript σ and τ identify whether the wave is normal stress wave or shear wave. The subscript $*$ implies that the wave has both normal and shear stress and hence the associated wave curve is an umbilic line. In the (σ,τ) plane, QPE and SRF are the umbilic lines. The curve PT is the locus of points at which the shear wave speed is identical to the normal shock wave speed. The curve RN is the locus of points at which the wave speed of the normal stress is identical to the shear shock wave speed. The curve TM is the locus of points at which the shear and normal shock wave speeds are identical. Finally, the point L is the point at which the shock wave speed for the shock from point B to L is identical to the normal wave speed for the normal stress at L. One can show that

$$\sigma_f - \sigma_B = 2(\sigma_L - \sigma_f) ,$$

where σ_f is the stress at the inflection point.

If point B is at a different place, the wave curves will have a different geometry. One could also consider wave curves for other cases. What is interesting is that the wave patterns are, if one considers the uncoupled 1x1 systems, simple and are given in Fig. 3 but become complicated if one considers the 2x2 system. Also, the shock wave for the 1x1 systems shown in Fig. 3 is the Lax shock [4], meaning that

$$c^+ \leq V \leq c^- ,$$

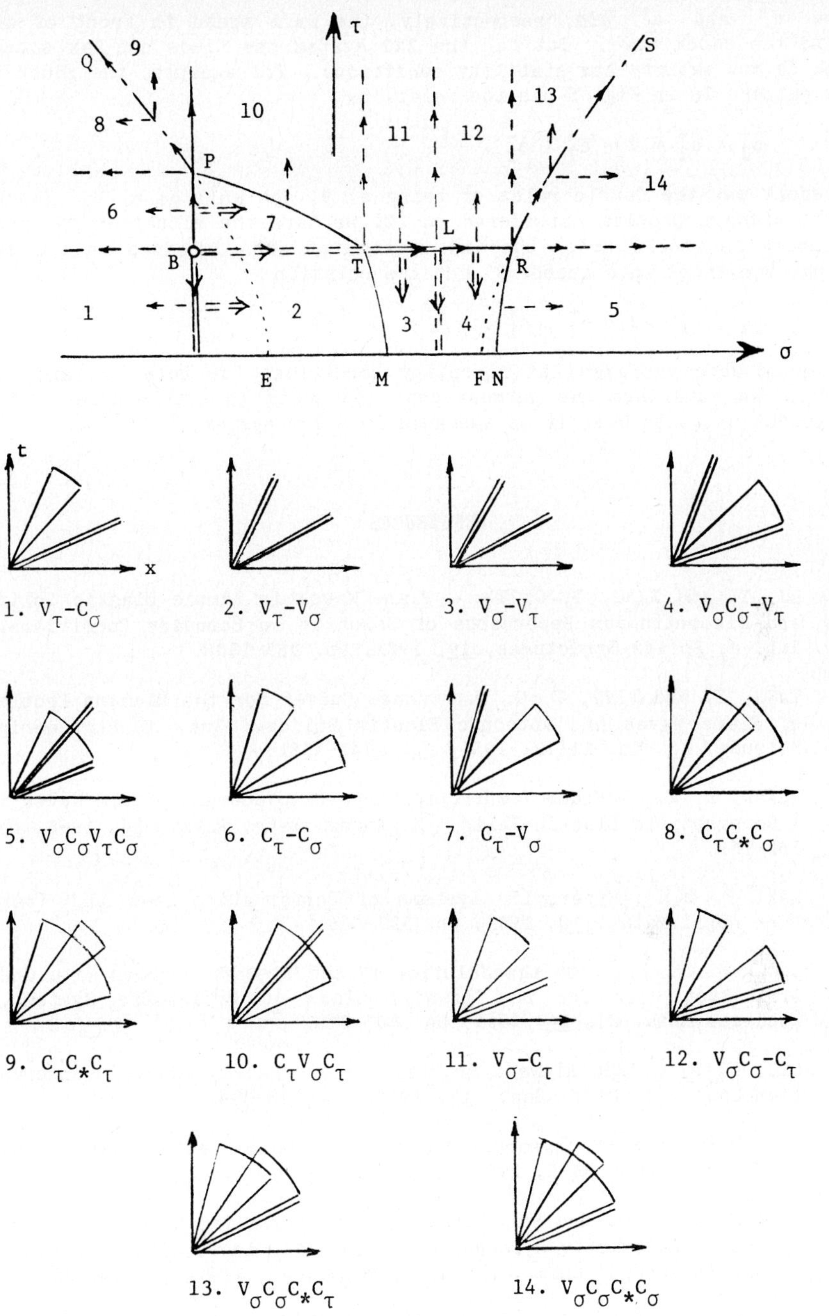

Fig. 5 Wave curves and wave patterns for Case I^+ for the Riemann problem for a fixed initial condition given by the point B in the (σ, τ) plane with arbitrary boundary condition.

where c^+ and c^- are, respectively, the wave speed in front of and behind the shock wave. But for the 2x2 system one finds non-Lax shocks which do not satisfy Lax stability conditions. For example, the shock in wave pattern 10 in Fig. 5 has the relations

$$c_3^+ < c_1^+ = V = c_3^- < c_1^- .$$

The shock has the double roles of being a V_1 as well as a V_3 shock. In the Riemann problem considered in [2] we have the situation in which the shock is stable under the 1x1 system but when imbedded in the 2x2 system, the shock wave speed V has the relation

$$c_3^- , c_3^+ < V < c_1^- , c_1^+ .$$

The shock wave violates Lax stability conditions for both V_1 and V_3 shocks. We have thus the paradox that the shock is stable under a 1x1 system but unstable when it is imbedded in a 2x2 system.

REFERENCES

[1] LI, Y. and TING, T. C. T.: 'Plane Waves in Simple Elastic Solids and Discontinuous Dependence of Solution on Boundary Conditions,' Int. J. Solids Structures, 19, 1983, pp. 989-1008.

[2] TANG, Z. and TING, T. C. T.: 'Wave Curves for the Riemann Problem of Plane Waves in Isotropic Elastic Solids,' Int. J. Engineering Science, 25, No. 11/12, 1987, pp. 1343-1381.

[3] BLAND, D. R.: 'Plane Isentropic Large Displacement Simple Waves in a Compressible Elastic Solid,' Z. Angew. Math. Phys., 16, 1965, pp. 752-769.

[4] LAX, P. D.: 'Hyperbolic Systems of Conservation Laws II,' Comm. Pure Appl. Math., 10, 1957, pp. 537-566.

[5] SMOLLER, J. A.: 'On the Solution of the Riemann Problem With General Step Data for an Extended Class of Hyperbolic Systems,' Michigan Math. J., 16, 1969, pp. 201-210.

[6] LIU, T. P.: 'The Riemann Problem for General Systems of Conservation Laws,' J. Diff. Eqs., 18, 1975, pp. 218-234.

[7] DAFERMOS, C. M.: 'Hyperbolic Systems of Conservation Laws,' Brown Univ. Rept. LCDS 83-5, 1983.

[8] SCHAEFFER, D. G. and SHEARER, M.: 'The Classification of 2x2 Systems of Non-Strictly Hyperbolic Conservation Laws, With Application to Oil Recovery,' Comm. Pure Appl. Math., 40, 1987, pp. 141-178.

[9] SHEARER, M., SCHAEFFER, D. G., MARCHESIN, D. and PAES-LEME, P. L.: 'Solution of the Riemann Problem for a Prototype 2x2 System of Non-Strictly Hyperbolic Conservation Laws,' Arch. Rat. Mech. Anal., 97, 1987, pp. 299-320.

[10] SCHAEFFER, D. G. and SHEARER, M.: 'Riemann Problems for Non-strictly Hyperbolic 2x2 Systems of Conservations Laws,' Trans. Amer. Math. Soc. 304, No. 1, 1987, pp. 267-306.

[11] LIU, T. P.: 'The Riemann Problem for General 2x2 Conservation Laws,' Trans. Amer. Math. Soc., 199, 1974, pp. 89-112.

[12] TRUESDELL, C. and NOLL, W.: 'The Nonlinear Field Theories of Mechanics,' Handbuch der Physik, III/3, Springer-Verlag, 1965.

[13] JEFFREY, A.: 'Quasilinear Hyperbolic Systems and Waves,' Pitman Pub., 1976.

RANDOM-CHOICE BASED HYBRID METHODS FOR ONE AND TWO DIMENSIONAL GAS DYNAMICS

E.F. Toro
College of Aeronautics, Cranfield Institute of Technology,
Cranfield, Beds, MK43 OAL, England

ABSTRACT

Two hybrid methods applicable to the unsteady Euler equations in one or more space dimensions are presented. Each method consists of a smooth solver for smooth parts of the flow and the Random Choice Method for discontinuities. The solution of the Riemann problem (approximate or exact) gives sufficient information for switching between the smooth solver and RCM. In 1-D the hybrid methods give very accurate results throughout the flow field. Discontinuities are of zero width. In the multidimensional case shock waves are compromised, but contact discontinuities are still of zero width.

1. INTRODUCTION

We are concerned with numerical methods for solving the unsteady Euler equations in one or more space dimensions. Emphasis is placed on accuracy for both the smooth part of the flow and the discontinuities (e.g. contact discontinuities, shock waves).

Traditional finite difference methods (Lax-Wendroff) are accurate for smooth flows but discontinuities are smeared over several computing zones; they also give rise to over/under-shoots followed by spurious oscillations that can lead to stability problems.

Modern finite difference techniques (e.g. high resolution methods) such as Roe's Method [1] are accurate in the smooth parts of the flow and discontinuities are more accurately represented than by traditional methods. Shocks are typically smeared over 3/4 zones and contacts over 6/7 zones. Contact discontinuities are more difficult to deal with when utilising these methods. Shock waves, unlike contacts, have a natural compression mechanism (converging characteristics) that helps their sharp numerical resolution. In some fields of application, such as reactive flow, the accurate representation of contact discontinuities is as important, if not more, than that of shock waves; for they carry discontinuities in temperature and energy.

The Random Choice Method (RCM) is capable of producing discontinuities with infinite resolution (zero width). This statement is true for the unsteady Euler equations in one space dimension, for the steady two-dimensional Euler equations (steady supersonic flow) and some other special unsteady multi-dimensional problems. RCM however, is inaccurate in smooth parts of the flow; randomness is a feature of the method (for details about RCM see Refs. 2-7). Randomness is not an important issue in homogeneous problems (no 'source' terms) but can be intolerable if problems of technological interest are to be solved.

In this report we present two hybrid methods for the unsteady Euler equations in one or more space dimensions. The idea is to identify the discontinuities (e.g. shocks, contacts) at every time step and deal with them via the Random Choice Method and utilise another method for the rest of the flow field. The resulting solution is accurate in smooth regions (unlike RCM) and discontinuities are of zero width (unlike all difference methods).

These hybrid methods are an attempt at combining the best features of available methods.

In Ref.8 we presented a hybrid method consisting of SORF (Second Order Random Flux) and RCM and illustrated applications to 1-D problems. In this report we extend the SORF/RCM hybrid method to 2-D problems and use updated monotonicity procedures for SORF as well as switching mechanisms. We also present another hybrid consisting of Roe's Method and RCM. ROE/RCM is applied to 1-D and 2-D problems. There is a clear distinction between 1-D and multidimensional problems. Not all 1-D features of the hybrid methods extend to the multidimensional case. In particular, shock waves must be compromised, i.e. solved by SORF or ROE. This is in part due to space operator splitting, which is the procedure considered here to solve multidimensional problems (see Ref.9).

The rest of this report is organised as follows: §2 contains the Euler equations and discusses the associated 1-D Riemann problem; in §3 we briefly review the component methods to be used in the construction of the hybrid methods; in §4 we describe the SORF/RCM and ROE/RCM hybrid methods for 1-D problems; in §5 we present extensions of the hybrid methods to two or more space dimensions; in §6 we draw some conclusions and point out aspects of possible further development.

2. EULER EQUATIONS

We are interested in hyperbolic conservation laws

$$U_t + F(U)_x + G(U)_y = 0 \qquad (1)$$

where $U = U(t,x,y)$ with t denoting time and x and y denoting space. U,F and G are vectors; subscripts denote partial differentiation. For the Euler equations the vectors U,F and G are

$$U = \begin{bmatrix} \rho \\ \rho u \\ \rho v \\ E \end{bmatrix}, \quad F = \begin{bmatrix} \rho u \\ \rho u^2 + p \\ \rho uv \\ u(E + p) \end{bmatrix}, \quad G = \begin{bmatrix} \rho v \\ \rho uv \\ \rho v^2 + p \\ v(E + p) \end{bmatrix} \qquad (2)$$

where ρ is density $V = (u,v)$ is velocity, p is pressure and E is total energy given by

$$E = \tfrac{1}{2}\rho(u^2 + v^2) + \rho e \qquad (3)$$

with e denoting the specific internal energy. For closure, an equation of state is used. Here we take the ideal gas case with

$$p = (\gamma - 1)\rho e \qquad (4)$$

where γ is the ratio of specific heats.

The non-linear system of partial differential equations (1) - (4) can be solved analytically or in closed form only for very special circumstances. An interesting special case is the Riemann problem, that is system (1) in 1-D with piece-wise constant date for initial condition. The corresponding 1-D initial value problem can be solved exactly, but not in closed form, not even for the simple equation of state (4).

The 1-D Riemann problem is an important subproblem that can be utilised locally for finding global solutions to the general multidimensional initial value problem for system (1) - (4). The solution for the 1-D unsteady Euler equations can be represented as in Fig.1. There are three waves. The left and right waves can be either shock or rarefaction waves and the middle wave is always a contact discontinuity. Hence there are four possible wave patterns. Cavitation is not considered.

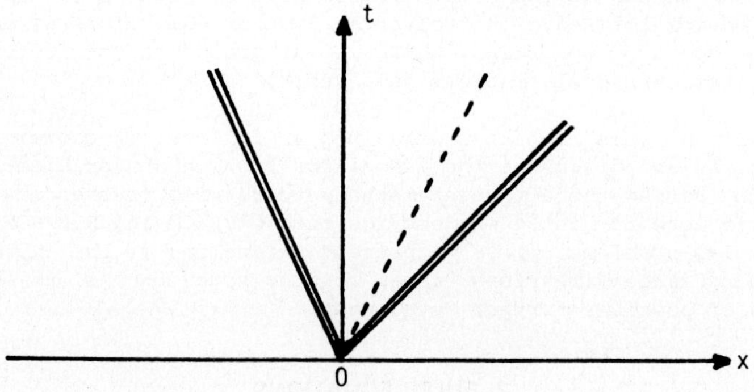

Fig.1: Solution of the Riemann problem for the 1-D unsteady Euler equations

Modern numerical methods for systems of type (1) use the solution of the 1-D Riemann problem. Thus an important aspect is the efficiency with which the solution is found. Finite-difference type methods use approximations to the exact solution. The Random Choice Method uses the exact solution. An efficient exact Riemann solver can be found in Ref.7.

3. REVIEW OF METHODS USED

The basic techniques used in this paper are: The Random Choice Method (RCM), the Second Order Random Flux (SORF) and the Flux Difference Splitting Method due to Roe (ROE).

3.1 The Random Choice Method (RCM)

This method is based on an existence proof by Glimm [2] and was first successfully implemented by Chorin [3]. The method has been further developed (e.g. Refs.4-7) to provide an efficient computational technique which is directly applicable to the unsteady 1-D Euler equations.

RCM approximates general data by piece-wise constant functions so that for a sufficiently small time the global initial value problem can be replaced by a set of Riemann problems RP(i,i+1). Fig.2 illustrates the two Riemann problems affecting cell i.

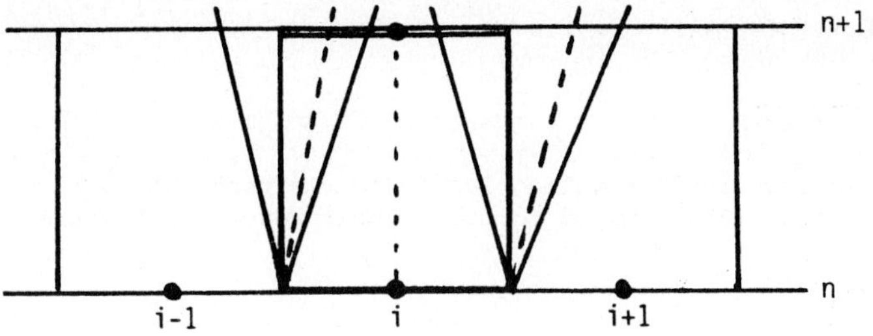

Fig.2: Riemann problems RP(i-1,i) and RP(i,i+1) affecting cell i

The solution at the grid point i at time level n+1 is obtained via

$$U_i^{n+1} = \bar{U}_i(Q_i) \tag{5}$$

where

$$Q_i = (x_i + \Theta_n \Delta x, t_n + \Delta T) \tag{6}$$

with $\Theta_n \varepsilon [0,1]$ a member of a pseudo-random sequence. In (5) $\bar{U}_i$ denotes the solution of the Riemann problems RP(i-1,i) and RP(i,i+1) inside cell i of length Δx, at time level n+1. Q_i is a "random" position inside cell i. This is the non-staggered grid version of RCM. More details about RCM can be found in Refs.2-7.

The Random Choice Method uses the exact solution of the Riemann problem. Here we use the Riemann solver presented in Ref.7. As for the sequence $\{\Theta_n\}$ we use the van der Corput sequence VDC(k_1,k_2) as done by Colella [4].

The most important feature of RCM is that discontinuities are of zero-width (infinite resolution). A disadvantage of the method is the randomness present in the smooth parts of the flow.

3.2 The Second Order Random Flux Method (SORF)

The basic form of this method was presented by the author in Ref.10. The method uses

$$U_i^{n+1} = U_i^n - \frac{\Delta T}{\Delta x}(F_{i+\frac{1}{2}}^{n+\frac{1}{2}} - F_{i-\frac{1}{2}}^{n+\frac{1}{2}}) \tag{7}$$

to update the solution. The intercell flux $F_{i+\frac{1}{2}}^{n+\frac{1}{2}}$ is evaluated at the random-choice solution of the Riemann problem RP(i,i+1) where the sampling length Δx is centred at the intercell boundary. In Ref.8 the method was shown to be 2nd order accurate, for the linear advection equation.

The method produces overshoots and spurious oscillations near discontinuities. To remove these overshoots one can sample the fluxes over a restricted sampling interval of length $T\Delta x$. The original scheme [10] has T=1; T=0 gives the Godunov's scheme. This is similar in spirit to the monotonicity procedures employed in 2nd order Godunov-type schemes and flux limiters [11-12]. An interesting analogy with flux limiters is given by the relation

$$T = \frac{\nu}{1 - (1-\nu)B} \tag{8}$$

where T is the restricted sampling length (normalised) and B is a flux limiter. Equation (8) is valid for the linear advection equation

$$u_t + au_x = 0 \tag{9}$$

and $\nu = \Delta Ta/\Delta x$ is the Courant number (a is constant).

Generalisation of (8) to 2x2 linear systems is direct but for non-linear systems the situation is not yet clear to us, although use of (8) with some empiricism gives quite satisfactory results.

SORF as described, is applicable to the unsteady 2-D (or 3-D) Euler equations via operator splitting in space, i.e. the 2-D initial value problem (1) is replaced by a sequence of 1-D problems. The simplest form proceeds as follows.

(i) Solve $U_t + F_x = 0$ (x-sweep) with data U^n for a time step of size ΔT. Denote the solution by U^{nx}.

(ii) Solve $U_t + G_y = 0$ with data U^{nx} for a time step of size ΔT again. The resulting solution is the solution U^{n+1} to (1) with data U^n.

This procedure works quite well for methods that compromise discontinuities, but fails for methods that give zero-width discontinuities (e.g. RCM). Ref.9 is useful in this respect. For 3-D one proceeds in an analogous fashion.

3.3 Roe's Method

This method approximates the Jacobian Matrices A and B corresponding to the flux functions F and G in equation (1) by $\tilde{A}$ and $\tilde{B}$, which are constructed to safisfy some desirable properties. One property is

$$\tilde{A}\Delta U = \Delta F \tag{10}$$

where $\Delta U = U_{i+1} - U_i$ and $\Delta F = F_{i+1} - F_i$. Projection of ΔF onto the eigenvectors of $\tilde{A}$ gives

$$\Delta F = \sum_{j=1}^{4} \alpha_j \lambda_j e_j . \tag{11}$$

The corresponding 1st order upwind scheme for equation (7) is

$$U_\ell^{n+1} = U_\ell^n - \frac{\Delta T}{\Delta x} \sum_{j=1}^{4} \alpha_j \lambda_j e_j \qquad (12)$$

$$\ell = \begin{cases} i & \text{if} \quad \lambda_j < 0 \\ i+1 & \text{if} \quad \lambda_j > 0 \; . \end{cases}$$

In equation (11) λ_j are eigenvalues of $\tilde{A}$ (wave speeds), e_j are the associated right eigenvectors and α_j are the wave strengths. For the x-sweep these are

$$\lambda_1 = u - a \;, \quad \lambda_2 = u \;, \quad \lambda_3 = u \;, \quad \lambda_4 = u + a \qquad (13)$$

$$e_1 = \begin{bmatrix} 1 \\ u - a \\ v \\ H - ua \end{bmatrix}, \quad e_2 = \begin{bmatrix} 1 \\ u \\ v \\ \frac{1}{2}q^2 \end{bmatrix}, \quad e_3 = \begin{bmatrix} 0 \\ 0 \\ 1 \\ v \end{bmatrix}, \quad e_4 = \begin{bmatrix} 1 \\ u + a \\ v \\ H + ua \end{bmatrix} \qquad (14)$$

$$\begin{aligned} \alpha_1 &= \frac{1}{2\tilde{a}^2} (\Delta p - \tilde{\rho}\tilde{a}\Delta u) \\ \alpha_2 &= \Delta\rho - \Delta p/\tilde{a}^2 \\ \alpha_3 &= \rho\Delta v \\ \alpha_4 &= \frac{1}{2\tilde{a}^2} (\Delta p + \tilde{\rho}\tilde{a}\Delta u) \end{aligned} \qquad (15)$$

where $q^2 = u^2 + v^2$ and $\tilde{a}$, $\tilde{\rho}$ are the Roe average values. For details on the 1-D basic scheme consult Ref.1.

To the 1st order scheme (12) higher order corrections can be added, preserving monotonicity. This involves the use of flux limiters [1 and 11).

4. HYBRID METHODS IN ONE DIMENSION

The intention is to combine the methods described in sections 3.2 and 3.3 with RCM to produce zero-width discontinuities and accurate representation of the smooth parts of the flow. RCM is used only at 'large' discontinuities (may be once or twice per time step). In one space dimension we identify shock waves and contact discontinuities from the solutions of the sequence of Riemann problems RP(i,i+1) at every time step n. If cell i is transversed by a discontinuity (see Fig.2) then the solution at time level n+1 is found from equation (5). Elsewhere one uses another method (SORF or ROE).

4.1 The ROERCM Hybrid Method

This method results by combining Roe's method (§4.3) and RCM (§4.1) in the manner just described. In order to assess the performance of the methods we consider a shock tube problem with data

$$\rho_\ell = 1.0, \; u_\ell = 0, \; p_\ell = 1.0, \; \rho_r = 0.125, \; u_r = 0, \; p_r = 0.1, \; \gamma = 1.4 \, . \qquad (16)$$

In all computed results shown we took a Courant number coefficient C = 0.4, and 100 grid points on a tube of length 1.0. The initial discontinuity is at $x_o = 0.5$.

Fig.3 shows the computed solutions and the exact solution using Roe's method alone after 140 time steps. By current standards this is an accurate solution to this problem. The shock is resolved within 3/4 zones (see velocity plot). The contact discontinuity is resolved within 7/8 zones (see energy plot).

Fig.4 shows the computed result using the ROERCM Hybrid Method. Notice that the shock wave and contact discontinuity are absolutely sharp, as we expected.

4.2 <u>The SORFRCM Hybrid Method</u>

This hybrid method was presented in Ref.8 for 1-D problems. The present version contains some improvements. Fig.5 shows results for problem (16) with data at t = 0. Compare with Fig.4.

5. HYBRID METHODS IN TWO SPACE DIMENSIONS

In order to test the performance of the hybrid methods against an exact result we used the shock tube problem (16) with the initial discontinuity placed at an angle α to the computing x-y grid. Then, along the normal direction of propagation of the resulting waves the problem is one-dimensional and thus the exact solution can be used again.

An important remark here is that since RCM fails in 2-D via operator splitting we do not expect to preserve all the good features of the hybrids that hold in 1-D problems. To illustrate this we show a computed solution using RCM alone for data at an angle $\alpha = 45°$. The results are shown in Fig.6. They are unacceptable. The oscillations are caused by the shock wave.

The hybrids in 2-D will therefore compromise shock waves. They will be treated by either ROE or SORF. Contact discontinuities however, will be absolutely sharp as in 1-D problems.

Fig.7 shows the solution obtained by ROERCM for the mesh 100x100 after 180 time steps. Notice that start up errors give inaccuracies near the tail of the rarefaction and just in front of the contact discontinuity (see internal energy plot). Fig.8 shows corresponding results obtained by the SORFRCM method.

6. CONCLUSIONS AND FURTHER DEVELOPMENTS

Two hybrid methods which are directly applicable to 1-D Gas Dynamics have been presented. Results are accurate in smooth parts and discontinuities are of zero width. Two or three dimensional extensions are implemented via space operator splitting. Shock waves must be compromised, but contact discontinuities are of zero width as in 1-D problems.

Further developments are possible for the SORFRCM hybrid. One aspect is monotonicity for SORF. Another is use of approximate Riemann solvers for SORF to increase computational efficiency. Some preliminary tests show that a two rarefaction approximation is sufficiently good. Further experience from applications to problems of scientific and engineering interest would also be valuable.

ACKNOWLEDGEMENTS

The author is grateful to Professors JF Clarke and PL Roe for useful discussions concerning the present work, which was partially supported by RARDE and AWRE.

REFERENCES

[1] Roe, P.L. 1981. "Approximate Riemann Solvers, Parameters, Vectors and Difference Schemes". J.Comp. Phys. 43 pp 357-372.

[2] Glimm, J. 1965. "Solutions in the large for non-linear hyperbolic systems of equations". Comm. Pure Appl. Maths. 18, 697.

[3] Chorin, A.J. 1976. "Random Choice Solution of Hyperbolic Systems". J. Comput. Phys. 22, 517.

[4] Colella, P. 1982. "Glimm's Method for Gas Dynamics". SIAM J. Sci. Stat. Comput. Vol.3, No.1, p.76.

[5] Sod, G.A. 1977. "A numerical study of a converging cylindrical shock". J. Fluid Mech. 83, 785.

[6] Gottlieb, J.J. 1987. "Staggered and non-staggered grids with variable node spacing for the Random Choice Method". Paper presented at the Second International Meeting on Random Choice Methods in Gas Dynamics. College of Aeronautics, Cranfield Institute of Technology, England, July 20-24 1987. (Submitted to J. Comput. Phys.).

[7] Toro, E.F. 1987. "A fast Riemann solver with constant covolume applied to the Random Choice Method". CoA Report 8719, Oct.1987, College of Aeronautics, Cranfield Institute of Technology, England.

[8] Toro, E.F. and Roe, P.L. 1987. "A hybridised Higher Order Random Choice Method for Quasi-Linear Hyperbolic Systems". Proc. 16th International Symposium on Shock Tubes and Waves. July 26-30, 1987, Aachen, W.Germany , pp.701 - 708. H.Cronig (Editor).

[9] Roe, P.L. 1987. "Discontinuous Solutions to Hyperbolic Systems under Operator Splitting". ICASE Report 87-64, NASA Langley Research Center, Hampton, VA3665, 1987.

[10] Toro, E.F. 1986. "A new numerical technique for quasi-linear hyperbolic systems of conservation laws". CoA Report 8620, College of Aeronautics, Cranfield Institute of Technology, England, December 1986.

[11] Sweby, P.K. 1985. "High resolution TVD schemes using flux limiters". Lectures in Applied Mathematics, Vol.22, pp. 289-309.

[12] Ben-Artzi, M. and Falcovitz, J. 1984. "A second-order Godunov-type scheme for Compressible Fluid Dynamics. J.Comput. Phys.55,1-32(1984).

[13] Simmonds, L.G. and Toro, E.F. 1988. "Development of numerical techniques for computing detonations in solids modelled by the reactive Euler equations". CoA Report NFP8801, College of Aeronautics, Cranfield Institute of Technology, Cranfield, England.

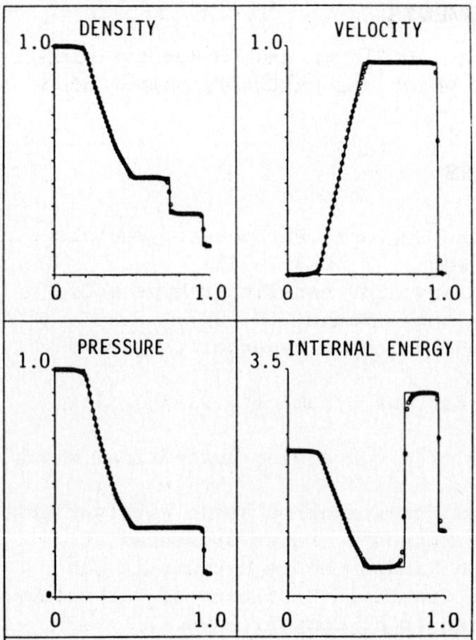

FIG. 3 : COMPUTED (SYMBOL) AND EXACT (LINE) SOLUTIONS. ROE METHOD USED.

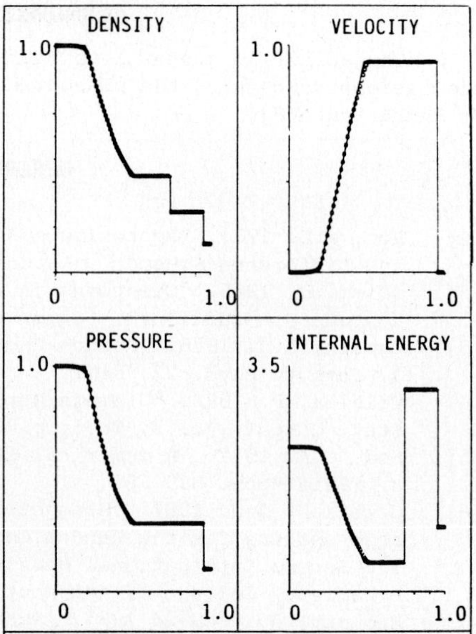

FIG. 4 : COMPUTED (SYMBOL) AND EXACT (LINE) SOLUTIONS. HYBRID ROE/RCM USED.

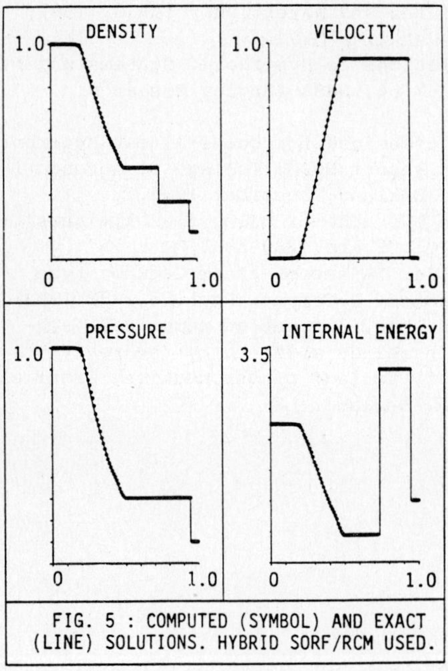

FIG. 5 : COMPUTED (SYMBOL) AND EXACT (LINE) SOLUTIONS. HYBRID SORF/RCM USED.

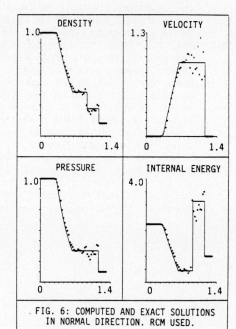

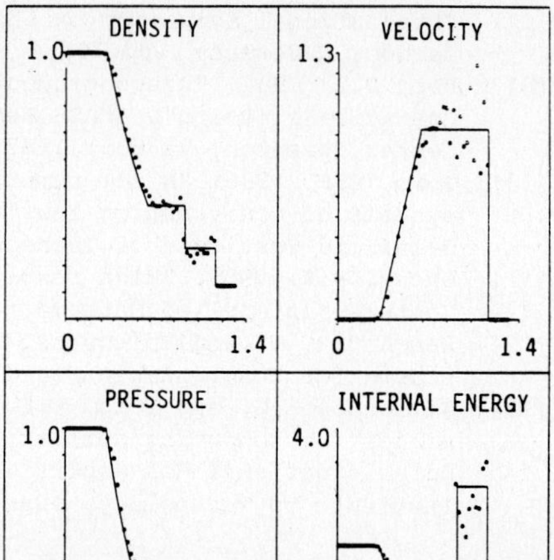

FIG. 6: COMPUTED AND EXACT SOLUTIONS IN NORMAL DIRECTION. RCM USED.

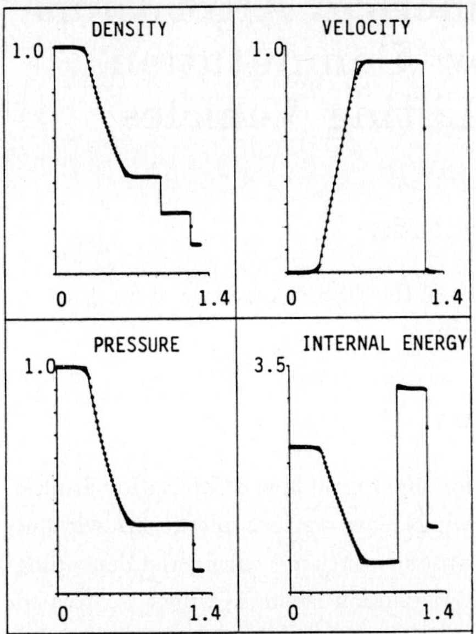

FIG. 7: COMPUTED AND EXACT SOLUTIONS IN NORMAL DIRECTION. ROE/RCM USED.

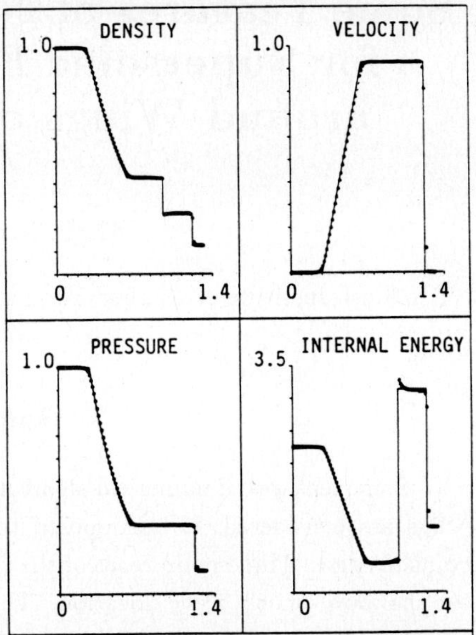

FIG. 8: COMPUTED AND EXACT SOLUTIONS IN NORMAL DIRECTION. SORF/RCM USED.

Some Features of Numerical Algorithms for Supersonic Flow Computation around Wings of Lifting Vehicles

G. P. Voskresensky

Keldysh Institute of Applied Mathematics of the USSR Academy of Science
Moscow

Summary

Some features of numerical algorithms for supersonic flow computation around wings are considered. It is supposed that wings have various planforms without cranked edges. Three main cases of the flow around wings are considered depending on the bow shock wave location. The flow domain around wings is divided into subdomains. In each subdomain there is a local coordinate system, a local computational mesh and separate algorithms. All algorithms are based on the inviscid gas model. Principal features of the algorithms and ways of their combination for the whole wing in various cases are described. Some computational examples are presented.

I

Wings of lifting vehicles have various planforms. Let us consider profiled wings without cranked edges. Three main cases can be isolated according to the nature of supersonic flow around them:

a) wings with a shock wave attached to the leading edge (supersonic leading edge) (Fig. 1)

b) wings with a shock wave attached to the apex but detached from the leading edge (subsonic leading edge) (Fig. 2)

c) wings with a shock wave completely detached from the leading edge (Fig. 3)

Case a) appears, as a rule, on sharp–edged trapezoidal wings. The tip edge of those wings is usually parallel to the axis $0X$. The flow on the upper side of the wing can be that of compression or expansion depending on the freestream Machnumber, angle of attack, the sweep angle and the opening angle of the airfoil nose. There is always compression flow below the wing.

In case b) the wing has a rectilinear or curvilinear leading edge and a sharp–nosed profile. The tip edge, as in the previous case can be parallel to axis $0X$. Detachment of the bow shock wave from the leading edge is determined by freestream Machnumber and the sweep angle.

In case c) the wing may have any planform with a curvilinear or rectilinear leading edge. The wings airfoil has a blunt or a sharp nose and the opening angle of the sharp nose is overcritical.

Computational meshes must be introduced for the finite–difference method of the flow computation. The construction of the mesh based on the universal coordinate system for all cases leads to unjustified difficulties. It is expedient to divide the flow domain around the wing into subdomains according to geometrical parts of the wing such as: fore part, tip–side part, central part. Local coordinate systems and local computational meshes are introduced in all subdomains isolated by some characteristic surfaces; it is more convenient to compute the flow around seperate parts of the wing.

In all subdomains the surfaces of the wing and the shock wave become coordinate surfaces and therefore the physical domains of the solution with complicated geometry are mapped on simple rectilinear computational subdomains. For each subdomain there are separate problems and corresponding algorithms based on the flow features. The general algorithm consists of these particular algorithms and depends on the computational case. The solution of the whole problem must be continuous.

Case c) is the most complicated. Let us first consider case c) as it is also the most common. Three subdomains can be isolated here. At first the Cartesian coordinate system is used with axis $0X$ directed along the central chord of the wing (Fig. 3). A local curvilinear coordinate system is introduced in each subdomain. Some parameters for a convenient description of the wing geometry, including aerodynamical deformation, can be introduced into the coordinate transformation. The subdomains are separated by space type surfaces. Characteristic cones emanating from the surfaces of this type will touch them only with their apexes. From this consideration the conclusion is drawn that on the space type surfaces there are no boundary conditions.

The first subdomain corresponds to the front part of the wing (Fig. 3). It is restricted above and below by wing and shock wave surfaces and downstream and from the tip–side by the space type surfaces Π and Q .

The second subdomain corresponds to the wingtip section. It is also restricted by wing and shock wave surfaces and on the sides of the upper and lower parts of the wing by space type surfaces Q and P .

The third subdomain consists of two parts — upper and lower. It covers the central part of the wing down the flow behind surface Π up to the trailing edge.

The same subdivision is preserved for case a). In case b) there is the second subdomain only, which is restricted on the side by a symmetry plane.

II

The solution of the whole problem is based on the inviscid gas model. For case c) this solution combines a pseudo–unsteady approach for the elliptic–hyperbolical problem in the first subdomain and a steady–state approach for the hyperbolical problems in the second and third subdomain. Gasdynamic functions on surfaces Π and Q are obtained by the solution of the first subdomain problem. The functions on surfaces Q are initial in the second subdomain and the functions on surface Π in the third subdomain.

Unsteady gasdynamic equations are utilized in the first subdomain. A curvilinear coordinate system is introduced by transformation:

$$t = \tau\,; \quad x = x(\tau,\xi,\eta,\zeta)\,; \quad y = y(\tau,\xi,\eta,\zeta)\,; z = z(\tau,\xi,\eta,\zeta).$$

Wing surface $G = G(\tau,\eta,\zeta)$ and shock wave surface $F = F(\tau,\eta,\zeta)$ become then coordinate surfaces. In new coordinates these equations are $\xi = 0$ and $\xi = 1$, respectively. The calculation domain is now enclosed in a rectangle:

$$0 \le \xi \le 1\,; \quad -\pi/2 \le \eta \le \pi/2\,; \quad 0 \le \zeta \le 1.$$

Surfaces Π above and below the wing have the equations $\eta = \pm\pi/2$. Plane $X0Y$ and surface Q have equations $\zeta = 0$ and $\zeta = 1$, respectively. The wing surface is approximated by local cubic splines in a time–dependent steady–state process. The final computational mesh is also generated in this process.

There is a steady flow in the second and third subdomain. The coordinate transformation is simpler here. In the second subdomain variables are now enclosed in rectangle:

$$0 \le \tau \le 1\,; \quad 0 \le \xi \le 1\,; \quad -\pi/2 \le \eta \le \pi/2,$$

where τ is one space coordinate. Surface Q has the equation $\tau = 0$ and surface P the equations $\eta = \pm\pi/2$.

In the third subdomain the variables alteration is:

$$0 \leq \tau \leq 1; \quad 0 \leq \xi \leq 1; \quad 0 \leq \zeta \leq 1.$$

Surface Π has equation $\tau = 0$ here, plane $X0Y$ has equation $\zeta = 0$ and surface P has equation $\zeta = 1$.

Computation in the second and third subdomain is carried out along axis 0τ simultaneously. On surface P which separates these subdomains the functions are smoothly completed from the inside of the second subdomain. In all subdomains numerical algorithms with implicit second–order finite–differences schemes are used [1].

III

Let us consider features of the first subdomain algorithm for case a). In this case the shock wave is attached to the leading edge when the velocity component normal to the leading edge is greater than sound velocity and the sum of the angle of attack and semi–angle of the airfoil nose opening is smaller then the critical deflection angle. In this case the mutual influence of the flow fields above and below the wing is absent. The flow domains are restricted by the shock wave or the characteristic surface above. This surface appears when the angle of attack is greater than the semi–angle of the airfoil nose opening.

There are two variants in the problem formulation. The first variant is for weakly swept wings ($\chi < 30°$). Here gasdynamic functions in the first subdomain on the leading edge are calculated asssuming that the flow around each element of the leading edge is the same as the flow around a sliding wedge. For calculation the freestream vector is decomposed into two components: normal and tangent to the leading edge. In compression region with the shock waves the functions are computed using relations for the flow around a wedge. In expansion region within a characteristic surface the functions are computed using Prandtl–Meyer relations. The functions on the leading edge represent initial data on surface Π for computations in the second and third subdomain.

In the second variant — for highly swept wings, it is convenient to use the algorithm of the third subdomain only. Here axis 0τ coincides with axis $0X$ and the initial data are on surface $x = x_0$ near the apex of the wing. The presence of the self–similar flow around the conical surface near the apex of the wing makes it possible to apply the self–similar stationing principle along coordinate 0τ for the flow computation on surface $x = x_0$.

In this case (a) the problem of the supersonic flow around the wingtip side takes a special place (Fig. 4). The solution of this problem represents initial data on surface Q_1 for the flow–field calculation of the downstream tip–side of the wing which is done by the algorithm of the third subdomain. The formulation of this problem is based on the assumption that the surface of the wingtip side is conical and the flow around it is self–similar. The flow under consideration and also the solution domain are separated from the freestream by the conical surface of a bow shock wave F, with the apex in point 0 and by characteristic surfaces L_u, L_l emanating from the same point. The self–similar flow makes it possible to apply the self–similar stationing principle with the algorithm of the second domain to the solution of that problem.

The flow around the tip–side of the wing is very complicated. The tip–edge vortices with large gradients of gasdynamic functions are generated here. This requires a careful choice of the computational grid and of local clustering of computational points.

IV

The algorithm of the second subdomain is used in case b). Axis 0τ coincides here with axis $0X$ and the initial data on surface $x = x_0$ are obtained in the same way as in the second variant of case a). For highly swept wings there are vortices on the leading edge such as on the tip–edge of the previous case.

V

Computation examples

According to the complete set of algorithms some examples were computed [1–3]. In Fig. 5 the wing of the elliptical planform with aspect ratio $\lambda = 2.56$ and thickness ratio of the blunted nose airfoil $\bar{c} = 0.3$ are shown. The boundaries of all subdomains are drawn as heavy lines. The bow shock wave forms corresponding to airfoils are shown here for $M_\infty = 2$ and $\alpha = 5°$.

The pressure distribution at $M_\infty = 2\,;3.5$, $\alpha = 5°$ along the two cross-sections of the same wing on the lower (solid lines) and on the upper (dashed lines) surfaces are shown in Fig. 6.

The pressure distribution on the lower (solid lines) and upper (dashed lines) surfaces of the delta wing with a detached shock wave is shown in Fig. 7. There is a sweep angle $\chi = 70°$. On plane XOY the symmetrical airfoil is generated by arcs of a circle. The thickness ratio of the airfoil is $\bar{c} = 0.038$. The cross–sections of the wing are ellipses. The freestream parameters are $M_\infty = 1.5$, $\alpha = 3°$. The algorithm of the second subdomain is used. The stream–surfaces cross–section cut by plane $x = 0.7$ (the length of the central chord is $x = 1$) for the same wing is shown in Fig. 8. On the upper surface a vortex–structure with tangential discontinuty is likely to develop. In Fig. 9 diagrams are given of aerodynamical loading Δp for a delta wing with sharp leading edges and an attached shock wave. The biconvex airfoil is generated by arcs of a circle with thickness ratio $\bar{c} = 0.05$ in the upper and $\bar{c} = 0.02$ in the lower parts. The sweep angle is $\chi = 60°$ and the freestream parameters are $M_\infty = 4$, $\alpha = 5°$. The algorithm of the third subdomain is used.

The pressure distribution p/p_∞ in plane Q_1 for the nose tip–side of the rectangular wing is given in Fig 10. The nose of the wing in this case is represented by a wedge and the wingtip surface is an elliptical semi–cone surface. The opening angle of the wedge is $\beta = 20°$. The pressure distribution on the lower surface is a solid line and on the upper surface is a dashed line. The freestream parameters are $M_\infty = 2$, $\alpha = 5°$. The clustering of the computational points in the tip region allows to approximate large gradients of functions.

References

[1] Voskresensky G. P., *Numerical method of calculation for supersonic flow around wings of lifting vehicles*, *Zh. vychisl. mat. i mat. fiz. V.24*, No. 6, 1984, 900–915.

[2] Voskresensky G. P., Lutskii A. E., Orlova M. G., *Supersonic flow around tip sides of wing with wedge–shaped airfoils and shock wave attached to leading edge*, *Prepr. No. 62 Keldysh Inst. Appl. Math. USSR Acad. Sci.*, Moscow, 1985.

[3] Voskresensky G. P., Ivanov O. V., Stebunov V. A., *Numerical modelling of supersonic flow around delta wings with airfoils and shock wave detached from leading edge*, *Prepr. No. 43 Keldysh Inst. Appl. Math. USSR Acad. Sci.*, Moscow, 1984.

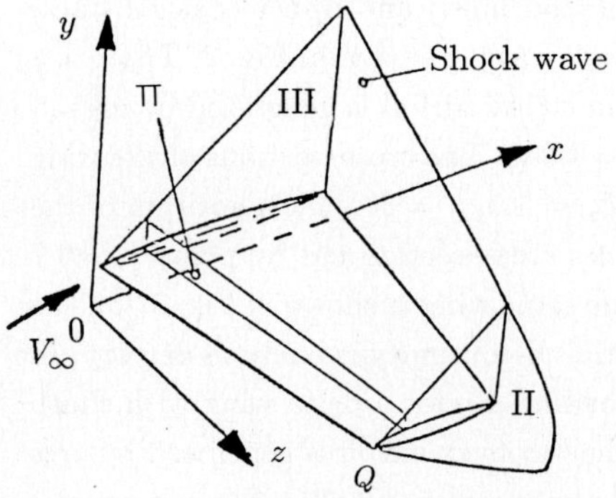

Fig. 1

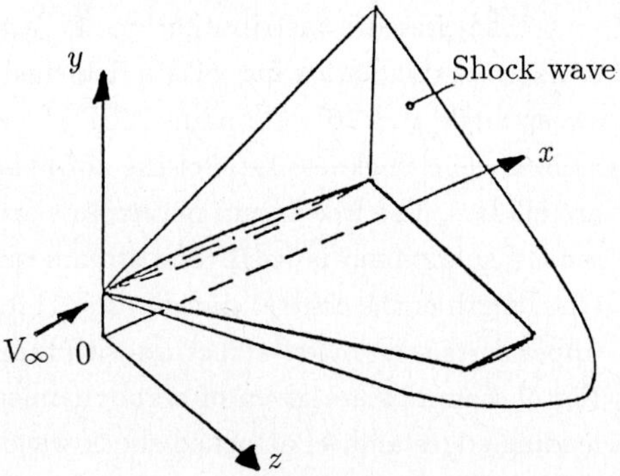

Fig. 2

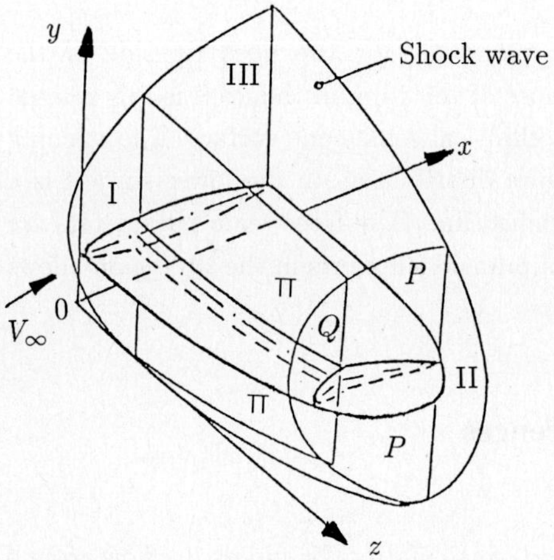

Fig. 3

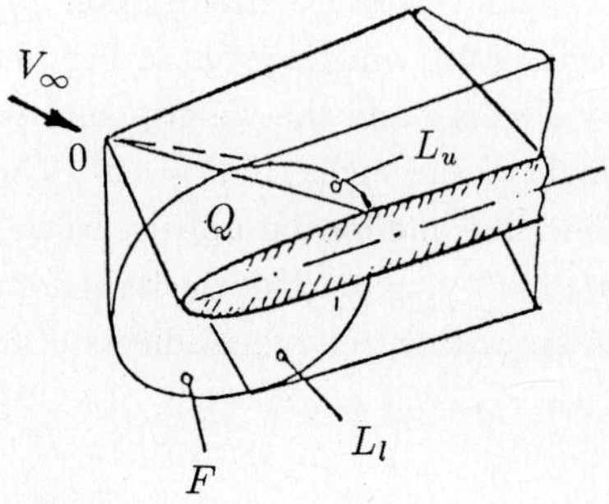

Fig. 4

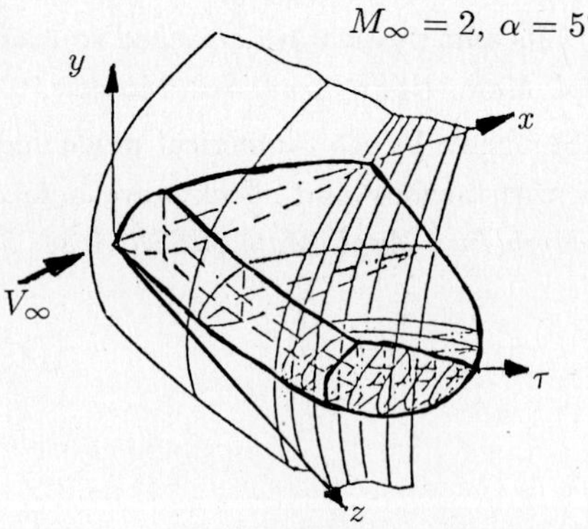

Fig. 5

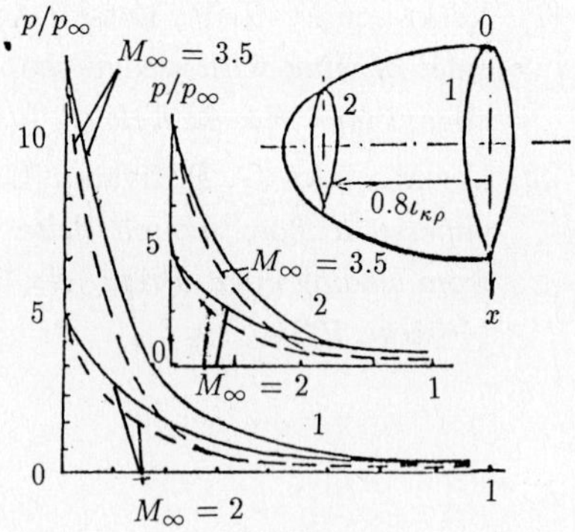

Fig. 6

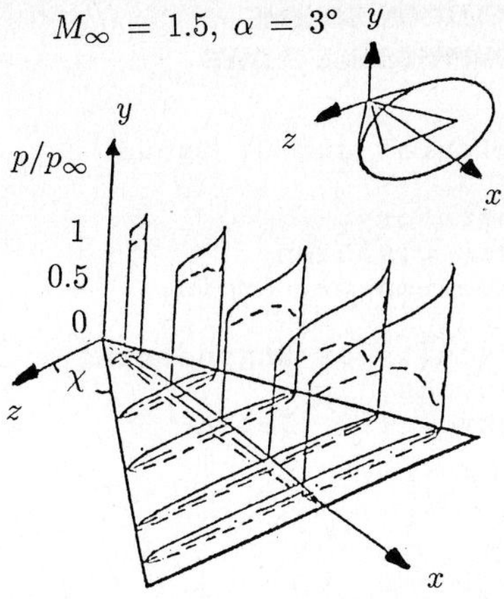

Fig. 7

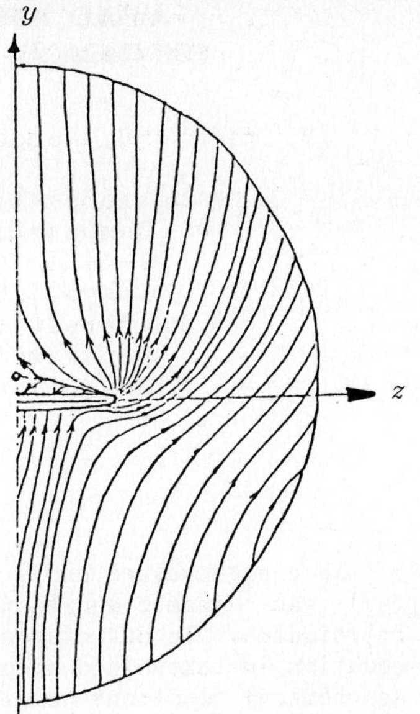

Fig. 8

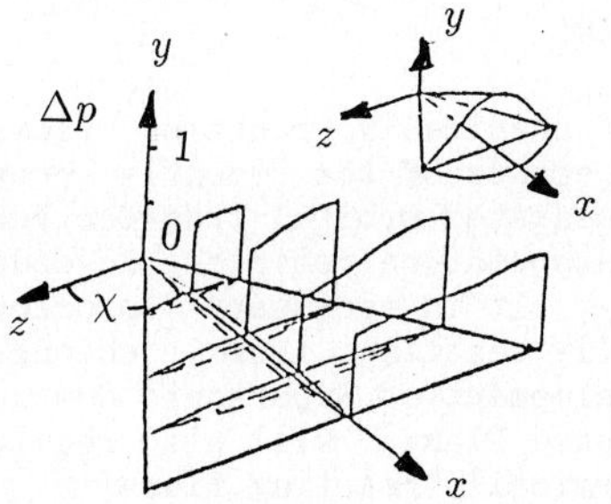

Fig. 9

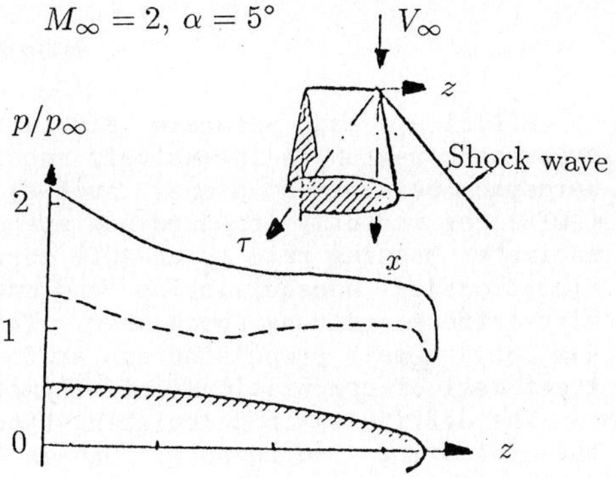

Fig. 10

A FULLY IMPLICIT HIGH-RESOLUTION SCHEME FOR CHEMICALLY REACTING COMPRESSIBLE FLOWS

Y. Wada*, H. Kubota**, T. Ishiguro* and S. Ogawa*

* National Aerospace Laboratory,
Computational Sciences Division,
7-44-1 Jindaiji-higashi-machi, Chofu,
Tokyo, 182, Japan

** University of Tokyo, Faculty of Engineering,
Department of Aeronautics,
Hongo, Bunkyo-ku, Tokyo, 113, Japan

SUMMARY

The eigenvalues and eigenvectors are analytically derived for general real gas dynamic equations in three-dimensional generalized curvilinear coordinates. In our diagonalizing formulation, the total mass conservation equation is taken into account, and arbitrary nonequilibrium effects, such as chemical reactions or vibrational nonequilibrium, can be treated in the same fashion. Making use of this diagonalization, we construct a fully implicit and high resolution scheme for chemically reacting real gases. Numerical results of two-dimensional shock-induced combustion problem show the efficiency and high resolution of our scheme.

1. INTRODUCTION

Efficient and accurate simulation of chemically reacting flows in hypersonic region is intensively required because of the recently proposed aerospace development plans, such as Aeroassisted Orbital Transfer Vehicle (AOTV), or recently proposed new space transportation vehicles. In order to estimate heating rate to an AOTV correctly, it is necessary to calculate vibrationally nonequilibrium and chemically reacting flows, accompanied with extremely strong shock wave. The development of Supersonic Combustion Ram Jet, a main propulsion system for a Space Plane, will also require a great deal of computational effort with chemically reacting flows.

The difficulty in calculating hypersonic reacting flow is mainly due to the following two reasons. One is the stiffness of system of equations. Chemically reacting flows can contain very different characteristic time scales, that of fluid or chemical reactions, even at the same time and same point. Such a disparity of time scales is referred to as stiffness, which causes numerical instability in time integration of the equations. Now the only effective means to avoid a numerical instability from the stiffness is an implicit treatment of chemical source terms[1]. In the past few years, several researchers[1,2,3] have calculated various nonequilibrium flows, using this approach, with great success. But the implicit treatment of chemical source terms is equivalent to advancing each state quantity at its own characteristic rate. So it seems to us that unsteady simulation of chemically reacting flows is essentially difficult and remains to be a future subject.

The second difficulty is the treatment of strong shock wave, which is a common problem with a perfect gas. Recently, several modern shock capturing methods are devised for a perfect gas, based on exact or approximate Riemann solver [4,5,6]. These first order upwinding schemes can be converted to the higher-order schemes under the restriction of Harten's TVD sufficiency conditions[7,8,9]. Various calculations for a perfect gas show the excellent accuracy, high resolution and robustness of the higher order TVD schemes[8,9]. Many approximate Riemann solvers make use of the eigenvalues and eigenvectors of the gas-dynamic matrices. So, when applying such higher-resolution schemes to real gases, those of real gas-dynamic matrices, for themselves, are needed. Eberhardt and Brown [10] have analytically obtained the eigenvalues and eigenvectors of the fully coupled system in Cartesian coordinate, including species conservation equations in addition to the gas dynamic equations. They have calculated reacting flows using a first-order TVD scheme. Very recently, Yee and Shinn [11,12] have extended their formulation to general curvilinear coordinates, and have solved shock ignition problem by the use of symmetric TVD schemes. However both of their basic equations do not include total mass conservation equation. Thus, the total mass is not conserved, especially when chemical source terms are treated implicitly. Park[13] suggests that the equations of the total mass and the all the species continuity should be solved together for flows in which dominant species do not exist.

In this paper, we present a new diagonalizing formulation of general real gas-dynamic matrices, which has the following favorable properties :

(1) Total mass conservation equation is included.
(2) Chemical reaction and vibrational nonequilibrium can be treated in the same fashion.
(3) Matrix multiplications are so simple that the increase in number of additional nonequilibrium equations does not violently increase the operational counts.
(4) This is a natural extension of Warming, Beam and Hyett's perfect gas-dynamic matrices diagonalization[14], which is a special case of our formulation.

Our diagonalization makes it possible for general nonequilibrium flows to construct finite difference schemes based on characteristic relations, or to simplify the inversion work of block-tridiagonal systems that arise in an implicit factored scheme. Numerical results of two-dimensional shock-induced combustion problem show the efficiency and high resolution of the scheme based on our diagonalization.

2. GOVERNING EQUATIONS OF GENERAL REAL GAS

There are two main differences between the governing equations of a perfect gas and that of general real gas. A general real gas needs the additional equations which represent nonequilibrium effects, such as chemical reaction or vibrational relaxation. Any of these equations is written in the weak conservation form :

$$\frac{\partial \rho f_i}{\partial t} + \frac{\partial \rho f_i u_k}{\partial x_k} = s_i \qquad \cdots (1)$$

where f_i is a physical property per unit mass, and s_i is its corresponding source term. If this equation represents the effect of chemical reaction, the property f_i is the mass fraction Y_i of species i, and s_i is the species production term. On the other hand, if this equation represents vibrational nonequilibrium effects, the property f_i and s_i are vibrational energy per unit mass e_{vibi} and the vibrational relaxation term, respectively. The second difference is that the internal energy contains additional terms for the case of general real gas. For a perfect gas, the internal, only translational and rotational, energy is proportional to the translational temperature T. On the other hand, the general real gas contains additional equilibrium and nonequilibrium vibrational energy and heat of formation. The resulting internal energy of a general real gas is given as :

$$e = \rho \Sigma e_{vib} + \rho \Sigma Y_i [\int Cp_i dT + \Delta HF_i] - P. \quad \cdots (2)$$

where ΔHF_i is the heat of formation for species i and Cp_i is a specific heat at constant pressure. Without radiation or lasing, there is no need to introduce the source term in this energy conservation equation, since the flow itself does not exchange energy with anything else, even when chemical reaction or vibrational relaxation takes place. However there are no simple relations between internal energy and translational temperature, nor between internal energy and pressure P. Therefore, for real gases, the pressure P is a complex function of density, internal energy and f_i ;

$$P = P(\rho, e, f_1, \cdots f_n). \quad \cdots (3)$$

Taking account of these differences between perfect and real gases, the three-dimensional Euler equations for general real gas , with n additional nonequilibrium equations, are given in the form :

$$\frac{\partial q}{\partial t} + \frac{\partial E}{\partial x} + \frac{\partial F}{\partial y} + \frac{\partial G}{\partial z} = S \quad \cdots (4)$$

where

$$q = \begin{bmatrix} \rho \\ \rho u \\ \rho v \\ \rho w \\ E \\ \rho f_1 \\ \rho f_2 \\ \cdot \\ \rho f_n \end{bmatrix} \quad E = \begin{bmatrix} \rho u \\ \rho uu + P \\ \rho vu \\ \rho wu \\ (E+P) u \\ \rho f_1 u \\ \rho f_2 u \\ \cdot \\ \rho f_n u \end{bmatrix} \quad F = \begin{bmatrix} \rho v \\ \rho uv \\ \rho vv + P \\ \rho wv \\ (E+P) v \\ \rho f_1 v \\ \rho f_2 v \\ \cdot \\ \rho f_n v \end{bmatrix} \quad G = \begin{bmatrix} \rho w \\ \rho uw \\ \rho vw \\ \rho ww + P \\ (E+P) w \\ \rho f_1 w \\ \rho f_2 w \\ \cdot \\ \rho f_n w \end{bmatrix} \quad S = \begin{bmatrix} 0 \\ 0 \\ 0 \\ 0 \\ 0 \\ s_1 \\ s_2 \\ \cdot \\ s_n \end{bmatrix}$$

The above equations are transformed into generalized curvilinear coordinate system in the form :

$$\frac{\partial \hat{q}}{\partial \tau} + \frac{\partial \hat{E}}{\partial \xi} + \frac{\partial \hat{F}}{\partial \eta} + \frac{\partial \hat{G}}{\partial \zeta} = \hat{S} \quad \cdots (5)$$

where

$$\hat{q} = q/J \qquad \hat{G} = (\zeta_t q + \zeta_x E + \zeta_y F + \zeta_z G)/J$$

$$\hat{E} = (\xi_t q + \xi_x E + \xi_y F + \xi_z G)/J \qquad \hat{S} = S/J$$

$$\hat{F} = (\eta_t q + \eta_x E + \eta_y F + \eta_z G)/J.$$

As the above system of equations includes the total mass continuity equation, we can conserve the total mass, which would violently suffer from numerical errors if the global continuity equation was calculated as the sum of all the species continuity equations[10].

3. DIAGONALIZATION OF GENERAL REAL GAS MATRICES

Every modern shock capturing method makes use of the hyperbolic properties of the Euler equations. This hyperbolicity comes from the real eigenvalues and the corresponding eigenvectors of the flux Jacobian matrices(gas-dynamic matrices). The flux Jacobian matrices P of the general real gas are given in the form :

$$P = k_t I + k_x A + k_y B + k_z C \qquad \cdots (6)$$

where

$$A = \frac{\partial E}{\partial q} \qquad B = \frac{\partial F}{\partial q} \qquad C = \frac{\partial G}{\partial q}$$

and $k = \xi, \eta$ or ζ for $P = \hat{A}, \hat{B}$ or $\hat{C}$, respectively.

The eigenvalues and eigenvectors of the gas-dynamic matrices P are obtained by diagonalization of P ;

$$P \equiv T \Lambda T^{-1} \qquad \cdots (7)$$

where Λ is the diagonal matrix of eigenvalues of P. T and T^{-1} are matrices representing its right eigenvectors and left eigenvectors, respectively. The form of the diagonalization is somewhat arbitrary. Here we choose the diagonalizing formulation so as to be identical to that of reference [14] in the case of a perfect gas. In this diagonalizing procedure, we assume no special relation among pressure, density, internal energy and f_i. This means that, in our formulation, an arbitrary equation of state for a gas can be treated, and any additional equation representing nonequilibrium effects can be included.

In the Appendix, we present the form of matrix T, Λ, T^{-1}. These matrices may look complex. However, in practical computation, only a multiplication of a vector and the matrix T or T^{-1} is needed. These results are written in a simple form[15], and the increase in number of additional nonequilibrium equations does not violently increase the operational counts.

4. NUMERICAL SCHEME

Euler implicit scheme is generally given as

$$[\text{Implicit Operator}]\,\Delta\hat{q} = \Delta t\,(-\Delta\hat{E} - \Delta\hat{F} - \Delta\hat{G} + \hat{S})$$
$$\hat{q}^{n+1} = \hat{q}^{n} + \Delta q\,. \qquad \cdots (8)$$

where $\Delta\hat{E} \equiv \hat{E}_{i+1/2} - \hat{E}_{i-1/2}$,etc. The form of a numerical flux and an implicit operator characterizes its scheme. In this paper, Harten-Yee's implicit TVD scheme, with a modified treatment of metrics[9], is used.

The implicit operator is approximately factorized to the multiplication of point and alternative direction implicit operators as follows;

$$[\text{Implicit Operator}] =$$
$$[I - hH]\, T_\xi\, [I + h\,\partial_\xi \Lambda + h D_\xi]\, O\, [I + h\,\partial_\eta \Lambda + h D_\eta]\, P\, [I + h\,\partial_\zeta \Lambda + h D_\zeta]\, T_\zeta^{-1}\,. \qquad \cdots (9)$$

Here, $h = \Delta t$, $O = T_\xi^{-1} T_\eta$, $P = T_\eta^{-1} T_\zeta$. and H is a Jacobian matrix of chemical source terms[1]. The implicit operators for convective terms are diagonalized so that it becomes the same as that of the diagonalized Beam-Warming scheme [16] except for the numerical viscosity,

$$D_\xi = \nabla [\mathrm{diag}\{-\max(\psi(\Lambda^m_{i+1/2})\}]\Delta \quad , \text{etc.}$$

Here, $\psi(z)$ is an absolute function modified to satisfy entropy condition.
The numerical flux, for example $\hat{E}_{i+1/2}$, is given as

$$\hat{E}_{i+1/2} = \frac{1}{2}[\hat{E}_i + \hat{E}_{i+1} + (T_\xi \Phi)_{i+1/2}]. \qquad \cdots (10)$$

where the element of Φ denoted by ϕ^m is

$$\phi^m_{i+1/2} = \frac{1}{J_{i+1/2}}[\frac{1}{2}\psi(\Lambda^m_{i+1/2})(g^m_i + g^m_{i+1}) - \{\psi(\Lambda^m + \gamma^m)\alpha^m\}_{i+1/2}].$$

The adjustment quantities, for high accuracy, g and γ are given as

$$g^m_i = \mathrm{minmod}[\alpha^m_{i+1/2}, \alpha^m_{i-1/2}],$$

$$\gamma^m_{i+1/2} = \frac{1}{2}\psi(\Lambda^m_{i+1/2}) \begin{cases} (g^m_{i+1} - g^m_i)/\alpha^m_{i+1/2} & \alpha^m_{i+1/2} \neq 0 \\ 0 & \alpha^m_{i+1/2} = 0. \end{cases}$$

$\alpha^m_{i+1/2}$ is defined on the basis of $T^{-1}_{i+1/2}(q_{i+1} - q_i)$, which is $T^{-1}_{i+1/2}(q_{i+1} - q_i)/J_{i+1/2}$ in Harten-Yee's original TVD scheme. Here, an attention should be paid to the evaluation of the average-state at i+1/2. For a perfect gas, Roe has wisely devised an average state which satisfies the "property U". However, this condition cannot be uniquely satisfied for the general real gas. In this paper, we use a simple average, or a Roe's average which would recover "Property U" for a perfect gas. Construction of a more accurate and efficient Riemann solver, for the general real gas, will require further research. The above formulations have been presented in three-dimensional form and two-dimensional form is easily obtained as a subset.

5. SAMPLE CALCULATIONS AND DISCUSSIONS

We calculate two-dimensional shock-induced combustion problem(Fig.1). A high temperature supersonic mixture gas of N_2, H_2, O_2 is ignited by the oblique shock waves, which are caused by the wedges located at the center and on the walls through the duct. The inlet conditions are P=1atm, T=1050K, and M=4. We employ Westbrook's H_2-O_2 chemical reaction model[17], which takes accounts of 17 reactions and 9 species(table 1). Here, we solve Navier-Stokes equations, using 120 x 40 grid system of the half computational domain. Figures 2-4 show the contours of temperature, mass fraction of H_2O and OH, respectively.

A steady state is obtained after the first 2000 time steps, which takes about 190 sec with FACOM VP-200. The same calculation using a perfect gas takes about 110 sec. This indicates the efficiency of our scheme. However, for the reacting flow, we cannot use a CFL number larger than the order of one. This is probably due to the strong nonlinearity of chemical source terms. We think that a more careful treatment of source terms is needed.

6. CONCLUSION

The eigenvalues and eigenvectors are analytically derived for general real gas dynamic equations including total mass continuity equation in generalized curvilinear coordinates. We construct fully implicit upwind schemes for the general real gas using our diagonalizing formulation. Sample calculations show the efficiency and high resolution of our scheme for chemically reacting compressible flows.

REFERENCES

[1] Bussing,T.R.A. and Murman,E.M., "A Finite Volume Method for The Calculation of Compressible Chemically Reacting Flows," AIAA Paper 85-0331, January 1985.

[2] Eklund,D.R. and Hassan.H.A., "The Efficient Calculation of Chemically Reacting Flow," AIAA Paper 86-0563, January 1986.

[3] Wada, Y., Yamaguchi, M. and Kubota, H., "Numerical Investigation of Nozzle Shape Effects on CO_2 Gas-Dynamic Laser Performance," AIAA Paper 87-1452, June 1987.

[4] Godunov, S.K., "A Finite Difference Method for the Numerical Computation of Discontinuous Solutions of the Equations of Fluid Dynamics," Mat. Sb. 47, pp.357-393, 1959.

[5] Osher,S. and Solomon,F., "Upwinded Schemes for Hyperbolic System of Conservation Laws," Math. Comp., Vol.38, pp.339-377,1981.

[6] Roe, P.L., "Approximate Riemann Solvers,Parameter Vectors, and Difference Schemes," J. Comp. Phys. Vol.43, pp.357-372,1981.

[7] Harten, A., "High Resolution Schemes for Hyperbolic Conservation Laws," J. Comp. Phys. Vol.49, pp.357-393.

[8] Chakravarthy, S. R. and Osher, S., " A New Class of high Accuracy TVD Schemes for hyperbolic Conservation Laws," AIAA 85-0363, January 1985.

[9] Takakura,Y., Ishiguro,T. and Ogawa,S.,"On the Recent DIfference Schemes for the Three-Dimensional Difference Schemes," AIAA Paper 87-1151, Jun 1987.

[10] Eberhardt, S. and Brown, K., "A Shock Capturing Technique for Hypersonic, Chemically Relaxing Flows," AIAA Paper 86-0231, January 1986.

[11] Yee,H.C., and Shinn, Judy L., "Semi-Implicit and Fully Implicit Shock-Capturing Methods for Hyperbolic Conservation Laws with Stiff Source Terms," AIAA Paper 87-1116CP, June 1987.

[12] Shinn, Judy L., Yee, H.C. and Uenishi, K., "Extension of a Semi-Implicit and Fully Implicit Shock-Capturing Algorithm for 3-D Fully Coupled, Chemically Reacting Flows in Generalized Coordinates," AIAA Paper 87-1577, June 1987.

[13] Park,C., "On Convergence of Computation of Chemically Reacting Flows," AIAA 85-0247, January 1985.

[14] Warming,R.F.,Beam,M.and Hyett,B.J., "Diagonalization and Simultaneous Symmetrization of the Gas-Dynamic Matrices," Math. of Comp., Vol 29, Num. 132, October 1975, pp.1037-1045.

[15] Wada,Y., Kubota,H., Ogawa,S. and Ishiguro,T. ,"A New Diagonalizing Formulation of General Real Gas-Dynamic Matrices with a New Class of TVD Schemes," National Fluid Dynamics Congress July, 1988, to be published.

[16] Pulliam, T., H., "Recent Improvements in Efficiency, Accuracy, and Convergence for Implicit Approximate Factorization Algorithms," AIAA Paper 85-0360, Jan., 1985.

[17] Westbrook,C.,K.,"Hydrogen Oxidation Kinetics in Gaseous Detonations," Combustion Science and Technology, Vol.29, pp.67-81, 1982.

Table 1 Westbrook's Chemical Reactions[17]

(1)	$H + O_2 \rightleftharpoons O + OH,$	(2)	$H_2 + O \rightleftharpoons H + OH,$
(3)	$H_2O + O \rightleftharpoons OH + OH,$	(4)	$H_2O + H \rightleftharpoons H_2 + OH,$
(5)	$H_2O_2 + OH \rightleftharpoons H_2O + HO_2,$	(6)	$H_2O + M \rightleftharpoons H + OH + M,$
(7)	$H + O_2 + M \rightleftharpoons HO_2 + M,$	(8)	$HO_2 + O \rightleftharpoons O_2 + OH,$
(9)	$HO_2 + H \rightleftharpoons OH + OH,$	(10)	$HO_2 + H \rightleftharpoons H_2 + O_2,$
(11)	$HO_2 + OH \rightleftharpoons H_2O + O_2,$	(12)	$H_2O_2 + O_2 \rightleftharpoons HO_2 + HO_2,$
(13)	$H_2O_2 + M \rightleftharpoons OH + OH + M,$	(14)	$H_2O_2 + H \rightleftharpoons HO_2 + H_2,$
(15)	$O + H + M \rightleftharpoons OH + M,$	(16)	$O_2 + M \rightleftharpoons O + O + M,$
(17)	$H_2 + M \rightleftharpoons H + H + M$		

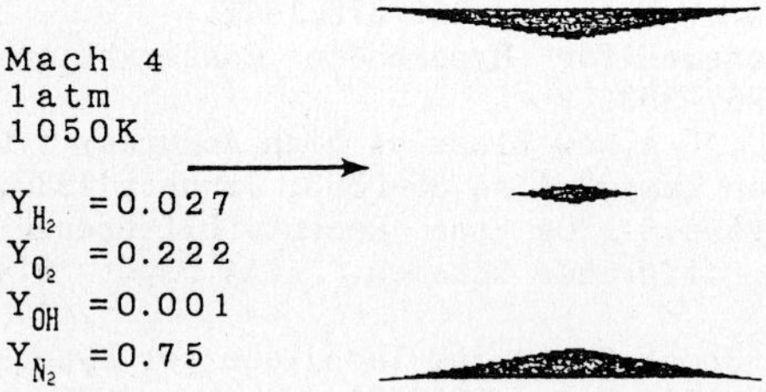

Fig.1 Shock-Induced Combustion Problem and Inflow Conditions

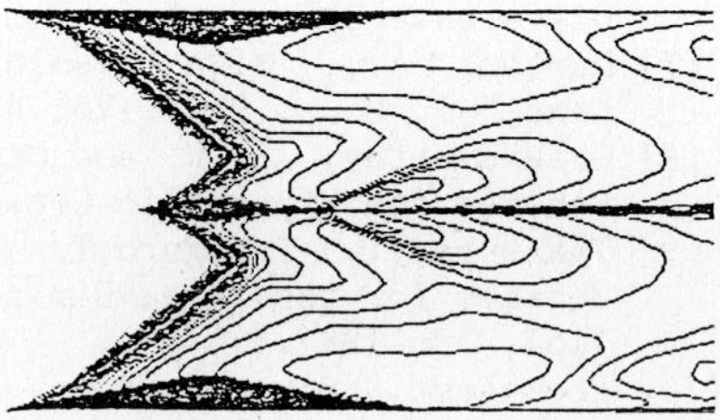

Fig.3 Contours of H_2O Mass Fraction (Maximum Y_{H_2O}=0.213)

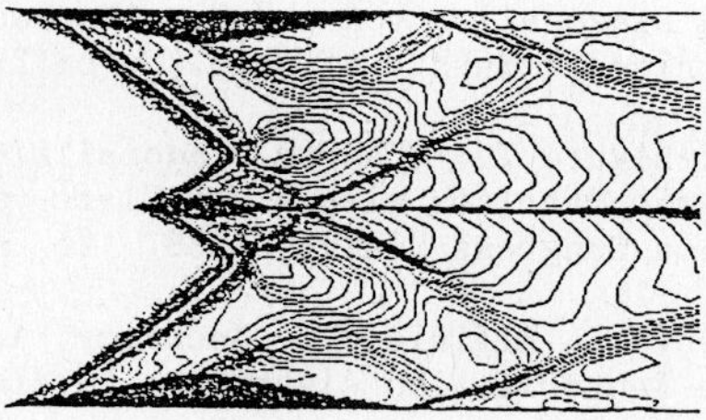

Fig.2 Temperature Contours (Maximum Temperature=3402K)

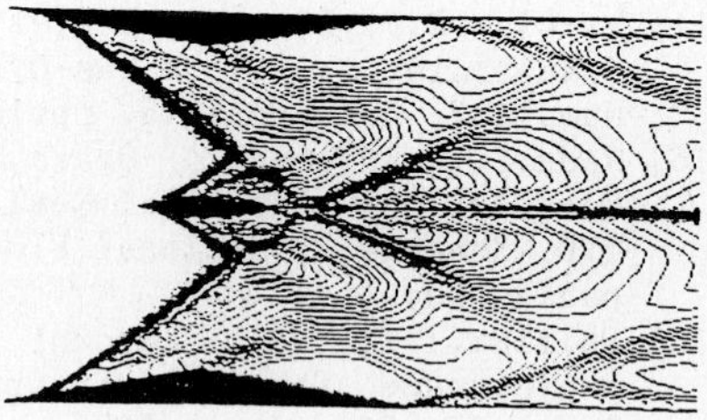

Fig.4 Contours of OH Mass Fraction (Maximum Y_{OH}=0.038)

APPENDIX

1. Common Notations

$$E \equiv e+\rho q^2/2$$
$$b \equiv e+P$$
$$(\tilde{\ }) \equiv (\ )/\sqrt{k_x^2+k_y^2+k_z^2}$$
$$\theta \equiv k_x u+k_y v+k_z w$$
$$q^2 \equiv u^2+v^2+w^2$$
$$c^2 = ab+\rho g$$

$$a \equiv \frac{1}{\rho}\frac{\partial P}{\partial e}\bigg|_{\rho,\, f_i},\qquad g \equiv \frac{1}{\rho}\frac{\partial P}{\partial \rho}\bigg|_{f_i,\, e},\qquad d_i \equiv \frac{1}{\rho}\frac{\partial P}{\partial f_i}\bigg|_{\rho,\, f_{j\neq i},\, e}$$

These notations are for three-dimension. When k_z and w are set to be zero, you get those of two-dimension.

2. Three-Dimensional Matrices

2.1 Eigenvalue Matrix

$$\Lambda \equiv \mathrm{diag}\,(U,\ U,\ U,\ U+C,\ U-C,\ \ U,\ U,\ \cdots,\ U)$$

where

$$U = k_t+k_x u+k_y v+k_z w \qquad C = \sqrt{k_x^2+k_y^2+k_z^2}\cdot c$$

2.2 Relation Between T^{-1} and T of the Form $T_k^{-1}T_l$

$$\left(\begin{array}{ccccc|c}
m_1 & m_2 & m_3 & -m_4/\sqrt{2} & m_4/\sqrt{2} & \\
-m_2 & m_1 & m_4 & m_3/\sqrt{2} & -m_3/\sqrt{2} & \\
-m_3 & -m_4 & m_1 & -m_2/\sqrt{2} & m_2/\sqrt{2} & O \\
m_4/\sqrt{2} & -m_3/\sqrt{2} & m_2/\sqrt{2} & (1+m_1)/2 & (1-m_1)/2 & \\
-m_4/\sqrt{2} & m_3/\sqrt{2} & -m_2/\sqrt{2} & (1-m_1)/2 & (1+m_1)/2 & \\
\hline
 & & O & & & \begin{matrix} 1 & & & O \\ & 1 & & \\ & & \cdot & \\ O & & \cdot & \\ & & & 1 \end{matrix}
\end{array}\right)$$

where

$$m_1 \equiv \tilde{k}_x\cdot\tilde{l}_x+\tilde{k}_y\cdot\tilde{l}_y+\tilde{k}_z\cdot\tilde{l}_z,\qquad m_2 \equiv \tilde{k}_x\cdot\tilde{l}_y-\tilde{k}_y\cdot\tilde{l}_x$$
$$m_3 \equiv \tilde{k}_x\cdot\tilde{l}_z-\tilde{k}_z\cdot\tilde{l}_x,\qquad m_4 \equiv \tilde{k}_y\cdot\tilde{l}_z-\tilde{k}_z\cdot\tilde{l}_y$$

2.3 Right Eigenvector Matrix T

$\widetilde{kx}$	$\widetilde{ky}$	$\widetilde{kz}$	$\frac{\rho}{\sqrt{2}\,c}$	$\frac{\rho}{\sqrt{2}\,c}$	0	0	...	0
$\widetilde{kx}u$	$\widetilde{ky}u - kz\rho$	$\widetilde{kz}u + \widetilde{ky}\rho$	$\frac{\rho\,(u+\widetilde{kx}c)}{\sqrt{2}\,c}$	$\frac{\rho\,(u-\widetilde{kx}c)}{\sqrt{2}\,c}$	0	0	...	0
$\widetilde{kx}v + \widetilde{kz}\rho$	$\widetilde{ky}v$	$\widetilde{kz}v - \widetilde{kx}\rho$	$\frac{\rho\,(v+\widetilde{ky}c)}{\sqrt{2}\,c}$	$\frac{\rho\,(v-\widetilde{ky}c)}{\sqrt{2}\,c}$	0	0	...	0
$\widetilde{kx}w - \widetilde{ky}\rho$	$\widetilde{ky}w + kx\rho$	$\widetilde{kz}w$	$\frac{\rho\,(w+\widetilde{kz}c)}{\sqrt{2}\,c}$	$\frac{\rho\,(w-\widetilde{kz}c)}{\sqrt{2}\,c}$	0	0	...	0
$\widetilde{kx}\,(\frac{q^2}{2} - \frac{g}{a})$ $+\rho\,(\widetilde{kz}v-\widetilde{ky}w)$	$\widetilde{ky}\,(\frac{q^2}{2} - \frac{g}{a})$ $+\rho\,(\widetilde{kx}w-\widetilde{kz}u)$	$\widetilde{kz}\,(\frac{q^2}{2} - \frac{g}{a})$ $+\rho\,(\widetilde{ky}u-\widetilde{kx}v)$	$\frac{\rho\,(\frac{q^2}{2} + \frac{b}{\rho} + \widetilde{\theta}c)}{\sqrt{2}\,c}$	$\frac{\rho\,(\frac{q^2}{2} + \frac{b}{\rho} - \widetilde{\theta}c)}{\sqrt{2}\,c}$	$-\frac{d1}{a}$	$-\frac{d2}{a}$	...	$-\frac{dn}{a}$
$\widetilde{kx}f1$	$\widetilde{ky}f1$	$\widetilde{kz}f1$	$\frac{\rho f1}{\sqrt{2}\,c}$	$\frac{\rho f1}{\sqrt{2}\,c}$	ρ		O	
$\widetilde{kx}f2$	$\widetilde{ky}f2$	$\widetilde{kz}f2$	$\frac{\rho f2}{\sqrt{2}\,c}$	$\frac{\rho f2}{\sqrt{2}\,c}$		ρ		
...	...	...	...	...	O		...	
$\widetilde{kx}fn$	$\widetilde{ky}fn$	$\widetilde{kz}fn$	$\frac{\rho fn}{\sqrt{2}\,c}$	$\frac{\rho fn}{\sqrt{2}\,c}$				ρ

2.4 Left Eigenvector Matrix T^{-1}

$hh \cdot \widetilde{kx} - \frac{1}{\rho}(\widetilde{kz}v - \widetilde{ky}w)$	$\frac{a\rho}{c^2}\widetilde{kx}u$	$\frac{a\rho}{c^2}\widetilde{kx}v + \frac{\widetilde{kz}}{\rho}$	$\frac{a\rho}{c^2}\widetilde{kx}w - \frac{\widetilde{ky}}{\rho}$	$-\frac{a\rho}{c^2}\widetilde{kx}$	$-\frac{d1kx}{c^2}$	$-\frac{d2kx}{c^2}$	$\cdots$	$-\frac{dnkx}{c^2}$
$hh \cdot \widetilde{ky} - \frac{1}{\rho}(\widetilde{kx}w - \widetilde{kz}u)$	$\frac{a\rho}{c^2}\widetilde{ky}u - \frac{\widetilde{kz}}{\rho}$	$\frac{a\rho}{c^2}\widetilde{ky}v$	$\frac{a\rho}{c^2}\widetilde{ky}w + \frac{\widetilde{kx}}{\rho}$	$-\frac{a\rho}{c^2}\widetilde{ky}$	$-\frac{d1ky}{c^2}$	$-\frac{d2ky}{c^2}$	$\cdots$	$-\frac{dnky}{c^2}$
$hh \cdot \widetilde{kz} - \frac{1}{\rho}(\widetilde{ky}u - \widetilde{kx}v)$	$\frac{a\rho}{c^2}\widetilde{kz}u + \frac{\widetilde{ky}}{\rho}$	$\frac{a\rho}{c^2}\widetilde{kz}v - \frac{\widetilde{kx}}{\rho}$	$\frac{a\rho}{c^2}\widetilde{kz}w$	$-\frac{a\rho}{c^2}\widetilde{kz}$	$-\frac{d1kz}{c^2}$	$-\frac{d2kz}{c^2}$	$\cdots$	$-\frac{dnkz}{c^2}$
$\frac{a\rho\frac{q^2}{2} - \widetilde{\theta}c + \rho g - \Sigma fidi}{\sqrt{2}\,\rho c}$	$-\frac{a\rho u - \widetilde{kx}c}{\sqrt{2}\,\rho c}$	$-\frac{a\rho v - \widetilde{ky}c}{\sqrt{2}\,\rho c}$	$-\frac{a\rho w - \widetilde{kz}c}{\sqrt{2}\,\rho c}$	$\frac{a}{\sqrt{2}\,c}$	$\frac{d1}{\sqrt{2}\,\rho c}$	$\frac{d2}{\sqrt{2}\,\rho c}$	$\cdots$	$\frac{dn}{\sqrt{2}\,\rho c}$
$\frac{a\rho\frac{q^2}{2} + \widetilde{\theta}c + \rho g - \Sigma fidi}{\sqrt{2}\,\rho c}$	$-\frac{a\rho u + \widetilde{kx}c}{\sqrt{2}\,\rho c}$	$-\frac{a\rho v + \widetilde{ky}c}{\sqrt{2}\,\rho c}$	$-\frac{a\rho w + \widetilde{kz}c}{\sqrt{2}\,\rho c}$	$\frac{a}{\sqrt{2}\,c}$	$\frac{d1}{\sqrt{2}\,\rho c}$	$\frac{d2}{\sqrt{2}\,\rho c}$	$\cdots$	$\frac{dn}{\sqrt{2}\,\rho c}$
$-\frac{f1}{\rho}$	0	0	0	0	$\frac{1}{\rho}$		O	
$-\frac{f2}{\rho}$	0	0	0	0		$\frac{1}{\rho}$		
$\vdots$	$\vdots$	$\vdots$	$\vdots$	$\vdots$			$\ddots$	
$-\frac{fn}{\rho}$	0	0	0	0	O			$\frac{1}{\rho}$

where

$$hh \equiv \frac{ab}{c^2}\left(1 - \frac{q^2}{2}\cdot\frac{\rho}{b}\right) + \frac{\Sigma fidi}{c^2}$$

3. Two-Dimensional Matrices

3.1 Eigenvalue Matrix

$$\Lambda \equiv \mathrm{diag}(U,\ U,\ U+C,\ U-C,\ \ U,\ U,\ \cdots,\ U)$$

where

$$U = k_t + k_x u + k_y v \qquad C = \sqrt{k_x^2 + k_y^2}\cdot c$$

3.2 Relation Between T^{-1} and T of the Form $T_k^{-1}T_l$

1	0	0	0				
0	m_1	$-m_2/\sqrt{2}$	$m_2/\sqrt{2}$				
0	$m_2/\sqrt{2}$	$(1+m_1)/2$	$(1-m_1)/2$				
0	$-m_2/\sqrt{2}$	$(1-m_1)/2$	$(1+m_1)/2$				
				1	0	0	0
				0	1	0	0
				0	0	$\cdot$ $\cdot$	0
				0	0	0	1

where $m_1 \equiv \widetilde{kx}\cdot\widetilde{lx}+\widetilde{ky}\cdot\widetilde{ly}$, $\quad m_2 \equiv \widetilde{kx}\cdot\widetilde{ly}-\widetilde{ky}\cdot\widetilde{lx}$

3.3 Right Eigenvector Matrix T

1	0	$\frac{\rho}{\sqrt{2}\,c}$	$\frac{\rho}{\sqrt{2}\,c}$	0	0	$\cdots$	0
u	$\widetilde{ky}\rho$	$\frac{\rho(u+\widetilde{kx}c)}{\sqrt{2}\,c}$	$\frac{\rho(u-\widetilde{kx}c)}{\sqrt{2}\,c}$	0	0	$\cdots$	0
v	$-\widetilde{kx}\rho$	$\frac{\rho(v+\widetilde{ky}c)}{\sqrt{2}\,c}$	$\frac{\rho(v-\widetilde{ky}c)}{\sqrt{2}\,c}$	0	0	$\cdots$	0
$\frac{q^2}{2}-\frac{g}{a}$	$\rho(\widetilde{ky}u-\widetilde{kx}v)$	$\frac{\rho(\frac{q^2}{2}+\frac{b}{\rho}+\tilde{\theta}c)}{\sqrt{2}\,c}$	$\frac{\rho(\frac{q^2}{2}+\frac{b}{\rho}-\tilde{\theta}c)}{\sqrt{2}\,c}$	$-\frac{d1}{a}$	$-\frac{d2}{a}$	$\cdots$	$-\frac{dn}{a}$
f1	0	$\frac{\rho f1}{\sqrt{2}\,c}$	$\frac{\rho f1}{\sqrt{2}\,c}$	ρ	0	0	0
f2	0	$\frac{\rho f2}{\sqrt{2}\,c}$	$\frac{\rho f2}{\sqrt{2}\,c}$	0	ρ	0	0
$\vdots$	$\vdots$	$\vdots$	$\vdots$	0	0	$\ddots$	0
fn	0	$\frac{\rho fn}{\sqrt{2}\,c}$	$\frac{\rho fn}{\sqrt{2}\,c}$	0	0	0	ρ

3.4 Left Eigenvector Matrix T^{-1}

hh	$\frac{a\rho}{c^2}u$	$\frac{a\rho}{c^2}v$	$-\frac{a\rho}{c^2}$	$-\frac{d1}{c^2}$	$-\frac{d2}{c^2}$	$\cdots$	$-\frac{dn}{c^2}$
$-\frac{1}{\rho}(\widetilde{ky}u-\widetilde{kx}v)$	$\frac{\widetilde{ky}}{\rho}$	$-\frac{\widetilde{kx}}{\rho}$	0	0	0	$\cdots$	0
$\frac{a\rho\frac{q^2}{2}-\theta c+\rho g-\Sigma fidi}{\sqrt{2}\,\rho c}$	$-\frac{a\rho u-\widetilde{kx}c}{\sqrt{2}\,\rho c}$	$-\frac{a\rho v-\widetilde{ky}c}{\sqrt{2}\,\rho c}$	$\frac{a}{\sqrt{2}\,c}$	$\frac{d1}{\sqrt{2}\,\rho c}$	$\frac{d2}{\sqrt{2}\,\rho c}$	$\cdots$	$\frac{dn}{\sqrt{2}\,\rho c}$
$\frac{a\rho\frac{q^2}{2}+\theta c+\rho g-\Sigma fidi}{\sqrt{2}\,\rho c}$	$-\frac{a\rho u+\widetilde{kx}c}{\sqrt{2}\,\rho c}$	$-\frac{a\rho v+\widetilde{ky}c}{\sqrt{2}\,\rho c}$	$\frac{a}{\sqrt{2}\,c}$	$\frac{d1}{\sqrt{2}\,\rho c}$	$\frac{d2}{\sqrt{2}\,\rho c}$	$\cdots$	$\frac{dn}{\sqrt{2}\,\rho c}$
$-\frac{f1}{\rho}$	0	0	0	$\frac{1}{\rho}$			
$-\frac{f2}{\rho}$	0	0	0		$\frac{1}{\rho}$		
$\vdots$	$\vdots$	$\vdots$	$\vdots$			$\ddots$	
$-\frac{fn}{\rho}$	0	0	0				$\frac{1}{\rho}$

where

$$hh \equiv \frac{ab}{c^2}\left(1-\frac{q^2}{2}\cdot\frac{\rho}{b}\right) + \frac{\Sigma fidi}{c^2}$$

STABILITY OF SEMI-DISCRETE APPROXIMATIONS FOR HYPERBOLIC INITIAL-BOUNDARY-VALUE PROBLEMS: STATIONARY MODES

ROBERT F. WARMING AND RICHARD M. BEAM

NASA Ames Research Center
Moffett Field, CA 94035, USA

SUMMARY

Spatially discrete difference approximations of hyperbolic initial-boundary-value problems (IBVPs) require *numerical boundary conditions* in addition to the analytical boundary conditions specified for the differential equations. Improper treatment of a numerical boundary condition can cause instability of the discrete IBVP even though the approximation is stable for the pure initial-value or Cauchy problem. In the discrete IBVP stability literature there exists a small class of discrete approximations which are called *borderline* cases. For nondissipative approximations, borderline cases are unstable according to the theory of the Gustafsson, Kreiss, and Sundström (GKS) but they may be Lax-Richtmyer stable or unstable in the L_2 norm on a finite domain. We show that borderline approximations can be characterized by the presence of a stationary mode for the finite domain problem. A stationary mode has the property that it does not decay with time and a nontrivial stationary mode leads to algebraic growth of the solution norm with mesh refinement. We give an analytical condition which makes it easy to detect a stationary mode and we investigate several examples of numerical boundary conditions corresponding to borderline cases.

1. INTRODUCTION

In this paper we consider the stability of spatially discrete approximations to hyperbolic IBVPs. For simplicity we consider the stability of approximations to the IBVP for the model hyperbolic equation

$$\frac{\partial u}{\partial t} = c\frac{\partial u}{\partial x}, \quad 0 \leq x \leq L, \quad t > 0 \tag{1.1}$$

where c is a real constant. One must specify initial data at $t = 0$, and the IBVP is well-posed if an *analytical* boundary condition is prescribed at $x = L$

$$u(L,t) = g(t) \quad \text{for} \quad c > 0. \tag{1.2}$$

A semi-discrete approximation of (1.1) is obtained by dividing the spatial interval into J subintervals of length Δx where $J\Delta x = L$, $x = x_j = j\Delta x$ and approximating the spatial derivative u_x by a difference quotient. As a prototype approximation we replace u_x by a second-order-accurate central-difference quotient to obtain a system of ordinary differential equations (ODEs)

$$\frac{du_j}{dt} = \frac{c}{2\Delta x}(u_{j+1} - u_{j-1}), \quad j = 1, 2, \cdots, J-1 \tag{1.3}$$

where $u_j = u_j(t)$ denotes the approximation to $u(x,t)$. The right boundary ($x = L$) is advanced by using the analytical boundary condition (1.2). For the stability analysis we assume that the boundary condition is homogeneous, i.e., $g(t) = 0$, and for the semi-discrete problem we write $u_J = 0$.

A complication in completing the approximation is the fact that more boundary conditions are required for the semi-discrete approximation than are specified for the partial differential equation. If we apply (1.3) at the left boundary ($j = 0$), then the difference stencil protrudes one point to the left of the boundary. It is clear that that a numerical boundary condition is required. For example, at the left boundary ($j = 0$) we can change from a centered approximation to a one-sided approximation of u_x:

$$\frac{du_0}{dt} = \frac{c}{\Delta x}[-\alpha u_2 + (1+2\alpha)u_1 - (1+\alpha)u_0] \tag{1.4}$$

where α is a parameter. Any procedure, e.g., (1.4), used to provide a numerical boundary condition is called a *numerical boundary scheme* (NBS). In the stability analysis of this paper it is convenient to express the NBS as a space extrapolation formula. The NBS (1.4) is equivalent to

$$q(E)u_{-1} = 0, \quad \text{where} \quad q(E) = (E-1)^2(2\alpha E - 1) \tag{1.5}$$

and the shift operator E is defined by

$$Eu_j = u_{j+1}. \tag{1.6}$$

The system of ODEs (1.3) together with the analytical boundary condition $u_J = 0$ and the NBS (1.4) can be written in vector-matrix form as

$$\frac{d\mathbf{u}(t)}{dt} = A\mathbf{u}(t) \tag{1.7}$$

where $\mathbf{u}$ is a J-component vector and A is a $J \times J$ matrix. The essential element in the stability of the semi-discrete approximation (1.7) is the behavior of the solution at a fixed time as the spatial mesh is refined. Consequently, one must consider an infinite sequence of ODE systems of dimension J where $J \to \infty$ as $\Delta x \to 0$.

For the semi-discrete approximation (1.3) with *periodic* boundary conditions the matrix $A = A_p$ is a skew-symmetric circulant matrix. Consequently the eigenvalues of A_p are pure imaginary and the semi-discrete approximation is said to be *nondissipative*. The analysis of stationary modes for dissipative approximations will be given in a subsequent paper.

Stability of a semi-discrete approximation with homogeneous boundary data means that there exists an estimate of the solution in terms of the initial data. For example, the semi-discrete approximation represented by the sequence of ODE's (1.7) is Lax-Richtmyer stable if there exists a constant $K > 0$ such that for any initial condition $\mathbf{u}(0)$

$$||\mathbf{u}(t)|| \leq K||\mathbf{u}(0)|| \tag{1.8}$$

for all $J > 0$, $J\Delta x = L$ and for all t, $0 \leq t \leq T$ with T fixed. In this paper the symbol $||\cdot||$ denotes the discrete L_2 norm defined by

$$||\mathbf{u}(t)||_2 = \left(\sum_{j=0}^{J-1} u_j^2 \Delta x\right)^{1/2}. \tag{1.9}$$

Two methods for carrying out a stability analysis are the energy method and the normal mode analysis. The normal mode analysis is an eigenvalue analysis. If we look for a solution of (1.7) of the form $\mathbf{u}(t) = e^{st}\boldsymbol{\phi}$, then we obtain $A\boldsymbol{\phi} = s\boldsymbol{\phi}$. But this is just the eigenvalue-eigenvector problem for the matrix A where $\boldsymbol{\phi}$ is the eigenvector and s is the eigenvalue. The practical problem of implementing stability tests on the eigenvalues is that the normal mode analysis for a discrete hyperbolic IBVP on a finite domain is, in general, analytically intractable.

The intractability can easily be demonstrated by the normal mode analysis of the ODE system (1.7). The components ϕ_j of the eigenvector $\boldsymbol{\phi}$ and the normalized eigenvalue $\hat{s} = (\Delta x/c)s$ are given by

$$\phi_j = a[\kappa^j - (-\kappa^2)^J(-1/\kappa)^j], \qquad 2\hat{s} = \kappa - \frac{1}{\kappa} \tag{1.10a,b}$$

where a is an arbitrary nonzero constant and κ is a root of the characteristic equation

$$q(\kappa) - (-\kappa^2)^{J+1} q(-1/\kappa) = 0. \tag{1.10c}$$

The polynomial $q(\kappa)$ depends solely on the NBS, i.e., $q(\kappa)$ is the polynomial associated with the NBS written as an extrapolation formula. For example, the polynomial $q(\kappa)$ for the NBS (1.5) is

$$q(\kappa) = (\kappa - 1)^2 (2\alpha\kappa - 1). \tag{1.11}$$

Since $J\Delta x = L$, the degree of (1.10c) increases as the spatial mesh increment Δx decreases. In general, one cannot solve for the roots of (1.10c) which accounts for the analytic intractability of the normal mode analysis on a finite-domain.

Although the Lax-Richtmyer condition (1.8) is a conventional stability definition, there is no known algebraic test to check the stability condition for discrete hyperbolic IBVPs on a finite domain. A more complicated stability definition is used in the theory developed by Gustafsson, Kreiss, and Sundström (GKS) [1]. Strikwerda [4] has extended the GKS theory to semi-discrete approximations. The advantage of the GKS theory accrues from the fact that a finite-domain problem with two boundaries is divided into a Cauchy problem and two quarter-plane problems each of which can be analyzed separately by a normal mode analysis. The analogues of (1.10) for the right-quarter plane problem are

$$\phi_j = a\kappa^j, \qquad 2\hat{s} = \kappa - \frac{1}{\kappa} \tag{1.12a,b}$$

where κ is a root of

$$q(\kappa) = 0. \tag{1.12c}$$

This is the same polynomial $q(\kappa)$ that appears in (1.10c). The roots of (1.12c) are easily found since $q(\kappa)$ is of low degree. Algebraic tests based on the roots of $q(\kappa)$ and the corresponding eigenvalues $\hat{s}$ provide necessary and sufficient conditions for GKS stability.

The connection between the normal mode analysis for the finite-domain problem and the normal mode analysis for the quarter-plane problem is rather obscure. In a recent paper [6] we derived asymptotic estimates of the eigenvalues for the finite-domain problem. These estimates were used to relate the normal mode analysis of the finite-domain problem and the GKS quarter-plane analysis. In order to derive the asymptotic estimates for the roots of (1.10c), we assumed that particular roots can be identified for each J and we write $\kappa = \kappa(J)$. Furthermore, there is no loss of generality in assuming that $|\kappa(J)| \leq 1$. We showed that the roots of the characteristic equation (1.10c) can be divided into three distinct classes according to the asymptotic behavior of $|\kappa(J)|^J$ in the limit $J \to \infty$. For $|\kappa(J)| < 1$ there are only two possibilites:

$$(I): \quad \lim_{J\to\infty} |\kappa(J)|^J = 0, \qquad (II): \quad \lim_{J\to\infty} |\kappa(J)|^J = \text{constant} > 0. \tag{1.13}$$

For roots in class (I), it is clear that (1.10c) reduces to the quarter-plane equation (1.12c) as $J \to \infty$. Consequently, a root in class (I) becomes a root of the quarter-plane polynomial (1.12c) in the limit $J \to \infty$. The eigenvalues corresponding to the κ's of class (II) are ***benign*** in the stability analysis in the sense that they satisfy the inequality $\Re(\hat{s}) \leq \text{const}/J$ (see [6]).

2. STATIONARY MODES

In addition to (I) and (II), there is a third class of roots

$$(III): \qquad |\kappa(J)| = 1. \tag{2.1}$$

If κ is in class (III) and is independent of J, then κ remains fixed on the unit circle for all J. For this to happen it is obvious from (1.10c) that the polynomials $q(\kappa)$ and $q(-1/\kappa)$ must have a common factor. This common factor leads to identical roots for both the quarter plane polynomial $q(\kappa)$ and the finite-domain characteristic equation (1.10c). These roots are fixed (independent of J) on the unit circle and from (1.10b) one obtains $\Re(\hat{s}) = 0$ with $\Im(\hat{s}) =$ fixed. Consequently, there is a *stationary* mode, i.e., a mode with κ and $\hat{s}$ independent of J.

If there is a stationary mode for the finite domain problem, the GKS perturbation test will always indicate the presence of a GKS generalized eigenvalue and consequently the semi-discrete approximation is GKS unstable. The existence of a GKS generalized eigenvalue is easily proved. Since $|\kappa| = 1$ and both κ and $-1/\kappa$ are roots of $q(\kappa) = 0$, we need two perturbation tests, i.e.,

$$\kappa_a^* = (1-\epsilon)e^{i\theta}, \quad \kappa_b^* = -(1-\epsilon)e^{-i\theta} \tag{2.2}$$

where κ_a^* and κ_b^* denote the perturbations of κ and $-1/\kappa$ inside the unit circle. But from (1.10b) it follows that $\Re[\hat{s}(\kappa_a^*)] = -\Re[\hat{s}(\kappa_b^*)] \neq 0$ and consequently there is a generalized eigenvalue $\Re(\hat{s}^*) > 0$ corresponding to either κ or $-1/\kappa$.

The importance of a stationary mode is the following. Gustafsson et al. [2] have proved that if the Cauchy stability requirement of the GKS theory is replaced by a more stringent energy estimate, then GKS stability implies Lax-Richtmyer stability in the L_2 norm. There is a small number of known examples showing that Lax-Richtmyer stability in the L_2 norm does not imply GKS stability. These examples all involve what are called *borderline* cases. We show that borderline cases have a stationary mode for the finite-domain problem. The GKS quarter plane analysis cannot detect whether or not a particular mode is stationary for the finite domain problem. However, from our analysis, stationary modes are easy to detect since they occur if and only if $q(\kappa)$ and $q(-1/\kappa)$ have a common factor.

From the point of view of an eigenvalue analysis, a semi-discrete approximation with a stationary mode must be treated separately since any instability derives not from an eigenvalue with a positive real part but from the algebraic growth (as $J \to \infty$) of the norm of the solution.

3. EXAMPLES OF STATIONARY MODES

In this section we examine several examples of stationary modes arising from various NBSs for semi-discrete approximations. We analyze the first example in detail but give only a summary of the stationary mode analysis for the subsequent examples. All of the examples are GKS unstable. Examples 1 and 2 are Lax-Richtmyer unstable but examples 3 and 4 are Lax-Richtmyer stable. In each of the examples we follow the convention of having the boundary condition of interest on the left boundary, i.e., the GKS analysis is done for a right quarter-plane problem.

EXAMPLE 1 Our first example of a semi-discrete approximation with a stationary mode is NBS (1.4) or equivalently (1.5) for $\alpha = -1/2$. The polynomial (1.11) becomes

$$q(\kappa) = -(\kappa^2-1)(\kappa-1), \quad q(-1/\kappa) = -(\kappa^2-1)(\kappa+1)\kappa^3. \tag{3.1}$$

The polynomials $q(\kappa)$ and $q(-1/\kappa)$ have the common factor (κ^2-1) and consequently there is a stationary mode. The characteristic equation (1.10c) has the roots

$$\kappa = \pm 1 \tag{3.2}$$

independent of J and, from (1.10b), $\hat{s} = 0$. According to a GKS stability analysis, $\alpha = -1/2$ is the *borderline* case between stability ($\alpha > -1/2$) and instability ($\alpha < -1/2$). For $\alpha = -1/2$ there is a stationary mode and consequently this borderline case is GKS unstable.

The NBS (1.4) with $\alpha = -1/2$ is

$$\frac{du_0}{dt} = \frac{c}{2\Delta x}(u_2 - u_0). \tag{3.3}$$

As an aid to interpreting the solution of a semi-discrete approximation with (3.3) as the NBS, it is useful to consider zeroth-order extrapolation

$$u_0(t) = u_1(t) \tag{3.4}$$

as an NBS. Differentiation of (3.4) with respect to time yields

$$\frac{du_0}{dt} = \frac{du_1}{dt}. \tag{3.5}$$

By replacing du_1/dt in the above equation by the interior approximation (1.3) evaluated at $j = 1$, one obtains the NBS (3.3). If we integrate (3.5) from 0 to t then

$$u_0(t) - u_1(t) = u_0(0) - u_1(0) = \text{constant}. \tag{3.6}$$

Hence the NBS (3.3) is equivalent to zeroth-order space extrapolation *if* the initial data is reset at the outflow boundary so that

$$u_0(0) - u_1(0) = 0. \tag{3.7}$$

A semi-discrete approximation with zeroth-order extrapolation (3.4) as the NBS is both GKS stable and Lax-Richtmyer stable. However, differentiation of the NBS (3.4) yields an approximation which is both GKS unstable and Lax-Richmyer unstable. Next we show that the Lax-Richtmyer instability is due to the presence of a stationary mode.

The stationary eigenvector corresponding to the stationary roots (3.2) is found by substitution of either $\kappa = 1$ or -1 into (1.10a) to obtain

$$\phi_j = a[1 - (-1)^{j+J}], \quad j = 0,1,,\cdots,J-1 \tag{3.8}$$

where a is an arbitrary nonzero constant. The L_2 norm of the eigenvector (3.8) is

$$\|\boldsymbol{\phi}\|_2 = \left(\sum_{j=0}^{J-1} \phi_j^2 \Delta x\right)^{1/2} = \begin{cases} |a|\sqrt{2L}, & J \text{ even} \\ |a|\sqrt{2L(J+1)/J}, & J \text{ odd} \end{cases} \tag{3.9}$$

where we have used $J\Delta x = L$.

The matrix A corresponding to the ODE system (1.3) with (3.3) as the NBS and $u_J = 0$ as the analytical boundary condition is $A = (c/\Delta x)\widehat{A}$ where

$$\widehat{A} = \frac{1}{2}\begin{bmatrix} -1 & 0 & 1 & & & \\ -1 & 0 & 1 & & O & \\ & \cdot & \cdot & \cdot & & \\ & & \cdot & \cdot & \cdot & \\ & & & \cdot & \cdot & \cdot \\ & O & & -1 & 0 & 1 \\ & & & & -1 & 0 \end{bmatrix}. \tag{3.10}$$

Since the first two rows are equal it is obvious that the matrix $\widehat{A}$ has an eigenvalue $\hat{s} = 0$. The eigenvector (3.8) is the right eigenvector $\boldsymbol{\phi}$ of the matrix (3.10) corresponding to $\hat{s} = 0$. The left eigenvector $\boldsymbol{\xi}$ corresponding to $\hat{s} = 0$ is easily found by inspection of the matrix (3.10) to be

$$\boldsymbol{\xi}^T = [-1,1,0,\cdots,0]. \tag{3.11}$$

If we choose $a = 1/2$ in (3.8), then the inner product of $\boldsymbol{\xi}$ and $\boldsymbol{\phi}$ is

$$\boldsymbol{\xi}^T\boldsymbol{\phi} = \pm 1 \tag{3.12}$$

where the plus sign is used if J is even and the minus sign if J is odd.

The matrix (3.10) has a complete set of eigenvectors and consequently the general solution of the ODE system (1.7) can be written as

$$\mathbf{u}(t) = \sum_{\ell=0}^{J-1} \beta_\ell e^{s_\ell t} \boldsymbol{\phi}(\ell) \tag{3.13}$$

where $\boldsymbol{\phi}(\ell)$ denotes the eigenvector corresponding to the eigenvalue s_ℓ. We denote the stationary eigenvector (3.8) by $\boldsymbol{\phi}(0)$ and rewrite (3.13) as

$$\mathbf{u}(t) = \beta_0 \boldsymbol{\phi}(0) + \sum_{\ell=1}^{J-1} \beta_\ell e^{s_\ell t} \boldsymbol{\phi}(\ell). \tag{3.14}$$

The sum on the right hand side of (3.14) consists of the eigenvalues and eigenvectors for zeroth-order space extrapolation (3.4) as the NBS.

For given initial data, the coefficient β_0 associated with the stationary eigenvector $\boldsymbol{\phi}(0)$ is determined by taking the inner product of the left eigenvector $\boldsymbol{\xi}$ given by (3.11) with $\mathbf{u}(0)$ given by (3.14):

$$\boldsymbol{\xi}^T \mathbf{u}(0) = [-u_0(0) + u_1(0)] = \pm\beta_0 \tag{3.15}$$

where the plus sign corresponds to J even and the minus sign to J odd. In obtaining (3.15) we have made use of the fact that the right and left eigenvectors of a matrix are orthogonal. Formula (3.15) for the coefficient β_0 is consistent with our earlier assertion that the NBS (3.3) is equivalent to zeroth-order space extrapolation if the initial data is reset at the outfow boundary so that $u_0(0) - u_1(0) = 0$.

One can show analytically that $\Re(s_\ell) < 0$ for all the modes except the stationary mode, and consequently the asymptotic solution ($t \to \infty$) is from (3.14)

$$\mathbf{u}(t) \approx \beta_0 \boldsymbol{\phi}(0), \qquad t \to \infty. \tag{3.16}$$

The L_2 norm of (3.16) for J even is

$$\|\mathbf{u}(t)\|_2 \approx |\beta_0| \|\boldsymbol{\phi}(0)\|_2 \approx |\beta_0| \sqrt{\frac{L}{2}} = |u_0(0) - u_1(0)| \sqrt{\frac{L}{2}}, \qquad t \to \infty \tag{3.17}$$

where we have used (3.9) with $a = 1/2$ and (3.15).

The Lax-Richtmyer instability resulting from the NBS (3.3) arises from the presence of the stationary mode. We illustrate this on the finite domain $0 \le x \le L$ by choosing the initial data

$$u_j(0) = \begin{cases} 1 & \text{for} \quad j = 0 \\ 0 & \text{for} \quad j > 0 \end{cases} \tag{3.18}$$

where $J\Delta x = L$. The L_2 norm of this initial data is

$$\|\mathbf{u}(0)\|_2 = \Delta x^{1/2} = \sqrt{\frac{L}{J}}. \tag{3.19}$$

For the initial data (3.18) with J even the L_2 norm of the asymptotic solution (3.17) is

$$\|\mathbf{u}(t)\|_2 \approx \sqrt{\frac{L}{2}}, \qquad t \to \infty. \tag{3.20}$$

From (3.20) and (3.19) it follows

$$\|\mathbf{u}(t)\|_2 \approx \frac{\sqrt{J}}{\sqrt{2}} \|\mathbf{u}(0)\|_2, \quad t \to \infty. \tag{3.21}$$

Consequently the L_2 norm of the solution is not uniformly bounded on $0 \le t \le T$ (with T large) for the initial data (3.18) and the semi-discrete approximation is Lax-Richtmyer unstable.

If the initial data is smooth, then the semi-discrete approximation converges although the global order of accuracy drops to first order. If we assume that the initial data is *smooth*, then

$$u_0(0) - u_1(0) = f(0) - f(\Delta x) = -\Delta x \frac{\partial f(0)}{\partial x} + O(\Delta x^2). \tag{3.22}$$

In this case (3.17) becomes

$$\|\mathbf{u}(t)\|_2 \approx \left|\frac{\partial f(0)}{\partial x}\right| \sqrt{\frac{L}{2}} \Delta x = O(\Delta x), \quad t \to \infty. \tag{3.23}$$

EXAMPLE 2a In this section the model hyperbolic equation is

$$\frac{\partial u}{\partial t} = -c\frac{\partial u}{\partial x}, \qquad c > 0, \qquad 0 \le x \le L, \qquad t > 0, \tag{3.24}$$

i.e., the analytical BC is specified at the left boundary, $x = 0$. The spatial derivative u_x is approximated by a centered approximation and the PDE (3.24) is replaced by the system of ODEs

$$\frac{du_j}{dt} = \frac{c}{2\Delta x}(u_{j-1} - u_{j+1}), \quad j = 1, 2, \cdots, J-1. \tag{3.25}$$

The homogeneous analytical boundary condition is

$$u_0 = 0. \tag{3.26}$$

The NBS is

$$\frac{du_J}{dt} = \frac{c}{\Delta x}(u_{J-1} - u_J). \tag{3.27}$$

The semi-discrete approximation (3.25) with boundary conditions (3.26) and (3.27) is both Lax-Richtmyer and GKS stable. In example 1 we showed that differentiation with respect to time of a stable NBS, i.e., zeroth-order extrapolation, resulted in an unstable semi-discrete approximation. In this section we show that differentiation of the analytical boundary condition (3.26) with respect to time, i.e.,

$$\frac{du_0}{dt} = 0 \tag{3.28}$$

leads to an unstable approximation. The source of the instability is the introduction of a stationary mode. It should be noted that this example differs from the other examples in this paper since the stationary mode is introduced by an improper modification of the analytical boundary condition, i.e., replacement of (3.26) by (3.28), rather than an improper choice of the NBS.

It is easy to show by evaluating (3.25) at $j = 0$ that (3.28) is equivalent to the extrapolation formula

$$h(E)u_{-1} = 0, \quad \text{where} \quad h(E) = 1 - E^2. \tag{3.29}$$

Since $h(\kappa) = 1 - \kappa^2$ it follows directly that

$$h(-1/\kappa) = -h(\kappa)/\kappa^2 \tag{3.30}$$

as one can easily verify. Therefore, $h(\kappa)$ and $h(-1/\kappa)$ have a common factor, namely $h(\kappa)$ itself, and consequently there is a stationary mode. The roots of $h(\kappa) = 0$ are $\kappa = \pm 1$. The stationary mode has the eigenvalue $\hat{s} = 0$ and the corresponding eigenvector is

$$\phi_j = a, \qquad j = 0, 1, 2, \cdots, J \tag{3.31}$$

i.e., a vector with constant elements.

Since all the modes except the stationary mode are damped for large time t, the asymptotic solution is

$$\mathbf{u}(t) \approx \beta_0 \boldsymbol{\phi}(0) \quad t \to \infty \tag{3.32}$$

where $\beta_0 = u_0(0)$ and the elements of $\boldsymbol{\phi}(0)$ are given by (3.31). For the initial data (3.18) the asymptotic solution is

$$\|\mathbf{u}(t)\|_2 \approx \sqrt{L} = \sqrt{J}\|\mathbf{u}(0)\|_2 \quad t \to \infty. \tag{3.33}$$

Consequently the norm of the solution is not uniformly bounded for $0 \leq t \leq T$ (with T large) for the initial data (3.18) and the semi-discrete approximation is Lax-Richtmyer unstable.

EXAMPLE 2b We return to the model equation (1.1) with the NBS on the left boundary, $x = 0$. If we choose the *overspecified* Dirichlet condition

$$u_0 = 0 \tag{3.34}$$

as the NBS, the resulting semi-discrete approximation is both Lax-Richtmyer and GKS stable. However if we differentiate (3.34) with respect to time

$$\frac{du_0}{dt} = 0 \tag{3.35}$$

the resulting approximation is both Lax-Richtmyer unstable and GKS unstable.

The NBS (3.35) is equivalent to the extrapolation formula

$$q(E)u_{-1} = 0, \quad \text{where} \quad q(E) = 1 - E^2. \tag{3.36}$$

The polynomial $q(\kappa) = 1 - \kappa^2$, and therefore,

$$q(-1/\kappa) = -q(\kappa)/\kappa^2. \tag{3.37}$$

Consequently, there is a stationary mode. For this example and the following examples, 3 and 4, all of the roots of the characteristic equation are in class (III), i.e., $|\kappa(J)| = 1$ and all the eigenvalues are pure imaginary. Consequently, the normal modes for the finite domain problem can be found analytically. For J even the stationary mode has a repeated eigenvalue $\hat{s} = 0$. For sufficiently large time the asymptotic solution is

$$u_j(t) \approx \begin{cases} \dfrac{ct}{L}, & j = \text{odd} \\ 0, & j = \text{even} \end{cases}, \quad t \to \infty. \tag{3.38}$$

For the initial data (3.18), the norm of the asymptotic solution is

$$\|\mathbf{u}(t)\|_2 \approx \frac{\sqrt{J}}{\sqrt{2}} \frac{|c|t}{L} \|\mathbf{u}(0)\|_2, \quad t \to \infty. \tag{3.39}$$

The Lax-Richtmyer instability arises from the factor $\sqrt{J}$ and not from the linear growth in t.

EXAMPLE 3 Consider the inconsistent NBS

$$u_{-1} = -u_1 \tag{3.40}$$

which can be written as

$$q(E)u_{-1} = 0, \quad \text{where} \quad q(E) = 1 + E^2. \tag{3.41}$$

This NBS was used by Trefethen [5] for the leap frog scheme as an example of a fully discrete approximation to (1.1) that is Lax-Richtmyer stable and GKS unstable. As outlined below the same stability results are obtained when the NBS (3.40) is used for the semi-discrete approximation (1.3).

Since $q(\kappa) = 1 + \kappa^2$, one has

$$q(-1/\kappa) = q(\kappa)/\kappa^2 \tag{3.42}$$

and consequently, there is a stationary mode. From (1.10c) one obtains

$$q(\kappa)[1 + (-\kappa^2)^J] = 0. \tag{3.43}$$

The roots of the equation (3.43) are determined from

$$q(\kappa) = 1 + \kappa^2 = 0, \quad \text{and} \quad 1 + (-\kappa^2)^J = 0. \tag{3.44a,b}$$

For the stationary mode the roots of $q(\kappa) = 0$ are $\kappa = \pm i$ and the corresponding eigenvalues are $\hat{s} = \pm i$. However from (1.10b) the roots $\kappa = \pm i$ both lead to trivial stationary eigenvectors for the finite domain problem.

For this example the eigensolutions of the finite domain problem can be found analytically. The remaining roots of the characteristic equation can be determined from (3.44b) by using the roots of unity formula. The corresponding eigenvalues are pure imaginary. In particular, for J odd the eigenvalues are

$$\hat{s}_\ell = i \sin\left(\frac{2\ell\pi}{J}\right), \quad \ell = 1, 2, \cdots, J \quad (J \text{ odd}). \tag{3.45}$$

Here we have the rather amazing result that for J odd, the eigenvalues are analytically identical to those of the circulant matrix associated with the spatially periodic problem. The eigenvectors are, of course, different.

One can show by the energy method that the semi-discrete approximation is Lax-Richtmyer stable. In particular

$$||\mathbf{u}(t)||_2 \leq \sqrt{2}||\mathbf{u}(0)||_2. \tag{3.46}$$

On the other hand the approximation is GKS unstable because there is a stationary mode.

We briefly outline the GKS perturbation analysis. For $\kappa = i$, the eigenvalue is $\hat{s} = i$ where obviously $\Re(\hat{s}) = 0$ and hence we must check to see if there is a GKS generalized eigensolution. Let κ^* denote a perturbation of $\kappa = i$ which is inside the unit circle, i.e.,

$$\kappa = e^{i\theta}(1 - \epsilon), \qquad \theta = \frac{\pi}{2} + \epsilon, \quad \epsilon > 0. \tag{3.47}$$

By inserting (3.47) into (1.10b), we find

$$\Re(\hat{s}^*) = \epsilon^2 + O(\epsilon^3). \tag{3.48}$$

Since the perturbation of $\Re(\hat{s})$ is positive there is a GKS generalized eigenvalue and semi-discrete approximation is GKS unstable. The fact that an ϵ perturbation of κ yields an ϵ^2 perturbation in $\Re(\hat{s})$ indicates a *weaker* type of GKS instability than for a conventional GKS generalized eigenvalue where the perturbation in κ and $\Re(\hat{s})$ are of the same order.

EXAMPLE 4 Our last example is due to Gustafsson[3]. The wave equation, $u_{tt} = u_{xx}$, written as a first-order system is

$$\frac{\partial \mathbf{w}}{\partial t} = B\frac{\partial \mathbf{w}}{\partial x}, \qquad \mathbf{w} = \begin{bmatrix} u \\ v \end{bmatrix}, \qquad B = \begin{bmatrix} 0 & 1 \\ 1 & 0 \end{bmatrix}. \tag{3.49}$$

The initial condition is $\mathbf{w}(x,0) = \mathbf{f}(x)$ and the boundary conditions are

$$u(0,t) = g_1(t), \qquad u(1,t) = g_2(t). \tag{3.50}$$

A semi-discrete approximation is

$$\frac{d\mathbf{w}_j}{dt} = \frac{B}{2\Delta x}(\mathbf{w}_{j+1} - \mathbf{w}_{j-1}). \tag{3.51}$$

The boundary conditions (3.50) are assumed to be homogeneous:

$$u_0 = 0, \qquad u_J = 0. \tag{3.52a,b}$$

The NBSs are

$$\frac{dv_0}{dt} = \frac{u_1 - u_0}{\Delta x}, \qquad \frac{dv_J}{dt} = \frac{u_J - u_{J-1}}{\Delta x}. \tag{3.53a,b}$$

Gustafsson [3] used this semi-discrete approximation as an example of a problem that is Lax-Richtmyer stable for homogeneous boundary data but unstable for highly oscillatory non-zero boundary data. We demonstrate that the approximation is GKS unstable because of the presence of a a stationary mode.

One can show that the NBSs (3.53a,b) are equivalent to the extrapolation formulas

$$h(E)u_{-1} = 0, \qquad p(E)u_{J+1} = 0 \tag{3.54a,b}$$

where

$$h(E) = 1 + E^2, \qquad p(E) = 1 + E^{-2}. \tag{3.55a,b}$$

In deriving (3.54) we have used (3.52). The polynomials $h(\kappa)$ and $p(\kappa)$ are

$$h(\kappa) = 1 + \kappa^2, \qquad p(\kappa) = 1 + 1/\kappa^2 \tag{3.56a,b}$$

and, therefore,

$$h(-1/\kappa) = h(\kappa)/\kappa^2, \qquad p(-1/\kappa) = p(\kappa)\kappa^2. \tag{3.57a,b}$$

Futhermore,

$$h(-\kappa) = h(\kappa), \qquad p(\kappa) = h(1/\kappa). \tag{3.58a,b}$$

As a consequence of (3.57) and (3.58) there is a stationary mode for the finite domain problem.

There is a close connection between this example and previous example. In fact, the example of this section can be written as two uncoupled semi-discrete problems of the form given by example 3. The NBS (3.41) is inconsistent but the analogous NBSs (3.54) only appear to be inconsistent, i.e., they are actually consistent.

As in the example of the previous section the eigensolutions of the finite domain problem can be determined analytically. We look for a solution of (3.51) of the form

$$\mathbf{w}_j = e^{st}\hat{\mathbf{w}}_j. \tag{3.59}$$

The components $\hat{\mathbf{w}}_j$ of the eigenvector $\hat{\mathbf{w}}$ for the finite domain problem are given by

$$\hat{\mathbf{w}}_j = a\kappa^j \boldsymbol{\phi}_+ + b(-1/\kappa)^j \boldsymbol{\phi}_+ + c(-\kappa)^j \boldsymbol{\phi}_- + d(1/\kappa)^j \boldsymbol{\phi}_- \tag{3.60}$$

where

$$\boldsymbol{\phi}_+ = \begin{bmatrix} 1 \\ 1 \end{bmatrix}, \qquad \boldsymbol{\phi} = \begin{bmatrix} 1 \\ -1 \end{bmatrix}. \tag{3.61a,b}$$

The constants a, b, c, d are determined by inserting (3.60) into the boundary conditions (3.52) and (3.54). One obtains a homogeneous linear system and a nontrivial solution exists if and only if

$$h(\kappa) = 0, \qquad \kappa^{2J} = 1. \tag{3.62a,b}$$

The polynomial $p(\kappa)$ does not appear here because of identity (3.58b).

We first consider $h(\kappa) = 0$ where $h(\kappa)$ is defined by (3.56a). The roots are $\kappa = \pm i$ and for J even these roots also satisfy (3.62b). For J odd one can show that the roots $\kappa = \pm i$ lead to trivial functions. From (3.62b) κ is a root of unity and hence

$$\kappa = e^{i\pi\ell/J}, \quad \ell = 1, 2, \cdots, 2J. \tag{3.63}$$

The associated eigenvalues are

$$\hat{s}_\ell = i \sin\left(\frac{\ell\pi}{J}\right), \quad \ell = 1, 2, \cdots, 2J \tag{3.64}$$

where $\hat{s} = s\Delta x$. For J odd, the eigenvalues (3.64) are identical to the eigenvalues of the circulant matrix associated with the IVP. Furthermore there are no eigenvalues $\hat{s} = \pm i$. But for J even there are are two distinct eigenvalues $\hat{s} = \pm i$.

One can prove by the energy method that the semi-discrete approximation of this section is Lax-Richtmyer stable. But the approximation is GKS unstable because there is a stationary mode. The GKS perturbation analysis is identical to the analysis at the end of example 3, i.e., an ϵ perturbation in κ yields an ϵ^2 perturbation in $\Re(\hat{s})$.

4. CONCLUSIONS

Stationary modes for semi-discrete approximations are easy to detect because $q(\kappa)$ and $q(-1/\kappa)$ have a common factor. For *simple* approximations the κ roots in class (III) defined by (2.1) can be determined analytically and consequently so can the normal modes for the finite domain problem. This includes, of course, stationary modes for nondissipative approximations where $|\kappa| = 1$ independent of J.

If there is a stationary mode for a nondissipative approximation on a finite domain problem, then the GKS perturbation test will always indicate the presence of GKS generalized eigenvalue. If in the GKS perturbation test the perturbation in κ is the same order as the perturbation in $\Re(\hat{s})$, then the approximation is Lax-Richtmyer unstable in the L_2 norm. On the other hand if the the perturbation of κ results in a higher order perturbation in $\Re(\hat{s})$, then there is a trivial stationary mode and the approximation is Lax-Richtmyer stable.

If a semi-discrete approximation with a stationary mode is Lax-Richtmyer unstable, then the solution exhibits algebraic growth as the mesh is refined. This is a weak type of instability and the approximation can be thought of as only *marginally* unstable since the approximation converges in the L_2 norm for smooth initial and boundary data on a finite domain.

REFERENCES

[1] B. Gustafsson, H.-O. Kreiss, A. Sundström: "Stability theory of difference approximations for mixed initial boundary value problems. II," Mathematics of Computation, 26, 649-686 (1972).

[2] B. Gustafsson: private communication from preliminary manuscript by B. Gustafsson, H.-O. Kreiss, and J. Oliger (1987).

[3] B. Gustafsson: " Stability analysis for initial-boundary value problems," Lecture Notes in Math., 1005, 128-141, Springer-Verlag, Berlin, (1982).

[4] J.C. Strikwerda: " Initial boundary value problems for the method of lines," J. Comput. Physics, 34, 94-107 (1980).

[5] L. N. Trefethen: "Instability of difference models for hyperbolic initial boundary value problems," Comm. Pure Appl. Math. 37, 329-367 (1984).

[6] R. F. Warming and R. M. Beam: "Stability of semi-discrete approximations for hyperbolic initial-boundary-value problems: asymptotic estimates," Proceedings of the International Symposium on Computational Fluid Dynamics, Sydney, Australia, North-Holland, Amsterdam (1987).

SOME SUPRACONVERGENT SCHEMES FOR HYPERBOLIC EQUATIONS ON IRREGULAR GRIDS

Burton Wendroff and Andrew B. White, Jr.

Los Alamos National Laboratory
Los Alamos, New Mexico 87545, USA

SUMMARY

An analysis of the truncation error for finite difference schemes frequently shows an apparent loss of accuracy when a nonuniform grid is used. Some schemes exhibit the phenomenon of supraconvergence, that is, there is no loss of accuracy in the global error. We show that this is the case for smooth solutions of the color equation for an upstream conservative scheme, for two versions of the Lax-Wendroff scheme, and for a variant of the von Neumann-Richtmyer scheme for gas dynamics, if the latter three are stable.

1. INTRODUCTION

Finite difference equations seem to work best on uniform grids, but even in one dimension one might be forced into using an irregular grid. A typical example of this situation occurs when materials with very disparate densities occur side-by-side; for example, at an air water interface. If Lagrangian coordinates are used, with mass taking the place of length, then equal mass cells would require roughly 1000 times as many cells in the water as in the air. One could change the grid size gradually but rapidly near the interface so that there is a grid gradient rather than a severe jump. This helps but does not remove the error. Giles and Thompkins [1] analyze the wave propagation properties of a grid gradient. Noh [4] shows the bad things that can happen with an exponentially varying grid when an infinite strength shock is sent through it. In any event, there is definitely a point to considering difference equations on totally nonuniform grids.

It is very clear that the local truncation error of most difference schemes suffers if the grid is nonuniform. These local errors for many difference schemes have been examined by Turkel [6], and by Pike [5]. The point of our presentation here is that the *global* error is sometimes better behaved than the local error would indicate, a property that has been called *supraconvergence*. This is well-known for finite element methods, but is somewhat of a surprise for finite difference methods. This phenomenon was observed by Manteuffel and White [3] for a cell-centered scheme for $u'' = f$ which is inconsistent on an arbitrary grid but maintains global second order accuracy. This was extended by Kreiss *et al* [2] to higher order equations and other difference methods, not all of which have this so-called supraconvergence property. Numerov's method, for example, fails to maintain its fourth order accuracy on some grids.

We show that supraconvergence obtains for smooth solutions of the color equation for an upstream conservative scheme, for an edge-centered version of the Lax-Wendroff scheme (with mild mesh restrictions), for a cell-centered Lax-Wendroff scheme, and for a variant of the von Neumann-Richtmer scheme for gas dynamics, provided that the latter three are stable.

2. AN UPSTREAM CONSERVATIVE SCHEME

In [5] Pike gives an example of a conservative scheme which is inconsistent on a nonuniform grid yet which compares favorably with a first-order method. For the differential equation

$$u_t + u_x = 0$$

the difference equation is

$$LU \equiv \frac{U_i^{n+1} - U_i^n}{\Delta t} + \frac{U_i^n - U_{i-1}^n}{h_i} = 0,$$

where U_i^n is the grid function at the cell edge (x_i, t^n), and $h_{i-1/2} = x_i - x_{i-1}$, $h_i = \dfrac{h_{i+1/2} + h_{i-1/2}}{2}$. This conserves $\Sigma h_i U_i$.

The truncation error is

$$Lu = \frac{h_{i-1/2} - h_{i+1/2}}{2h_i} u_{x,i} + O(h, \Delta t)$$

and therefore, in general, the scheme is inconsistent. In fact, it is *first-order accurate* if u(x,t) is smooth. To see this, note that

$$\frac{h_{i-1/2} - h_{i+1/2}}{2h_i} u_{x,i} = \frac{h_{i-1/2} u_{x,i-1} - h_{i+1/2} u_{x,i}}{2h_i} + O(h).$$

Defining the corrected error

$$e_i = u_i - U_i + \frac{1}{2} h_{i+1/2} u_{x,i}$$

we see that

$$Le = O(h, \Delta t)$$

since $u_{x,i}$ is a smooth function of t. This scheme is monotone and stable if $\Delta t / h_i \le 1$, in which case both e and $u - U$ are $O(h)$.

3. LAX-WENDROFF

3.1. An Edge-Centered Version

Pike also considered an edge-centered version of the Lax-Wendroff scheme. With the same notation as above the difference equation is

$$LU = \frac{U_i^{n+1} - U_i^n}{\Delta t} + \frac{U_{i+1}^n - U_{i-1}^n}{2h_i} - \frac{\Delta t}{2h_i}\left[\frac{U_{i+1}^n - U_i^n}{h_{i+1/2}} - \frac{U_i^n - U_{i-1}^n}{h_{i-1/2}}\right] = 0.$$

The truncation error is

$$Lu = \frac{1}{2} u_{xx,i}(h_{i+1/2} - h_{i-1/2}) + O(h^2, \Delta t^2).$$

Note that for the low order part of the truncation error we can write

$$\frac{1}{2}u_{xx,i}(h_{i+1/2}-h_{i-1/2})+O(h^2,\Delta t^2)=\frac{T_{i+1/2}-T_{i-1/2}}{2h_i}+O(h^2),$$

where

$$T_{i+1/2}=\frac{1}{2}u_{xx,i+1/2}h_{i+1/2}^2.$$

Now consider the following difference equation:

$$\frac{\delta_{i+1}-\delta_{i-1}}{2h_i}-\frac{\Delta t}{2h_i}\left[\frac{\delta_{i+1}-\delta_i}{h_{i+1/2}}-\frac{\delta_i-\delta_{i-1}}{h_{i-1/2}}\right]=\frac{T_{i+1/2}-T_{i-1/2}}{2h_i}. \tag{1}$$

We are going to show that under mild mesh restrictions this difference equation has a solution δ which is smooth in t and is $O(h^2)$. From this it follows that

$$L(u-U-\delta)=O(h^2,\Delta t^2).$$

The rest hinges on the stability of the difference equation $LU=r$. If it is stable with the same grid restriction in some norm then the *global error* in that norm will be *second order*.

Note that a solution of the following is clearly a solution of eq. (1):

$$\delta_{i+1}+\delta_i-\nu_{i+1/2}(\delta_{i+1}-\delta_i)=T_{i+1/2},$$

where $\nu_{i+1/2}=\dfrac{\Delta t}{h_{i+1/2}}$. Marching to the left from some N, taking $\delta_N=0$, we see that

$$\delta_i=\sum_{l=i+1}^{N}(-1)^{l-1}\left[\prod_{k=i+1}^{l-1}\frac{1-\nu_{k+1/2}}{1+\nu_{k+1/2}}\right]\frac{T_{l-1/2}}{1+\nu_{l-1/2}}.$$

Thus, δ will indeed be $O(h^2)$ if there exists ρ such that

$$\left|\prod_{k=r}^{s}\frac{1-\nu_{k+1/2}}{1+\nu_{k+1/2}}\right|\le\rho^{s-r},\quad 0<\rho<1,$$

for $s-r$ sufficiently large.

Let

$$h_{\min}=\min_i h_i.$$

We require the

<u>*Main Mesh Condition*</u>: Assume there exists θ, $0<\theta<1$, such that

$$0<\theta\le\frac{\Delta t}{h_{\min}}\le 1.$$

For example, if the cell lengths are alternately h and h^2, for $0<h<1$, and if

$$\frac{1}{2}\le\frac{\Delta t}{h^2}\le 1,$$

then since half the cells have length h^2,

$$\left|\prod_{k=r}^{s}\frac{1-\nu_{k+1/2}}{1+\nu_{k+1/2}}\right|\le\left(\frac{1}{2}\right)^{\frac{s-r}{2}}=\rho^{s-r},\ \rho=\sqrt{\frac{1}{2}}.$$

In general, it is sufficient (but not necessary) that the fraction of intervals which have a bounded ratio with the smallest interval remain bounded from below as the mesh is refined. Of course, this only permits a correction which cancels the bad part of the truncation error. The actual global error depends on the stability of the scheme.

We still have to show that the divided time difference of the correction δ is $O(h^2)$. This is so, since

$$\frac{T_{l+1/2}^{n+1} - T_{l+1/2}^{n}}{\Delta t} = \frac{1}{2} h_{i+1/2}^2 \frac{u_{xx,l+1}(t+\Delta t) - u_{xx,l+1}(t)}{\Delta t}$$

and we are assuming that u(x,t) is sufficiently smooth.

3.2. A Cell-Centered Version

We consider the particular solution $u(x,t) = (x-t)^2$. For a cell-centered Lax-Wendroff the full error equation is

$$Le = \frac{e_{i+1/2}^{k+1} - e_{i+1/2}^{k}}{\Delta t} + \frac{e_{i+1}^{*} - e_{i}^{*}}{h_{i+1/2}} - \frac{\Delta t}{2h_{i+1/2}} \left\{ \frac{e_{i+3/2} - e_{i+1/2}}{h_{i+1}} - \frac{e_{i+1/2} - e_{i-1/2}}{h_i} \right\}$$

$$= \frac{1}{4}(h_{i+3/2} - h_{i-1/2})$$

$$- \frac{\Delta t}{4} \left\{ \frac{h_{i+3/4} - 2h_{i+1/2} + h_{i-1/2}}{h_{i+1/2}} \right\}$$

where

$$e_{i+1}^{*} = \frac{h_{i+3/2} e_{i+1/2} + h_{i+1/2} e_{i+3/2}}{h_{i+3/2} + h_{i+1/2}} .$$

Let $d_{i+1/2} = e_{i+1/2} - \frac{1}{4} h_{i+1/2}^2$. Since

$$\frac{h_{i+3/2} h_{i+1/2}^2 + h_{i+1/2} h_{i+3/2}^2}{h_{i+3/2} + h_{i+1/2}} = h_{i+3/2} h_{i+1/2}$$

and

$$\frac{h_{i+3/2}^2 - h_{i+1/2}^2}{h_{i+3/2} + h_{i+1/2}} = h_{i+3/2} - h_{i+1/2}$$

then

$$Ld = 0 .$$

In general, a similar analysis shows that for sufficiently smooth solutions there is a correction which removes the low-order truncation terms.

4. VON NEUMANN-RICHTMYER

We consider the classic von Neumann-Richtmyer method without viscosity. The gas-dynamic equations in Lagrangian form are

$$v_t - u_x = 0,$$

$$u_t + p_x = 0,$$

$$e_t + p v_t = 0,$$

$$p = p(v,e).$$

The velocities u are set at the cell corners, and the volumes and energies v and e are set at the cell centers see Fig. 1.1. Thus, the difference equations are

$$\frac{V_{i+1/2}^{n+1/2} - V_{i+1/2}^{n-1/2}}{\Delta t^n} - \frac{U_{i+1}^n - U_i^n}{h_{i+1/2}} = 0, \tag{2a}$$

$$\frac{U_i^n - U_i^{n-1}}{\Delta t^{n-1/2}} + \frac{p(V_{i+1/2}^{n-1/2}, E_{i+1/2}^{n-1/2}) - p(V_{i-1/2}^{n-1/2}, E_{i-1/2}^{n-1/2})}{h_i} = 0, \tag{2b}$$

$$\frac{E_{i+1/2}^{n+1/2} - E_{i+1/2}^{n-1/2}}{\Delta t^n} \tag{2c}$$

$$+ \frac{\Delta t^{n-1/2} p(V_{i+1/2}^{n+1/2}, E_{i+1/2}^{n+1/2}) + \Delta t^{n+1/2} p(V_{i+1/2}^{n-1/2}, E_{n+1/2}^{n-1/2})}{\Delta t^n} \frac{V_{i+1/2}^{n+1/2} - V_{i+1/2}^{n-1/2}}{\Delta t^n} = 0.$$

where

$$h_{i+1/2} = x_i - x_{i-1}, \; h_i = \frac{1}{2}(h_{i+1/2} + h_{i-1/2}),$$

and

$$\Delta t^{n+1/2} = t^{n+1} - t^n, \; \Delta t^n = \frac{1}{2}(\Delta t^{n+1/2} + \Delta t^{n-1/2}).$$

Now we need to look at the detailed truncation error to see if the appropriate corrections can be made.

In eqs. (2a) and (2c) it is the uncentered time difference that contributes to the low order truncation error. Specifically, the form is

$$\frac{f(t^{n+1/2}) - f(t^{n-1/2})}{\Delta t^n} = f_t + \frac{1}{8} f_{tt} \frac{(\Delta t^{n+1/2})^2 - (\Delta t^{n-1/2})^2}{\Delta t^n} + O(\Delta^2),$$

where the functions on the right are evaluated at t^n. The second term on the right can be rewritten as

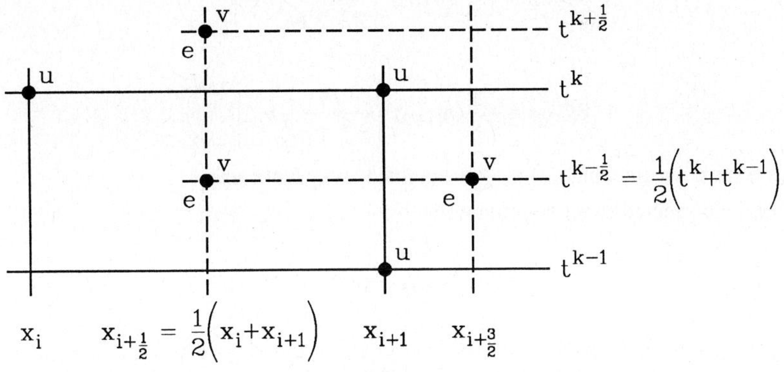

Fig. 1.1. Staggered grid.

$$\frac{\frac{1}{8}(\Delta t^{n+1/2})^2 f_{tt}^{n+1/2} - \frac{1}{8}(\Delta t^{n-1/2})^2 f_{tt}^{n-1/2}}{\Delta t^n} + O(\Delta^2).$$

Thus there is a second-order correction $\bar{f}$ to f which satisfies

$$\frac{\bar{f}^{n+1/2} - \bar{f}^{n-1/2}}{\Delta t^n} = f_t + O(\Delta^2),$$

where

$$\bar{f}^{n+1/2} = f(t^{n+1/2}) - \frac{1}{8}(\Delta t^{n+1/2})^2 f_{tt}(t^{n+1/2}).$$

It follows immediately that, replacing f by v, the truncation error for eq. (2a) is $O(\Delta^2)$ for $\bar{\mathrm{v}}$, that is,

$$\frac{\bar{\mathrm{v}}_{i+1/2}^{n+1/2} - \bar{\mathrm{v}}_{i+1/2}^{n-1/2}}{\Delta t^n} - \frac{u_{i+1}^n - u_i^n}{h_{i+1/2}} = O(\Delta^2).$$

Since the average p used in eq. (2c) is already second-order accurate it follows in the same way that eq. (2c) is also second-order for $\bar{e}$ and $\bar{\mathrm{v}}$.

There remains the momentum equation (2b) to deal with. We first note that for any smooth second-order perturbations α and β

$$p(\bar{\mathrm{v}}_+ + \alpha_+, \bar{e}_+ + \beta_+) - p(\bar{\mathrm{v}}_- + \alpha_-, \bar{e}_- + \beta_-) = p(\bar{\mathrm{v}}_+, \bar{e}_+) - p(\bar{\mathrm{v}}_-, \bar{e}_-) +$$
$$\frac{\partial p}{\partial \mathrm{v}}(\alpha_+ - \alpha_-) + \frac{\partial p}{\partial e}(\beta_+ - \beta_-) + h_i O(\Delta^2),$$

where $\alpha_+ = \alpha_{i+1/2}$, etc. The low order part of the truncation error of the off centered $\partial p/\partial x$ in eq. (2b) is

$$\frac{1}{8} p_{xx} \frac{(h_{i+1/2})^2 - (h_{i-1/2})^2}{h_i}.$$

In order to cancel it we can take the corrections to be

$$\alpha_{i+1/2}^{n-1/2} \equiv -\frac{1}{8}(h_{i+1/2})^2 \left[\frac{p_\mathrm{v}}{p_\mathrm{v}^2 + p_e^2} p_{xx}\right]_{i+1/2}^{n-1/2}$$

and

$$\beta_{i+1/2}^{n-1/2} \equiv -\frac{1}{8}(h_{i+1/2})^2 \left[\frac{p_e}{p_\mathrm{v}^2 + p_e^2} p_{xx}\right]_{i+1/2}^{n-1/2},$$

where we must assume that $p_\mathrm{v}^2 + p_e^2$ is bounded away from zero.

For the corrected exact solution given by

$$\hat{\mathrm{v}} \equiv \bar{\mathrm{v}} + \alpha,$$
$$\hat{e} \equiv \bar{e} + \beta,$$

and

$$\hat{u} \equiv u,$$

the local truncation error is $O(\Delta^2)$. Since these differ from the exact solution itself by $O(\Delta^2)$, stability would imply second-order accuracy.

REFERENCES

1. M. Giles and W.T. Thompkins, Jr., "Interval reflection due to a nonuniform grid," in *Advances in Computer methods for Partial Differential Equations V*, ed. R. Vichnevetsky and R.S. Stepleman, pp. 322-328, IMACS, 1984.
2. H.-O. Kreiss, T.A. Manteuffel, B. Swartz, B. Wendroff, and A.B. White, Jr., "Supraconvergent schemes on irregular grids," *Mathematics of Computation*, vol. 47, pp. 537-554, 1987.
3. T.A. Manteuffel and A.B. White, Jr., "The numerical solution of second-order boundary value problems on nonuniform meshes," *Mathematics of Computation*, vol. 47, pp. 537-554, 1987.
4. W.F. Noh, "Errors for calculation of strong shocks using an artificial viscosity and an artificial heat flux," *Journal of Computational Physics*, vol. 72, no. 1, pp. 78-120, 1987.
5. J. Pike, "Grid adaptive algorithms for the solution of the Euler equations on irregular grids," *Journal of Computational Physics*, vol. 71, pp. 194-223, 1987.
6. E. Turkel, "Accuracy of schemes with nonuniform meshes for compressible fluid flows," *Applied Numerical Mathematics*, vol. 2, pp. 529-550, 1986.

THE HOMOGENEOUS HOMENTROPIC COMPRESSION OR EXPANSION- A TEST CASE FOR ANALYZING SOD'S OPERATOR-SPLITTING

H. Westenberger, J. Ballmann
Lehr- und Forschungsgebiet Mechanik
RWTH Aachen
D-5100 Aachen, West Germany

SUMMARY

The paper shows that the choice of dependent variables in conservation laws with source terms may strongly influence the quality of numerical solutions using operator-splitting. A further result is that the splitting may violate the second principle of thermodynamics. For quasi one-dimensional flows, closed form solutions were derived for a family of variable cross-sectional functions. A numerical test problem is formulated including this analytical solution. Combinations of the operator-splitting with Godunov's Method and with the Random Choice Method are discussed by means of this example. Reasonable results were obtained by both methods but special attention must be paid to the entropy.

INTRODUCTION

The quasi one-dimensional flow of a non-viscous and non heat conducting gas is described by a hyperbolic system of conservation laws with a source term. Among the different methods of integrating a system of homogeneous conservation laws Godunov's Method and the Random Choice Method are well established. These methods can be used to integrate a non-homogeneous system applying Sod's splitting technique [4]. Because of it's conceptional simplicity this technique is in common use. Moreover, new developments exist to integrate conservation laws with source terms [1-3].

THE OPERATOR-SPLITTING

A hyperbolic system of conservation laws with a source term is given by

$$\underline{u}_{,t} + \underline{F}(\underline{u})_{,r} = \underline{Q}(\underline{u},r) \quad . \qquad (1)$$

where $\underline{F}$ is the flux, $\underline{Q}$ is the source term, $\underline{u} \in \mathbb{R}^k$ and $r \geq 0$, $t \geq 0$. The integration of equation (1) over the domain $G = [\, r_{m-\frac{1}{2}} , r_{m+\frac{1}{2}} \,] \times [\, t_n , t_{n+1} \,]$ in space and time leads to

$$\int_{r_{m-\frac{1}{2}}}^{r_{m+\frac{1}{2}}} \underline{u}\,(r,t_{n+1})\,dr \;-\; \int_{r_{m+\frac{1}{2}}}^{r_{m+\frac{1}{2}}} \underline{u}\,(r,t_n)\,dr \;+\; \int_{t_n}^{t_{n+1}} \underline{F}\,(\underline{u}(r_{m+\frac{1}{2}},t))\,dt$$

$$-\int_{t_n}^{t_{n+1}} \underline{F}\,(\underline{u}(r_{m-\frac{1}{2}},t))\,dt \;=\; \iint_G \underline{Q}\,dr\,dt \quad . \qquad (2)$$

Evaluating the integrals by the approximations

$$\underline{u}_{m,n} = \frac{1}{\Delta r}\int_{r_{m-\frac{1}{2}}}^{r_{m+\frac{1}{2}}} \underline{u}\,(r,t_n)\,dr \;, \qquad \underline{g}_{m+\frac{1}{2},n} = \frac{1}{\Delta t}\int_{t_n}^{t_{n+1}} \underline{F}\,(\underline{u}\,(r_{m+\frac{1}{2}},t_n))\,dt\,,$$

$$\underline{Q}_{m,n} = \frac{1}{\Delta r\,\Delta t}\iint_G \underline{Q}\,dr\,dt \;, \quad \Delta r = r_{m+\frac{1}{2}} - r_{m-\frac{1}{2}} \;, \quad \Delta t = t_{n+1} - t_n$$

one obtains the discrete formulation of (2) :

$$\underline{u}_{m,n+1} = \underline{u}_{m,n} - \lambda\left(\underline{g}_{m+\frac{1}{2},n} - \underline{g}_{m-\frac{1}{2},n}\right) + \Delta t\,\underline{Q}_{m,n} \;, \quad \lambda = \frac{\Delta t}{\Delta r} \;. \qquad (3)$$

We introduce the abbreviation $\{\underline{u}\}_n := \{\underline{u}_{m,n}\}_{m \in \mathbb{N}}$. The solution operator $H : \{\underline{u}\}_n \to \{\underline{u}\}_{n+1}$ is defined by (3) with $\underline{Q}_{m,n} = \underline{0}$. The concept of operator-splitting is to approximate (1) by the two simpler equations

$$\underline{u}_{,t} + \underline{F}(\underline{u})_{,r} = \underline{0} \qquad (4)$$

and

$$\underline{u}_{,t} = \underline{Q}(\,\underline{u},\, r) \qquad (5)$$

to be solved separately. The differential equation (5) is usually integrated by explicit Euler integration. The corresponding solution operator S has the form

$$S^{(\varepsilon)}\underline{u}_{m,n} = \underline{u}_{m,n} + \varepsilon\,\Delta t\,\underline{Q}(\,\underline{u}_{m,n}, r_m) \;, \qquad \varepsilon = \tfrac{1}{2} \text{ or } \varepsilon = 1 \;. \qquad (6)$$

In order to approximate the solution of (1) three different combinations of the operators H and S will be considered :

$$\{\underline{u}\}_{n+1} = \begin{cases} S^1\,H\,\{\underline{u}\}_n & \text{variant 1} \\ S^{\frac{1}{2}}\,H\,S^{\frac{1}{2}}\,\{\underline{u}\}_n & \text{variant 2} \\ (\,S^1 + H - Id\,)\{\underline{u}\}_n & \text{variant 3}\;. \end{cases}$$

Therein, variant 3 can be derived by choosing a special approximation of the integral of $\underline{Q}(\underline{u}, r)$ over the domain G, setting $\underline{Q}_{m,n} = \underline{Q}(\underline{u}^*, r^*)$ with $r^* = r_m$, $\underline{u}^* = \underline{u}_{m,n}$. Id is the identity. Introduction of $r^* = r_m$, $\underline{u}^* = \underline{u}_{m,n} - \lambda(\underline{g}_{m+\frac{1}{2},n} - \underline{g}_{m-\frac{1}{2},n})$ into $\underline{Q}(\underline{u}^*, r^*)$ leads to variant 1.

THE OPERATOR-SPLITTING FOR THE GASDYNAMIC EQUATIONS FOR QUASI ONE-DIMENSIONAL FLOWS

THE INFLUENCE OF THE FORMULATION

The gasdynamic equations for quasi one-dimensional flows (cross section $A(r)$) of gases with constant specific heats can be described in the form (1). But the set of unknown functions is not unique, since different dependent variables can be chosen. Here, two different formulations will be considered :

	conservative variable $\underline{u}$	flux $\underline{F}$	source term $\underline{Q}$
(*) :	$\begin{pmatrix} \rho \\ m \\ \rho e \end{pmatrix}$	$\begin{pmatrix} v\rho \\ vm + p \\ v(\rho e + p) \end{pmatrix}$	$-\frac{A_r}{A}\begin{pmatrix} v\rho \\ vm \\ v(\rho e + p) \end{pmatrix}$,
(**) :	$\begin{pmatrix} \rho A \\ mA \\ \rho e A \end{pmatrix}$	$\begin{pmatrix} v\rho A \\ (vm + p)A \\ v(\rho e + p)A \end{pmatrix}$	$A_r \begin{pmatrix} 0 \\ p \\ 0 \end{pmatrix}$

with ρ density, v velocity, $m = \rho v$, e specific total energy, p pressure. In order to compare the influence of the different formulations we evaluate the operators H and S for a trivial solution ($p \equiv$const., $\rho \equiv$const., $v \equiv 0$). Version (*) gives

$$\text{flux} : \begin{pmatrix} 0 \\ p \\ 0 \end{pmatrix}, \quad \text{flux difference} : \underline{g}_{m+\frac{1}{2},n} - \underline{g}_{m-\frac{1}{2},n} = \underline{0}, \quad \text{source term} : \underline{0}.$$

This means that H = Id and S = Id for trivial initial conditions, and all three variants of operator- splitting reproduce the trivial solution. Version (**) gives under the same conditions

$$\text{flux} : \begin{pmatrix} 0 \\ pA \\ 0 \end{pmatrix}, \quad \text{flux difference}: \begin{pmatrix} 0 \\ p\triangle A \\ 0 \end{pmatrix}, \quad \text{source term} : \begin{pmatrix} 0 \\ A_p \\ 0 \end{pmatrix},$$

where $\triangle A := A(r_{m+\frac{1}{2}}) - A(r_{m-\frac{1}{2}})$. Hence, it follows $H \neq Id$ and $S \neq Id$. The solution operator H for example keeps ρ and e unchanged while the momentum, the pressure and the velocity field are changed. To obtain the trivial solution for the full time step the disturbances of the steps H and S must balance each other. Normally, this is satisfied only by variant 3 and $A_r = \triangle A / \triangle r$.

ENTROPY AND INTEGRATION OF THE SOURCE TERM

Simple considerations for an ideal gas with the entropy s and

$$\frac{s-s_0}{R_m} = \frac{1}{\varkappa-1} \ln\left(\frac{p}{p_0}\left(\frac{\rho_0}{\rho}\right)^{\varkappa}\right).$$

where R_m and $\varkappa$ are the gas constant and the ratio of specific heats, show that the operator S may diminish the entropy, e. g. S step for version (**) yields

$$\begin{pmatrix} \rho A \\ m A \\ \rho e A \end{pmatrix}_1 = S \begin{pmatrix} \rho A \\ m A \\ \rho e A \end{pmatrix}_0 = \begin{pmatrix} \rho A \\ m A \\ \rho e A \end{pmatrix}_0 + \Delta t\, A_r \begin{pmatrix} 0 \\ p \\ 0 \end{pmatrix}_0 .$$

Density and total energy are not changed but

$$m_1 = m_0 + \Delta t \frac{A_r}{A} p_0 \neq m_0 .$$

Whenever A_r and v_0 are of the same sign then $|m_1| > |m_0|$ and, because of

$$p_1 = (\varkappa-1)\left(e_1 - \frac{1}{2}\frac{m_1^2}{\rho_1}\right) \neq p_0 ,$$

it is easy to see that $s_1 < s_0$.

THE HOMOGENEOUS HOMENTROPIC COMPRESSION OR EXPANSION

THEOREM : In a tube with variable cross section, $A(r) > 0$ for $r > 0$, $A(0) < \infty$, $A \in C^2$, a non-trivial solution of the conservation laws with constant entropy and thermodynamic variables only depending on time is to be found in the domain $(r,t) \in \mathbb{R}_+ \times [0,T]$ with the initial conditions $v(r_0,0) = v_0$ and the boundary condition $v(0,t) = 0$. Then, such a solution exists only for cross-sectional functions

$$A(r) = \gamma r^{\nu}, \quad \gamma > 0, \quad \nu \geq 0 \quad \text{and for} \quad \frac{v_0}{r_0} t > -1$$

and can be given in the form

$$\rho(t) = \rho(0)\left(1 + \frac{v_0}{r_0} t\right)^{-(1+\nu)}, \qquad v(r,t) = \frac{v_0}{1 + \frac{v_0}{r_0} t} \frac{r}{r_0} .$$

Proof : The gasdynamic equations for isentropic flow reduce to

$$\frac{\partial}{\partial t}(\rho A) + \frac{\partial}{\partial r}(\rho v A) = 0 , \tag{7}$$

$$\frac{\partial}{\partial t}(\rho v A) + \frac{\partial}{\partial r}((\rho v^2 + p) A) = \frac{\partial A}{\partial r} p . \tag{8}$$

Assumed a solution $\rho(t)$ exists, then from equation (7) it follows

$$A(r) \frac{d\rho}{dt} =: A(r)\, \Omega(t) = - \rho(t) \frac{\partial (v(r,t) A(r))}{\partial r} . \tag{9}$$

With the prepositions $A(0) < \infty$, $v(0,t) = 0$ integration of (9) and insertion of equation (8) yield

$$\frac{d^2\rho}{dt^2} - \frac{1}{\rho}\left(\frac{d\rho}{dt}\right)^2\left(2 - \frac{1}{A^2(r)}\frac{\partial A(r)}{\partial r}\int_0^r A(x)\,dx\right) = 0 \cdot \tag{10}$$

A solution $\rho(t)$ can only be found if the term $2 - \frac{1}{A^2(r)}\frac{\partial A(r)}{\partial r}\int_0^r A(x)\,dx$ is independent of r. This is equivalent to finding a solution f of the ODE

$$\frac{f(r)\,f''(r)}{(f'(r))^2} = 2 - q := h \tag{11}$$

where $f(r) := \int_0^r A(x)\,dx$ and q is a constant and the initial condition $f(0) = 0$ is satisfied. (11) can be integrated to

$$f^{-h} f' = \beta \quad , \tag{12}$$

where β is a constant. The integration of (12) gives reasonable results only for $h \in [0,1)$, i. e.

$h = 0$: $\quad A(r) = \text{const.}$,

$h \in (0,1)$: $\quad A(r) = \text{const.} \cdot r^{\frac{h}{1-h}}$.

Up to now it was shown that only one family of cross-sectional functions allows a solution. Now it must be shown that in fact a solution exists. Using the result for $A(r)$, equation (10) can be integrated twice yielding

$$\frac{1}{h-1}\left(\rho(t)^{h-1} - \rho(0)^{h-1}\right) = \alpha t \quad .$$

The constant α is determined from the initial conditions as

$$\alpha = -v_0 \rho_0^{\,h-1} \frac{A(r_0)}{\int_0^{r_0} A(x)\,dx} \quad .$$

The final solution

$$\rho_h(t) = \rho_0\left(1 + \frac{v_0}{r_0} t\right)^{\frac{1}{h-1}} \quad , h \in [0,1), \ \frac{v_0}{r_0} t > -1 \tag{13}$$

depends on the parameter h and is related to corresponding cross-sectional functions $A_h(r)$,

$$A_h(r) = \gamma r^{\frac{h}{1-h}} \quad , h \in [0,1) \quad , \gamma > 0 \text{ arbitrary} . \tag{14}$$

By setting $\nu := \frac{h}{1-h} \in [0,\infty)$ the proof is completed .

Remark 1 : In the solution family the special cases of plane ($\nu = 0$), cylindrical ($\nu = 1$), and spherical ($\nu = 2$) homogeneous compression or expansion are contained.

Remark 2 : By integration of equation (9) it can be shown that $v_\nu(r,t)$ does not really depend on ν :

$$v_\nu(r,t) = -\frac{\Omega(t)}{\rho(t)} \frac{1}{A_\nu(r)} \int_0^r A_\nu(x)\,dx = \frac{v_0}{1 + \frac{v_0}{r_0}} \frac{r}{r_0} .$$

Remark 3 : The particle paths, which can be given by the integration of the differential equation $\frac{dR}{dt} = v(R,t)$, are all straight lines through the point ($r = 0$, $t = -\frac{r_0}{v_0}$). It is obvious that for $v_0 < 0$ the particle paths converge (compression case) and for $v_0 > 0$ they diverge (expansion case).

Remark 4 : The $C_{\pm 1}$ - characteristics can be found by integration of

$$\frac{dC}{dt} = v(C,t) \pm a(t) ,$$

where a is the velocity of sound. Assume $v_0 \neq 0$: then, with $\varepsilon = \frac{1}{2}(\nu+1)(\varkappa-1)$ the C_{-1} - characteristic through the point ($r = r_0$, $t = 0$) is given by

$$C(t) = \frac{r_0}{r_0 + v_0 t} \left(\frac{a_0}{\varepsilon v_0} \left(\left(1 + \frac{v_0}{r_0} t\right)^{-\varepsilon} - 1 \right) + 1 \right) . \tag{15}$$

If $\frac{\varepsilon v_0}{a_0} < 1$, the characteristic line reaches the centre $r = 0$ at the time

$$t = \frac{r_0}{v_0} \left(\left(1 - \frac{\varepsilon v_0}{a_0}\right)^{-1/\varepsilon} - 1 \right) .$$

A NUMERICAL TEST CASE

A piston, compressing with constant velocity v_0 a spherical volume for time $t < 0$, is suddenly stopped at $t = 0$ at the position $r = r_0$. The initial values and boundary conditions for this problem are

$$p(r,0) = p_0 , \quad \rho(r,0) = \rho_0 , \quad A(r) = r$$

$$v(r,0) = r \frac{v_0}{r_0} , \quad v_0 = -\frac{1}{2} \frac{a_0}{\varkappa - 1} , \quad r \in [0, r_0] ,$$

$$v(r_0,t) = v(0,t) = 0 , \quad t > 0 .$$

The stopping of the piston produces a rarefaction wave. The head of this wave is given by the above mentioned C_{-1} - characteristic (15) at the interface connecting the domains of rarefaction and homogeneous compression (Figure 1).

Godunov's Method, applied to the formulation (*), gives good results in all of the three variants of splitting (Figure 2). Within the zone of homogeneous compression the results agree very well with the analytical solution. For the formulation (**) only the splitting H+S-Id gives comparable results, whereas the other variants give worse results for this formulation as could be expected from the above made considerations. Looking at the entropy, which is of special interest in the analysis of this paper, we notice a diminishing of entropy during the reflection of the rarefaction wave at the centre (Figure 2c). Another non-physical effect is the production of entropy by the centered rarefaction wave starting from the stopped piston. There, the increase of entropy is of the same order of magnitude as the jump of entropy across the shock, which has developed at the tail of the rarefaction wave. The same effect is observed in plane flows. Therefore, it is not caused by the splitting but by the averaging process involved in Godunov's Method. From a physical point of view this is similar to a mixing process which increases entropy. Hence, in zones of isentropic flow with strong gradients entropy can numerically grow.

The test problem was also solved by the Random Choice Method using the first variant of splitting. The computation was repeated several times with different random numbers. The expected values are called the result of the method. The level of pressure agrees very well to the results by Godunov's Method but it's distribution is rough (Figure 2). During the reflection of the rarefaction wave at the centre numerical oscillations occur. The entropy is diminished beneath the centre whereas the centered wave at the piston produces no entropy numerically because of the absence of averaging processes.

CONCLUSIONS

Sod's method represents a splitting up of the integration of conservation laws with a source term into the integration of a homogeneous conservation law and the time integration of the source term. Different forms of combination can be chosen. Furthermore, different formulations of the conservation laws lead to different splittings. Both influences the quality of the numerical results. As a minimum requirement on the method, trivial solutions should be reproduced.

For the case of quasi one-dimensional gas flows with variable cross-sectional areas two formulations were compared, one with density, momentum and total energy as dependent variables and the other with the same variables multiplied by the cross-sectional function. The first formulation satisfies the above mentioned minimum requirement with all of the three variants of splitting, whereas only one combination satisfies the condition for the second formulation. Further, it was shown that the splitting influences the entropy numerically.

An analytical solution for the gasdynamic equations with area variation, where the thermodynamic state depends only on time, exists for special cross-sectional functions. It was found that for these solutions all particle paths are running through one point. Those solutions exist for plane, cylindrical or spherical quasi one-dimensional flows.

Godunov's Method and the Random Choice Method with operator-splitting were tested with an example, where in a subdomain such an analytical solution is given. The Random Choice Method shows small numerical oscillations near the centre whereas the results of Godunov's Method agree very well with the analytical solution if an appropriate combination of formulation and splitting is used. It was confirmed that entropy can be reduced numerically by operator-splitting whereas in Godunov's Method an increase of entropy may arise from the averaging.

REFERENCES

[1] GLAZ, H. M., LIU, T.-P. : "The asymptotic analysis of wave interactions and numerical calculations of transonic nozzle flow", Adv. Appl. Math., 5 (1984) pp. 111-146.

[2] GLIMM, J., MARSHALL, G., PLOHR, B. : "A generalised Riemann problem for quasi one-dimensional gas flows", Adv. Appl. Math., 5 (1984) pp. 1-30.

[3] ROE, P. L. : "Upwind differencing schemes for hyperbolic conservation laws with source terms", in : CARASSO, C., RAVIART, P. A., SERRE D. (eds.) : "Nonlinear hyperbolic problems", Springer Verlag, Proceedings St. Etienne 1986.

[4] SOD, G. A.: "A numerical study of a converging cylindrical shock", J. Fluid Mech., 83 (1977) pp. 785-794.

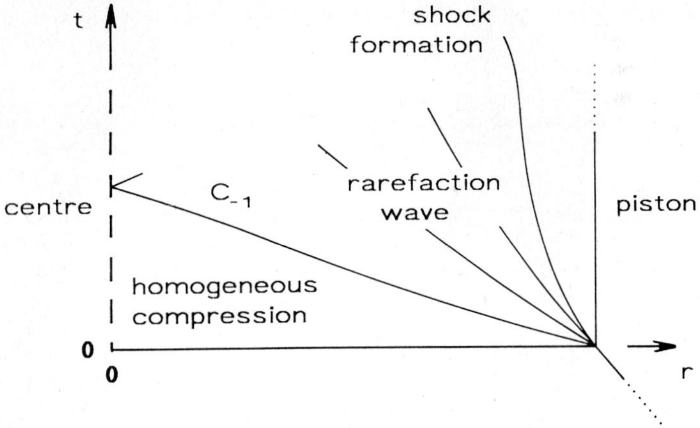

Fig. 1 The homogeneous compression with a suddenly stopped piston: wave field qualitatively

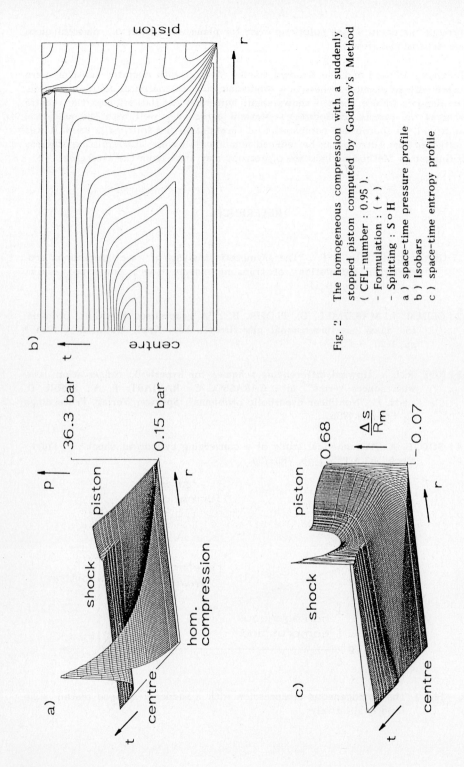

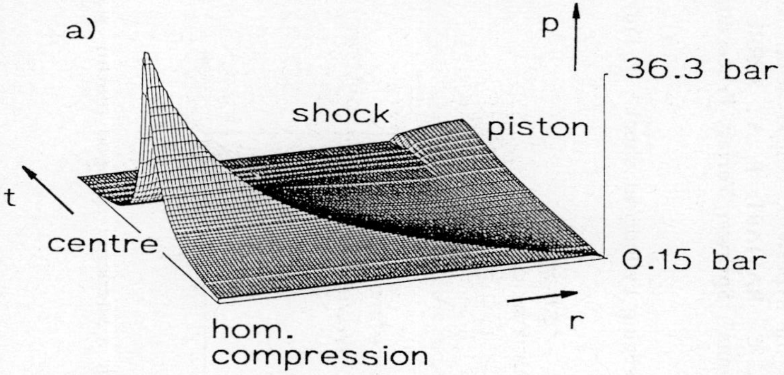

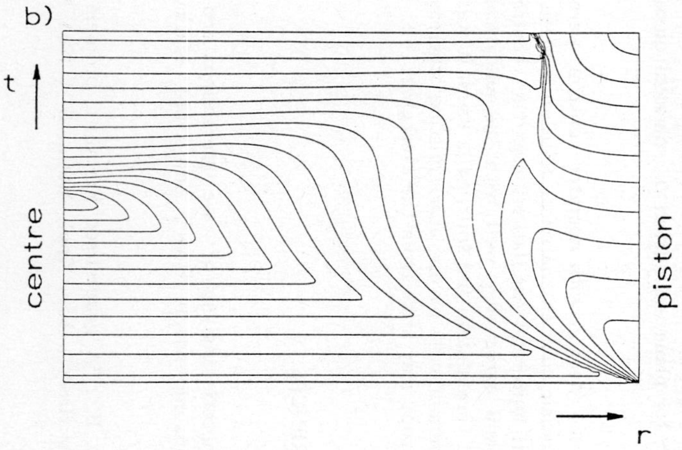

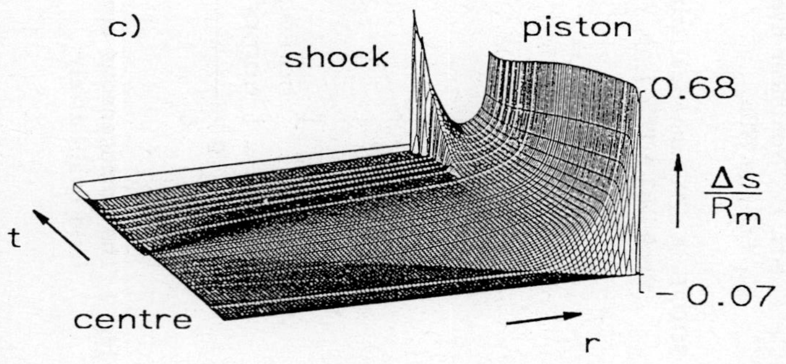

Fig. 2 The homogeneous compression with a suddenly stopped piston computed by Godunov's Method (CFL-number : 0.95).

- Formulation : (*)
- Splitting : S ∘ H

a) space-time pressure profile
b) Isobars
c) space-time entropy profile

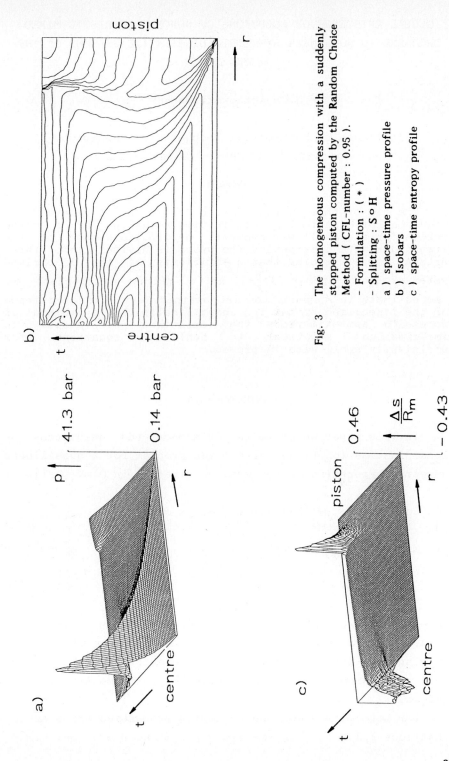

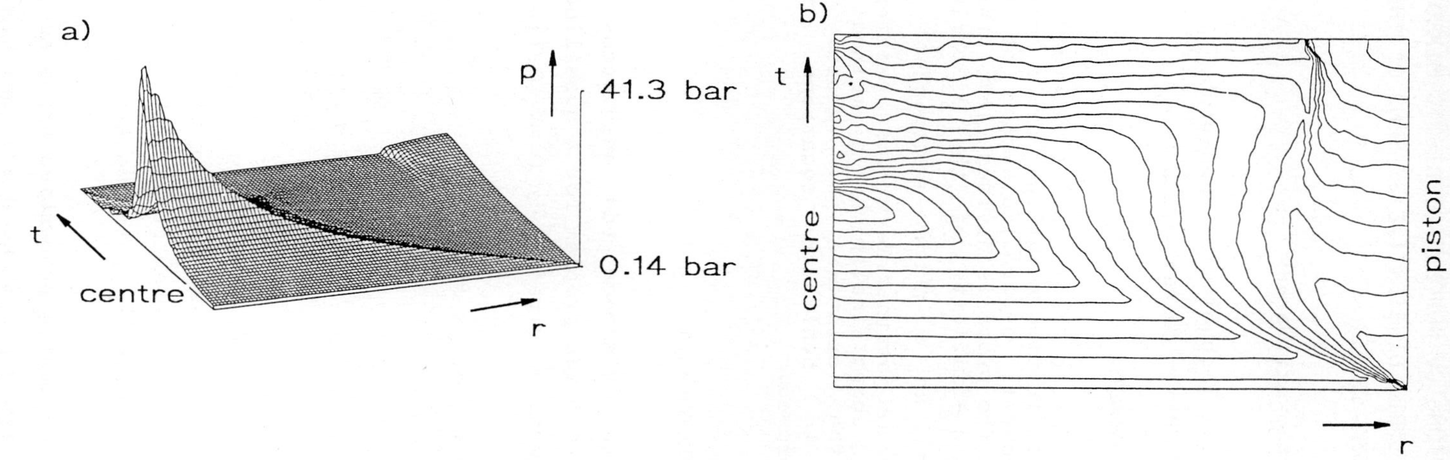

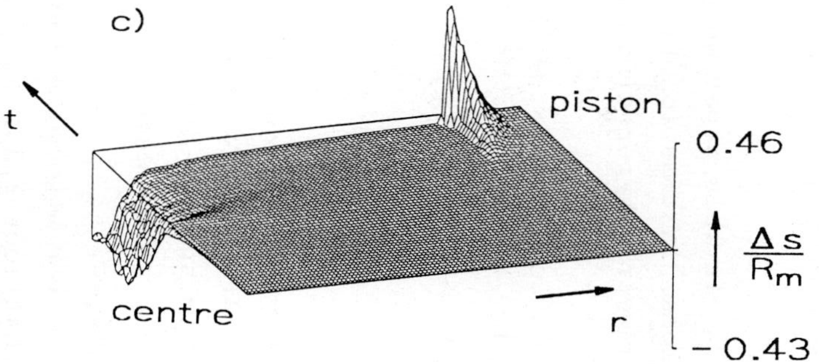

Fig. 3 The homogeneous compression with a suddenly stopped piston computed by the Random Choice Method (CFL-number : 0.95).
- Formulation : (*)
- Splitting : S ∘ H
a) space-time pressure profile
b) Isobars
c) space-time entropy profile

GLOBAL EXISTENCE OF SOLUTIONS OF NONCHARACTERISTIC MIXED PROBLEMS TO NONLINEAR SYMMETRIC DISSIPATIVE SYSTEMS OF THE FIRST ORDER

W.M.Zajaczkowski,Warsaw.

Institute of Fundamental Technological Research,
Swietokrzyska 21,00-049 Warsaw,Poland

SUMMARY

The aim of this paper is to show the existence and uniqueness of classical,global in time solutions for a mixed problem for a quasilinear symmetric hyperbolic system of the first order with dissipation.The proof is divided into three parts.At first assuming that coefficients of the linearized system are in H^l-spaces $(l>\frac{n}{2}+1)$ an a priori independent on time estimate in H^l spaces too is obtained.Next the existence for the linearized system is shown.Finally by a method of successive approximations the existence of global in time,classical solutions to nonlinear equations for sufficiently small data is proved.

1. INTRODUCTION

In this paper we prove the existence and uniqueness of global in time solutions for a mixed problem for a quasilinear symmetric hyperbolic system with dissipation in $\Omega^T=\Omega\times[0,T]$:

$$Lu\equiv E(t,x,u)u_t+\sum_{i=1}^{n} A_i(t,x,u)u_{x_i}+B(t,x,u)u=F(t,x) \quad \text{in } \Omega^T,$$

$$M(t,x',u)u|_{\partial\Omega}=g(t,x') \quad \text{on } \partial\Omega^T, \quad (1)$$

$$u|_{t=0}=u_0(x) \quad \text{in } \Omega,$$

where $t\in[0,T]$, $x=(x_1,\dots,x_n)\in\Omega\subset\mathbb{R}^n$, $x'\in\partial\Omega$, $u=(u_1,\dots,u_m)\in G\subset\mathbb{R}^m$, $E(t,x,u)$, $A_i(t,x,u)$, $i=1,\dots,n$, $B(t,x,u)$ are real mxm matrices.The state space G arises because physical quantities such as the density or temperature should always be positive.Assume that $u_0\in G_0\subset\subset G$.

For $t\in[0,T]$, $x\in\Omega$, $u\in G$ the following properties are assumed:
1.Matrices E, A_i, $i=1,\dots,n$, are symmetric,E and B are uniformly

positively definite, so

$$Eu\cdot u\geq\alpha_o u^2,\ Bu\cdot u\geq\beta_o u^2, \tag{2}$$

where α_o,β_o are positive constants and dot denotes the scalar product in $\mathbb{R}^m$.

2. Let $A_{\underline{n}}=A\cdot\underline{n}$, where $\underline{n}$ is the unit outward vector normal to the boundary $\partial\Omega$ and $A=(A_1,\dots,A_n)$. The matrix $A_{\underline{n}}$ is symmetric, so it has eigenvalues and a complete ortonormal system of eigenvectors in $\mathbb{R}^m$:

$$-A_{\underline{n}}\gamma_\mu=\lambda_\mu\gamma_\mu,\quad \mu=1,\dots,m. \tag{3}$$

Assume that in a neighbourhood of the boundary the eigenvalues are disjoined from zero, so there exists a constant c_o such that

$$\min_{\mu\in\{1,\dots,m\}} |\lambda_\mu|\geq c_o>0. \tag{4}$$

Let λ^+_μ, $\mu=1,\dots,l$, be the positive eigenvalues of $-A_{\underline{n}}$ and γ^+_μ corresponding them eigenvectors. Similarly by $\lambda^-_\mu,\gamma^-_\mu$, $\mu=l+1,\dots,m$, we denote the negative eigenvalues and corresponding them eigenvectors.

3. Assume the following form of the matrix M

$$\begin{aligned} M(t.x',u)=&\sum_{\mu,\nu=1}^{l}\alpha_{\mu\nu}(t,x',u)\gamma^+_\mu(t,x',u)\gamma^+_\nu(t,x',u)\\ &+\sum_{\mu=1}^{l}\sum_{\nu=l+1}^{m}\beta_{\mu\nu}(t,x',u)\gamma^+_\mu(t,x',u)\gamma^-_\nu(t,x',u), \end{aligned} \tag{5}$$

where matrices $\{\alpha_{\mu\nu}\},\{\beta_{\mu\nu}\}$ are such that

$$\max_{\partial\Omega^T\times G}|\alpha^{-1}_{\mu\nu}|\leq\delta^{-1},\quad \max_{\partial\Omega^T\times G}|\beta_{\mu\nu}|\leq\delta^{-1}_1, \tag{6}$$

where δ_o,δ_1 are constants.

4. Assume that eigenvalues are bounded

$$\begin{aligned} &\max_{\nu\in\{1,\dots,l\}}\ \max_{\partial\Omega^T\times G}\lambda^+_\nu(t,x',u)\leq c_1,\\ &\max_{\nu\in\{l+1,\dots,m\}}\ \max_{\partial\Omega^T\times G}|\lambda^-_\nu(t,x',u)|\leq c'_1, \end{aligned} \tag{7}$$

where c_1,c'_1 are constants.

The proof of theorem of the existence of solutions to problem (1) is the following. At first we prove the existence of solutions to linearized problem (1). Hence it is disjoined into two problems: Cauchy and boundary (see [6]). Having found a priori estimates to these problems (Lemmas 1,2,[7,8]) the existence of weak solutions is shown [2], then we prove that the weak solution is strong [3] and finally we show that the strong solution is sufficiently regular.(Theorem 1, see [5]).Now obtaining a priori estimates for nonlinear system (1) (see Lemmas 1,2,also,[7,8]) we prove the existence of solutions to nonlinear problem (1) by the method of successive approximations [7,8](Theorem 2).We have to mention that the existence of local in time solutions of Cauchy problem to nonlinear system $(1)_1$ was proved [4].Using that system $(1)_1$ is dissipative we are able to prove the existence of global in time solutions for sufficiently small data functions.The solutions are classical to (1).

2. NOTATION

Let $H=H(t,x,u(x,t))$.By D_t, D_x we denote the total derivatives: $D_t H=H_t+H_u u_t$, $D_x H=H_x+H_u u_x$, where by H_t, H_x, H_u (or $H_{,t}$, $\partial_t^k H$, and so on) are denoted partial derivatives.

By $|\ |$ we denote the Euclidean norm of a vector or of a matrix.At first we introduce the following notations ($l,k\in\mathbb{Z}_+$, $p\in\mathbb{R}$, $p>1$)

$$\|g\|_{C^l(QxG)}=|g|_{l,QxG}\ ,\ \sup_{QxG}|g| = |g|_{QxG}\ ,\ \|u\|_{Lp(Q)} = \|u\|_{p,Q}\ ,$$

$$\|u\|_{H^l(Q)}=\|u\|_{l,2,Q}\ ,\text{where } Q=\Omega \text{ or } Q=\Omega^T \text{ or } Q=\Omega\times\mathbb{R}^1_+ \text{ and so on.}$$

Now we define the space $\Gamma^l_o(\Omega)$ with the norm

$$\|u\|_{\Gamma^l_o(\Omega)}\equiv|u|_{l,o,\Omega}=\Big(\sum_{|\nu|\le l}\int_\Omega|D^\nu_{t,x}u|^2dx\Big)^{1/2}$$

where

$$D^\nu_{t,x}=D_t^{\nu_o}D_{x_1}^{\nu_1}\dots D_{x_n}^{\nu_n},\ \nu_o+\nu_1+\dots+\nu_n=|\nu|.$$

and also the following space

$\Pi^l_o(\Omega^T)=\bigcap_{i=o}^{l} L^{l-i}_\infty(0,T;H^i(\Omega))$ with the norm

$\|u\|_{\Pi^l_o(\Omega^T)}=|u|_{l,o,\infty,\Omega^T}$ .The same spaces can be introduced for functions defined on $\partial\Omega$.

Let $V\subset\mathbb{R}^{\nu}$. $C^{\infty}_{(o)}(V)(C^{\infty}_{(o)o}(V\times\mathbb{R}^{1}))$ is the set of restrictions to V $(V\times\mathbb{R}^{1})$ of functions $C^{\infty}_{o}(\mathbb{R}^{\nu})(C^{\infty}_{o}(\mathbb{R}^{\nu}\times\mathbb{R}^{1}))$.

Finally we define the scalar produkt of functions defined on $D\subset\mathbb{R}^{k}$ by $\int_D u(x)v(x)dx=(u,v)_D$.

3. EXISTENCE OF SOLUTIONS TO (1)

To prove the existence of solutions of problem (1) we have to find suitable a priori estimates. To do this we linearize problem (1) replacing u by v, which is treated as a given functions, in coefficients of (1). Let us denote the new problem by (1)'. Multiplying $(1)'_1$ by u, integrating the result over Ω and using assumptions 1÷4 one has

Lemma 1.

Let the assumptions 1÷4 be satisfied. Let $v\in G$, $\tilde{L}\in C^{1}(\Omega^{t}\times G)$ $(\tilde{L}=(E,A_1,\dots,A_n,B))$, $v\in W^{1}_{\infty}(\Omega^{t})$, $(c_o+c_1)\delta_o^{-2}\delta_1^{-2}\le \frac{1}{4}c_o$,

$$|E_t|_{\Omega^{t}\times G}+|\operatorname{div} A|_{\Omega^{t}\times G} < \frac{1}{4}\beta_o,\ |E_v|_{\Omega^{t}\times G}|v_t|_{\Omega^{t}}+|A_v|_{\Omega^{t}\times G}|v_x|_{\Omega^{t}}<\frac{1}{4}\beta_o,$$

$$|E|_{\Omega^{t}\times G}\le c_3.$$

Then for functions u in (1)' such that $u\in H^{1}(\Omega^{t})\cap L_{\infty}(0,t;L_2(\Omega))\cap L_2(\Omega^{t})\cap L_2(\partial\Omega^{t})$ the following estimate holds

$$\alpha_o\int_{\Omega}u^2+\frac{c_o}{2}\int_{\partial\Omega^{t}}u^2+\frac{\beta_o}{2}\int_{\Omega^{t}}u^2\le c_2\int_{\partial\Omega^{t}}|Mu|^2+\frac{2}{\beta_o}\int_{\Omega^{t}}|Lu|^2+c_3\int_{\Omega}|u_o|^2, \quad (8)$$

where c_2, c_3 are constants and $c_2=2(c_o+c_1)\delta_o^{-2}$.

To obtain further estimates for solutions to (1)' we have to consider it in a half space $\Omega=\mathbb{R}^{n}_{+}=\{x:x_1>0\}$. Looking for solutions in H^{s} spaces $(s\ge2)$ we have to add the following compatibility conditions

$$D^{\alpha}_{x},D^{\beta}_{t}[M(t,x',v(x',t))u(x',t)-g(t,x')]=0 \text{ at } x_1=0, t=0, \quad (9)$$
$$|\alpha|+\beta\le s-2,$$

where v is a solution to problem (1)' where u and v are replaced by v and w, respectively, and $u|_{t=o}=v|_{t=o}=w|_{t=o}=u_o$. Then

derivates $D^{\sigma}_{t}u, D^{\sigma}_{t}v$, $|\sigma|\le s-2$, at $x_1=0$, $t=0$ are calculated from (1)', so (9) depends on u_o and its derivatives only.

By differentiating (1)' we obtain the following system of problems in $\Omega_1 \equiv \mathbb{R}^n_+$:

$$LD^{\sigma}_{t,x'}u = D^{\sigma}_{t,x'}Lu + (LD^{\sigma}_{t,x'}u - D^{\sigma}_{t,x'}Lu) \quad \text{in } \Omega^T_1,$$

$$MD^{\sigma}_{t,x'}u = D^{\sigma}_{t,x'}Mu + (MD^{\sigma}_{t,x'}u - D^{\sigma}_{t,x'}Mu) \quad \text{on } \partial\Omega^T_1, \qquad (10)$$

$$D^{\sigma}_{t,x'}u|_{t=0} = D^{\sigma}_{t,x'}u|_{t=0} \quad \text{in } \Omega_1,$$

where the right-hand side of $(10)_3$ must be calculated from equations (1)'.

Apply Lemma 1 to problem (10) and use the noncharacteristic boundary conditions which imply that

$$u_{x_1} = A_1^{-1}[F - Eu_t - \sum_{i=2}^{n} A_i u_{x_i} - Bu] \qquad (11)$$

can be calculated because det $A_1 \neq 0$. Then by theorems of embedding and some inductive considerations one gets

Lemma 2

Let the assumptions of Lemma 1 and the compatibility conditions (9) be satisfied. Let $\tilde{L}, M \in C^l(\Omega^t \times G) \cap \Pi^l_o(\Omega^t)$, $v \in G \cap \Pi^l_o(\Omega^t)$, $l > \frac{n}{2} + 1$, $a_1 = \sup_t |A_1^{-1}|_{\Omega^t \times G}$, $a_2 = \sup_t |\partial^1_{t,x,v}\tilde{L}|_{l-1,\Omega^t \times G}$,

$a_3 = \sup_t |\partial^1_{t,x}v|_{l-1,o,\infty,\Omega^t}$, $a_4 = \sup_t |\tilde{L}|_{l,\Omega^t \times G}$,

$b_1 = \sup |\partial^1_{t,x',v}M|_{l-1,\partial\Omega^t \times G}$. Assume the restriction

$$ca_2^2\, p_{1s}(a_1^2 a_4^2)p_{2s}(a_3^2) \leq \frac{1}{4}\beta_o, \qquad (12)$$

where $s \in \mathbb{Z}_+$ and p_{1s}, p_{2s} are some polynomials. Let $Lu \in H^s(\Omega^t) \cap \Gamma^{s-1}_o(\Omega)$, $Mu \in H^s(\partial\Omega^t)$, $u|_{t=0} \in \Gamma^s_o(\Omega)$.

Then there exists polynomials $p_{i,s} = p_{i,s}(a_1, a_2, a_3, b_1)$, $i = 0,1,2,3$, such that for $u \in \Gamma^s_o(\Omega) \cap H^s(\Omega^t) \cap H^s(\partial\Omega^t)$ in (1)' the following estimate holds

$$\alpha_o|u|^2_{s,o,\Omega} + \beta_o\|u\|^2_{s,2,\Omega^t} + c_o\|u\|^2_{s,2,\partial\Omega^t} \leq p_{o,s}|u|^2_{s,o,\Omega}\Big|_{t=0}$$

$$+ p_{1,s}\|Mu\|^2_{s,2,\partial\Omega^t} + p_{2,s}\|Lu\|^2_{s,2,\Omega^t} + p_{3,s}|Lu|^2_{s-1,o,\Omega}, \qquad (13)$$

where $s\in\mathbb{Z}_+$, $s\le 1$, $l>\frac{n}{2}+1$ and $p_{i,s}$, $i=1,2,3$, depend at least linearly on $\sup_t |\partial_{t,x}\bar{L}|_{l-1,\Omega^t\times G} + \sup_t |\partial_v L|_{l-1,\Omega^t\times G} |\partial_{t,x} v|_{l-1,0,\infty,\Omega^t}$,

where $\bar{L}=(\tilde{L},M)$.

Having proved estimates (8),(13) we are able to prove the existence of solutions to linear problem (1)' in the half space $x_1>0$. The existence in a bounded domain can be obtained immediately by a suitable partition of unity. Let $\varphi\in C^\infty_{(0)}(\mathbb{R}^1_+)$, $\varphi=\varphi(x_1)$ and $\varphi=1$ in a neighbourhood of $x_1=0$. Then we introduce a function $\tilde{u}=\tilde{u}(x,t)$, $t>0$, determined by $\tilde{u}|_{t=0}=u_0\varphi$ in such a way that if $u_0\in H^s(\Omega)$, then $\tilde{u}\in H^{s+1/2}(\Omega^t)$ and $\|\tilde{u}\|_{s+1/2,2,\Omega^t}\le c\|u_0\|_{s,2,\Omega}$.
Introducing $w=u-\tilde{u}$ we see that w is a solution to the problem

$$\begin{aligned} &Lw=F-L\tilde{u}\equiv\tilde{F} && \text{in } \Omega_1^T, \\ &Mw=g-M\tilde{u}\equiv\tilde{g} && \text{on } \partial\Omega_1^T, \\ &w|_{t=0}=u_0(1-\varphi) && \text{in } \Omega_1. \end{aligned} \tag{14}$$

Knowing that $w|_{t=0}=0$ near $x_1=0$ we introduce new quantities w_1' and w_2 by splitting (see[6])

$$w=X(t)w_1'+w_2, \tag{15}$$

where $X\in C_0^\infty(-\delta,\delta)$, $X=1$ in a neighbourhood of $t=0$ and δ is sufficiently small (support of X depends on support of φ (see[6])) and maximal speed of propagation of considered system (1) (see [1])). Then problem (14) can be disjoined into the Cauchy problem

$$\begin{aligned} &Lw_1'=\tilde{F}_1, && x_1>0,\ 0<t<T, \\ &w_1'|_{t=0}=u_0(1-\varphi), && x_1>0,\ t=0, \end{aligned} \tag{16}$$

and mixed problem

$$\begin{aligned} &Lw_2=\tilde{F}-X\tilde{F}_1-w_1'\partial_t X\equiv F_2, && x_1>0,\ 0<t<T, \\ &Mw_2=\tilde{g}, && x_1=0,\ 0<t<T, \\ &w_2|_{t=0}=0, && x_1>0,\ t=0, \end{aligned} \tag{17}$$

where as compatibility conditions for solutions in H^s we assume $\tilde{F}_1$ such that

$$D_t^\alpha F_2|_{t=0}=0 \quad \text{for } \alpha\le s-1. \tag{18}$$

Assume that $\tilde{u}_0$ is an extension of $u_0(1-\varphi)$ such that $\tilde{u}_0|_{t=0}=$ $=u_0(1-\varphi)$ and put $w_1=w_1'-\tilde{u}_0$. Then instead of (16) one has

$$\begin{aligned} &Lw_1=\tilde{F}_1-L\tilde{u}_0\equiv F_1, && x_1>0,\ 0<t<T, \\ &w_1|_{t=0}=0, && x_1>0,\ t=0. \end{aligned} \tag{19}$$

By the compatibility conditions (9)' such that $D_t^\alpha\tilde{g}=0$ at $t=0$ for

$\alpha \le s-1$ and (18), the functions $\tilde{G}$ and $\tilde{F}_2$ can be prolongated by zero for $t<0$ in the sense of H^s spaces. Therefore we replace (17) by the following boundary problem

$$\begin{aligned} &Lw_2=F_2, && x_1>0, t\in\mathbb{R}^1, \\ &Mw_2=\tilde{g}, && x_1=0, t\in\mathbb{R}^1. \end{aligned} \tag{20}$$

To prove the existence of solutions we need adjoint operators and adjoint boundary conditions. Therefore we consider the following identities

$$(Lu,v)_{\mathbb{R}^n\times[0,T]}=(u,L^*v)_{\mathbb{R}^n\times[0,T]}, \quad \forall\ u,v\in C_0^\infty(\mathbb{R}^n\times[0,T]) \tag{21}$$

for the Cauchy problem, and

$$\begin{aligned} &(Lu,v)_{\mathbb{R}^n_+\times\mathbb{R}^1}+(A_1\alpha^{-1}Mu,v)_{\mathbb{R}^{n-1}\times\mathbb{R}^1}=(u,L^*v)_{\mathbb{R}^n_+\times\mathbb{R}^1} \\ &+(u,A_1\alpha^{*-1}M^*v)_{\mathbb{R}^{n-1}\times\mathbb{R}^1}, \end{aligned} \tag{22}$$

$\forall u,v\in C^\infty_{(0)0}(\mathbb{R}^n_+\times\mathbb{R}^1)$, for the boundary problem (20), where

$$L^*=-E\partial_t-\sum_{i-1}^{n}A_i\partial_{x_i}+(B^*-E_t-\operatorname{div} A), \tag{23}$$

$$M^*=\sum_{\mu,\nu=l+1}^{m}\alpha^*_{\mu\nu}\gamma^-_\mu\gamma^-_\nu+\sum_{\mu=l+1}^{m}\sum_{\nu=1}^{l}\beta^*_{\mu\nu}\gamma^-_\mu\gamma^+ \tag{24}$$

and

$$\sum_{\sigma=1}^{n}\lambda_\mu\alpha^{-1}_{\mu\sigma}\beta_{\sigma\nu}=\sum_{\sigma=l+1}^{m}\lambda_\nu\alpha^{*-1}_{\nu\sigma}\beta^*_{\sigma\mu}, \mu=1,\dots,l, \nu=l+1,\dots,m. \tag{25}$$

From Lemmas 1 and 2 we get a priori estimates (8) and (13) for solutions of problems (19), (20) and their adjoint. Then by the well known methods (see[3]) we prove the existence of strong solutions of these problems in L_2 spaces (see[6]). Then by the Friedrichs' mollifiers technics we solve the regularity problem for solutions of (19) and (20). Knowing that $u=\tilde{u}+X(w_1+\tilde{u}_0)+w_2$ and using a suitable partition of unity we obtain the following result for the linearized problem (1)' in a bounded domain.

Theorem 1

Let the assumptions of Lemma 2 be satisfied and $\partial\Omega\in C^{s}$. Then there exists a unique solution to problem (1)' such that $u\in\Pi_{o}^{s}(\Omega^{t})\cap H^{s}(\Omega^{t})\cap H^{s}(\partial\Omega^{t})$, $s\le l$, $l>\frac{n}{2}+1$ and estimate (13) is valid

To prove the existence of solutions to problem (1) we use the following method of successive approximations

$$L(\overset{m-1}{u})\overset{m}{u}\equiv E(t,x,\overset{m-1}{u})\overset{m}{u}_t+\sum_{i=1}^{n}A_i(t,x,\overset{m-1}{u})\overset{m}{u}_{x_i}+B(t,x,\overset{m-1}{u})\overset{m}{u}=$$

$$=F(t,x) \quad \text{in } \Omega^{t},$$

$$M(t,x',\overset{m-1}{u})\overset{m}{u}=g(t,x') \quad \text{on } \partial\Omega^{t}, \qquad (26)$$

$$\overset{m}{u}|_{t=0}=u_o(x) \quad \text{in } \Omega,$$

where $m=1,2,\dots,\overset{o}{u}=u_o$. Introduce the set $Q(N)=\{u\in G\subset\mathbb{R}^{m}:|u|_{l,o,\infty,\Omega^{t}}\le N,\ l>\frac{n}{2}+1\}$. Let $\overset{m-1}{u}\in Q(N)$, $m=1,2,\dots$, and the assumptions of Theorem 1 be satisfied for $\overset{m-1}{u}\in Q(N)$. Then by Theorem 1 there exist a unique solution of (26) such that $\overset{m}{u}\in\Pi_{o}^{l}(\Omega^{t})\cap H^{l}(\Omega^{t})\cap H^{l}(\partial\Omega^{t})$ and

$$\|\overset{m}{u}\|_{l,\Omega^{t}}\equiv|\overset{m}{u}|_{l,o,\Omega}+\|\overset{m}{u}\|_{l,2,\Omega^{t}}+\|\overset{m}{u}\|_{l,2,\partial\Omega^{t}}\le N,\ \forall t\in\mathbb{R}_{+}^{1} \qquad (27)$$

if

$$p_{o,l}|u|^{2}_{l,o,\Omega}|_{t=o}+p_{1,l}(N)\|g\|^{2}_{l,2,\partial\Omega^{t}}+p_{2,l}(N)\|F\|^{2}_{l,2,\Omega^{t}}$$
$$+p_{3,l}(N)|F|^{2}_{l-1,o,\Omega}\le N, \qquad (28)$$

where the last inequality holds for sufficiently small u_o, F, g, N and a_2, b_1.

To show convergence of sequence $\{\overset{m}{u}\}$ we consider the following problem

$$L(\overset{m-1}{u})\overset{m}{U}=-L'(\tilde{u}^{m-1})\overset{m-1}{U},$$
$$M(\overset{m-1}{u})\overset{m}{U}=-M'(\tilde{u}_{o}^{m-1})\overset{m-1}{u}\ \overset{m-1}{U}, \qquad (29)$$
$$\overset{m}{U}|_{t=o}=0,\ m\ge1,\quad \overset{o}{U}=u_o(x),$$

where $\overset{m}{U}=\overset{m}{u}-\overset{m-1}{u}$, $L(\overset{m-1}{u})-L(\overset{m-1}{u})=L'(\tilde{u}^{m-1})(\overset{m-1}{u}-\overset{m-2}{u})$, $M(\overset{m-1}{u})-M(\overset{m-2}{u})=M'(\tilde{u}^{m-1})(\overset{m-1}{u}-\overset{m-2}{u})$, prime denotes derivative with respect to u, and $\tilde{u}^{m-1}\in[\overset{m-1}{u},\overset{m-2}{u}]$. Then by Lemma 2 we have

$$\|\overset{m}{U}\|_{s,\Omega^{t}}\le h(\|\overset{m-1}{u}\|_{l,\Omega^{t}},\|\overset{m-2}{u}\|_{l,\Omega^{t}})\|\overset{m-1}{u}\|_{l,\Omega^{t}}\|\overset{m-1}{\bar{U}}\|_{s,\Omega^{t}} \qquad (30)$$

where $s<l$. Assume that N is such that

$$h(N,N)N<1. \tag{31}$$

Then the sequence $\{\overset{m}{u}\}$ converges strongly in $L_\infty(0,t;\Gamma_0^s(\Omega))\cap H^s(\Omega^t)\cap H^s(\partial\Omega^t)$, $s<l$, $l>\frac{n}{2}+1$. Hence by (27) and well known arguments (equations (1) are satisfied in the classical way because $\Pi_0^l(\Omega^t)\subset C^1(\Omega^t)$ for $l>\frac{n}{2}+1$ (see [1], Ch. 3, Th. 10.4)) one obtains

Theorem 2

Let the assumptions of Theorem 1 be satisfied and $s=l$. Let data u_0, F, g and coefficients of (1) be such that (28) and (31) are satisfied. Then there exists a solution u to the problem (1) such that $u\in L_\infty(0,t;\Gamma_0^l(\Omega))\cap H^l(\Omega^t)\cap H^l(\partial\Omega^t)$, $t\in\mathbb{R}^1_+$, $\|u\|_{l,\Omega^t}\le N$ and (1) are satisfied classically.

We proved the global existence because we have not any restrictions on time.

Finally we prove uniqueness. Assume that u^1, u^2 are two solutions to problem (1). Then $U=u^1-u^2$ is a solution to the problem

$$L(u^1)U=-(L(u^1)-L(u^2))u^2\equiv -L'(\tilde u)u^2U \qquad \text{in } \Omega^t,$$

$$M(u^1)U=-(M(u^1)-M(u^2))u^2\equiv -M'(\tilde u)u^2U \qquad \text{on } \partial\Omega^t,$$

$$U|_{t=0}=0 \qquad \text{in } \Omega,$$

where $\tilde u\in[u^1,u^2]$. Then by Lemma 1 one has

$$\alpha_0\|U\|^2_{2,\Omega}+\frac{c_0}{2}\|U\|^2_{2,\partial\Omega^t}+\frac{\beta_0}{2}\|U\|^2_{2,\Omega^t}\le c_2\|M'(u_2)u^2U\|^2_{2,\partial\Omega^t}$$

$$+\frac{2}{\beta_0}\|L'(\tilde u)u^2U\|^2_{2,\Omega^t}.$$

Assuming that

$$\frac{2}{\beta_0}\|L'(\tilde u)u^2U\|^2_{2,\Omega^t}\le\frac{2}{\beta_0}|\tilde L'(\tilde u)|^2_{\Omega^t\times G}(|u^2|^2_{\Omega^t}+|\partial^1_{t,x}u^2|^2_{\Omega^t})\|U\|^2_{2,\Omega^t}$$

$$\le\frac{\beta_0}{4}\|U\|^2_{2,\Omega^t}, \tag{32}$$

$$c_2\|M'(\tilde u)u^2U\|^2_{2,\partial\Omega^t}\le c_2|M'(\tilde u)|^2_{\partial\Omega^t\times G}|u^2|_{\partial\Omega^t}\|U\|^2_{2,\partial\Omega^t}\le\frac{c_0}{4}\|U\|^2_{2,\partial\Omega^t},$$

one has

Theorem 3

Let the assumptions of Lemma 1 be satisfied, $v=u$, $\tilde L$, M, be boundedly differentiable with respect to u, (32) be satisfied. Then for solutions of (1) in $C^1(\bar\Omega^t)$ we have uniqueness.

REFERENCES

[1] Besov,O.V.,Ilyin,V.P.,Nikolskij,S.M.:"Integral representations of functions and theorems of imbedding",Moscow 1975 (in Russian).

[2] Friedrichs,K.O.:"Symmetric positive linear differential equations",Comm.Pure Appl.Math.11(1958)pp.333-418.

[3] Lax,P.D.,Phillips,R.S.:"Local boundary conditions for dissipative symmetric linear differential operators", Comm.Pure Appl.Math.13(1960)pp.427-455.

[4] Majda,A.:"Compressible fluid flow and systems of conservation laws in several space variables",Appl.Math.Sc-53, Springer,New York 1984.

[5] Nagumo,M.:"Lectures on modern theory of partial differential equations",Moscow 1967 (in Russian).

[6] Rauch,J.:"L_2 is a continuable initial condition for Kreiss' mixed problems",Comm.Pure Appl.Math.25 (1972) pp.265-285.

[7] Zajaczkowski,W.M.:"Initial boundary-value problems for nonlinear symmetric systems of first order",Selected problems of modern continuum theory",ed.Kosinski,W.,..., Pitagora Bologna 1987 pp.195-205.

[8] Zajaczkowski,W.M.:"Noncharacteristic mixed problems for nonlinear symmetric hyperbolic system (to be published).

STABILITY OF INITIAL BOUNDARY VALUE PROBLEMS FOR HYPERBOLIC SYSTEMS

M. Ziółko

AGH, Institute of Automatic Control

al. Mickiewicza 30, 30-058 Kraków, Poland

SUMMARY

The energy integral method was used to prove the asymptotic stability for a set of first order linear equations with constant coefficients. The main problem consists in finding a norm which would enable to establish sufficient conditions for boundary and interior stability. These conditions are also necessary if considered system has some additional properties.

THE HYPERBOLIC SYSTEMS

There are various approaches to the theory of hyperbolic differential equations in which the energy integral method is used. At the beginning of 20-th century this method was applied to prove the uniqueness and to establish the existence of the solutions of hyperbolic partial differential equations. Friedrichs in his work [1] quoted papers dealing with these problems. Secondly, the energy integral method was used by Gunzburger [2] in 1977 and Layton [4] in 1983 to establish the stability of Galerkin method.

Consider the initial boundary value problem

$$\frac{\partial y}{\partial t} + \Lambda \frac{\partial y}{\partial x} = D\, y \qquad y_{(x,0)} = y_{0(x)} \tag{1}$$

for $y_{(x,t)} \in R^n$, $0 \le x \le 1$, $0 \le t \le T$. Λ and D are nxn real and constant matrices. Without loss of generality Λ may be taken to be diagonal

$$\Lambda = \begin{bmatrix} \Lambda^- & 0 \\ 0 & \Lambda^+ \end{bmatrix} \qquad \begin{aligned} \Lambda^- &= \mathrm{diag}(\lambda_1, \lambda_2, \ldots, \lambda_r) < 0 \\ \Lambda^+ &= \mathrm{diag}(\lambda_{r+1}, \ldots, \lambda_n) > 0, \end{aligned}$$

Well-posed linear homogeneous boundary data can always be written as

$$y^{+}_{(0,t)} = S_0\, y^{-}_{(0,t)}$$
$$y^{-}_{(1,t)} = S_1\, y^{+}_{(1,t)} \qquad (2)$$

where vector $y = [y^{-T}\ y^{+T}]^T$ is the partition of y corresponding to the partition of Λ. S_0 and S_1 are matrices with dimensions $(n-r)\times r$ and $r\times(n-r)$, respectively. The initial conditions are C^1 continuous and satisfy

$$y^{+}_{0(0)} = S_0\, y^{-}_{0(0)}$$
$$y^{-}_{0(1)} = S_1\, y^{+}_{0(1)} . \qquad (3)$$

STABILITY OF HYPERBOLIC PROBLEMS

For a positive definite matrix G, the energy functional is defined as

$$E_{(t)} = \|y\|^2_G = \int_0^1 y^T G\, y\, dx \qquad (4)$$

where $\|y\|^2_G$ denotes the second power of $(L^2_{(0,1)})^n$ norm. In this work the asymptotic stability is considered. By definition, problem (1) (2) is stable if $\|y\|_G \to 0$ as $t\to\infty$. By use of Leibnitz pattern the time rate of change of the energy functional is given by

$$\frac{dE}{dt} = \int_0^1 y^T(D^T G + GD)\, y\, dx - y^T \Lambda G\, y\Big|_0^1 + \int_0^1 y^T(G\Lambda - \Lambda G)\,\frac{\partial y}{\partial x}\, dx . \qquad (5)$$

If we assume additionally that matrix G is also diagonal, then $G\Lambda=\Lambda G$ and the last segment become equal to zero. Only then is it possible to obtain useful results. Assumption, that matrix G is diagonal, sets additional restrictions for matrices D, S_0 and S_1. It requires a certain "symmetry" of the boundary value problem. Taking into account that $G\Lambda=\Lambda G$, and substituting boundary conditions (2) to equation (5) we obtain

$$\frac{dE}{dt} = \frac{dE^i}{dt} + \frac{dE^b}{dt} \qquad (6)$$

where

$$\frac{dE^{i}}{dt} = \int_{0}^{1} y^{T}(D^{T}G + GD)\, y\, dx \tag{6a}$$

$$\frac{dE^{b}}{dt} = y^{-T}_{(0,t)}(G^{-}\Lambda^{-}+S_{0}^{T}G^{+}\Lambda^{+}S_{0})y^{-}_{(0,t)} - y^{+T}_{(1,t)}(G^{+}\Lambda^{+}+S_{1}^{T}G^{-}\Lambda^{-}S_{1})y^{+}_{(1,t)} . \tag{6b}$$

G^{-} and G^{+} are the partitions of G, similarly like Λ^{-} and Λ^{+} are the partitions of matrix Λ.

Definition 1.
The initial boundary value problem (1)(2) is interior stable (resp. boundary stable) if there is a positive definite diagonal matrix G, independent of initial conditions, such that the inequality $dE^{i}/dt < 0$ (resp. $dE^{b}/dt < 0$) holds for every $t \geq 0$ and $y \neq 0$ (resp. $y^{-}_{(0,t)} \neq 0$ and $y^{+}_{(1,t)} \neq 0$).

Theorem 1.
If problem (1)(2) is interior stable then all real parts of eigenvalues of matrix D are negative.

Proof.
If problem (1)(2) is interior stable then $\int_0^1 y^{T}(GD + D^{T}G)y\, dx < 0$ for every $t \geq 0$, including the arbitrary taken $y = y_{0(x)}$. It follows that $GD + D^{T}G < 0$. This condition can be fulfilled by a positive definite matrix G only if all real parts of eigenvalues of matrix D are negative. ∎

Theorem 2.
If all eigenvalues of matrix $D + D^{T}$ are negative, then problem (1)(2) is interior stable.

Proof.
If $D + D^{T} < 0$ then we can take G equal to the unit matrix and obtain $dE^{i}/dt < 0$ for every $y \neq 0$. ∎

If we assume additionally that matrix D is symmetric, then theorem 2 would gives not only the sufficient (with G equal to the unit matrix), but also the necessary condition for interior stability. This idea can be developed for a much larger class of matrices D.

Definition 2.

A matrix D is symmetrizable by a nonsingular matrix K if $KD = D^TK$.

Theorem 3.

Assume that matrix D is symmetrizable by a diagonal positive definite matrix, e.g. $K^2D = D^TK^2$. Then problem (1)(2) is interior stable if and only if all eigenvalues of matrix D are negative.

Proof.

For a symmetrizable matrix D we have $KDK^{-1} = K^{-1}D^TK$. It follows that KDK^{-1} is a symmetric matrix and all eigenvalues of matrix D are real. Substituting $G = K^2$ and $y = K^{-1}\eta$ we obtain

$$y^T(GD + D^TG)\,y = 2\,\eta^T KDK^{-1}\eta.$$

This quadratic form is negative definite if and only if all eigenvalues of matrix D are negative. ■

If matrix D is symmetric or symmetrizable by a positive definite matrix, it is possible to obtain useful stability conditions. For this reason symmetrizability of equations of classical physics were tested. Many years ago, Fridrichs observed that almost all equations, linear and linearized, can be transformed to a symmetrical form. For example it is easy to verify that symmetrizers are positive definite for the equations of electric RLC and RLGC networks, linear gas pipeline and heat exchangers.

Lemma 1.

If S_o and S_1 are matrices with dimensions $m\times r$ and $r\times m$ respectivelly, then all nonzero eigenvalues of matrix S_oS_1 are identical with nonzero eigenvalues of matrix S_1S_o, that is

$$\det(S_oS_1 - \lambda I) = \lambda^{m-r}\det(S_1S_o - \lambda I). ■$$

Theorem 4.

If problem (1)(2) is boundary stable then all modules of

eigenvalues of matrices S_0S_1 and S_1S_0 are less than 1.

Proof.

If $dE^b/dt < 0$ for every nonzero solution of equation (1) then we obtain inequalities for quadratic forms

$$S_0^T G^+ \Lambda^+ S_0 < -G^- \Lambda^-$$
$$S_1^T (-G^- \Lambda^-) S_1 < G^+ \Lambda^+ .$$

Substituting $G^+\Lambda^+$ by $S_1^T(-G^-\Lambda^-)S_1$ in the first inequality we obtain

$$(S_1S_0)^T(-G^-\Lambda^-)S_1S_0 - (-G^-\Lambda^-) < 0 .$$

This inequality condition can be fulfilled by a positive definite matrix $-G^-\Lambda^-$ only if all modules of eigenvalues of matrix S_1S_0 are less than 1. It follows from lemma 1 that all eigenvalues of matrix S_0S_1 are also less than 1.■

Theorem 5.

If there exists $g > 0$ such that $\|S_0\| < g$ and $\|S_1\| < g^{-1}$ then problem (1)(2) is boundary stable. $\|\cdot\|$ denotes the Hilbert (spectral) matrix norm.

Proof.

We can take $g = g_-/g_+$, where g_- and g_+ are positive constants. From the assumptions of this theorem and the definition of Hilbert norm it follows that

$$\frac{y^{-T} S_0^T S_0 y^-}{y^{-T} y^-} < \left(\frac{g_-}{g_+}\right)^2$$

$$\frac{y^{+T} S_1^T S_1 y^+}{y^{+T} y^+} < \left(\frac{g_+}{g_-}\right)^2$$

for every $y^- \neq 0$ and $y^+ \neq 0$. If we now take $-G^-\Lambda^- = g_-^2 I_n$ and $G^+\Lambda^+ = g_+^2 I_{n-r}$ we finally obtain $dE^b/dt < 0$.■

Theorem 6.

Assume that the square matrix $S = \begin{pmatrix} 0 & S_1 \\ S_0 & 0 \end{pmatrix}$ is symmetrizable by a diagonal positive definite matrix. Then problem (1)(2) is boundary stable if and only if all eigenvalues of matrix S_0S_1, or equivalently S_1S_0, are less than 1.

Proof.

Condition (6b) for the boundary stability can be written in matrix notation

$$y^T(S^T|G\Lambda|S - |G\Lambda|)y < 0 \qquad (7)$$

where $y = \begin{pmatrix} y^{-T}_{(0,t)} & y^{+T}_{(1,t)} \end{pmatrix}^T$ and $|\cdot|$ denotes the absolute value of all elements of matrix, i.e. $|G\Lambda| = \begin{pmatrix} -G^-\Lambda^- & 0 \\ 0 & G^+\Lambda^+ \end{pmatrix}$.

For a symmetrizable matrix S we have $KSK^{-1} = K^{-1}S^TK$, that is KSK^{-1} is a symmetric matrix and it follows that all eigenvalues of matrix S are real. Substituting $|G\Lambda|=K^2$ and $\eta=Ky$ we obtain condition (7) for the boundary stability in the form

$$\frac{\eta^T(KSK^{-1})^2\eta}{\eta^T\eta} < 1 .$$

This Rayleigh's quotient is less than 1 for arbitrary $\eta\neq 0$, if and only if all modules of eigenvalues of the matrix S are less than 1. If λ denotes eigenvalue of matrix S and $y = \begin{pmatrix} y_1 \\ y_2 \end{pmatrix}$ its block eigenvector then

$$\begin{pmatrix} -\lambda I & S_1 \\ S_0 & -\lambda I \end{pmatrix} \begin{pmatrix} y_1 \\ y_2 \end{pmatrix} = 0 .$$

It follows that

$$S_0S_1y_2 = \lambda^2 y_2$$
$$S_1S_0y_1 = \lambda^2 y_1$$

i.e. λ^2 is an eigenvalue of matrices S_0S_1 and S_1S_0. ■

The theorem 6 enable to test the boundary stability only for the cases when there are positive definite diagonal matrices G^+ and G^- such that

$$S_0^T G^+\Lambda^+ = -G^-\Lambda^-S_1 . \qquad (8)$$

This condition can be fulfilled only for the cases when each element of matrix S_0^T and the element of matrix S_1 situated in the same place are either equal zero or have the same sign. There are important cases of the initial boundary value problems which violate this constraints. Nevertheless it is possible to extend the theorem 6 over the cases where some elements of matrix S_0^T have other sign then corresponding elements of matrix S_1.

Theorem 7.

Problem (1)(2) is boundary stable if there exists a diagonal, nonsingular matrix K such that $KS = S^T K$, and all modules of eigenvalues of matrix $K^{-1}|K|S$, or equivalently $S|K|K^{-1}$, are less than 1.

Proof.

We can take $|G\Lambda| = |K| = JKJ$ where J is a complex diagonal matrix, consisting of only two elements: 1 or $-i$. The matrices $K^{-0,5}$ and J are complex, but their product $K^{-0,5}J$ is a real diagonal matrix. After the transformation $K^{-0,5}J\eta = y$, condition (7) for the boundary stability takes form

$$\eta^T(K^{-0,5}JS^TJKJSJK^{-0,5} - I)\eta < 0 \qquad (9)$$

If $KS = S^T K$ holds, then a real matrix $K^{0,5}JSJK^{-0,5}$ is symmetric. Therefore inequality (9) is satisfied for every nonzero real vector η if all modules of eigenvalues (which are real) of a complex matrix JSJ are less than 1. Now, if y denotes a eigenvector of the matrix JSJ and λ its eigenvalue, then $(JSJ-\lambda I)y=0$. It follows that $(J^2S-\lambda I)Jy=0$ and $(SJ^2-\lambda I)J^{-1}y=0$, that is matrices JSJ, $J^2S=K^{-1}|K|S$ and $SJ^2=S|K|K^{-1}$ have the same eigenvalues.■

STABILITY OF TWO EQUATIONS SYSTEM

Consider the system which consists of only two equations

$$\frac{\partial y}{\partial t} + \begin{pmatrix} \lambda^- & \\ & \lambda^+ \end{pmatrix} \frac{\partial y}{\partial x} = Dy \qquad (10)$$

where $\lambda^- < 0$ and $\lambda^+ > 0$ are speeds of return and progressive wave, respectively. Without lost of generality the energy functional can be define as

$$E_{(t)} = \int_0^1 y^T \begin{pmatrix} 1 & 0 \\ 0 & g \end{pmatrix} y \, dx \, . \qquad (11)$$

The time rate of change of the energy functional can be split into two components

$$\frac{dE^i}{dt} = \int_0^1 y^T \begin{pmatrix} 2d_{11} & d_{12}+gd_{21} \\ d_{12}+gd_{21} & 2gd_{22} \end{pmatrix} y \, dx \tag{11a}$$

$$\frac{dE^b}{dt} = (\lambda^- + g\lambda^+ S_0^2) y^{-2}_{(0,t)} - (g\lambda^+ + \lambda^- S_1^2) y^{+2}_{(1,t)} \tag{11b}$$

where d_{ij} are elements of matrix D. Boundary conditions (2) are set by scalars S_0 and S_1.

Applying theorem 3 we obtain the conclusion that symmetrizable system (10) is interior stable if, and only if $trD < 0$ and $detD > 0$. Symmetrizable matrix D fulfills either condition $0 < d_{12}/d_{21} = g$ or $d_{12} = d_{21} = 0$ (then $g=1$). Because the system under consideration is simple, formula (11a) can be analyzed in a straightforward way.

Theorem 8.

If $d_{11} < 0$, $d_{22} < 0$ and $detD > 0$ then and only then exists a diagonal matrix G such that problem (10)(2) is interior stable and parameter g fulfills either conditions

$$\frac{\left(\sqrt{detD} - \sqrt{d_{11}d_{22}}\right)^2}{d_{21}^2} < g < \frac{\left(\sqrt{detD} + \sqrt{d_{11}d_{22}}\right)^2}{d_{21}^2} \quad \text{for } d_{21} \neq 0 \tag{12a}$$

$$\text{or} \quad g > \frac{d_{12}}{4\, d_{11} d_{22}} \quad \text{for } d_{21} = 0. \blacksquare \tag{12b}$$

The conclusions obtained from theorem 8 are more valuable then conclusions from theorem 3. We obtain all admissible values for g and additionally the conditions for stability if matrix D is nonsymmetrizable (then $d_{12}d_{21} \le 0$). However, there are some other cases when matrix D fulfills necessary conditions for the stability but does not exist a diagonal matrix G such that $dE^i/dt < 0$. All these cases are described by inequalities $trD < 0$, $detD > 0$ and $d_{11}d_{22} \le 0$.

Theorem 6 can be applied to establish the boundary stability either for the cases $S_0S_1 > 0$ or $S_1 = S_2 = 0$. Then matrix S is symmetrizable by a positive definite diagonal matrix. A case $S_0S_1 < 0$ can be verified by theorem 7. For both these cases we obtain the necessary and sufficient condition for the boundary stability: $|S_0S_1| < 1$. Then the positive and monotonically decreasing energy functional (11) is defined either by $g = |\lambda^- S_1 / \lambda^+ S_0|$ or $g > 0$ if $S_0 = S_1 = 0$. However, the case when only one of the elements either S_1 or S_0 equals zero, can not be established on this way.

Theorem 9.

If $|S_0 S_1| < 1$ then and only then there exists a diagonal matrix G such that problem (1)(2) is boundary stable and parameter g fulfills inequalities

$$\frac{-\lambda^-}{\lambda^+} S_1^2 < g < \frac{-\lambda^-}{\lambda^+} \frac{1}{S_0^2} . \qquad (13)$$

Proof.

If $dE^b/dt < 0$ for arbitrary solution of the hyperbolic equation (10) then

$$g\lambda^+ S_0^2 < -\lambda^-$$
$$-\lambda^- S_1^2 < g\lambda^+.$$

The both sides of these inequalities are nonnegative, and therefore we can multiply them to obtain

$$S_0^2 S_1^2 < 1 .$$

To prove the sufficient condition, we conclude from inequality

$$\frac{-\lambda^-}{\lambda^+} S_1^2 < \frac{-\lambda^-}{\lambda^+} \frac{1}{S_0^2}$$

that there exists g such that

$$\frac{-\lambda^-}{\lambda^+} S_1^2 < g < \frac{-\lambda^-}{\lambda^+} \frac{1}{S_0^2} . \qquad \blacksquare$$

We obtained all admissible values for g if problem (10)(2) is boundary stable. Moreover, we proved the stability for a "nonsymmetric" case when $S_0 S_1 = 0$ but $S_1 + S_0 \neq 0$.

If problem (10)(2) is interior and boundary stable, it is not always possible to find such g that both formulas (11a) and (11b) would be negative simultaneously. Some additional conditions are then and only then fulfilled

$$\frac{\left(\sqrt{\det D} - \sqrt{d_{11} d_{22}}\right)^2}{d_{21}^2} S_o^2 \leq \frac{-\lambda^-}{\lambda^+} \leq \frac{\left(\sqrt{\det D} + \sqrt{d_{11} d_{22}}\right)^2}{d_{21}^2} \frac{1}{S_1^2} .$$

These conditions are obtained from (12a) and (13) under the assumption that $d_{21} \neq 0$. For the case when $d_{21} = 0$ we obtain from inequalities (12b) and (13)

$$\frac{-\lambda^-}{\lambda^+} \geq \frac{d_{12}^2 \, S_o^2}{4 \, d_{11} d_{22}} .$$

REFERENCES

[1] Fridrichs K.O: "Symmetric Hyperbolic Linear Differential Equations", Comm. Pure Appl. Math., 17 (1954) pp. 345-392.

[2] Gunnzburger Max.D: "On the stability of Galerkin methods for initial-boundary value problems for hyperbolic systems", Mathematics of Computation, 31 (1977) pp. 661-675.

[3] Kreiss H.-O: "Initial boundary value problems for hyperbolic Systems", Comm. Pure Appl. Math., 23 (1970) pp. 277-298.

[4] Layton W.J: "Stable Galerkin methods for hyperbolic systems", SIAM J.Numer. Anal., 20 (1983) pp. 221-233.

List of Participants and Authors(*)

Alber, H. D., Dr., Mathematisches Institut A, Universität Stuttgart, Pfaffenwaldring 57, 7000 Stuttgart 80, *FRG*

Andersen, G. R., Mathematics and Physics Branch Dept. of the Army, US Army Research Development and Standardization Group (UK), 'Edison House', 223 Old Marylebone Rd., London NW1 5TH, *Great Britain*

* **Andersson, H. I.**, Norwegian Institute of Technology, Trondheim, *Norway*

Ansorge, R., Prof. Dr., Institut für Angewandte Mathematik, Universität Hamburg, Bundesstr. 55, 2000 Hamburg 13, *FRG*

* **Arminjon, P.**, Prof. Dr., Dépt. de Mathématiques et Statistiques, Université de Montréal, C.P. 6128 Succursale A, Montréal, Québec H3C 3J7, *Canada*

Arrenbrecht, W., Dr., Lehrstuhl für Mechanik, RWTH Aachen, Templergraben 64, 5100 Aachen, *FRG*

Bäcker, M., Fachbereich Mathematik, Universität Kaiserslautern, Postfach 3049, 6750 Kaiserslautern, *FRG*

* **Ballmann, J.**, Prof. Dr.-Ing, Lehr- und Forschungsgebiet Mechanik, RWTH Aachen, Templergraben 64, 5100 Aachen, *FRG*

Barbry, H., Dr., C.E.A., Centre de Limeil, B.P. 27, 94190 Villeneuve St. Georges, *France*

Barley, J. J., Prof. Dr., Dept. of Mathematics, University of Reading, Whiteknights, P.O. Box 220, Reading RG6 2AH, *Great Britain*

Barth, T. J., Prof. Dr., CFD Branch, NASA Ames, M.S. 202a-1, Moffett Field, CA 94035, *USA*

* **Beam, R. M.**, Prof. Dr., NASA Ames Research Center, Moffett Field, CA 94035, *USA*

* **Becker, K.**, MBB - UT, TE 212, Hünefeldstr. 1-5, 2800 Bremen 1, *FRG*

Benetschik, H., Dr., Institut für Strahlenantriebe und Turboarbeitsmaschinen, RWTH Aachen, Templergraben 55, 5100 Aachen, *FRG*

Berger, H., Mathematisches Institut A, Universität Stuttgart, Pfaffenwaldring 57, 7000 Stuttgart 80, *FRG*

* **Billet, G.**, O.N.E.R.A., 29, Avenue de la Division Leclerc, 92320 Chatillon Cedex, *France*

* **Binniger, B.**, Dr., Aerodynamisches Institut, RWTH Aachen, Templergraben 55, 5100 Aachen, *FRG*

Bisbos, C. D., Dr., Institute of Steel Structures, University of Thessaloniki, Kleiton 10, Thessaloniki, Greece

* **Bourdel, F.**, Dr., Groupe d'Analyse Numérique, O.N.E.R.A.-C.E.R.T., 2, Ave Edouard Belin - B.P. 4025, 31055 Toulouse Cedex, *France*

Bourgeat, A., Prof. Dr., Equipe d'Analyse Numérique, Université de Saint-Etienne, 23, Rue du Docteur Paul Michelon, 42023 Saint-Etienne Cedex 2, *France*

Braess, D., Prof. Dr., Institut für Mathematik, Ruhr-Universität Bochum, Universitätsstr. 150, Geb. NA, 4630 Bochum, *FRG*

* **Brakhagen, F.**, Dr., Gesellschaft für Mathematik und Datenverarbeitung mbH, Schloß Birlinghoven, Postfach 1240, 5205 Sankt Augustin 1, *FRG*

Braun, M., Prof. Dr.-Ing., Fachbereich 7 — Maschinenbau Fachgebiet Mechanik, Universität Duisburg, Lotharstr. 1, 4100 Duisburg 1, *FRG*

Brawer, R., Dr., Seminar für Angewandte Mathematik, ETH Zentrum, 8092 Zürich, *Switzerland*

* **Brio, M.**, Prof. Dr., Dept. of Mathematics, Building #89, University of Arizona, Tucson, AZ 85721, *USA*

* **Cahouet, J.**, Prof. Dr., Research Branch, Laboratoire National d'Hydraulique, 6, Quai Watier, B.P. 49, 78401 Chatou Cedex, *France*

Carasso, C., Prof. Dr., Equipe d'Analyse Numerique, Université de Saint–Etienne, 23, Rue du Docteur Paul Michelon, 42023 Saint–Etienne Cedex, *France*

Carstens, V., Dr., Inst. für Aeroelastik, DFVLR — AVA, Bunsenstraße 10, 3400 Göttingen, *FRG*

* **Causon, D. M.**, Dr., Dept. of Mathematics and Physics, John Dalton Faculty of Science and Engineering, John Dalton Building, Chester Street, Manchester M1 5GD, *Great Britain*

Chakravarthy, S., Prof. Dr., Rockwell Science Center, P.O. Box 1085, Thousand Oaks, CA 91360, *USA*

Chaput, E., Dr., Aerospatiale, Route de Verneuil, B.P. No 2, 78133 Les Mureaux Cedex, *France*

Chen, G., Dr., Courant Institute of Mathematical Science, New York University, 251 Mercer Street, New York, NY 10012, *USA*

* **Childs, P. N.**, Prof. Dr., Oxford University Computing Laboratory, 8–11 Keble Road, Oxford OX1 3QD, *Great Britain*

* **Christiansen, S.**, Prof. Dr., Laboratory of Applied Mathematical Physics, Technical University of Denmark, 2800 Lyngby, *Denmark*

Colombeau, J. F., Prof. Dr., U.E.R. de Mathématiques et d'Informatique, Université de Bordeaux 1, 351, Cours de la Liberation, 33405 Talence Cedex, *France*

* **Coquel, F.**, Prof. Dr., Research Branch, Laboratoire National d'Hydraulique, 6, Quai Watier, B.P. 49, 78401 Chatou Cedex, *France*

Cordova, J. Q., Dr., NASA Ames Research Center, M.S. 202a–1, Moffett Field, CA 94035, *USA*

Croisille, J. P., Dr., O.N.E.R.A., 29, Ave. Division Leclerc, B.P. 72, 92320 Chatillon Cedex, *France*

* **Dadone, A.**, Prof. Dr., University of Bari, Via Re David 200, 70125 Bari, *Italy*

* **Dang Dinh Ang**, Dept. of Mathematics, Dal Hoc Tong Hop, Ho Chi Minh City, Vietnam

* **Degond, P.**, Prof. Dr., Centre de Mathématiques Appliquées, Ecole Polytechnique, 91128 Palaiseau Cedex, *France*

* **Delorme, Ph.**, Dr., O.N.E.R.A., Division de l'Energétique, B.P. 72, 92320 Chatillon Cedex, *France*

* **De Luca, P.**, Dr., Centre d'Etudes de Gramat, 46500 Gramat, *France*

* **Deshpande, S. M.**, Prof. Dr., Dept. of Aerospace Engineering, Indian Institute of Science, Bangalore 560012, *India*

* **Dervieux, A.**, INRIA, Sophia–Antipolis 1 et 2, 2004, Route des Lucioles, 06565 Valbonne Cedex, *France*

Diederich, J., Dr., Schlachthofstraße 38, 4690 Herne 2, *FRG*

Dörfler, W., Institut für Angewandte Mathematik, SFB 256, Universität Bonn, Wegelerstr. 6, 5300 Bonn 1, *FRG*

Dohmen, L., Dr., IES GmbH, Bastionstraße 11–19, 5170 Jülich, *FRG*

Donato, A. A., Prof. Dr., Dipartimento di Matematica, Università di Messina, Contrada Papardo, Salita Sperone 31, 98010 Sant'Agata, Messina, *Italy*

* **Dubois, F.**, Prof. Dr., Aerospatiale, SDT–STMI, B.P. 96, 78133 Les Mureaux Cedex, *France*

Dziuk, G., Prof. Dr., Inst. für Angewandte Mathematik, Universität Bonn, Wegelerstr. 6, 5300 Bonn, *FRG*

Egnesund, L., Dr., Dept. of Scientific Computing, University of Uppsala, Foermansgatan 17, 72466 Vasteras, *Sweden*

Einfeldt, B., Dr., Inst. für Geometrie und Praktische Mathematik, RWTH Aachen, Templergraben 55, 5100 Aachen, *FRG*

* **Eliasson, P.**, Prof. Dr., FFA, The Aeronautical Research Institute of *Sweden*, P.O. Box 11021, 161 11 Bromma, *Sweden*

Enander, R., Dr., Dept. of Scientific Computing, University of Uppsala, Sturegatan 4B 2tr, 75223 Uppsala, *Sweden*

* **Engelbrecht, J.**, Prof. Dr., Institute of Cybernetics, Estonian Academy of Sciences, 200108 Tallinn, Estonia, *USSR*

Engels, H., Prof. Dr., Institut für Geometrie und Praktische Mathematik, RWTH Aachen, Templergraben 55, 5100 Aachen, *FRG*

* **Fabrizio, M.**, Prof. Dr., Istituto di Matematica, Università di Bologna, Piazza Porta San Donato, Bologna, *Italy*

Favini, B., Prof. Dr., Dipartimento di Meccanica e Aeronautica, Facolta di Ingegneria, Via Eudossiana 18, 00195 Roma, *Italy*

Feistauer, M., Prof. Dr., Faculty of Mathematics and Physics, Charles University Prague, Sokolovska 83, 18600 Prague 8, *ČSSR*

* **Fernandez, G.**, Prof. Dr., INRIA, Sophia–Antipolis, 06560 Valbonne Cedex, *France*

* **Fezoui, L.**, INRIA, Sophia–Antipolis 1 et 2, 2004, Route des Lucioles, 06565 Valbonne Cedex, *France*

Finckenstein, Graf K. von, Prof. Dr., Fachbereich Mathematik, TH Darmstadt, Schloßgartenstr. 7, 6100 Darmstadt, *FRG*

* **Fogwell, Th. W.**, Prof. Dr., Gesellschaft für Mathematik und Datenverarbeitung mbH, Schloß Birlinghoven,Postfach 1240, 5205 Sankt Augustin 1, *FRG*

Frehse, J., Prof. Dr., Inst. für Angewandte Mathematik, Universität Bonn, Behringstr. 6, 5300 Bonn, *FRG*

* **Freistühler, H.**, Dr., Institut für Mathematik, RWTH Aachen, Templergraben 55, 5100 Aachen, *FRG*

Frühauf, H. H., Dr.–Ing., Institut für Raumfahrtsysteme, Universität Stuttgart, Pfaffenwaldring 31, 7000 Stuttgart 80, *FRG*

Fusco, D., Prof. Dr., Dipartimento di Matematica, Universita di Messina, Contrada Papardo, Salita Sperone 31, 98010 Sant'Agata – Messina, *Italy*

Gibbons, J., Dr., Dept. of Mathematics, Imperial College of Science and Technology, University of London, 180 Queen's Gate, London, SW7 2BZ, Great Britain

Gilquin, H., Equipe d'Analyse Numerique, 23, Rue du Docteur Paul Michelon, 42023 Saint–Etienne Cedex, *France*

* **Gimse, T.**, Prof. Dr., Dept. of Mathematics, University of Oslo, P.O. Box 1053, 0316 Oslo, *Norway*

* **Glimm, J.**, Prof. Dr., Courant Institute of Mathematical Sciences, New York University, 251 Mercer Street, New York, NY 10012, *USA*

* **Goldberg, M.**, Prof. Dr., Dept. of Mathematics, Technion , Haifa 32000, Israel

Gowda, V., Dr., INRIA, Domaine de Voluceau – Rocquencourt, B.P. 105, 78153 Le Chesnay, *France*

* **Greenberg, J. M.**, Prof. Dr., Dept. of Mathematics, University of Maryland, 5542 Suffield Court, Catonsville, MD 21228, *USA*

Grönig, H., Prof. Dr., Stoßwellenlabor, RWTH Aachen, Templergraben 55, 5100 Aachen, *FRG*

* **Gustafsson, B.**, Prof. Dr., Dept. of Scientific Computing, Uppsala University, Sturegatan 4B 2tr, 752 23 Uppsala, *Sweden*

* **Hackbusch, W.**, Prof. Dr., Institut für Informatik und Praktische Mathematik, Christian–Albrechts–Universität zu Kiel, Olshausenstr. 40, 2300 Kiel 1, *FRG*

* **Hänel, D.**, Dr., Aerodynamisches Institut, RWTH Aachen, Templergraben 55, 5100 Aachen, *FRG*

* **Hagemann, S.**, Inst. für Informatik und Praktische Mathematik, Christian–Albrechts–Universität zu Kiel, Olshausenstr. 40, 2300 Kiel 1, *FRG*

Hagstrom, Th., Dr., Dept. of Applied Mathematics and Statistics, State University of New York at Stony Brook, Stony Brook, NY 11794, *USA*

Halpern, L., Dr., Centre de Mathématiques Appliquées, Ecole Polytechnique, 91128 Palaiseau Cedex, *France*

Hanche–Olsen, H., Prof. Dr., Division of Mathematical Sciences, Norwegian Institute of Technology, Oystein Moylas v. 23, 7031 Trondheim–NTH, *Norway*

* **Hanyga, A.**, Prof. Dr., Institute of Geophysics, Polish Academy of Sciences, ul. Pasteura 3, 00973 Warsaw, *Poland*

* **Harabetian, E.**, Prof. Dr., Dept. of Mathematics, University of Michigan, 3220 Angell Hall, Ann Arbor, MI 48109–1003, *USA*

* **Henke, H.**, Dr., MBB GmbH Bremen, TE 234, Hünefeldstraße 1–5, 2800 Bremen, *FRG*

Hettich, R., Prof. Dr., Fachbereich IV – Mathematik, Universität Trier, Postfach 3825, 5500 Trier, *FRG*

* **Holden, H.**, Prof. Dr., Institute of Mathematics, University of Trondheim, 7034 Trondheim–NTH, *Norway*

* **Holden, L.**, Dr., Norsk Regnesentral, Postboks 114, Blindern, 0314 Oslo 3, *Norway*

Holing, K., Statoil Trondheim, Postuttak, 7004 Trondheim, *Norway*

* **Hornung, K.**, Prof. Dr., Fakultät für Luft– und Raumfahrttechnik, Universität der Bundeswehr München, Werner–Heisenberg–Weg 39, 8014 Neubiberg, *FRG*

* **Hsiao, L.**, Prof. Dr., Dept. of Mathematics, University of Washington, Seattle, WA 98195, *USA*

* **Hunter, J. K.**, Dr., Dept. of Mathematics, Colorado State University, Fort Collins, CO 80523, *USA*

* **Isaacson, E. L.**, Dept. of Mathematics, University of Wyoming, Laramie, WY 82071, *USA*

* **Ishiguro, T.**, Prof. Dr., Computational Sciences Division, National Aerospace Laboratory, 7–44–1 Jindaiji–higashi–machi, Chofu, Tokyo, 182, *Japan*

Jaffre, J., Prof. Dr., INRIA, Domaine de Voluceau – Rocquencourt, B.P. 105, 78153 Le Chesnay Cedex, *France*

James, F., Dr., Centre de Mathématiques Appliquées, Ecole Polytechnique, 91128 Palaiseau Cedex, *France*

Jeffrey, A., Prof. Dr., Dept. of Engineering, University Newcastle upon Tyne, Stephenson Building, Claremont Road, Newcastle upon Tyne, NE1 7RU, *Great Britain*

Jeltsch, R., Prof. Dr., Institut für Geometrie und Praktische Mathematik, RWTH Aachen, Templergraben 55, 5100 Aachen, *FRG*

* **Jeschke, M.**, Aerodynamisches Institut, RWTH Aachen, Templergraben 55, 5100 Aachen, *FRG*

Johnson, C., Prof. Dr., Dept. of Mathematics, Chalmers University of Technology, 41296 Goeteborg, *Sweden*

Jongen, H., Prof. Dr., Lehrstuhl C für Mathematik, RWTH Aachen, Templergraben 55, 5100 Aachen, *FRG*

Jourdren, H., Dr., C.E.A., Centre de Limeil, B.P. 27, 94190 Villeneuve St. Georges, *France*

* **Klein, R.**, Institut für Allgemeine Mechanik, RWTH Aachen, Templergraben 64, 5100 Aachen, *FRG*

* **Klingenberg, C.**, Dr., Institut für Angewandte Mathematik, Universität Heidelberg, Im Neuenheimer Feld 294, 6900 Heidelberg, *FRG*

* **Klopfer, G. H.**, Prof. Dr., NEAR Inc., Mountain View, CA 94043, *USA*

Kocaaydin, S., Lehr– und Forschungsgebiet Mechanik, RWTH Aachen, Templergraben 64, 5100 Aachen, *FRG*

* **Koren, B.**, Dr., Center for Mathematics and Computer Science, P.O. Box 4079, 1009 AB Amsterdam, *Netherlands*

* **Kosiński, S.**, Dr., Inst. Inzynierii Budowlanej I–32, Politechnika Łodzka, Al. Politechniki 6, 93–590 Łódź 40, *Poland*

* **Kosiński, W.**, Dr., Institute of Fundamental Technological Research, Polish Academy of Sciences, ul. Świętokrzyska 21, 00–049 Warsaw, *Poland*

* **Kozel, K.**, Prof. Dr., Dept. of Computational Techniques and Informatics, TU Prague, Suchbátarova 4, 166 07 Prague 6, *ČSSR*

* **Kröner, D.**, Dr., Institut für Angewandte Mathematik, Universität Heidelberg, Im Neuenheimer Feld 294, 6900 Heidelberg, *FRG*

Kroll, N., Institut für Entwurfsaerodynamik, DFVLR e. V. Am Flughafen, 3300 Braunschweig, *FRG*

* **Kubota, H.**, Prof. Dr., Faculty of Engineering, Dept. of Aeronautics, University of Tokyo, Hongo, Bunkyo–ku, Tokyo, 113, *Japan*

Küpper, T., Prof. Dr., Institut für Angewandte Mathematik, Universität Hannover, Welfengarten 1, 3000 Hannover 1, *FRG*

Kunik, M., Mathematisches Institut A, Universität Stuttgart, Pfaffenwaldring 57, 7000 Stuttgart 80, *FRG*

* **Lar'kin, N. A.**, Prof. Dr., Inst. of Theoretical and Applied Mechanics, 630090 Novosibirsk - 90, *USSR*

* **Larrouturou, B.**, Prof. Dr., INRIA, Sophia–Antipolis, 06560 Valbonne Cedex, *France*

Lazareff, M., Prof. Dr., O.N.E.R.A. 10AT3, 29, Ave de la Division Leclerc, 92320 Chatillon Cedex, *France*

* **Le Floch, P.**, Dr., Centre de Mathématiques Appliquées, Ecole Polytechnique, 91128 Palaiseau Cedex, *France*

* **Le Roux, A. Y.**, Prof. Dr., U.E.R. Mathématiques, Université de Bordeaux 1, 355 Cours de la Liberation, 33405 Talence Cedex, *France*

Le Veque, R. J., Prof. Dr., Dept. of Mathematics GN–50, University of Washington, Seattle, WA 98195, *USA*

Leclercq, M. P., Dr., Avions Marcel–Dassault–Berguet Aviation, 78, Quai Marcel Dassault, 92214 Saint–Cloud Cedex, *France*

Leutloff, D., Dr. Ing., Institut für Mechanik, TH Darmstadt, Bert Brecht Straße 5, 6107 Reinheim 3, *FRG*

Lorentz, R., Dr., Gesellschaft für Mathematik und Datenverarbeitung mbH, Schloß Birlinghoven, Postfach 1240, 5205 Sankt Augustin 1, *FRG*

Luh, Y., Gesellschaft für Mathematik und Datenverarbeitung mbH, Schloß Birlinghoven, Postfach 1240, 5205 Sankt Augustin 1, *FRG*

* **Mandal, J. C.**, Prof. Dr., Dept. of Aerospace Engineering, Indian Institute of Science, Bangalore 560012, *India*

Mao, D.–K., Prof. Dr., Dept. of Mathematics, University of California, 405 Hilgard Avenue, Los Angeles, CA 90024–1555, *USA*

* **Marchesin, D.**, Prof. Dr., Instituto de Matemática Pura e Aplicada, Pontificia Univ. Catholica, Estrada d. Castorina 110, Rio de Janeiro 22453, *Brazil*

* **Marshall, G.**, Prof. Dr., EPFL, GASOV Group, 1015 Lausanne, Switzerland

Martensen, E., Prof. Dr., Mathematisches Institut II, Universität Karlsruhe, Englerstr. 2, 7500 Karlsruhe 1, *FRG*

Maslov, V. P., Prof. Dr., Moscow Institute of Electronic Machinebuilding, B. Vusovsky 3/12, 109 028 Moscow, *USSR*

* **Mazet, P. A.**, Prof. Dr., O.N.E.R.A. - C.E.R.T., B.P. 4025, 31055 Toulouse Cedex, *France*

* **Mertens, J.**, Dr., MBB - UT, TE 212, Hünefeldstr. 1–5, 2800 Bremen, *FRG*

Mönig, R., Dr., Institut für Strahlenantriebe und Turboarbeitsmaschinen, RWTH Aachen, Templergraben 55, 5100 Aachen, *FRG*

* **Montagné, J.–L.**, Prof. Dr., Division DAT1, O.N.E.R.A., 29, Avenue de la Division Leclerc, 92320 Chatillon Cedex, *France*

* **Moretti, G.**, Prof. Dr., G.M.A.F. Inc., P.O. Box 184, Freeport, NY 11520, *USA*

* **Morton, K.W.**, Prof. Dr., Oxford University Computing Laboratory, 8–11 Keble Road, Oxford OX1 3QD, *Great Britain*

Müller, B., Dr., DFVLR — AVA, SM — TS, Bunsenstr. 10, 3400 Göttingen, *FRG*

* **Munz, C. D.**, Dr., Institut für Reaktortechnik und Neutronenphysik, Kernforschungszentrum Karlsruhe, Postfach 3640, 7500 Karlsruhe, *FRG*

* **Mustieles, F. J.**, Dr., Centre de Mathématiques Appliquées, Ecole Polytechnique, 91128 Palaiseau Cedex, *France*

Nastase, A., Prof. Dr., Lehrgebiet Aerodynamik des Fluges, RWTH Aachen, Templergraben 55, 5100 Aachen, *FRG*

* **Niclot, B.**, Centre Technique, Citroen, DAT/CSI, Route de Gisy, 78140 Velizy–Villacoublay Cedex, *France*

* **Nguyen Van Nhac**, Dr., Dept. of Mathematics, Faculty of Nuclear Engineering, TU Prague, Trojanova 13, 120 00 Prague 2, *ČSSR*

* **Ogawa, S.**, Prof. Dr., Computational Sciences Division, National Aerospace Laboratory, 7–44–1 Jindaiji–higashi–machi, Chofu, Tokyo, 182, *Japan*

* **Osher, S. J.**, Prof. Dr., Dept. of Mathematics, University of California, Los Angeles, CA 90024, *USA*

Paert, E., Dr., Dept. of Scientific Computing, University of Uppsala, Sturegatan 4B 2tr, 75223 Uppsala, *Sweden*

Panagiotopoulos, P. D., Prof. Dr., RWTH Aachen, Templergraben 55, 5100 Aachen, *FRG*

* **Pandolfi, M.**, Prof. Dr., Dipartimento di Ingegneria Aeronautica e Spaziale, Politecnico di Torino, Corso Duca degli Abruzzi 24, 10129 Torino, *Italy*

Pennisi, S., Dr., Dipartimento di Matematica, Citta Universitaria, Viale A Doria N. 6, Catania, *Italy*

* **Peradzyński, Z.**, Dr., Institute of Fundamental Technological Research, Polish Academy of Sciences, ul. Świętokrzyska 21, 00–049 Warsaw, *Poland*

* **Pfitzner, M.**, Dr., MBB GmbH, Postfach 801169, 8000 München 80, *FRG*

* **Pham Ngoc Dinh, A.**, Dr., Dépt. de Mathématiques et d'Informatique, Université de Orleans, B.P. 67–59, 45067 Orleans - Cedex 2, *France*

* **Plohr, B. J.**, Prof. Dr., Computer Sciences Dept., University of Wisconsin, Madison, WI 53706, *USA*

Pöppe, C., Dr., SFB 123, Universität Heidelberg, Im Neuenheimer Feld 294, 6900 Heidelberg, *FRG*

Pohl, B., Institut für Geometrie und Praktische Mathematik, RWTH Aachen, Templergraben 55, 5100 Aachen, *FRG*

Priestley, A., Dr., Dept. of Mathematics, University of Reading, Whiteknights, P.O. Box 220, Reading RG6 2AH, *Great Britain*

Racke, R., Dr., Inst. für Angewandte Mathematik, Universität Bonn, Wegelerstraße 10, 5300 Bonn 1, *FRG*

Rannacher, R., Prof. Dr., Fachbereich Angewandte Mathematik und Informatik, Universität des Saarlandes, 6600 Saarbrücken, *FRG*

Rascle, M., Prof. Dr., Laboratoire de Mathématiques, Université de Nice (U.A. CNRS No.168), Parc Valrose, 06034 Nice, *France*

Raviart, P. A., Prof. Dr., Centre de Mathématiques Appliquées, Ecole Polytechnique, 91128 Palaiseau Cedex, *France*

Reutter, F., Prof. Dr., Lütticherstraße 238, 5100 Aachen, *FRG*

Risebro, N .H., Dr., Institute of Mathematics, University of Oslo, P.O. Box 1053, Blindern, 0316 Oslo 3, *Norway*

* **Rizzi, A.**, Prof. Dr., FFA, The Aeronautical Research Institute of Sweden, P.O. Box 11021, 161 11 Bromma, *Sweden*

Romstedt, P., Dr., Gesellschaft für Reaktorsicherheit, Forschungsgelände, 8046 Garching, *FRG*

* **Rostand, P.**, Prof. Dr., INRIA–Menusin, Domaine de Voluceau – Rocquencourt, B.P. 105, 78153 Le Chesnay Cedex, *France*

Rozhdestvensky, B. L., Dr., Keldysh Institute of Applied Mathematics, Academy of Sciences of USSR, Miusskaja Sq.4, 125047 Moscow, *USSR*

Rusanov, V. V., Prof. Dr., Keldysh Institute of Applied Mathematics, Academy of Sciences of USSR, Miusskaya Sq. 4, 125047 Moscow A – 47, *USSR*

* **Schick, P.**, Dr., Fakultät für Luft– und Raumfahrttechnik, Universität der Bundeswehr München, Werner–Heisenberg–Weg 39, 8014 Neubiberg, *FRG*

Schlechtriem, S., Lehr– und Forschungsgebiet Mechanik, RWTH Aachen, Templergraben 64, 5100 Aachen, *FRG*

* **Schmidt, L.**, Dr., Institut für Angewandte Mathematik, Universität Karlsruhe, Englerstraße 2, 7500 Karlsruhe, *FRG*

Schneider, M., Prof. Dr., Mathematisches Institut I, Universität Karlsruhe, Englerstraße 2, 7500 Karlsruhe 1, *FRG*

* **Schöffel, St.U.**, Dr., Mechanische Verfahrenstechnik und Strömungsmechanik, Fachbereich Maschinenwesen,Universität Kaiserslautern,Postfach 3049,6750 Kaiserslautern, *FRG*

Schulte, M., Dr., Fachbereich Mathematik, WE03, Freie Universität Berlin, Arnimallee 2–6, 1000 Berlin 33, *FRG*

Schwarz, M., WWU Münster, D 208, Stadtlohnweg 11, 4400 Münster, *FRG*

Schwenzfeger, K.J., Prof. Dr., Institut I für Mathematik und Rechneranwendung,Fakultät für LRT, Universität der Bundeswehr, Georgenstr. 24, 8021 Sauerlach, *FRG*

Serre, D., Prof. Dr., Ecole Normale Superieure de Lyon, 46, Allée d'Italie, 69364 Lyon Cedex 07, *France*

* **Sethian, J.**, Prof. Dr., Dept. of Mathematics, University of California, Berkeley, CA 94720, *USA*

* **Shapiro, R. A.**, Dr., Computational Fluid Dynamics Laboratory, Massachusetts Institute of Technology, 77 Massachusetts Avenue, Cambridge, MA 02146, *USA*

Smit, J. H., Prof. Dr., Dept. of Mathematics, University of Stellenbosch, 7600 Stellenbosch, *South–Africa*

* **Sommerfeld, M.**, Dr., Lehrstuhl für Strömungsmechanik, Friedrich–Alexander Universität Erlangen-Nürnberg, Egerlandstr. 13, 8520 Erlangen, *FRG*

Sonar, Th., Inst. f. Entwurfsaerodynamik, DFVLR, Am Flughafen, 3300 Braunschweig, *FRG*

Song, J., Universität Bonn, An der Ohligsmühle 29, 5300 Bonn 1, *FRG*

Sowa, J., Dr., Dept. of Scientific Computing, University of Uppsala, Sturegatan 4B 2tr, 75223 Uppsala, *Sweden*

* **Staat, M.**, Dr.–Ing., Lehr– und Forschungsgebiet Mechanik, RWTH Aachen, Templergraben 64, 5100 Aachen, *FRG*

Steffen, B., Dr., Zentralinstitut für Mathematik, Kernforschungsanlage Jülich, Postfach 1913, 5170 Jülich, *FRG*

* **Steve, H.**, INRIA, Sophia–Antipolis 1 et 2, 2004, Route des Lucioles, 06565 Valbonne Cedex, *France*

* **Stoufflet, B.**, AMD–BA, DGT–DEA, B.P. 300, 78, Quai Marcel Dassault, 92214 Saint–Cloud, *France*

* **Straškraba, I.**, Prof. Dr., Československa Akademie Ved Matematicky Ustav, Žitná 25, Praha 1, 115 67, *ČSSR*

* **Sweby, P. K.**, Prof. Dr., Dept. of Mathematics, University of Reading, Whiteknights, P.O. Box 220, Reading RG6 2AH, *Great Britain*

Szepessy, A., Prof. Dr., Dept. of Mathematics, Chalmers University of Technology and University of Goeteborg, Sven Hultins Gata 6, 41727 Goeteborg, *Sweden*

Szmolyan, P., Prof. Dr., Dept. of Mathematics and Statistics, University of Maryland, Baltimore County Campus, Baltimore, MD 21228, *USA*

* **Tadmor, E.**, Prof. Dr., School of Mathematical Sciences, Tel Aviv University, Tel Aviv 69928, Israel

* **Temple, B.**, Prof. Dr., Dept. of Mathematics, University of California, Davis, CA 95616, *USA*

* **Ting, T. C. T.**, Prof. Dr., Dept. of Civil Engineering Mechanics and Metallurgy (M/C 246), University of Illinois at Chicago, Box 4348, Chicago, IL 60680, *USA*

* **Toro, E. F.**, Dr., College of Aeronautics, Cranfield Institute of Technology, Cranfield, Beds MK43 QAL, *Great Britain*

Trangenstein, J. A., Prof. Dr., Lawrence Livermore National Laboratory, University of California, Post Box 808, Livermore, CA 94550, *USA*

Turchak, L. I., Prof. Dr., Computational Fluid Dynamics Dept., USSR Academy of Sciences Computing Centre, 40 Vavilova, 117333 Moscow, *USSR*

Tveito, A., Prof. Dr., Institute of Informatics, University of Oslo, Postboks 1080, Blindern, 0316 Oslo 3, *Norway*

Varnhorn, W., Dr., TH Darmstadt, Schloßgartenstraße 7, 6100 Darmstadt, *FRG*

* **Vavřincová, M.**, Prof. Dr., Dept. of Computational Techniques and Informatics, TU Prague, Suchbátarova 4, 16607 Prague 6, *ČSSR*

Vecchi, I., Dr., Institut für Angewandte Mathematik, Universität Heidelberg, Im Neuenheimer Feld 294, 6900 Heidelberg, *FRG*

Vila, J.–P., Prof. Dr., Cemagref, Domaine Universitaire, B.P. 76, 38402 Saint Martin d'Heres, *France*

* **Vinokur, M.**, Prof. Dr., Sterling Software, Palo Alto, CA 94030, *USA*

* **Voskresensky, G.P.**, Prof. Dr., Keldysh Institute of Applied Mathematics, USSR Academy of Sciences, Miusskaya Sq. 4, 125047 Moscow, *USSR*

* **Wada, Y.**, Prof. Dr., Computational Sciences Division, National Aerospace Laboratory, 7–44–1 Jindaiji–higashi–machi, Chofu, Tokyo, 182, *Japan*

Wagner, D. H., Prof. Dr., Dept. of Mathematics, University of Houston, University Park, 4800 Calhoun Road, Houston ,TX 77004, *USA*

Walter, A., Fachbereich Mathematik, Universität des Saarlandes, Im Stadtwald, 6600 Saarbrücken 1, *FRG*

Walter, J., Prof. Dr., Institut für Mathematik, RWTH Aachen, Templergraben 55, 5100 Aachen, *FRG*

* **Warming, R. F.**, Prof. Dr., CFD Branch MS 202a–1, NASA Ames Research Center, Moffett Field, CA 94035, *USA*

Warnecke, G., Dr., Mathematisches Institut A, Universität Stuttgart, Pfaffenwaldring 57, 7000 Stuttgart 80, *FRG*

Wegner, W., Institut für Aeroelastik, DFVLR – AVA, Bunsenstraße 10, 3400 Göttingen, *FRG*

Weiland, C., Dr.–Ing., Kommunikationssysteme und Antriebe KT 225, MBB GmbH, Postfach 801169, 8000 München 80, *FRG*

* **Wendroff, B.**, Prof. Dr., Theoretical Division MSB 28484, Los Alamos National Laboratory, Los Alamos, NM 87545, *USA*

Werner, K. D., Institut für Geometrie und Praktische Mathematik, RWTH Aachen, Templergraben 55, 5100 Aachen, *FRG*

Werner, W., Dr., Gesellschaft für Reaktorsicherheit (GRS) mbH, Forschungsgelände, Postfach, 8046 Garching, *FRG*

Wesolowski, Z., Prof. Dr., Institut für Physik, Max–Planck–Institut für Metallforschung, Postfach 80 06 65, 7000 Stuttgart, *FRG*

* **Westenberger, H.**, Dr., Lehr- und Forschungsgebiet Mechanik, RWTH Aachen, Templergraben 64, 5100 Aachen, *FRG*

* **White, A. B., Jr.**, Prof. Dr., Los Alamos National Laboratory, Los Alamos, NM 87545, *USA*

Wiemer, P., Dr., Rheinmetall GmbH, Postfach 6609, 4000 Düsseldorf 30, *FRG*

Wirz, H. J., Dr., II. Institut für Mechanik, TU Berlin, Straße des 17. Juni, 1000 Berlin, *FRG*

Witsch, K., Prof. Dr., Angewandte Mathematik, Universität Düsseldorf, Universitätsstr. 1, 4000 Düsseldorf, *FRG*

Woodward, P. R., Prof. Dr., Astronomy Dept., University of Minnesota, 116 Church St. S.E., Minneapolis, MN 55455, *USA*

* **Yee, H. C.**, Prof. Dr., NASA Ames Research Center, Mail Stop 202a–1, Moffett Field, CA 94035, *USA*

* **Zajaczkowski, W.M.**, Prof. Dr., Institute of Fundamental Technological Research, Świętokrzyska 21, 00–048 Warsaw, *Poland*

Zeller, H., Prof. Dr., Aerodynamisches Institut, RWTH Aachen, Templergraben 55, 5100 Aachen, *FRG*

Zeller, R., Dr., Cray–Research GmbH, Perhamerstr. 31, 8021 Sauerlach, *FRG*

Zhang, T., Prof. Dr., Institute of Mathematics, Academia Sinica, Beijing, *China*

Zhu, Y.–L., Prof. Dr., Stochastische Mathematische Modelle,SFB 123, Universität Heidelberg, Im Neuenheimer Feld 294, 6900 Heidelberg, *FRG*

* **Ziółko, M.**, Dr., Institute of Automatic Control, University of Mining and Metallurgy, al. Mickiewicza 30, 30 – 059 Kraków, *Poland*

Support and Sponsorship Acknowledgements

Scientific Committee

Y. L. Zhu, Beijing, Z. Wesolowski, Warsaw, C. Weiland, München, B. van Leer, Ann Arbor, Y. Shokin, Krasnoyarsk, P. A. Raviart, Palaiseau, M. Pandolfi, Torino, S. Osher, Los Angeles, O. Oleinik, Moscow, T. P. Liu, Maryland, C. Klingenberg, Heidelberg, R. Jeltsch, Aachen, A. Jeffrey, Newcastle uponTyne, J. D. Hoffman, West–Lafayette, B. Gustafsson, Uppsala, A.A. Donato, Messina, C. Dafermos, Providence, C. Carasso, St. Etienne, J. Ballmann, Aachen.

Host Organization

Rheinisch Westfälische Technische Hochschule Aachen

Supporting Organizations — Assistance Gratefully Acknowledged

Control Data GmbH, Düsseldorf
Cray Research GmbH, München
Deutsche Forschungsgemeinschaft
Diehl Gmbh & Co., Röthenbach
Digital Equipment GmbH, Köln
FAHO Gesellschaft von Freunden der Aachener Hochschule, Düsseldorf
IBM Deutschland GmbH, Stuttgart
Mathematisch Naturwissenschaftliche Fakultät der RWTH
Mayersche Buchhandlung, Aachen
Ministerium für Wissenschaft und Forschung des Landes Nordrhein–Westfalen
Office of Naval Research Branch, Office, London, England
Rheinmetall GmbH, Düsseldorf
Stadt Aachen
US Air Force EOARD
US Army European Research Office, London
Wegmann GmbH & Co., Kassel